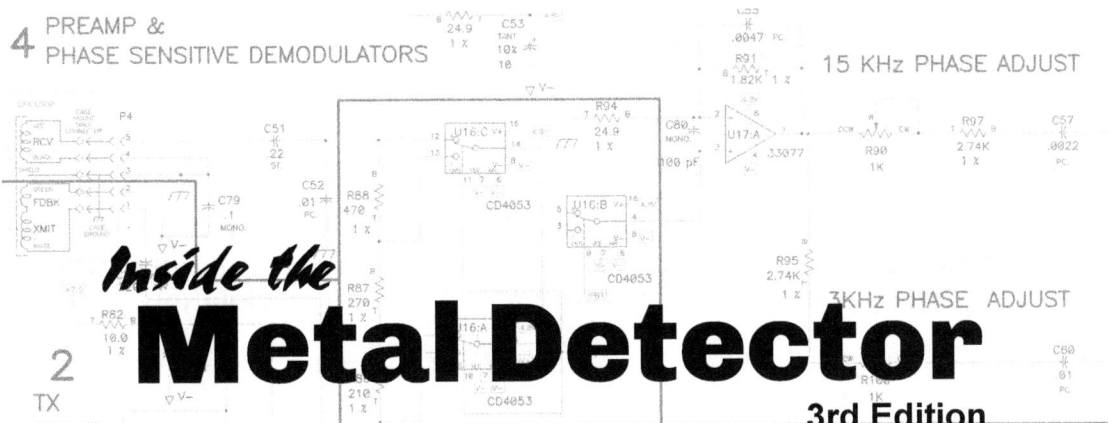

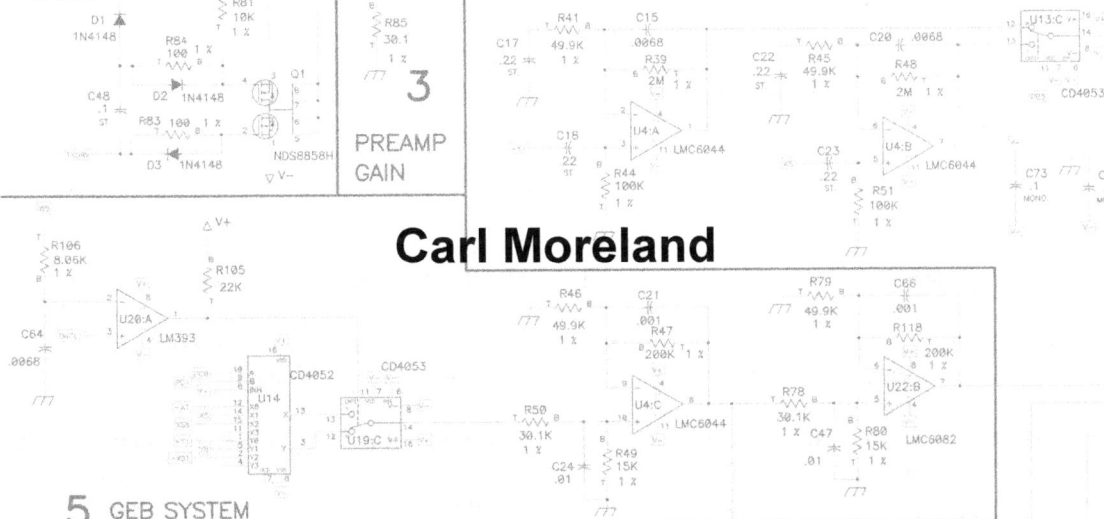

Inside the
Metal Detector

3rd Edition

Carl Moreland

Geotech Press
Republic, Washington

Published by:
Geotech Press
Republic, WA 99166
geotech1.com/itmd

Cover: Minelab Equinox coil X-ray, courtesy Pav_DoK
Background schematic: White's *DFX*

Many trademarked names are mentioned throughout this book. The trademarks are held by their respective companies.

The circuits presented in this book are for experimental purposes only. Some designs may be covered under active patents and may not be developed, either commercially or for personal use, without permission from the patent holder. Some of the projects might also fall under government regulation for radio emissions. It is the responsibility of the user to ensure adherence to these regulations.

Some studies have indicated that devices which emit radio frequencies may be linked to certain health problems, including disruption of pacemakers. Although there has been no indication that metal detectors are among these devices, health safety is the responsibility of the user.

Neither the author nor the publisher of this book shall be held responsible, whether legally, financially, or any other way, for any damages incurred by the use of the contents of this book.

Finally, this book represents years of hard work with relatively little payback due to its specialized topic and small reader market. Please respect the effort spent bringing you this book; do not upload it to the Internet for others to get for free. *Thank you.*

This book was written in Adobe FrameMaker in New Times Roman 10pt.
Line art was drawn in Microsoft Visio. Photographs were taken with a camera.

ISBN 978-0-9858342-3-4

Quick Look

Part 1: Fundamentals

Part 2: Coils

Part 3: Proximity Methods

Part 4: Induction Balance

Part 5: Pulse Induction

Part 6: Advanced Methods

Part 7: Leftovers

Part 8: Appendices

M-SCOPE

*If it is metallic
the M-Scope will find it*

FISHER

Table of Contents

Part 6: Advanced Methods

Random Images

Introduction

"In science one tries to tell people, in such a way as to be understood by everyone, something that no one ever knew before. But in poetry, it's exactly the opposite."
— *Paul Dirac* (1902–1984)

Throughout history people have hidden or lost vast amounts of treasure, and were followed by other people who would try to find it. Early treasure hunters tended to be either wealthy or well-financed and the process was often expensive and sometimes dangerous. Much of the early European exploration of the Americas was essentially done by treasure hunters.

The development of the metal detector transformed treasure hunting and brought it to the common people who could hunt for lost valuables as a weekend hobby. The hobby of metal detecting barely existed in the 1950s when surplus military mine detectors were just about the only metal locator available. Twenty years later, inexpensive transistorized metal detectors could be found just about everywhere. And the magnetometer — a close cousin of metal detectors — has been responsible for the vast majority of shipwreck discoveries since the 1960s.

The metal detecting hobby experienced its "boom" period a number of years ago but even nowadays it remains a popular hobby amongst all ages. It is relatively inexpensive to get started; an entry level detector can be purchased for around $100. That doesn't get you much, either in performance or quality, and many newcomers make the mistake of buying a cheap discount-store detector only to get discouraged and quit. Although some high-end models can cost well over $1000, a good quality entry-level detector can be found for $200-$300[1].

In the early years, metal detectors had to be tuned, balanced, adjusted, and tweaked continuously during use. On some models, the user really had to know exactly what all the knobs did, and how they interacted. But many of the newest models feature turn-on-and-go operation, automatic ground tracking, and "target identification," so today's user can get reasonable results with no knowledge about how the instrument actually works.

Some models now have such an array of features that it can be extremely confusing for the beginner deciding which machine to purchase. Regardless of whether a particular design is all-analog or microprocessor-controlled; has a simple tonal response or an LCD display; or has a single control knob or an array of control buttons; it must abide by the same basic electromagnetic principles as any other metal detector. The addition of more complexity, for the most part, provides more bells and whistles with little substantial gain in depth. Those bells & whistles, though, can often be more valuable than raw depth.

Even though metal detectors are not especially expensive and many detectorists own more than one model, the engineers amongst us experience an irresistible urge to develop our own homebrew detectors, to both understand the basic principles and to try and match or even exceed the achievements of the commercial models. This book is especially aimed at those kind of people.

But it is also for those who want a deeper understanding of how detectors work so they can get the most out of the units they hunt with. The most successful detectorists are those who have a good understanding of what all those controls are doing. How does discrimination work? What, exactly, is happen-

---

1. Ironically, this has been true for the last 50 years. In 1975, a White's *Coinmaster IV* was $239. That comes to $1530 in 2024 dollars, making today's detectors seem like bargains, especially when comparing the capabilities of, say, a Nokta *Simplex* with the *Coinmaster IV*.

ing when you ground balance a detector? How do different targets respond? What are the advantages and disadvantages of different coils? Even if you never build a detector circuit this book will help answer those questions, and more. But this book does not cover detecting techniques or the need to do proper research; many other books are available that attempt to do that.

Finding Technical Information

In the realm of consumer electronics, it's remarkable just how little information has been published on metal detector technology. With other electronic technologies—radio, television, computers, remote control, etc.—there not only exist vast amounts of information about the theory, but often full schematics of consumer products are easily obtainable. Not so with metal detectors. Companies rarely will provide schematics, and often take steps to prevent reverse-engineering, such as potting circuits in epoxy, grinding off IC part numbers, or using tamper-protected microcontrollers.

But there have been a few books on detector technology. In 1927, a book with the strange title "Modern Divining Rods" by R. J. Santschi offered what was probably the first technical book on metal detectors. Most detectors of that era were the orthogonal two-box locators, such as those sold by Fisher Labs, and all had tube-based circuits. Santschi not only covered the basic types of locators of his era but did a masterful job with technical analyses and included a number of full schematics. Ironically, the book does not cover divining rods.

A book in 1969 by Dr. Arnold Kortejarvi, "Official Handbook of Metal Detectors," included a chapter on "How Detectors Function." Also in 1969, treasure hunter E. S. "Rocky" LeGaye produced "The Electronic Metal Detector Handbook," of which a majority is dedicated to technical descriptions of detectors. He included both BFO and induction balance, though not PI which was in its infancy and not well-known. LeGaye provided little in the way of actual schematics. Most other books on detector technology are more focused on do-it-yourself projects, such as "Metal Detector Projects" by Charles Rakes, "Metal Locators" by Traister and Traister, "Proximity Sensors and Metal Locators" by John Potter, and "How to Build Your Own Metal & Treasure Locators" by F.G. Rayer. Although these books are chocked full of project circuits, they offer little in the way of theory and detailed technical description.

Besides these few books, there have been a fair number of articles published in various electronics hobby magazines describing metal detector projects. Although a few of these articles made decent attempts at explaining theory, most of the projects have been variations of basic designs and most do not cover operational theory very well. Many of these magazine articles are listed in Appendix D.

Perhaps the most valuable source of technical information are the patents that have been filed by detector manufacturers. A patent is usually very thorough in describing all the details of the invention and, in the case of metal detectors, sometimes include schematics. However, patents tend to be written rather obscurely, often in intentionally obfuscated legalese, perhaps to confuse the reader as to exactly what new method they have come up with. Appendix E contains a fairly comprehensive list of US metal detector patents and also what to watch out for when walking through a patent minefield.

Finally, the Internet is slowly accumulating technical information on metal detectors. Several web sites and forums focus largely on the technological aspects and many schematics and articles are now available if you dig around a little. Again, Appendix D lists some of these web sites and forums.

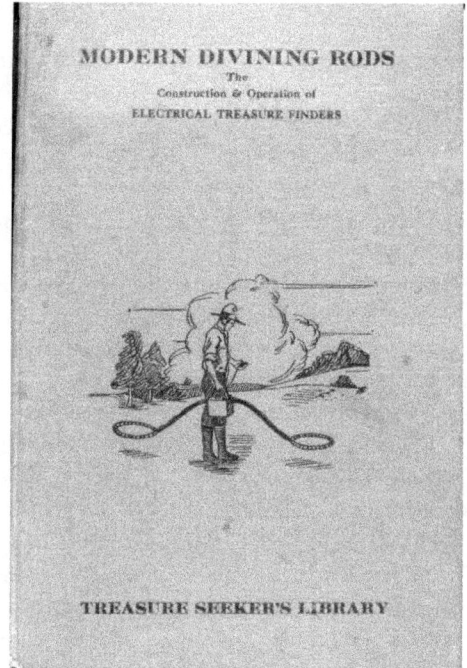

Building Detectors

The first two editions of this book included numerous detector projects that were thoroughly documented including schematics, parts list, PCB artwork, and construction and adjustment procedures. This edition does not have ready-to-build projects, rather it has a number of "designs." Schematics are presented and all further details are intended to be off-loaded to a supporting web site where it is much easier to update the information and provide support to experimenters. Most of the designs in this book are "new" and do not replicate the projects in the first two editions.

These designs are intended to teach basic concepts. They are not intended to be comprehensive circuits that you can build and get results comparable to commercial designs. They also are presented with some ICs (especially opamps and logic chips) lacking any overt power supply connections. It should be obvious from the schematic which power supplies should be connected. Furthermore, it is good practice to place bypassing capacitors on each IC supply pin, something that is omitted from the schematics to maintain clarity.

For those interested in building circuits, unless you are an experienced electronics hobbyist it is not advised that you jump straight to an advanced design. You might end up with something that does not work and you will not have a clue as to why. One of the things this book does not attempt to do is teach basic electronics. It is assumed you know Ohm's Law, those of Kirchhoff, transistor basics, opamps, filtering, and trigonometry. There are many books and on-line videos that teach these concepts. A metal detector project is a frustrating way to learn basic electronics so unless you have a good grasp of the concepts it is better to start off with simpler circuits.

As in using a metal detector, building one requires patience. Some chapters include small side experiments that are useful in demonstrating key concepts. Some projects are slowly developed from a simple circuit to a more complex and complete version. Don't try to jump ahead and build the complex version if you don't have a thorough understanding of the concepts taught in prior projects; use the simpler versions as stepping stones.

What will you need to build these circuits? There are two areas to consider: the hardware needed to build the circuit itself, and the test equipment needed to debug it when things go wrong[2]. In the old days, Veroboard or perf board was used with point-to-point soldered wiring. It was a slow process. Today, there are two easy methods: plug-in breadboards, and custom PC boards. Figure 1 shows a plug-in breadboard which happens to be hosting a metal detector circuit: my first bipolar pulse induction design. Breadboards offer a wonderful method of building quick circuits that are easy to modify. When done, you can rip everything up and re-use the parts in another design. Breadboards are not good for high-frequency circuits but work just fine for most metal detector designs.

Fig. 1: **Breadboard**

For permanent solutions (something you can stuff in a box and actually use) Veroboard or perfboard is still an option but custom PC boards are even better. Fifteen years ago iron-on transfers would be a preferred option; they are easy to make at home and have a professional appearance. Today, small-quantity multi-layer manufactured board prices have fallen so low that it is silly to consider anything else[3]. A big advantage in using a PC board is the ability to use surface-mount components and shrink

---

2. Oh, yes, things *will* go wrong.

3. At the time this was written, small-quantity 2-layer PCBs could be bought for less than $5 each.

Fig. 2: **PC board comparisons**

the size of the board dramatically. Figure 2 shows the same pulse induction design implemented using through-hole components (left) and surface-mount components (right).

Many of the designs in this book have a PC board layout available. Some are a singled-sided PCB while others use a 2-layer or 4-layer board. All use through-hole components. The rationale is that these circuits are intended to be teaching circuits and having the ability to easily swap component values is very useful. Through-hole components allow you to use IC sockets and even pin sockets for passive devices. Not only can you swap components easily but you can reuse most components in multiple designs. The choice of board layers is a matter of economy. While any of these project can be implemented on a single-layer board, doing so imposes severe restrictions on the layout which can compromise performance. With even a 4-layer PCB being reasonably cheap[4] it no longer makes much sense to limit ourselves to bad layouts.

Many times people try to build a complete circuit and then, when it doesn't work, debug it. The best approach is to build the circuit in stages and test the stages as you go. Generally start with the power supplies; they have to work for anything else to work. Then move to the clocking circuitry, which also controls the operation of other blocks. Next would be the transmitter, then move through the receiver blocks until you reach the speaker.

Experimentation and testing requires test equipment. It is not reasonable to expect that you can build circuits without encountering occasional problems that need debugging, and some of the circuits in this book require adjustment settings that can only be done with the proper equipment.

First and foremost is a good digital multimeter (DMM or DVM). Besides the normal voltage, current, and resistance, a DMM that can measure capacitance and frequency will be of great value, and generally costs less than $50. If you are also designing and building coils then an LCR meter is another must-have. Preferably look for one that supports 10kHz and measures inductance down to the micro-henries[5]. There are a few inexpensive LCR meters and meter kits available online but I am unfamiliar with any of them.

---

4. To expand on footnote #2, a 150x100mm (6"x4") 4-layer PCB costs $42 for (5) pieces, or $8.40 each. Use one, sell the others to friends & family.

5. My personal favorite is the DER EE DE-5000 which runs about $100.

Secondly is an oscilloscope[6]. These used to be fairly expensive, but have come way down in price, especially in the used market[7]. For metal detectors, you can get by with a basic 2-channel model with a bandwidth as low as 20MHz. These can be had for as little as $50 used[8], generally without probes. New probes can be found on the Internet for as little as $15 each.

Finally, you will of course need components. It's getting more difficult to find local parts houses[9], but far easier to find everything on the Internet. Digikey, Mouser, and Jameco are popular sources in the US; Farnell and Maplin in the UK and Europe. Large "kits" can be found on eBay, such as 1/4 watt resistor kits and capacitor kits. Also other bulk components like diodes and transistors. Be careful buying semiconductors (especially opamps) on eBay and Ali Express as fake devices have become a big problem[10].

Through-hole construction is, by far, the easiest for most home brewers, but through-hole components are slowly getting obsoleted and many new chips are released only in surface mount. However, there are adapter boards available that convert SMT chips to through-hole. The designs in this book tend to use older devices that are still easily found in through-hole packages. Through-hole construction also offers the opportunity to use sockets for components, allowing you to easily swap components in a design to see if one opamp is better than another, or to easily change resistor or capacitor values. It also allows you to recycle the more expensive components in the next circuit.

Designing and building metal detectors is a finicky task. It is quite easy to get a metal detector to operate fairly well on the bench, and many times novice experimenters are excited to discover how easily they can achieve good depth, or discriminate targets. The excitement then evaporates when they put the detector to the ground and see the depth vanish, or find that it produces an audio catastrophe. My rule of thumb is this: once you have a circuit working on the bench, you are 5% of the way done. The remaining 95% is to make it work well in the field.

Acknowledgements

ITMD began as two independent efforts — of myself and George Overton — that turned into a collaboration after meeting on the *Geotech* forums. Although this edition is a solo effort it benefits greatly from the prior contributions of George and in his generous time reviewing new material.

I have also benefited greatly from the online forum discussions that have occurred over the years. It started with a PI Tech forum on Findmall where Eric Foster was generously helping people understand pulse induction techniques. Eric was therefore not only a pioneer in pulse induction technology, but also a pioneer in the on-line proliferation of metal detector knowledge. The *Geotech* web site and forums were created out of my desire to expand his efforts. It is because of this that I continue to dedicate the third edition of this book to Eric. Sadly, Eric passed away in 2022 and his presence is sorely missed.

Third Edition Notes

This is the 3rd edition of *ITMD*. The first edition was one of those efforts where George and I had lofty intentions but tired fingers, so a lot of material got cut just to get the damned thing finished. And in the push to get it published, we overlooked a number of errors, ergo the second edition.

Besides a bunch of corrections (thanks to our astute readers), the second edition cut out the chapters on Long Range Locators and the Pistol Detector. LRLs in general are make-believe devices and don't really belong in a book that deals with real physics; they were included in the first edition because we thought it was important factual information that had never been published and needed to be. For

---

6. Do not think for a minute you will be successful building metal detector circuits without an oscilloscope, A common problem on the *Geotech* forums is people asking for help because they don't have an oscope.

7. Dominated by eBay. In the US there is also Craigslist where I often see oscopes, and Facebook Marketplace.

8. Be very very very careful in buying used test equipment. Many units are sold "as-is" with no warranty and no returns; try to make sure it is listed as working, with a return policy if it does not.

9. No one repairs electronic circuits any more.

10. I once bought some TL071 opamps on eBay, which is a single opamp in an 8-pin package. What I received was labeled a TL071 but was actually some kind of dual opamp.

those interested, the Pistol Detector chapter has been released as a stand-alone PDF, free for download on the *Geotech* web site. The LRL chapter is intended to get rolled into a book of its own.

The third edition of *ITMD* is greatly expanded[11] and has far more technical and theoretical content. Pretty much everything from the first edition has been rewritten and expanded upon, and there is a lot of brand-new material. Despite careful proof reading, I expect there will be a couple of errors that need correcting. Because the book is published on-demand, corrections will be made as needed, the book will be "revision numbered" similar to software, and a list of errata posted on the *Geotech* web site for those who don't want to buy a newer edition just for minor corrections. It is also intended that new material will be added over time.

Additional Resources

As stated previously, this edition of *ITMD* does not include additional support information for the various designs beyond the schematic and design description. Parts list, PCB artwork, and construction and adjustment procedures, and source code will be available on *Geotech*. Start at *geotech1.com/itmd*. This is also the starting point for any other information regarding this book, including errata and interactive discussions.

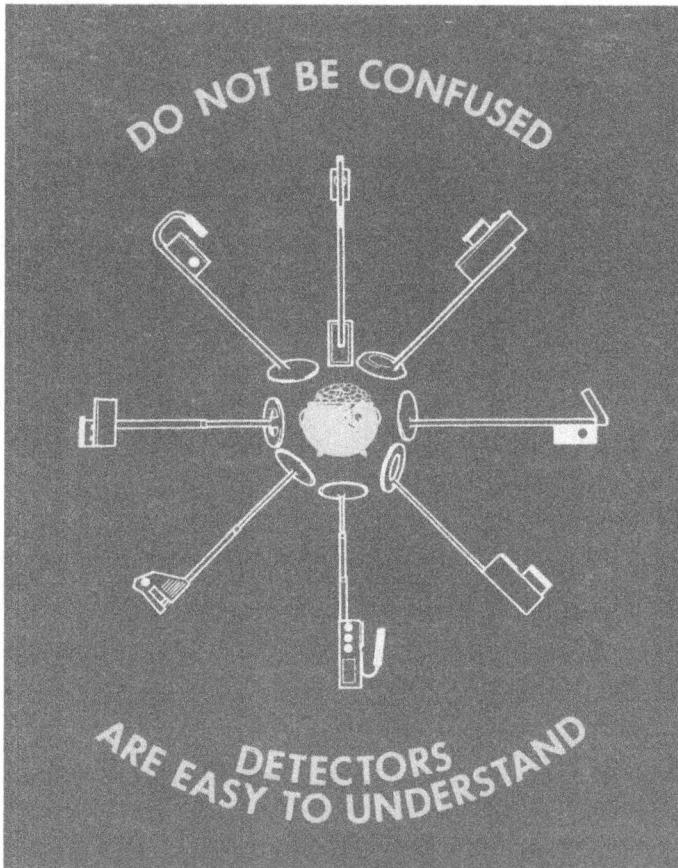

11. The 1st & 2nd editions were 6"x9"; the 3rd edition is 7"x10", a 30% increase!

PART 1
Fundamentals

SCIENTIFICAMERICAN

USING AN INDUCTION BALANCE TO LOCATE UNEXPLODED SHELLS BURIED IN AN ERSTWHILE BATTLEFIELD IN FRANCE. (See Page 425)

Vol. CXIII. No. 20
November 13, 1915

Munn & Co., Inc., Publishers
New York, N. Y.

Price 10 Cents
$3.00 A Year

History

"History is a set of lies agreed upon."

— *Napoleon Bonaparte*

The nineteenth century saw rapid development in the new frontier of electrical science. One area in particular — the relationship between electrical currents and magnetic fields — became the foundation for the development of the metal detector. Although we could dive right in to metal detector circuits, it is interesting to look back just a little bit further, to the scientists who developed some of the underlying principles that enabled this invention, and to the applications that have resulted. Also, the hobby of metal detecting has seen many companies and technologies come and go — we'll take a quick look at some of the ones that have made a lasting impact.

The People

The information on our historical heros will be brief. The reader is urged to seek out more detailed biographies on these fascinating people.

1.1: Hans Christian Øersted

In 1820, Øersted (1777-1851), quite accidentally noticed that the current through a wire caused a nearby compass needle to deflect. He subsequently investigated and determined that the current was producing a magnetic field. His findings sparked widespread interest in electrodynamics.

1.2: Michael Faraday

One person in particular, Michael Faraday (1791-1867), was instrumental in discoveries involving electricity and magnetism. With little more than a primary education, Faraday worked his way up to be the director of the laboratory of the Royal Institute of London. Although not a theorist like many of his peers, Faraday had a keen insight and his experimental methods won him acclaim. It is in his honor that the unit of capacitance is called the *farad*.

One of Faraday's greatest discoveries was the principle of *induction*. It had already been shown that an electric current through a wire produced a magnetic field and therefore Faraday reasoned that a magnetic field should produce an electric current. He showed this to be true, and his discovery directly led to the inventions of the electric generator and the electric motor.

What Faraday found was that when a wire moves through a magnetic field an electric current is produced in the wire. It is also true that when a wire is brought near a *changing* magnetic field a current is produced. The electric current that is developed in the wire will also result in a counter-magnetic field around the wire. This is the very principle used in metal detectors.

1.3: Heinrich Wilhelm Dove

Heinrich Wilhelm Dove's (1803-1879) primary interest was in meteorology but he contributed to various other scientific areas. In 1841 he published an invention called the "differential inductor," basically a 4-coil induction balance. This was effectively the first induction metal detector. Dove used charged Leyden jars to inject a transient current through the primary coils, essentially making it a one-pulse PI design. He would hold the wires of the balanced secondary and noted that when metal was introduced he would receive a shock.

1.4: David Hughes

David Hughes (1831-1900) was primarily a professor of music but also worked in experimental physics, winning numerous awards for his work. He likely discovered wireless radio transmission long before Guglielmo Marconi but Hughes' weakness was his inability to analyze and describe the science and math behind his experiments. In the realm of metal detecting, Hughes worked with and improved on Dove's induction balance. Hughes came up with a 3-coil induction balance which was the forerunner of the modern concentric design used in many T/R and VLF style metal detectors. His induction balance was initially used to investigate the conductive properties of ore samples.

1.5: Alexander Graham Bell

Without a doubt, David Hughes experimented with the effects that metal targets had on his induction-balanced systems. It's very possible that other scientists, perhaps someone we haven't mentioned, did similar experiments. It's even possible that someone intentionally built a crude portable metal detector using this method. But it was a medical emergency that spurred what is considered to be the first applied hand-held metal detector.

In 1881, United States President James Garfield lay dying from an assassin's bullet lodged in his back. Using the horribly unsanitary technique of bare fingers, doctors attempted to remove the bullet but were unable to locate it. The bullet was thought to be critically close to the liver and doctors were personally unwilling to probe too deeply and risk killing the President in order to find the bullet.

Inventor Alexander Graham Bell (1847-1922) decided to volunteer his efforts. On his initial visit to the White House, he found that a Hughes Induction Balance had been sent by a Mr. George Hopkins of New York. Bell took Hughes' idea and merged it with some of the same techniques he developed for the telephone. Bell was successful in building a metal detector, and it could easily detect a bullet held in a clenched fist. He even tested the device on at least one Civil War veteran and showed that it could successfully locate an embedded bullet.

The drawing in Figure 1-1 (from Harper's Weekly) shows Bell and an assistant using the detector on Garfield. Notice in the lower right corner there is a wire going out the door. The (incomplete) inset in

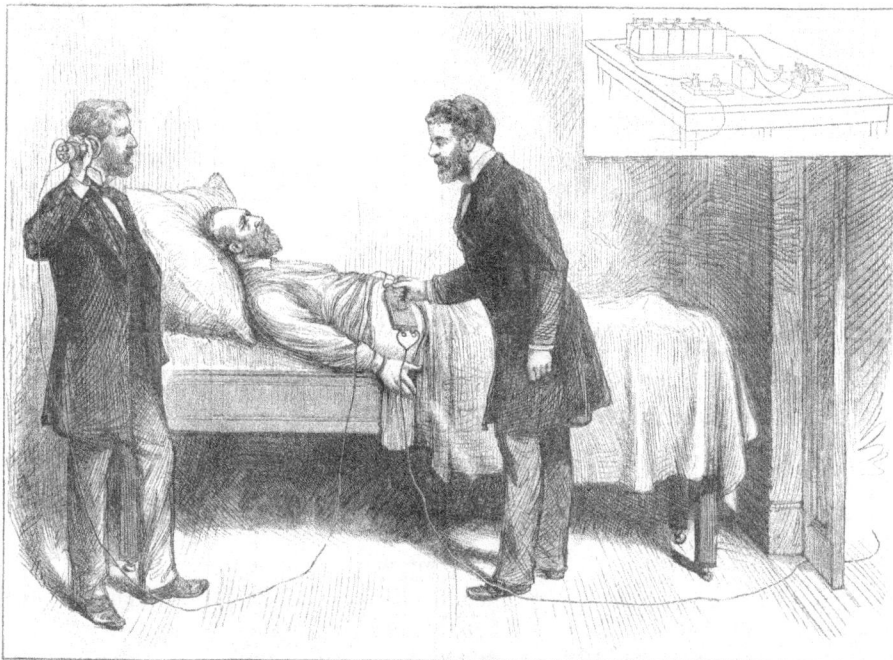

THE WOUNDED PRESIDENT—ASCERTAINING THE LOCATION OF THE BULLET—FROM A SKETCH BY W. SHINGLE.—[SEE PAGE 565.]

Fig. 1-1: **Alexander Graham Bell using his metal detector on U.S. President James Garfield (1881)**

the upper right corner shows the clapper circuit that functioned as the oscillator. The clapper made such a loud racket that it had to be placed in another room so the assistant could listen for faint signals.

When they first tried to use it on Garfield they got readings everywhere. They soon realized Garfield was laying on one of the earliest spring-coil mattresses. More attempts failed because, it turned out, the bullet lodged in Garfield was too deep and he died shortly afterward from an infection caused by the unsanitary efforts of the doctors. It also turned out that the bullet was not in a critical location and Garfield would have survived if they had simply dressed the bullet wound. Bell was distraught over his failure to locate the bullet, and wrote to his wife:

> I feel much disturbed by the result of the Autopsy of the President. It is now rendered quite certain why it was that the result of the experiment with the Induction Balance was "not satisfactory" as I stated in my report -- for the bullet was not in any part of the area explored. This is most mortifying to me and I can hardly bear to think of it -- for I feel that now the finger of scorn will be pointed at the Induction Balance and at me -- and all the hard work I have gone through -- seems thrown away. I feel all the more mortified -- because I feel that I have really accomplished a great work -- and have devised an apparatus that will be of inestimable use in surgery -- but this mistake will re-act against its introduction.

Bell's metal detector design work included experiments with coils which achieve induction balance by partially overlapping them. This same exact method is still used in DD-style coils today. Figure 1-2 shows his original drawings. The top drawing shows a 4-coil IB and the lower one is a 2-coil "OO" type[1] IB; notice the ability to slide the two coils until balance is achieved.

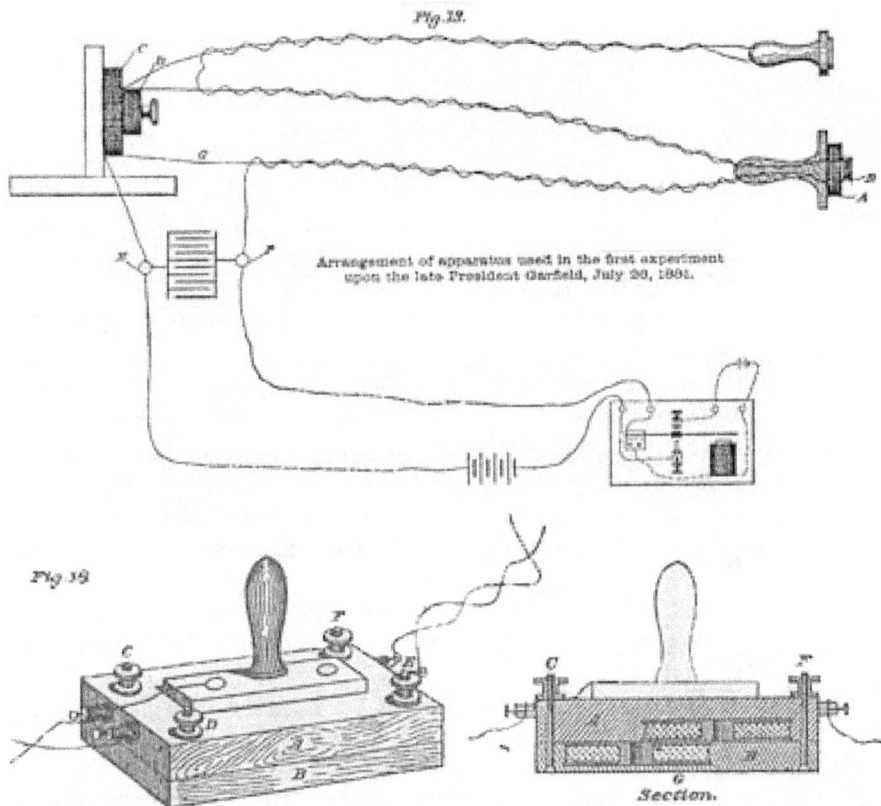

Fig. 1-2: Drawings by Alexander Graham Bell (1881)

1. See 6.2.

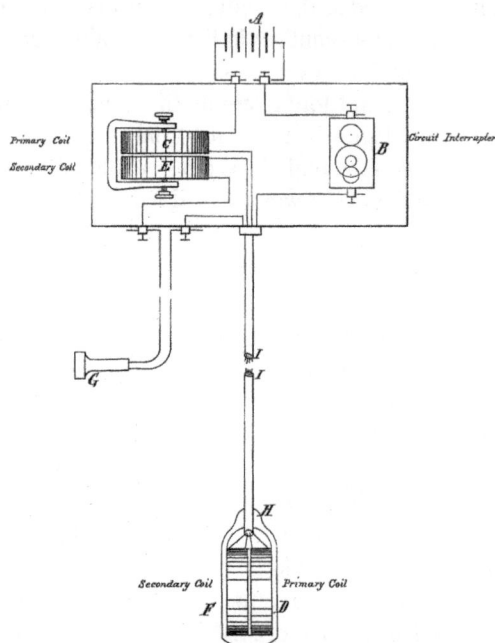

Fig. 1-3: **McEvoy's underwater metal detector (1882)**

1.6: Other Early Inventors

Bell's work with metal detectors is known only because it revolved around a high-profile case. Bell never received — and probably never filed for — a patent for his detector but we can look at the patent records to see the work of other people who might otherwise be unknown.

The first U.S. patent we find (269,439) is from 1882, *Apparatus for Finding Torpedoes*, by Charles Ambrose McEvoy. Interestingly, the patent states that the invention was also patented in England (5,581) in 1881. This puts McEvoy's work coincident with Bell's, so he may have had a working detector prior to Bell. McEvoy's patent illustrates a coil assembly that is lowered into the water (Figure 1-3), so this would probably make it the first underwater detector.

The next patent we run across (1,126,027) is from 1915, *Apparatus for Detecting Pipe-leads or Other Metallic Masses Embedded in Masonry*, by Max Jullig. The interesting things to note from this patent are two illustrations; one shows the coils arranged in an orthogonal configuration, and another shows the schematic equivalence of those coils drawn in a figure-8 style. See Figure 1-4.

A patent from 1924 (1,492,300) is *Means For Electro Aviatic Proof and Measuring of the Distance of Electric Conductive Bodies* by Heinrich Lowy of Austria. This patent describes a detector for ore bodies carried by an airship, such as a Zeppelin. It is interesting in that the design is a type of off-resonance detector and it also describes the possibility of using a secondary oscillator to form a BFO detector.

Shirl Herr was issued a patent (1,679,339) in 1928 for a *Hidden-Metal Detector*. This detector uses an orthogonal coil configuration but, unlike other orthogonal designs of the time, the transmit coil was fully contained within the receive coil.

Two other early patents are *Method Of and Apparatus For Locating Terrestrial Conductive Bodies* (1,812,392 in 1931) by Theodor Zuschlag, and *Electrical Apparatus for Locating Bodies Having Anomalous Electrical Admittances* (1936, 2,066,135) by William Barret and Randolph Mayer. The former has a rather bizarre receive coil setup and the latter shows the detector as a traditional two-box design.

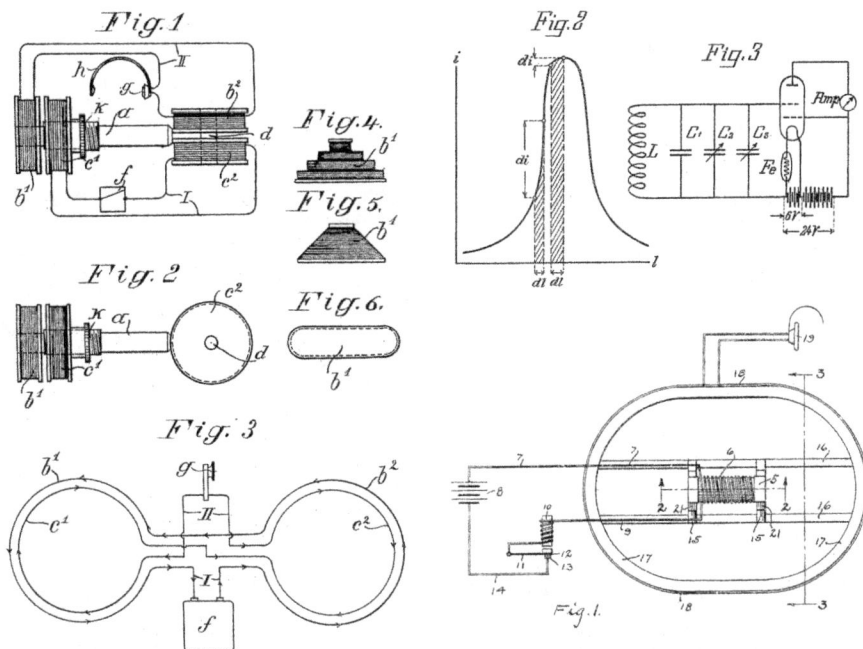

Fig. 1-4: **Max Jullig patent; shows orthogonal coils, plus a figure-8**

1.7: Gerhard Fisher

It was apparently not until 1931 when the first portable metal detectors were widely marketed based on an orthogonal coil design invented in 1925 by Gerhard Fisher[2] (1899-1988). In the 1920s he was working with radio navigation equipment for Naval aircraft when he noticed that the equipment would occasionally register small errors in the directional reading. Subsequent investigation led Fisher to the discovery that it was due to large metal objects, such as buildings.

He received U.S. Patent 2,066,561 for his "Metalloscope" in 1937. Though this was a year after the Barret and Mayer two-box patent above, Fisher actually filed his patent in 1933, a year prior to Barret and Mayer. It is often stated that Fisher received the first metal detector patent but even discounting the Barret/Mayer patent this is an erroneous claim.

In 1931 Fisher left his Naval work and founded Fisher Research Laboratory for the purpose of developing his own commercial detectors. His earliest designs were two-box locators in which the coils are inductively balanced in an orthogonal configuration. Figure 1-5 shows an early Fisher, the Model *47B*, which is constructed entirely of wood. In the 1950s and 60s, Fisher *T-10* and *T-20* models were very popular in the emerging hobby of metal detecting. Fisher is considered by many to be the father of the modern metal detector and, today, the Fisher brand lives on as part of First Texas Products.

1.8: Later Contributors

Many other people have been instrumental either in the development of detectors or in the promotion of the industry. Two of the earliest promoters in the hobby market were Robert Gardiner and Kenneth White, both of whom began in the 1950s. Gardiner developed and sold a number of innovative designs including an early discriminating detector. White started out producing the *Oremaster* Geiger counter during the uranium craze of the 1950s but switched to metal detectors in 1959 when the US government stopped buying uranium. White's Electronics was a prominent name in hobby detectors until they closed operations in 2020.

Bill Mahan, a developer of BFO detectors and founder of D-Tex, was a prominent figure in the early days of treasure hunting. Charles Garrett likewise began with BFOs in the early 60s, and Garrett

---

2. Fischer was his German name; he adopted Fisher as an American.

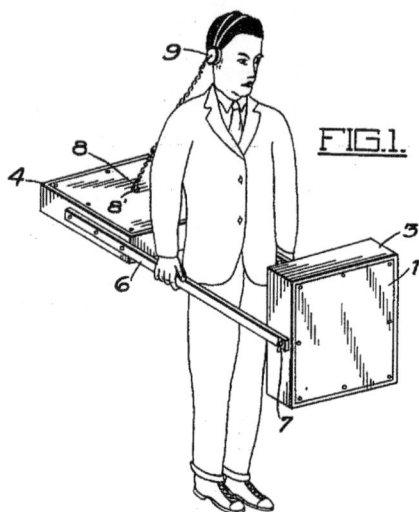

Fig. 1-5: **Fisher patent drawing (1937); author with Fisher Model 47B**

Electronics is now one of the largest detector companies in the world, especially in the security sector. Both men were strong ambassadors for the hobby.

It is uncertain who invented basic target discrimination in metal detectors. The method widely used for discrimination — determining phase using synchronous demodulation — is easily found in patents back to the early 1960s and even before, though largely applied to other types of instruments. Basic TR discrimination first showed up in hobby detectors in the early-mid 1970s, in the Technos *Phase Read-out Gradiometer* and in models produced by White's and others.

George Payne is widely recognized as one of the most significant contributors to VLF metal detector technology, having invented the ground canceling motion discriminator, tonal target ID, and visual target ID. In pulse induction, Eric Foster was a leading developer from the 1960s and some of his designs were instrumental in the discovery and recovery of Spanish shipwreck treasure, such as with Mel Fisher and the *Atocha*. Bruce Candy has made tremendous contributions to both multifrequency VLF and PI technologies which are found in Minelab detectors. David Johnson was a prolific developer who designed some of the most popular detector models for Fisher, Tesoro, and White's.

Other noted contributors to metal detector design include Bob Podhrasky (Garrett), Jack Gifford (C&G, Tesoro), Rick Maulding (White's), Mark Rowan (White's), John Earle (Compass, White's), Alan Hametta (A.H. Electronics), Dick Hirschi (White's, Compass, Teknetics), and Jim Karbowski (loop design), among countless others.

The Applications

The first known application of a metal detector, as described in the section on Alexander Bell, was medical in nature. Let's look at some other areas where metal detectors are used.

1.9: Military Use

Even before Fisher began building metal detectors, they were being used by military forces searching for unexploded ordinance. A Popular Mechanics article from Feb, 1916 shows two operators using a detector, described as an induction balance design, to search a French field for unexploded ordinance during World War I. Likely this is in reference to the Maxwell bridge bomb detector that M.C. Gutton of France experimented with.

By the second World War mine detectors were in widespread use. The SCR-625 was developed by Hazeltine Corp. in 1941[3] and introduced to troops in 1942. During Vietnam, the AN/PSS-11 mine

BURIED SHELLS FOUND BY INDUCTION BALANCE

This Instrument Detects the Presence of Shells or Other Metal Buried in the Soil, Thereby Enabling the Farmer to Remove Them before Tilling a Field

Fig. 1-6: **UXO detection in WW1 (Popular Mechanics, 1916)**

detector was popular. It used an unusual 4-over-1 induction-balanced coil scheme. Before the wide-spread availability of commercial detectors treasure hunters used surplus mine detectors which are still widely available on the used-surplus market, though they are now only useful as collector items.

1.10: Commercial & Industrial Use

In manufacturing, metal detectors can be used in robotic assembly as position sensors for accurately placing objects. In the food industry, detectors are used to ensure no foreign metal objects get into the final product. In logging, detectors look for embedded metal objects such as nails that could damage cutting tools.

Security is a rapidly growing market due to theft and threats to safety. Walk-through metal detectors are widely employed in high-risk security areas such as airports, government buildings, and prisons. Most walk-throughs are pulse induction designs but newer tech-

Fig. 1-7: **Early Walk-Through (Radio News, 1926)**

---

3. See US2451596, which is the patent for the SCR-625 search coil and possibly the first concentric design. The inventor is Harold Wheeler, also known for the Wheeler formula used to calculate coil inductance.

nologies such as terrahertz imagers that can "see" non-metallic weapons are gaining popularity. Businesses are using walk-through detectors in an effort to stem theft, both from customers and employees. This application has been around longer than you might think — the April, 1926 issue of Radio News describes an early walk-through detector being used at a company in Germany to prevent employee theft; see Figure 1-7.

1.11: Treasure Hunting

There are numerous magazine articles throughout the 1920s, 30s, and 40s covering the application of new metal detection technology to both treasure hunting and ore prospecting. Many of these applications use devices which were quite large and required two people to operate. During these early years of electronic detection treasure hunting was largely relegated to the few who could either buy or build the instruments and could then figure out how to make use of them.

It was not until post-World War II that the availability of handheld surplus mine detectors brought treasure hunting to a wider audience. Even though the mine detectors were handheld and a one-man operation they were still heavy and unwieldy. It was companies like Detectron and Fisher who, in the early 1950s, introduced more reasonable consumer detectors, although still using vacuum tubes.

With transistorized circuitry in the 1960s low-cost lightweight metal detectors began to emerge specifically for the consumer market. Unlike Fisher's design 30 years prior, most of these units were simple BFO detectors which could only discern between ferrous and non-ferrous targets. There were some TR machines but they

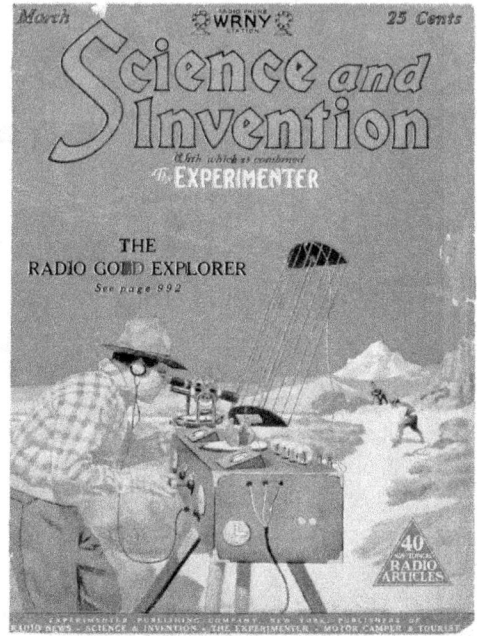

Fig. 1-8: **Science & Invention March 1926**

were generally more expensive and, while exhibiting some impressive air test depths, had severe problems with mineralized ground. With some of the cheaper detectors selling well below $100[4], treasure hunting became a hobby for the masses.

Initially, the major players in the consumer market were D-Tex, Garrett, Relco, and White's. Because the BFO was such a simple design, it was not long before new start-ups entered the market. Most of these companies did not have the research and development teams needed to produce new innovations that competition ultimately demands and so many of them were quickly overtaken by the more established manufacturers.

The introduction of integrated circuitry, in particular the operational amplifier (opamp), made the TR design much more attractive to develop. Like the BFO, early TR detectors could only differentiate between ferrous and non-ferrous targets. But in 1974 another breakthrough technology was introduced by Technos called the *Phase Readout Gradiometer*[5] which was a TR discriminator. This new circuitry looked at the phase shift of the received signal and, using a threshold level set by the user, either signaled the target as good or ignored it as undesirable. The discovery that led to this new feature was that different metals — aluminum, nickel, copper, gold, silver, etc. — exhibited slightly different phase

---

4. A $49 detector in 1965 is equivalent to $491 in 2024 dollars. So, not all that cheap.

5. The PRG wasn't really the first development of a TR discriminator. The patent for the PRG is US3826973 which was filed in 1973. Several other patents predate this one, notably US3020470 was filed in 1943 and awarded in 1962, so the idea of phase discrimination clearly dates back to the 1940s. However, the PRG was the probably the first discriminator in the hobby metal detecting market. It was very expensive ($850, or $5,400 in 2024 dollars), and only about 50 were made.

shifts in the return signal due to differing conductivities. By measuring this phase shift the detector could discriminate between targets. This new feature was a two-edged sword, however, as some desirable targets — rings and nickels — appeared to have a similar response as aluminum pull tabs. Thus discriminating out the "trash" could also mean losing some of the treasure. However, in areas with heavy concentrations of buried bottle caps and pull tabs the trade-off was well worth it, especially in the days when there were so many silver coins to find.

TR detectors normally operate at a 100kHz or less, called the low frequency range (LF), while BFOs often operated above 100kHz. The lower frequencies yielded better soil penetration. By lowering the frequency even further to just a few kilohertz — called the very low frequency range (VLF) — designers found that they could achieve one more desired result, that of ground cancelation. With both BFO and TR detectors ground mineralization creates a false signal. Ground mineralization in most places is fairly consistent over a given search area so the user can retune the detector once the search coil is lowered near the ground. However, any variation in the height of the search coil above the ground will result in a change in signal, a very annoying phenomenon.

Some detectors up to the mid-70s included both a ground-cancel mode and a discrimination mode, but the two modes could not operate at the same time. The next technological breakthrough occurred in 1977 when Bounty Hunter introduced the world's first VLF motion discriminator, the *Red Baron*. Analog filter techniques allowed users to both ground cancel and discriminate at the same time, as long as the search loop remained in motion. In the case of the *Red Baron* and other first-generation motion discriminators, the search loop had to be moved quickly, with almost a whip-like motion. Later developments (lead by Fisher and Tesoro) allowed slower loop motion.

The Technologies

1.12: BFO

The beat-frequency oscillator was invented in 1901 by Reginald Fessenden. Its early use included the demodulation of radio signals, especially the on-off keyed modulation of radio-telegraph signals. Since the BFO was invented before the vacuum tube amplifier, the earliest BFOs were implemented with a motor-driven mechanical wheel.

Possibly the earliest application of the BFO to metal detecting was in 1924 when Daniel Chilson invented the *Chilson bridge*. A treasure hunt in Panama using a Chilson bridge device was reported by the New York Times in 1927, whereby pirate treasure was recovered. This may be the first documented case of finding treasure with an electronic locator.

While induction balance was the dominant technology throughout the first half of the 20th century, BFO surged during the 1960s and 70s when transistors made for simple and cheap designs. Today, the BFO has completely vanished from the market. Although it has run its course commercially, the simplicity of the BFO makes it a good beginning project. Also, the BFO belongs to a larger family of *proximity detectors* which still show up in some pinpointer designs.

1.13: IB

Induction balance was used in the very first metal detectors and continues to be the dominant technology today. The earliest designs predated vacuum tubes and instead used mechanical clappers to create an AC transmit signal (Figure 1-2). In the 1920s and 1930s, commercially-produced detectors using vacuum tube circuitry became commonly available.

The move to transistors briefly gave the BFO detector the advantage as their mono coil and simpler circuitry made them far cheaper to build. But ultimately the BFO could not perform as well as an IB detector and the move to integrated circuits (especially opamps) shifted the advantage back to IB. Furthermore, the addition of synchronous demodulation gave IB detectors capabilities in canceling ground and discriminating targets that BFO could not match.

Although the fundamental operational theory of the induction balance has not changed in over 140 years the details have changed tremendously. The newest IB detectors are computerized, can ignore ground mineralization, and offer both tonal and visual target identification. Not to mention a much lighter weight and more ergonomic design.

1.14: PI

A couple of early papers to describe pulse induction technology were "A Method of Detecting a Mass of Non-Ferrous Metal Located at Depth in the Earth" by J.H. Wescott in 1955 and "A Pulsed Bomb Locator" by F.B. Johnson in 1956. But other papers (even back to 1884[6]) showed research into magnetic field step responses and a 1942 patent (US2278506) shows a pulse induction metal detector using a motor driven cam to actuate the TX and RX switches.

Possibly the earliest commercial PI detector was designed by Eric Foster at Oxford University and produced by Elsec in the 1960s. Foster also produced his own models in the Goldscan series and other models mostly tailored to gold prospecting. In the 1980s White's brought PI to the masses with the PI-1000 and the Surfmaster PI. Other manufacturers quickly followed.

In 1995 Minelab introduced the *SD2000* PI detector for prospecting. It began a long run of innovative PI designs (mostly from Bruce Candy) that currently dominate the prospecting market. Minelab, along with other companies, also manufactures PI-based mine detectors for military and humanitarian demining.

The Companies

There have been countless detector companies in the last 100 years, especially in the broader scope of military and industrial applications. This book primarily focuses on hobby detectors and even in that market companies came and went in the blink of an eye. As mentioned before, the transition of technology from BFO to IB and subsequent technological developments left many of the early hobby-oriented companies unable to compete, and the brief history of the hobby detector is littered with names that have come and gone. The companies considered here were long-lived, had a broad impact on the hobby market, or brought unique developments to metal detecting.

1.15: Fisher

Fisher Research Labs (FRL) was one of the earliest commercial manufacturers of metal detectors, created in 1931 by Gerhard Fisher. Initially it focused on two-box style locators, but in the early 1960s produced a popular TR model called the *T-10*. The tube-based *T-10* was redesigned into the transistorized *T-20*, and other handheld models followed. To my knowledge, Fisher never made a BFO detector. During the 1950s and 60s Fisher also made geiger counters and magnetometers and eventually branched out into utility locators including line tracers and acoustic leak listeners.

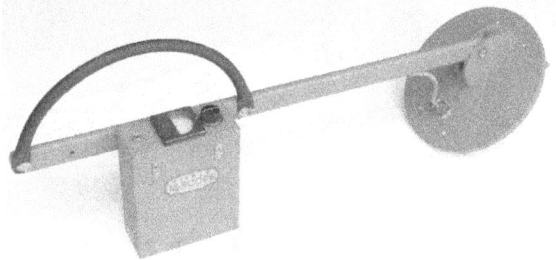

Fig. 1-9: **Fisher *T-10***

By 1978 Fisher joined the motion-VLF market with the *550*-series detectors[7] which had a strikingly attractive gold design. As with many similar C-rod designs, they were heavy and hard on the wrist for all-day hunting. In 1985, the *1210X* model was released with a new S-rod design and was followed by several more *1200*-series models[8]. These included the *1280x*, a

Fig. 1-10: **Fisher *555-D***

6. "On the Induction of Electric Currents in Cylindrical and Spherical Conductors," Horace Lamb

7. Partly designed by Jack Gifford.

scuba-rated VLF detector that, due to its low frequency of 2.4kHz, was somewhat usable in salt water.

In 1991 Fisher produced a 2-frequency detector called the *CZ-6*. This was a few months after Minelab released the *Sovereign*, their first 2-frequency detector, so while Minelab edged out Fisher in the first-to-market it is clear that Minelab and Fisher were developing 2-frequency technology at the same time. The CZ series was expanded to include the *CZ-7*, a model with digital target ID and LCD readout, and the *CZ-21*, another Scuba-rated model.

Fig. 1-11: **Fisher *F75***

Gerhard Fisher retired in 1967 and died in 1988. Fisher continued to serve the hobby market though they became a weaker competitor over time. In 2006 Fisher was purchased by First Texas Products who revamped the hobby product line with new models including the highly successful *F75*.

1.16: White's

White's Electronics was founded in 1950 by Kenneth White in Sweet Home, Oregon beginning in the basement of his furniture store. At that time uranium prospecting was popular and White decided he could build a better Geiger Counter. For several years he produced the *Oremaster* geiger counter in several models. By 1960 the U.S. government stopped buying uranium from private parties which killed the uranium prospecting boom.

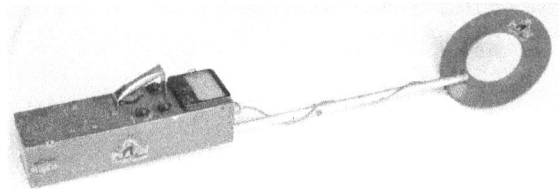

Fig. 1-12: **White's *M-60***

White's was already transitioning to metal detectors, having produced its first BFO model in 1958 called the *M58*. Throughout the 1960s White's made various models of *Goldmaster* detectors in BFO and eventually TR technologies, starting with vacuum tube circuits and migrating to transistorized circuits. In the early 1970s, White's produced a popular TR series called the *Coinmaster*, and the *Coinmaster IV* in particular sold in large numbers.

Fig. 1-13: **White's *6dB***

White's was also an early adopter of VLF technology, producing the *Coinmaster V* in 1975. When George Payne invented VLF motion discrimination in 1977, White's was quick to jump on board and made the *Coinmaster 6dB* (1977) and *6000/D* (1978). The *6000* went through several iterations over the next 27 years and was hugely popular for the company. It culminated in the *XL Pro*, the last all-analog detector White's produced.

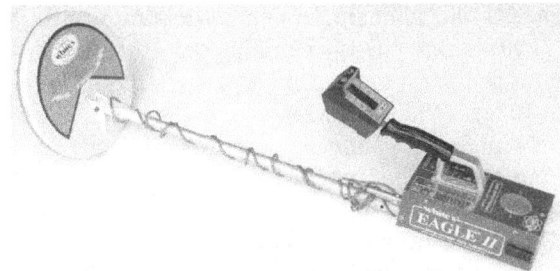

Fig. 1-14: **White's *Eagle II***

White's introduced the first microprocessor-based detector in the *Eagle* model. The *Eagle 2* followed, then the *Spectrum*, *Spectrum Eagle*, and *Spectrum XLT* (1994). The *XLT* was also highly popular. In 2001 they released the *DFX* (dual-frequency XLT) and eventually the 3-frequency *V3* in 2008.

---

8. Many of the 1200-series models were designed by David Johnson.

During the 1980s and 90s White's was perhaps the dominant metal detector brand but by the 2000s they were fading as Minelab began rising in popularity. A key element in their demise was the refusal to adapt to the changing landscape of Internet-based sales and big-box stores, insisting instead to stay with a distributor/dealer sales model. In 2020 White's shut down and sold their assets to Garrett.

1.17: Garrett

Charles Garrett was unimpressed with early BFO detectors and their excessive drift. He thought he could design a better detector, and in 1962 started Garrett Electronics. Garrett did what he set out to do, and for almost 20 years produced the best BFO detectors on the market. His designs found favor with popular authors Karl von Mueller and Roy Lagal, whose books often featured Garrett models.

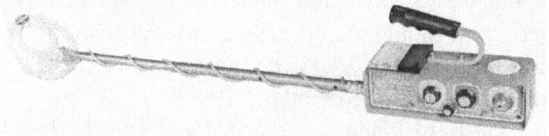

Fig. 1-15: **Garrett *Master Hunter* BFO**

Garrett successfully made the transition to TR and VLF designs, and their *Groundhog* models of VLF/TR-discriminators (1975) were popular due to their light weight and good depth. The *Master Hunter ADS-III* and later versions were VLF motion discriminators popular with cache and relic hunters. The *GTI* series (1996) added modern features like target ID and LCD displays.

Fig. 1-16: **Garrett *Groundhog***

In 1984 Garrett entered the security detection market with both wands and walk-throughs. Their *PD6500* became a very popular model that is often seen in government buildings and large sports venues. At least for a while, security metal detector sales were a majority of Garrett's business.

In the 2000s Garrett struck triple-gold with three detectors. First, the *Ace 250* (2006) became the low-cost leader with features normally found only in more expensive detectors. It spent a decade or more as the most popular entry-level model. Then Garrett introduced the *Pro-Pointer* model pinpointer (2008) which instantly became the standard for competitors to beat. Finally, the *AT-Pro* (2010) took a good-performing detector design with popular features and put it into a completely waterproof package, submersible to 10 feet. While this wasn't the first for such a detector (Garrett *AT-4*, White's *Beachhunter*, Tesoro *Tigershark*, etc.), it was the first time a waterproof detector included all the features normally found in a typical land detector, including a keypad and LCD interface. The *AT-Pro* became a sensation and had competitors scrambling to come up with a their own waterproof models, something that is not as easy at it seems.

Charles Garrett died in 2015 but the company is still family-owned. Unlike what happened with Tesoro, Garrett maintains an active engineering department and continues to innovate. In fact, Bob Podhrasky, an engineer who joined Garrett in 1970 and was

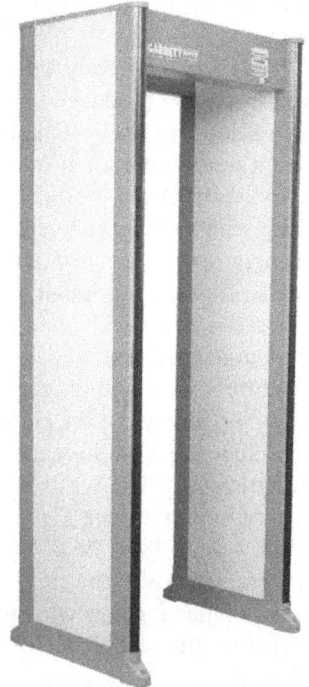

Fig. 1-17: **Garrett *PD6500***

responsible for much of their TR and VLF developments, is still managing the technical side of the business. That gives Podhrasky over 50 years at one detector company, surely a record, and is likely a strong reason for Garrett's longevity.

1.18: Bounty Hunter

Pacific Northwest Instruments was formed in 1970 in Klamath Falls, Oregon. The first models were the *Bounty Hunter I, II,* and *III* which were BFO designs. Eventually, the company adopted "Bounty Hunter" as their brand name and they produced very popular TR and TR-discriminators[9] and even a unique BFO-TR hybrid model called the *Outlaw*.

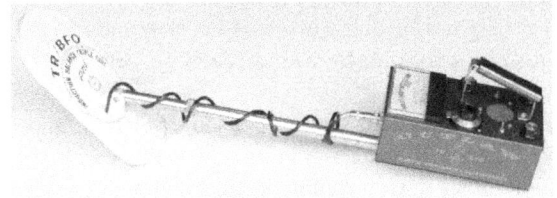
Fig. 1-18: **Bounty Hunter** *Outlaw*

In the 1970s it was sold to Space Data Corporation and moved to Tempe, Arizona. In 1976 George Payne left White's to work for Bounty Hunter, where he developed the VLF motion discriminator and Bounty Hunter released it as the *Red Baron.* This was a major milestone in metal detector design and for a year or two Bounty Hunter sold all the *Red Barons* they could possibly build while other companies scrambled to compete.

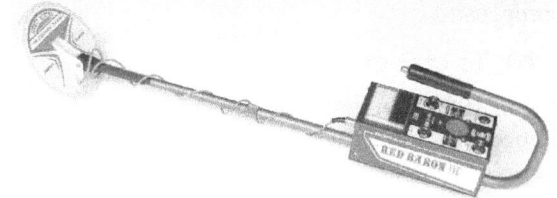
Fig. 1-19: **Bounty Hunter** *Red Baron*

Metal detectors were not a primary business for Space Data and after Payne left they did not invest much in development. In 1979 Bounty Hunter was sold to Ray Smith who then sold it in to Teknetics[10] and the company moved back to Oregon (Lebanon) until Teknetics went bankrupt. Both the Bounty Hunter and Teknetics brands were purchased by Techna[11] in El Paso where Bounty Hunter detectors continued to be produced. However, quality was dismal and the Bounty Hunter brand became poorly regarded in the metal detecting community.

In 1990 Techna sold Bounty Hunter and Teknetics to First Texas Products, also in El Paso. FTP corrected the quality problems, introduced many new and redesigned products, and then aggressively marketed the Bounty Hunter line to big-box stores and online outlets. For many years Bounty Hunter was the largest-selling (volume-wise) metal detector brand.

Fig. 1-20: **Bounty Hunter Lone Star Pro**

1.19: Compass

Compass was founded in 1970 when Don Dykstra left White's Electronics and joined up with Ron Mack, who became Compass' president. Compass (of Forest Grove, Oregon) was an early innovator in TR and TR discrimination, and the *Yukon* series (especially the *77B*) was very popular in the 1970s.

Fig. 1-21: **Compass 77B**

The later *Magnum* series combined VLF with a TR-discriminators. The *Coin Magnum*, in particular, was a unique no-motion VLF discriminator. Compass also produced more conventional VLF motion discriminators with the *Challenger* series, and its final and very capable line was the *Scanner* series. Many of the *Challenger* and *Scanner* models still

---

9. Several of their popular TR-discriminators were design by Jack Gifford.

10. Ironically, Teknetics was partly created and owned by George Payne.

11. Techna was owned by John E. Turner, the "JET" in the Jetco brand of detectors. Turner also created Ranger Security Detectors which manufacturers walk-through type models.

---

bring premium prices on the used market, and their early TR detectors (like the *77B*) remain a specialty unit in some arsenals because their high operating frequency was effective in ignoring nails.

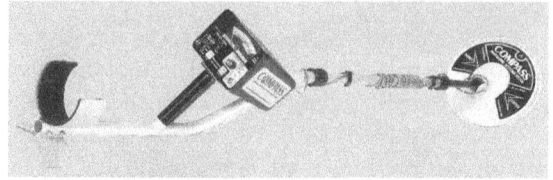

Fig. 1-22: **Compass Coin Scanner**

Compass had some very competitive and well-respected detectors in its day. But, as with so many other detector companies, mismanagement caused its downfall. The company was already struggling in 1994 when the factory mysteriously burned to the ground and brought Compass to an abrupt end.

1.20: Teknetics

In 1983 George Payne left White's (for the second time) and — with Gary Morris, Bill Wend, and Sloan Smith — founded Teknetics in Lebanon, Oregon, just 15 minutes from the White's facility in Sweet Home. Right away Teknetics produced some innovative detectors, with their initial products introducing metered target identi-fication and the first LCD in the *9000* model.

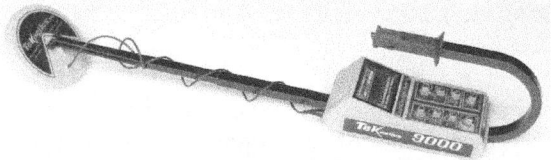

Fig. 1-23: **Teknetics 9000**

Teknetics also included tone ID in the *Mark 1* model which is still highly prized by Teknetics fans. As with Compass, poor management lead to its demise. Teknetics and Bounty Hunter were bought out of bankruptcy by Techna of El Paso, Texas.

1.21: C&G/Tesoro

While Bounty Hunter was based in Tempe, Arizona they hired an engineer named Jack Gif-ford. Jack and fellow employee Ray Crum left Bounty Hunter in 1976 and founded a detector company called C&G Technologies. C&G pro-duced the "Cat" line of VLF detectors culminat-ing with the *Wildcat* model. The C&G models had good discrimination and were well-liked but the use of a coaxial search coil design resulted in less-than-stellar depth.

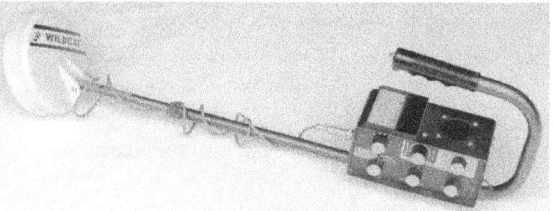

Fig. 1-24: **C&G Wildcat**

Gifford parted ways with Crum and, after brief stops at Fisher and again at Bounty Hunter, in 1980 he created Tesoro Electronics in Prescott, Arizona. Gifford focused on so-called "2-filter" VLF discrim-inators which reduced by half the number of analog filter stages and did not require the "coil whipping" motion needed for decent depth with the 4-filter designs. Tesoro models were known for their outstand-ing discrimination and became favorites with British field hunters and American relic hunters.

Early Tesoro models (*Deep Search*) were a typical C-rod configuration, but they quickly migrated to a more ergonomic post design. In 1996 Tesoro struck ergonomics gold when they produced the *MicroMax* S-rod design at only 2.5 pounds (1.14 kg). To do so, they moved the cir-cuitry to surface mount technology, one of the first companies to do so.

Tesoro detectors gained a reputation for extreme lightness and exceptional performance.

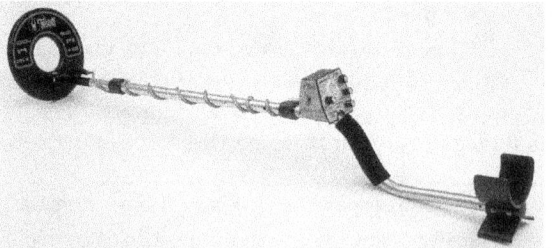

Fig. 1-25: **Tesoro Bandido uMax**

However, most models were "beep & dig;" they could discriminate, but did not perform target ID. There was a model with tone ID (*Golden*) and even a few with displays (*Toltec*, *Cortes*, *DeLeon*) but while those technologies were universally adopted by other companies they never seemed to be favored by Tesoro. Tesoro was also different in their distribution, preferring to sell through individual dealers instead of mail order companies like Kellyco.

Jack Gifford retired in 2004 and died in 2015. His sons James and Vince continued the company but James departed in 2010. Whether it was a lack of passion or some other issue, after Jack's death Tesoro simply coasted on the designs they had. In a last-ditch effort, they began selling through Kellyco but by then other companies had far surpassed them in performance and features and had equaled Tesoro's ergonomics. Though still favored by some of the old-timers, it was not enough to keep the business going and in 2018 they ceased operations.

1.22: Minelab

Minelab of Australia was founded in 1985 after Bruce Candy was asked to design a metal detector capable of handling the harsh ironstone soils found in Australia. Their initial detector models — beginning with the *Goldseeker 15000* — were specifically designed just for that purpose and did it well.

Fig. 1-26: **Minelab Goldseekers 15000**

In 1991 Minelab introduced the first multifrequency hobby detector, the *Sovereign*, which transformed the hobby as much as the Bounty Hunter *Red Baron* did 14 years prior. The Sovereign was renowned for both depth and the ability to deal with the harshest soils and became immensely popular with relic hunters. Minelab has continued to dominate the multifrequency market with the *Explorer* series, *Etrac*, *CTX3030*, *Equinox* and *Vanquish* models, and the *Manticore*.

In 1995 Minelab released the *SD2000*, their first PI design for prospecting. It was an instant

Fig. 1-27: **Minelab Sovereign**

hit and ushered in a market niche that Minelab has almost uniquely owned since then. The *SD* series was followed by the *GP/GPX* series and the *GPZ 7000* model. Minelab PI detectors have sold phenomenally well in the emerging gold prospecting markets, especially in Africa.

1.23: XP

XP Detectors of France was started by Allain Loubet in 1998. Loubet is both a detectorist and an electrical engineer and was not satisfied with the available detectors at the time. His first designs (*Adventis*, *GoldMaxx*) were good and later versions featured the first integrated wireless headphones. But it was the introduction of the *Deus* that made XP a formidable competitor. The

Fig. 1-28: **XP Deus**

Deus was the first detector to use a wireless coil which resulted in a very sleek and lightweight design. As a bonus it had one of the fastest recovery speeds on the market. In 2021 XP released the *Deus II*, a waterproof multifrequency detector that is still wireless.

1.24: Nokta

Turkish metal detectorist Muzaffer Önlek started Makro (2001) and Nokta (2003). Originally, Makro did the product development and Nokta did the product distribution. Önlek's younger brother, Mehmet Önlek, joined the company soon after to run the Nokta side of the business. A few years later, Mehmet began producing his own models under the Nokta label and the two brands eventually began overlapping and competing with each other.

In 2014, Muzaffer decided to depart the metal detector business and sold out his share to Mehmet, who then combined the companies into Nokta-Makro. In 2023 the name was shortened to just Nokta. Although a relative newcomer to the global market, Nokta has made a name for itself

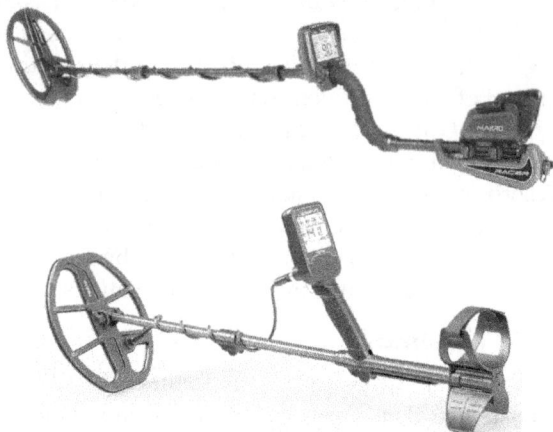

Fig. 1-29: **Makro Racer & Nokta Legend**

by producing good performing detectors with excellent feature sets at low prices. In a relatively short time they have managed to put out models that easily compete with the best on the market, currently culminating with their *Legend* multifrequency model.

1.25: Others

There are other detector companies, some still active and many that have come and gone. Some notables are:

- Detectron — An early (1950s) competitor to Fisher with TR designs. Detectron is still in business and sells a two-box utility detector.
- Metrotech — A 1960s producer of the popular and sensitive *Model 220* TR detector.
- Gardiner Electronics — Beginning in the 1950s Bob Gardiner produced a variety of models using BFO, TR, and off-resonance.
- Jetco — A prominent name in mail-order and department store detectors in the 1970s. Most of them were BFO (the most popular being the *Treasure Hawk*) but there were a couple of TR units.
- Relco — Another prominent mail-order name with ads in practically every male-oriented magazine of the 1960s and 70s. All BFO.
- A.H. Electronics — Allen Hametta produced the *A.H. Pro* and other models, all of which were off-resonance designs. Hametta likely produced the first commercial pinpointer.
- Wilson-Neuman — Paul Wilson and Chuck Neuman produced VLF detectors in the 1970s and (with just Wilson) the 1980s. Popular models were the *Daytona* and the *ATD*, and were known for exceptional all-metal depth.
- C.Scope — A British company founded in 1975 and still in business. Over the years it has produced TR, VLF, and PI designs and, despite not being a top-tier name, has remained competitive. A specialty seems to be non-motion VLF detectors.
- Discovery Electronics — Founded by ex-White's employees including George Payne, Discovery made the *Baron* series of detectors which could be upgraded with plug-in modules. Discovery also made the *TF-900* cache locator, which they also produced for both White's and C.Scope.
- DetectorPro — Started in 1996 by Gary Storm, DetectorPro specialized in diver detectors in which all the electronics are inside the headphone, called the *HeadHunter* series. DetectorPro no longer makes metal detectors but focuses on headphones and other accessories.
- D-Tex — Founded by Bill Mahan and produced early BFO models. The company was later resurrected for a short while by Bill Mahan Jr. and produced some TR and VLF models that were carried over from the demise of Gold Mountain.

- Gold Mountain — Founded by Phil Storck, Gold Mountain produced some well-received TR and VLF models that were designed by Bill Mahan Jr.
- AKA — A Russian company currently making several cutting-edge VLF and multifrequency models.
- Rutus — A Polish company currently making several cutting-edge VLF and multifrequency models.
- Heathkit — Heathkit made do-it-yourself kits for all kinds of electronics (even televisions) and had a few metal detector designs, including off-resonance, TR, and VLF.
- Nautilus — In the 1990s Jerry Tyndell produced a line of VLF detectors especially popular with US Civil War relic hunters. They were notable for their extreme detection depth.

ANNUAL EXPERIMENTERS' NUMBER

August

Science *and* Invention

FORMERLY
ELECTRICAL
EXPERIMENTER

25 cents

HOW TO BUILD
A BURIED TREASURE
FINDER
See Page 335

TELEVISION   RADIO'S GREATEST MAGAZINE   RADIOVISION

FEBRUARY
25 Cents
Over 200
Illustrations

RADIO
NEWS

WRNY

RADIO PROSPECTING
SEE PAGE 716

EXPERIMENTER PUBLISHING COMPANY, 230 FIFTH AVENUE, NEW YORK

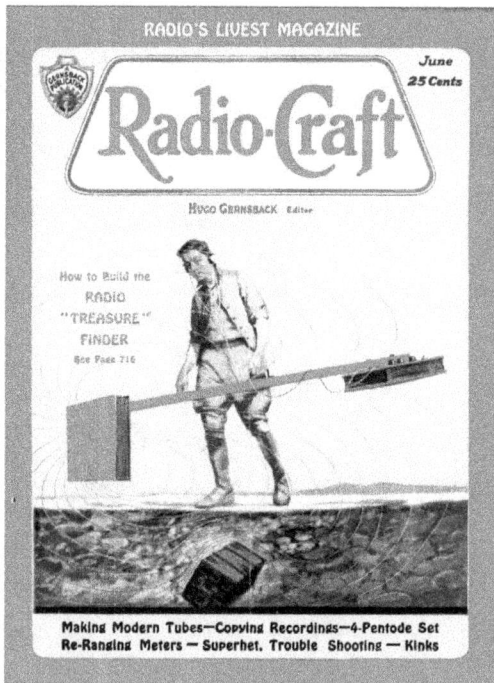

RADIO'S LIVEST MAGAZINE

June
25 Cents

Radio-Craft

HUGO GERNSBACK *Editor*

How to Build the
RADIO
"TREASURE"
FINDER
See Page 716

Making Modern Tubes—Copying Recordings—4-Pentode Set
Re-Ranging Meters — Superhet. Trouble Shooting — Kinks

RADIO'S LIVEST MAGAZINE

July
25 Cents
Canada 30¢

Radio-Craft

HUGO GERNSBACK *Editor*

Building and Operating
the New
"TREASURE"
FINDER
See Page 8

New Tube Data—Copper-Oxide Rectifiers—Building an Auto-Radio Set
Point-to-Point Capacity Testing—How to Make a Good Cutting Head

2

Types of Detectors

"It was all too complicated and, where it is too complicated,
it meant that someone was trying to fool you."

— *Terry Pratchett (The Fifth Elephant)*

In writing a book on metal detector technology, we inevitably need to deal with the terminology surrounding the different "types" of metal detectors. We've all heard of "VLF" detectors but what, exactly, does that mean? What does TR, IB, PI, and BFO really mean? What is a multifrequency machine? Or a time-domain detector? Many of these terms have been misused over the years, sometimes for marketing purposes.

We will divide detectors into three broad categories: proximity effect, induction balance, and pulse induction. In the previous editions of this book, the term *frequency shift* was used instead of proximity effect, but frequency shifting is only one of the proximity effects that can occur; there is also power loading. Beyond these categories, we will also consider continuous-time versus discrete-time designs, and frequency domain versus time domain.

Proximity Effect

Proximity effect detectors are usually characterized by having a single loop in the search head; it oscillates at a particular frequency and targets cause a measurable change in the oscillator itself. The change is either in the frequency of oscillation, or the power loading of the circuit, or both.

2.1: BFO

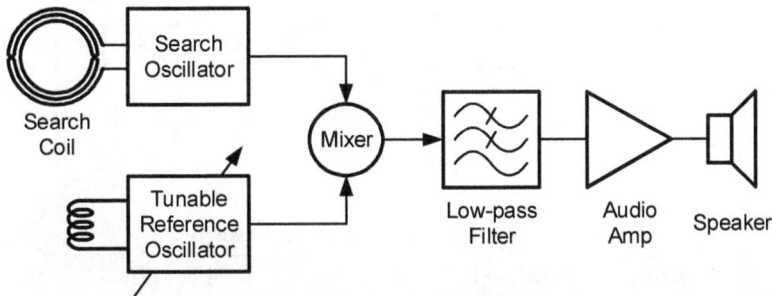

Fig. 2-1: **BFO Block Diagram**

The easiest term to deal with is BFO: *Beat Frequency Oscillator*. This refers to a particular type of circuit design in which two similar oscillators are fed into a mixer and the frequency difference (the "beat" frequency) is used to drive an audio stage. It sounds simple, and it is; a basic BFO can be built with just 3 transistors.

BFO detectors had their heyday in the 1960s and 70s during the very early years of the metal detecting hobby. At about the same time, transistors replaced vacuum tubes and the BFO became dirt-simple and mud-cheap; anyone and everyone could easily produce BFO detectors. Companies like Garrett and White's developed fairly sophisticated designs (by BFO standards) with crystal oscillators to minimize drift. In 1982, Garrett was perhaps the last major company with a BFO, and by then it had some rudimentary form of target discrimination.

BFO is a term that is pretty solidly defined and has not been excessively abused by the detecting community. We will look at BFO designs in detail in Chapter 10.

2.2: Energy Theft

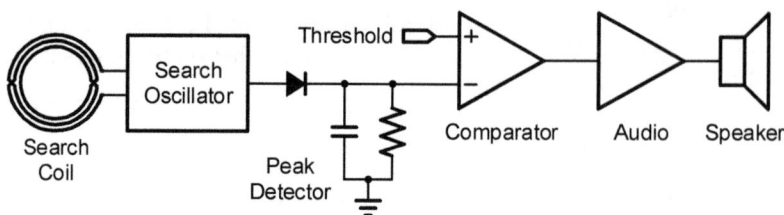

Fig. 2-2: **Energy-Theft Block Diagram**

In a BFO detector a target changes the Q of the search oscillator. This has two major effects: it causes the frequency of the oscillator to shift slightly, and it causes a change in the oscillation amplitude. While the traditional BFO design looks solely at the frequency shift effect, it is possible to design a circuit that considers only the amplitude shift that occurs. This can be done without a second oscillator so it can't be called a BFO even though it performs much like one. If it uses a free-running oscillator it is often called an "energy theft" method, owing to the fact that target eddies steal energy from the oscillator. Sometimes it is referred to as a "loaded oscillator" method because the oscillator driver is loaded by the target.

While the BFO has vanished from the market, the energy-theft design lives on. Many popular pin-pointers including the Garrett *Pro-Pointer*, the older Minelab *Pro-Find*, and White's *Bullseye II* use/used this circuit. Some handheld security wands do as well. In these applications where discrimination isn't needed and performance demands are modest, the loaded oscillator offers simplicity and low-power consumption.

2.3: Off-Resonance

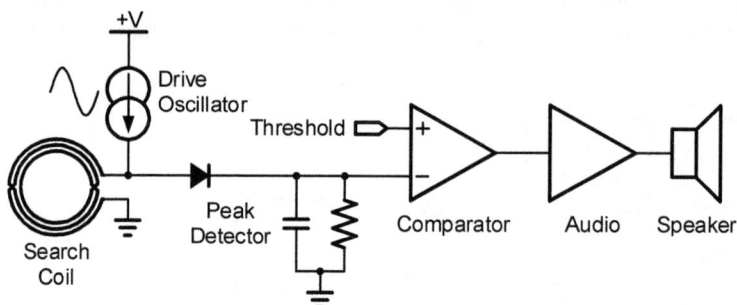

Fig. 2-3: **Off-Resonance Block Diagram**

In the proximity detectors described so far the transmit oscillator is allowed to free-run at its resonant frequency. Targets then cause a shift both in frequency and amplitude of the oscillation. It is possible to forcefully drive the LC tank at a frequency that is different than its resonant frequency; this is called an *off-resonance* design. It was successfully marketed in the 1970s by A.H. Electronics as the *A.H. Pro* detector.

Since the LC tank is driven at a fixed frequency a proximate target no longer causes a frequency shift, only an amplitude shift. If the drive frequency is higher than the resonant frequency, then higher conducting metals cause a rise in amplitude and lower conducting metals or ferrous metals cause the amplitude to drop. For some particular metal conductivity, the amplitude doesn't change. It is possible to select the frequency drive to alter the metal conductivity break point, and this forms a method of discrimination. An off-resonance design is presented in Chapter 11.

2.4: PLL

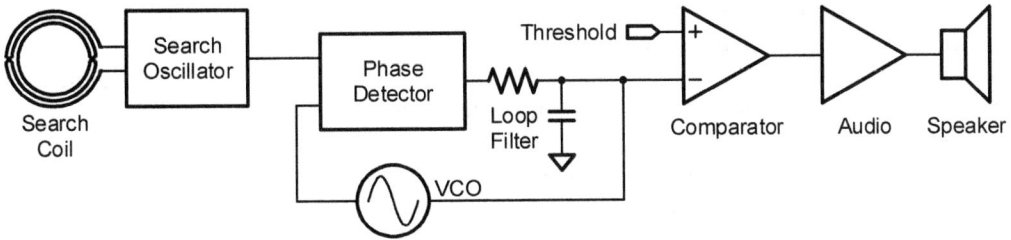

Fig. 2-4: **PLL Block Diagram**

The BFO looks at the frequency shift effect, comparing the transmit oscillator to a stable reference oscillator. However, the frequency shift can be extracted without the use of a reference oscillator. For example, the search oscillator can drive a *phase-locked loop* (PLL), and the drive voltage to the PLL-VCO (voltage-controlled oscillator) can be used as a target indicator. This is often referred to simply as a "PLL" design.

It is possible that the PLL design has never made it into a production metal detector. It showed up in a 1970s Signetics application note for their NE565 PLL chip and over the years has created interest amongst experimenters. While one could argue that the VCO in the PLL is actually a second oscillator, there still is no "beat frequency" effect taking place, so it is not a BFO. A PLL design is presented in Chapter 12.

Induction Balance

Induction balance (IB) means that there is a transmit coil and a receive coil arranged in an induction-balanced state. This is the oldest type of metal detector, dating back to the 1880s in the works of David Hughes and Alexander Graham Bell.

2.5: IB

Again, IB stands for *Induction Balance*. It is really not a type of metal detector, but rather a type of coil design where the TX and RX coils are arranged in an induction balance. However, people commonly use IB in detector jargon to encompass TR and VLF detectors, which practically always use an IB coil. Keep in mind, though, that PI detectors[1] can be designed to use an IB coil (and some require it), though no one ever seems to refer to them as an IB "type" of detector. Although it was the mainstay of Fisher Research Labs from the 1930s onward and was commonly found in military mine detectors starting in the 1940s, IB didn't hit its stride in the hobby/treasure market until the 1970s. It is now the dominant technique. There are many ways to design IB coils, and Chapter 6 explores many of these.

2.6: TR

TR stands for *Transmit-Receive*. It refers to the search coil design as well as the electronics. In its most basic meaning, a TR detector continuously transmits a signal and simply looks for a response on

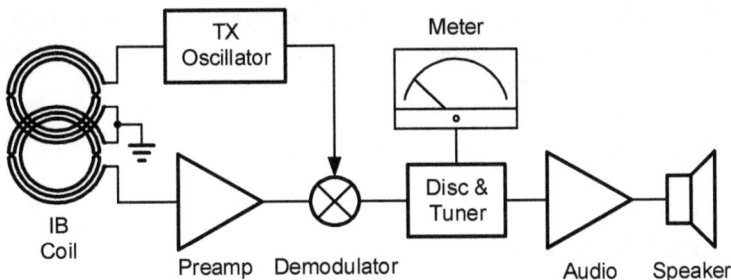

Fig. 2-5: **TR Block Diagram**

---

1. And even the lowly BFO

the receive coil caused by an imbalance from a nearby target. Technically, TR applies to most detectors all the way back to Bell (1881) but the term only became widespread in the 1960s and 70s. The earliest TR detectors were designed with vacuum tubes and instead of using a sinusoid for the transmit waveform they transmitted a periodic burst of sine waves. The period of the burst was in the audio range so when the RX signal was applied to a simple diode detector the result could be heard directly, greatly simplifying the hardware.

In the era of transistors TR designs became more complex and could transmit continuous sine waves and create audio by other means. TR models coexisted with BFO models and, for a short period in the mid-70s, TR detectors began to dominate the market. TR detectors tended to have better sensitivity and could discriminate but were more expensive (mostly due to the more complex coil) and worked poorly in mineralized ground, but so did BFOs.

The earliest TR detectors with burst transmitters could not phase-demodulate the received signal and therefore could not discriminate. Later models added discrimination by synchronously demodulating the received signal. Most TR detectors ran at somewhat higher frequencies, from 30kHz to 100kHz or more.

Nowadays some people reserve the term TR to mean two-box locators. The last two primary two-box locators produced — the Fisher[2] *Gemini 3* and the White's *TM808* (now defunct) — run at 82kHz and 6.592kHz respectively. The *Gemini 3* is designed more like a traditional TR detector, but the *TM808* is a standard VLF design with quadrature demodulators.

2.7: RF

RF, or *Radio-Frequency*, is a term that has almost exclusively been applied to two-box detectors, and mostly by a relatively small number of people. However, it's out there in literature, so it's included here. Those who refer to two-box locators as RF detectors tend to reserve the term "TR" for pancake-coil detectors.

2.8: VLF

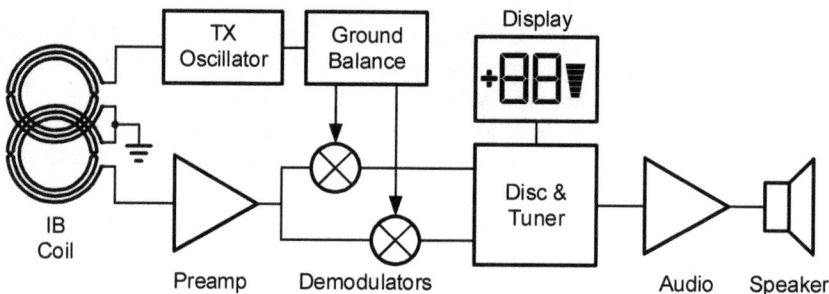

Fig. 2-6: **VLF Block Diagram**

VLF is an acronym for *Very Low Frequency*, and is defined by worldwide standards to be the frequency band from 3kHz to 30kHz. In metal detectors, it was first used to describe a new class of TR detectors that operated in or near this frequency band. Even so, there are many metal detectors that are considered to be VLF that operate outside of the VLF band. Some examples are:

- Compass *Relic Magnum*: 2 kHz
- Compass *AU52*: 52 kHz
- Fisher *1280x*: 2.4 kHz
- Fisher *Gold Bug 2*: 71 kHz
- Minelab *Eureka*: 6.5/20/60 kHz
- White's *Coinmaster V*: 1.75 kHz
- White's *V3*: 2.5/7.5/22.5 kHz

---

2. Fisher also makes an identical *TW-6* model for the utility locator market.

The *Eureka* and the *V3* provide examples of detectors that operate at multiple frequencies, both inside and outside of the VLF band.

If the term VLF in detector-land does not strictly mean operation in the 3-30 kHz band, then what is considered a distinguishing feature of a VLF detector? At the same time manufacturers made the shift to low frequencies (and some of the very first "VLF" designs were under 3 kHz), synchronous demodulation was becoming the norm in receiver design which allowed the design to "cancel ground." So practically all[3] of the early VLF detectors were ground-canceling as well. This ability increased detection depth in moderate ground and opened up severely mineralized ground that previously couldn't be hunted.

The ability to ground cancel is not found in so-called TR detectors, and is always found in so-called VLF detectors regardless of their actual operating frequency. So for all practical purposes, we'll consider a VLF detector to be one that can ground balance[4]. However, other than the addition of ground filters there really isn't much that separates a VLF from a TR design, and most VLFs instantly become a TR when operated in static pinpoint mode.

2.9: VLF/TR

This is not really a type of detector, but rather a detector that includes two modes. The first ground-canceling VLF detectors had only a single-channel synchronous demodulator that required the user to simultaneously adjust both the "ground" setting and the "tuner" setting as they pumped the coil up and down. It was a bit tricky to get it right and even trickier to maintain it while hunting. Furthermore, the use of a single demodulator meant that you could choose ground cancel or discrimination, but not at the same time. So manufacturers included two modes: a VLF ground cancel mode and a TR discriminate mode. Both modes usually ran at the same VLF frequency, they just used the demodulator in different ways.

Motion-mode VLF designs (starting with the Bounty Hunter *Red Baron*) could simultaneously cancel ground *and* discriminate, but required a very fast coil motion (coil "whipping") to detect targets while canceling ground. Recognizing that many people might prefer to turn off the ground cancel in mild soil, manufacturers continued the practice of including a non-GB TR mode. Eventually motion VLF got to the point where it worked with very slow coil motion and the TR mode was dropped.

2.10: Multifrequency

The term *multifrequency* (MF) can technically apply to either VLF or PI type designs, but we normally only associate it with VLF. In a PI design, the frequency is referred to as the *pulse rate* and the pulse rate has less to do with performance than does the TX pulse width, the exact shape of the pulse, or the RX sample timing. PI detectors that use multiple TX pulse timings are called *multi-pulse* or *multi-period* detectors. This section concerns VLF multifrequency.

Most single frequency detectors transmit a sinusoidal coil current. To date, multifrequency detectors never transmit sinusoids[5]; instead the coil is driven with a digital voltage (square wave, or multi-rectangular waveform) which produces a triangular or multi-ramp coil current. The transmit waveform then contains energy at two (or more) frequencies.

In the case of the Fisher *CZs* and White's *DFX* and *V3*, the receive signal is split into the individual frequency components by band-pass filters, demodulated, and processed. The Fisher and White's designs do the filtering and demodulation in analog circuitry. The Minelab *Equinox* does it digitally inside the microprocessor and instead of channelization filters more likely uses narrowband demodulation. In the case of Minelab's various BBS and FBS models, two frequencies are transmitted in sequential fashion and demodulation is done temporally; there is no need for band-pass filters. Regardless of the design approach, the information from two frequencies allows simultaneous ground balancing and

---

3. A very few of the early VLFs did not have ground cancel; they were simply TR detectors operating in the VLF band.
4. To add a little confusion, many PI detectors can also ground balance.
5. Except, see US5642050 & US5654638, but this never made it to a product.

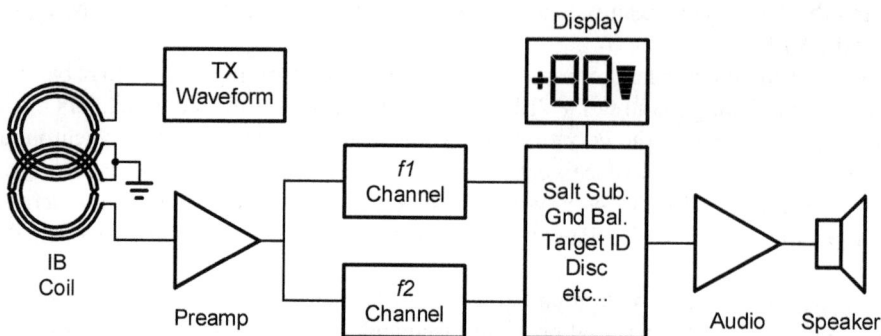

Fig. 2-7: **Multifrequency**

salt cancelation, plus provides more information about the target. The block diagram in Figure 2-7 closely represents the DFX, with each "Channel" consisting of a channel filter, I/Q demodulator, and post filter; see Chapter 24 for more details of this and other architectures.

Pulse Induction

Pulse Induction (PI) has become a poor name for a class of detectors, and we'll see why as things unfold. But for now it is used to describe a class of detectors which transmit periodic pulses of energy and process the received signal using non-continuous demodulators.

2.11: Basic PI

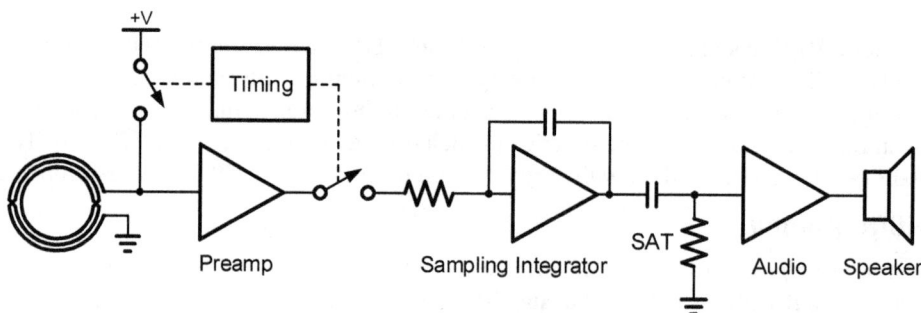

Fig. 2-8: **PI Block Diagram**

Everything described so far relies on a continuous waveform[6], almost always sinusoidal. Furthermore, all proximity and IB -type detectors transmit a signal while simultaneously receiving a signal or evaluating the effects of a target. PI detectors are characterized by a time-sequential transmit and receive. That is, they are not performed at the same time. The transmitter sends out a pulsed current, after which the receiver looks at the resulting target signal, often using the same coil[7]. Many people have likened this to a pulsed radar system, where an antenna sends out a pulsed signal and then looks for reflections from targets. It's not a perfect analogy, but not a terrible one, either.

2.12: Ground Balance PI

Early commercial PI detectors were simple designs which used a mono coil and operated only in "all-metal" mode. Later designs added the ability to ground balance, still using a mono coil, and added a rudimentary ability to do some amount of discrimination. While most PIs are designed for use with a mono coil, it is possible to instead use separate TX and RX coils, and they can be induction-balanced[8]

---

6. Ignoring the burst-mode TR designs.

7. Since the PI detector transmits and receives, could we call it a TR type? No, we don't want to encourage that now.

though it's not necessary. The Garrett *Sea Hunter MkII* uses separate coils in a non-IB arrangement, while Minelab PI detectors can use mono or DD coils.

2.13: Other PI Developments

A basic PI has a transmit pulse followed by a delayed receive sample. Target and ground performance are very dependent on the transmit pulse width and the delay and width of the sample pulse. A *multi-period PI* transmits two or more TX pulse widths followed by their own RX samples which feed separate channels. The multiple channels can be combined for ground balance and to provide better response over a wider range of targets. The Minelab *SD* series (*SD2000 - SD2200*) were the first hobby multi-period PIs.

A further enhancement to the multi-period concept is to use different drive voltages for the different TX pulse widths, where a short pulse gets a higher drive voltage than a long pulse. The reasons for doing this will be presented in Chapter 25. These can be called *multi-voltage PI*, with the understanding that they are also multi-period. The Minelab *GPX* series represent multi-voltage designs; Minelab has coined the term *Dual Voltage Technology* (DVT) since they use two TX pulse widths with two voltage drives.

A curious design that has confused quite a few people is the Minelab *GPZ7000*. It is technically a *Constant Current PI* (CCPI) detector whereby, instead of a truncated exponential or ramp, the transmit signal is quickly driven to a high current level and held there. This could be done as a punctuated bipolar pulse as in Figure 2-10b, or as a continuous bipolar square wave as in Figure 2-9c. The *GPZ* uses a continuous square wave; that is, the coil current slams back and forth between a positive and negative current. This makes it a continuous time transmitter. The receiver demods are (presumably[9]) discrete time, and the signal processing would most likely be time domain. Many people have considered the *GPZ* to be a VLF design due to the continuous time transmitter, but it's really a variant of PI.

The latest PI developments include analyzing the receive signal not only after the TX pulse, but also during the TX pulse. This absolutely requires the use of an IB coil. Furthermore, some of these developments use a transmit signal that is not easily defined as a pulse but rather some other kind of discrete (non-continuous) signal. In fact, such a detector was marketed in the 1990s — the Fisher *Impulse* — which used a bipolar asymmetrical triangle waveform instead of a pulse. Yet it was marketed as a "PI."

In the end, there are quite a few detectors that are clearly PI. But the trend is toward designs that are not clearly PI and not clearly VLF, but rather something in between. At some point the detecting community[10] will figure out a term to call these designs, whether that term is technically correct or not, and the confusion is sure to continue.

Other Jargon

The main purpose of this chapter is to clear up the meanings of various terms that are loosely tossed about in the discussion of metal detectors. Even though we now understand (sort of) what "TR" and "VLF" mean, there are other terms that go beyond the basic types of detectors.

2.14: Continuous-Time vs Discrete-Time: *Transmitters*

A *continuous-time*[11] (CT) transmitter is one that is continuously transmitting a signal. That is, the transmitted signal has no "off" time. A *discrete-time* (DT) transmitter has a series of active (transient) waveforms punctuated by regions of zero current.

Most CT transmitters have sinusoidal coil currents. This includes pretty much every proximity-type detector (BFO, etc) by necessity, since they use free-running oscillators. Most of the later TR

---

8. Never call a PI an IB, even if it is.

9. Apologies to the reader, I did not strip and analyze an $8,500 *GPZ* for this book.

10. More likely, some company's marketing department.

11. Also known as *continuous-wave*.

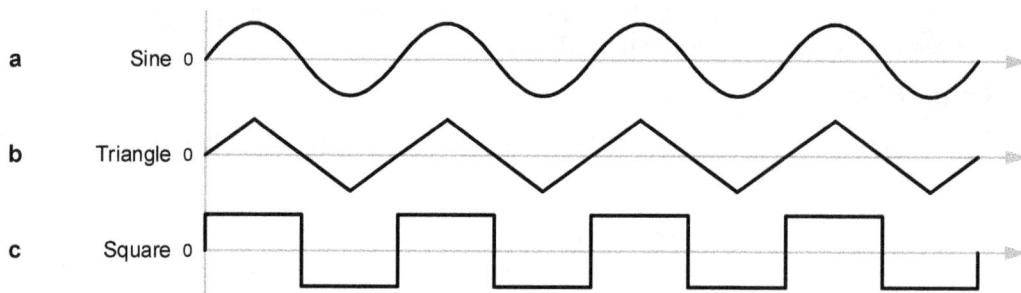

Fig. 2-9: **Continuous-Time Signals**

detectors and most of the modern VLF single-frequency detectors also use sinusoidal transmitters. Older TR detectors used a burst transmit signal and multifrequency models (and perhaps a few single-frequency models) use triangular or ramped coil currents. Finally, the Minelab *GPZ7000* uses a square wave current. Figure 2-9 shows a few examples of continuous-time signals.

Almost all PI detectors use DT transmitters. A typical design might have a 100μs current pulse repeated at a rate of 1kHz, meaning that 90% of the signal has zero current. Newer PI detectors use bipolar pulsing (both positive and negative current pulses) but they are still transmitting discretely. Most PIs don't transmit a pure pulse, but rather a truncated exponential or a sawtooth. As mentioned before, the Fisher *Impulse* had an odd dual-ramp waveform. There is also a method called "half-sine" and "truncated half-sine."

Figure 2-10 shows examples of discrete-time signals. Notice that the square wave can be considered either a continuous or a discrete signal. In the CT case, it switches between a positive and negative value[12]; in the DT case, it switches between some value and zero, or 'off.'

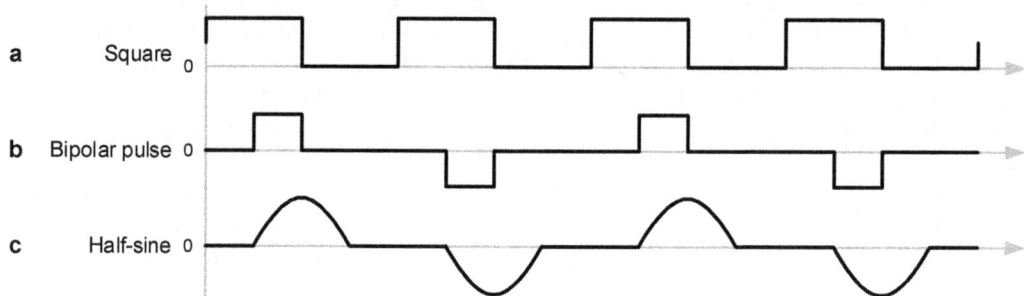

Fig. 2-10: **Discrete-Time Signals**

2.15: Continuous-Time vs Discrete-Time: *Demodulators*

Most modern detectors demodulate the received signal to determine target behavior. Whether the detector uses a single demodulator (in the case of a simple TR or PI) or a pair of quadrature demodulators (as with most VLFs) they are always clocked synchronously with the transmit signal. Proximity-type detectors don't have demodulators, but their signal processing is continuous.

Like the transmitter, the demodulators can be clocked either in continuous-time or discrete-time. It might seem that this would simply follow the method (CT or DT) of the transmitter, but not necessarily, at least in the case of the CT transmitter. For a DT transmitter, the demodulators are always[13] DT as well. This means the received signal is demodulated using short discrete "sample windows." Figure 2-11 shows an example of the transmitter and demodulator timing for a typical PI detector. On the demod waveform, a "+1" means an additive sample is taken, a "−1" means a subtractive sample is taken, and "0" means the demodulator is off.

---

12. The Minelab *GPZ7000* uses a continuous-time square wave.

13. At least I am not aware of any detector that has a DT transmitter and CT demods. Readers will be kind enough to correct me, no doubt.

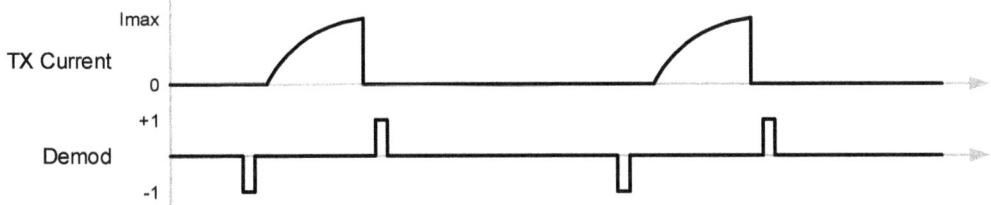

Fig. 2-11: **Discrete-Time Demodulation**

Most standard VLF detectors that have a CT sinusoidal transmitter use CT demodulators. The normal demodulation scheme has an in-phase (I) demodulator for reactive signals and a quadrature-phase (Q) demodulator for resistive signals[14]. Figure 2-12 shows a typical VLF setup; note that the demod waveforms are always sampling the received signal, there is no "off" time as there was with the PI demodulator.

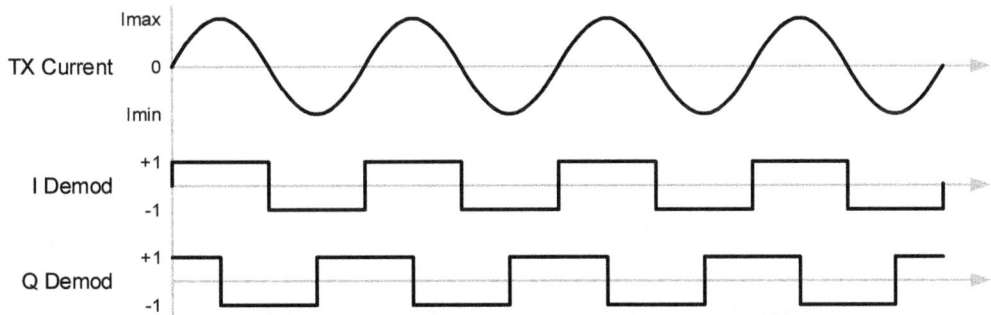

Fig. 2-12: **Continuous-Time Demodulation**

There is no reason why a detector that uses a CT transmitter must also use CT demodulators. The example VLF scheme in Figure 2-12 could be modified so that both demodulators have a positive and negative sample windows punctuated with dead time; see Figure 2-13. This particular demodulation method is called "Tayloe" mixing, named for inventor Dan Tayloe. It does not appear to be used in any metal detector designs, so why bring it up? First, it provides a simple example of DT demodulation in an otherwise CT system. Second, it is an introduction to an alternate style of quadrature demodulation that might produce performance benefits in a metal detector; we will consider this in Chapter 16.

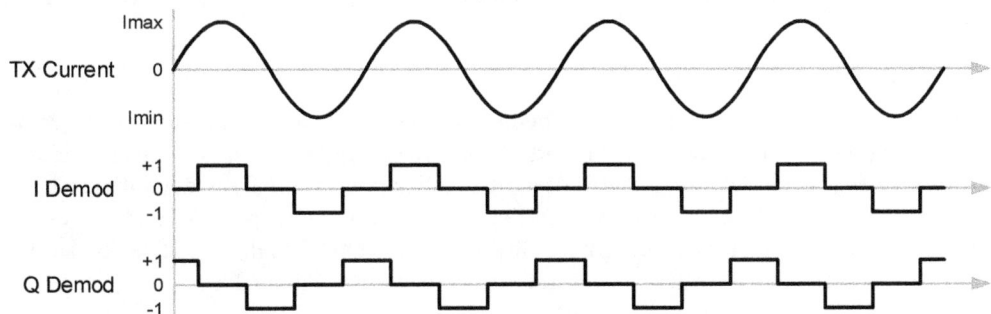

Fig. 2-13: **CT Transmit / DT Demodulation**

While metal detectors don't use Tayloe demodulation, there are metal detectors that have CT transmitters with DT demodulators. Two in particular are the Minelab BBS & FBS multifrequency detectors, and the Minelab *GPZ7000*. We will cover these methods in later chapters.

---

14. Depending on the exact design of the transmitter and receiver, these functions might be reversed.

2.16: Continuous-Time vs Discrete-Time: *Signal Processing*

After the receive signal is demodulated, the detector further processes the demodulation signals for the purpose of noise filtering, ground balance, discrimination, audio generation, and so forth. In the old days, this was done using analog circuitry in a continuous-time manner. The CT nature of the signal processing allowed every nuance of the target response to be heard and many detectorists became very skillful at discriminating "between the ears" even before tone ID became available.

Practically all[15] modern metal detector designs now perform the vast amount of signal processing in the digital domain. Shortly after the demodulators the signals are sampled and digitized by an analog-to-digital converter (ADC), a process that converts the CT analog signals into DT digital signals. In fact, "discrete time" is a formal term used in electrical engineering to refer to sampled data systems.

After the ADC all further processing takes place in the digital domain. Digital signal processing (DSP) is almost always performed in a microprocessor or microcontroller. Done right, DSP can be very powerful in producing higher performance detectors than can ever be achieved with analog processing. Done wrong, it leaves detectorists yearning for the good ol' days of the all-analog designs.

2.17: Frequency Domain vs Time Domain

Two other terms that are thrown about rather carelessly are *frequency domain* (FD) and *time domain* (TD). They are somewhat related to continuous-time and discrete-time in that detectors which use both a CT transmitter and CT demodulation tend to be frequency domain, and detectors that use both a DT transmitter and DT demodulation tend to be time domain[16].

So just what is FD and TD? A detector which demodulates a single-frequency into reactive and resistive baseband signals for the purpose of determining the signal's amplitude and phase characteristics is a frequency domain design. This is almost always done using CT demodulators, but a DT Tayloe demodulator (and perhaps other DT demod designs) could also be used in a FD design.

A detector which demodulates a broadband signal using DT demodulators for the purpose of determining the signal's amplitude and decay characteristics is a time domain design. Such a detector usually has a DT transmitter, but a CT transmitter could also be used in a TD design, as is the case with the Minelab *GPZ* or Tarsacci *MDT 8000*.

Because most detectors now perform the majority of signal processing in a microprocessor where it is effectively hidden from curious book-writers, it is not always obvious exactly how the detector is processing the demodulated signals. But how the demodulators are clocked (assuming they are also not in software) usually offers a pretty good idea as to what is going on in DSP. The section on Hybrid Designs (2.20) will discuss some specific examples of detectors where CT, DT, FD, and TD are all mixed up.

2.18: Synchronous Demodulation

In all the prior examples, regardless of whether the transmitter and demodulators are continuous time or discrete time in nature, the demodulators clocks are synchronous with the transmit signal. This is called *synchronous demodulation*[17] and is used by practically every modern metal detector design[18]. The outputs of the demodulators are baseband signals with just enough bandwidth (10-50Hz) to respond to targets passing under the coil. These slow near-DC signals are then digitized by an ADC and further processed in DSP to determine e.g. target ID and depth. See Figure 2-14.

---

15. The last modern all-analog continuous-time detector products I'm aware of were the Troy *Shadow X3/X5*. On the PI side, the all-analog Fisher *Impulse/AQ* is currently in production.

16. DT = discrete time, and TD = time domain; similar but different.

17. The term *synchronous demodulation* is commonly used in metal detectors; the first motion-discrimination metal detector, the Bounty Hunter *Red Baron*, coined the term *Synchronous Phase Discrimination* (SPD). In radio-land it's called a *direct conversion receiver* (DCR) or *zero-IF* (ZIF).

18. An example of a modern design that does not use synchronous demodulation is the Fisher *Gemini*, which uses a non-synchronous low-IF demodulation because the transmitter and receiver are physically isolated.

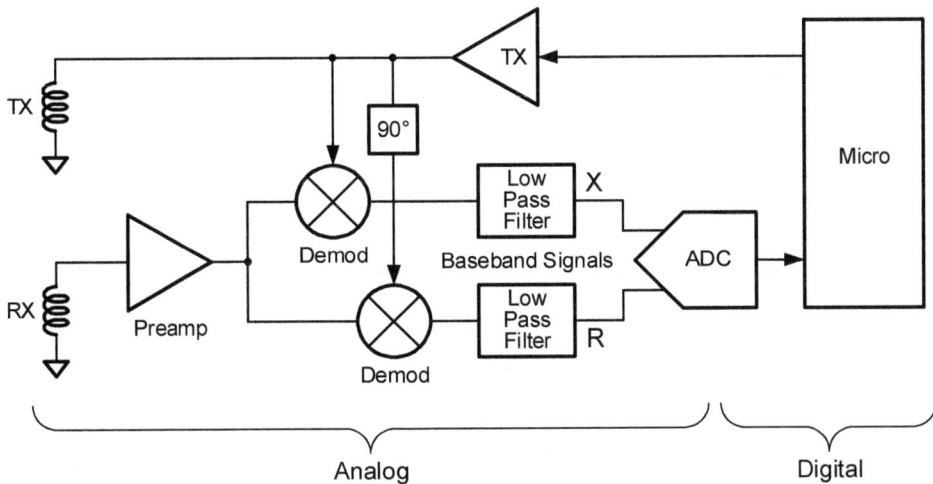

Fig. 2-14: **Analog Demodulation**

Synchronous demodulation is used in metal detectors because the transmitter and receiver are conveniently located in the same circuitry making the method simple and cheap. Even multifrequency designs use synchronous demodulation, generally with each frequency channel clocked at its designated fundamental frequency but always in sync with the transmitter. Synchronous demodulation became popular in the mid-1970s with the introduction of TR discrimination and (separately) VLF ground canceling models. Before that, most IB designs used an amplitude-modulated transmit signal and an envelope detector in the receiver.

2.19: Direct Sampling

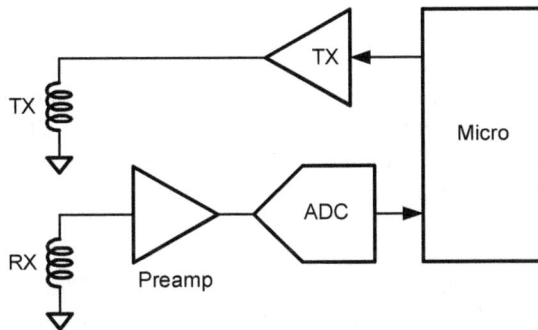

Fig. 2-15: **Direct Sampling Receiver**

Some detectors have successfully utilized a *direct sampling receiver* (DSR) design instead. In direct sampling[19], the analog demodulators and filters are eliminated and the ADC is placed on the output of the preamp, as shown in Figure 2-15. Obviously this simplifies the circuit design, but demands a much higher performing ADC because, instead of digitizing two lazy baseband signals, the ADC must now digitize the full-speed RX signal at a rate at least four times faster than the highest frequency of interest and with higher precision than a baseband ADC.

With direct sampling, the output of the ADC is still an RF signal and must be digitally demodulated to produce the required baseband signals. So by using a higher performance ADC the demodulation and baseband filtering is moved into the digital domain, which can make for a more flexible design as digital processing is easier to alter than analog components. At the time of this book, single-frequency detectors using direct sampling are the Minelab *X-Terra* series, White's *Prizm 6T* (a.k.a. *Coin GT*),

---

19. In radio-land it's also called RF sampling, or IF sampling if done in an IF stage.

Minelab *Go Find* series, and the XP *Deus*. The Minelab *Equinox* and *Vanquish* models, the XP *Deus II*, and the Nokta *Legend* are examples of direct-sampling multifrequency detectors, and the (newer) Bounty Hunter pinpointer is an example of direct-sampling PI.

2.20: Hybrid Designs

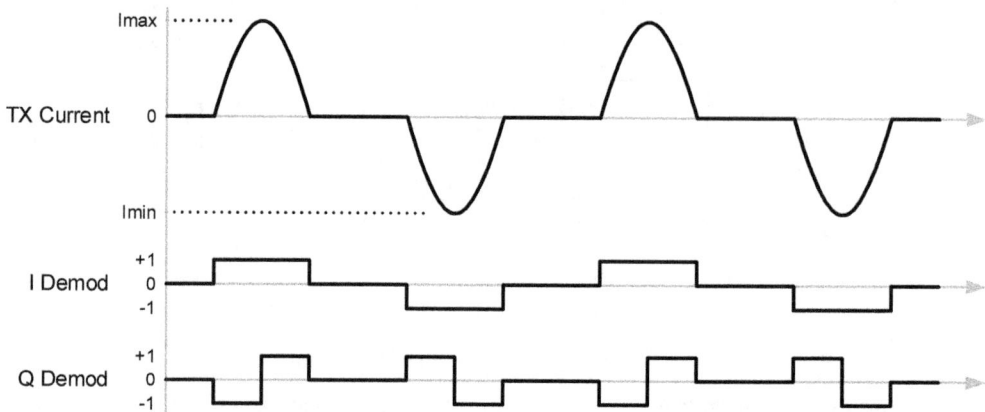

Fig. 2-16: **Hybrid: Half-Sine**

Let's go deeper down the rabbit hole and look at some hybrid possibilities. A method that was used by Barringer Research in the 1970's is called *half-sine* (HS)[20]. The TX waveform for this method is shown in Figure 2-16. Obviously this is a sine wave that has been split apart with some zero-time between the positive and negative half-sine pulses. A DT waveform, no question. And you would definitely need to use DT demods to look at target responses. However, all of this discrete-time TX/RX can still produce a FD result. Each half-sine pulse can be quadrature demodulated to extract reactive & resistive signals that indicate amplitude and phase.

But that's not all. Notice that each half-sine comes to a sudden dead stop followed by zero current. That looks a lot like... a PI. So not only can you sample the half-sines and extract FD information, you can also extract PI-like information and do parallel DT processing. The VLF-like responses and the PI-like responses occur continuously and simultaneously, making this a true hybrid method, This and other hybrid techniques are covered in Chapter 26.

2.21: The Future

It is apparent that while "Proximity Effect" aptly covers BFO and the like, newer designs have rendered the categories "Induction Balance" and "Pulse Induction" too narrow in scope. I was tempted to use the categories "Continuous Time" and "Discrete Time" but, again, designs like the *Explorer* and *GPZ7000* still don't fit neatly into either. So I stuck with the familiar for now, with the caveat that future editions of this book may move in a different direction.

What we see from all this is that one day there could well be an IB design using a MF CT VLF TX combined with MF-THS and PI and processed in DT using both FD and TD methods, with a PLL thrown in for good measure. What would we call it? Probably "confusing."

---

20. See patents US3020471 & US4506225

Magnetics & Induction

"I am busy just now again on electro-magnetism,
and think I have got hold of a good thing, but can't say."

— *Michael Faraday*

Understanding metal detectors begins with understanding magnetics. This chapter has a number of simple experiments that demonstrate the behavior of magnetic fields, and goes on to explain the principle of induction. All of this is done in preparation for a detailed discussion of eddy currents in the next chapter, which is a critical concept in understanding how metal detectors work.

Magnetics

3.1: Static Magnetic Fields

Metal detectors operate via the principle of *induction*. Induction is the method of coupling two circuits though an *alternating magnetic field*. To clarify, we'll step back in time and look at some of the very same experiments mentioned in the *History* chapter. Let's start with basic magnetic fields first.

When you pass a direct current (DC) through a wire a *static* magnetic field develops around the wire[1]. This is a very simple experiment that is often demonstrated in grade school science class, and one that anyone can do at home.

Experiment 3-1: Demonstrate that a current through a wire produces a magnetic field.

Required: Battery (C-cell is fine); a length of small-gauge wire (18 in. or 1/2 m will do), and a needle-style magnetic compass.

Figure 3-1 illustrates the setup. Place the compass on a wooden table, and form the wire into a large circular loop. While holding the battery, *briefly*[2] short the wire across its terminal. If you move the loop close to the compass, you will see the needle jump. Try to determine which orientation of the wire, as compared to the normal direction of the needle, causes the greatest deflection.

What happens when the wire is held above the compass? What happens when the wire is held below the compass? Perpendicular to the compass? What happens when the battery polarity is reversed?

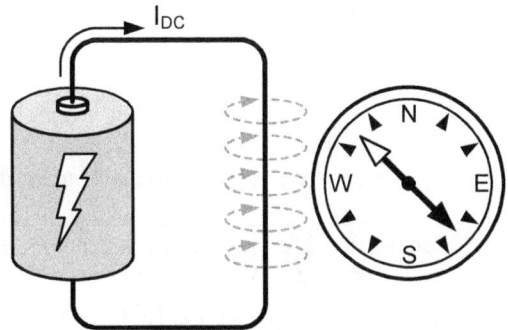

Fig. 3-1: **Magnetic Field Experiment**

---

1. As discovered by Øersted in 1820.
2. *Briefly* is the operative word here. The wire will get quite warm if you hold it on too long. "Too long" is about a second.

If you wrap the wire in a long tight coil (Figure 3-2) you can enhance the magnetic field, because the coil "focuses" the field though its center. Each turn of wire adds more "flux" to the field. It turns out that the strength of the magnetic field depends on the current through the coil (I), the dimensions of the coil (length *l* and radius *r*), the number of turns of wire (N), and the material used as the core of the coil. The maximum magnetic field strength of such a coil is:

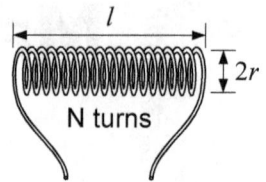

Fig. 3-2: **Basic Coil**

$$B = \frac{\mu NI}{\sqrt{4r^2 + l^2}} \quad \text{(Teslas)} \qquad \qquad \text{Eq 3-1}$$

where μ is the *permeability* of the core material the coil is wrapped around. The term B is *magnetic flux density* in units of *Teslas* (T), and Equation 3-1 gives the flux density (field strength) exactly at the center point *inside* the coil. The lines of flux that make up the magnetic field exit one end of the coil, wrap around the outside, and enter the other end of the coil to form a complete path. The flux density reduces slightly as you move from the center towards the ends of the coil, and is weaker around the outside of the coil because it is not confined and can spread out. Figure 3-3 illustrates. Keep in mind that the *total flux* through the center of the coil is exactly the same as the total flux returning around the outside of the coil even though their *flux densities* are different. Also, the magnetic flux lines have direction, and the flux external to the coil has a direction opposite to the flux inside the coil. This will be important in the design of metal detector coils.

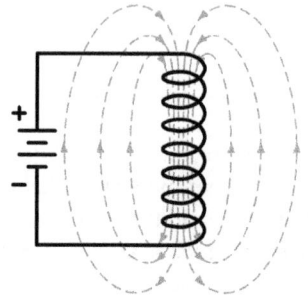

Fig. 3-3: **Coil Field**

So what is "permeability" in the equation above? It is the magnetic equivalence of conductivity[3] for current. A material that has a higher permeability is more "conducive" to the presence of magnetic flux. The coil in Figure 3-2 appears to have no core material other than air, and the permeability of air is $\mu = 4\pi \times 10^{-7}$ Henries/meter (H/m). Air is often considered to be the norm by which other permeabilities are compared[4] and is given the special symbol μ_0:

$$\mu(air) = \mu_0 = 4\pi \times 10^{-7} \text{ H/m}$$

If we were to wrap our coil around a steel core, such as a nail, then the magnetic field would be enhanced because steel has a much higher μ than air, about 100 times higher. A material's permeability as compared to air is called *relative permeability* (μ_r) so for steel $\mu_r = 100$:

$$\mu(steel) = \mu_r \mu_0 = 100 \cdot 4\pi \times 10^{-7} = 400\pi \times 10^{-7} \text{ H/m}$$

It is the iron content in steel that gives it a high permeability, and pure iron is even higher; 99.8% pure iron has a $\mu_r = 5000$. Certain alloys are made to have a $\mu_r = 1,000,000$ or more. A coil wrapped around a nail forms a simple electromagnet, also often demonstrated in elementary school science class.

Experiment 3-2: Demonstrate that winding the wire into a coil increases the magnetic field strength.

Required: Battery (C-cell is fine); a length of small-gauge wire (36 in. or 1 m will do), a needle-style magnetic compass, a wooden pencil, and a large nail.

Wrap the middle 18 inches (1/2m) of wire around the pencil to form a coil with 9-inch "pigtails" on each end. See Figure 3-4. For a normal-sized #2 pencil, you should end

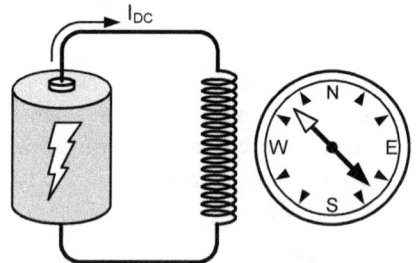

Fig. 3-4: **Coil Experiment**

---

3. Conductivity is, of course, the reciprocal of resistance.

4. Actually a vacuum is the norm, but air is almost identical.

---

up with about 15 turns. You may remove the pencil; the coil will probably have a tendency to unspring slightly, that is nothing to worry about.

Repeat the actions in Experiment 3-1, holding the coil close to the compass. Try different orientations of the coil relative to the at-rest needle, such as on top of, and along the edge of, the compass. Slide the nail inside the coil and repeat the experiment.

In performing these last two experiments you should have noticed a difference in how the orientation of the wire (or coil) affects the deflection of the compass needle. In the case of the straight wire, the magnetic field "wraps around" the wire, as shown in Figure 3-1. It is always true that the magnetic field attempts to "wrap around" the flow of current, so when we wind the wire into a coil, the magnetic field tries to wrap around each individual coil winding, and ends up focused through the center on the coil and "wrapping" back around the outside of the coil, like we saw in Figure 3-3.

Let's go back to Equation 3-1 which gave us the strength of the magnetic flux density B at the center of a coil. If the coil is much longer than its radius (long & skinny), then $l \gg r$ and this simplifies to:

$$B = \frac{\mu NI}{l} \text{ (T)} \qquad\qquad \text{Eq 3-2}$$

But metal detector coils are not long & skinny, they are short and large, where $r \gg l$:

$$B = \frac{\mu NI}{2r} \text{ (T)} \qquad\qquad \text{Eq 3-3}$$

Again, this is the field strength exactly in the center. Assuming the strength of the magnetic field is an important issue in designing a metal detector[5], what we see from this is that for a given coil size (radius r) we can either increase the number of turns of wire, increase the current through the coil, or use a core material with higher permeability. If you've looked at a lot of detector coils, you will notice that the core material is always air, or some material (plastic, fiberglass, resin, etc.) that has a $\mu_r = 1$ (that is, equal to air). So we can assume that choosing a different μ is not a viable option or companies would be doing that. This leaves us with only the number of turns and current; these two parameters are often combined into a single term called *ampere-turns* which is simply $N \times I$. It also appears that large radius coils should have a weaker field but we'll see later that this depends on the distance from the coil.

3.2: Alternating Magnetic Fields

So far, we have only considered the effect of a DC current[6] through a wire. As long as the current is constant, the magnetic field will be constant. If the current is varied, then the magnetic field will vary as well. An alternating current (AC) in a wire produces an *alternating* magnetic field around the wire. That is, as the current continuously reverses direction, the polarity of the magnetic field reverses, too.

As with the DC experiments, if we wind the wire up into a coil we can intensify either effect. Pass an AC current through the coil and we get a stronger AC magnetic field. Also, again, a coil with a high permeability core (like iron) will have a stronger magnetic field than a coil with a low permeability core (like air).

Experiment 3-3: Demonstrate that an AC current[7] produces a changing magnetic field.

Required: The coil, compass, and nail from Experiment 3-2, plus a variable signal generator and a 100Ω resistor.

Instead of driving the coil with a battery, we will now drive it with an oscillator. Use a signal gener-

---

5. And we haven't really established this yet, have we? Be patient.
6. "DC" means "direct current", so "DC current" means "direct current current", which is redundantly repetitive. But in electronics, we also use the term "DC voltage", so DC has become synonymous with "constant".
7. Likewise, "AC current" is redundant, so just pretend AC means "alternating". Usually, AC refers to a periodic (repeating) signal, most often a simple sinusoid.

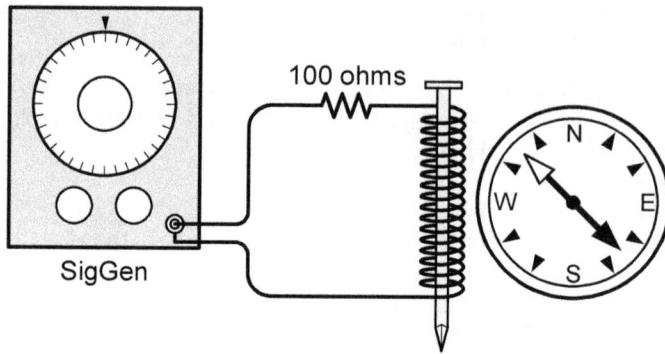

Fig. 3-5: **AC Test**

ator that can go down to about 1 Hz; a sine wave is preferable but a square wave will work. Since the generator probably has a more limited current drive than the battery, you may need to use the nail as a core for the coil.

With the generator set to oscillate at about 1 Hz or so, if you place the coil near the side of the compass (try the north or south position, with the coil aligned east-west), you should see the needle move back and forth slightly. As you slowly increase the generator's frequency, the needle will move faster, but at some point mechanical inertia will begin to limit its movement. It's possible, if you can develop a strong enough magnetic field, to get just the right frequency so that the needle starts spinning in one direction; this is a crude motor.

If a current through a wire can create a magnetic field, an obvious question to ask is, can a magnetic field create a current in a wire? The answer is yes, but only in the case of a *changing* magnetic field. In other words, a wire or other conductor (generally as part of some closed-loop circuit) placed in an alternating magnetic field generates an electromotive force (EMF) in the wire which, in turn, can produce an AC current in the wire. Clamp-on ammeters use this principle to determine the current in household wiring without having to actually cut the wire. The AC magnetic field surrounding the wire couples into the metal clamp and the EMF can be measured.

It might seem that placing a wire in a static magnetic field would generate a DC current, but not so[8]. To generate a current with a static field, you have to move the wire *through* the field, for which you get a *transient*[9] EMF (and transient current). If you move the wire through the field in a periodic manner, then you will get a periodic (AC) current. This is how a generator (alternator) works. But a wire sitting motionless in a static magnetic field will get you zilch.

Curiously, an alternating magnetic field gives rise to an alternating electric field (and vice-versa). Although it's really not important to this discussion, a changing magnetic field produces an electric field which can move electric charge and, conversely, electric charges in motion (whether they are inside a wire or not) create a magnetic field. So they mutually support one another, and the combination of the two is known as an *electromagnetic* (EM) *field*. The EM field is how radio signals travel through space with seemingly no means of support.

The reason for bringing this up is because we will often refer to the alternating magnetic field as an "EM field", and others do so as well. But metal detectors only make use of the magnetic field portion, and the electric field part does nothing for us. So when you see the term "EM field" applied to metal detectors, it really means the alternating magnetic field[10].

---

8. That would be nice, because you could use it to generate free electricity.

9. A transient signal (voltage or current) is not constant, nor is it periodic, so it generally is not considered DC or AC.

10. A static magnetic field is not considered an electromagnetic field, even when it is produced by an "electromagnet".

3.3: Coil Fields

The previous two experiments showed that it is possible to intensify the magnetic field by winding wire into a multi-turn coil, and this is exactly what metal detectors use. Instead of a solenoidal coil (long and skinny) most detectors use large-diameter "skinny donut" coils — there really is no standard term for these kinds of coils. The diameter, number of turns, and wire gauge are usually chosen for best performance for a particular design; there is no "one size fits all" coil design. We will consider all this in Chapter 8, but for now it is useful to complete our look at magnetics to see what a coil field actually looks like.

Figure 3-3 shows, in a crude drawing, how the magnetic field behaves for a solenoidal coil. Magnetic fields are invisible but can be photographed by their effect on iron filings. This only works with

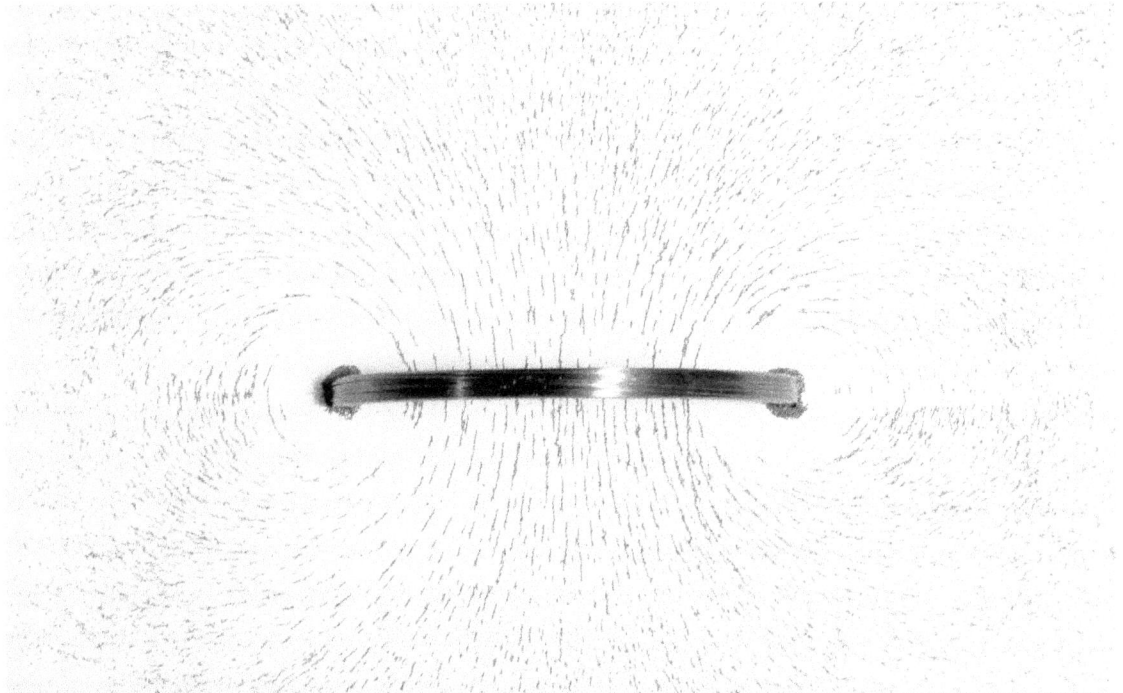

Fig. 3-6: **Magnetic Field**

Fig. 3-7: **FEM Magnetic Field**

DC magnetic fields, not AC, but an AC field will have the same "shape" of flux lines. Figure 3-6 shows the magnetic field for a 4" (10cm) round coil. Notice that the space close to the windings is devoid of filings; that's because the field strength in this region pulls the filings all the way to the coil. See the *Explore* section at the end of this chapter for details on how to do this experiment yourself.

There are more modern ways to find how magnetic fields behave. Computer programs that use a method known as *finite element modeling* (FEM) can, for any coil design, calculate and display the magnetic field. FEM software is commonly used in the design of motors and generators. Figure 3-7 shows an FEM plot for the same coil as before.

FEM software shows not only the pattern of the magnetic field but can also show the magnetic flux density as color gradients. This book does a poor job of showing color so the flux density is shown as gray-shaded regions. Notice how the flux density at shallow depths is stronger at the coil windings than at the center axis. The mathematics involved in calculating the three-dimensional field is extremely complicated involving elliptic integrals[11], which is why FEM analysis is commonly used.

In metal detectors we are mostly concerned with the *on-axis* field strength; that is, the strength of the B-field directly along the center axis of the coil. This is because we generally recognize the maximum target response is along this axis. Finding the total response of the three-dimensional field is not necessary, and we can just focus on the on-axis strength which is much easier to deal with.

Equation 3-3 gives the field strength exactly in the center of the coil, and the field strength diminishes as you move along the axis away from the coil. At a point on the axis a distance d from the plane of the coil the magnetic field strength is given by[12]:

$$B = \frac{1}{2} \cdot \frac{\mu N I r^2}{\sqrt{r^2 + d^2}^3} \ (T) \qquad\qquad \text{Eq 3-4}$$

where, again,

- r is the radius of the coil
- N is the number of turns
- I is the coil current
- μ is the permeability, assumed to be μ_0

This is a more generalized form of Equation 3-3 and for d=0 (exactly in the center of the coil) it simplifies to Equation 3-3. In cases where d >> r Equation 3-4 becomes

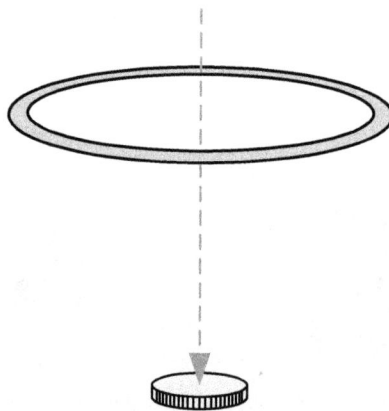

Fig. 3-8: **On-Axis Field**

11. C.A. Bartberger, *The Magnetic Field of a Plane Circular Loop*, Journal of Applied Physics, Vol. 21 November 1950, p1108-1114.

12. This equation is commonly derived in any college *electromagnetics* textbook. My favorite is Kraus & Carver.

$$B = \frac{1}{2} \cdot \frac{\mu N I r^2}{d^3} \ (T) \qquad\qquad \text{Eq 3-5}$$

This suggests that as we get farther away from the coil, the field strength becomes roughly proportional to $1/d^3$.

Example: Calculate the on-axis *relative* flux density versus depth for an arbitrary coil radius.

Solution: Let B_0 be the flux density at the center of the coil, given by Equation 3-3. The relative on-axis flux density is the ratio of B/B_0 given by dividing Equation 3-4 by Equation 3-3:

$$\frac{B}{B_0} = \left(\frac{r}{\sqrt{r^2 + d^2}} \right)^3 \qquad\qquad \text{Eq 3-6}$$

The depth is 0 at the center of the coil, and then we can consider a depth of r/2, r, 2r, 2.5r, and so forth. The relative field strength vs depth is:

depth	B/Bo	r^3/d^3
0	1.000	—
0.5r	0.7155	—
r	0.3536	1.0000
1.5r	0.1707	0.2963
2r	0.0894	0.1250
2.5r	0.0512	0.0640
3r	0.0316	0.0370
4r	0.0143	0.0156
5r	0.0075	0.0080
6r	0.0044	0.0046
8r	0.0019	0.0020
12r	0.0006	0.0006
16r	0.0002	0.0002

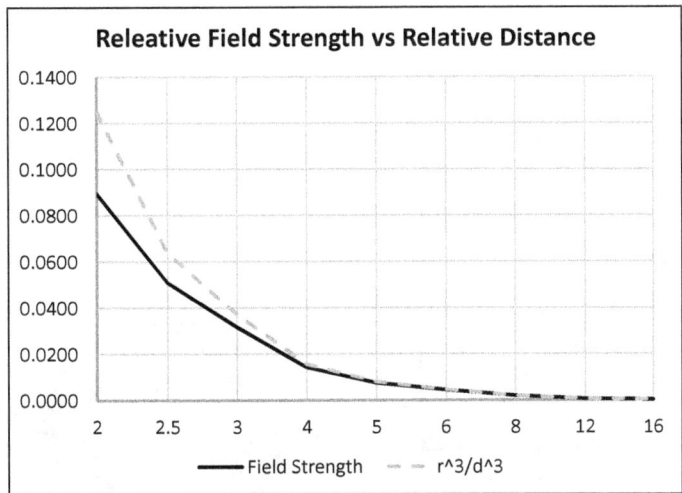

Fig. 3-9: **Magnetic Field vs Depth**

What this says is that — for a given current, number of turns, and core permeability — the field rapidly diminishes with distance and beyond about twice the coil diameter (4r) it approaches a rate of r^3/d^3 as shown in the last column, and is even more apparent in the graph. That is to say, if you double the depth, the magnetic field strength will reduce to $(1/2)^3$, or 1/8th.

In the example above the same numbers hold true regardless of the coil diameter, current, or number of turns. That is, for *any* circular coil, the *relative* field strength at one diameter of depth will be $0.0894 \cdot B_0$. How diameter, turns, and current affect practical depth will be further explored in Chapter 8.

Induction

If we take two coils and place them close to each other, then driving one coil with AC will create the EM field, which will *induce* an AC voltage (EMF) in the other coil. This creates a *transformer* (Figure 3-10). This is the stuff that was figured out by Michael Faraday. It's called *induction*, because something (a voltage) is getting *induced*.

What magic is doing the inducing? With wires it's electron motion, where the electron motion (current i_1) in one coil produces a magnetic field, and where that same magnetic field cutting across conductors in a second coil produces an electromotive force (EMF) that causes electrons to move, and results in a secondary current (i_2). In transformers, the driven coil is called the *primary* coil, and

Fig. 3-10: **Inductive Coupling**

the induced coil is called the *secondary* coil. In metal detectors, the driven coil is called the *transmit* (TX) coil, and the induced coil is called the *receive* (RX) coil. As we will see, metal detectors operate very much like transformers, but with an additional layer of complication.

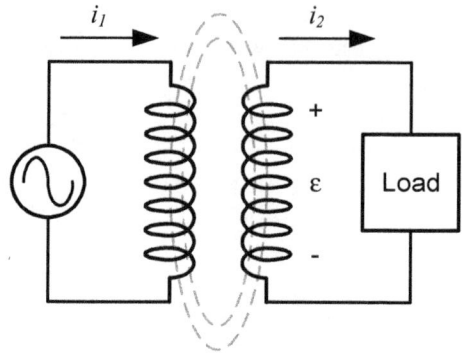

Experiment 3-4: Demonstrate the principle of induction.

Required: Small-gauge magnet wire, a signal generator, an oscilloscope.

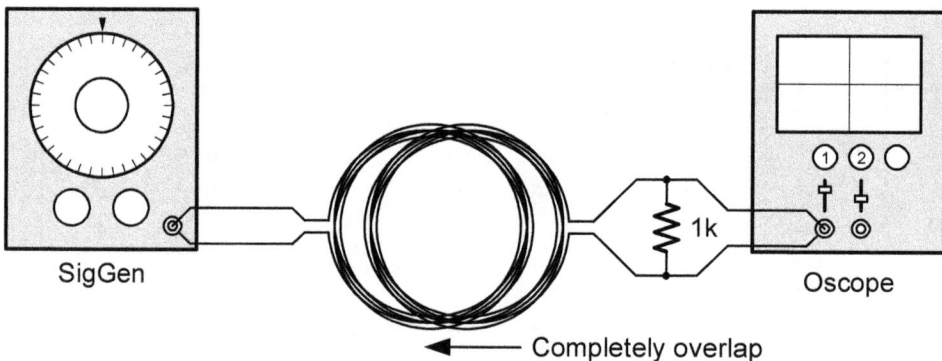

← Completely overlap

Fig. 3-11: **Induction Experiment**

Wind two coils of wire from small-gauge magnet wire (such as 26-28AWG), 50-turns each, about 4 inches (10cm) in diameter. For each coil leave ~10 inch (25cm) pigtails. Connect one of the coils to the signal generator and the other coil to the oscilloscope with a 1kΩ parallel load resistor. Arrange everything as shown in Figure 3-11, with one coil laying directly on top of the other. For the signal generator, a sine wave at 10kHz works well; set the output amplitude to fairly high level. Observe the signal amplitude on the oscilloscope.

Now place a sheet of paper or card stock between the two coils. Does this have any effect? Replace the paper with aluminum foil. What happens now? Finally, try a piece of steel sheet metal[13].

---

13. Perhaps a steel cookie pan.

3.4: Induction Balance

The above experiment shows that the EM field produced by one coil will induce electron motion in another coil, thereby generating a voltage. Again, this is called a transformer. Normally, transformers are designed for maximum coupling between the primary and secondary coils in order to get the highest efficiency possible. Usually they include an iron-based core material to help maximize efficiency through a higher mutual inductance. With metal detectors, we actually want the opposite: we want to minimize the *direct* coupling between the primary and secondary coils; in fact, we want none at all, if possible. That way, when a target enters the field the only signal on the secondary side is due to the indirect coupling via the target.

In Experiment 3-4 the secondary coil was laid directly on top of the primary, so that they were coaxially aligned. In this manner the magnetic field from the primary coil coupled with the secondary coil uniformly. If we slide the secondary coil sideways so that they overlap only partially, the secondary coil will begin to couple with a portion of the magnetic field on the interior of the primary coil and, simultaneously, with a portion of the magnetic field on the exterior of the primary coil. Figure 3-12 illustrates. Since the magnetic field on the outside of the primary coil is opposite in direction to the interior field, there will be some cancelation in field coupling to the secondary coil. As we continue to slide the secondary coil over, at some point there will be a position where the opposing fluxes coupling to the secondary coil exactly cancel, and there will be no overall coupling between the primary and secondary coils. This state is called *induction balance*. Keep in mind that even though the secondary coil is in a state of zero-coupling this is still effectively a transformer.

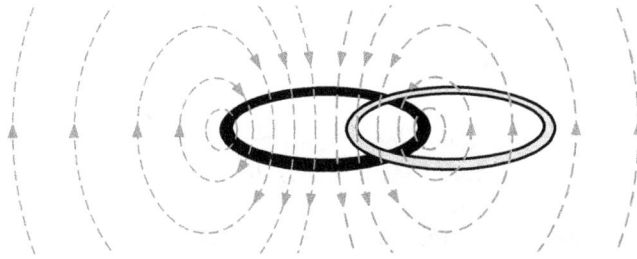

Fig. 3-12: **Induction-Balanced Coils**

As mentioned before, we call the primary coil the transmit (TX) coil and the secondary coil the receive (RX) coil. Figure 3-13 shows how the coupling varies as the position of RX coil is moved in relation to the TX coil, from off to the far left (1); to the left-side null (2); to maximum coupling (3); to the right-side null (4); and finally to the far right side (5).

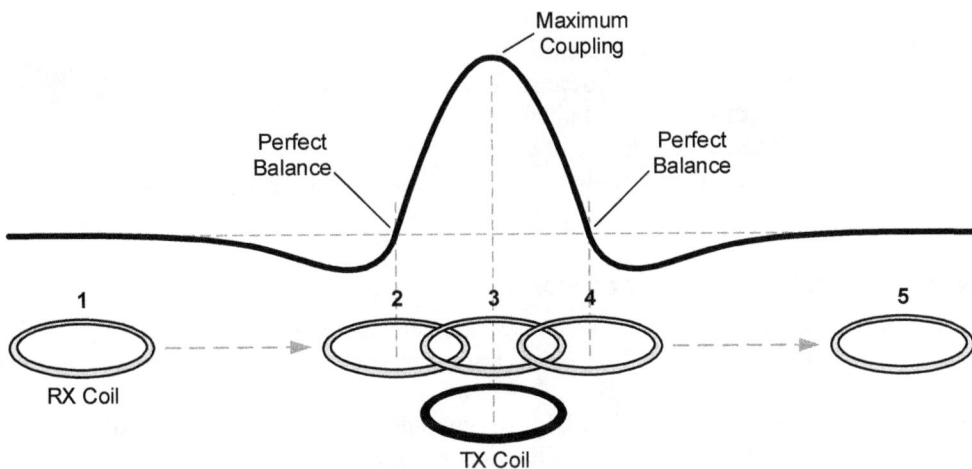

Fig. 3-13: **Coupling vs. Overlap**

Outside of the null points the coupling has a negative polarity compared to the area of maximum coupling. This can be seen on an oscope by triggering with the TX signal and noting the RX signal is inverted. Also notice that right at the points of null, the slope of the coupling is at a maximum. This makes positioning the coils for a null difficult as it is the most sensitive point to movement. Later, we'll look at how to deal with this problem.

Experiment 3-5: Demonstrate induction balance.

Required: The setup from Experiment 3-4.

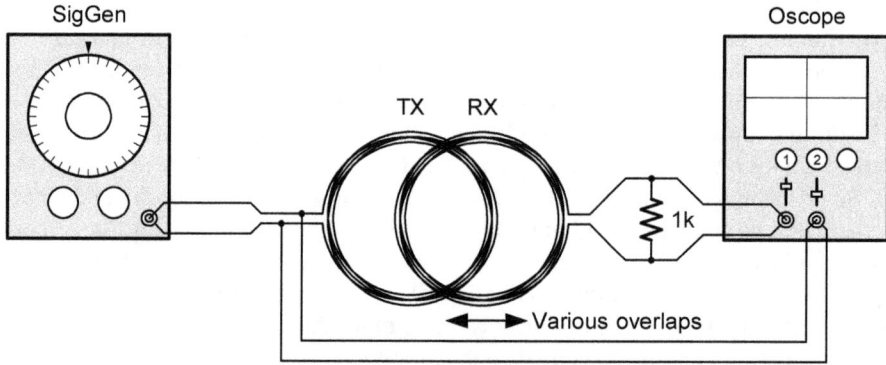

Fig. 3-14: **Induction Balance Test**

Use the same setup as in Experiment 3-4, but also connect the second channel of the oscope to the transmit (TX) coil as shown.

Place the receive (RX) coil directly on top of the TX coil. Adjust the oscilloscope until the signal amplitude reasonably fills the screen. Begin sliding the RX coil off the TX coil. What happens to the signal? Try to achieve zero coupling; is it possible? (Increase the Ch1 sensitivity of the oscilloscope as the signal decreases.) Continue sliding the second coil beyond this point; what happens to the signal?

Return the second coil to the position where there is minimal coupling. Now introduce a large metal target close to the coils. What happens to the signal?

Experiment 3-5 demonstrates the basics of an induction balance metal detector. Alexander Graham Bell built this same arrangement in 1881, using a mechanical "interrupter" for the signal generator and a telephone earpiece in place of the oscilloscope. The interrupter was noisy and had to be placed in an adjacent room. The coils were placed in a pair of sliding wooden blocks to permit adjustment of the induction null.

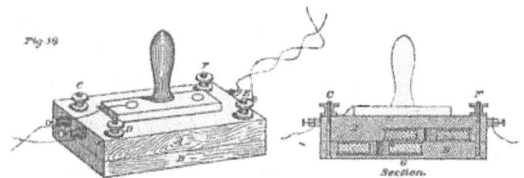

Fig. 3-15: **Bell's IB Coil**

Project: Alexander Bell's Detector

In honor of Alexander Bell's achievement at building the first applied metal detector, we will make our first detector project a replication of his effort. Using the same coils from Experiment 3-5, build the following circuit:

The relay can be most any 5V SPDT or DPDT relay; I used a PR16-5V-540-2C from CUI Devices. A 10µF capacitor is used to slow down the oscillation and give the "NO" pole a chance to make good contact. The "speaker" should be an earbud or some other high-impedance element; an ordinary 8-ohm speaker will not work. I used an AT-1220-TT-R transducer from PUI Audio. A 50Ω speaker may also

Inside the Metal Detector

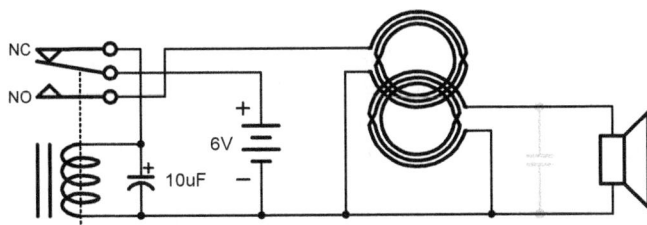

Fig. 3-16: **Bell Metal Detector**

work. Depending on the speaker element, a parallel capacitor may help produce a cleaner audio. In the spirit of Bell, do everything with point-to-point wiring; there were no PCBs or breadboards in 1881.

When the battery is connected the relay will begin oscillating as the 'NC' contacts energize the coil, but as soon as it's energized the contacts are opened and it turns back off once the capacitor discharges enough. During the brief energized state, the 'NO' contacts apply the battery voltage directly to the TX coil. This arrangement is basically the same as Bell's design.

When you listen to the earbud, you will hear a strong buzzing sound if the coils are not balanced. As the coils are balanced, the buzzing sound becomes fainter but never completely vanishes. The relay itself makes a lot of noise which can make it hard to hear the receiver sound, which is why an earbud is best. This is also why Bell had to put the interrupter in a separate room. With no metal targets near the coils, adjust the overlap until the receiver audio sound reaches a minimum. Then, moving a metal target over the coils should produce a slight increase in the audio response. In my case, I could clearly hear a large silver coin at an inch or so but a 58 caliber Minnie was barely detectable even inside the overlap.

As it is shown, this configuration produces a similar transmit signal as a PI design (see Chapter 19). Bell added a "condenser" (capacitor) across the TX coil which causes the signal at turn-off to "ring" for a longer duration and provide an extended audio signal.

Explore

The main contents of the chapters in this book are meant to teach the art of metal detector design and sometimes gloss over details to keep the Big Picture in focus[14]. Likewise, many of the designs in this book are imperfect and are meant to teach concepts, not to set you up with polished designs ready to manufacture. The designs also tend to be on the lean side of complexity in order to keep them simple to understand and easy to build and debug. Many of the chapters in this book end with a section called *Explore*. This is where extra material is placed with the intent to

- Clarify or expand on some topic presented earlier in the chapter
- Present oddball variations of previous examples
- Show additional enhancements or variants of project circuits
- Talk about weird stuff

In some cases *Explore* will point you in directions without giving full details of what you might find. That's why it is called *Explore*: you get to seek and discover.

3.5: Compass Experiments

In the three experiments that use a compass, the type of compass matters. Many better compasses are dampened with liquid and these are slow to respond, both to a transient event (Experiments 3-1 & 3-2) or to a periodic event (Experiment 3-3). In the transient experiments it may be difficult to see a dampened compass deflect, and a periodic magnetic field may cause barely perceptible oscillations in the compass needle. In Experiments 3-2 and 3-3 winding a lot more turns of wire around the nail (say, 100) will help.

For these experiments a cheap plastic compass is preferred over a quality one. These can be found on eBay for a few dollars or from science supply houses.

---

14. The classic "forest-for-the-trees" idiomatic failure.

3.6: Visualizing Magnetic Fields

Figure 3-6 is a photo of the effects of a magnetic field on iron filings sprinkled around a coil. This makes for a fun experiment although it does require a beefy power supply to pull it off. The coil used had a diameter of 4"/10cm with 96 turns of 24 AWG bonded magnet wire. Other sizes and turns can be used with the following guidance: a smaller coil will produce a higher flux density for a given turns and current; more turns will increase flux density for a given size and current. If your power supply is limited to, say, 1 amp then a smaller coil with the same or more turns will produce better results.

Use a large sheet of stiff cardboard and cut two holes exactly at the diameter of the coil. If your coil is wound on a bobbin using bonded wire (as was mine) then the holes will be rectangular; if your coil is hand-wound and tied or taped, then round holes are more appropriate. Also cut a slot between the holes. See Figure 3-17 for help visualizing this. Slide the coil through the slot; carefully center the coil and glue it in place perpendicular to the cardboard. Hot melt works well for this. Finally, cover the cardboard with two sheets of ordinary white printer paper, notching the edge of the sheets where the coil passes through the cardboard. With the sheets abutting each other along the center of the cardboard[15], attach them to the cardboard. Double-sided tape is useful for this.

Lay the whole assembly flat and carefully level it — it is important that the board be as flat as possible, with no bowing or twisting; this allows the iron filings freedom to properly align with no influence from gravity. Liberally sprinkle iron filings on the paper, trying to get even coverage everywhere. Something like a salt shaker is useful for this. Iron filings are easily found on eBay or science supply houses.

Attach a DC power supply and run a large current through the coil. Figure 3-6 was achieved with about 6 amps. With current flowing, some gentle taps to the assembly will coax the iron filings to align. The coil can get quite hot with this much current so **be brief**. You may want to turn on the current for a few seconds, tap the board, then turn the current off. While off, the filings will retain their position. Wait a minute for the coil to cool off, then repeat until the field pattern looks good. If you don't like the results, pour the filings back into the container and try again. In Figure 3-6 there is a notable lack of filings in the immediate areas around the coil. Here the field is strong enough that the filings get pulled all the way to the coil winding.

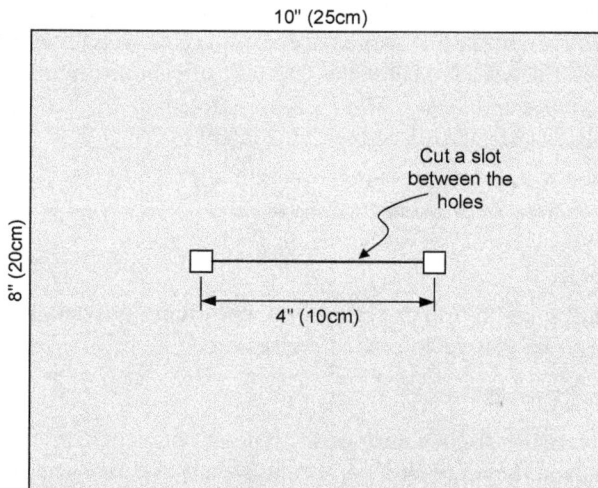

Fig. 3-17: **Cardboard Holder**

---

15. If you look closely at Figure 3-6, you will see a faint line to the left side of the coil where the paper sheets were joined.

Target Responses

"To be sure of hitting the target, shoot first,
and, whatever you hit, call it the target."

— Ashleigh Brilliant

Metal detectors generally invoke two types of target responses: magnetic, and eddy. The term "target" is not specific to just buried items but also the matrix in which they are buried. As we will see, ground minerals, sea water, and even the motion of the coil through the Earth's magnetic field produce a response. Simultaneously, individual targets produce a response that can be entirely magnetic, entirely eddy, or a mixture of the two. The mechanisms for the two responses are completely different so we will consider them separately and, in the next chapter, look at more complex combinations of signals.

Magnetic Responses

All materials respond to magnetic fields in some way. Five behaviors exhibited are diamagnetism, paramagnetism, ferromagnetism, ferrimagnetism and antiferromagnetism[1]. For the metal detectorist, only ferromagnetism and ferrimagnetism are of interest. In both of these behaviors elements can be magnetized, either temporarily or permanently. Examples of ferromagnetic materials are iron, nickel, and cobalt which, not coincidentally, are the dominant elements used to make permanent magnets. Magnetite is the important ferrimagnetic mineral in metal detection. For the purpose of metal detection we will lump these two behaviors into one term: the *magnetic response*.

4.1: Magnetic Lag

Exploring the magnetic response begins with a little bit of backpedaling. In the previous chapter the magnetic field produced (*transmitted*) by a coil was expressed as a "B-field," which is actually the magnetic flux *density* within a given material matrix. The *driving magnetic field* is called the "H-field" and for static fields is related by the equation

$$B = \mu H \qquad\qquad\qquad \text{Eq 4-1}$$

where μ is the permeability of the medium. In air (all we have considered thus far) the permeability is just μ_0, and in many other materials it is expressed as μ_0 times a relative term ($\mu = \mu_r \mu_0$) so B and H differ only by a constant. But for ferromagnetic materials in a *changing* magnetic field (H) the permeability can vary in real time with the resultant magnetic flux density (B) within the material. The result is

---

0° Phase Reference

Metal detectors typically measure the amplitude and phase of the receive signal to determine characteristics of the target. A phase measurement needs a reference point for comparison, and throughout this book the 0° phase reference is the *transmitted magnetic field*. Thus all phase-domain responses we discuss will be with respect to a sinusoidal TX field. When we discuss time domain responses, all time delays will also be with respect to the TX magnetic field, whatever waveform that field happens to have.

---

1. The exact nature of these 5 behaviors is easily found elsewhere and the details are not particularly important to metal detecting.

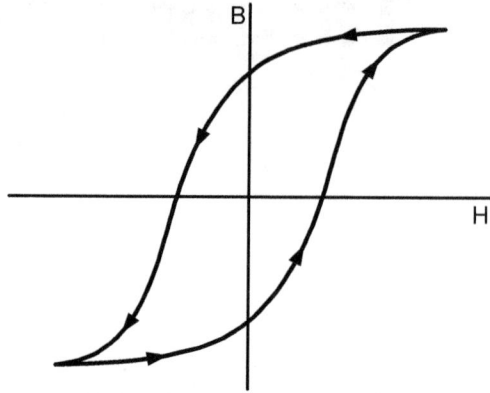

Fig. 4-1: **Magnetic Hysteresis Loop**

that the material's magnetic flux density exhibits hysteresis and when the applied magnetic field H is swept back and forth between positive and negative extremes (as is the case with metal detectors) the material flux density B traces out a hysteresis loop as shown in Figure 4-1.

At any point on the curve the material's instantaneous μ is the slope of the B-H curve. The hysteresis curve shows that if the H-field gets strong enough, the B-field exhibits a saturation limit where additional applied H-field produces little additional B-field. This is the point where the μ of the material falls to μ_0, the same as air. Metal detectors don't have a strong enough H-field to drive a magnetic material into saturation[2] so the B-H curves for typical magnetic targets look more like those in Figure 4-2.

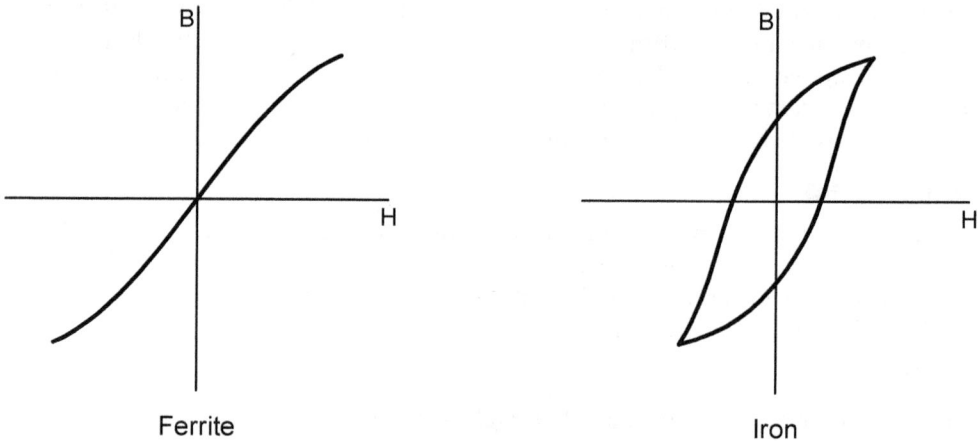

Ferrite

Iron

Fig. 4-2: **Hysteresis Curves**

The first curve in Figure 4-2 has no hysteresis gap so that B follows H with no lag. This would be typical of pure magnetite (or a perfect ferrite); it will distort the metal detector's transmit field but only in an instantaneous manner. The result is that the receive coil[3] will see a signal that is perfectly in phase with the transmit signal; that is, a 0° signal at some amplitude, though it may be slightly distorted. The second curve has hysteresis which results in a magnetic "lag" in B with respect to H. This is common with many small iron targets such as nails. Like ferrite, these targets will distort the TX field but the distortion will also have a delay. The result at the RX coil is a distorted signal that is no longer in phase with the TX signal, shown in Figure 4-3.

---

2. But fluxgate magnetometers do.

3. We assume the receive coil is normally in a state of perfect induction balance with the transmit coil. That is, lacking any kind of target disturbance, the receive coil signal has zero amplitude.

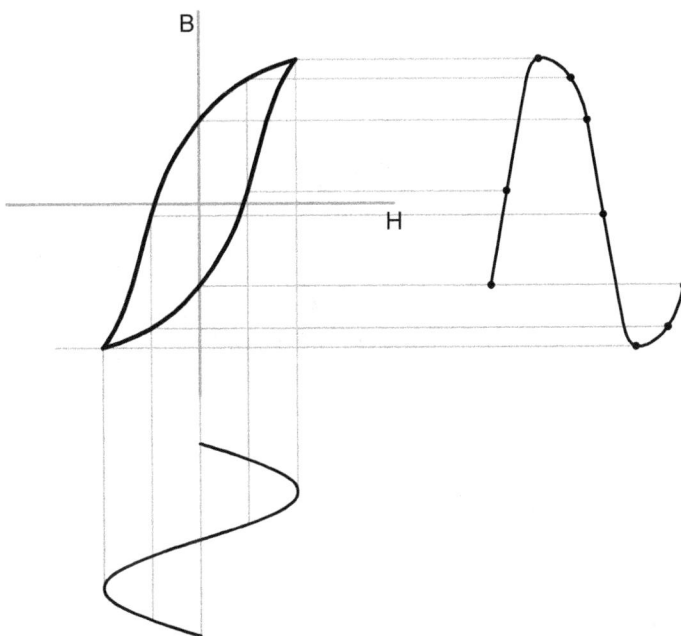

Fig. 4-3: **Hysteresis Response**

The amount of phase shift depends on the magnetic lag which, for many iron targets, is fairly independent of frequency, except when magnetic viscosity effects are present. The reason for this is that the magnetization of iron is a static effect; that is, if you apply a magnetic field to a piece of iron it becomes magnetized and will remain magnetized even if the field is removed. A reverse field is required to demagnetize the iron, and the frequency at which this process occurs largely doesn't matter, at least within the realm of frequencies at which metal detectors typically operate.

Magnetic lag is proportional to the spread in a B-H curve and the amount of spread represents real work done in magnetizing and demagnetizing the magnetic material. The energy lost in this process is dissipated as heat and is commonly referred to as *magnetic loss*. Another term, *magnetic loss angle*, refers to the phase angle between a sinusoidal driving H-field and the resultant B-field. This is of interest to us. Ideal magnetite has a loss angle of 0° but, in reality, even pure ferrite isn't perfect and can exhibit a phase shift of up to a couple of degrees. In testing and calibrating metals detectors, a ferrite material is often used to produce a near-zero magnetic phase response, though the selection of ferrite should be made with care as some[4] produce a greater lag than others.

Materials with a gap in the B-H curve produce a greater loss angle. Typical soil values range from a few degrees up to 15° in extreme cases. Iron targets have a much higher loss angle, usually in the area of 40-60°. The lag between the B-field and the H-field can never be greater than 90°, so any purely magnetic response will be limited to a phase shift between 0° and 90°.

Experiment 4-1: Look at a magnetic response.

Required: The setup from Experiment 3-5, plus a ferrite rod[5] and a silver dollar[6].

---

4. So-called "soft" ferrites typically have low lag while "hard" ferrites have greater lag. Hard and soft have nothing to do with the feel of the ferrite, they all feel hard.

5. Or any piece of ferrite, including toroid or a pot core.

6. Or any large, thick, (preferably) silver coin.

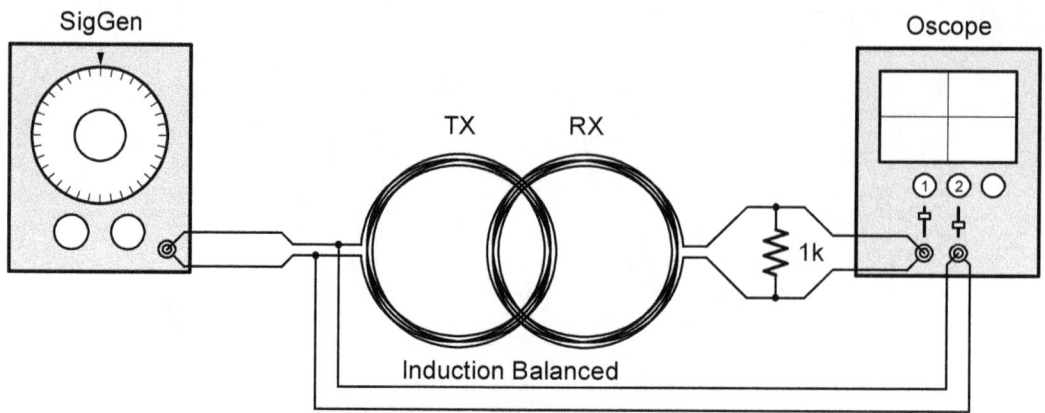

Fig. 4-4: **Magnetic Response Test**

Using the same setup as in Experiment 3-5, set the signal generator to produce a sine wave at 10kHz at the highest peak voltage possible without distortion. Adjust Ch2 on the oscilloscope until the TX signal reasonably fills the screen. Place the RX coil exactly on top of the TX coil and verify that the RX signal is in-phase with the TX signal. There is a 50% chance that it will be 180° out-of-phase which means the RX coil is connected backwards. The easiest remedy is to simply flip it over.

Set Ch1 (the RX signal) to 100X the resolution (volts/div) of the TX channel. Slide the receive coil over the transmit coil until you achieve induction balance (it will be approximately the position shown above). At this point the RX signal should be negligible and will be a fairly flat (but perhaps noisy) line across the screen.

Place the ferrite inside the overlapped area of the coils taking care not to touch either coil or otherwise upset the induction balance. What happens to the RX signal? (You may need to increase the resolution even more.) Remove the ferrite and introduce the silver dollar; what happens to the signal? Try other frequencies and see how the ferrite response behaves, both in terms of phase and amplitude.

You should see that the ferrite RX response is in-phase with the TX signal while the silver dollar response is closer to 180° out-of-phase. The plots in Figure 4-5 are taken from a digital oscilloscope with averaging enabled.

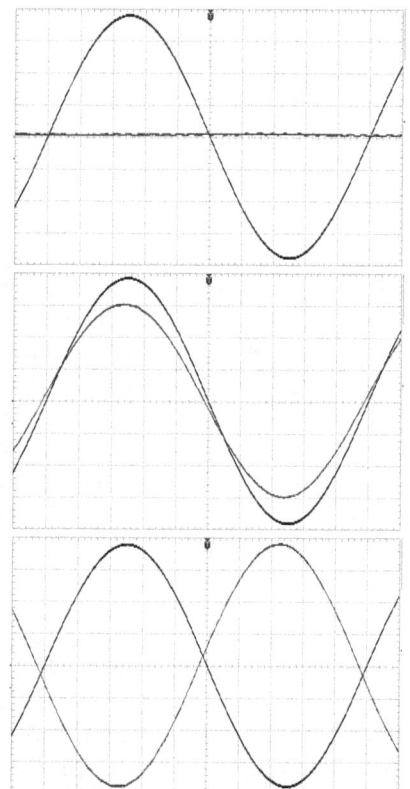

Fig. 4-5: **Phase Responses: No Target, Ferrite, Silver Dollar**

4.2: Ground vs Iron

Let's back up a little and revisit magnetite. Magnetite is an oxide of iron (Fe_3O_4) and is commonly used in making high-permeability cores for inductors and RF transformers. Magnetite is also a common mineral in soil; generally, the term "mineralized soil" or "mineralization" refers to soil containing magnetite. How a metal detector responds to magnetite is important and modern designs go to great pains to try to "ignore" mineralization. We will look at specific methods in future chapters but, broadly speaking, ground mineralization has two characteristics that make it unique from other magnetic responses and other target responses in general.

As mentioned earlier, pure magnetite has a near-zero phase response. Soil mineralization is not always so ideal, but is still typically found to be in the range of about 0° to perhaps 5-6° with extreme soils even higher. Practically no targets fall in this range so metal detectors take advantage of this fairly narrow range of exclusivity to recognize it as a ground signal and implement some manner of ground compensation.

The second characteristic of ground mineralization that differs from normal targets is that targets are highly localized responses while mineralization is widely dispersed throughout the soil and in many cases is relatively homogeneous. This means that the signal at the receive coil will have a sharp narrow response for targets but a slowly changing response for mineralization. Applying a high-pass filter to the receive signal attenuates the ground signal with little effect on targets.

Other magnetic responses — mostly iron targets — produce a significantly faster response at a higher phase angle than ground so these techniques cannot be used to eliminate them, assuming that is a desired goal. Instead, most metal detectors employ classical discrimination techniques based on phase response to eliminate undesired targets. The following table lists a few small iron targets plus ferrite and a "hot" rock with their phase responses from 1kHz to 100kHz.

	1kHz	2kHz	5kHz	10kHz	20kHz	50kHz	100kHz
Ferrite	0°	0°	0°	0°	0°	0°	0.2°
Red volcanic rock	3.7°	4.1°	5.0°	5.9°	6.6°	7.8°	8.9°
16d nail on-end	30.4°	34.0°	39.5°	43.5°	45.0°	51.3°	51.3°
Square nail on-end	26.6°	29.6°	32.6°	35.1°	36.2°	36.6°	36.9°
Square nail flat	13.4°	16.4°	23.2°	31.6°	44.1°	69.2°	91.7°
3/8" Washer	68.0°	90.0°	112.1°	122.1°	131.0°	142.0°	149.7°

Table 4-1: **Ferrous Responses vs Frequency**

4.3: Hot Rocks

Many soils have a reasonably homogeneous distribution of mineralization within a given area large enough to accommodate an average coil sweep. That is, a typical coil swing will not normally result in large changes in mineralization. There are, however, exceptions to this. In cases where the soil has been disturbed there can be abrupt transitions in mineralization phase. An example might be a trench line where utilities are buried and covered over and the resulting soil disturbance alters the mineralization compared to undisturbed soil. This is not uncommon in urban settings.

Another exception are so-called *hot rocks*. These are mineralized rocks which have a different loss angle than the surrounding soil matrix. They are also highly localized responses and so in terms of response speed they look more like a target than ground. The term *hot rock* is a general term to describe rocks that fall into the typical range of soil responses but differ from the immediately surrounding soil matrix. Ideally, a soil matrix containing pure magnetite would have a 0° phase, but in reality there are effects that usually push the ground phase a few degrees above 0. Therefore it is possible for a hot rock to have a phase angle that is either higher or lower than the matrix. When specifically differentiating these two cases, rocks with a phase angle higher than the matrix are still called hot rocks, but rocks with a phase angle lower than the matrix are called *cold rocks*[7].

Hot rocks are usually not very noticeable in a typical silent-search type metal detector. Where they are most annoying is with threshold-based detectors, especially a first derivative type detector typically used for gold prospecting. And, to make matters worse, it is often the areas where gold is found that prospectors encounter more severe soil conditions containing hot rocks. In a worst-case scenario, the

---

7. Some people use the term *negative hot rocks* to refer to cold rocks, and *positive hot rocks* for regular hot rocks.

soil matrix itself is highly variable and contains both hot rocks and cold rocks, creating miserable conditions for many metal detectors. This situation can be found in stream beds where the rocks are from many different upstream locations. Using a common single-frequency gold detector in such conditions can result in rapid machine-gun like responses that make detecting nuggets almost impossible.

4.4: Viscous Magnetic Response

There is a final variation of magnetic response called *viscous remanence magnetization* or VRM. So far we have explored magnetite which has a near-zero phase lag and does not retain residual magnetism after an applied field is removed. We also considered iron targets which have a much higher loss angle and can retain residual static magnetism.

A third consideration is viscous magnetic material. This is characterized by its ability to be temporarily magnetized but when the driving field is removed the material's magnetization decays back to zero or some low-level value. That is, it has a tendency to hold the magnetization for a very brief time before returning to a demagnetized state. The magnetic decay has a power law response[8] independent of drive frequency, meaning that it will have a phase response that can increase with frequency. Maghemite is the predominant mineral with VRM and is another iron oxide (Fe_2O_3).

Most hobby metal detectors are designed to only deal with the more mundane soil conditions and generally that's all a single-frequency IB-type detector can do. To tackle both instantaneous and remanence responses at the same time demands either a multifrequency IB detector or a PI detector. We will explore this more in later chapters.

4.5: Earth Field

A final consideration in magnetic responses is the magnetic field of the Earth. In a hobby metal detector the search coil is moved back and forth, and a coil moving through a magnetic field will experience an induced EMF. The Earth's magnetic field can be sufficient to produce a decently strong EMF, though it varies with location. At the equator the effect is minimal, while at the Earth's poles[9] it is more pronounced. This phenomenon is usually dealt with in the design of the demodulators so we will save the solutions for later.

Eddy Responses

Table 4-1 shows that ferrite and small iron targets produce phase responses between 0° and 90°. Experiment 4-1 offered a glimpse at another response mechanism caused by the silver dollar whereby the response was closer to 180°, far beyond that of magnetic delay. We will now consider what causes the silver dollar signal.

Experiment 3-4 showed that a changing magnetic field induces an EMF in a secondary coil, and the EMF can be seen as a voltage on an oscilloscope. If the coil is connected to some sort of useful circuit then the EMF will push electrons and produce a current. What if, instead of a secondary coil connected to a circuit, we introduce a disconnected piece of metal to a changing magnetic field? There will still be an EMF and it will still try to push the electrons around, but when you have a disconnected piece of metal the electrons have nowhere to go. So they do something really odd: they just go around in circles, something called *eddy currents* (Figure 4-6). If you have a *really* strong EM field you can get really high eddy currents, enough to make the metal heat up. An induction cooktop generates enough eddy currents in a metal pot or pan to cook food; without the pot/pan, no heat is produced. Foundries also use induction crucibles to melt steel.

---

8. The time domain response is $f(t) = k \cdot t^{-\alpha}$ where α is ideally assumed to be close to 1; that is, $f(t) \cong t^{-1}$.

9. While it is perfectly reasonable to assume that there is not much opportunity for metal detecting at the poles, metal detectors have been employed in Antarctica to look for meteorites. When I worked at White's we provided GMTs for such an expedition, and they were successful not only in locating meteorites but also a buried food cache left by a 1950s expedition.

So with a metal detector we drive an AC current through a coil which produces an alternating EM field around the coil. When a metal target is near the coil, the EM field induces an EMF in the target, per Faraday's Law of Induction, which generates eddy currents. How is that useful? It turns out — per Lenz's Law — that the direction of the induced eddy currents are such that they create a target magnetic field that *opposes* the incident magnetic field (Figure 4-6).

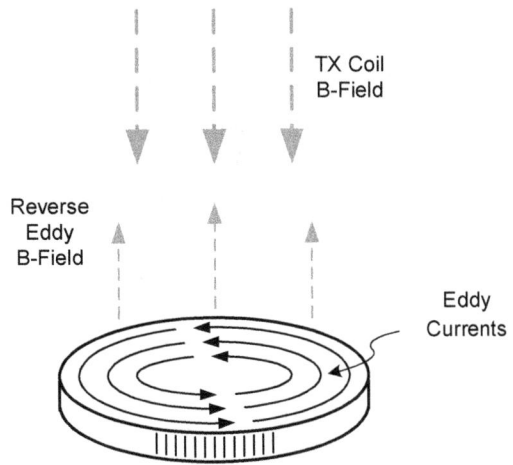

Fig. 4-6: **Eddy Currents & Field**

This reverse EM field gives us something to look for in order to detect the presence of a metal target. The problem is, the target field is extremely weak compared to the one that was produced by the primary (TX) coil. If we introduce our secondary (RX) coil right on top of the TX coil (position 3 in Figure 3-13) the resulting induced RX signal will be swamped by the direct-coupled TX signal, and the target signal will be impossible to distinguish[10]. What we need to do is eliminate the presence of the TX field *at the RX coil* so that the RX coil "sees" only the target field. Creating this state of induction balance was the purpose of Experiment 3-5.

With the RX coil positioned so there is minimal direct coupling to the TX field we can now detect the presence of a metal target by its eddy-induced reverse field. This reverse target field inductively couples with the receive coil and generates the signal that says "you've found something." You can think of this reverse target field either as separate from the transmit field or as a *distortion* of the transmit field — it has been described both ways by various authors, and either way of thinking is fine. Technically, it is a distortion of the TX field but superposition allows us to treat it as a separate entity.

In Chapter 3 the induction-balanced coil arrangement was described as a transformer, which it is. When a target is introduced the system of the TX coil, the target, and the RX coil become a *double transformer*. For one transformer the TX coil is the primary and the target is the secondary. For the other transformer the target is the primary and the RX coil is the secondary. See Figure 4-7.

If an induced eddy current in a metal target creates a magnetic field that exactly opposes the incident magnetic field then that implies a 180° phase shift. The phase shift actually depends on some parameters of the target and only a perfect conductor (a *superconductor*) gives a perfect 180° phase shift. Everything else produces a lesser phase shift, and it turns out that variations

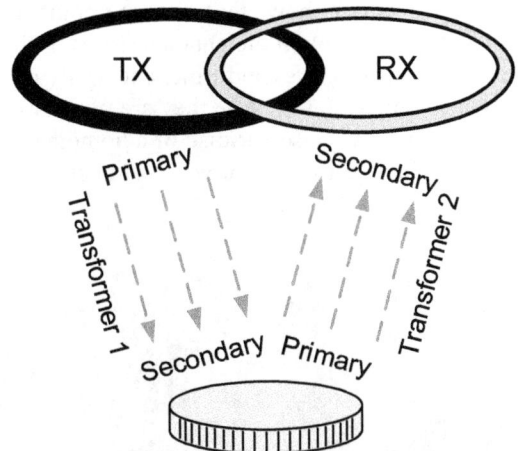

Fig. 4-7: **Double Transformer**

in the total phase shift is how we can discriminate between different targets. Thicker, higher conductive targets, like Big Silver Coins, are closer to 180°. Thinner, lower conductive targets, like aluminum foil, are closer to 90°. Ironically, while the magnetic lag response is limited to the range of 0° to 90°, the eddy response is limited to the range of 90° to 180°. If you read this and think, "Oh good, an easy way to tell ferrous from non-ferrous," you will soon see that reality is not so black-and-white.

---

10. It's like trying to hear someone whisper at a rock concert.

Experiment 4-2: Demonstrate target phase shifts.

Required: Everything from Experiment 3-5, various targets.

Again use the same set-up as in Experiment 3-5. Now introduce silver, copper, and cupronickel coins. Try a small (2"/5cm) square of household aluminum foil. Pay attention to the resulting amplitudes and phase shifts for each target.

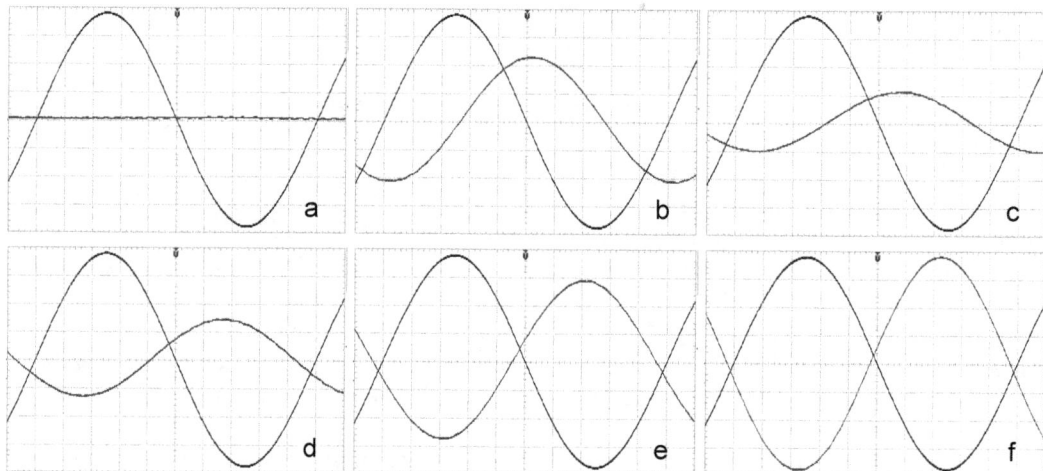

Fig. 4-8: **Target Phase Shift Test**

Figure 4-8 shows the results for (a) no target, (b) foil, a US (c) nickel, (d) cent, (e) quarter, and (f) silver dollar.

Experiment 4-2 shows that different coins produce different phase shifts. The phase of the eddy currents depends a lot on the characteristics of the target including the metal alloy, the shape, and the physical dimensions. We can further exemplify by looking at coin responses in more detail. Coins make for a good comparison because they are popular detecting targets and simplify the discussion by being almost universally a circular disc of a homogeneous metal alloy. Furthermore, we'll use pre-clad US coinage[11] which, for over 100 years, almost never changed in either size or alloy. Here are our initial targets:

Fig. 4-9: **US "Classic" Coins**

Besides a long-term consistency, US coinage has the advantage of having four different-sized coins of identical alloy (silver), two coins of close dimensions but different alloys (cent & dime), and a coin that is larger and thicker than these latter two but with a different alloy still (cupronickel). These all serve to effectively illustrate how dimensions and alloy each have strong influence on the eddy response.

---

11. The coins shown in Figure 4-9 represent my favorite US designs, pinnacles of US coin artistry before the unfortunate move toward depicting dead presidents.

Coin	Value	Slang alloy	Actual Alloy	Diameter	Thickness[1]
Cent	1¢	Copper	95%Cu 5%SnZn	19mm	1.55mm
Nickel	5¢	Nickel	75%Cu 25%Ni	21.2mm	1.95mm
Dime	10¢	Silver	90%Ag 10%Cu	17.9mm	1.35mm
Quarter	25¢	Silver	90%Ag 10%Cu	24.3mm	1.75mm
Half	50¢	Silver	90%Ag 10%Cu	30.6mm	2.15mm
Dollar	$1	Silver	90%Ag 10%Cu	38.1mm	2.4mm

Table 4-2: **US "Classic" Coinage**

1. Thicknesses are nominal; actual thicknesses vary depending on the relief.

For the time being, the Induction Balance Test setup used in Experiment 4-2 will suffice. The resulting phase shifts at 10kHz[12] are:

	Phase
Cent	166°
Nickel	123°
Dime	169°
Quarter	174 °
Half	175°
Dollar	177°

Table 4-3: **Measured Coin Phases**

Despite having the same metal conductivity, the phase shifts of the four silver coins increase as both target size and thickness increase. The cent is slightly lower than the dime despite being slightly larger and thicker. The nickel shows that, even with a larger diameter and thickness than the cent or the dime, it has a lower phase shift yet. From this we can deduce that the phase shift is due to a combination of the alloy properties, plus the size and thickness of the object[13]. To understand what is going on we need to look at the physics of how eddy currents are generated.

4.6: The Physics of Eddy Currents

The transmit coil produces a magnetic field with flux density B. At some distance from the coil, a coin presents itself as a target to that B-field and now we need to consider the physics of that interaction. According to Faraday's Law, a changing magnetic field imposed on a conductor induces an electromotive force that is

$$\varepsilon = -\frac{d\Phi}{dt}$$ Eq 4-2

The term Φ[14] is the *total magnetic flux* which intersects the surface area of the coin. For a perpendicular field[15] it is equal to the magnetic flux density B times the surface area of the coin, or

$$\Phi = B \cdot A$$ Eq 4-3

12. 10kHz was commonly used in Chapter 3 and will continue to be our "standard" frequency.

13. And, perhaps, the shape, should we explore targets other than round flat discs.

14. Note that Φ (capital phi) is used for flux; we will use ϕ (little phi) for target phase shift.

15. That is, the magnetic flux is perpendicular and the coin is horizontal, or flat.

That is, Φ is simply the total amount of B that actually hits the target. For a given B field strength, a bigger target will see a larger Φ and produce more ε. This is why large targets are easier to detect than small targets, everything else being equal.

Equation 4-2 states that, somewhere, there is an electromotive force (EMF) created in or on the target. EMF is measured in volts so a reasonable question might be, "Where can I measure this voltage?" The answer is: *you can't*. The EMF is generated in a closed loop that encircles the incident magnetic line of flux that intersects the surface. This loop can be modeled (for now) as a voltage (the EMF) and a resistance (the circular path through the metal). If we take this to the simplest lumped-element model we have just a volt-

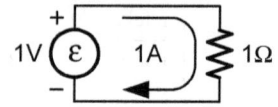

Fig. 4-10: **Single-Element Lumped Model**

age source and a resistor. To simplify the discussion, let's say the EMF is 1 volt and the total metal resistance is 1Ω. This model tells us that a current of 1A will be generated. This is the eddy current.

We can progressively create a more accurate model by breaking down the EMF and the resistance into smaller elements. Figure 4-11 shows the target modeled with two elements and four elements. The model consists of a single loop of elements because the eddy currents flow in circles. The increased elements proportionately divide the EMF and the resistance into smaller values but the resulting induced current is still 1A. Imagine that we continue this to 1000 elements; that is, 1000 voltage sources, each with 1mV of EMF, and each with $1m\Omega$ of series resistance. The overall current is still 1A. This can continue down to the level of individual electrons which is ultimately where the induction effect occurs. At that point, the model would have billions of elements of nearly-zero volts and nearly-zero ohms but with the overall effect of still creating a 1 amp eddy current. We can't actually measure a voltage — it's dispersed to an infinitesimally small level — nor can we measure the current directly; all we can do is indirectly measure the effects.

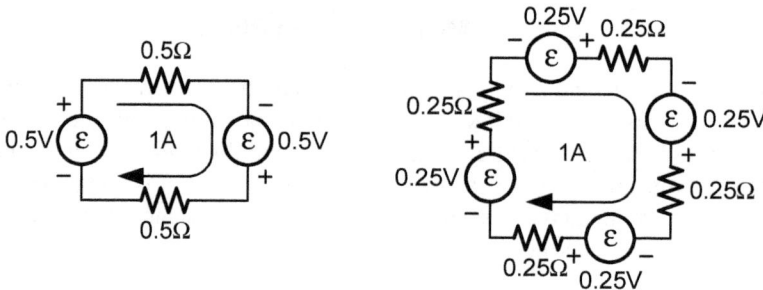

Fig. 4-11: **2x & 4x Lumped Models**

The resistance in the above discussion is due to the finite conductivity of the target's metal. It is reasonable to assume that a high-conductivity metal (like pure silver) will produce a stronger eddy current than a low-conductivity metal (like lead), given the exact same size and shape. This is a little bit of a simplification because there is more to the metal than just resistance when it comes to AC inductive effects. This additional complication is, in fact, what gives rise to the varying phase shifts, which is covered in subsequent sections.

Getting back to the effects of the transmit field on a target, Equation 4-3 gives us the total flux when the target is perfectly normal (perpendicular) to the field. For a coin which is angled to the direction of the field (Figure 4-12) the total flux is

$$\Phi = B \cdot A \cos\theta \qquad\qquad \text{Eq 4-4}$$

where θ is the tilt angle of the coin in the field. This shows that a target which is tilted relative to the magnetic field lines will generate weaker eddy currents and be more difficult to detect. Metal detectors which attempt to estimate depth do so assuming the targets are flat. A tilted target will be shallower than the metal detector's depth reading. Keep that in mind when you plunge a digger into the soil.

As the angle approaches 90° the magnetic flux cutting through the surface of the coin approaches zero and the coin theoretically cannot be detected. In reality, the edge of the coin can also support eddy currents though the response will be weak. If the coin is close enough to the coil then the motion of the coil over the coin can create vertical responses on either side of center and you will hear a double response. More about this in the next chapter.

It is apparent from this investigation that the optimum condition for creating maximum target eddy currents is a transmit coil parallel to the ground, a target whose maximum surface area is parallel to the coil, and the target

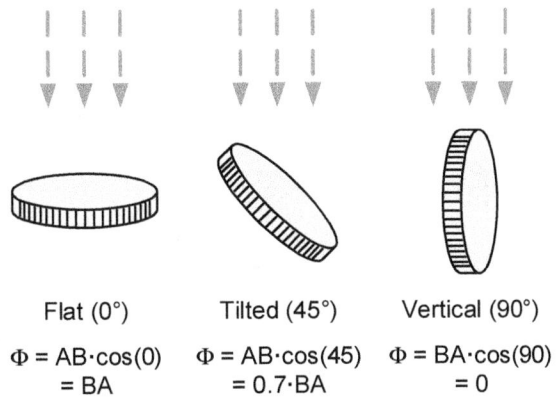

Flat (0°)	Tilted (45°)	Vertical (90°)
$\Phi = AB \cdot \cos(0)$	$\Phi = AB \cdot \cos(45)$	$\Phi = BA \cdot \cos(90)$
$= BA$	$= 0.7 \cdot BA$	$= 0$

Fig. 4-12: **Incident Field vs Target Angle**

positioned directly under the center axis of the transmit coil. The eddy currents then create a reverse magnetic field which induces a voltage in the receive coil, also according to Equation 4-2. The same conditions for creating maximum target eddies also apply to generating a maximum signal at the receive coil; namely, the receive coil is parallel to the target and the target is under the center axis of the receive coil.

4.7: Eddy Depth Distribution

Consider a coin in a perpendicular magnetic field. The incident field creates a surface eddy current[16] which produces a reverse magnetic field. This reverse field partially (but not completely) cancels the incident field so that a reduced incident field continues to penetrate into the interior of the coin. This reduced field then generates a deeper eddy current which is weaker than the surface eddy. The deeper eddy further reduces the incident field and the effect continues until either the penetrating incident field is reduced to zero, or it reaches the far side of the coin.

To better illustrate this let's consider a coin sliced into several separate layers as shown in Figure 4-13. The reverse field of the top layer opposes the incident magnetic field and weakens the amount that makes it to the second layer, producing a weaker eddy current at that layer. This effect continues with additional depth resulting in an ever-diminishing incident field and resulting eddy current at each layer. The figure above splits the coin into four layers but, like the model in Figure 4-11 which can be subdivided infinitely, the layers tend toward an infinite number that approach zero thickness.

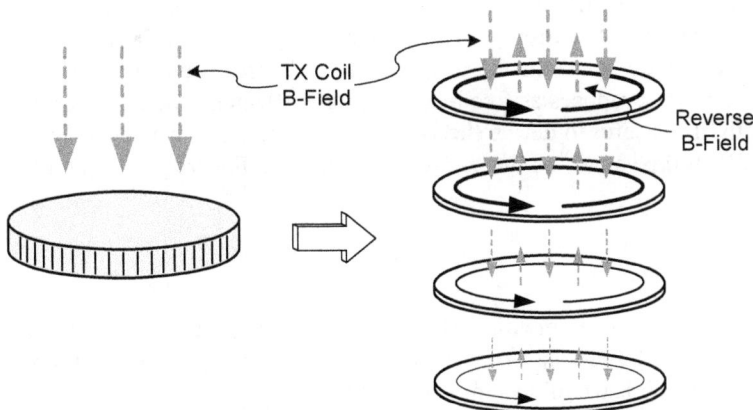

Fig. 4-13: **Eddy Currents in a Layered View**

16. For illustrative purposes we'll draw it as a single eddy current path.

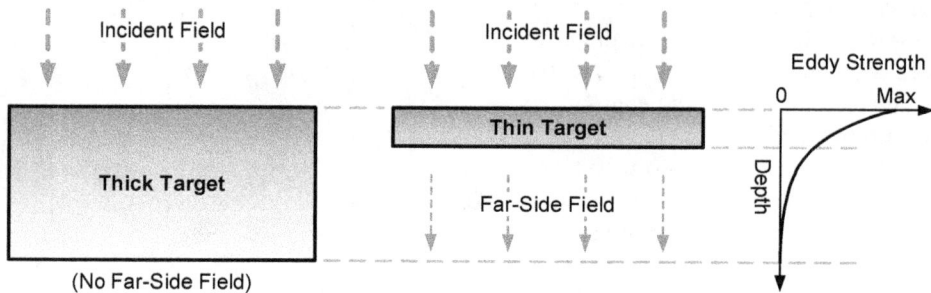

Fig. 4-14: **Eddy Reduction vs Depth**

The diminishing eddy currents with depth is exponential in nature so if the target is thick enough the deeper eddy currents will diminish effectively to zero. Figure 4-14 illustrates this in another way by looking at the side view of the target. A thick target may have near-zero eddy currents on the far surface which means the incident magnetic field has been fully attenuated and there is no remaining field on the far side. A thinner target will still have substantial eddy currents on the far surface, meaning that not all of the incident field was attenuated and there will be some remaining incident field on the far side[17]. To use light as an analogy, the thick target would be opaque and the thinner target would be translucent.

4.8: Skin Effect

The process described in the last section is known as *skin effect*. It is present whenever an AC current flows through any conductor. In an ordinary wire conducting a DC current the current density is uniform throughout the cross-section of the wire. But when conducting an AC current, skin effect causes a reduction in the current density at the center of the wire in the same way our coin sees a reduction in eddy currents at depth. We will see in a future chapter a special type of wire that is made to counter this effect.

Eddy currents are created by the EMF produced by the incident magnetic field which, from Equation 4-2, involves the derivative of the field waveform. Assuming that the incident magnetic field is a sine wave with amplitude B_0 — that is, $B(t) = B_0 \sin(\omega t)$ — the EMF will have the form[18]

$$\varepsilon(t) = -\frac{d\Phi}{dt} = -A \cdot \frac{d}{dt} B_0 \sin(\omega t) = -A B_0 \omega \cdot \cos(\omega t) \qquad \text{Eq 4-5}$$

where A is the area of the target.

Equation 4-5 shows that EMF increases with frequency and we assume the eddy currents will also increase. A stronger surface eddy current will produce a stronger reverse-field, and the stronger reverse-field in turn more effectively weakens the deeper eddy currents. The result is that the penetration of eddy currents into the target reduces as frequency increases. Because the eddy currents decrease exponentially with depth, it is useful to define the *skin depth* as the point where the interior eddy currents have reduced to 37% of the value[19] of the surface eddy current. For frequencies typical of metal detectors skin depth is

$$\delta = \sqrt{\frac{2}{\sigma \omega \mu}} \qquad \text{Eq 4-6}$$

where σ is the conductivity of the metal, μ is the permeability of the metal, and ω is the radian frequency of the magnetic field. For non-ferrous metals the permeability is roughly μ_0 (ferrous metals may be substantially higher) so it is useful to simplify this to

17. Not just weaker, but also phase-shifted. Remember, the TX field must be changing, e.g., sinusoidal.

18. It is assumed that $\Phi(t) = A \cdot B(t)$ which is valid for non-ferrous metals.

19. 37% is e^{-1}, the classic one-tau decay of an exponential.

$$\delta = 503.3 \times \sqrt{\frac{1}{\sigma f}}$$

Eq 4-7

As we see, the skin depth is inversely proportional to the square root of the frequency.

Example: Calculate the skin depth for silver and lead at 1kHz, 10kHz, and 100kHz.

Solution: The conductivity of silver and lead are 6.29×10^7 and 4.54×10^6 S/m respectively, and $\mu = \mu_0 = 4\pi \times 10^{-7}$ H/m.

	1kHz	10kHz	100kHz
Silver	2.01mm	0.63mm	0.20mm
Lead	7.47mm	2.36mm	0.75mm

Table 4-4: **Skin Depth Examples**

The example shows that eddy currents can penetrate deeper in a low conductor like lead versus a high conductor like silver. It also shows that penetration decreases with increasing frequency. For any target the eddy current at five skin depths is e^{-5} or 0.67% the value of the surface current, or a reduction of 99.3%. For all practical purposes, 5 skin depths is considered the total penetration depth of the eddy currents.

4.9: Eddy Surface Distribution

So far eddy currents have been vaguely described as circular currents. An eddy current arises from a magnetic field intersecting a conductive surface and we have been depicting the B-field from the TX coil as several vectors (Figure 4-6). In reality, it is a continuum of field lines that intersect the surface of the target but for illustration we use discrete vectors. There are a couple of ways to visualize how the eddies behave.

Suppose seven discrete vectors represent the incident field on a coin, and suppose each vector gives rise to an eddy current. The resulting surface eddies would look like those in Figure 4-15a. While all the eddies flow counterclockwise, we see that the center eddy opposes each of the other six eddies where they meet.

If the seven eddies are equal then the interior eddy currents will cancel while a perimeter eddy current is sustained. Even though there is a continuum of induction across the surface of the coin the resultant eddy flow occurs at the perimeter of the coin. This is the simplistic view of eddy behavior; now let's look at a more physical approach similar to the eddy depth analysis.

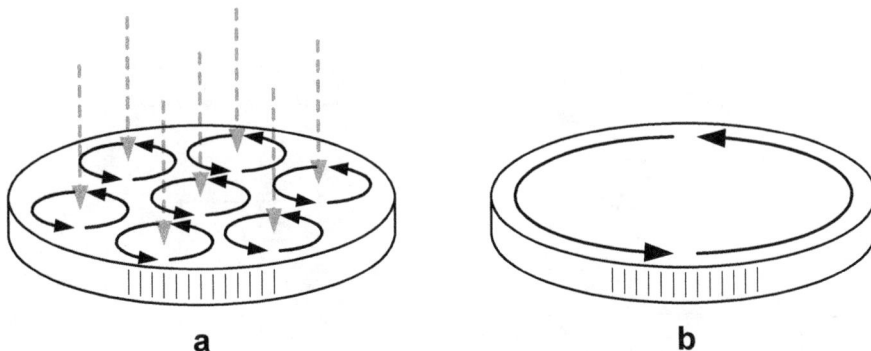

a   b

Fig. 4-15: **Eddy Surface Distribution**

Suppose that, instead of creating small individual eddies as shown in Figure 4-15, the incident B-field creates a concentric eddy flow as shown in Figure 4-16. While shown as individual vectors, the eddy flow is actually continuous across the surface of the coin but for illustration we will consider it as discrete concentric currents.

Imagine that the coin is divided into individual concentric rings (plus a center core) and for each piece there is an eddy current (Figure 4-17a). Suppose we also imagine that this can be viewed on-edge as if the coin is cut in half (Figure 4-17b); eddy currents are then represented with a 'dot' (eddy current coming out of the page) and an 'x' (eddy current going into the page[20]). Each of these currents creates a corresponding circular B-field that helps form the overall target reverse B-field we keep talking about.

Let's start with the outer ring in Figure 4-17c; the eddy flow creates a B-field[21] that encircles the ring as shown. On the inside of the ring the B-field opposes the incident TX field. This means that the second ring (d) will see a weakened incident field resulting in a weakened eddy current. Likewise, additional flux lines from the outer ring will also oppose the TX field seen by the other rings (e and f), reducing their eddies.

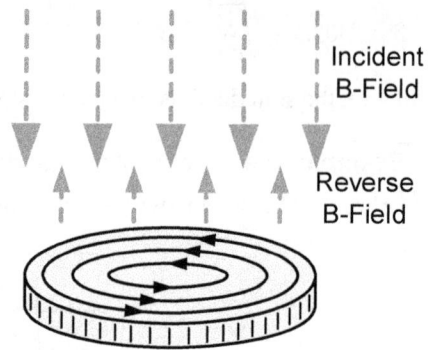

Fig. 4-16: **Simplistic View**

The B-field from the second ring (d) additionally opposes the TX field presented to the third ring (e) and the core (f). However, it *adds* to the TX field presented to the first ring which will make the outer ring eddy current *stronger*. The flux created by the other inner eddies (in e and f) likewise boost the outermost eddies.

The net result from this is that the B-fields from the concentric eddy currents progressively weaken the interior currents while at the same time strengthen the outermost currents. The overall effect is that the eddies push themselves to the perimeter of the coin. Recall from the discussion of skin effect that the weakening of eddies versus depth was exponential in nature and that after five skin depths the eddy current was effectively zero. The exact same mechanism is at play across the surface of the coin with the eddies getting exponentially weaker as we move from the perimeter of the coin to the center. By the time we are five skin depths from the perimeter the eddies are effectively zero.

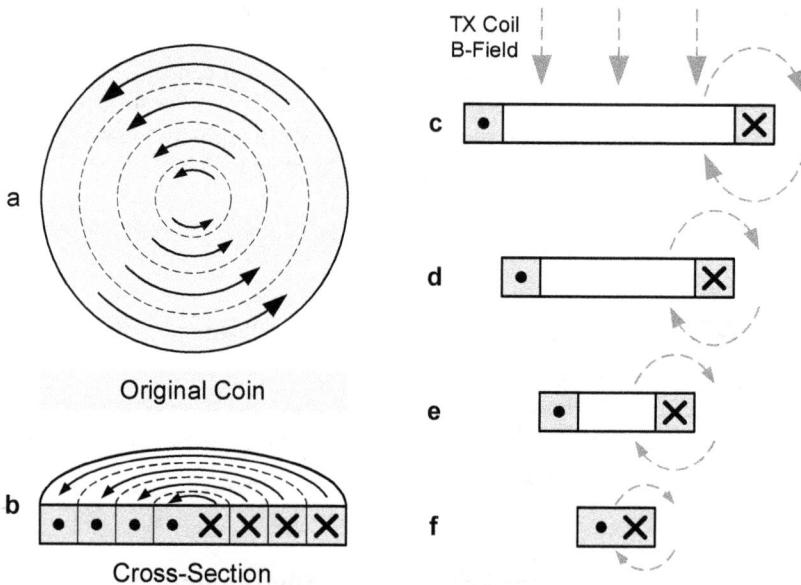

Fig. 4-17: **Concentric Ring Model of Eddies**

20. In vector math the dots represent the point of the arrow and the X represents the tail of the arrow.

21. This is but a representation of a single flux line. See Figure 3-6 for a more complete picture.

Fig. 4-18: **Solid vs Hollow Coin**

If all this is correct then a hollowed-out coin should have about same response as a solid coin, in both amplitude and phase. In comparing the two coins in Figure 4-18, this turns out to be true.

What we've seen with both the depth and surface analyses is that eddy current crowding is both exponential versus depth and exponential versus the radius. The resulting overall eddy distribution is shown in Figure 4-19 for two cases, both of which are depicted as the cross section of a coin: (a) represents a coin with a high conductivity and (b) represents a low conductivity coin. The high conductivity coin has a lower skin depth so eddies are pushed very close to the perimeter and also do not penetrate very deeply. In the low conductivity coin eddies penetrate even to the far surface of the coin and are also dispersed across more of the top surface.

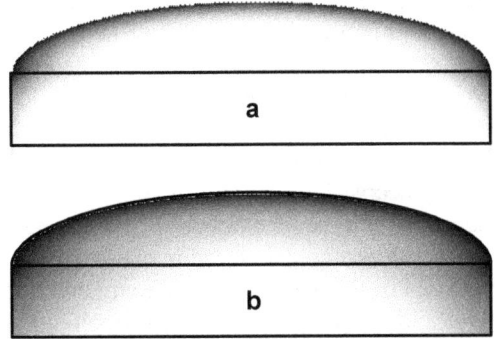

Fig. 4-19: **Eddy Distribution**

4.10: Target Field Phase

Suppose we have a perfect *superconductor*[22]; that is, $\sigma = \infty$. Regardless of frequency, the skin depth is zero which means that all eddy current generation is limited to the surface. This also means that none of the incident magnetic field penetrates the target. Therefore the target's reverse magnetic field is exactly equal in amplitude and exactly 180° out-of-phase with the incident magnetic field. In terms of an easily understood visual, a superconductor looks like a perfect mirror to the incident field and at the surface reflects it back 100%. On the far side of the superconductor target there is no magnetic field to be seen.

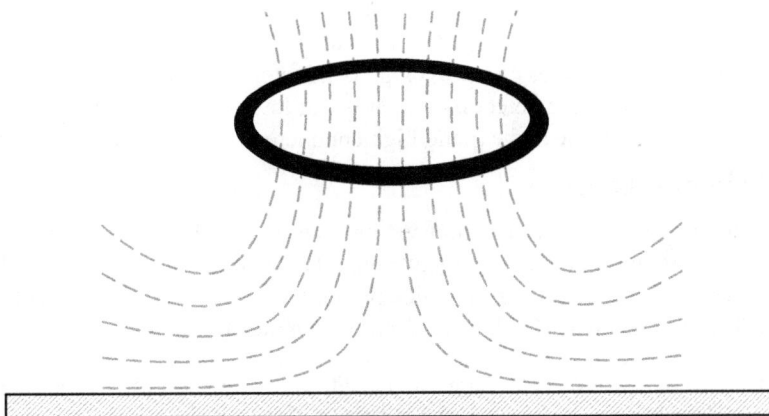

Fig. 4-20: **Superconductor**

22. Superconductors really exist, but rarely at room temperature. You're unlikely to find one with a metal detector, but they make for a good hypothetical example. For a practical example of a target with exceptionally high conductivity you will see "Atocha bars" mentioned throughout the book. An Atocha bar is an 80 pound (36 kg) bar of silver. 962 were recovered from the Atocha shipwreck.

For a second example, consider a 5mm thick silver plate in a 10kHz magnetic field. The thickness (5mm) is roughly 8 times the skin depth (0.63mm) so the incident field absorption is 99.96% and only 0.04% makes it to the far side. In engineering we call this "roughly zero" so that at the far side of the silver plate there is essentially no magnetic field. The silver plate is behaving much like the superconductor and appears to be a near-perfect mirror to the incident magnetic field. That is, the reverse field at the surface is very close to 180° compared to the TX field[23].

This is a rather simplistic view of the physics. In reality, the eddy current at the very top surface of the plate is at 0° with respect to the EMF. Just beneath the top surface, the eddy current experiences a phase lag due to the inductive and resistive effects of the metal, and this phase lag continues linearly through the thickness of the plate. The phase shift of the eddy current at any given depth is 1 radian per skin depth[24]. That is, the eddy currents at the surface are at 0°, the eddy currents at 1δ deep are 1 radian (or about 57°), the eddy currents at 2δ deep are 2 radians (or about 114°), and so forth. Again, all with respect to the surface EMF, not the incident B-field.

It seems confusing that the surface eddies are at 0° and at 3δ deep they are almost 180°. We are looking for an explanation as to why the EMF-to-eddy phase shift is close to 90° for the thick silver example. The answer is that while deeper eddy currents have progressively more phase shift (even beyond 180°) they are also exponentially weaker and contribute to the surface phase shift in an exponentially decreasing amount. Therefore the total phase shift seen at the surface (in the form of the eddy-induced B-field) never exceeds 90° compared to the EMF.

It is interesting that the eddy phase shift of 1 radian per skin depth has no dependency on any other properties of the metal. Therefore any metal target that is 5δ thick will have an overall EMF-to-eddy shift of almost 90° whether it is silver, aluminum, or lead. This leads to the conclusion that even though silver coins are considered high conductors, a coin of any other metal could mimic a silver coin if it is thick enough[25].

To illustrate how skin depth affects the target response Table 4-5 shows the phase response for the US dime, quarter, and dollar coins at 1kHz, 10kHz, and 100kHz.

	1kHz	10kHz	100kHz
Dime	134.0°	169.4°	179.9°
Quarter	142.3°	174.4°	179.99°
Dollar	152.7°	178.0°	179.999°

Table 4-5: **Phase Shifts vs Frequency**

Notice that at 100kHz all three coins are crowded together at 180°. Experienced detectorists know that detectors which run at high frequencies compress high conductors toward the 180° end of the VDI scale while expanding low conductors, but detectors which run at low frequencies compress low conductors toward the 90° area of the VDI scale and expand high conductors. More on this in Chapters 16 and 24.

4.11: Target Field Amplitude

So far our discussion of skin depth has focused on its effect on the phase of the target field, especially versus frequency. But phase isn't the only attribute that changes with frequency; the amplitude of the target field also changes. Consider a 2mm-thick pure silver coin in an exceptionally low frequency (100Hz) TX field. The skin depth at 100Hz is 6.35mm which means that a large amount of the incident

---

23. By necessity, this conversation covers the phase shift of eddies both with respect to the incident TX field and also with respect to the induced EMF. It is always a difference of exactly 90°.

24. This is derived in many papers and books concerning eddy current *non-destructive testing*, a popular way of looking for defects in metal parts. An excellent book is *Introduction to Electromagnetic Non-Destructive Test Methods* by Hugo Libby.

25. Detectorists commonly refer to "low conductors" and "high conductors" with respect to the phase (or decay) response of targets. This is clearly a misnomer as the metal conductivity is only part of the response. Since I'm also guilty in this abuse of terminology, nothing more will be said about it.

Inside the Metal Detector

field (about 73%) completely penetrates the coin and continues on the other side; therefore, the strength on the "reflected" field is low (about 27%). If the frequency is increased to 1kHz, the skin depth is now 2mm and more of the incident field is reflected (~73%) with less seen at the other side. At 10kHz the skin depth is 0.63mm and the reflection is stronger (96%), and at 100kHz the skin depth is 0.2mm and virtually all of the incident field is reflected. Figure 4-21 illustrates what this looks like.

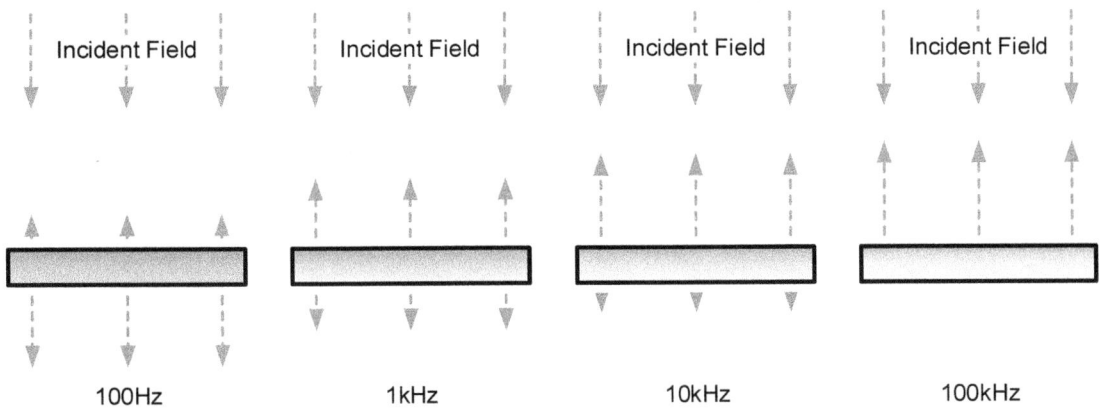

Fig. 4-21: **Target Amplitude vs Frequency**

The above analysis is not very rigorous because the math is very involved and depends on the size of the coin as well as thickness and metal conductivity. But as the incident field is increased from some low frequency to some high frequency, both the amplitude and the phase of the reflected field also increases. Different coins of different sizes, thicknesses, and conductivities will have different specific amplitude and phase responses but they all follow the same trend. Since a higher frequency offers a higher reflected amplitude it seems to suggest that a metal detector should operate at the higher frequency, but we will see in Chapter 16 why this is not so.

4.12: RL Target Model

In explaining how eddy currents are physically generated we proposed a target model consisting of an EMF generator and a resistor. This model obviously does not account for the frequency-dependent phase shift that occurs, so when a model doesn't work it is time to refine the model. Previous examples show that with increasing frequency the phase of

Fig. 4-22: **RL Model**

the reverse magnetic signal increases toward 90° with respect to the EMF. This can be equated to a composite eddy current that increases toward 90° with respect to the EMF as frequency increases. A model that replicates this behavior is a simple RL circuit in Figure 4-22.

In this model the inductor is the EMF generator; an incident magnetic field will produce an EMF in an inductor exactly per Equation 4-2. The combined inductor and resistor also model the frequency dependency of the eddy phase response. At low frequencies the resistor dominates and the loop current is almost in-phase with the EMF, which is 90° with respect to the TX field. At high frequencies the inductor dominates the response and the loop current approaches 90° with respect to the EMF, or 180° with respect to the incident B-field. Since the reverse field has the same phase as the RL loop current, the overall result is a reverse B-field phase response between 90° and 180° from the incident field.

If we were to use this model in an actual Spice simulation we would need to choose values for R and L. Suppose we want to model a coin that has a 140° phase shift at 10kHz. Subtracting the 90° shift due to the EMF, the EMF-to-eddy phase shift becomes

$$\phi = \arctan\left(\frac{\omega L}{R}\right) = 50° \qquad \text{Eq 4-8}$$

or L/R = 19µs. For phase shift it doesn't matter what the actual selected values are so the easiest solution is to choose 1Ω for the resistor and 19µH for the inductor. In a simulation, the target inductor is mutually coupled to the TX inductor, and then again to the RX inductor[26]. The strength of the response can be adjusted in the mutual coupling parameters. Figure 4-23 shows how to set up a simulation — although the target RL is an isolated magnetic circuit it will need to be grounded to satisfy the Spice requirement that every node has a DC path to ground. Because the RL model is limited to the response range of 90° to 180° it is not valid for modeling magnetic (ferrous) responses.

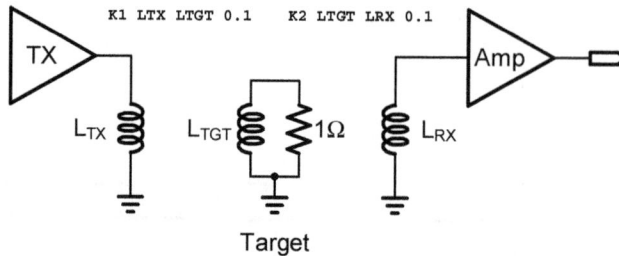

Target

Fig. 4-23: Sim Setup For Using Target Model

Note that L/R is actually a time constant; it is the *characteristic time constant* of the target (often called the *target tau*) and is independent of frequency[27]. Equation 4-8 can be rewritten as

$$\phi = \arctan(\omega\tau)$$ Eq 4-9

This can be rearranged to extract the target tau from a measured phase response at a known frequency:

$$\tau = \frac{\tan\phi}{2\pi f}$$ Eq 4-10

Again, ϕ is the EMF-to-eddy phase shift and does not include the 90° EMF shift.

Spice simulations allow us to look at signals that we normally can't see in a real circuit, like the eddy current response of a target. It is important to remember that we still can't see the EMF of the coil (it is infinitely distributed) and that the voltage across L_{TGT} represents the eddy current response, not the EMF. Also, when looking at the RX voltage waveform it will have yet another EMF derivative due to the RX coil that has not been covered. See Chapter 14 for this discussion.

4.13: Wire Loop Tau

Let's link this new target model back to the eddy current physics that have been explored. It was shown in Figure 4-15 that the generated eddy currents push themselves to the perimeter of the target. For a coin, this means that all the effective eddy current flows in just an outer ring of metal. Before we try to determine exactly what that means let's first consider a single turn shorted coil of ordinary wire. The inductance of a single-turn loop is[28]

$$L = \mu r\left(\ln\left(\frac{8r}{r_w}\right) - 2\right)$$ Eq 4-11

where µ is the permeability of air, r is the average radius of the loop, and r_w is the cross-sectional radius of the wire. This wire ring also has physical resistance which is the resistor in the RL model. The resistance can be calculated as

---

26. Again, forming a double transformer system.

27. Not exactly; tau can vary with frequency due to the skin effect altering the effective inductance. For now we'll ignore this.

28. This simplified formula was proposed by F. W. Grover in *Inductance Calculations*, 1946. A rigid solution using Maxwell's equations involves elliptic integrals. We are thankful for Mr. Grover's solution.

$$R = \frac{1}{\sigma} \cdot \frac{l}{a}$$
Eq 4-12

where σ is the conductivity[29] of the metal, l is the length (circumference) of the loop, and a is the cross-sectional area of the wire. Dividing L by R gives

$$\tau = \mu\sigma \cdot \frac{r_w^2}{2} \cdot \left(\ln\left(\frac{8r}{r_w}\right) - 2 \right)$$
Eq 4-13

Example: Calculate the tau for a shorted loop of 18 AWG copper wire with a diameter of 50mm (2").

Solution: Let's calculate inductance and resistance separately to get an idea of their magnitudes. The loop radius is 25mm and the wire radius is about 0.5mm. The core of the loop is air so $\mu = \mu_0$. Inductance is therefore

$$L = \mu_0 \cdot 0.025 \left(\ln\left(\frac{8 \cdot 0.025}{0.0005}\right) - 2 \right) = 125.4\text{nH}$$

The conductivity of copper is 5.96×10^7 S/m and the circumference is 157mm so the resistance is

$$R = \frac{1}{5.96 \times 10^7} \cdot \left(\frac{0.157}{\pi \cdot 0.0005^2} \right) = 3.35\text{m}\Omega$$

Therefore, the tau is

$$\tau = \frac{L}{R} = \frac{125.4\text{nH}}{3.35\text{m}\Omega} = 37.4\mu\text{s}$$

We can check this result by constructing this wire loop and measuring the phase response at, say, 10kHz. The result should be

$$\phi = \arctan(\omega\tau) = 66.9°$$

The actual phase measurement is 67.4°, or a tau of 38.3µs. This is in reasonable agreement.

A loop of wire was used in the prior example because it has definitive characteristics: there is a definite average loop radius, a definite wire radius, and the loop is pure copper so conductivity is also definite. This makes calculating inductance and resistance easy and the loop of wire is well-behaved in a verification test.

Coins are a bit more iffy. How close to the perimeter do the eddy currents flow? Although the current crowds the perimeter it also has an exponential distribution (Figure 4-19) that makes both the average surface radius and the conduction radius difficult. What is the exact conductivity of the alloy? Some coins are not homogeneous; modern US clad coinage has a core of pure copper sandwiched between cupronickel layers. Notice that the wire example used μ_0 for permeability; that's because the core of a single-turn loop of wire is air. For a coin where the eddy current flows along the perimeter in the same manner as the wire, the core of the "coin loop" is the rest of the metal in the coin. Whether we can use μ_0 depends on the alloy. Many coin alloys[30] have a relative permeability of 1 but those that have a high iron or nickel content (and therefore tend to be magnetic) may have a much higher permeability.

All of this makes it difficult to mathematically calculate the tau of a coin. Applying the method used in the prior example and using dimensions determined by the predicted eddy distributions of Figure 4-19 will usually gives results that are roughly in the ballpark. But the best way to determine the tau of a coin is to just measure it.

---

29. Many textbooks use $R = \rho \cdot l / a$ where $\rho = 1/\sigma$ and is the resistivity. Same results.

30. All silver and gold alloys, most copper alloys, and some cupronickel alloys.

4.14: Salt Water

Salt water is electrically conductive so it will support the creation of eddy currents just like any other conductor. While there is a fair amount of variation, seawater has a typical conductivity of 4 S/m so at our three previously used frequencies the skin depth is:

	1kHz	10kHz	100kHz
δ	7.96m	2.52m	796mm

Table 4-6: **Saltwater Skin Depths**

The question now becomes, what is the phase shift caused by seawater? As with a metal target, phase would normally depend on the thickness (depth) of the material compared to skin depth and salt water can extend very deeply into the sand. But seawater has such a large skin depth that it easily exceeds the practical reach of the transmit field. For example, at 10kHz the 1δ depth is 2.5m; at that depth a typical TX field is close to vanishing[31]. With coins and other metal targets, we assume that the thickness of the target is small enough that the flux density of the incident magnetic field is the same across that small distance. When salt water is the target, that clearly is not the case.

Size also plays a crucial role in the response of salt water. So far we have considered the TX magnetic flux lines to be parallel and 90° to the surface of the target (e.g. Figure 4-6). Salt water encompasses a much larger volume of the TX field including flux reversals. See Figure 4-24. The resulting eddy current flows are far more complex than what we've seen in coins.

With most detectors salt water ends up with a phase shift very close to 90°, including the contribution from the EMF derivative. This is about the same as some of the thinnest pieces of aluminum foil, such as a paper gum wrapper that has a foil laminate. On a discriminating detector, foil can be eliminated but when hunting in salt water the signal is continuous and often large in amplitude, reducing sensitivity to desired targets and/or making the detector very noisy. This issue plus solutions are explored in later chapters.

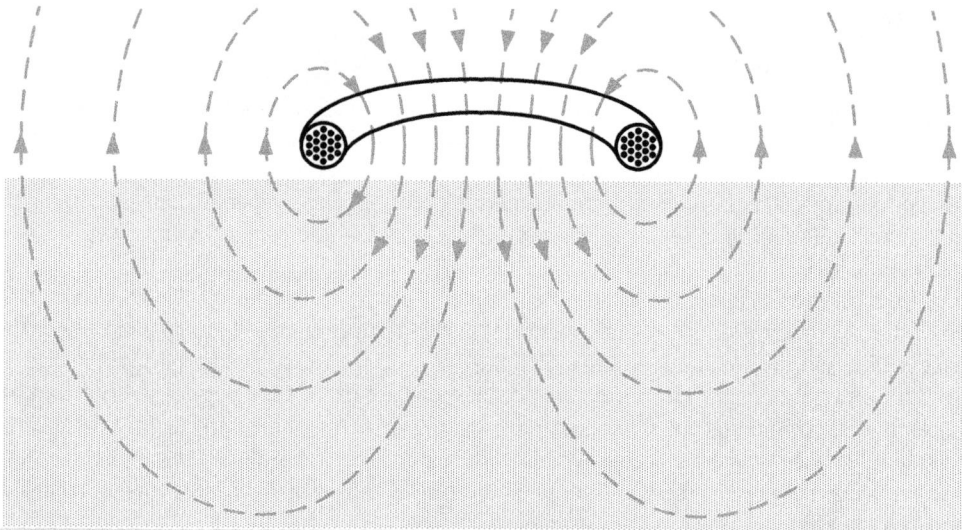

Fig. 4-24: **Salt Volume**

---

31. From Equation 3-6, a 25cm coil has lost 99.988% of B_0 at 2.5m.

Transmitter Effects

When a metal target is introduced to an alternating magnetic field eddy currents are produced in the target. The eddies produce an opposing magnetic field that can be detected by an induction balance arrangement, but there may be other effects to the system depending on the design of the circuitry.

Ideally, if the transmit coil current is a single frequency sinusoid then the coil can be resonated with a capacitor and whatever power is require to create the magnetic field is returned to the circuit when the magnetic field collapses. That is, an ideal (resonated) sinusoidal transmitter should consume zero power. Obviously this won't be the case as circuit non-idealities (such as coil resistance) will cause power losses. But the amount of power consumed is constant over time, at least in the absence of a target.

The eddy currents produced in a target are real currents, and due to the metal resistance they dissipate real power in the form of heat[32]. This power is delivered to the target from the transmit coil via the magnetic field and ultimately results in a noticeable power loss in the transmit circuit. So even if a metal detector doesn't have a receive coil to detect the eddy field it is still possible to detect the presence of metal by carefully monitoring the transmit circuit power losses. While typically not as sensitive as using a receive coil, this method is still in use today in popular pinpointer products and we'll look at such a design in Chapter 9.

Another effect of targets on the transmitter is a frequency shift. Metal detectors almost always use an air-core transmit coil where the permeability is μ_0, that of air. The introduction of a target with higher permeability — say, ferrite minerals or iron — will increase the inductance of the transmit coil. If that inductance is a determining part of the transmit frequency then the frequency will change. In most cases, a rising inductance creates a frequency drop.

Eddy targets can also produce a shift in frequency, but not because of a change in permeability. Most non-ferrous metals have a μ_r close to 1, the same as air. Instead, the eddy currents steal energy from the transmit coil, making the coil appear as if it has a lower inductance. A lower inductance generally creates a rise in the frequency. See Figure 4-25.

The effect of targets (whether ferrous/magnetic or non-ferrous/eddy) on transmit frequency is called *frequency pulling* and is normally only found in free-running (non-driven) transmit circuits, such as a Colpitts or Hartley oscillator. Older designs often used free-running oscillators; some, like BFO, work entirely on the concept of frequency shift. Others, like TR, ignore or even compensate for frequency pulling.

Most modern detectors use a driven transmitter design; that is, the coil is driven with a voltage or a current at a frequency determined elsewhere in the circuit, such as a microprocessor. In this case, a target will still affect the apparent inductance of the coil but the coil no longer determines the transmit frequency. The frequency remains stable but, depending on the circuit design, the inductance shift may

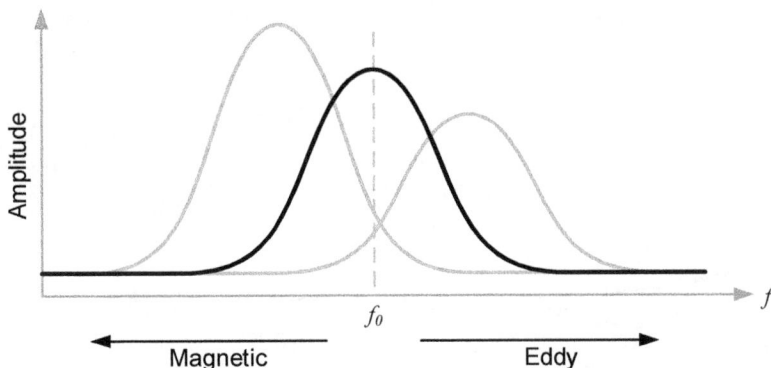

Fig. 4-25: **Target Effects on TX Oscillator**

---

32. Sometimes enough to cook dinner, or even melt steel, as mentioned before.

---

still slightly distort the transmit waveform and/or affect the efficiency of the transmitter if, for example, an LC tank is being used to recycle energy. Waveform distortion can affect the TX current zero-crossing which is typically used in VLF receivers to trigger the demods, and timing shifts due to the distortion can affect ground balance.

The design of the transmitter also determines how much of an effect eddy targets have on the power loss in the transmit circuit. With a voltage driven coil (such as a square wave drive) the effect may be minimal, whereas with a current driven coil it may be more pronounced. Some commercial detectors designs compensate for this loss to minimize the effect on whatever ground compensation methods are used.

Experiment 4-3: Determine target effects on the transmitter.

Use an LCR meter to measure the inductance of the RX coil from previous experiments. If the meter has several frequency options, select 10kHz or the next highest frequency available. Measure the inductance with no metal nearby. Place the ferrite in the center and see what happens to the inductance. Remove the ferrite and place the silver dollar in the center.

	Air	Ferrite	Silver Dollar
L	1816µH	1827µH	1796µH

The table above shows the results for an example coil. As expected the ferrite causes an increase in the inductance while the coin causes a decrease. If you place an oscilloscope on the coil while measuring its inductance you will also see that the ferrite slightly decreases the LCR drive amplitude while the coin increases it. The amplitude change is very slight and requires zooming into the signal peaks to see the effect.

Explore

4.15: Alloy Conductivities

It is tempting to assume that an alloy of two metals will have a conductivity somewhere between the conductivities of the two pure metals. Cupronickel is a very popular alloy in coinage throughout the world because it is durable and cost-effective[33]. Figure 4-26 shows a graph of copper-nickel alloys from 100% copper to 100% nickel[34]. Pure copper has the highest conductivity as expected, but even

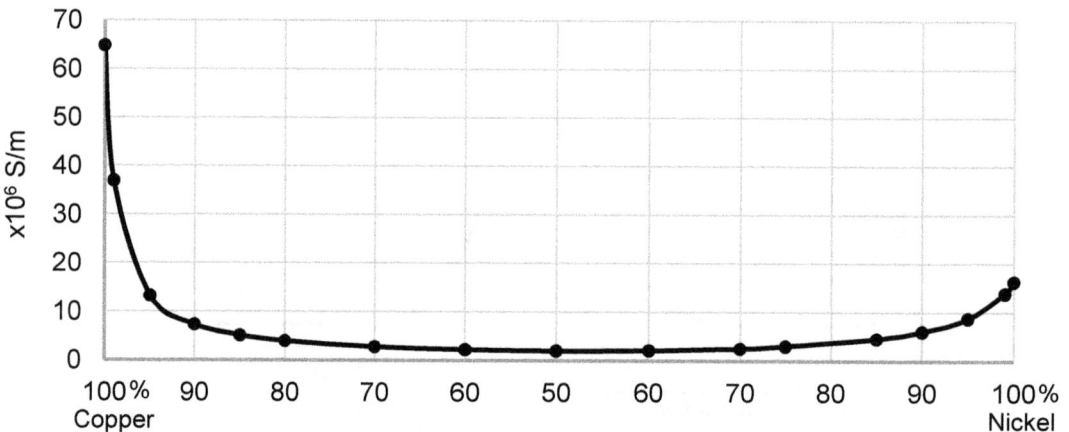

Fig. 4-26: **Conductivity of Copper-Nickel Alloys**

---

33. Or was. As of 2023, it cost 11.5¢ to manufacture a 5¢ US nickel.

Inside the Metal Detector

pure nickel has a higher conductivity than most of the alloy range. This is true with most coin alloys. Notice that a mere 1% nickel content cuts the conductivity (from pure copper) almost in half.

4.16: Practical Test Targets

In building and testing metal detectors you will need a variety of test targets. The most obvious is a good selection of coins. Sometimes you will need to look at RX signals with little or no amplification and the response of a single coin may prove to be too weak. This can be overcome by simply using a larger target but we still want the larger target to exactly mimic the behavior of its smaller cousin. In the case of coins, the way to do that is to glue seven identical coins to a piece of cardboard to increase the effective target size, as shown in Figure 4-27.

Fig. 4-27: **Making the coins look larger**

It is important that the coins do not touch so that the eddy response is limited to each individual coin. Even then, if the coins are spaced closely the eddy B-fields will interact and alter the collective response tau to be slightly lower than the individual coin. With proper spacing, the overall response is the same as an individual coin, only stronger. The same trick can be applied to ferrite and many other types of targets.

If you want a finely graduated collection of test targets then coins may not be the best choice as there will be large gaps in the taus. Household aluminum foil can be used for this. We commonly think of aluminum foil as being a low conductor with a low tau. But the foil can be stacked in layers to increase the thickness which increases the tau. Figure 4-28 shows a collection of foil squares that are

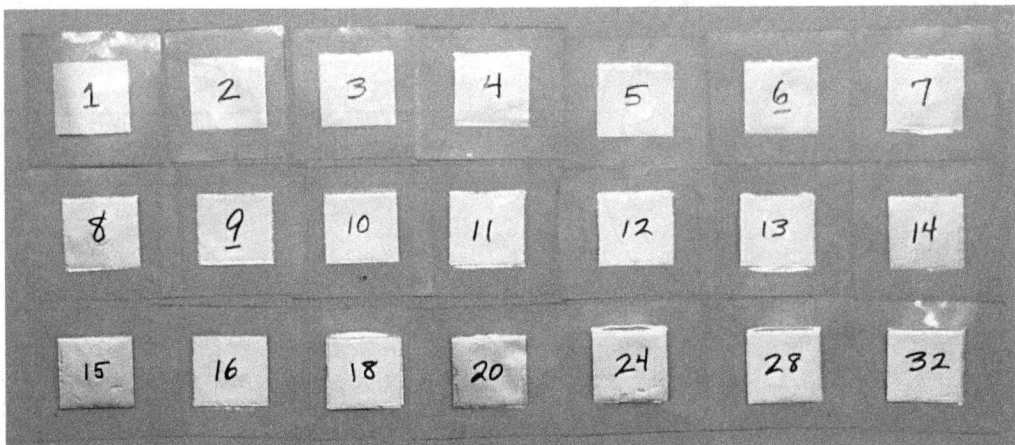

Fig. 4-28: **Foil Target Standards**

---

34. Data is from the CRC Handbook of Chemistry and Physics, 87th Ed.

25mm (1") square and range from 1x thick to 32x thick. Each one is sandwiched in clear packing tape to keep the layers compressed. The taus range from 2µs to 44µs.

With foil targets you still "get what you get" in terms of tau. If you want a specific tau (say, exactly 10µs) you can alter the dimensions and stack thickness of the foil but it's still a bit of a guessing game to achieve a certain tau. Another way to make a test target is to literally replicate Figure 4-22. Recall that we modeled coin targets as a parallel RL circuit and as an example used a shorted loop of 18 AWG wire to simulate a target. The inductance of the loop of wire and the resistance of the wire determined the tau.

We can extend this concept by adding more turns to the loop to increase inductance, or adding a physical resistor to increase resistance. One of the issues with this kind of target is that the wire itself can be detected if the wire gauge is large enough. As an example, the 18 AWG shorted loop of wire had a measured tau of 38.3µs but if the loop is broken it still has a tau of 4.4µs. Therefore 18 AWG is too large to simulate a low tau target.

While a smaller gauge wire will solve this problem, the higher intrinsic resistance will limit the maximum tau that can be achieved. Adding more turns can help; the wire resistance increases linearly while the inductance increases by roughly N^2, therefore the maximum tau will be roughly proportional to the number of turns. Still, even a very small wire gauge (say, 30 AWG) can be detected by a high-frequency gold detector so, depending on the desired tau, it may be necessary to use different wire gauges. Fortunately high-tau targets are less important in high-frequency detectors (usually used for nugget hunting) so this is not necessarily a big deal. So it may be necessary to make high-tau wands with a lower gauge wire (say, 26-30 AWG) with the understanding that they won't be accurate with high-frequency detectors. Low-tau wands can be made with higher gauge wire (say, 30-40 AWG) and they will still be accurate at low frequencies, but cannot produce high taus.

Figure 4-29 shows an RL target wand design. The coil can be any size[35] and can be one turn or several turns as needed. A physical resistor is added to accurately create any lower tau desired. The resistor can be, say, an 0402 chip resistor (which is small enough not to be detected) soldered directly to the ends of the coil. If a leaded resistor is used then it needs to be mounted away from the coil so it is not detected as a target. A long handle not only makes this possible but also serves as convenient way to wave the target in tests. It is imperative to use twisted pair along the handle from the coil to the resistor to prevent this from becoming part of the inductance loop. Finally, using a potentiometer as the resistor creates an easily variable tau target that can be used, for example, to find the target null of a PI detector.

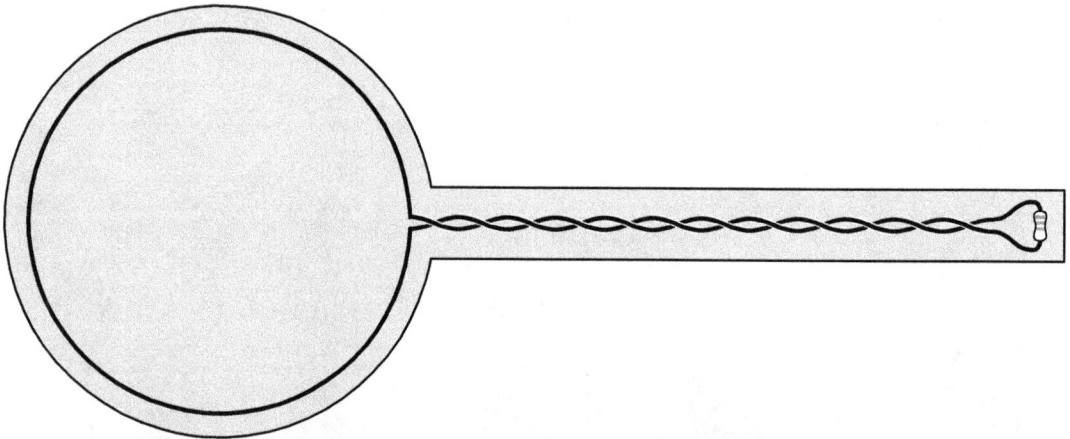

Fig. 4-29: **RL Wand Target**

---

35. A diameter of 2"/5cm to 4"/10cm is a good range; use a small diameter when testing with small search coils. Both inductance and resistance increase linearly with diameter so tau versus diameter is mostly a wash. However, a larger diameter produces a stronger RX response.

4.17: Measuring Target Taus

Several times in this chapter a target was stated to have a certain tau. Target tau can be measured in two ways, both which require jumping far ahead in the book to understand and build. The first is to use a VLF-style circuit to measure the target's phase response at a particular frequency (10kHz being our favorite median frequency) and then calculating the tau using Equation 4-10. For this a circuit like Figure 16-32 will work. Since the phase can be measured at the outputs of the X and R demodulators any circuitry beyond those points is not really needed. It is important that the demod timing be properly calibrated for accurate phase measurements.

An alternative to using a homebrew circuit is to simply tap into the demod outputs of any good analog VLF metal detector. Again, be sure that the circuit is properly calibrated and measure the actual TX frequency to use in Equation 4-10. Newer digital detectors that have analog demods can also be used but they might do target rotation in software so you may need to measure the phase of ferrite and manually calibrate each target measurement to that of ferrite.

A final alternative is to use a phase meter like the HP 3575A. You will still need a transmit signal but this can be a simple signal generator, perhaps boosted by an amplifier. An astute designer could take a VLF circuit (such as in Figure 16-32) and create a complete phase readout device, similar to a modern detector except that instead of a VDI number you present the measured phase.

A second method of measuring target taus is by looking at the step response using a pulse induction (PI) design. As with the VLF approach you do not need to build an entire PI system, just the transmitter and the receiver preamp. Even the demodulator is not needed. Most PI designs run the TX pulse at 1-3 kHz and the pulse width at around 100μs. A critical requirement for measuring taus is that the TX pulse width must be about five times the tau you want to measure, and that's beyond the time required to flat-top the coil current. For big silver coins the pulse width might be as high as 2ms and a practical pulse rate might be 50Hz. Also, faster flat-topping of the TX current is desirable so a higher-than-normal series resistance can be preferable. This will reduce the peak coil current somewhat but for a tau-tester we are not all that interested in achieving more depth.

We want to graph the exponential target decay in a log-linear plot which produces a straight line and then find the slope of the line. Few oscilloscopes have the option of selecting log axes[36] so the easiest solution is to save the waveform data, import it into a spreadsheet (e.g. Excel), and create a log-linear graph with the data. Because the coil and preamp will contribute to the decay curve it is important to first save a no-target waveform and then subtract this data from the target waveform data.

Figure 4-30 shows the results for a US silver quarter. A trendline has been added which matches the decay slope of the target. We can then use two points on the trendline to calculate the tau. The trendline intersects the Y-axis ($t = 0$μs) at about 850mV and is 42mV at $t = 439$μs. The exponential equation is

$$42\text{mV} = 850\text{mV} \cdot e^{-439\mu s/\tau} \hspace{4cm} \text{Eq 4-14}$$

Solving results in $\tau = 146$μs.

You will notice that the initial portion of the decay curve is very non-linear and at around 146μs it becomes linear. Recalling the discussion in Section 4.7, eddy currents penetrate through the depth of the coin but this does not happen instantaneously; initially the eddy currents are limited to the surface of the coin. Also, as discussed in Section 4.9, the overall eddy flow tends to be pushed to the perimeter of the coin by its own magnetic field. Again, this takes time and the eddies start out distributed across the face of the coin. The process where eddies go from their initial even surface distribution to their eventual distribution at the perimeter and at five skin depths is called *diffusion* and takes about 1τ. The initial non-linear part of the curve is exactly that process in action. The steeper initial slope (~15μs) suggests a lower tau because the eddies are still spread out over a broader part of the surface making the effective radius appear smaller and also limited to the surface making the target appear thinner. Also note the "rattiness" of the data at the far end of the curve[37]. This occurs because the logarithm increasingly expands the resolution as the data approaches zero.

---

36. The PicoScope PC-based oscilloscopes apparently can do this.

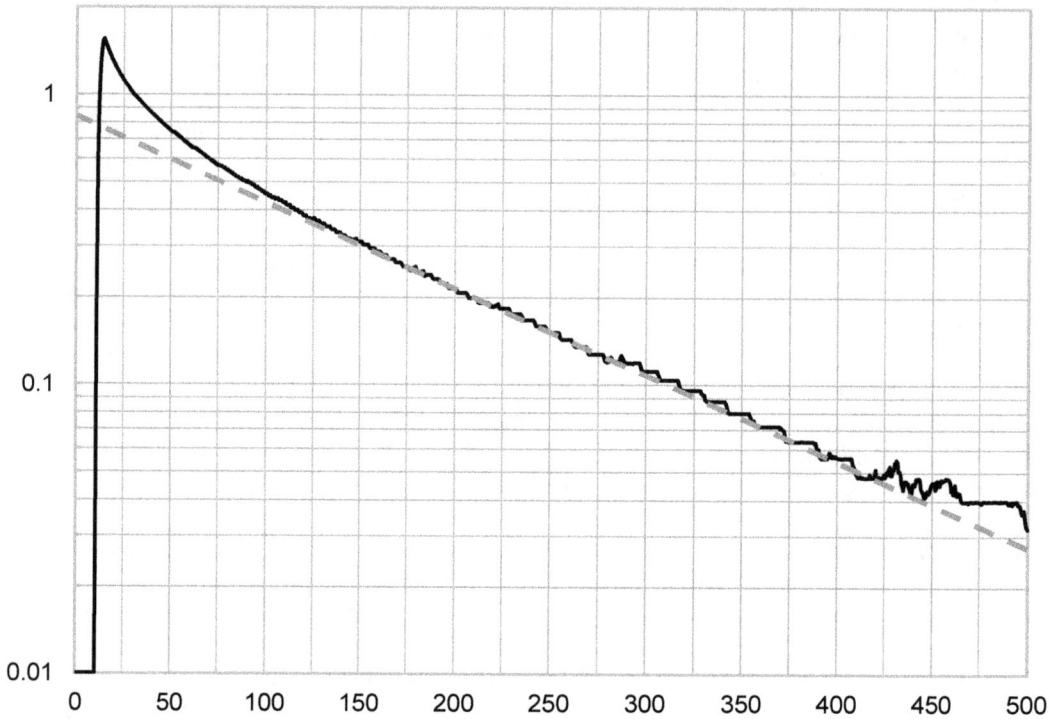

Fig. 4-30: **Excel Plot of a US Quarter Response**

The measured taus for the coins in Figure 4-9 are listed in Table 4-7 along with the 10kHz phase angles calculated from the taus per Equation 4-9. Compare these calculated phase angles to the measured phase angles (using the HP 3575A) in Table 4-3.

	Tau	Phase
Cent	66.1µs	166°
Nickel	10.2µs	123°
Dime	80.8µs	169°
Quarter	146.5µs	174°
Half	184.8µs	175°
Dollar	310.1µs	177°

Table 4-7: **Measured Coin Taus**

Obviously this technique requires a digital oscilloscope with the ability to save waveform data. If you lack a digital oscilloscope then there is one more trick that will make this work on an analog oscilloscope. A log amp can be added to the preamp to convert the exponential decay into a linear slope[38] much like Figure 4-30. This method cannot compensate for the decay contribution of the coil and preamp so extra attention is needed in their design. It may still be necessary to look later into the decay curve to ensure you are looking only at target decay. The VLF phase measurement method presented earlier may prove to be an easier design and can be used with any oscilloscope.

---

37. The data was taken with averaging set to 64. You can easily see the LSB steps of the ADC used in the digital oscilloscope and if you look closely you can also see periodic DNL errors.

38. This method was successfully developed on the *Geotech* forums by denizen 'green' under the thread called "Target Response Tester." His threads on measuring time domain taus were valuable discussions.

Inside the Metal Detector

4.18: Fun With Eddy Currents

It's been mentioned that if eddy currents are strong enough they can heat metal. Induction cooktops use them to cook food and furnace crucibles use them to melt steel. Eddy currents are weird and interesting, so what other applications are they used for?

Eddy currents can also stop things. In metal detectors eddy currents are generated by a varying magnetic field imposed on a metal target. But a moving piece of metal in a static magnetic field will also generate eddy currents. Some exercise cycles use a large aluminum disc for the "wheel" and a pair of magnets for the resistance. The magnets — on either side of the aluminum wheel but not touching— generate eddy currents in the wheel when the wheel is in motion. The reverse magnetic fields created by the eddies act against the magnets to create drag[39]. As the magnets are moved closer to the surface of the aluminum wheel the increasing magnetic field creates higher eddies which increases the pedaling resistance.

Some roller coasters use the same effect to stop the train at the end of the ride. For safety, the brakes on a roller coaster must be fail-safe so magnets are placed on the track such that metal blades on the coaster cars passes between them. As

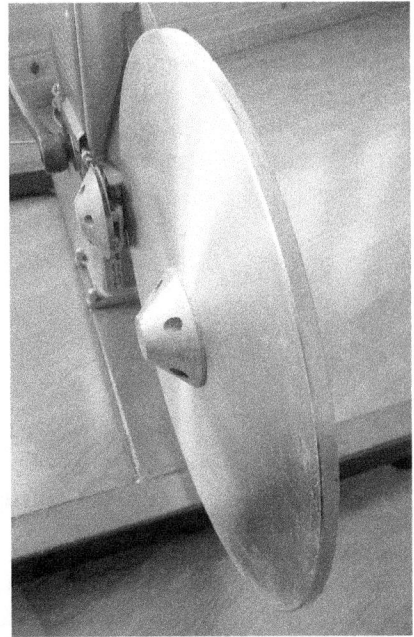

Fig. 4-31: **Spin Bike Eddy Wheel**

with the exercise cycle, strong eddy currents are created in the metal blades and the resulting magnetic field acts against the static magnets to stop the train.

Eddy currents can also move things. Science museums often have a "ring launcher" whereby you press a button and an aluminum ring is propelled vertically into the air, usually constrained by a thin rod. The propelling force is performed by eddy currents. At rest, the ring sits on a hidden coil which is charged with a high current, generating a magnetic field. When the current is suddenly turned off the collapsing magnetic field generates a strong eddy current in the ring and its field opposes the field of the coil, propelling the ring upwards. The design of a ring launcher is pretty much identical to the transmitter design in a pulse induction detector.

Militaries have experimented with "coil guns" whereby a series of coils are energized and progressively turned off (like the ring launcher) as a projectile passes through, thereby accelerating the projectile to very high speeds. Some roller coasters are now using a similar method. Instead of climbing a large hill for gravity propulsion, they can launch from a flat dead standstill using an electromagnetic launcher[40]. A massively scaled-up version of this concept has been deployed on the *USS Gerald Ford* aircraft carrier, whereby the traditional steam-powered catapults used to launch aircraft have been replaced by electromagnetic launchers. They would be impractical if not for the on-board nuclear power plant.

In the movie *The Hunt for Red October* the fictional *Red October* submarine used a propulsion system referred to as a "caterpillar drive." This part of the movie is non-fiction; there is a real technique known as magnetohydrodynamics (MHD) which uses magnetic fields to propel a conductive fluid (like seawater) via the induced eddy currents. MHD drives have been built (the Japanese have achieved a speed of 8 knots) but they have poor efficiency. Conversely, a moving conductive fluid through a magnetic field can generate electricity. This has been observed with ocean tidal action at river outlets.

---

39. Throughout this chapter we have ignored the fact that any conductor carrying a current (including eddies) and placed in a magnetic field experiences a force (called the *Lorentz force*) due to the interaction of the fields. If you're thinking this could provide a way to make coins pop out of the ground, sorry, you won't be able to carry enough batteries. Read about the "Master Magnet" on the next page.

40. I once rode the "Rock-n-Roller Coaster" at Disney Studios that uses this method. Very impressive acceleration.

Eddy currents can deform things. There is a technique called "magneforming" whereby pulsed coils are used to deform metal via eddy current forces. An impressively dangerous application of this method can be found in "coin shrinkers." A coil is placed around the perimeter of the coin and connected to a bank of capacitors. The capacitors are charged to a high voltage and then the energy is suddenly dumped into the coil. The resulting eddy currents generated in the coin produces radial forces[41] which causes the coin to shrink in diameter but increase in thickness. It also causes the coil to literally explode,[42] making this a very dangerous endeavor requiring a blast shield.

Fig. 4-32: **Shrunken US Quarter (the smaller one)**

Finally, can eddy currents attract things? In the September, 1962 edition of Popular Electronics magazine there is an article called *The Master Magnet* which is touted as being able to pick up non-ferrous objects including silver and aluminum. In one configuration it could pick up six silver half dollars. However, it is a difficult project and runs on household AC drawing many amps of current so it is not the kind of project for making a battery-operated portable device. But the concept might be reducible to something portable that could pick up gold nuggets up to a few grains. Although it may lack any practical application it might win a few bets.

41. Figure 4-17 helps explain how coin shrinking works.

42. This is when you know you've completely maximized your PI transmitter.

5

Complex Responses

"For every complex problem, there's a
solution that is simple, neat, and wrong."

— *H.L. Mencken*

The previous chapter looked at magnetic responses and eddy responses and noted that the former can never be more than 90° and the latter can never be less than 90°. In a utopian world it would be quite easy to distinguish ferrous from non-ferrous targets, but any detectorist who has dug more than two holes knows that iron targets commonly produce "non-ferrous" responses. We also viewed eddy responses from the perspective of a flat coin in a vertical incident field — this might always be valid for a vending machine coin reader but targets in the ground are far more variable. This section will consider those variables, and also look at non-sinusoidal systems.

Problem Targets

5.1: Composite Responses — Iron

Iron[1] is the bane of metal detecting because it is widespread and variable. Since it is a metal it can produce an eddy response as well as a magnetic response. We call this a *composite* response. Depending on the alloy it can have a wide range of magnetic phase responses, and depending on size and shape it can have a wide range of eddy responses. The result is that an iron target can have a composite phase response that lands just about anywhere.

In general, small iron targets like nails have little eddy response and so their predominant magnetic response makes them ID as a ferrous target on most detectors. Large iron items — especially flat steel (bottle caps) and discs (washers) — have very strong eddy responses and will often identify as non-ferrous targets. As with real non-ferrous targets, the eddy response portion of iron will vary with frequency making the same target respond a little differently to different detectors. To make matters worse, the response phase can vary with depth as, for example, the magnetic response may fall off faster with depth than does the eddy response, or as the reactive soil signal masks the target's reactive response, making a deeper target look more non-ferrous than if it were shallow. And, finally, to make matters almost miserable, the response of iron targets can vary dynamically even as the coil sweeps over them, sometimes making one target appear as two or three targets.

The common steel bottle cap is a good example of the composite response. When lying flat (horizontal) and centered under a perpendicular field its response is dominated by eddy currents and many detectors will mis-identify it as a coin[2]. But, being steel, it can also have a strong magnetic response under certain conditions. An obvious condition is if the bottle cap is vertical where the incident field is unable to cut through the flat surface and generate any substantial eddy currents. Figure 5-1 illustrates.

Unlike most coins iron decomposes in soil. As it does so, iron oxide leaches into the soil and alters both the magnetic and eddy components of the target's response. Therefore old rusty iron can have a significantly different response than its new counterpart. It can make the target more deeply detectable, something many people call the *halo effect*. More on this in Chapter 28.

---

1. By iron we mean any ferrous target, but mostly iron.
2. Due to its high permeability (μ) a bottle cap can identify as a high conductive coin even though it is fairly thin. See Equation 4-6.

TX B-Field

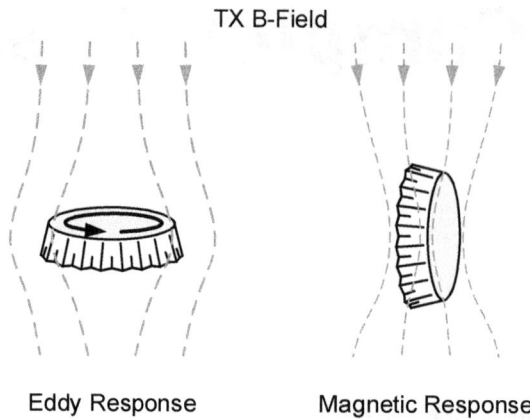

Eddy Response   Magnetic Response

Fig. 5-1: **Bottle Cap Response vs Orientation**

Experiment 5-1: Demonstrate that a bottle cap can have either a ferrous or non-ferrous response depending on orientation.

Required: A steel bottle cap; a metal detector that can respond to both ferrous and non-ferrous targets.

Set up the detector so it does not discriminate any targets, and preferably so it will have distinctly different ferrous and non-ferrous tones[3]. The easiest way to perform this test is to lay the detector on a (non-metal) table. Most metal detectors are motion-based, especially in modes which incorporate target identification, so the bottle cap will need to be in motion. However, don't sweep the bottle cap *across* the coil as you might normally do; rather, bob the bottle cap directly into and away from the center of the coil.

Hold the bottle cap flat-wise to the coil and bob up & down. You should hear/see a non-ferrous response. The target ID will depend on the detector model as it can vary considerably. Turn the bottle cap edge-wise and bob up & down. You should hear/see a ferrous response.

Finally, with the bottle cap flat-wise to the coil, sweep it across the coil. How does the detector respond?

In the final step of sweeping the bottle cap across the coil, most detectors will call it a coin of some type though it may differ from the "bobbing" response. But what the detector actually sees as the bottle cap moves across the coil is more complex. Throughout the last chapter we considered that the TX field is comprised of parallel vertical lines of flux (e.g. Figure 5-1). But that is not the case, reality looks like Figure 5-2. While the flux lines are vertical through the center of the coil and directly below, they diverge outward and eventually circle around the outside of the coil. This means the lines of flux are, at some places, horizontal.

As the coil is swept over a target, the target can potentially see an initial horizontal field, fol-

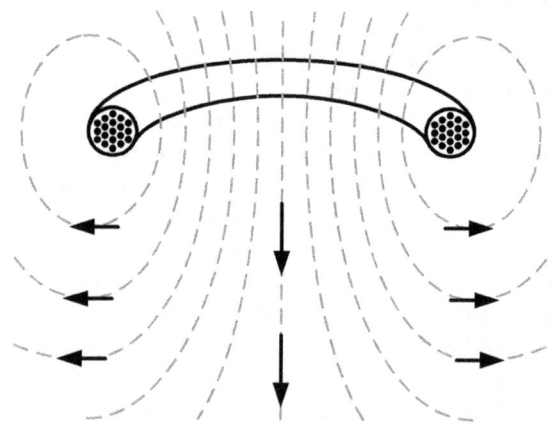

Fig. 5-2: **Realistic TX Flux Lines**

3. Some detectors can do this, some cannot.

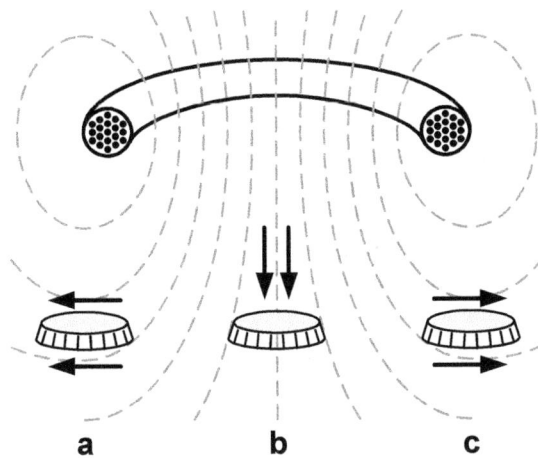

Fig. 5-3: **Bottle Cap Response**

lowed by a vertical field as the coil is directly over the target, followed by a second horizontal field. Experiment 5-1 demonstrated that a field normal (90°) to the surface of a bottle cap produces a non-ferrous response, while a field parallel to the surface produces a ferrous response.

Figure 5-3 shows what happens with a bottle cap at different positions under the coil. At position 'a' the field flux is oriented horizontally so the bottle cap response is magnetic. When the bottle cap is directly under the center of the coil (position 'b') the flux is vertical and an eddy response dominates. At position 'c' there is again a magnetic response. Thus if a coil is swept over a bottle cap we would expect to hear a quick succession of iron-coin-iron response from the detector. Many detectors, though, process the target ID at the peak of the signal which, with bottle caps, is usually the eddy response, resulting in some kind of a coin ID. Note that if the bottle cap happened to be buried in a vertical position we would expect a coin-iron-coin response, but instead would probably hear either two coins if the detector has a super-fast response, or one coin that is difficult to pinpoint.

Composite responses are not limited to undesirable items. Many coins are now produced using a ferrous alloy which behaves much like a bottle cap so the moment you figure out a way to eliminate bottle caps you may also eliminate these coins. Many of them are "high-dollar" coins and therefore desirable such as the Canadian $1 (Loonie) and $2 (Toonie). The trend toward making coins with ferrous alloys (Figure 5-4) will continue as long as the prices of copper and nickel continue to climb.

Fig. 5-4: **Ferrous Coins**

5.2: Positional Quirks

A phenomenon with similarities to the bottle cap concerns tilted targets and nails. We've already seen (Figure 4-12) that a tilted coin reduces the induced eddy currents resulting in a weaker return signal. But the dispersion of the TX field creates an additional effect (see Figure 5-5). Maximum eddy currents are generated when the incident field is 90° to the surface of the coin and for a tilted target this now occurs off-center to the coil (d). A vertically-oriented coin might have the strongest response at the edge of the coil or just beyond. While a moderately tilted coin will not produce a double response a vertical coin usually will. All of this makes pinpointing the location of the target difficult but performing an "X-sweep" will often help clarify the situation.

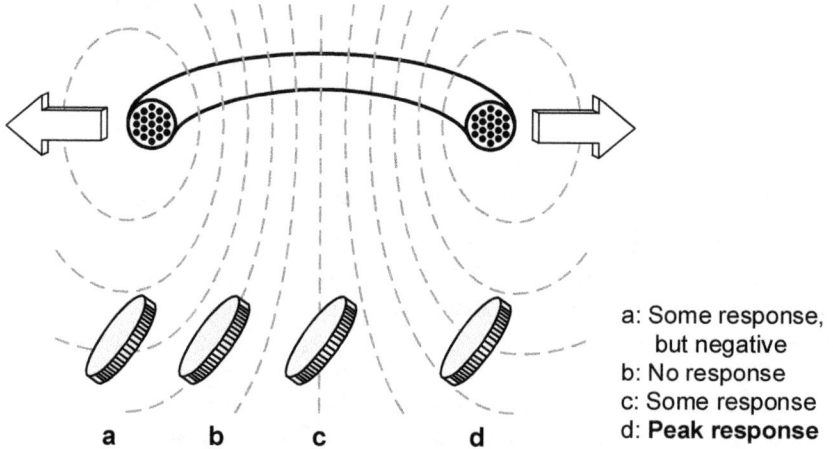

a: Some response,
   but negative
b: No response
c: Some response
d: **Peak response**

a   b   c   d

Fig. 5-5: **Tilted Coin Response**

Nails have little cross-sectional surface area regardless of orientation and predominantly have a magnetic response no matter what[4]. However, the same effect as in Figure 5-3 is in play and can produce different responses depending on orientation and position under the coil. Figures 5-6 and 5-7 show two common situations: a vertical nail and a longitudinal[5] nail. A nail is most susceptible to flux when the nail is aligned with the flux lines. The vertical nail will therefore produce the greatest response at position 'b' and little response at 'a' or 'c.' The longitudinal nail will do the opposite, producing a weak response at 'b' but strong responses at 'a' and 'c,' resulting in a double-beep on most detectors.

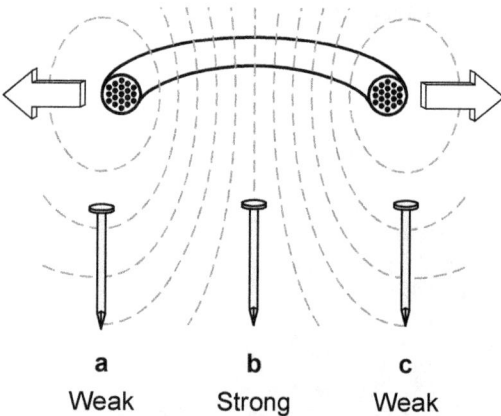

a   b   c
Weak   Strong   Weak

Fig. 5-6: **Vertical Nail**

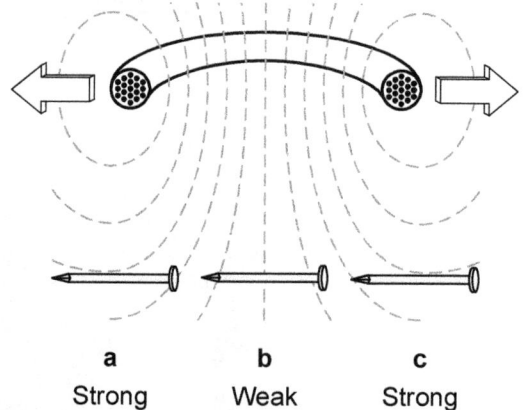

a   b   c
Strong   Weak   Strong

Fig. 5-7: **Longitudinal Nail**

---

4. An exception I've seen is with a circa 1970 Bounty Hunter BFO which runs at 1MHz and tends to produce a non-ferrous response on nails. That is, at 1MHz the eddy response exceeds the magnetic response.

5. Longitudinal meaning: if the coil is swept east-west, the nail is also aligned east-west. Also called in-line.

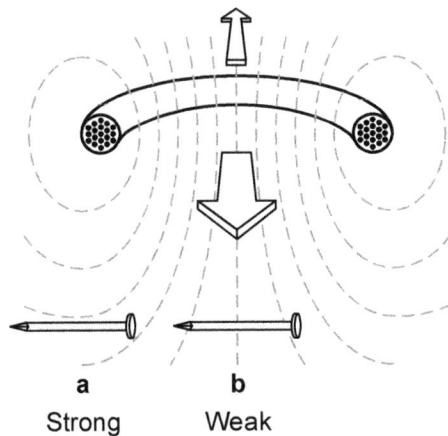

Fig. 5-8: **Transverse Nail**

For the transverse-oriented[6] nail (Figure 5-8), when the coil is swept directly over the center of the nail (b) the nail is never aligned with any flux lines so it will have a minimal response. But if the coil is moved off-center (a) the nail will begin to respond and have a maximum response near the edges of the coil, again making a difficult job of pinpointing. Note the sweep direction of the coil as compared to Figures 5-6 and 5-7.

5.3: Composite Responses — Co-Located Targets

A composite response can also be produced by two (or more) co-located targets. Consider a silver coin with a piece of foil buried exactly above it. The coin produces a high conductor response and the foil has a low conductor response. The detector sees both responses at the same time resulting in a composite response that is somewhere between that of the coin and the foil. The "somewhere" depends on the solo phase responses of the two targets and their relative strengths. But it can also depend on the detector, especially its operating frequency. For example, at 1kHz the foil will have a very weak response that pulls down the coin response only slightly. At 100kHz the foil response will be more significant and so will the composite error.

Now suppose there is a nail buried exactly above the silver coin. The coin produces an eddy response, the nail produces a magnetic response, and the detector again sees both responses mixed together. As with the foil, the nail's ferrous response will pull the composite response below that of the silver coin alone. Depending on the size of the nail, the depth difference, and the detector's operating frequency the error could be significant but probably will still end up as a non-ferrous reading. Detectability then depends on the discrimination setting because, for example, the composite response could look like a common pull tab.

But suppose instead the coin is a low conductor, like a US nickel. Now the composite response may get pulled into the ferrous region and be rejected if any level of ferrous discrimination is being used. It is not uncommon to see videos of tests showing a detector able to detect a coin under a nail. "Iron see-through" is a term that is sometimes used to describe a detector's ability to detect coins through nails. Generally there is no iron see-through, just a composite response that depends on the targets they chose for the test, their positions and depth, and a discrimination setting that does not reject the composite response.

So far we have not considered the orientation of the nail. If the nail is longitudinal to the sweep direction (left scenario, Figure 5-9) then the nail will have peak responses at 'a' and 'c' but a minimal response at 'b' where the coin has a maximum response (recall Figure 5-7). Therefore a detector with a fast recovery speed (or a large coil) may be able to largely isolate the coin. This is a case where exceptional recovery speed is mistaken for iron see-through. When the nail is transverse to the sweep direc-

---

6. Transverse meaning: if the coil is swept east-west, the nail is aligned north-south. Also called perpendicular.

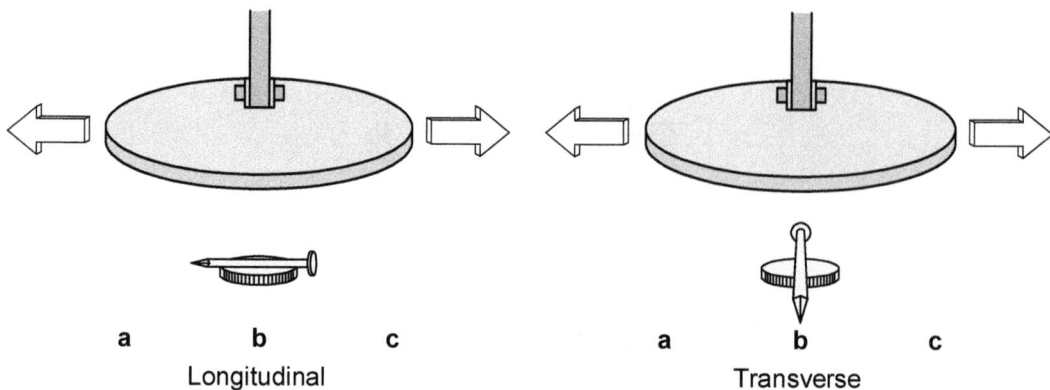

a	b	c		a	b	c
	Longitudinal				Transverse	

Fig. 5-9: **Coin + Nail Composite Response**

tion (right scenario, Figure 5-9) then both the nail and coin responses will peak at 'b,' although in this orientation the nail response tends to be weaker. In any case, you will get a composite response and no amount of recovery speed will help. If the coil is moved slightly ahead or behind the targets then the nail response will increase while the coin response decreases, thereby shifting the composite response more to the ferrous range. X-sweeping a target and moving the coil ahead or behind the peak response, and even perhaps lifting the coil a few inches, can often help clarify a difficult situation but it takes close attention and experience to make this effective.

Some detectors, notably old TR machines that ran 100kHz or more, actually have trouble detecting nails. These detectors are still valued for their ability to hunt in a "carpet of nails" and pick out coins even though they have no ability to ground balance and their detection depth is limited to just a few inches. A multifrequency detector may have better success at figuring out co-located targets than a single frequency detector because it can see the composite response at more than one frequency. This will usually vary the dominance of one target versus the other, although whether that helps depends on how the signals are processed. For example, the White's *DFX* and *V3* had a "correlate" mode whereby targets were rejected if the responses at different frequencies did not agree, as would be the case with a composite response. However, it's an effective way to implement bottle cap rejection.

5.4: Adjacent Targets

If two targets are buried with a slight horizontal separation (say, 2 inches/50mm) then a "fast" detector might be able to pick them apart into separate signals. This is often called "target separation." Many people test detectors for this quality as it is an important factor in picking out good targets amongst trash. The detector attribute responsible for good target separation is commonly referred to as "target response speed" or "target recovery speed." These two terms actually have slightly different meanings. Target response speed is how fast a target's response is presented[7] as the coil is swept over it. Target recovery speed refers to how quickly the detector can recover from one target and present a response to a second target. Before target separation became a concern detector designs often had an intentionally lengthened response (*beeeeeep*) which precluded having a fast recovery speed. Newer designs have a much faster response speed (*zip*) which allows for fast recovery as well. The two terms usually go hand-in-hand but for clarity we are concerned with recovery speed.

Target separation is obviously important when hunting in high trash areas. A detector with a fast recovery speed allows for a faster coil sweep which lets you cover more ground. However, it is often the case that a wicked fast recovery comes with the penalty of reduced depth. That is, you can go deep, or you can have fast recovery, but not both at the same time. This is a matter of how the post-demodulation filters are designed. Fast recovery might also be more susceptible to EMI.

---

7. Almost always referring to the target's audio response.

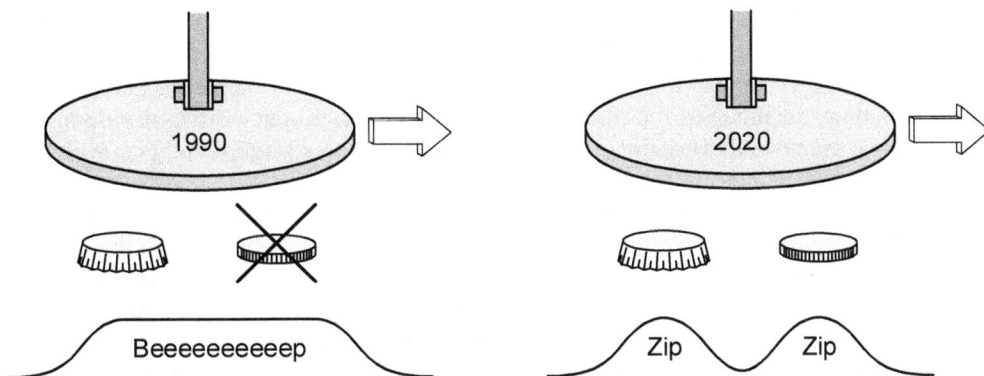

Fig. 5-10: **Recovery Speed**

5.5: Difficult Metals

We've discussed targets that we can detect, including ferrous targets which we generally don't want to detect. There are some metal targets that we do want to detect but often cannot. Stainless steel is an alloy which contains iron and chromium (plus other components). The overall behavior of a stainless steel alloy is very dependent on the alloy; one type of stainless steel can be highly ferromagnetic while another can be highly diamagnetic[8]. You can see this with stainless utensils and cutlery whereby some are attracted to a magnet and some are not.

In some cases stainless steel can be practically invisible to a metal detector because the alloy has a high resistance to eddy currents. Since there are not a lot of valuable stainless targets — perhaps some high-end watches or a nice Scuba knife — this might be considered a benefit. However, in hobby detecting there are also few junk stainless targets so the benefit is minimal. The problem with detecting stainless steel mostly affects security metal detectors[9] where there is a need to find stainless knives and, in prisons, razors and needles. The best approach is to use a very high frequency.

Titanium is another metal that is exceptionally difficult to detect, again due to high resistance. It, and stainless steel, are commonly used in the medical industry both for surgical instruments and implants. Metal detectors are sometimes used in hospitals in varying capacities[10] and need to be able to detect these difficult targets.

5.6: Composite Responses — Multiple Domain Targets

Our eddy target examples have so far been coins which are fairly well-behaved. Other targets may be more complicated. Suppose a common aluminum "Coke" can is run over by a lawn mower which shreds the can into small mangled pieces, and those pieces then work their way into the soil. These pieces are affectionately known as "can slaw" to detectorists. Because they are chopped, bent, and twisted, multiple different faces with different effective thicknesses and sizes can be presented to the incident field all at the same time.

Fig. 5-11: **Can Slaw**

These different faces will create their own eddy currents so that a single metal target may have several different responses happening at the same time. Such a target is said to have *multiple domains* due to the multiple response taus, as opposed to coins which are mostly single domain.

---

8. The same is true with coins containing nickel.

9. See Chapter 23 for a brief discussion of security detectors.

10. I once assisted a major medical company in the development of a metal detector for ORs.

In the realm of desirable targets two categories of multiple domain targets stand out: jewelry and gold nuggets. Except for a simple plain ring, jewelry is usually made up of various bits and pieces put together. Examples are a ring with a setting, or a necklace made of lots of links plus a hasp. The various pieces will sometimes be made of different alloys, and will often produce different response phases. Gold nuggets have tremendous variation in surface texture with lots of lumps and pits, with the added excitement that no two are alike.

While each eddy domain will have its own response, the response of the overall target will be some combination of the domains. But a multiple domain response can vary depending on the orientation of the target. Usually the overall response is still dominated by an "effective" conductivity, thickness, size, and shape so that, for example, a sub-gram gold nugget will always appear to be at the low end of the eddy phase range and a nugget the size of a tennis ball will have a much higher phase response regardless of all the lumps and pits. But some targets like can slaw can produce a lot of variability in the phase response even as the direction of the coil sweep over the target is changed. This can make for "jumpy" target ID numbers.

It is not just oddly shaped targets that can have a multiple-domain response. Some coins are "bi-metal," meaning they consist of more than a single metal alloy. In the US, clad coinage consists of a pure copper core and an outer thin shell of copper-nickel alloy. Clad coins have a changing skin depth as the eddies progress through the layers but they appear to mostly have a single domain response. Other coins have a core of one metal bonded to an outer ring of another metal, so these types of coins can produce a multiple-domain response. The worst of these coins combine a ferrous alloy with a low-conductivity non-ferrous alloy, making them very difficult to keep in the non-ferrous response region.

Fig. 5-12: **Bi-Metal Coins**

5.7: Size & Shape Effects

An incident magnetic field normal to the surface of a metal target causes eddy currents flow in a circular motion and the eddies tend to push themselves to the perimeter of the target. We have seen in Chapter 4 that larger thicker targets made of high-conductive metal have a higher phase response and thinner smaller targets made of low-conductive metal have a lower phase response. It is also generally true that larger targets have a stronger response but whether thicker targets or high-conductive metal produces a stronger response can depend on the detector.

Some targets may have a large mass of metal but are difficult to detect because they are made up of many small pieces that do not support large eddy currents. A notorious example is chain jewelry whereby

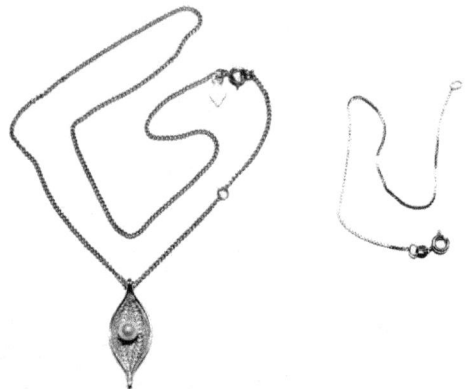

Fig. 5-13: **Necklace & Bracelet**

the chain is made of many small links. Each link creates a small eddy current and a reverse field, and while all those individual reverse fields add up the sum total is not nearly as strong as you would see from a single lump of metal of equivalent weight.

This issue is common with chain jewelry but also applies to coin caches. It may seem that a mason jar full of coins would produce a very strong signal that could be easily detected but that may not be the case. For an experiment I once (around 2010) buried a 3-pound (1.4 kg) plastic tub of silver US quarters

Fig. 5-14: **Cache Test**

(220 coins) at a depth of 24 inches (61 cm). I had a variety of VLF and PI detectors with coils up to 24" in diameter, yet none of them could detect the cache. However, a solid 20-ounce silver bar (567g) was detectable at 24 inches.

Another issue that can affect detectivity is the "integrity" of the target. We know that eddies are pushed to the perimeter of a coin so it stands to reason that a metal ring should make for a good target. And, in fact, rings of the jewelry variety make dandy targets. A ring behaves as a single-turn inductor and very efficiently supports an eddy current. But what happens when the ring is broken? The broken ring still supports eddy currents, but the eddies can no longer circulate around the complete ring. We can visualize in Figure 5-15a that they now circulate in small areas bounded by the inner and outer edges. In reality, they push themselves to the edges (as explained in Chapter 4) to create the overall eddy path shown in Figure 5-15b. The outer eddy path creates a B-field vector coming out of the page and the inner eddy path creates a B-field vector going into the page, which means that the two largely cancel each other. The broken ring therefore has an overall response that is a fraction of its unbroken response.

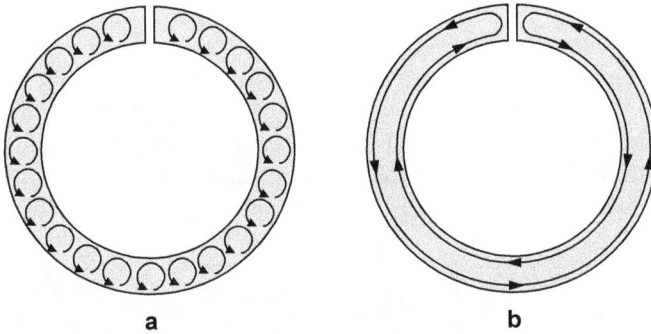

a   b

Fig. 5-15: **Broken Ring Response**

Time Domain Responses

This chapter and the last have thus far focused on target responses when the transmit signal is a sinusoid. Most detectors which utilize a sinusoidal transmitter process the receive signal in the frequency domain, meaning the amplitude and phase responses are extracted for a given frequency. Induction occurs for any dB/dt so a sinusoid isn't necessary, and we will look at a few other types of transmit waveforms.

5.8: Ramp Response

Normally, signal processing considers responses to sinusoids, steps, and impulses. We will soon look at the step response, but with metal detectors it is a little easier to start with the ramp response. This is not just busy work on our way to something useful; understanding the ramp response is also valuable because (currently) all multifrequency designs use multi-ramp TX current waveforms (triangle waves and similar).

Consider an ideal transmit coil (pure inductor) driven by an ideal voltage source. The current is

$$i_L = \frac{1}{L}\int_0^t v(t)\,dt \qquad\qquad \text{Eq 5-1}$$

At t < 0 the voltage and current are both zero. At t = 0 the voltage steps from 0 to V volts. The current is now

$$i_L = \frac{V}{L}t \qquad\qquad \text{Eq 5-2}$$

That is, the current is a linear ramp with a slope V/L. It will theoretically continue ramping to infinity though, obviously, the power supply will run out of gas long before infinity.

Example: At t = 0 a 1-volt battery is suddenly applied to a 1mH inductor. What is the resulting current?

Solution: At t = 0 the current is zero. Afterwards, it is a linear ramp with a slope of 1V/1mH, which is 1000A/s = 1A/ms = 1mA/µs.

This infinitely ramping current creates an infinitely ramping magnetic field. Again, in a practical metal detector design we don't generate an infinite ramp; instead, we might drive the coil with a square wave voltage which produces a triangle wave current and magnetic field:

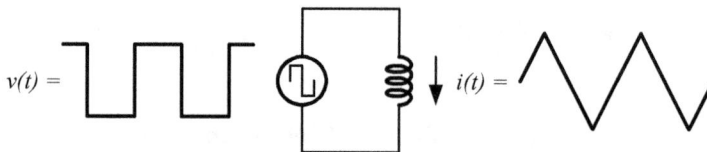

Fig. 5-16: **Practical Ramp-Mode Transmitter**

A bipolar square wave voltage with a peak-to-peak amplitude of V_{pp} will create a peak-to-peak coil current of

$$I_{pp} = \frac{V_{pp}}{L}\times\frac{T}{2} \qquad\qquad \text{Eq 5-3}$$

where T is the waveform period.

Example: A 1mH coil is driven with a 10 Vpp square wave at 10kHz. What is the resulting coil current?

Solution: The coil current is a triangle wave with a p-p amplitude of $10V/1mH \times 50\mu s = 500mA$.

Though not commonly used outside of multifrequency designs, a triangle-wave current can make for a practical single-frequency design, especially a variable frequency design. The next step is to determine what the target responses look like. For magnetic targets, we first consider ideal ferrite which has no magnetic lag. As with the sinusoid, ferrite produces a perfectly "in-phase" distortion of the TX signal. That is, the secondary target B-field is a slightly distorted triangle wave with no delay as in Figure 5-17a.

For magnetic targets with magnetic lag (nails and such) we have to look at the B-H curve to determine the response. Since the driving H-field is now a perfect triangle wave we can draw this onto the B-H plot and directly produce the secondary target B-field that is produced (Figure 5-17b). Like the sinusoidal case, the secondary field is distorted and delayed. This is but one B-H curve for a certain target, and different ferrous targets will result in a myriad of response possibilities. But the target B-field curve above is fairly indicative of what to expect.

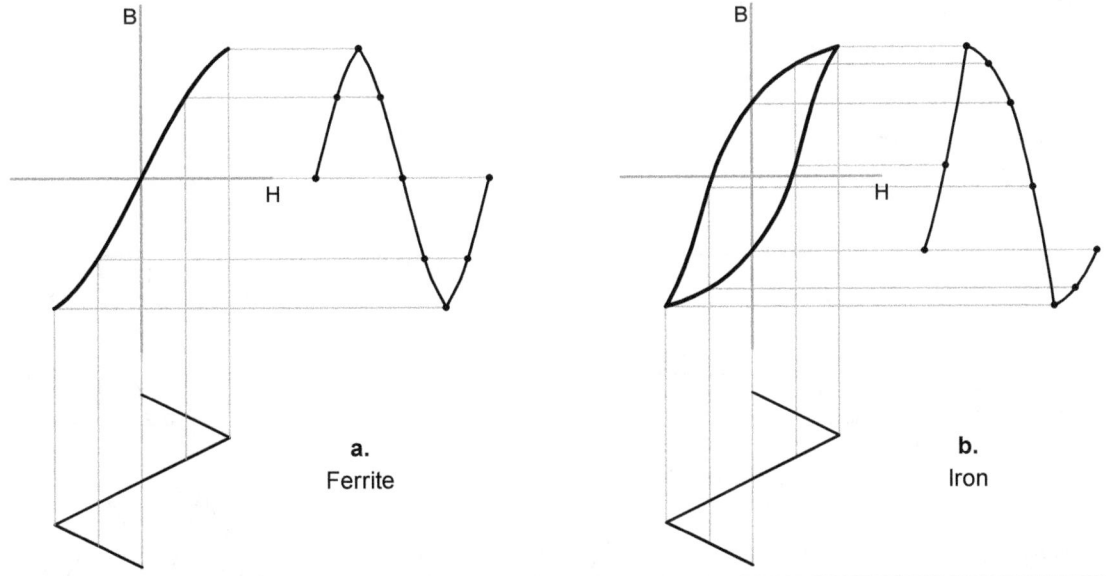

a.
Ferrite

b.
Iron

Fig. 5-17: **B-H Curve Response**

The eddy response is a bit more straightforward. The induced target EMF is the (negative) derivative of the incident B-field (per Equation 4-2) so if the TX field is a triangle wave the EMF will be a square wave. The strength (amplitude) of the EMF is proportional to the slew rate of the incident field meaning that, just like the sinusoidal case, a stronger EMF (and hence a stronger target response) can be achieved by increasing either the amplitude or the frequency of the TX field.

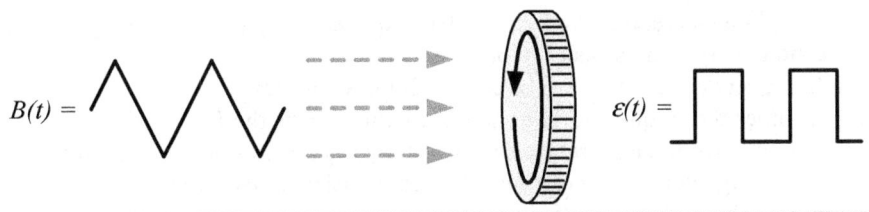

$B(t) =$ $\varepsilon(t) =$

Fig. 5-18: **Triangle-Induced EMF**

When considering sinusoidal eddy activity we determined that a typical eddy target can be modeled as an RL circuit which accounts for the phase shift behavior. The same RL model is valid for any waveform so for a square wave EMF the eddy current is an alternating exponential:

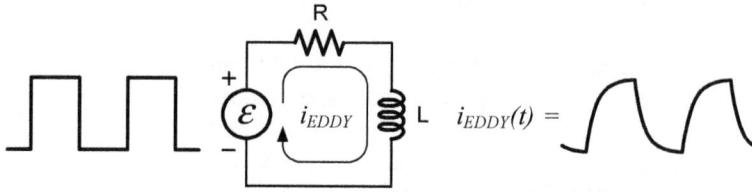

Fig. 5-19: **Eddy Exponential Response**

The time constant (τ) of the exponential eddy current is the L/R of the target model and is the characteristic time constant of the target itself. With a sinusoidal excitation the target's τ results in a phase shift of the secondary field which depends on the transmit frequency and the τ has to be calculated by Equation 4-10. With a triangle excitation the tau is directly evident in the secondary waveform.

Since the target tau is constant the eddy response will be a consistent exponential. However, the triangle waveform frequency determines the ease at which the RX circuit can determine what that tau is. As examples, let's compare a low tau target with a high tau target at both a low frequency and high frequency, with the assumption that the TX voltage drive is identical at both frequencies. If we could directly measure the EMF and eddy current responses they would look like this:

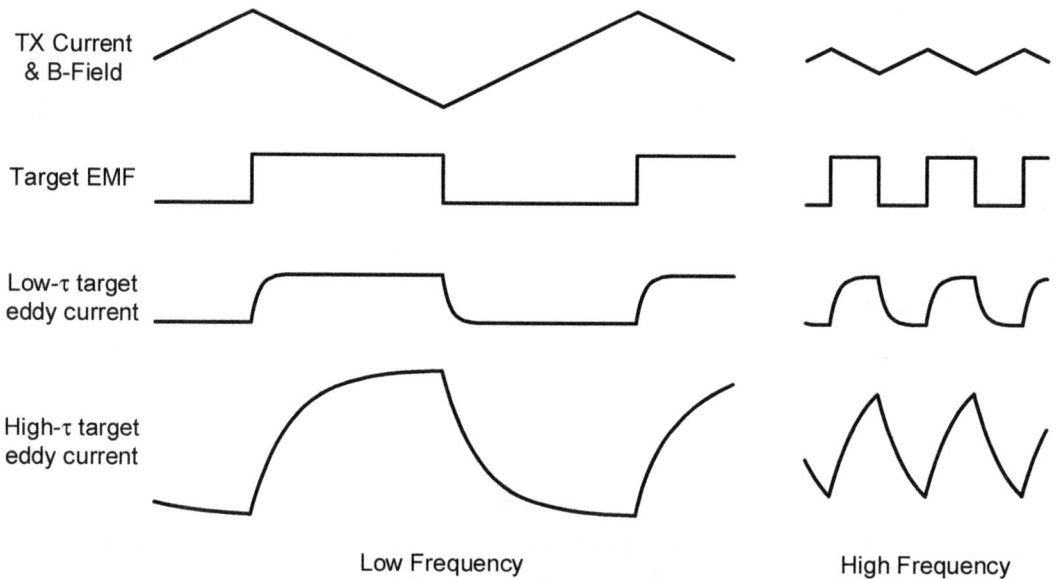

Fig. 5-20: **Eddy Responses vs Tau & Frequency**

At a low frequency the low-tau target is difficult to measure as it is approaching a square wave response. A receiver with typical demodulator timing will not be able to measure the early and brief exponential. The high-tau target has a more complete exponential (shown with a higher amplitude since it has a higher conductivity) that is easier to process.

On the high frequency side the TX is shown with the same slew rate since the voltage drive is the same, but at four times the frequency. The same slew rate means the EMF will have the same amplitude. Even though the low-tau target response is basically the same as before it comprises a more complete exponential during each half-period that makes it easier to determine a target ID using typical demodulator timing. The high-tau target is almost a triangle wave which, again, makes it difficult to resolve a target ID since very little of the exponential is available to analyze.

For complex responses (including composite responses and positional quirks) the same issues found with sinusoidal excitations apply for any other excitation. Because a composite ramp response can potentially consist of a distorted and delayed ramp (magnetic response) plus an exponential (eddy response) it is even more difficult to analyze than a composite sinusoidal response.

At the beginning of this section on ramp responses it was stated that all multifrequency designs use ramp (or multi-ramp) TX waveforms. That is true, and it is also true that the target signal is comprised of exponential responses, but it does not necessarily mean they process the RX signals as exponentials. Continuous multifrequency detectors typically use narrowband circuitry to split the frequency components and signals are then processed as sinusoids. Sequential multifrequency designs (Minelab's BBS & FBS) do appear to directly demodulate the exponential responses. In the single frequency realm the only detector that, to my knowledge, directly processes exponential responses is the Tarsacci *MDT 8000*[11].

5.9: Step Response

Target step responses are a little more complicated than ramp responses. For a transmitter to have a perfect step in the magnetic field requires a perfect step in the transmit coil current, which requires an infinite applied coil voltage. This isn't possible, but it is possible to approach a decent step current and some detectors (notably PI) do this. For this discussion we'll assume two things: the current steps from −Ip to +Ip in a non-zero time of Δt, and the step is actually a repeating square wave. The coil voltage drive required to achieve this is an alternating-polarity high-voltage pulse. Figure 5-21 illustrates.

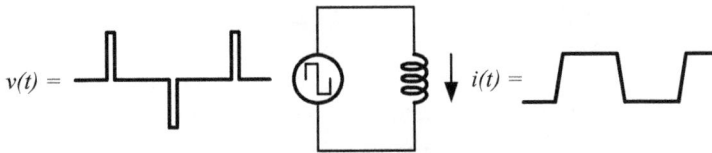

Fig. 5-21: **Square Wave TX**

The required amplitude of the voltage pulses is V = L·di/dt. As an example, if we were to switch from −500mA to +500mA in 1μs (di/dt = 1A/μs) in a 1mH inductor then the voltage pulses would need to be ±1000V. While this doesn't appear to be very realizable (and perhaps this particular example is pushing the practical limits) such a design is possible and is used in the Minelab *GPZ7000*.

Starting again with magnetic targets, and in particular ferrite, a step in the incident magnetic field produces a step in the ferrite response, with some distortion imposed in the transition slews. But it essentially looks like the same square wave. A magnetic target with hysteresis (small iron) creates a delay in the transition slews in addition to distorting them. But it essentially looks like a delayed square wave.

Eddy targets are more interesting. In this scenario the resultant TX field alternates from one DC value to another (equal but negative) DC value. As we know from the discussion on induction in Chapter 3, a DC magnetic field does not induce an EMF, but the transition slews do. The resulting induced EMF is a pulse train shown in Figure 5-22.

By now you may have noticed a conspicuous relationship between the induced EMF waveform ε(t) and the transmit voltage drive: they are the same basic waveforms but with opposite polarities. This is true if the TX coil is an ideal inductor. We have not considered that the coil invariably has real resis-

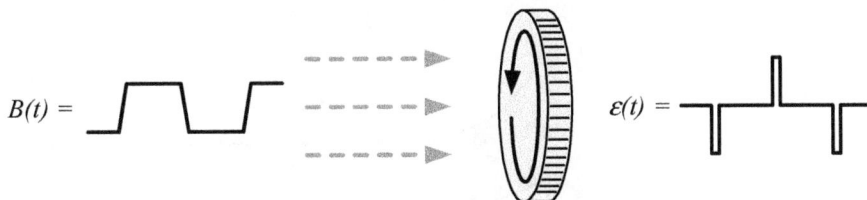

Fig. 5-22: **Square Wave Target EMF**

11. See patent US10969512 for an excellent description of what Mr. Gargov has done.

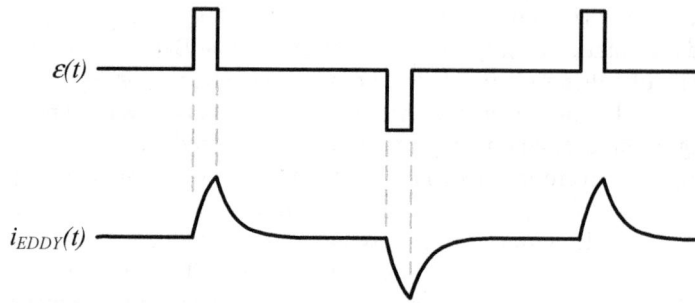

Fig. 5-23: **Square Wave Target Eddy Currents**

tance, or that the transmit circuit has non-linearities. But when exploring new ideas, assuming the target EMF has the same (but inverted) waveform as the TX voltage drive is a reasonable first-order approximation.

Because the induced EMF is a brief impulse instead of a sustained voltage, the resultant eddy currents will also be some kind of brief transient. We know from the ramp response that a sustained step in the EMF produces a sustained exponential response. If the EMF instead steps up to a value and then immediately back to zero, we should expect the eddy currents to exponentially rise to a peak current and then decay exponentially to zero.

The pulsed EMF has an amplitude and pulse width that depends on the transition slew of the incident B-field. Assuming the peak-to-peak strength of the B-field is constant, increasing the slew rate (dB/dt) produces a proportionately higher peak EMF while simultaneously reducing the pulse width by a proportional amount. A higher EMF provides more drive for the creation of eddy currents while a wider pulse width allows more time for the exponential eddies to build up. One might then ask which is better: a higher peak EMF, or a wider pulse width?

Let's take a look at two different target taus at two different slew rates. Figure 5-24 shows $\varepsilon_1(t)$ with a wider EMF with a lower peak voltage due to a particular dB/dt step. The next two waveforms are

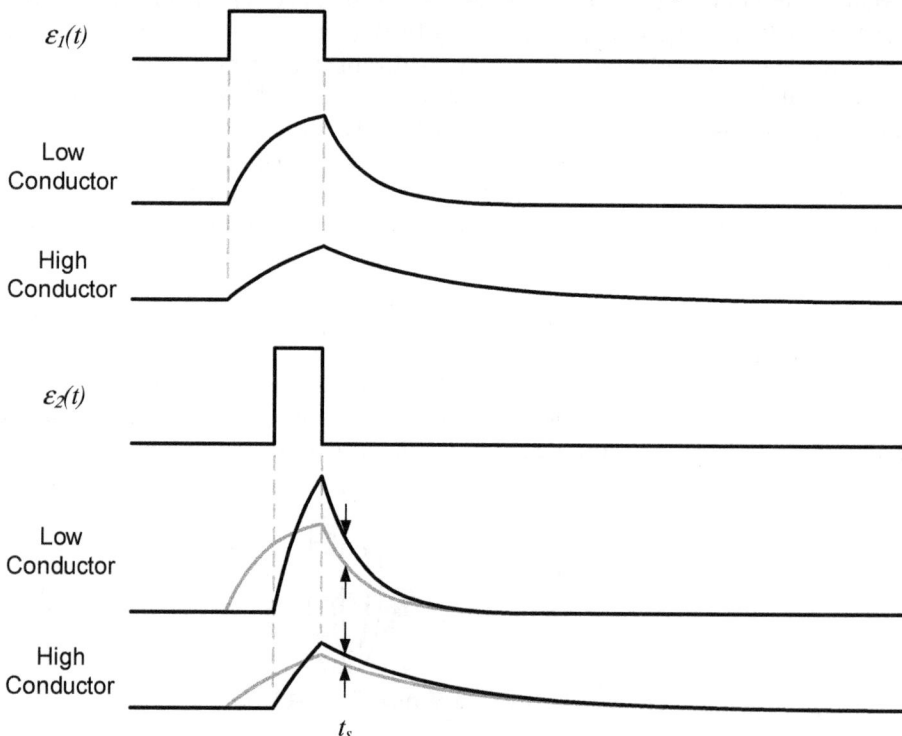

Fig. 5-24: **Slew Rate Variation**

the eddy current responses for a low conductor and a high(er) conductor. During the EMF pulse eddies build up exponentially. For the faster target the current has a noticeable exponential curve and reaches a higher amplitude while the slower target has a more linear-looking curve that reaches a lower amplitude.

If the slew rate (dB/dt) is doubled then the EMF ($\varepsilon_2(t)$) pulse width will be halved while its amplitude doubles. Both targets will experience twice the drive level for eddy generation but for half the time. The low conductor reaches a substantially higher eddy current amplitude while the high conductor shows a marginal improvement. The $\varepsilon_2(t)$ responses are shown with the $\varepsilon_1(t)$ responses superimposed in gray to illustrate the differences. They are time-aligned at the falling edges of $\varepsilon(t)$ because this is usually the starting point at which the detector receiver's sample delay begins. If the sample point is at t_s and we assume the eddy signal produces a proportional RX signal at the detector then the low conductor will exhibit a substantial signal improvement while the high conductor's improvement is more modest.

This shows that a faster slew rate is better but only up to a point, which depends on the fastest target tau you are trying to find. A good rule of thumb is that the slew transition should be no slower than one-fifth the fastest target tau desired. That is, if you are looking for 5μs targets, anything faster than a 1μs slew will yield little additional target signal. This also generally holds for the turn-off slew rate in traditional PI detectors.

5.10: Exponential Response

In a passive PI detector the turn-on portion of the transmit pulse is often an exponential and the turn-off portion is (approximately) a step. For PI detectors that use a mono coil the step response is the only viable portion of the signal and the exponential portion is ignored. With PI detectors that can use an induction-balanced coil (such as a Minelab with a DD coil) the step response is still the primary means of detecting targets (especially small gold) but the exponential response is often used as a secondary signal for discriminating iron since it contains both resistive and reactive components.

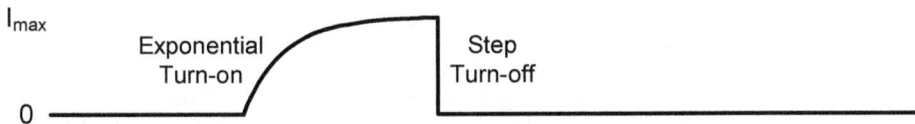

Fig. 5-25: **Typical PI Transmit Current**

A second (unintentional) use of exponential excitation occurs in a square wave driven coil commonly used in a multifrequency detector (including its single frequency modes). Figure 5-26 shows that a square wave applied to a coil produces a triangle-wave current but ignores the parasitic resistances of the coil and switches that drive the coil. The total resistance is usually on the order or 1-10 ohms, but at low frequencies the resistance distorts the triangle ramps into exponentials.

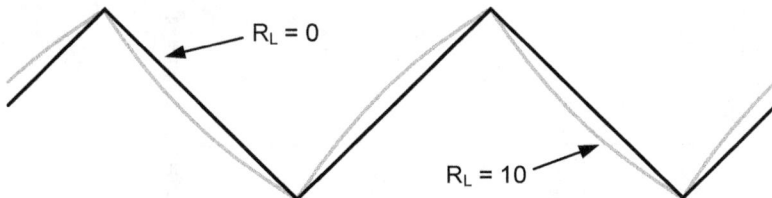

Fig. 5-26: **Triangle Wave Distortion**

The same mechanism (coil/switch resistance) is responsible for both the exponential turn-on in a PI waveform and the exponential distortion in a triangle waveform, so in both cases the exponential is not intentionally created, rather it is a by-product. Because of these cases it can be useful to understand how targets respond. However, exponential responses are significantly more complicated than the sinusoidal and linear responses dealt with so far and they are presented in painful detail in Appendix C.

Explore

5.11: Testing Recovery Speed

Since about 2010 recovery speed has become an important performance metric for detectors. Old coins are not being replenished and the easy ones have been found so detectorists are now trying to pick out good targets in high-trash areas that were passed over before. Detectors are often tested for recovery speed and many videos are produced demonstrating this, often using a coin and one or two nails. As with the co-located scenario, nail orientation matters. If the nail is longitudinal (left scenario, Figure 5-27) then one of its double responses may end up exactly on top of the coin response. While this can certainly happen in the field, for test purposes it is a poor arrangement as it does not properly evaluate recovery speed but rather demonstrates a composite response. The transverse nail (right scenario, Figure 5-27) has a weaker response but this arrangement keeps the target responses centered over the respective targets. The weaker nail response can be compensated for by using a larger nail.

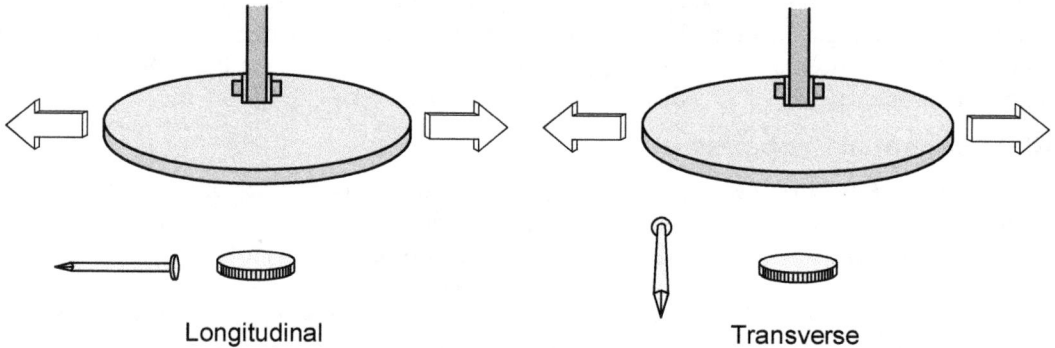

Longitudinal   Transverse

Fig. 5-27: **Coin + Nail Adjacent Response**

5.12: The Monte Nail Board

An especially popular test method for target recovery is to use the "Monte nail board." Shown in Figure 5-28, the nail board represents a real hunt situation whereby metal detectorist Monte Berry was hunting in a Utah ghost town littered with nails and spied a surface coin lying amongst some nails — an

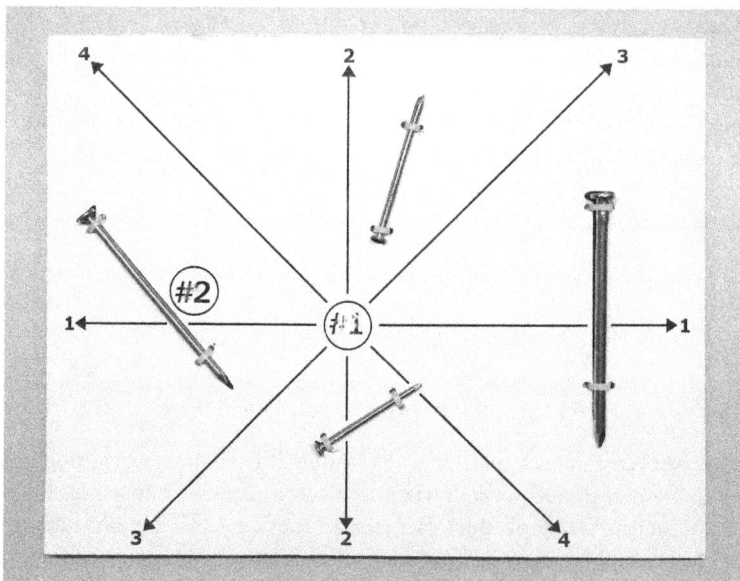

Fig. 5-28: **Monte Nail Board**

Indian Head cent at position #1. Several detectorists tried their luck but got no response; the discriminated nail responses were masking the coin response. Monte reconstructed the scenario into a test board.

While the nail board is a valid situation (it really happened) it simultaneously presents multiple problems to the detector that may exceed the majority of what most people encounter. It is also subject to improper test methods that people sometimes use to post videos on the Internet showing how a particular detector excels with the nail board. This usually involves super-short and super-fast sweeps that tend to isolate the coin response from the nails instead of using a broader, slower sweep as normally used in hunting. This isn't the fault of the nail board itself but rather the lack of controls in how it is used. Furthermore, there is no control for coil height over the board, and the board only represents two distinct target locations within the fixed nail array with no option to vary spacing. We'd prefer a test method that is more controlled.

5.13: Simple Test Jig

Figure 5-29 shows a recovery speed test jig which offers more control of the test variables. It is made of dimensional "2x4" lumber[12] glued together so it is cheap and easy to make. There are two tar-

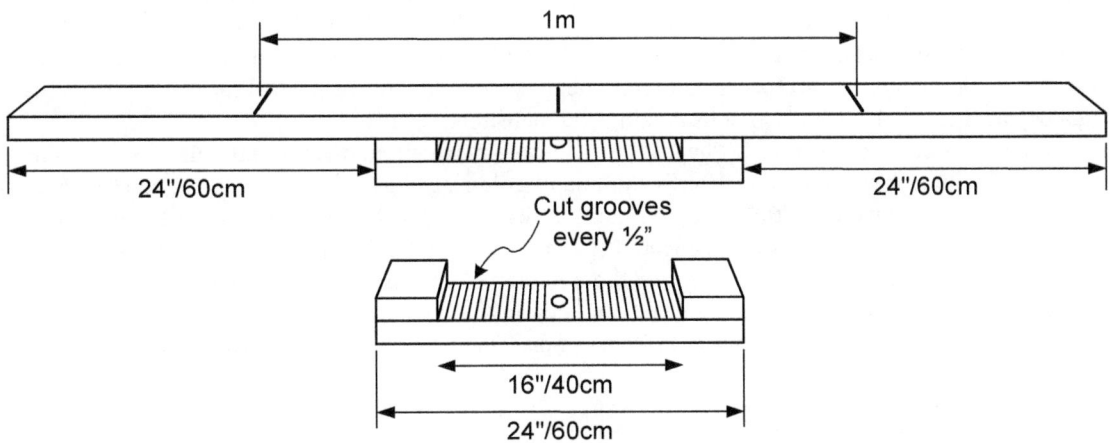

Fig. 5-29: **Recovery Speed Test Jig**

12. In the US, a 2x4 is actually 1.5" x 3.5". It seems that even in metric countries the term "2-by-4" is still used, although the finished dimensions may be closer to 45mm x 95mm which is a bit larger than the US 2x4.

get "tiers" which are 3" and 6" below the top surface. My preference is to place a coin (I use a silver US quarter) between two transverse nails (I use 16d common nails). The tiers have slight notches cut every ½" so the nails don't roll around during the test. Marks are added to the top board that are ½ m from the center.

The detector should be set up in a normal search mode with no discrimination. If it has tone ID then enable that to the highest possible resolution and if it has "iron grunt[13]" enable that as well. If it has a recovery speed setting then place that on its fastest setting.

Starting with the nails spaced, say, 6" from the center of the coin, glide the coil back and forth across the top surface. Use a metronome[14] set to 1 click per second and use the clicks as a cue to reverse coil direction at the ends of the sweep, exactly where the ½ m marks are located. All of this gives you a very repeatable 1m/s sweep rate at a fixed height above the targets.

As the coil sweeps over the nails and coin you should hear three distinct responses from the detector (bonk — beep — bonk) if it has tone ID and an iron grunt, or something similar otherwise. The reason for placing nails on both sides of the coin is to get the same response regardless of sweep direction. Move the nails closer to the coin and the triple responses will be closer together. At some point you will no longer be able to discern three distinct responses as they start to blend together into a single response. The last spacing at which you can still make out all three responses is the minimum recovery spacing.

The jig offers two tiers for testing at two depths[15]. You can also place the coin on one tier and the nails on the other tier to test for staggered-depth testing. With both the coin and nails at a deeper depth target separation becomes more difficult — it's an issue of geometry. With the coin deeper than the nails separation is also more difficult, this time due to relative target strengths.

The test jig can also be used for testing composite responses; for example, placing foil or a nail on top of a coin. You still get the benefit of a repeatable depth and sweep rate. It is useful to compare multiple detectors to see how much, say, a 16d transverse nail pulls down the target ID of a quarter. Finally, the test jig can be used with the Monte nail board, again for the purpose of getting the depth and sweep rate under control.

5.14: Gold versus Aluminum

A widely desired improvement to current metal detector designs would be the ability to distinguish gold from aluminum. From the last two chapters we can surmise that it is easily possible for gold and aluminum targets to share the exact same tau, making it unlikely that the dominant phase response will ever be able to distinguish between gold and aluminum.

In a forum discussion, it was pointed out that eddy current methods are used in industry to sort aluminum from valuable metals like gold and silver, raising the possibility that detectors might do the same. However, industrial sorters use eddy currents solely to create a Lorentz force[16] in the metals to push them sideways on a conveyor. Aluminum, being very light, will be pushed farther, thereby creating a physical separation. This is not usable in metal detectors.

Some users claim to be able to "hear" the difference between gold rings and pull tabs. Assuming they are correct, a possible explanation is that many gold rings have a single domain eddy response while many pull tabs have multiple eddy domains, at least one for the ring and one for the beaver tail. Some detectors may produce a very nuanced difference in the audio that a trained ear can discern.

It may be useful to develop a way for the detector electronics to distinguish a single domain response from a multiple domain response. But keep in mind that while most coins and simple gold bands are single domain, many other gold items (more elaborate rings, chains, earrings, nuggets, etc) have multiple eddy domains.

---

13. Or some unique iron audio feature.

14. Or a metronome app on your phone.

15. Add more tiers for more depths. However, once you establish recovery speed for, say, 3" the recovery speeds at other depths will change mostly due to geometry, not detector performance, though coil size makes a difference.

16. See section 4.18.

PART 2
Coils

BFO ALL-PURPOSE SERIES

DESIRABLE SEARCHCOIL ATTACHMENTS — PRICES — INFORMATION
STATE MODEL AND SERIAL NUMBER WHEN ORDERING FOR ANY DETECTOR

3½-inch Single Coil. Hot response to small coins, nuggets, *etc.* Ore identification. Good extra coil. *Model No. 22064.* **$39.95.**

6½-inch Single Coil. Popular size coil for coin and relic hunting and building searching. Good all-around use. *Model No. 22066.* **$39.95.**

3½x6½-inch Dual Coil. Independent operation. Saves carrying and changing two coils. Most popular size. *Model No. 22069.* **$89.95.**

5x12-inch Dual Coil. Very popular sized coil for coins, caches, relics, *etc.* Provides the best spread between coil sizes. Good response to coins as well as to deeper objects. *Model No. 22071.* **$89.95.**

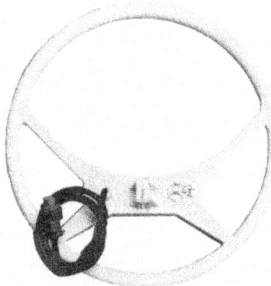

12-inch Single Coil. Very good single coil for caches, relics, large size coins, *etc.* Economical, lightweight, makes a popular added accessory to go with small coils. *Model No. 22068.* **$39.95.**

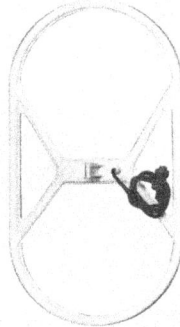

13x24-inch Deepseeking Coil. Light, produces depth, fast ground coverage. Economical. Most popular size for small or large caches, veins, relics, *etc.* Makes good addition to any combination of BFO coils. All-Purpose except coins. *Model No. 22072.* **$89.95.**

5-inch Underwater Coil with fifty feet of waterproofed extension cable. Can be lowered 50 feet to detect metallic objects as small as coins. An excellent coil for deep work. *Model No. 22023.* **$89.95.**

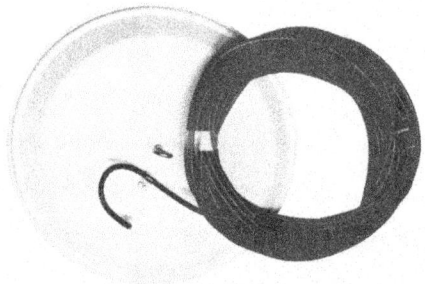

12-inch Underwater Coil with fifty feet of waterproofed extension cable. Can be lowered 50 feet to detect large metallic objects. Its heavier weight makes it ideal in swiftly running water. Not for objects smaller than baseball size. *Model No. 22024.* **$99.95.**

C
H
A
P
T
E
R

6

Coil Types

"Nothing is so simple that it cannot be misunderstood."
— *Freeman Teague*

This chapter will present an overview of search coil designs that are commonly used in metal detectors. The search coil is, at a glance, an incredibly simple circuit but one of the most underestimated elements in the design of a metal detector. The nuances are difficult to get right and a poorly designed or manufactured coil can severely handicap an otherwise excellent detector design. Chapters 7 and 8 will further explore some of those nuances.

There is some confusion when talking about a "coil" as to whether it refers to the total "search coil" or one of the "coils" contained in the "search coil." Some people call the search coil the "loop[1]" or the "search head," or even the "antenna[2]." When talking about the complete search coil in this book we will use the term "search coil," or use a specific type such as "mono coil" or "concentric coil." We will reserve the word "coil" (by itself) for individual coils of wire and will furthermore use the terms "TX coil" and "RX coil" to denote specific individual coils.

Planar Coils

A *planar* coil is simply one where the one or more windings lie in a flat plane. When the search coil has two or more windings it is often referred to as *coplanar*. Sometimes the coils overlap each other so that technically they are not coplanar, but that generally does not disqualify them from the club.

6.1: Mono Coil

The most simplistic metal detectors (BFO and PI) can use the most simplistic type of coil: a single winding[3] usually referred to as a *mono* coil. Often a mono coil is obvious by the design of the housing, such as the PI coil (Eric Foster *BeachScan*) in Figure 6-1.

In the case of a BFO, the mono coil is part of a self-oscillating circuit and a target induces a frequency shift in the circuit. There is no need for a separate receive coil. In the case of a PI, a mono coil acts as both transmitter and receiver in sequential time. A mono coil seems like it would be dirt-simple and to some extent it is. But with advanced PI designs it may be the limiting factor to performance and achieving that slight performance edge can be exceptionally difficult.

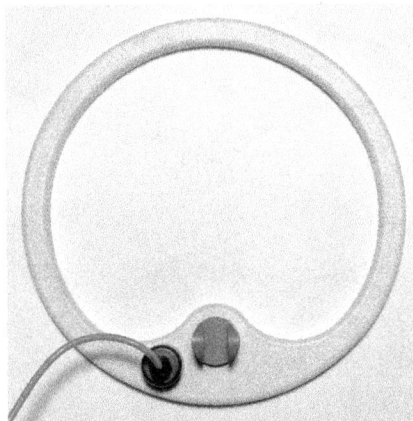

Fig. 6-1: **Example Mono Coil**

---

1. When I worked at White's, the women in the "Loop Room" often reminded me of the difference between coils and loops: *"That's not a coil, it's a loop!"*

2. For the purist, an antenna is a transducer that transmits/receives far-field electromagnetic (radio) waves. The search coil deals with near-field magnetic fields only so is not really an antenna. See 28.1.

3. A "single winding" does not imply a single turn of wire; a single winding may have, say, 100 turns of wire.

6.2: Double-O Coil

Induction balance designs require two or more coils, at least one each for transmit and receive. There are numerous ways to set up an induction-balanced coil system, many of which have been used by metal detector manufacturers over the years. The *Double-O*[4] coil consists of two round coils which are partially overlapped to achieve induction balance. In Chapter 3, the Double-O (OO) coil was seen in Experiment 3-4 and Experiment 3-5, as well as the Bell detector project.

The OO coil was used by Alexander Graham Bell in his second attempt to locate President Garfield's bullet (1881) and probably predates Bell. Figure 6-2 illustrates this, with the transmit (TX) coil in black and the receive (RX) coil in gray. The RX coil is partially overlapped with the TX coil so that a portion of the inner field of the TX coil goes through the RX coil, and a portion of the outer field of the TX coil also goes through the RX coil. The inner and outer fields of the TX coil have opposite magnetic "polarities" (flux directions) so if the RX coil is precisely positioned it is possible to get the effects of the opposing field polarities to cancel, resulting in near-zero static coupling.

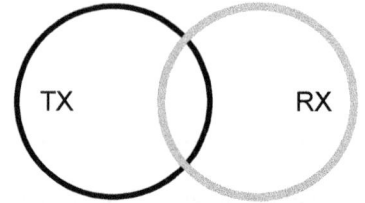

Fig. 6-2: **Double-O Coil**

In most cases both coils are the same size but that is not a requirement. Usually, the wire gauge and number of turns are different between transmit and receive in order to optimize each coil for its particular task. The coils also don't have to be round; they can be any shape, and the double-elliptical has become a popular variant over the past decade. The coplanar overlapped coil could also use square, rectangular, triangular, or any shaped coils you like, including mismatched shapes. Figure 6-3 shows a Double-O coil used on various Nexus models, a Double-0 "*SEF*" coil popularized by Detech, and the Tesoro *Cleansweep* coil using rectangular coils. A very popular variant is the Double-D coil which has sub-variants of its own and deserves its own section.

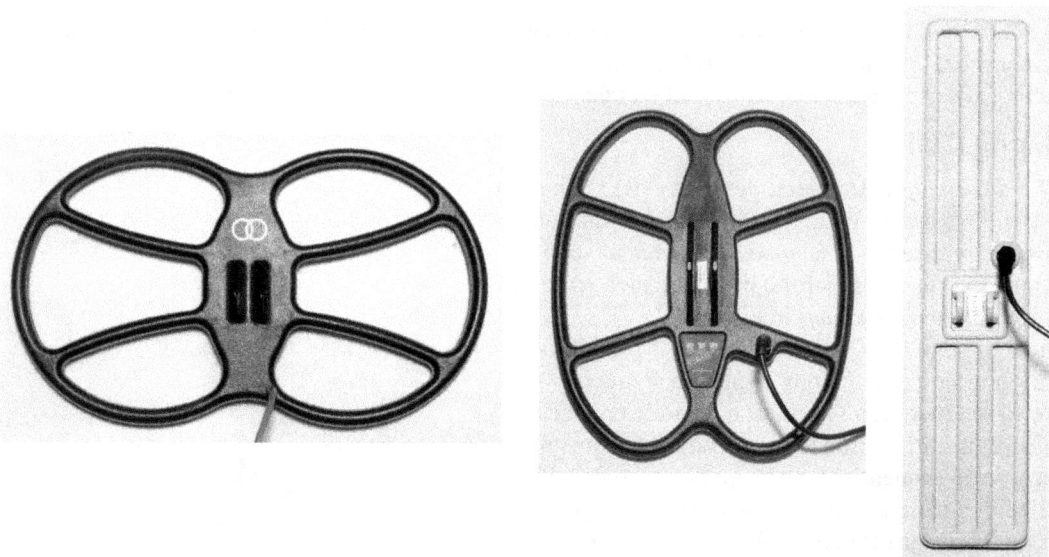

Fig. 6-3: **Example Coils: Nexus (OO), Detech *SEF* (00), and Tesoro *Cleansweep* (☐☐)**

---

4. Pronounced "double-oh", or just "d'oh" for Simpson coils.

Inside the Metal Detector

6.3: Double-D Coil

The *Double-D* (DD) coil[5] has become the search coil of choice for serious detectorists and is often the stock coil on higher-end detector models. "DD" is an appropriate name because the windings are literally shaped like "D"s placed back-to-back. They are overlapped just like the OO coil until the positive and negative transmit field coupled into the RX coil exactly balances. The result is a long narrow overlap section as shown in Figure 6-4.

In practically every case, the transmit and receive coil share the same shape. They are often designed so that the overall shape of the search coil is a round disc but some DD coils have an overall elliptical shape (Figure 6-5). As with any coil type, they may be wound with different gauge wire and different numbers of turn in order to optimize their respective characteristics. Usually, the transmit coil has fewer turns of heavier gauge wire than the receive coil.

The overlap region of the DD can be designed to be long and narrow (standard DD, Figure 6-4) or broader but shorter ("bowed" DD, Figure 6-6). Long-and-narrow offers a broader sweep range, but broad-and-short goes a little deeper and is easier to pinpoint with. It is often said that the DD coil has consistent depth across the length of the overlap region resulting in a search pattern like a "windshield wiper," as opposed to a conical pattern of e.g. a concentric coil. We'll look at coil detection patterns in the next chapter and see this is not true.

The DD was first seen in a 1938 U.S. patent awarded to Charles Hedden (US2129058). It was not seen in commercial detectors until Compass and White's began using them in the 1970s. Many people erroneously believe Compass — and engineer Don Dykstra in particular — invented the DD coil[6] but that was not the case. DD coils are widely regarded as the best configuration for highly mineralized soil so they are very popular in areas where other coils struggle. Figure 6-7 shows a photograph of a Minelab *Sovereign* DD coil which is a bowed DD as in Figure 6-6. A few things to note are: the transmit coil (left) is slightly larger than the RX coil; the TX coil appears thicker because it has a higher inductance and is a double winding, likely to reduce resistance; and there is a PCB in the coil which includes the RX preamp. Figure 6-7 also shows an 18" EXcelerator DD and Garrett DD as examples of the flat-overlapped DD and elliptical DD.

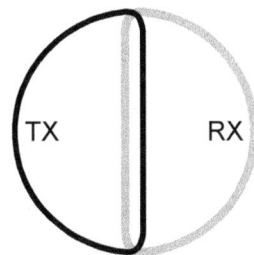

Fig. 6-4: **Double-D Coil**

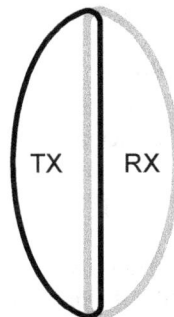

Fig. 6-5: **Elliptical DD**

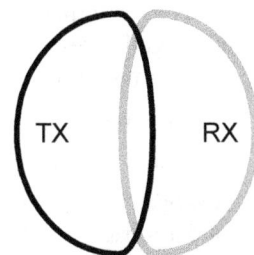

Fig. 6-6: **Bowed DD**

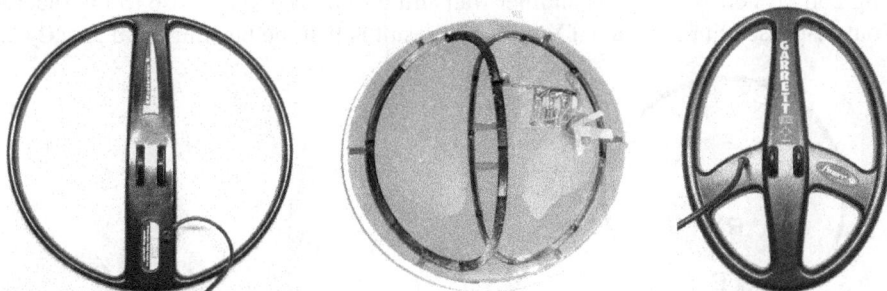

Fig. 6-7: **Example DD Coils: Straight Overlap, Bowed Overlap, Elliptical**

---

5. Also commonly called the *widescan* coil.

6. And that they are even called "DD" because of Don's initials.

6.4: Other Overlaps

It's easy to see that both the OO coil and the DD coil are basically the same general "overlap" design, just with coil windings of different shapes. And we also saw variants using ellipses and rectangles. The shapes of the coils produce variations in the behavior of the overall search coil which we'll consider later.

The overlapped coil design can further be expanded by adding additional coils. The Minelab *GPZ7000* uses a "DOD" coil which consists of a central transmit coil (the "O") and two "D" receiver coils on either side, connected out-of-phase. A diagram is shown in Figure 6-8. The DOD (or more appropriately an "oOo") first showed up in patent US3002262 in 1961 as part of a mine detector design.

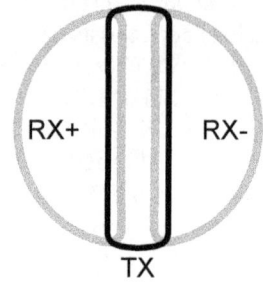

Fig. 6-8: **DOD Coil**

Another multi-coil overlap design is the 4-over-1 coil (4/1) which is found in the *AN/PSS-11*, a mine detector in widespread use during the Vietnam War (Figure 6-9). In this design a single TX coil is flanked by 4 RX coils. The overall result was not especially impressive in performance.

Both the DOD and 4/1 designs have RX coil pairs that are connected out-of-phase. This is similar to the figure-8 coil (coming up soon) and has similar benefits in that ground mineralization and far-field EMI are first-order canceled. Also like the figure-8, there is a dead zone between the coils but because the coils are aligned with

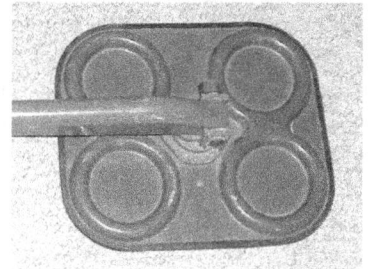

Fig. 6-9: **4-Over-1 Coil**

the direction of a normal sweep pattern the dead zone is temporal in nature. That is, a target gives a positive-null-negative response as the coil sweeps over, and there is no chance of a null-only response as there is with a figure-8 coil.

6.5: Concentric Coil

The workhorse search coil in the hobby detector market — especially in lower cost models — is the *concentric coil*, shown in Figure 6-10. Although not widely used in commercial detectors until the 1980s, it dates back to at least 1948 in a U.S. patent awarded to Harold Wheeler (US2451596). Another patent was awarded to White's (US4293816, 1981) for essentially the same design leading some people to assume that White's invented the coplanar-concentric coil. Just prior to the White's patent Compass was awarded a patent for a non-coplanar concentric design (US4255711, 1981).

Normally, a receive coil placed concentrically within the transmit coil will couple in a large amount of the transmit field and present a large continuous signal to the receive circuitry. To compensate for this a bucking coil[7] is added, which is another transmit coil placed very close to the receive coil but wired 180° out-of-phase with the main TX coil. The result is that the bucking field exactly cancels the

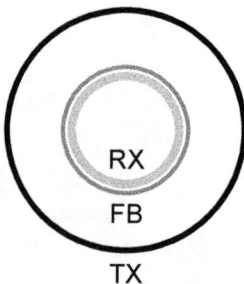

Fig. 6-10: **Concentric Coil**

Fig. 6-11: **White's Concentric**

7. Also called a *nulling* or *feedback* coil. The drawings label them with 'FB' for feedback.

main transmit field — only in the exact position of the receive coil — without much effect on overall target depth. The bucking coil can be placed just outside the receive coil, or just inside it, or even above it. In patent US4255711 Compass shows a transmit coil with a smaller bucking coil above the TX coil, and an even smaller receive coil below the TX coil[8].

Figure 6-10 shows the diagram for a standard concentric design; the main TX coil determines the overall search coil diameter, the RX coil is typically 1/2 the diameter of the TX coil, and the bucking coil is placed just outside the RX coil and often wound directly on top of it (radially speaking) in manufacturing. Figure 6-11 shows an example concentric coil made by White's. Although most RX coils are ½ the diameter of the TX coil as with the White's example, this is not a requirement. Figure 6-12 shows examples where the RX coil is smaller (Tesoro) and larger (Fisher) than the usual one-half, with the Tesoro coil being slightly oval. Nor do the coils need to be the same shape; Fisher made a concentric search coil with a round TX coil and an elliptical RX coil for the CZ series.

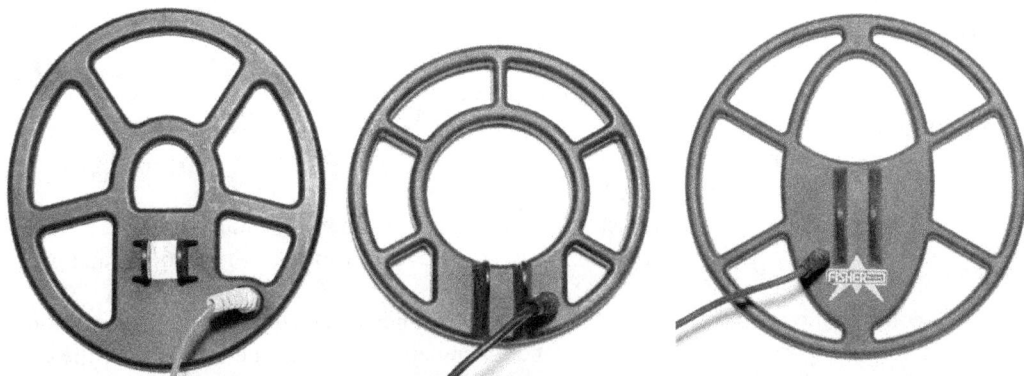

Fig. 6-12: **Tesoro & Fisher Variants; Fisher CZ**

One might assume that making the RX coil so much smaller than the TX coil will penalize depth. That is not the case. The size of the TX coil determines the strength of the magnetic field at depth. Coin-sized targets produce a small eddy field which couples as easily to a small RX coil as a large one so there is no practical depth penalty in making the RX coil small. An advantage of a small RX coil is a reduction in the strength of the ground signal which produces an overall SNR advantage.

The concentric has become the standard search coil for many detectors, primarily because they have excellent depth in low-to-moderate mineralization, have about the best pinpointing capability of all search coil types, and are relatively easy to manufacture.

6.6: Omega Coil

Figure 6-13 shows a coplanar coil design that was very popular in the 1970s and early 80s. It is called an "omega" coil because the lower portion of the TX coil looks like the Greek letter omega (Ω). It is also commonly referred to as a "4B" loop (a White's term) and was widely used in TR and VLF detectors from White's and Bounty Hunter. The earliest depiction of an omega coil was in a 1969 U.S. patent awarded to Robert Penland (US3471773).

A small part of the transmit coil is folded inward and the receive coil lies across this section. The folded portion produces a reverse transmit field through a portion of the receive coil that cancels the larger transmit field. As with overlapped coils, nulling is easily achieved by sliding the RX coil until the fields cancel. Figure 6-14 shows a photograph of a Bounty Hunter *Red Baron* search coil. Note the small epoxied circuit board near the top, used to trim the balance.

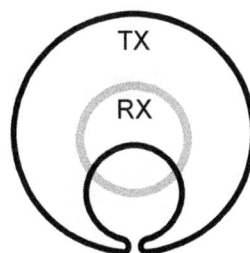

Fig. 6-13: **Omega Coil**

---

8. The Compass patent describes a non-coplanar concentric coil which should be in the *Non-Planar* section as it is really a coaxial coil. It is mentioned here because its operation is identical to the common coplanar concentric.

It seems that manufacturing a folded transmit coil would be extremely difficult. The same effect could be achieved by winding an ordinary transmit coil, then winding a separate reverse bucking coil and wiring it in series (but out-of-phase) with the transmit coil, similar to how a concentric bucking coil is wired. With the bucking coil now independent of the transmit coil, the size and number of turns could be better optimized for performance. The clarity of hindsight makes it hard to imagine why anyone would have ever manufactured an omega coil, at least in the way they were manufactured.

Fig. 6-14: **Red Baron Coil**

6.7: Figure-8 Coil

An interesting coplanar configuration is the "figure-8" coil. The earliest depiction of this coil arrangement is in a 1915 U.S. patent awarded to Max Jullig of Austria-Hungary (US1126027) where he used it as an "equivalent" diagram of an orthogonal configuration. There are several ways to make a figure-8 coil; in Figure 6-15, the RX coil is twisted into a figure-8 and placed inside the TX

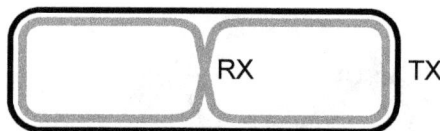

Fig. 6-15: **Figure-8 Coil**

coil. The transmit field couples equally in each half of the RX coil and the induced signals cancel. This is the approach used in the popular *Bigfoot* coil designed by Jim Karbowski (Figure 6-16) and produced by his company, Applied Creativity[9].

This coil configuration has two quirks. First, the RX coil has a detection null at the crossover, so it will have poor sensitivity right at the center. Second, the received signal at the front half of the coil will be 180° degrees out-of-phase with that of the rear half. This means that in a phase discriminating design proper target ID will work only in one of the halves unless the detector is specifically designed to handle opposing phase quadrants[10].

But the figure-8 also has some advantages. It does an excellent job of rejecting electromagnetic interference (EMI) because the RX coil sees the same amount of far-field EMI on the in-phase half as on the out-of-phase half. It also has excellent inherent ground rejection for the same reason: ground signals are canceled by each half. The inherent EMI rejection of the figure-8 also makes it a good choice for the lab bench, where test equipment and other indoor noise sources often make metal detector development difficult.

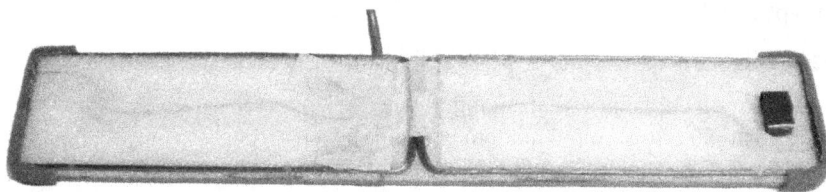

Fig. 6-16: **Inside the Bigfoot**

9. Jim died in 2007, and for a while the *Bigfoot* coils were contract-manufactured by Roy Van Epps in the defunct Discovery Electronics plant for Jimmy "Sierra" Normandi. Eventually this effort fizzled out. People often ask why no one picks up this opportunity as the *Bigfoot* has a strong reputation. I once visited Roy on a day he happened to be building Bigfoot coils; he told me that only half of the ones he builds work correctly, the rest are trashed. It is a difficult design to get right, and tedious and time-consuming to make.

10. Since the *Bigfoot* coil was specifically produced for White's detectors, White's in turn specifically designed some models to deal with the *Bigfoot's* quirky behavior.

It's possible to design a figure-8 so that the TX coil is twisted and placed in the middle of the RX coil (Figure 6-17). This will work but is inferior to the first method as it can no longer cancel EMI and ground signals, and it will still have the funny front-back target phase problem and the dead spot in the center.

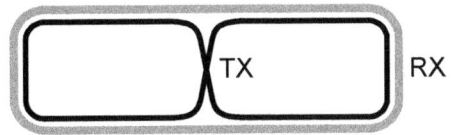

Fig. 6-17: **Alternate Figure-8 Coil**

Figure 6-18 shows an attempt to get the benefits of the figure-8 while improving target detection. The center portion of the search coil will detect targets correctly, and there are now two dead spots but they are moved closer to the ends of the coil. The outer lobes could be made even smaller by increasing their turns compared to center lobe, further improving the coil. As with most coil

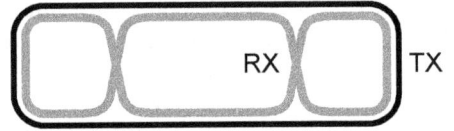

Fig. 6-18: **Improved Figure-8 Coil**

designs, there are potential improvements that could be made to the figure-8.

6.8: Butterfly Coil

The butterfly coil (Figure 6-19) appears to be identical to the figure-8 coil and it is with one exception: the direction of sweep. With the figure-8 a target passing under the upper RX lobe produces a positive response, under the lower RX lobe it produces a negative response, while a target swept through the crossover may not be detected at all. With the butterfly the target passes under each RX lobe regardless of where it passes under the coil and passes perpendicular to the crossover. While this makes it impossible to get a null response it still creates an odd positive-negative (or negative-positive) response that the detector will need to be specifically designed to handle. The closest commercial use of the butterfly coil in hobby detectors is the DOD coil on the Minelab *GPZ7000*. In security walk-through detectors the butterfly coil is common. See the *Explore* section of Chapter 23 for more information on how it is used in a walk-through.

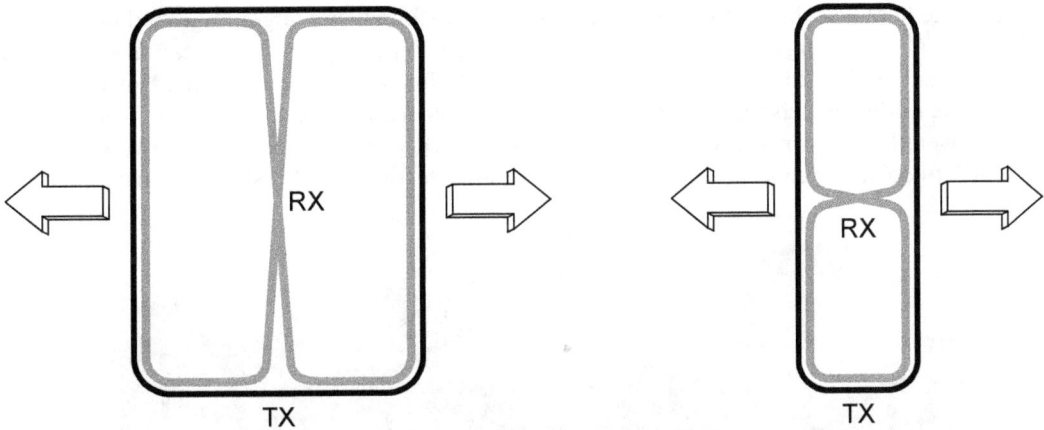

Fig. 6-19: **Butterfly vs Figure-8**

Non-Planar Coils

6.9: Coaxial Coil

The coils presented in the last section are often referred to as "coplanar" because the TX and RX coils lie in the same plane. Figure 6-20 shows another method of attaining induction balance in a coil. This is commonly called a *coaxial* coil because all the coils lie along the same center axis[11]. The left version (a) shows the TX coil sandwiched between two RX coils wired in opposition, such that the induced RX signals cancel each other. An alternative method (b) has the RX coil placed exactly between two equal TX coils that are wired in opposition so that their fields precisely cancel at the RX coil. In practice, no one uses the version on the right and subsequent discussions will ignore it.

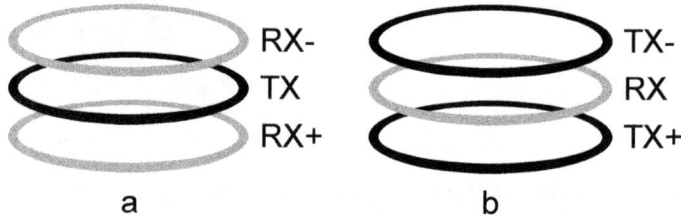

Fig. 6-20: **Coaxial Coils**

Variations of coaxial coils date back to the work of David Hughes and Alexander Bell in the late 1800s. In Bell's earliest attempt to locate Garfield's bullet he used a double-coaxial coil arrangement which was excessively complicated, at least in hindsight. In hobby detectors, the coaxial coil was used in the innovative *Phase Readout Gradiometer* made by Technos in the mid-1970s and also in early Garrett VLF machines and in detectors from C&G Technology (maybe others as well) in the late 1970s. The C&G units used a modified version of the left-hand stack where the receive coils were smaller than the transmit coil. This still achieves induction balance and, according to a conversation with Jack Gifford, was done to reduce the magnitude of the ground signal for a better target-to-ground ratio.

The coaxial coil arrangement usually carries a slight depth penalty because of partial signal cancellation between the RX+ coil and the RX– coil, especially for larger deeper targets. However, it does have a couple of distinct advantages. Because induction balance is achieved in the vertical direction it makes the overall coil less sensitive to metal targets on the edge of the stack. This allows a coaxial coil to get closer to objects such as metal fence poles without detecting them[12]. Since the coaxial has oppo-

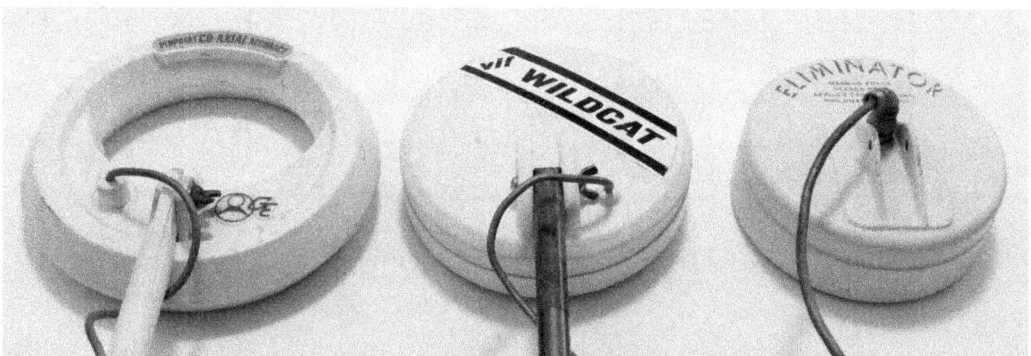

Fig. 6-21: **Example Coaxial Coils**

---

11. The concentric coil is also coaxial, but no one calls it a coaxial coil. And the coaxial coil is also concentric but, again, it's never called that.

12. While working for White's Electronics, I was asked by an engineer for the Eugene, Oregon city utilities if there was a way to detect right up to a chain-link fence without detecting the fence itself. White's engineer Dan Geyer built a one-off coaxial coil for their MXT, and loaned it to the utility company to try out. It worked so well they declined to give it back.

Inside the Metal Detector

site-polarity receive coils it is also exceptionally good at canceling EMI and so it does well
under power lines. Like the figure-8, it is a
good choice for lab bench development[13].

The last coaxial design produced was the
"Eliminator" aftermarket coil for White's
detectors, made by Jim Karbowski's company
Applied Creativity. Figure 6-21 shows coaxial
coils from Garrett, C&G, and Applied Creativity. Figure 6-22 shows an inside look at the
Eliminator. This is actually Karbowski's pattern design with notes written on it listing wire
gauges and numbers of turns.

Fig. 6-22: **Inside the *Eliminator***

Before leaving the coaxial section it is
worth revisiting the aforementioned Compass
concentric/coaxial design (US4255711).
Unlike most coaxial coils (Figure 6-20a)
where the TX coil is sandwiched between two
out-of-phase RX coils, the Compass design is
really a bucking-concentric coil where the
coils are spread apart vertically. Figure 6-23

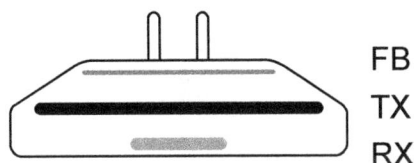

Fig. 6-23: **Compass Concentric-Coaxial Coil**

shows a side view of the coils. A possible benefit is that the bucking coil (FB) is farther removed from
ground effects so there could be a reduction in lift-off. Also, spacing the bucking coil away from the RX
coil should decrease capacitive coupling and improve the resistive null and bandwidth. Compass called
this a "Tri-Planar" coil and used it on the *Coin Magnum*.

6.10: Orthogonal Coil

Figure 6-24 shows an *orthogonal* arrangement which seems to be one of the earliest
methods of induction balance used in commercial detectors. A detector using this coil design
is often referred to as a *two-box locator*
because many of the designs[14] are literally two
boxes mounted on a pole. In the past it was
also commonly called an "RF" detector. The
receive coil is turned 90° to the transmit coil
and placed so that it lies exactly along the isomagnetic lines of the transmit field. In this
way, the receive coil has no magnetic flux cut-

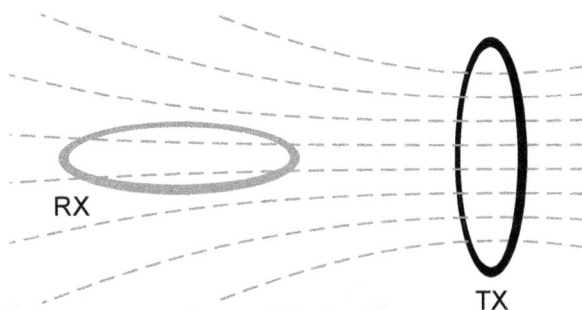

Fig. 6-24: **Orthogonal Coils**

ting *through* it[15] and therefore no voltage is induced; inductive coupling is theoretically zero. With
many two-box designs the tilt of the RX box is adjustable so the user can tweak the inductive null. In
some designs this is how the threshold is adjusted.

---

13. Also see the Tophat design in the *Explore* section of Chapter 19.

14. Including the modern-day Fisher Gemini.

15. Recall Equation 4-4; with $\theta = 90°$ the resulting $\Phi = 0$.

The earliest account of this coil arrangement is in Jullig's 1915 patent (US1126027). It began showing up on the earliest commercial detectors — including those sold by Fisher — and was popular throughout the 1930s and 40s, judging from both commercial designs and magazine construction projects. Figure 6-25 shows an early Fisher unit being used by Gerhard Fisher himself.

Because the coils are orthogonal it would seem that the TX magnetic field lines would not produce target eddy fields in a manner that are efficiently detected by the RX coil. Figure 6-26 shows that the curvature of the TX field becomes vertical beneath the RX coil and everything aligns. The spacing of the coils relative to their diameters determines optimal target depths. Larger spacing increases the optimal depth but does so with a weaker field, meaning that increases in depth only occur on larger targets.

Fig. 6-25: **Early Fisher Two-Box**

The two-box detector is, in fact, primarily a "big deep" target detector. It is only for deeper targets that the TX field lines have a substantial vertical component. Shallow targets will see TX flux lines that are almost horizontal and therefore produce almost no target vertical eddy response that can be detected by the RX coil.

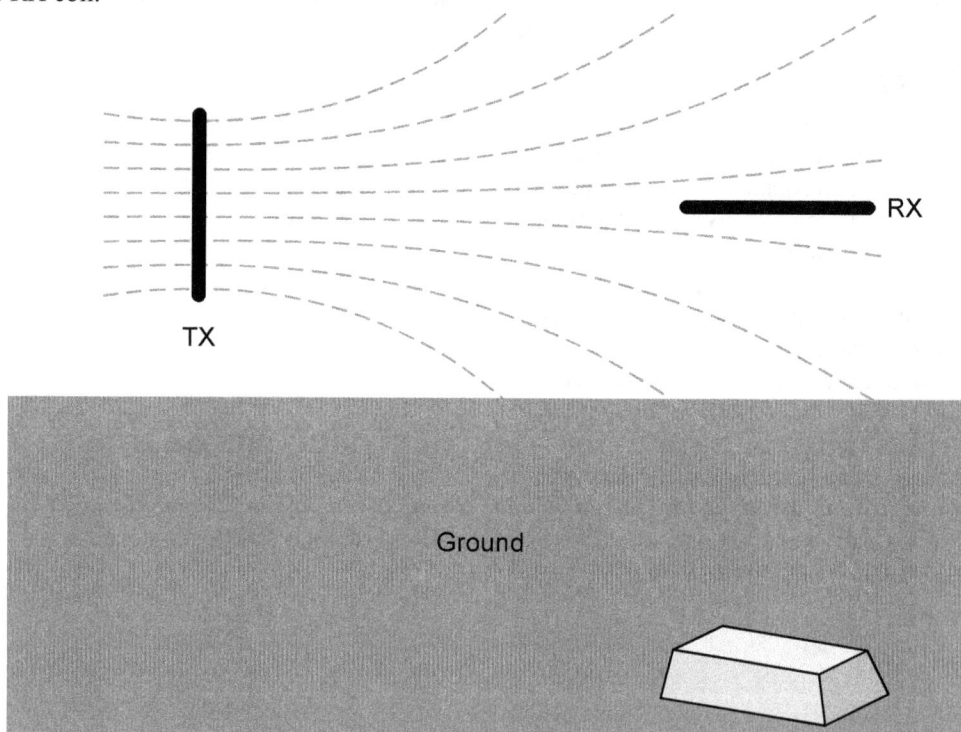

Fig. 6-26: **TX Field Curvature**

While this approach has been commonly used in large two-box locators, it has never found favor in hand-held detectors. The only example of its application in a hand-held design is a 1969 article in *Popular Electronics* (Figure 6-27). In this example, the search coil has one TX coil and two RX coils arranged orthogonally resulting in a very odd-looking design. Most likely the detection pattern is at least as odd.

A final example of a non-planar design is the short-lived "humpback" coil used on the Wilson-Neuman *Daytona* detectors of the early 1980s (Figure 6-28). This design placed the TX and RX coils in the normal concentric arrangement, but a portion of the TX and RX coils are placed in a vertical position in the "hump" to provide the bucking for induction balance. This design is described in patent US4345208 (1982) and includes the illustration on the left.

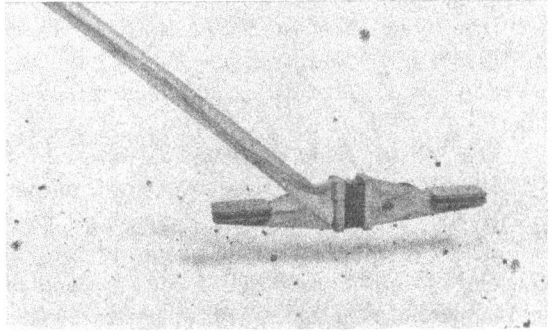

Fig. 6-27: **Orthogonal Search Head (Popular Electronics Feb 1969)**

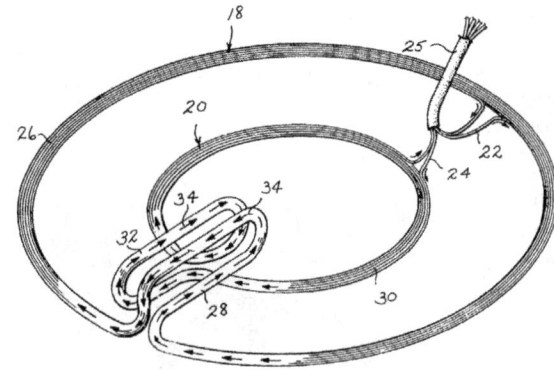

Fig. 6-28: *Daytona* **Humpback Coil**

The advantage of this design is that the bucking signal is perpendicular to the ground and has no effect on depth, plus it minimizes lift-off effect. The drawback is that it is more difficult to manufacture, and is ugly.

Dual Coils

There have been a few interesting cases of metal detectors with dual coils. The first was seen on various Garrett BFO's from the 1970s in which the search coil contained two independent mono coils that were selected by a toggle switch on the control box. See Figure 6-29. The coils could not be run simultaneously and because they were not perfectly matched the user had to readjust the tuner after switching over to the alternate coil. Still, this was a clever feature that users liked.

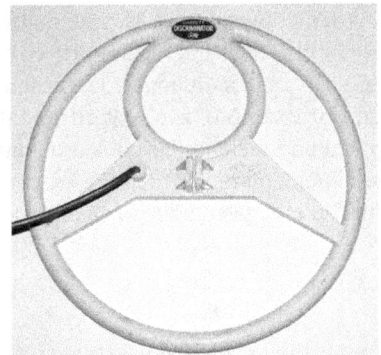

Fig. 6-29: **Garrett BFO Dual Coil**

Both Fisher & White's have used dual coils in PI detectors. The White's *Surfmaster PI* and *TDI* both used a "Dual Field" search coil which has an outer coil and an inner coil wired in series so that they are both energized at the same time.

The White's *Dual Field* coil is covered by a patent (US7994789) and an aftermarket coil company called Razorback[16] wanted to make something equivalent. To avoid the White's patent they came up with the "folded" coil concept shown in Figure 6-31. A coil is wound and twisted into a figure-8, then folded over to create, in effect, a small coil inside a larger one. The result is fairly effective, both at improved detection and avoiding the patent.

For metal detectors which normally use a mono coil (e.g. BFO and PI), developing a dual coil is not too difficult a task. But for IB designs (e.g. VLF) the solution is more difficult. There are already two coils — a TX and an RX — and the RX must be inductively balanced with the TX. Assuming we are trying to broaden the sensitivity of the coil to include both large/deep and small/shallow targets, adding another RX coil of potentially a different size requires that we now inductively balance both RX coils simultaneously.

The only commercially produced "dual" IB coil (that I'm aware of) is the Garrett "imaging" coil used on their *GTI*[17] models. The coil design is described in patent US5786696 and very lightly in a Garrett technical white paper. There are two receive coils: one about half the diameter of the TX coil, and the other at about ¾ the TX diameter. Both RX coils are individually induction-balanced, probably using traditional bucking coils. The signals from the two coils are fed to separate RX channels and compared to determine both target size and depth.

Fig. 6-30: **White's Dual PI Coil**

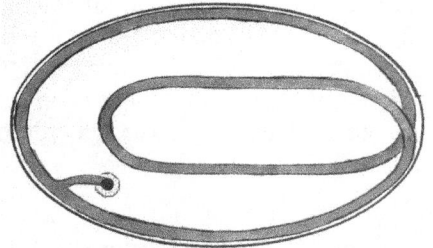

Fig. 6-31: **Miner John Folded Coil**

Fig. 6-32: **Garrett Imaging Coil**

Explore

6.11: Coil Sizes & Weights

When comparing coils, size matters. Coil sizes are stated in *diameter* for round coils and *length/width* for elliptical and rectangular coils. It is easy to compare two round coils by comparing their diameters; a 10" coil is obviously larger than an 8" coil. It gets difficult when one coil is elliptical or rectangular. For example, is a 6x9 elliptical coil larger or smaller than an 8" round coil? It turns out that elliptical and rectangular coils are equivalent in performance to a round coil that is closer to the *minimum* dimension rather than the *maximum* dimension. For this, I've empirically devised an equation to normalize elongated coils to an equivalent round coil size:

$$D_{eff} = \sqrt{\frac{2\,l^2 w^2}{l^2 + w^2}}$$

where l is the length and w is the width. The 6x9 example is effectively equivalent to a 7" round coil.

---

16. Razorback was purchased by Miner John Products; Figure 6-31 courtesy of Miner John.

17. GTI = Garrett Target Imaging.

Besides performance, coil size also affects the weight of the coil. And because the coil is situated at the end of a pole and is swung about, its weight plays a large role in the ergonomic quality of the overall detector. Even a well-balanced detector with a 2-pound/1kg coil can still wear a user down because of the stopping and starting torque required at the end of every swing. A lighter weight will always win. The weight of a coil has less to do with the coil type (concentric, DD, etc.) and more to do with the design of the windings (wire gauge and number of turns) and the construction methods. For example, the White's concentric coil in Figure 6-11 is made with a lightweight foam carrier; the Tesoro *Cleansweep* coil in Figure 6-3 is poured in solid slab of epoxy.

Table 6-1 lists a number of commercial coils with their dimensions (in inches) and their weights[18] (in ounces). D_{eff} is given for each coil and a "weight quality factor" Q is given as

$$Q = \frac{\text{Weight}}{\sqrt{D_{eff}}} \times \frac{1}{Q_0}$$

where Q_0 is a scaling factor so that a 12" coil that weighs 16 ounces has a Q of 1.00. This suggests that, ideally, weight should scale linearly with diameter-squared. That is, a coil will need to be four times the diameter to weigh twice as much. This is not an unfair scaling because, for a given detector design, the amount of wire used in a coil is roughly the same no matter the coil size[19]. Therefore with increasing diameter the increase in weight is primarily due to the plastic housing and whatever filler is used.

Personal note: Probably my all-time favorite coil is the Tesoro 8x9 concentric. I prefer a smaller coil for general hunting and the 8x9 is ideal; it is exceptionally lightweight, very thin, and has an open spoke design that makes pinpointing easy. At the other end of the table is the Tesoro *Cleansweep* which, as stated before, is a heavy slab of epoxy.

Coil	Length	Width	D_{eff}	Weight	Q
Teknetics 8" Conc.	8	8	8.00	10.75	0.82
Minelab GM-05	5	5	5.00	9.00	0.87
Tesoro 8x9 Conc.	9	8	8.46	12.00	0.89
Fisher F44 Conc.	9	5	6.18	10.25	0.89
Fisher F11 Conc.	11.5	6.5	8.00	12.25	0.94
XP Deus 9" DD	9	9	9.00	13.25	0.96
Troy Shadow 9" Conc.	9	9	9.00	13.50	0.97
Teknetics 7x11 DD	11	7	8.35	13.50	1.01
GoldQuest 11" Mono	11.5	11.5	11.50	16.00	1.02
Sunray 12" DD	12	12	12.00	17.00	1.06
Fisher Goldbug 3x6	6.5	3.25	4.11	10.00	1.07
Fisher Impulse AQ 12" Mono	12.5	12.5	12.50	17.50	1.07
Minelab E-Trac 11" DD	11	11	11.00	16.50	1.08
Minelab Equinox DD	10	10.5	10.24	16.00	1.08
Fisher 5x9 Conc.	9.5	5.5	6.73	13.00	1.08
White's Super-12	12	12	12.00	16.50	1.09
Tesoro 10x12 DD	12	10	10.86	16.75	1.10
Minelab V12X	11.5	8.5	9.67	15.75	1.10

Table 6-1: **Coil Dimensions, Weights, & Q-Factor**[1]

---

18. Be aware that the weight of a given model search coil can vary, especially for epoxy-poured construction.

19. As the coil gets larger it takes fewer windings to maintain the same inductance.

Coil	Length	Width	D_{eff}	Weight	Q
GoldQuest 8"	8.25	8.25	8.25	14.75	1.11
White's 950 Conc. (thin)	9.5	9.5	9.50	15.75	1.11
Garrett Viper	11	6	7.45	14.25	1.13
White's 10" DD	10	10	10.00	16.75	1.15
Nokta LG15	5.5	5.5	5.5	12.50	1.15
Teknetics 7x11 DD	11	7	8.35	15.50	1.16
White's 950 Conc. (Eclipse)	9.5	9.5	9.50	16.50	1.16
Garrett Viper	10.5	5.75	7.13	14.25	1.16
Minelab GM-10	9.5	6	7.17	14.50	1.17
Garrett 8x10.5 DD	10.5	8	9.00	16.50	1.19
Nokta LG28	10.5	10.5	10.5	17.75	1.19
XP GoldMaxx 9" DD	9	9	9.00	16.75	1.21
SEF 10x12 00	12	10	10.86	18.50	1.22
Fisher 8" Conc. (CZ)	8	8	8.00	16.00	1.22
Fisher 5x10 DD	10	5	6.32	14.25	1.23
Tesoro 10x12 Conc.	12	10	10.86	18.50	1.23
White's 9" Conc. (Prizm)	9	9	9.00	17.50	1.26
NEL Tornado 12" DD	12.5	12.5	12.50	21.25	1.30
Garrett Axiom 11DD	10.5	7.5	8.63	18.25	1.34
White's 6x10 DD (GMT)	10	6	7.28	16.75	1.34
Minelab Explorer XS	10	10	10	19.5	1.34
EXcelerator 14x10 DD	14	10	11.51	21.50	1.37
Makro Racer	11	7	8.35	18.75	1.40
Fisher 5x10 DD	10	5	6.32	16.50	1.42
SEF 12x15 00	15	12	13.25	24.75	1.47
Minelab 8" DD (BBS)	8	8	8.00	19.25	1.47
Fisher 10" Conc. (CZ)	10	10	10.00	23.00	1.57
Bigfoot	18	3.5	4.86	17.50	1.72
White's Bluemax 256	10	10	10.00	26.25	1.80
Garrett 10x14 DD	14	10	11.51	28.50	1.82
EXcelerator 18" DD	18	18	18.00	35.75	1.82
Fisher 15" DD	15	15	15.00	34.50	1.93
Tesoro Cleansweep	18	4	5.52	25.50	2.35

Table 6-1: **Coil Dimensions, Weights, & Q-Factor**[1]

1. All in imperial units, apologies to metric readers. Dimensions are rounded to the nearest ¼ inch; weights are rounded to the nearest ¼ ounce. Weight includes the entire coil cable but no other hardware, and no skid plate.

Coil Performance

> "What I am going to tell you about is what we teach our physics students in the third and fourth year of graduate school... It is my task to convince you not to turn away because you don't understand it. You see my physics students don't understand it... That is because I don't understand it. Nobody does."
>
> — *Richard P. Feynman*

The previous chapter covered most of the coil topologies in past and current use but did not discuss performance in any detail. We will now consider performance parameters, not only in terms of coil topologies but also in variations within a given topology. For the builder this will offer insight on how to design coils and for the user it offers objective comparisons that may help in selecting the right coil for the task.

Transmit

The first element in coil performance is the transmitted magnetic field. A stronger TX field will always create a stronger target response but the mechanics of this are not so clear. Calculating the on-axis field strength for a circular coil is pretty easy[1] but this does not offer a complete picture of the total field seen by a given target. And calculating the on-axis field strength for other coils is not so easy, such as a non-circular coil as commonly found in DD search coils. Or perhaps you want to find the angular direction of the flux lines at some arbitrary point in space such as when aligning orthogonal coils. Magnetic simulation software is available for this which uses finite element modeling (FEM) to create magnetic field vector plots for any imaginable coil geometry. These programs tend to be expensive and difficult to use. We will use an alternative method that anyone can do: empirical measurement. To do this a measuring tool is needed, something we will call a *Magnetic Field Probe* (MFP). Appendix A gives the theory and construction details for the MFP.

7.1: Using the Probe

A good starting point for measuring magnetic fields is with a simple circular coil as it is often used as a transmitter in a concentric topology. Current through the coil produces a magnetic field which was described in Chapter 3 in terms of the flux line pattern. The flux lines tell us the direction of the magnetic field and the relative strength of the field is implied in the density of the flux lines; but the flux lines don't tell us the actual field strength. This is the purpose of the magnetic field probe.

The field strength can be measured in two ways: the strength as measured with the probe aligned to the flux lines for true readings, and the strength as measured with the probe always per-

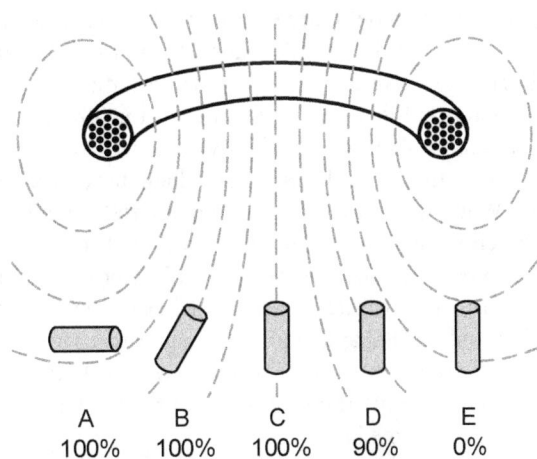

A	B	C	D	E
100%	100%	100%	90%	0%

Fig. 7-1: **Probe Alignment Error**

---

1. There is an excellent coil field calculator at https://www.accelinstruments.com/Magnetic/Coil-Calculator.html. It can calculate the field strength anywhere in the 3D space around the coil.

pendicular to the plane of the coil. Alignment with the flux lines (Figure 7-1: A, B, C) gives us the actual field strength as well as the flux direction at all points, but keeping the probe vertical (C, D, E) shows what a flat-buried coin would actually see. Figure 7-1 shows that at the center-line of the coil (C) there is no difference, but off-center (B, D) the error increases and near the fringes (A, E) the difference is huge as the field becomes horizontal and a vertically-oriented probe will produce no signal.

Measuring the true field is a more tedious job than measuring the perpendicular field because, at each measurement point, the proper angle of the probe must be found. For the most part we will make measurements on the perpendicular (or, later, with flat-oriented targets) but it is useful to make a comparison. Figure 7-2 shows the field measurements using both methods. The plot on the left shows the depth at which a flat-oriented coin will "see" certain field strengths using perpendicular measurements. The plot on the right shows that coins which are tilted to optimum angles can "see" a certain field strength at a somewhat greater depth because the magnetic field bends outward from the center line[2].

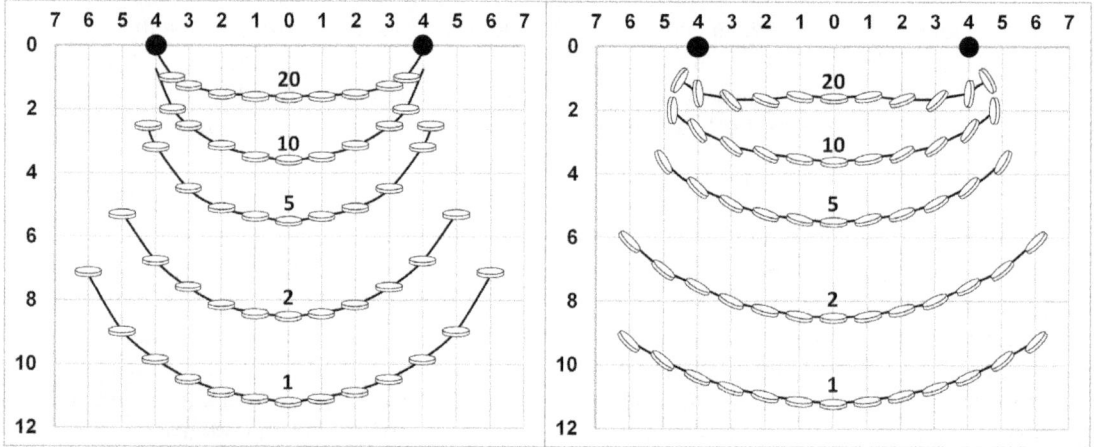

Fig. 7-2: **Field Measurements: Perpendicular vs Aligned Probe**

Instead of magnetic lines of flux[3], these field measurements show *isomagnetic* lines of field strength (flux density). That is, each line represents a constant level of field strength measurement, similar to the way elevation contour lines on a map represent a constant elevation. This is generally more useful in metal detecting than knowing the lines of flux. Each set of five isomagnetic lines represent the same magnetic field intensities and the lines are labeled for relative intensity[4]. That is, the line labeled "10" has a field strength that is 10 times stronger than the line labeled "1." The isomagnetic lines clearly show that the maximum depth for a given field strength occurs along the center axis of the coil. Keep in mind that these plots are for the TX field only and do not represent lines of target sensitivity. That will be shown later. Finally, the black dots at the top of the plots show where the TX coil cross section is; in this case, the coil has a diameter of 8 inches, centered at 0.

When we plot either the magnetic field flux lines or the isomagnetic lines we tend to do so in a vertical cross sectional view of the coil as if it is sliced through the middle. Circular coils have identical such plots regardless of the angle of the cross section; that is, from the front or from the side. The same is not true of non-circular coils. The next section will only consider circular coils, but if we were to consider oval or D-shaped coils, we might want to include "isoplots" for more than one coil cross section. Or, since the search coil is usually swept side-to-side, we might only consider a front-to-back cross section that presents the swept field profile. This will become a bigger issue when we consider target response isoplots later.

---

2. Recall Equation 4-4 where the "cosθ" term accounts for the angle of the coin with respect to the angle of the flux cutting through it.

3. Such as in Figure 3-6.

4. We could label the isolines with their actual magnetic field strengths in Teslas, but this is an exercise in relative comparisons.

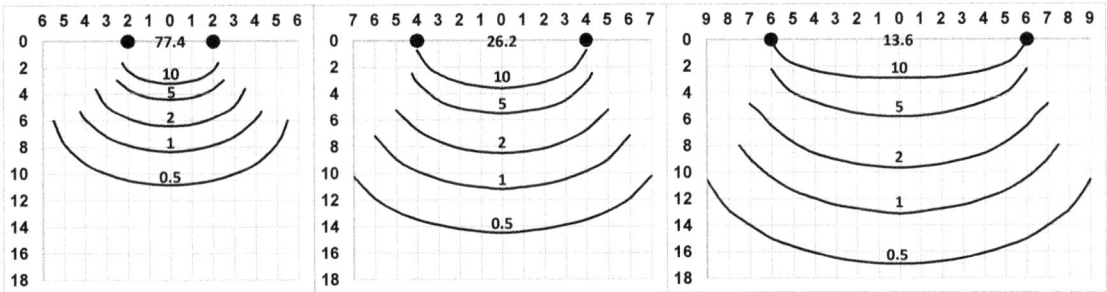

Fig. 7-3: **Isomagnetic Lines for 4, 8, and 12-inch coils (all 1mH)**

7.2: Comparative Measurements

Isomagnetic measurements can be made for coils of different diameters, number of turns, wire gauge, and peak TX current, and also at different frequencies. For circular coils, the results will have isomagnetic lines of similar overall shapes but with different levels of field strengths. Figure 7-3 shows the isomagnetic lines for 4, 8, and 12-inch coils, all 1mH and all driven at 100mAp-p. The relative peak field strengths at the centers of the coils is also given[5].

It is interesting to see that the isoline labeled "0.5" is deepest on the 12-inch coil (17 inches max) while it is 14.5 inches on the 8-inch coil and 10.8 inches on the 4-inch coil. This tells us that a larger coil can produce deeper magnetic fields. However, also notice that the isoline labeled "10" on the 12-inch coil has less depth than for the smaller coils; it is actually the 8-inch coil that is deepest for this particular field strength. Furthermore, the peak field at the center of the coil windings shows that the smallest coil excels here. All of this helps illustrate why a large coil is a better choice for large deep targets and why a small coil is a better choice for small shallow targets.

In Chapter 3 it was mentioned that the field strength of a coil is effectively determined by the "ampere-turns" — that is, the number of turns of wire multiplied by the coil current. It is time to put that concept to the test. Figure 7-4 shows a 44-turn 8-inch coil driven with a current of 100mAp-p (left), and an 88-turn 8-inch coil driven with 50mAp-p (right). In both cases, the ampere-turns is a constant 4400mA-turns so we should expect the transmit fields to be equal. From the iso plots in Figure 7-4 we see that this is true; the relative field strengths are almost exactly equal, with any variations are likely due to measurement error.

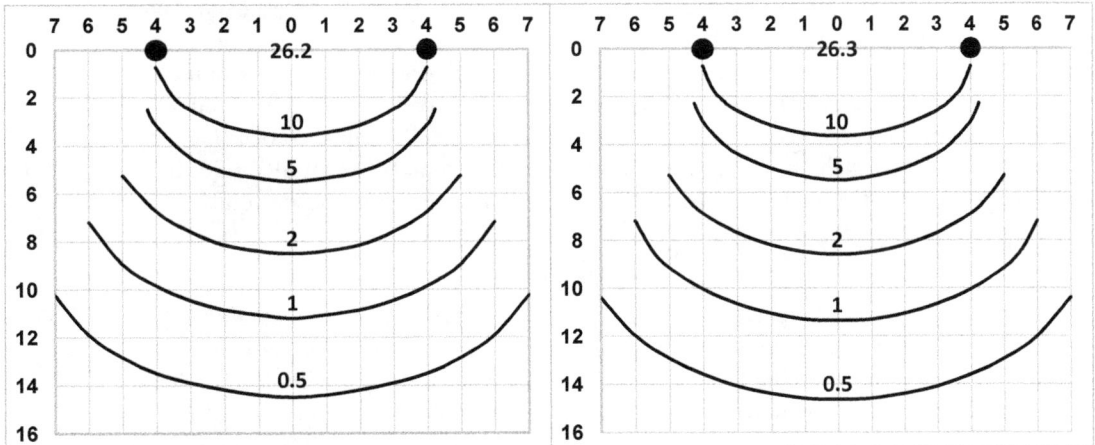

Fig. 7-4: **Field @ N = 44 I = 100mA (L) and N = 88 I = 50mA (R); d = 8 inches, f = 10kHz**

---

5. If you want actual field strength numbers in Teslas, use Equation 3-3 to calculate the field strength at the center of the winding, then scale using the relative numbers shown.

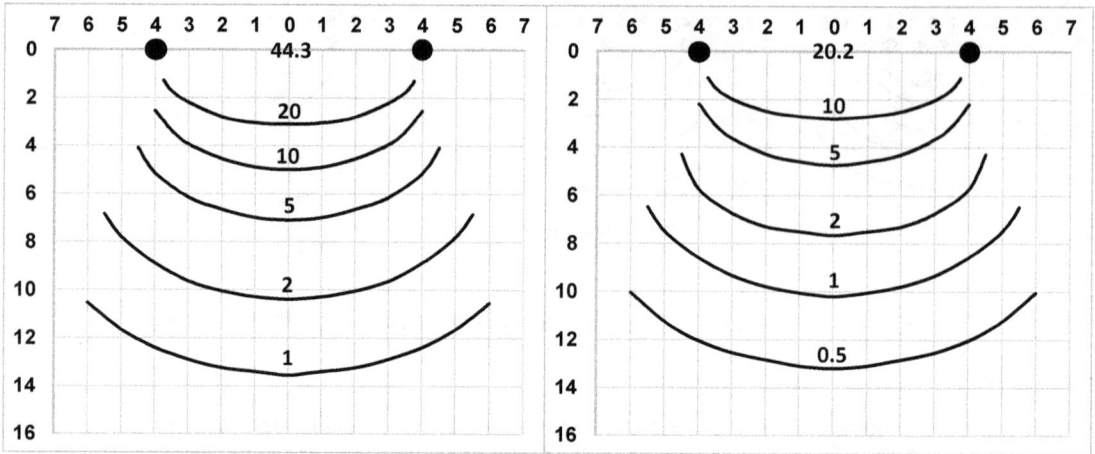

Fig. 7-5: Isomagnetic Lines for 1mH and 5mH coils (both 8 inches) @ 10vp-p

Let's now consider the response of coils with inductances of 1mH and 5mH. If each coil was driven with a constant current (say, 100mA as before) then the 5mH coil would easily have the deepest field contours. Instead, let's consider that we have a TX circuit design that can drive the coil at 10vp-p, and we want to decide whether to use a low-inductance or a high-inductance TX coil. Figure 7-5 shows the isolines for 8-inch coils of 1mH (left) and 5mH (right), each driven at 10kHz and 10vp-p. It is obvious that the lower inductance wins, by more than a factor of two. Although the 5mH coil has more turns (102 versus 44), the ampere-turns favors the lower inductance because the coil current is inversely proportional to the coil impedance, and that increases as a function of N^2. Therefore, ampere-turns is inversely proportional to N, with some additional loss due to the increased wire resistance.

This suggests that a 1-turn coil is the best solution, but then it is difficult to resonate (if that is desired) and also difficult to design a variety of coil sizes with consistent inductances (if that is important). Furthermore, an extremely low TX coil inductance will likely require a very low TX drive power supply to keep current consumption reasonable, which makes the drive circuitry more difficult to design. Historically, commercial VLF designs have had TX coils up to several millihenries and only dropped below 500μH in high frequency nugget detectors. Modern designs (especially multifrequency) have tended to settle around 500μH or less with some as low as 100μH. Due to the need for low capacitance, PI detectors have ranged from 200μH up to 1mH, with 300μH being especially popular.

As a final exercise, we could drive a coil with different frequencies and see if the field varies. Whether it does or not depends on the frequencies and the self-capacitance of the coil. At high frequencies the interwinding capacitance begins to shunt some of the coil current, which reduces the TX field. This will only occur as the drive frequency begins to approach the self-resonant frequency of the coil. As an example, the 8-inch 1mH coil used in the previous iso plots has an SRF of 390kHz so over the usual band of interest (1kHz-100kHz) the magnetic field frequency response is quite flat.

Receive

Most detectors utilize both a transmit and a receive coil arranged in an induction balance. While it's interesting and useful to measure the transmit field, what we really want to know is the performance of various coil designs in detecting targets. Different coil types usually have different detection patterns and even different variants within a type (like overlapped) can have different detection patterns.

7.3: Receive Coil Physics

Before we get into measuring detection patterns, let's first discuss the mechanics of how a receive coil reacts to magnetic fields. Chapter 4 explored the interaction of a magnetic field on a metal target, specifically that the magnetic field induces an EMF in the target that is uniformly distributed in a closed circular pattern. The target EMF generated is given by Equation 4-2. The EMF gives rise to an eddy current and, ultimately, the eddy current pushes itself toward the perimeter of the target. Therefore a metal target behaves as a single-turn closed loop.

A receive coil is a multi-turn loop of wire but the physics are the same. For a coil of wire with N turns and an area A the induced EMF from a target's magnetic field is

$$\varepsilon = -N \cdot \frac{d\Phi}{dt} = -NA \cdot \frac{dB}{dt} \cdot \cos\theta \qquad \text{Eq 7-1}$$

where θ is the angle between the lines of flux and the axis of the coil. If the lines of flux are aligned with the coil's axis then the cosine term is 1. The NA product is sometimes called the *area-turns* of the coil.

Equation 7-1 implies two things. One, the more turns the RX coil has the better; and two, the bigger the RX coil the better. Adding more turns to the RX coil does increase the induced EMF so it seems like this would be an effective way to get free depth. Since the TX coil, the target, and the RX coil form a double transformer, the ratio of the RX turns to the TX turns is sometimes called the *transformer gain*. In earlier days of metal detector design it was commonly believed that a high transformer gain was desirable, and often the RX coil would end up with 20-50mH of inductance for a TX coil of, say, 1mH.

And this works, if there is no ground mineralization and no EMI. However, the signal strengths induced by the target, the ground minerals, and EMI are all proportional to N so the effective target-to-noise ratio is independent of N. And a higher N means a heavier coil, so the current design wisdom is to use a low transformer gain[6] with a modest RX inductance closer to 1mH.

A larger RX coil only helps if the target field is uniform and completely aligned with the RX coil's axis. You guessed it, it is not. Figure 7-6 shows that the target field disperses outward and even curls around in an opposite direction. Therefore, making the RX coil larger will capture both the "forward" flux and the "return" flux, and the return flux will cancel the forward flux and weaken the response.

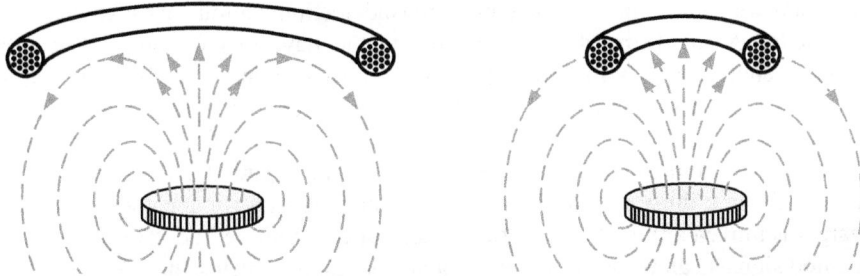

Fig. 7-6: **Target Field Interaction vs RX Coil Diameter**

This is one reason why small coils are more effective at finding small targets, and large coils for large targets. It is also why concentric coil designs have generally settled on an RX coil diameter that is around one-half the TX coil diameter[7]. With DD and 00 coils, the RX coil is usually the same size as the TX coil just for aesthetics; it could be made smaller to improve small target sensitivity.

7.4: Detection Patterns

As with the TX fields in the last section, a detection pattern is simply a plot of "contour" lines that represent a constant level of detection for a particular target. An example is shown in Figure 7-7. The lines represent the amount of voltage deflection measured at the output of an amplified demodulator. Since these voltages are highly dependent on the design of the test circuitry they are only useful in a relative manner. That is, we can use them to see where a target responds with equal strength across the detection space of the coil; we can see where a target strength falls by a factor of 10 (for example); and we can compare the relative target strengths between two coils. At the top of the plot are black and gray

---

6. The Sovereign coil is an example of a transformer gain less than 1, where the TX coil is 1mH and the RX coil is 400µH. And yet the Sovereign was no slouch on depth.

7. Although there are exceptions; see Figure 6-12.

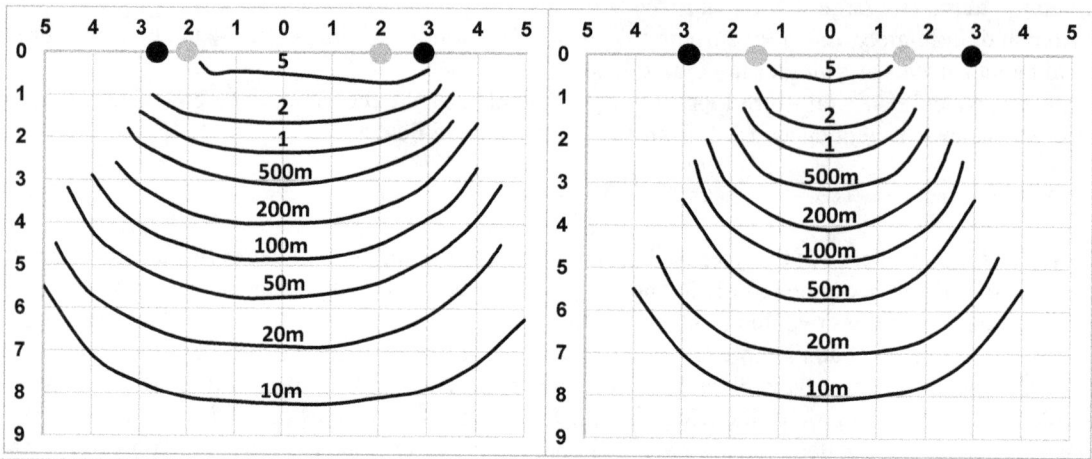

Fig. 7-7: **Coil Detection Pattern (Fisher F44 Ovoid)**

dots. As with the TX field plots, they represent the locations (respectively) of the TX and RX winding cross sections.

The lines of equal detection depth are called *isodetection* lines because they are isometric and represent detectability. They show how detection strength varies with position, with peak detectability along the center axis of the coil for a flat-buried coin. This means that a target at the fringes of maximum attainable depth will only be detected if the center of the coil passes over it; see Figure 7-8. Detectorists who understand this will heavily overlap their coil swings[8] if they want to maximize deep coverage of an area. It is also a contributing reason why places are never "hunted out."

For this exercise we want to measure and plot the isodetection contour lines which means we need a way to determine target signal strength. Most metal detectors have an audio output where the loudness depends on target strength, but it is difficult to produce a consistent and objective loudness level. Some detectors have a visual target strength indicator in the form of a depth meter but it tends to have a coarse resolution.

A finer level of measurement can be attained by directly measuring the analog signals produced by the demodulators[9]. Placing these signals on an oscilloscope allows us to accurately monitor the target signal as the target is moved around the detection space. One or both demodulator outputs can be monitored. Both demod signals can be used to produce a vector response using the XY-mode of the oscope,

Fig. 7-8: **Coverage at Depth**

---

8. Perhaps by more than the rule-of-thumb 50% overlap, depending on the coil.

9. Many newer detectors do not have analog demodulators but use direct sampling. These models cannot be used for this exercise.

and the length of the vector represents the target signal strength. However, it is easier to use just one demod voltage and look at its amplitude, as it alone will be proportional to the total vector response. And while most VLF detectors use the resistive channel as the target channel, for this exercise it really doesn't matter which demod voltage is monitored as long as you're consistent.

Using a commercial metal detector by probing its demodulator signals presents two limitations. First, you can only characterize the coils that work with that particular metal detector which means using a lot of different detectors to characterize a lot of different coils. The second limitation is that different detectors drive coils with different peak TX currents. Even though the depth pattern for a given coil on a given detector would be valid and it would be valid to compare patterns of different coils from different detectors, it would not be valid to compare their relative depths.

A partial solution is to design a test circuit that drives all the coils identically and has the same RX front-end (preamp and demodulator(s)). This assumes that all the coils we want to test will work optimally at the same frequency. Choosing the usual 10kHz will work with many coils but not all, and it still leaves some coil design variables unaccounted for. For example, we saw in the previous section that RX signal strength is proportional to the number of windings in the RX coil, so a search coil with 100 turns in the RX winding will appear to outperform a search coil that has an RX winding of 50 turns (but otherwise identical). However, in practical use the 100-turn coil may require a lower sensitivity setting to manage the increased noise level. The complete solution is to design a suite of coils that are consistent in their parameters (like inductances and overall size) so that the number of variables is reduced to only what we are interested in. That is a lot of work for so few book sales, so for this exercise we will use a number of commercial coils and measure them with a single test circuit, shown in Figure 7-9. All coils are driven at 10kHz and 10vp-p.

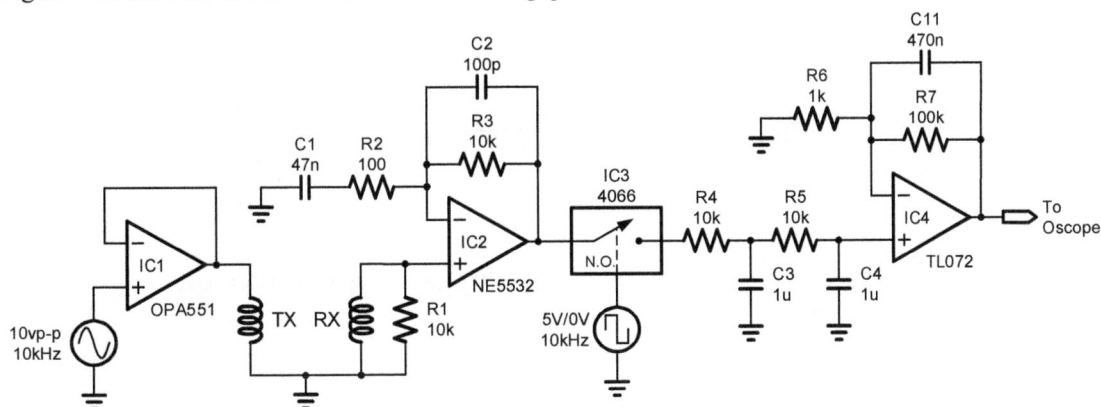

Fig. 7-9: **Test Circuit**

Now that we have that resolved, let's finalize the details of the test. The detection area under the coil is a three-dimensional space, so it would seem that we need to measure the isodetection lines within *all* this space. Instead, we will measure the isodetection pattern for a target which passes through the center of the coil in a normal sweep. Some coils (like round concentric and coaxial) have the same response patterns across the coil regardless of sweep direction (side-to-side, front-to-back). For these, a two-dimensional plot across one direction (front-to-back) will suffice. Other coils (oval, DD, 4B, figure-8, etc) have response patterns that depend on direction. For these, two plots will be recorded: a front-to-back plot and a side-to-side plot. Figure 7-10 illustrates.

For recording the isodetection lines the coil is placed against a metal-free platform which has a sheet of paper attached. Figure 7-11 illustrates the setup for a side-to-side plot. For a front-to-back plot the coil is rotated 90°. Finally, we'll standardize a couple of variables in the test. For the target we will use a US nickel, which represents a medium size lower-conductive coin. For the tilt of the coin we will use a flat tilt (0°) at all positions.

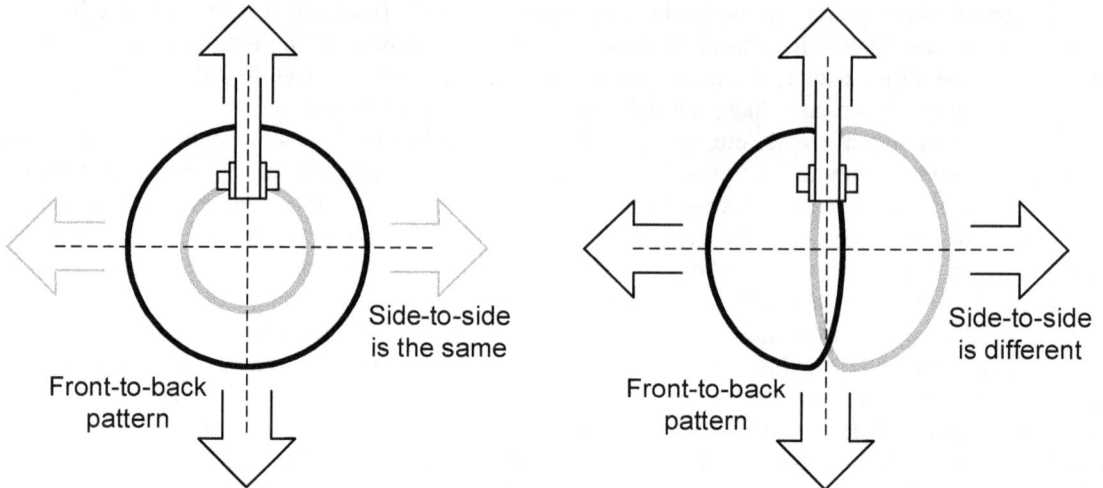
Fig. 7-10: **Directionality Examples**

Side-to-side
is the same

Front-to-back
pattern

Side-to-side
is different

Front-to-back
pattern

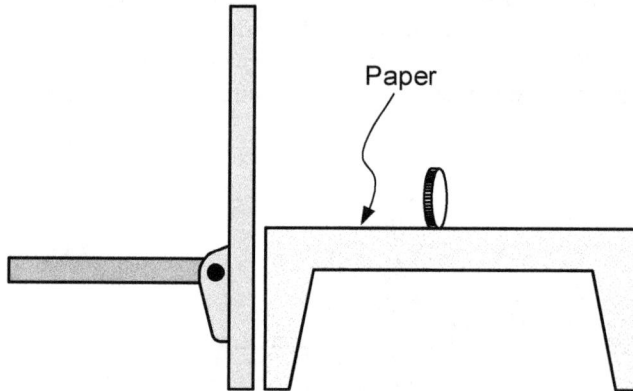
Paper

Fig. 7-11: **Pattern Measurement Setup (for side-to-side plot)**

7.5: Concentric Coils

We will jump right in and look at detection patterns of some commercial coils. Purely round concentrics (White's *950*, Fisher CZ-8) will have only a single side pattern plot whereas the Fisher CZ-10 (Figure 6-12) is highly asymmetric and will also include a front pattern plot. The Tesoro 9x8 is slightly asymmetric but not enough to matter.

The White's *950* (*DFX*, *MXT*) in Figure 7-12 is a very basic round concentric with an RX coil diameter of 50%. The Tesoro 9x8 in Figure 7-13 is slightly oval and has an RX coil diameter of 33%. The Fisher CZ-8 in Figure 7-14 gives us an example where the RX coil diameter is 60%. Ignoring the differences in depth, the Fisher coil has a flatter pattern for shallow targets while the Tesoro coil has a flatter pattern for deep targets, with the 950 in between. Notice also that the *BlueMax 350* (a 3.5 inch coil) has relatively poor deep-target response compared to the larger coils.

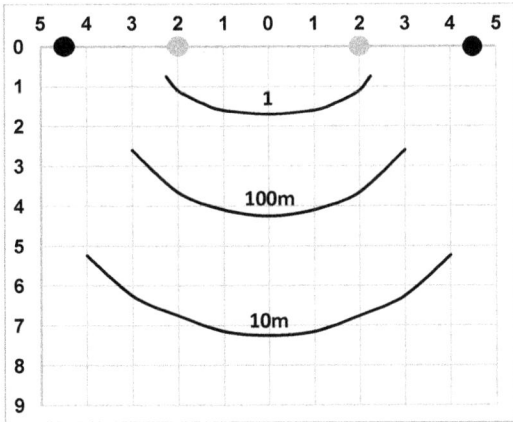

Fig. 7-12: **White's** *950*

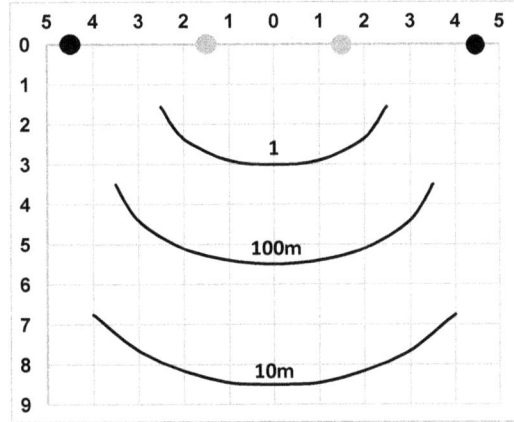

Fig. 7-13: **Tesoro 9x8**

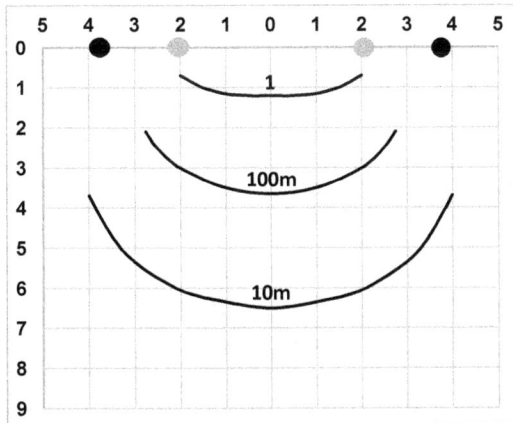

Fig. 7-14: **Fisher CZ 8-inch**

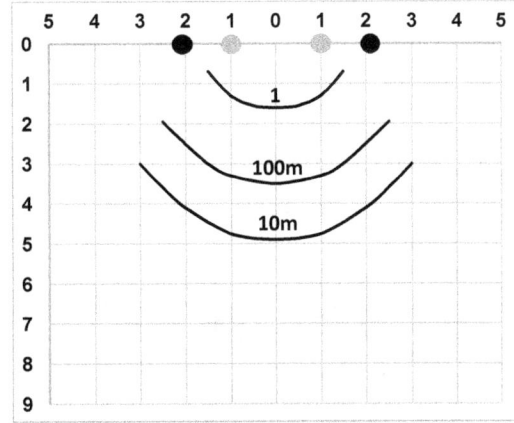

Fig. 7-15: **White's** *BlueMax 350*

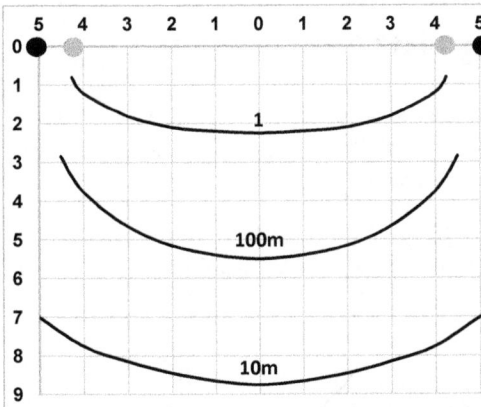

Fig. 7-16: **Fisher CZ 10-inch**

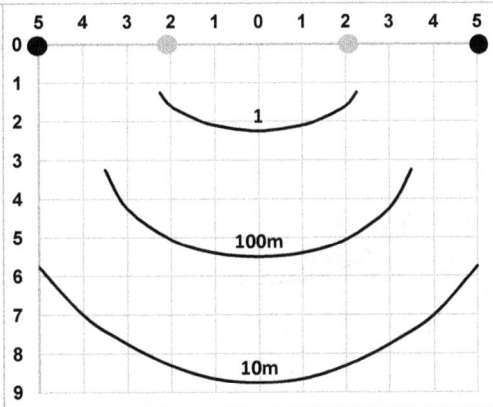

7.6: Overlapped Coils

The following plots show the detection patterns for several commercial overlapped coils. Some of these coils were pictured in the last chapter.

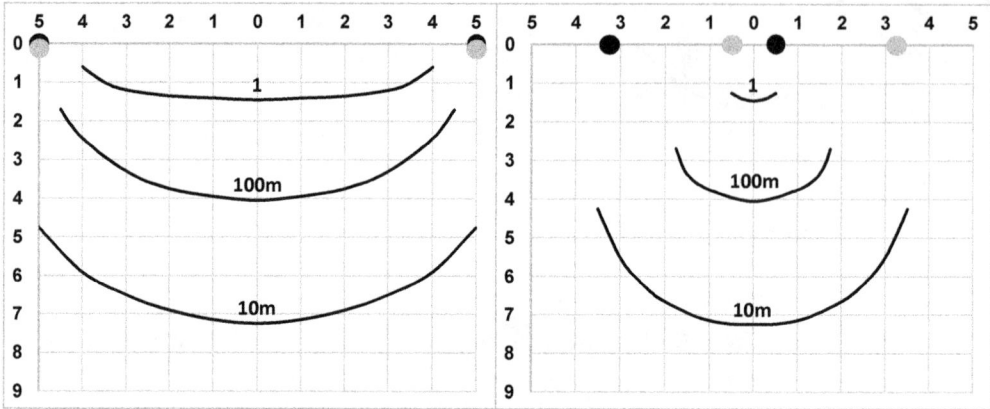

Fig. 7-17: **Fisher 7x11-inch DD**

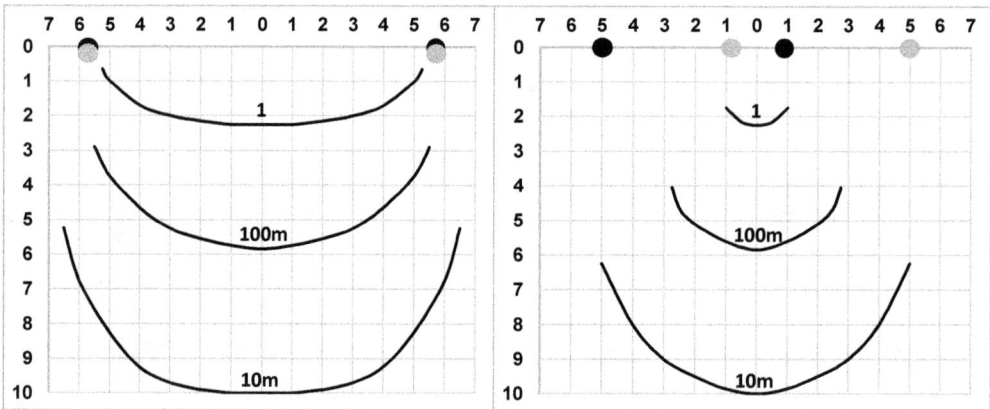

Fig. 7-18: **Tesoro 10x12 DD**

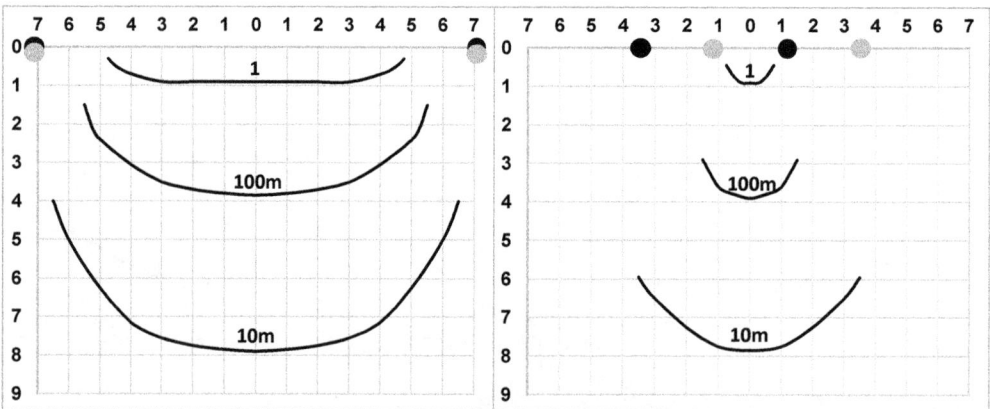

Fig. 7-19: **12x15 SEF**

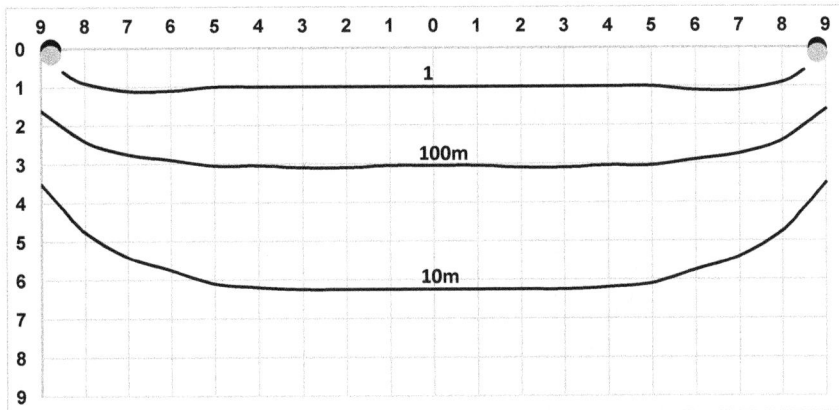

Fig. 7-20: **Tesoro *Cleansweep***

DD coils are often described as having a detection pattern akin to a "windshield wiper," that is, very narrow side-to-side and with consistent depth front-to-back. In fact, one manufacturer illustrates the detection pattern as shown in Figure 7-21. As we can see from actual measured plots, this is completely wrong. It is certainly true that the DD coil has a fairly narrow side-to-side response compared to a concentric, especially with targets close to the coil. And some coils (like the *Cleansweep*, and the *SEF* for very close targets) can exhibit a very consistent front-

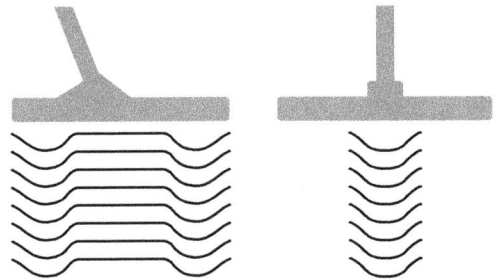

Fig. 7-21: **Mythical DD Pattern**

to-back response. But, for typical round DD coils, as the target moves farther from the coil the response pattern rapidly takes on a broader curvature, looking more and more like a concentric coil. Therefore, even a DD coil needs overlapping sweeps if you want to be sure of hitting the deeper fringe targets.

For a direct comparison, Figure 7-22 overlays the response patterns of the Fisher 7x11 DD (black) with the White's *950* concentric (gray). The DD's detection pattern exhibits the alleged "windshield wiper" coverage only for strong targets *very* close to the coil. At depth, the front-to-back pattern (left plot) closely mimics the concentric. The side-to-side response (right plot) is sharper for the DD and broader for the concentric, which should make the DD the better choice for target separation. A curiosity of these patterns is that the concentric beats the DD in sensitivity at shallower depths but loses this advantage for deeper targets. See 7.12 for an explanation.

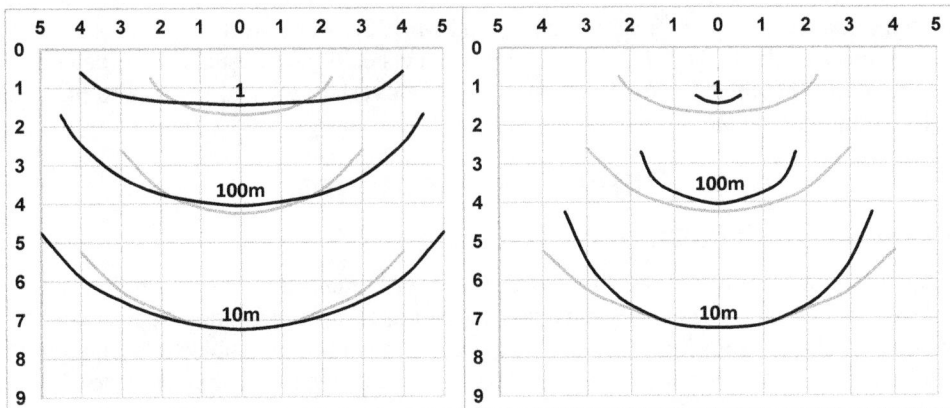

Fig. 7-22: **DD versus Concentric Patterns**

7.7: Coaxial Coils

Coaxial coils are relatively rare. They were used in production detectors for the early Garrett VLF and several C&G Technology models. In the aftermarket space, Jim Karbowski made the *Eliminator* for various White's detectors. See Figure 6-21 for a photo of all three.

Although each of these three used different design approaches we will only consider the Eliminator coil, for which all three coils (RX−, TX, and RX+) are 6 inches in diameter. Figure 7-23 shows the detection pattern. What is not shown is the edge sensitivity which exhibits a null for metal targets that span the entire thickness of the coil. This is a primary advantage of coaxial coil as it allows hunting very close to metal fence posts and the like.

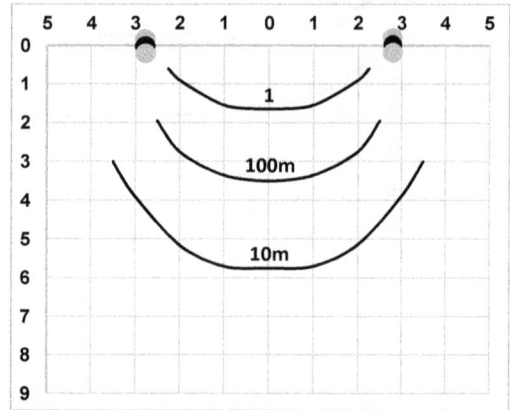

Fig. 7-23: **Applied Creativity *Eliminator***

7.8: Figure-8 Coils

The only figure-8 coil that was ever widely available was the *Bigfoot* coil for various White's detectors. Here are the detection patterns for it:

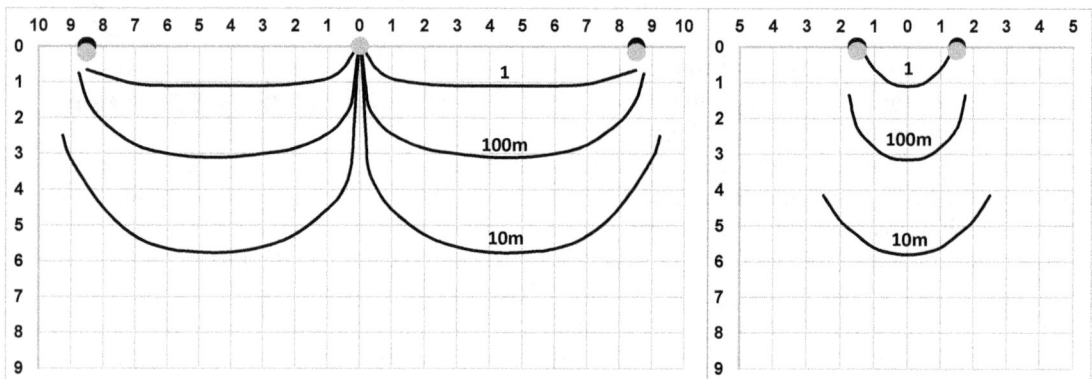

Fig. 7-24: ***Bigfoot* Coil**

Notice that the front-to-back pattern has a considerable notch in depth at the center of the coil. This is because the front half of the coil (toe) has a positive response and the rear half (heel) has a negative response. In the exact center there is no response. And while the shallowest targets otherwise get fairly consistent depth across the length of the coils, deeper targets exhibit the usual curved detection pattern of every other coil. This means the *Bigfoot* must also be overlapped, and by 75% to place the peak detection lobe on top of the previous sweep's center null.

Other Considerations

7.9: Mineralization & Lift-Off

Ground mineralization consists of iron oxides, generally magnetite (Fe_3O_4) but also maghemite (Fe_2O_3). As we saw in our basic experiments, iron concentrates a magnetic field so iron-based mineralization will distort the transmit field. The TX magnetic field is bi-directional; that is, the field extends upward above the coil with the same strength and pattern as downward. As the search coil is lowered to the ground, mineralization will start to compress the TX field, primarily on the lower (ground) side. We'll call this *ground compression* or *ground effect*. Figure 7-25 illustrates.

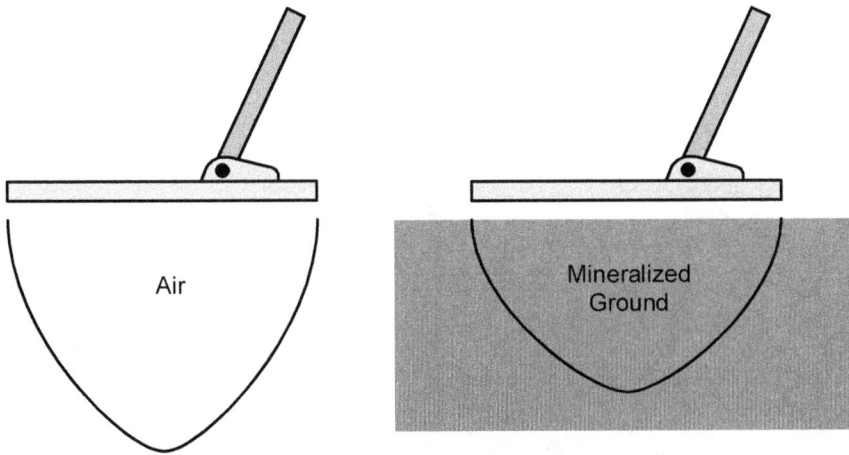

Fig. 7-25: **Ground Effect on the TX Field**

Ground compression occurs because mineralized ground has a higher permeability than air. Permeability represents how well a medium "conducts" a magnetic field so it would seem that the TX field would go deeper in mineralized ground. But the field naturally diverges (curves) around the windings and the higher permeability increases the curvature so that the flux takes an even shorter path, thereby compressing the overall field. Figure 7-26 shows an FEM analysis with the field over mineralized ground (black) versus the field in air (gray).

The vast majority of detectors use induction-balanced coils, where the coils are carefully arranged such that the RX coil is in a balanced (nulled) position with respect to the TX coil, and the residual (no target) RX signal is as close to zero as possible. This is done at the factory in complete absence of any target or ground effect. Obviously a metal target will upset this balance and create a legitimate target signal, as intended. The compression effect from mineralized ground can also be strong enough to upset the balance and great effort has been put into developing "ground balance" methods to minimize its effect. But ground balance schemes deal with the problem after the coil balance has been upset and a

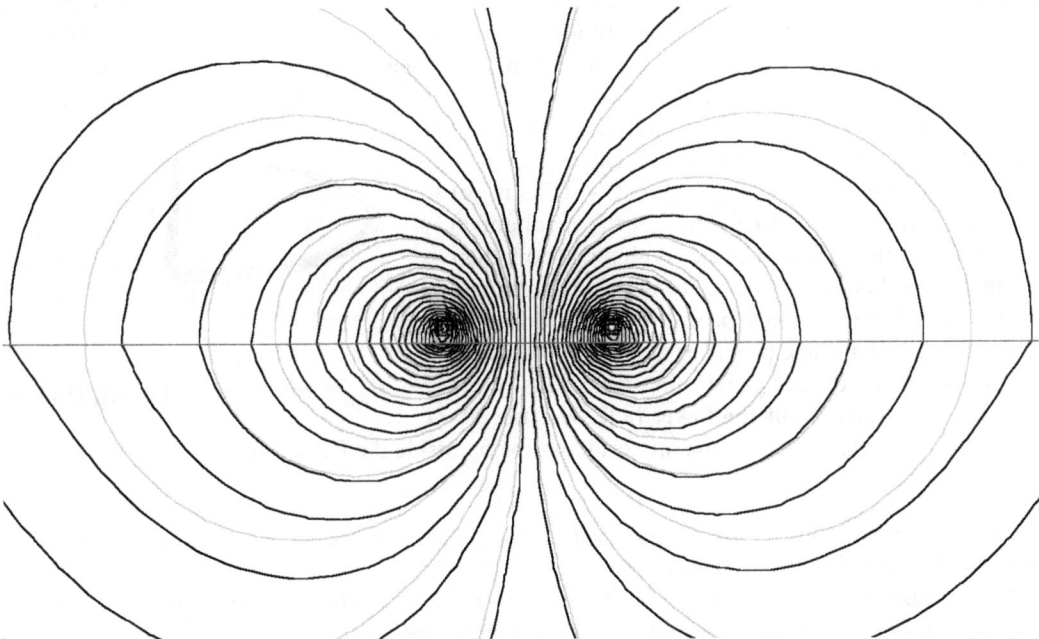

Fig. 7-26: **Field Compression**

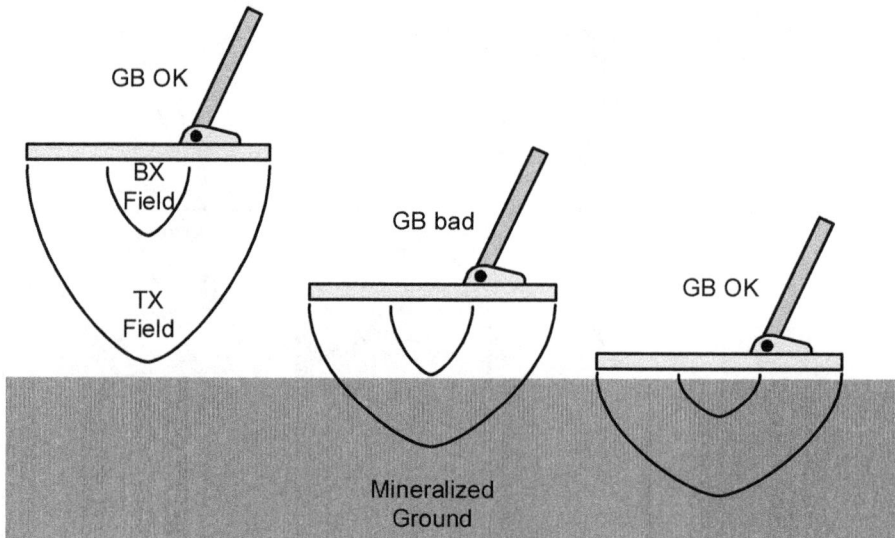

Fig. 7-27: **Concentric Ground Effect and Lift-Off**

ground signal has been produced. The most effective way to minimize ground effect is to minimize it at the coil, and some coils are better at this than others.

The imbalance caused by ground effect is worse for non-planar coils (coaxial, orthogonal) which is why most coils are planar. But even with planar coils the effect varies, both in causes and in severity. The concentric coil[10] is widely known to struggle in highly mineralized ground and is not difficult to get into an overload condition. The reason is there are two TX coils: the main TX coil and the bucking (feedback) TX coil. The main coil produces a very strong TX field and the bucking coil produces a weak TX field but is placed right next to the RX coil for a stronger effect.

As the search coil is lowered to the ground, the main TX field gets distorted first because it is stronger. This upsets the balance and the detector "sees" the ground. As the search coil gets even closer to the ground, the bucking field also gets distorted and can actually pull the system back toward a balanced state. In some cases, the bucking field distortion can outrun the main field distortion and cause an inversion in the induction imbalance. Any of this can create a weird positive-negative ground noise as the coil is moved up and down very close to the ground, or when swinging the coil over uneven ground. This effect is called *lift-off*[11] and can make manual ground balancing very difficult and or cause automatic ground balancing systems to behave erratically. Figure 7-27 illustrates[12].

The DD coil is widely known to do a better job handling mineralization than the concentric. But it is still subject to ground effect and even lift-off, though the mechanism is entirely different than the concentric. With a DD coil, there is only one TX coil but, unlike the concentric, it is not round; it tends to have a flat(ish) side, a broad rounded side, and two tight corners. Such a shape produces a TX field that is very non-uniform, with a deeper penetrating field density in the middle and more intense but shallower field densities at the

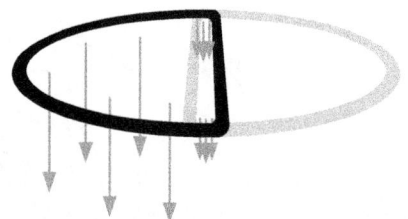

Fig. 7-28: **DD Coil Field Density**

tight corners, which usually have the misfortune of being inside the RX overlap.

---

10. For now, only the round concentric is considered. Other concentric shapes (elliptic, etc) have similar problems, plus other problems that will become apparent as we proceed.

11. Lift-off is a term borrowed from eddy current *non-destructive testing*, where it is used to describe a similar effect of response non-linearity versus probe distance from the material.

12. The "BX" field is that from the bucking (BX) coil. Also called the feedback (FB) coil in Chapter 6.

As the DD coil is lowered to the ground the non-uniform TX field is compressed by the mineralization at different rates resulting in a varying imbalance at the RX coil. The "tightness" of the corners and their spatial relationship with the RX coil determine the severity of lift-off. Even though the DD is not immune from ground effect, it almost always outperforms a concentric.

If not for the non-uniform TX field, it would seem that an overlap coil design could be immune from ground effect. This can be observed in a double-O coil which exhibits little ground effect. Except for Nexus, no detectors use the OO coil, most likely for aesthetic reasons. But the double-0 (oval) coil performs almost as well and has become fairly popular as an aftermarket coil.

Two coils that (theoretically) have zero ground effect are the figure-8 and the DOD. Both of these coil designs use a coplanar differential RX coil which geometrically cancels whatever ground field might be present. This assumes well-matched RX coils and reasonably uniform ground with the search coil held level to the ground. The worst coils for ground effect are the coaxial and orthogonal. While the coaxial has a differential RX coil, they are not coplanar and therefore do a poor job at canceling ground effect.

One coil not mentioned in all this is the mono coil. The mono coil is not part of an induction balance system so it is reasonable to assume there is no ground effect. That is true, except that a BFO has an obviously strong reaction to mineralization. This is not due to magnetic field distortion, but rather it is the mineralization's effect on the inductance of the search coil. That is to say, it is an entirely different mechanism (see Chapter 10). With a PI mono coil, the TX magnetic field is turned off just prior to RX sampling so even if it is heavily distorted by the ground the RX circuit will never see it.

7.10: Target Separation

Target separation is the ability to distinguish between two adjacent targets. As an example, suppose a 1914-D Lincoln cent is buried adjacent to a 1972 aluminum pull tab and they are in alignment with your coil sweep direction. Ideally, you would like to hear each target separately no matter how close together they are, and hopefully get separate tones or visual IDs as well.

For target separation we only need to consider the side-to-side detection patterns. As we saw in Figure 7-22 the side-to-side detection pattern for the DD coil is substantially narrower than that of the concentric, especially for shallower targets. For deep targets it still holds a diminishing advantage. Therefore, it stands to reason that the DD coil will be superior for target separation. This is true with a minor caveat. The DD coil also has a slightly negative response for shallow targets that fall just beyond the DD overlap, and this was not shown in the pattern plots. Figure 7-29 shows what this looks like the coil is swept over a shallow coin.

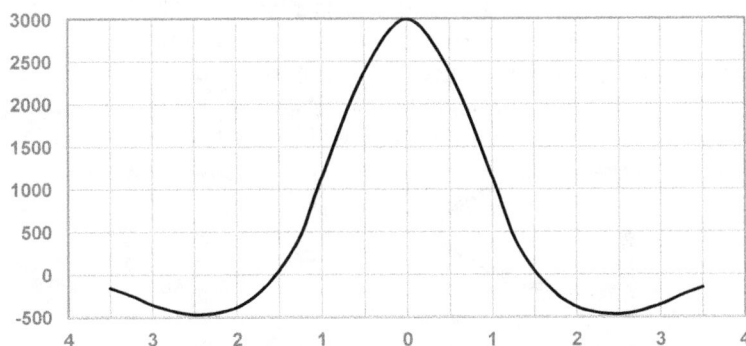

Fig. 7-29: **DD Close-Target Response Pattern**

The plot shows that a shallow target produces a 3000mV response (measured at the R demod) when it is under the center of the coil. The same target has a negative response that peaks at almost −500mV at about 2.5 inches from the center of the coil. What this means is that a shallow target in close proximity to a deeper target can completely nullify the deeper target's response. This is an example of *masking*, where an adjacent target (or hot rock) can alter or nullify another target's response.

For the DD, this effect vanishes after a couple of inches of depth and deeper targets only produce a positive response across the width of the coil. The concentric does not exhibit this issue but it tends to have a broader sweep response which still hurts target separation as the responses of close proximity targets tend to blend together.

7.11: EMI

Target detection causes an imbalance in an IB system and therefore depends on both the TX and RX coils. *Electromagnetic Interference* (EMI) is simply the coupling of outside RF signals to the RX coil and has no causality with the TX coil. In the most simplistic terms, coils with a single RX winding (mono, concentric, 00, DD, orthogonal) are subject to EMI pickup, whereas coils with differential RX windings (coaxial, figure-8, DOD) naturally cancel EMI. EMI pickup is dependent on the RX coil area so larger diameter coils tend to have worse EMI performance. This would seem to offer concentric coils an edge over DD coils because the concentric RX coil is generally smaller.

However, EMI pickup is also proportional to the number of turns of wire in the RX coil and for the same inductance a smaller coil has more turns. The smaller coil still wins as the product of the coil area and the number of turns ($N \times A$) favors the smaller coil for constant inductance. The area-turns product increases roughly by the square root of the coil diameter.

The RX coil is not the only element that determines whether a detector suffers from EMI. Some detectors fare better at EMI rejection than others. Good coil shielding plays a role, especially at suppressing higher frequency EMI. In the circuitry, narrow band filtering can help a lot, which means that a wideband design (multifrequency or PI) will typically have more EMI problems. The sampling rate of the demodulator can also be optimized for very specific EMI, such as power line (50/60 Hz) noise. Some detectors have a minor frequency adjustment (*frequency offset*) to help minimize interference, including detector-to-detector cross-talk interference.

7.12: Spacial Alignment

A final consideration is what we will call *spacial alignment*. This has to do with where each coil exhibits peak target coupling in the 3D detection space. Some coils designs are handicapped by how the RX coil is aligned in relation to the TX coil. The best aligned coil is the concentric, where the RX coil is both coplanar and coaxial to the TX coil. Assuming a flat-oriented target, the peak target eddy response occurs when the TX coil is exactly centered over the target, and at the exact same time the RX coil has maximum sensitivity to the reverse field produced by the target eddies. The stars align perfectly.

The OO, 00, and DD coils don't share this fortuitous alignment. The TX and RX coils are spaced laterally so that they do not appear directly over the target at the same time. This results in a geometric misalignment as the coil is swept over the target. This alone will reduce depth compared to a concentric of the same overall diameter as, when the target is under the center of the coil, the TX field incident on the target is something less than 90°, as is the target's field incident on the RX coil.

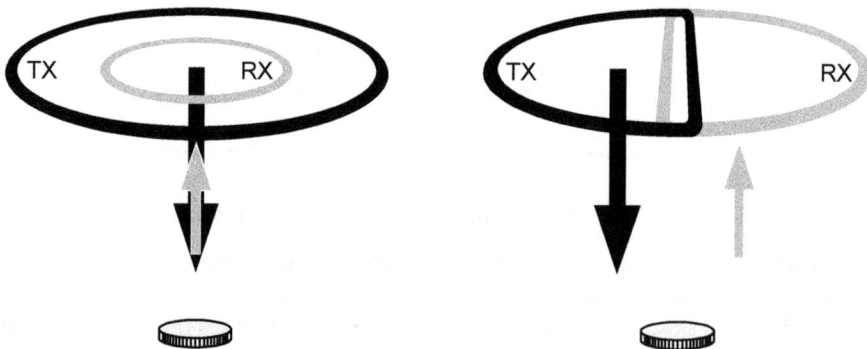

Fig. 7-30: **Concentric vs DD Spacial Alignment**

Figure 7-22 compares the detection patterns for a DD and a concentric. The concentric has better sensitivity for the shallowest target, while the two coils are roughly equal at depth. This illustrates that the geometric disadvantage of the DD is worst for shallow targets where the misalignment angles are greatest.

Another handicap of the DD coil is that, for the same overall diameter as a concentric, the TX winding is smaller and the RX winding is larger. In comparing the TX windings, we should expect the concentric to punch a little bit deeper than the DD. And in comparing the RX windings, we should expect the concentric to have a little better target field sensitivity. The DD remains popular because of its superior performance in harsh ground and in separating close targets.

The coaxial coil has the TX and RX coils spaced vertically with the RX coil placed closest to the ground. Therefore, the RX coil is optimally as close to the target as possible while the TX coil is spaced farther away, which results in a weakened flux density at the target.

A final example is the orthogonal coil where the peak TX and RX zones are in completely different directions. The TX coil is vertical so the transmit field is horizontal and any flux that couples to the target relies on the field diverging from horizontal and curving into the ground (Figure 6-26). Meanwhile, the RX coil is horizontal which is optimal for targets directly below it, but the peak response of a target may be well off its center line.

Explore

7.13: More Power

Detectorists often imagine that if they could only increase the detector's transmit power they could find deeper targets. That is true in general but it has a rapidly diminishing return, and the fantasized "doubling" of depth just isn't reasonable. Figure 7-31 shows the center-line depth results for a concentric coil driven at both 10vp-p and 20vp-p. As expected, doubling the TX voltage drive results in a doubling of the RX target voltage. That is, a 100mv target at 10vp-p becomes a 200mv target at 20vp-p.

While we can say that a strong target response (the 1v target) has its "depth" doubled — from ~1 inch to ~2 inches — the relative benefit diminishes with depth. For 10vp-p a 10mv target is just under 7 inches and doubling the voltage drive (and therefore the TX current) adds just under one extra inch. This is about a 13% depth increase for twice the "power[13]." And it gets depressingly worse with deeper targets.

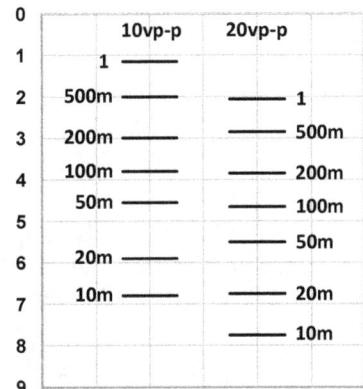

	10vp-p	20vp-p
0		
1	1	
2	500m	1
3	200m	500m
4	100m	200m
	50m	100m
5		50m
6	20m	
7	10m	20m
8		10m
9		

Fig. 7-31: **10vp-p vs 20vp-p**

7.14: The Math of Depth: Part 1

In considering only the transmit field, Equation 3-5 shows that the flux density diminishes at a rate roughly proportional to d^3. The ensuing example shows how the $1/d^3$ approximation compares to the exact given in Equation 3-6. Let's see how well Equation 3-6 fits the measured field strength for the 8-inch coil in Figure 7-3.

For this we take the center-line measurements and compare them to $k(r/(\sqrt{r^2 + d^2}))^3$ where k is a scaling factor and should equal the measurement at the center of the coil. With $k = 26.2$ the calculated curve is shown in Figure 7-32 along with the discrete measured points. The calculated curve is quite predictive and we should consider this an engineering success.

---

13. See 28.2 for a discussion of what "power" means.

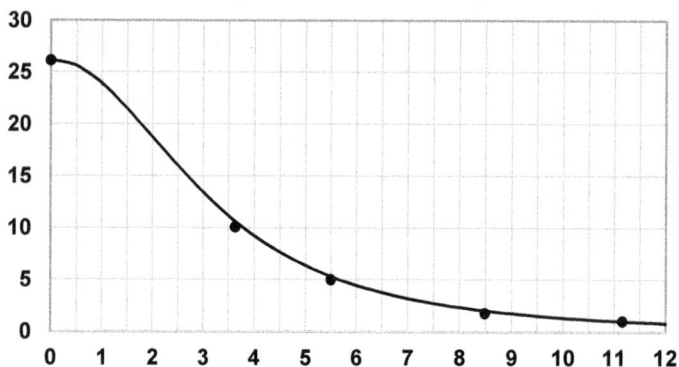

Fig. 7-32: **Calculated B-Field vs Depth**

7.15: The Math of Depth: Part 2

The same physics holds true for the induced target field. That is, the strength of the target's reverse B-field will diminish with distance from the target, eventually at a rate of $1/d^3$. With coin-sized targets, "eventually" happens very quickly. We saw in Equation 3-6 that the field attenuation for a coil hits the $1/d^3$ rate beyond about twice the diameter of the coil; this also holds true for a coin but instead using the coin's diameter, so beyond about 2 inches (5 cm) the coin's eddy field is dropping at a rate proportional to $1/d^3$.

This means the round-trip field strength should diminish at a rate proportional to d^6. Let's now consider the depth plots and see if they correspond to this prediction. For this we'll use the F44 coil (Figure 7-7) which has more detail, and instead of a simple $1/d^6$ reduction we'll use a slightly more complicated equation incorporating Equation 3-6:

$$S = k \cdot \left(\frac{r_c}{\sqrt{r_c^2 + d^2}} \right)^3 \left(\frac{r_t}{\sqrt{r_t^2 + d^2}} \right)^3 \qquad \text{Eq 7-2}$$

where S is the relative sensitivity, r_c is the coil radius, r_t is the target radius, d is the depth, and k is again a scaling factor. Since the F44 coil data is in inches we'll state the radii in inches as well: $r_c = 3$ inches and $r_t = 1/2$ inch. With $k = 240$ (empirically found to give a good curve fit), the equation is now

$$S = 240 \cdot \left(\frac{3}{\sqrt{3^2 + d^2}} \right)^3 \left(\frac{0.25}{\sqrt{0.25^2 + d^2}} \right)^3$$

which produces the graph in Figure 7-33. Center-line data points from Figure 7-7 are added for comparison.

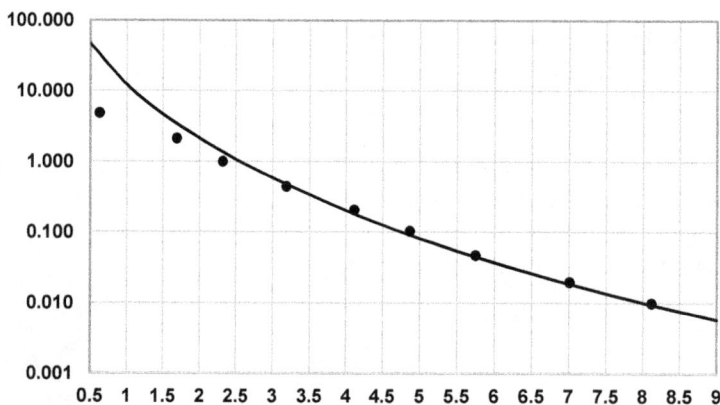

Fig. 7-33: **Calculated Relative Sensitivity vs Depth**

Inside the Metal Detector

Because of the wide range of values the Y axis is presented logarithmically. When $k = 240$ the calculated curve fits the data fairly well except when the target is extremely close to the coil. In that case, the math predicts a stronger response than observed. Continue reading the next section to find out why this happens.

7.16: RX Coil Size Revisited

Equation 7-1 shows how a changing magnetic field incident upon a coil induces an EMF. We assume the coil is normal to the target field so θ is 0 and the cosine term is 1. This tells us that the EMF is proportional to the area-turns ($N \cdot A$) of the RX coil. Next, let's assume that we want the RX coil to be a particular inductance; inductance is (roughly[14]) proportional to $N^2 A$. This means that N^2 is proportional to L/A and therefore the EMF is proportional to

$$\varepsilon \propto \sqrt{A \cdot L}$$
Eq 7-3

This suggests that when L is held constant a larger RX coil will produce a higher EMF, everything else being equal. But this assumes a homogeneous target field that is always as broad as the RX coil. If the target field is homogeneous but limited to the target area as in Figure 7-34a then the ideal coil diameter is roughly the same as the target diameter. The increased area of a larger RX coil does not couple any more flux and the reduced number of turns to maintain inductance is detrimental.

But the homogeneous field is a wrong assumption. For small targets the flux lines look more like that in Figure 7-34b. The field diverges and wraps around the target resulting in a diminishing vertical field as the distance (depth) increases. This shows that a larger coil now suffers a double penalty: the reduced number of turns as before, but now the increased RX coil area begins to see the reverse flux which actively cancels a portion of the forward flux (see also Figure 7-6). This is why small targets are often invisible to large coils and also why prospectors using mono coil PI detectors can detect the tiniest targets with the coil edge even though the target is invisible across the face of the coil.

It is also why the target measurements close to the coil in Figure 7-33 are lower than what the math predicts. The math assumes a homogeneous vertical field and does not account for field divergence or flux reversal. An equation that does account for these would be complex indeed and such analysis is better left for something like FEM software.

Whether you think of the target field as shown in (a) or (b) of Figure 7-34, the size of the RX coil should be appropriate for the expected size of the target. This is likely why some Tesoro concentric coils (intended for coin hunting) have smaller-than-usual RX coils.

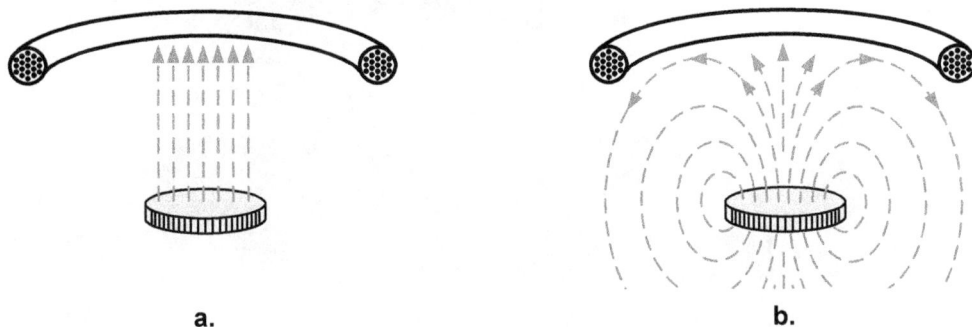

a.   b.

Fig. 7-34: **Homogeneous vs Actual Target Field**

---

14. See Appendix B for practical inductance equations.

Ore sample
containing
GOLD-COPPER
COBALT

5" CONCRETE BLOCK

Full scale
meter reading
on metal side
(RIGHT)

8

Building Coils

"In theory, there is no difference between theory and practice. In practice, there is."

— Yogi Berra

A key element in detector performance is the search coil. Designing and building a coil for a modest-performing detector is quite easy. But that same coil quickly becomes a limiting factor when attempts are made to significantly improve the performance of the detector. This chapter looks deeper into coil performance and at the trade-offs faced in designing them.

The design of a metal detector circuit and the design of its coil are not independent. By necessity, some of the information presented in this chapter relies on circuit design knowledge presented in later chapters. The alternatives were to either place this information near the end of the book — meaning the circuit chapters would be difficult to fully flesh out — or to disperse coil design information throughout the book, making it either difficult to find needed details or constantly repeating those details in each chapter.

Also, be sure to read this whole chapter no matter what coil you intend to build. Important information is spread around, so if you want to build a concentric coil don't just jump to that section and begin. Read everything first.

8.1: Coil Windings

The coil windings are the central elements in a coil design. It all seems so simple until you realize there are a lot of nuances at play. We'll start with the very basics and cover those nuances as we go.

8.1.1: Coil Parametrics

A coil is an inductor, one of the three fundamental passive elements in electronics. An inductor resists *changes* to current flowing through it, proportionally to its inductance L. An inductor is often used in circuits as a choke or a tuning element and, in those cases, the inductance is an important parameter.

Metal detector coils are used to generate and sense a magnetic field; that is, they are used as transducers. The inductance is not so much an important parameter in these functions. Rather the size, number of turns, current, frequency, and so forth matter more. However, transmit coils are often designed as part of a resonant circuit and receive coils are also often broadly tuned, so in many cases inductance does matter for other reasons. In mass production, keeping performance consistent (both amongst different units and in using different coils on a given unit) is achieved partly by maintaining fairly tight control over coil parameters, including inductance.

Besides inductance, a coil has a series resistance due to the resistance of the wire. The close proximity of adjacent windings also gives rise to *interwinding capacitance*[1]. Both resistance and capacitance are distributed throughout the coil but are often modeled as lumped parasitic elements (Figure 8-1). We'll see later that R_L or C_L can be critical in some coil designs, yet not matter much in others.

Measuring coil parametrics is not difficult. An inductance meter (usually an LCR meter) is the best way to measure inductance but beware of cheap and inaccurate meters. You want one that has

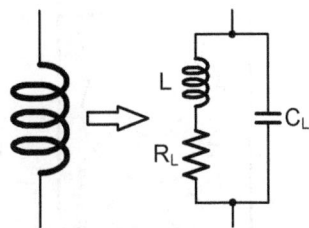

Fig. 8-1: **Coil Parametrics**

---

1. Also called *self-capacitance*.

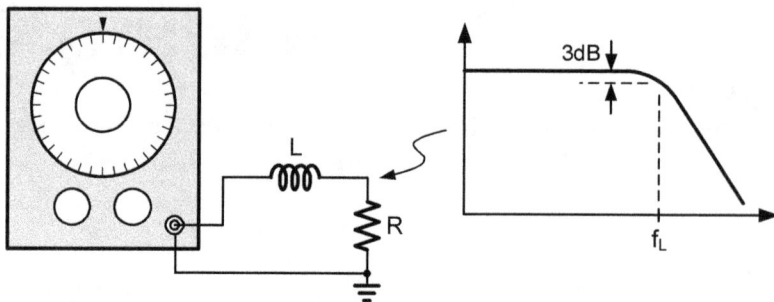

Fig. 8-2: **Measuring Inductance**

decent resolution in the microhenry range and preferably one that can measure at 100Hz, 1kHz, and 10kHz. The "DER EE" model DE-5000 is my preferred meter and can be found on eBay for about $100. As a bonus it has a 100kHz measurement setting.

If you don't have an inductance meter you can indirectly measure inductance using a simple R-L low-pass filter in Figure 8-2. Start with the frequency moderately low (100Hz or so) and increase it until the amplitude drops. Using the frequency f_L at which the amplitude is 70.7% (−3dB) of the passband amplitude, the inductance is calculated as

$$L = \frac{R}{2\pi f_L}$$

Eq 8-1

This is simple filter response math, widely described in electronics texts. R should be selected so that the pole frequency is much lower than the self-resonance frequency. 10kHz is usually a good target, but if you don't know the inductance to start with how do you select a resistor? You guess. Metal detector coils tend to be between 100μH and 25mH, so as a starting guess pick 2mH. This puts the resistor at about 120Ω. Measure the −3dB frequency and solve for L. In most cases a high coil inductance is accompanied by a large amount of wire resistance and this can produce measurement error. If 120Ω produces a −3dB frequency well below 10kHz, increase the resistor and try again. If it's well above 10kHz, decrease and repeat.

Coil (wire) resistance is easily measured with an ohmmeter except that many coils have only a few ohms of resistance and some ohmmeters (especially general-purpose multimeters) have significant errors when trying to measure such low values. LCR meters measure resistance, but be aware that sometimes the "resistance" is actually "impedance" so that when measuring an inductor you are really measuring $R + j\omega L$. You can easily see if this is the case by measuring R at two frequencies, and it is equally easy to derive the raw R from these measurements.

Capacitance is not easily measured directly. Because of the low coil resistance, a capacitance meter cannot measure the capacitance of a coil. We turn again to a simple filter measurement, this time a band-pass filter (Figure 8-3). At low frequencies the coil is a short circuit and at high frequencies the self-capacitance C_L of the coil becomes another short circuit. The peak response occurs at

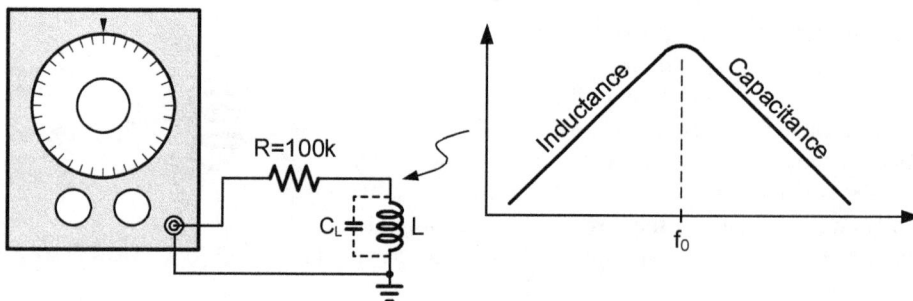

Fig. 8-3: **Measuring SRF & Self-Capacitance**

$$f_0 = \frac{1}{2\pi\sqrt{LC_L}}$$

Eq 8-2

and is known as the *self-resonant frequency* (SRF) of the coil. With a measured inductance value the capacitance is calculated as

$$C_L = \frac{1}{L(2\pi f_0)^2}$$

Eq 8-3

Typically metal detector coils have an SRF in the range of 50kHz to 1MHz and a self-capacitance on the order of 100-1000pF. Usually the higher the inductance (more windings), the higher the self-capacitance. Using the ~120Ω resistor suggested before will work but a higher resistor value (10-100 kΩ) will also work well.

There is another way to determine SRF. If the coil is energized with an impulse current it will ring at this frequency, something that happens in pulse induction designs and is cured with a damping resistor. Measure the ringing frequency f_0, then solve the above equation for C_L.

8.1.2: Wire

In electronics, a coil is a piece of wire wound (most often) in a circular shape. That is about as simple a concept as possible. But the type of wire and size make a big difference. Copper is the overwhelming choice for metal detector coils although, for a while, White's made their *950* concentric coil with an aluminum TX coil to reduce weight. All coils in this book assume the use of copper wire of some kind.

Wire is sized according to *gauge*, and there are two predominant gauge standards. The first is "American Wire Gauge" (AWG) which is an imperial standard of wire diameter that originated in England in the 1850s. The other is metric, where wire is specified simply by its diameter. The tables below list common sizes used in metal detector coils, along with resistance and weight per length. The table

American Wire Gauge					Metric			
AWG	Diameter (mils)	Resistance mΩ/ft	Weight lb/1000ft		mm	Diameter (mils)	Resistance mΩ/m	Weight g/m
18	40.3	6.51	4.92		1.0	39.4	22.38	7.05
19	35.9	8.21	3.90		0.9	35.4	27.64	5.71
20	32.0	10.4	3.09		0.8	31.5	34.97	4.51
21	28.5	13.1	2.45		0.71	27.9	44.39	3.55
22	25.4	16.5	1.94		0.6	23.6	62.20	2.53
23	22.6	20.8	1.54		0.56	22.1	71.38	2.20
24	20.1	26.2	1.22		0.5	19.7	89.52	1.76
25	17.9	33.0	0.970		0.45	17.7	110.5	1.43
26	15.9	41.6	0.769		0.4	15.8	139.9	1.13
27	14.2	52.5	0.610		0.355	14.0	177.7	0.887
28	12.6	66.2	0.484		0.315	12.4	225.5	0.700
29	11.3	83.4	0.384		0.28	11.0	285.6	0.552
30	10.0	105	0.304		0.25	9.84	358.4	0.440
31	8.93	133	0.241		0.224	8.82	446.4	0.353
32	7.95	167	0.191		0.2	7.87	559.6	0.282
33	7.08	211	0.152		0.18	7.09	691.0	0.228
34	6.31	266	0.120		0.16	6.30	875.7	0.180
35	5.62	335	0.095		0.14	5.51	1143	0.138
36	5.00	423	0.076		0.125	4.92	1433	0.110

on the left is for AWG, the table on the right is for metric wire. The two tables are aligned for roughly equivalent sizes with diameter in mils (milli-inch) given for both standards for easy comparison.

In the AWG system wire diameter is halved every 6 gauge sizes. This means resistance goes up by a factor of 4 but weight goes down by a factor of 4. Since the metal detector coil is usually placed at the end of a tubular rod and swung by the user, weight is a primary consideration. This would tend to argue for the use of the smallest and lightest wire possible, except that resistance and sometimes current capacity also play even more primary roles, especially on the transmit side.

Wire also comes in a variety of types. Solid/stranded and insulated/bare are commonly used to distinguish types of wire. It goes well beyond this, as there are many different insulations and even different ways to strand wire. In metal detectors, *magnet wire* is the predominant choice for manufacturing coils. Magnet wire is a solid-core copper wire with a thin coating of insulation, usually a polyester film. Furthermore, magnet wire can be obtained with a "bond" coating which is an additional thermoplastic adhesive that can be activated with either isopropyl alcohol or heat to adhesively bond all the turns of wire together into a rigid bundle. This is important in the manufacture of search coils as it provides a solid self-supporting structure that is more easily handled.

In a pinch, coils can be wound with ordinary insulated wire, such as PVC-jacketed wire. The insulation is substantially thicker than that of magnet wire, resulting in a lower interwinding capacitance (a good thing). But, given the same dimensions and number of windings, the increased wire spacing also has the effect of reducing inductance, which means additional turns are required to get back to the needed inductance value, assuming that is an important part of the design.

Teflon-jacketed wire has the same issues as PVC wire but Teflon has an advantage in that it has a relative permittivity lower than PVC[2], resulting in even lower interwinding capacitance. Teflon insulated wire is fairly expensive but has been used in some PI designs where a "fast" coil is crucial to overall performance.

Although the purpose of the transmit coil in a metal detector is to induce eddy currents in a nearby target, what we don't want to do is induce eddy currents in the coil wire itself. Induced eddy currents within the coil wire tend to force the resultant current flow to the outer surface of the wire which increases the apparent wire resistance, an effect that gets worse with higher frequency. A special type of wire called *litz*[3] was developed to specifically address the problem in high-frequency coils[4]. High-frequency VLF gold detectors and PI detectors often use litz wire in the TX coil to minimize power loss.

Litz wire is similar to ordinary stranded wire except each individual strand is completely insulated throughout the length of the wire, and the strands of wire are also twisted in a way that they all alternate between the middle and the surface of the bundle. The individually insulated wires ensure that each strand carries the same portion of the total current, and the woven twist ensures that each strand experiences the same effective magnetic field. An example of litz wire is "42/40" which means it has 42 strands of 40 AWG wire, which is roughly equivalent to a solid wire of 24 AWG.

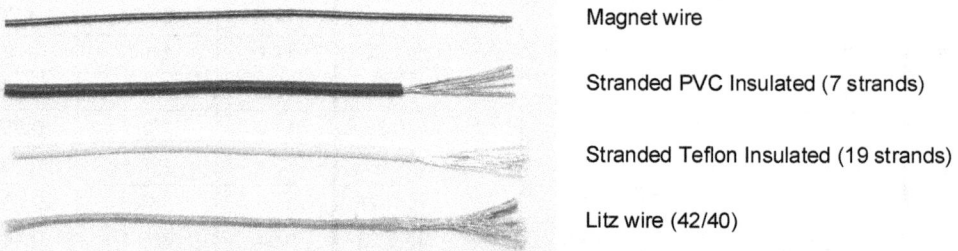

Magnet wire

Stranded PVC Insulated (7 strands)

Stranded Teflon Insulated (19 strands)

Litz wire (42/40)

Fig. 8-4: **Different Wires (all 24 AWG)**

2. PVC $\varepsilon_r = 3.0$, Teflon/PTFE $\varepsilon_r = 2.1$.

3. From the German word *litzendraht*, meaning "braided" or "woven."

4. Minelab's very first patent was on the use of litz wire in metal detector coils (US4890064, 1989) to reduce eddy current generation despite the fact that litz wire was invented almost 100 years earlier specifically to address this issue.

8.1.3: Coil Windings

The most common type of coil winding is called a *solenoidal* coil or a *helical* coil, as shown in Figure 8-5. Many RF circuit inductors and some relays (solenoids) use this type. It is an efficient coil structure with a low parasitic capacitance. It also works well with ferrite core rods which can improve performance in certain applications. Because it is more difficult to manufacture and results in a tall coil structure, it is practically never used in metal detectors with the exception of pinpointers.

A second type is the *spiral* coil. It shares the advantage of the solenoidal coil in that it has low parasitic capacitance. But it is a less efficient coil because the magnetic flux created by outer windings cut through the inner windings and partly cancel the effects of the inner windings; the inner windings do the same to the outer windings. Nevertheless, the spiral coil has one advantage: it is completely planar and can be created on any flat surface, including a PC board like that in Figure 8-6. The earliest PCB spiral coil was in a 1970s BFO detector sold in kit form by Radio Shack (USA). The Tesoro *Sandshark* also used a printed-spiral coil and was probably the only commercially produced detector to do so.

Many coils are a combination of solenoid and spiral, in that they are wound in several layers of tight helical windings. The layers are carefully laid down by machine so that each layer is tightly packed for highest efficiency. A metal detector coil is most often manufactured using a *scramble-wound* technique, whereby the windings are not laid down with any particular care. The result is simply a "bunched" coil of wire. This method produces more variation in the inductance value, but it is generally still within a couple percent of a nominal value. It also produces a higher and less repeatable inter-winding capacitance, but in many metal detector designs this is not a critical issue. Scramble-wound coils are fast and easy to produce, and the parametric variations are well within the needs of typical metal detectors.

High-performance metal detectors may require coils with much tighter tolerances to consistently meet their performance goals, or may require the lowest self-capacitance achievable. The latter is especially true in cutting-edge PI designs where speed is critical, or in high-frequency gold detectors. Consistency is better achieved with carefully produced windings (such as a spiral) instead of the randomness of scramble windings. Another method for achieving low self-capacitance is the *basket wound* coil (Figure 8-8). In a basket wound coil individual windings are staggered and spaced so wire-to-wire capacitance is minimized. The result is an overall low self-capacitance. Basket-wound coils were

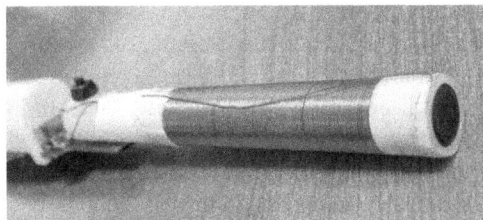

Fig. 8-5: **Solenoidal (Fisher F-Pulse)**

Fig. 8-6: **PCB Spiral**

Fig. 8-7: **Scramble Wound**

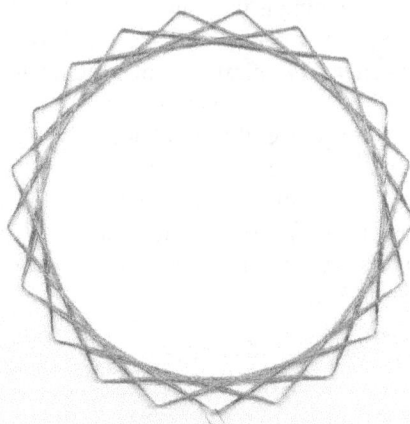

Fig. 8-8: **Basket Wound**

popular in older RF applications but in metal detectors have only appeared in home-brewed solutions due to the difficult nature of manufacturing.

In all of these topologies the wire used to make the coil can be any of the wire types mentioned in the last section. Besides the wire type and the basic winding topology, there is also shape, size, and number of turns to consider. These can vary from round to square, from 1/2-inch diameter to 2 meters, and from 1 turn to many 100s of turns. The possibilities are nearly infinite in number so it is important to have a good understanding of the underlying principles.

The following table compares scramble-wound coils using magnet wire (bonded and unbonded), PVC insulated, Teflon insulated, and litz wire, plus solenoidal, spiral, and basket wound coils using magnet wire only. The standards for this comparison are 24 AWG or equivalent wire gauge, 40 turns, and a nominal diameter of 10-inch|25cm. For a scramble-wound coil of magnet wire this produces a nominal inductance of about 1mH.

Type	Wire	L	R (DC)	R (100kHz)	SRF	B_0	B_{10}	Eff. %
Scramble	Magnet (bonded)	1074µH	2.79Ω	14.54Ω	345kHz	20.2µT	1.72µT	8.5
	Magnet (unbonded)	1091µH	2.79Ω	14.63Ω	383kHz	20.2µT	1.70µT	8.4
	PVC (7/32)	934µH	2.60Ω	4.55Ω	514kHz	20.2µT	1.80µT	8.9
	Teflon (19/36)	954µH	2.51Ω	6.86Ω	667kHz	20.2µT	1.78µT	8.8
	Litz (42/40)	1009µH	2.79Ω	3.19Ω	547kHz	20.2µT	1.78µT	8.6
Spiral	Magnet	860µH	2.91Ω	3.75Ω	1023kHz	20.2µT	1.74µT	8.6
Solenoidal	Magnet	822µH	2.83Ω	3.75Ω	1076kHz	19.1µT	1.54µT	8.1
Basket	Magnet	730µH	2.90Ω	4.02Ω	849kHz	20.8µT	1.46µT	7.0

Table 8-1: **Comparison of Wire Types and Winding Styles**

For scramble-wound coils, we see that wire with thinner insulation (magnet wire) makes a tighter bundle that gives a higher inductance[5]. The thickest insulation (PVC) has the lowest inductance. While the DC resistances are very close, the AC resistance at 100kHz is not. AC resistance is caused by cross-eddy currents inside the wire that push the current flow out toward the surface of the wire, in the same way that target eddies get pushed toward the perimeter of the target. Solid wire is the worst as seen with the magnet wire, and litz wire is almost unchanged over frequency. This is why litz is the best choice for TX coils in high frequency gold detectors. When litz is not available, even ordinary stranded wire shows a marked improvement over magnet wire due to its thicker insulation.

SRF is dominated by the interwinding capacitance which depends on the wire spacing. Magnet wire — with its thin insulation coating — has the highest capacitance and thus the lowest SRF. The bonded winding is the tightest and therefore has a higher capacitance than the unbonded winding. Wire with a heavier insulation (like PVC) is better, but Teflon insulation with its low dielectric constant is even better. The spiral and solenoidal basket windings show advantages in SRF and are more reproducible but are harder to manufacture. The basket winding, while better than the scramble windings, shows no advantage over the spiral and solenoidal windings and is exceptionally difficult to manufacture.

All coils are driven with a 10kHz sinusoid at a peak-to-peak current of 100mA. B_0 is the field strength at the planar center of the coil and B_{10} is the on-axis field strength measured at a distance of 1d (10-inch|25cm). Measurements are made with the Magnetic Field Probe (Appendix A). To see if these numbers are reasonable we can calculate the expected field strengths for a 25cm 40T coil. Repeating Equation 3-4:

$$B = \frac{1}{2} \cdot \frac{\mu N I r^2}{\sqrt{r^2 + d^2}^3}$$

Eq 8-4

---

5. It's a little odd that the non-bonded magnet wire coil has a slightly higher inductance than the bonded coil.

Inside the Metal Detector

and plugging in $\mu = \mu_0 = 1.257\mu H/m$, $r = 12.5$cm, $d = 25$cm, $N = 40$, and $I = 100$mA we get

$B_0 = 19.8\mu T$

$B_{10} = 1.80\mu T$

which are close to our measured values above.

There are a couple of notables regarding the measured field strengths. We would expect that 40 turns of wire at the same nominal diameter and with the same current should produce the same B_0 field strength, and that is exactly what happens, except for the solenoidal and basket coils. The solenoidal coil is easy to explain: the windings are not planar so some of them are farther away from the surface plane which reduces their contribution. With the basket coil, the effective diameter turned out to be a little less than 10-inches|25cm (Figure 8-8). Even so, the basket coil has the worst field efficiency, whereas most of the other coils are close in efficiency. If you are looking for good field efficiency and low self-capacitance, the spiral coil appears to be the best although — if inductance is important — it requires a few more turns to equal the nominal target of 1mH.

8.1.4: Winding Coils

Building a search coil always requires making one or more coil windings. This section assumes scramble-wound coils as they are the most common. In a production environment this is done with an automated winding machine where the wire is laid onto a form (or bobbin) with the exact number of turns needed. Simpler (and cheaper) methods are available for homemade coils.

One of the easiest (and most flexible) methods is to place nails or pegs on a board in the shape of the coil. The wire is then wrapped around the nails. The completed winding can be bonded or zip-tied and then removed. An advantage of this method is that it is easy to make a winding jig for any shape coil, even a continuous figure-8. It is also easy to apply zip ties in the case where you have non-bondable wire. Figure 8-9 shows a nail board for a 10-inch DD coil. The same jig can be used for winding the TX and RX coils.

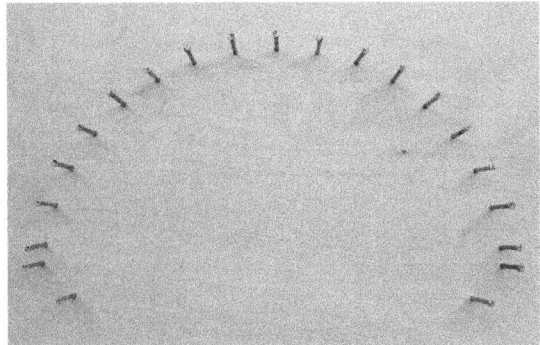

Fig. 8-9: **DD Nail Board**

Another option is to make a bobbin. This can be done with three layers of plywood where the center piece is cut to the shape of the winding and the outer pieces are sized slightly larger to constrain the winding. See Figure 8-10. Like the nail board method, the bobbin can be made for most any winding shape (although not a figure-8) but requires a bit more work to produce. However, a bobbin makes a neater and more consistent winding than a nail board.

Fig. 8-10: **Plywood Bobbin**

The bobbin shown in Figure 8-10 consists of an 8-inch center disc and two outer 9-inch discs. The discs are all made with 1/4-inch plywood. In the case where you want to use a bobbin to wind a variety of coils with different wire gauges and/or different turns, you may find it useful to make two or three center discs of different thicknesses. For example, if you have center discs that are 1/8-inch and 1/4-inch thick (with the same diameter) then you can wind a coil with a bundle width of 1/8-inch, 1/4-inch, or 3/8-inch. If you use wire that is alcohol-bondable then all discs should be finished with an oil-based polyurethane that resists the alcohol. If you plan to use non-bondable wire, then cut several notches around the perimeter so that zip ties can be applied before disassembling the bobbin.

8.1.5: Bondable Wire

By far the easiest way to make a rigid coil winding is to use bondable magnet wire. It is activated by simply squirting some isopropyl alcohol on it; the alcohol softens the bonding agent and then evaporates, leaving a rigid coil winding. Figure 8-11 shows a finished coil using bondable magnet wire, in fact wound on the bobbin shown in Figure 8-10. Compare to the non-bonded coil in Figure 8-7.

It is important to make sure all of the alcohol has evaporated before committing the coil to further assembly steps. Often the alcohol on the outer windings will evaporate and create an evaporation barrier for the inner windings. It is not easy to tell if this has happened as the coil will feel rigid, but may still be soft inside. The problem is that, over time, the inner windings can outgas the remaining alcohol and cause movement of the coil. In some coil designs (especially DD) this can shift the null significantly[6]. Isopropyl alcohol boils at a temperature of about 180°F (82°C) so the best solution is to bake the coil somewhat above the boiling point, perhaps 200°F (93°C), for around 30 minutes, maybe longer for extremely thick windings. This will ensure all of the alcohol is fully evaporated.

Most magnet wire that has a bond coat can also be bonded thermally. This is achieved by passing a high current (several amps) through the coil for a short time. The current heats up the wire to the melting point of the bond coat and fuses the windings together. The amount of current needed depends on the wire gauge. The advantage of this method is that it is much quicker than alcohol bonding (a few seconds and it's ready to use) but requires building a high current "coil zapper." This is a device that drives a high current through the coil and simultaneously monitors the resistance of the coil, which is a good indicator of temperature. When the temperature reaches the melting point of the bond coat, the current is held for a few seconds and then turned off.

Most magnet wire you are likely to run across (either on eBay or from the big e-houses like Mouser and DigiKey) is not bondable. Most likely, you will need to order it directly from a manufacturer[7]. For a few one-off coil designs bondable wire is more of a luxury. If you find the need to wind a lot of coils, it may be a worthwhile purchase.

8.1.6: Automated Winder

Hand-winding a coil up to one or two hundred turns is a little tedious but not difficult. Hand-winding a coil with a thousand turns, or winding 100 coils for a small production run, is just awful. This is where you might consider buying or making a coil winder. Inexpensive coil winders are available from the usual on-line sources. You want a stepper motor drive with enough torque to turn your largest (and heaviest) bobbin pulling the heaviest gauge wire, and one with a counter that displays the number of turns. The cheaper units use a foot switch to control the winder, where you have to watch the counter and stop at the desired turns. More expensive models are computer controlled; you enter the number of turns and it stops automatically.

In most cases a coil winder is used to wind magnet wire, and usually with alcohol bonding. For this you will want a wire feeder and tensioner which keeps the magnet wire at the right tension and prevents kinks. Tensioners are also available on-line. For alcohol bonding, the wire is usually passed over an alcohol wick which wets the wire as it is being wound. The wick is often placed at the outfeed of the coil tensioner.

---

6. At White's we had a terrible problem with this happening to the 4x6 DD "Shooter" coils. The windings were small and so they had a lot of turns, resulting in a thick wire bundle that was difficult to fully bake out.

7. Elektrisola, Essex, MWS, and Rea are the companies I am familiar with, there are surely more.

Fig. 8-12: **Inexpensive Coil Winder**

8.2: Other Elements

Building a coil is conceptually simple but the details matter. In some magazine projects coils are built with the windings attached to a piece of plywood. This is always an option but a professional-looking coil may not be a big step beyond plywood, either in difficulty or cost. Besides just the coil winding(s), the required components for a nice assembly are a coil shell, a cable and connector, shielding, and a way of stabilizing the coil windings. Before getting into specific designs let's cover these other items as they can be common amongst the designs.

8.2.1: Coil Shell

The coil shell is usually a plastic housing that is either injection-molded or vacuum-formed. In home-brew coils it is almost always the latter. Some people have gone to the extreme of building their own vacuum former to produce whatever coil shell they want, and some of them have ended up making and selling shells to other experimenters. Since these people come and go they will not be listed here; either the *Geotech* forums or eBay are good places to see what is available. It is important to procure a coil shell before you embark on winding any coils as the coils will need to be sized specifically for the shell.

Figure 8-13 shows a variety of vacuum-formed coil shells[8], including round, oval, rectangular, mono, concentric, DD, coaxial, open-center, spoked, and solid, from 6 inches to 18 inches. Coils tend to

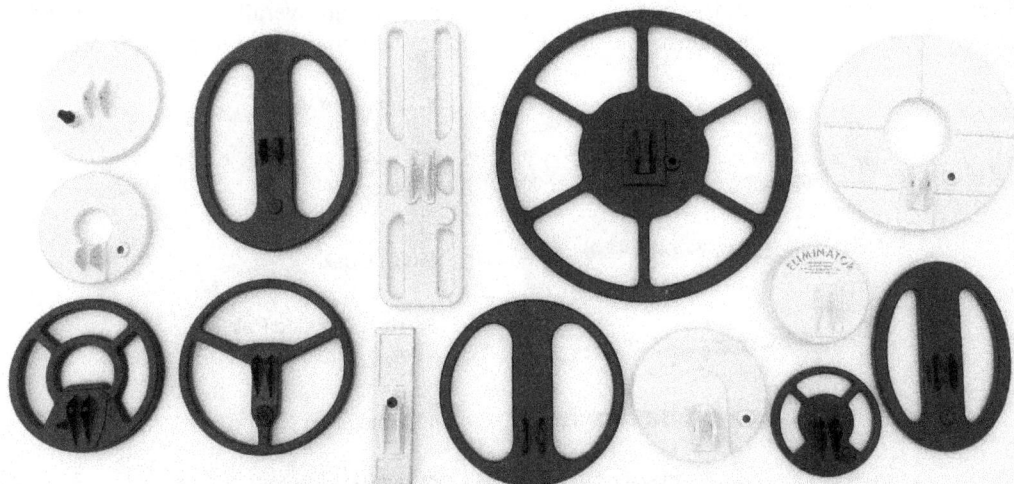

Fig. 8-13: **A Variety of Vacuum-Formed Coil Shells**

---

8. Many of the pictured shells came from Bill Hays who used to make them for hobbyists but passed away in 2017.

be black or white but they could be any color. White is best for extreme heat such as desert hunting and are also easier to see if shallow water hunting. The coil housing may consist of only a top shell or might also have a bottom shell. Coils are almost always assembled in the top shell. If there is a bottom shell then that is installed as a final step, if there is no bottom shell then the top shell is filled out with epoxy.

As mentioned above it is possible to simply use a piece of plywood to mount the coils on. This is crude but workable and will provide the necessary rigidity. Shielding and waterproofing can be difficult, though. Other clever homebrew solutions have been proposed in various magazine construction articles. See the Appendix D for a list of articles; some of these are available on the *Geotech* web site.

8.2.2: Potato Chipping

Vacuum formed coil shells have a tendency to "potato chip," or warp in a Pringles[9] sort of way. Usually it is slight but enough to be annoyingly evident, and applying a shield paint can make it worse. Figure 8-14 shows a particularly egregious example. When hot-melting the coils into the shells or making an epoxy pour you want to make sure the shell is held down and reasonably flat.

Fig. 8-14: **Potato Chipping**

This can be a bit of a challenge and may require that you create some kind of clamping jig to press the shell flat. Once the epoxy pour cures, your troubles are over.

If you have a warped shell and want to use it in a non-epoxy construction then that implies there will be a bottom shell, which might also be warped. In this case, you will need a clamping jig during the seam sealing process so that the overall assembly is held flat while the seam cement is applied and cured.

8.2.3: Cable

The cable and connector requirements depend on the type of coil. A BFO or PI with a simple mono coil can make use of a standard coaxial cable with a stranded center conductor for flexibility. Cable capacitance is not an issue for BFO but for high-performance PI low capacitance is critical. Coaxial cable is often specified with an impedance rating and a higher number (like 125Ω) will have less capacitance per meter than a lower number (like 50Ω). Cable resistance[10] can be a concern with PI coils as it contributes to the peak current limit and also to power loss. With PI coils also pay attention to the voltage breakdown rating of the cable.

In the case of an IB coil you will need at least three conductors and preferably four. Up through the 1980s it was common to see multi-conductor cable surrounded by an overall shield. As newer designs pushed sensitivity levels higher, most detectors began using dual coaxial cable which is often custom

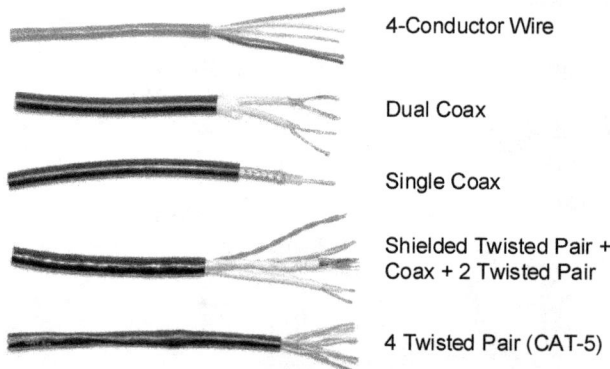

4-Conductor Wire

Dual Coax

Single Coax

Shielded Twisted Pair +
Coax + 2 Twisted Pair

4 Twisted Pair (CAT-5)

Fig. 8-15: **Example Cables**

---

9. Canned potato chips in the US.

10. The cable's impedance rating has nothing to do with ohmic resistance.

made. Off-the-shelf solutions can be found in cheap eBay S-video cables which are dual coax, or various companies like Belden or Vandamme make dual coax.

Cable capacitance is less of a concern in a narrowband VLF as the TX and RX are usually resonated with capacitors anyway. In multifrequency designs, capacitance might be an issue as it causes waveform errors that vary over frequency. Cable resistance is mostly a concern on the TX side and contributes to power loss, as well as phase error in a narrowband VLF and waveform distortion (and, ultimately, phase error) in a multifrequency.

The Minelab *X-Terra* detector series successfully used CAT-5 cable which has four sets of twisted pair. However, it is important to note that the *X-Terra* has the RX preamp embedded in the coil so that the RX twisted pair is driven with a low impedance. This helps prevent contamination of the RX signal by the TX twisted pair signal. But with two more twisted pair available to provide power (and, perhaps, control signals), placing the preamp in the coil is a worthy idea for allowing the use of cheap cable.

The design of a submersible detector imposes an additional requirement on the cable: the jacket must be waterproof. This seems easy as most cables have a jacket made of some kind of plastic. However, not all plastics are completely waterproof. PVC, which is a common material for cable jackets, often has micro holes which allow water to seep through the jacket. This is especially a problem in a diver detector which is exposed to higher water pressures at depth, but may also occur even with a surf detector. For a submersible design, use a cable with a polyurethane jacket.

Finding just the right cable can be a challenge. Cable is often sold in large rolls but can be found in per-meter or per-foot quantities. Besides the usual eBay/AliExpress, try wireandcableyourway.com or rfparts.com.

Where the cable enters the coil housing can be a point of high cable stress and also a place for potential water ingress. A popular strain relief cable gland is made by Heyco; Figure 8-16 shows two different sizes for different cable diameters; the smaller is a Heyco 3237 which works well with cables that are about 1/4-inch (6mm) in diameter. These include a soft rubber gasket that is compressed around the cable as the gland is tightened, which provides a watertight seal. It is still a good idea to pour some epoxy or apply some RTV sealant on the inside of the coil shell covering the gland and cable entry for extra protection.

Fig. 8-16: **Heyco Glands**

The length of the coil cable is typically 48-54 inches (120-140 cm) for a detector where the control housing is mounted at the top of the handle grip. If you intend to mount the control housing below the arm cup or will use a harness to body-mount the control housing, be sure to add enough cable length to accommodate this.

8.2.4: Connector

The choices for a cable connector are plentiful. A common and inexpensive solution used in many commercial detectors is the microphone-style (GX16) connector. These can be found with 2-8 pins (Figure 8-17a). Switchcraft's Slimline series were popular for many years but have become expensive

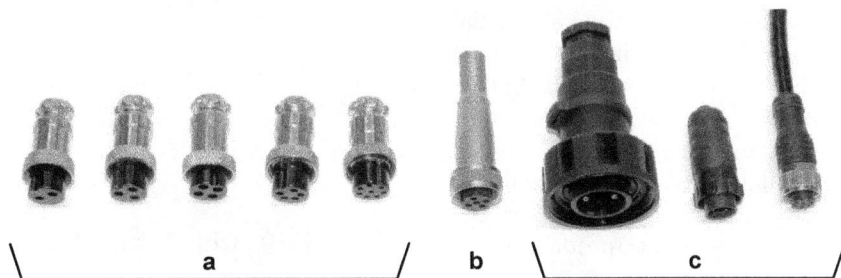

Fig. 8-17: **Example Connectors**

and hard-to-find (Figure 8-17b). If you need a waterproof connector the choices have grown considerably in the last several years. There are standard metric connectors (M12, M16) and also company-custom solutions, all with various numbers of pins (Figure 8-17c). As with cabling, pay attention to the voltage rating when selecting a PI connector. All-plastic connectors may present a challenge with shielding against external EMI or falsing in salt surf conditions, especially with a monopolar PI design. Sources for connectors are the usual big e-houses like Digikey, Mouser, and Farnell; also eBay, Amazon, AliExpress, or americanradiosupply.com.

8.2.5: Shielding

An otherwise well-designed and well-built coil will perform poorly if not adequately shielded. Shielding is the process of adding an electrically conductive "shell" around the coil windings. The shield (commonly called a *Faraday shield*) performs two major functions: it presents a consistent capacitive coupling environment to the coil windings, and it helps reduce external EMI noise.

Let's consider the first function and use a BFO coil as an example. Detection in a BFO (as well as other proximity designs) relies on small shifts in the TX frequency caused by target effects. However, a change in the capacitance of the coil can also create a frequency shift that sounds like a target. Part of the coil capacitance is the inter-winding capacitance of the coil itself which is normally not important as long as it is stable. However, ground capacitance is a proximity effect that alters the coil capacitance. It is mostly a nuisance in moist-to-wet grass. As the search coil is lowered to the ground, the total capacitance "seen" by the search oscillator is slightly altered by the distributed parallel coil-to-ground capacitance as shown in Figure 8-18. This change in capacitance will alter the search frequency and do so as the height of the coil above the ground varies, resulting in false signals.

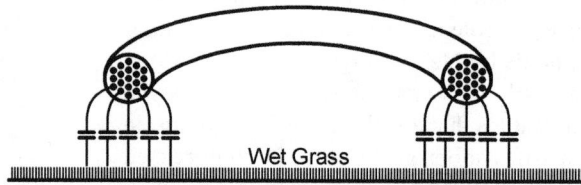

Wet Grass

Fig. 8-18: **Ground capacitance**

The addition of a grounded shield presents a consistent capacitive environment for the coil regardless of what happens outside the shield. This is less of a problem with IB transmitter designs, even those with a free-running TX oscillator, as small frequency shifts are largely ignored. On the RX side, variable ground capacitance can cause falsing from capacitively-coupled TX signals and also minor phase errors in target responses. PI designs can be sensitive to transient capacitance changes as it alters the coil damping and therefore the instantaneous decay, resulting in false signals.

The other function is the mitigation of EMI noise. EMI shows up in the forms of both magnetic field and electric field signals. There is little we can do to reduce the magnetic field EMI we receive[11] and attempting to do so can also reduce the magnetic field received from targets. But electric field EMI can be lowered with shielding. Besides electric field EMI, another situation arises when sweeping the coil over dry dusty ground where static electricity can build up on the plastic shell and discharge into the coil windings, causing noise bursts. Instead, the static build-up will discharge into the shield.

There have been several production methods of shielding coils. They generally involve either shielding the winding(s) individually or shielding the entire coil assembly as a whole. Shielding an individual winding is more likely to be seen in a mono coil for BFO or PI. This is done by directly wrapping the conductive shield around the winding. At first glance it seems to be completely counterproductive to put a

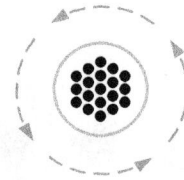

Fig. 8-19: **Coil & Shield Cross-Section**

---

11. Except for using particular "EMI-canceling" coil designs, like coaxial, DOD, or figure-8.

conductive cover around a coil intended to detect conductors. However, if the shield is thin, completely surrounds the cross section of the coil, but does not extend around the entire circumference of the coil then it has a minimal effect on magnetic fields. Figure 8-19 shows that the flux lines of the magnetic field (dashed arrows) are parallel to the shield surface (gray circle); eddy generation will therefore occur only along the circular length of the shield. To prevent an overall eddy current from circulating around the shield we simply leave a small break in the shield, usually at the location of the coil pigtail wires.

The easiest way to apply a wrapped shield is to use a conductive fabric tape like 3M's CN-3190, which is a copper-nickel metallized fabric tape available in different widths. The 3M brand tape can be expensive but cheap options are available in the usual eBay and AliExpress venues. A full metal tape like copper could be used but the metallized fabric tape is far easier to work with and lighter in weight. As a bonus, the acrylic adhesive used on the fabric tapes is, itself, conductive which helps produce a better shield. Figure 8-20 shows a coil that has metallized tape shielding; In this example, the gap in the shield tape is opposite the position of the wires.

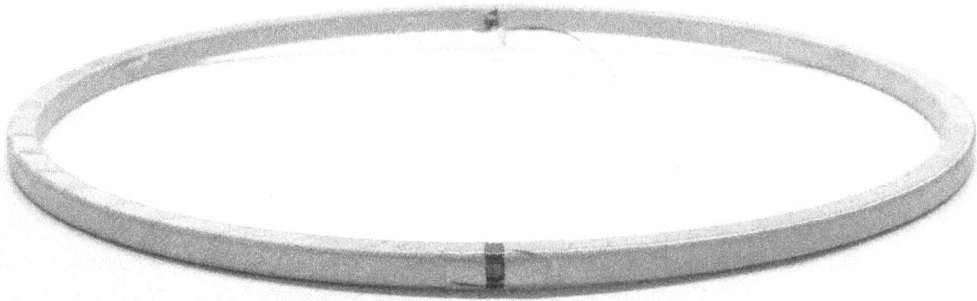

Fig. 8-20: **Wrapped Tape Shielding**

In high-volume production coils the shield is often a conductive paint applied to the inside of the plastic housing. This shields the entire assembly and is especially effective for more complex induction balance coils. For coils with an upper shell and a lower shell, both shell halves are painted[12] and connected with drain wires to a ground. Figure 8-21 shows a top shell that has been sprayed with conductive paint. Before the paint is applied, a shield drain wire is tacked down to the plastic with cement, then painted over, and then fully cemented.

With any IB coil design that has a shield-painted bottom shell there is a possibility that attaching the bottom shell will cause the null to shift slightly. This is usually more of a problem with concentrics than with DD coils. It is useful to test for this *before* the final null tweak is done. If the null shifts, make a note of how much, then intentionally mis-null the coil in the opposite direction by the same

Fig. 8-21: **Shield-Painted Top Shell (with drain wire)**

amount. When the bottom shell is attached the null will then shift to its desired minimum level.

---

12. Some low-end detectors only shield the bottom shell. This prevents wet grass and static discharge problems but is less effective at EMI reduction.

Some coils do not have a plastic bottom; the coils are placed in a top shell and then epoxy-potted, with the exposed epoxy forming the bottom (look ahead to Figure 8-26). In this case, the top shell is shield painted and attached with a drain wire, and the coils are placed in the shell and connected to the cable. A first epoxy-pour covers the coils but does not fill the shell. Once it cures, shield paint is applied directly to the epoxy and connected with another drain wire. A second epoxy pour covers the shield paint and fills out the shell.

In shielding the whole assembly, eddy currents will be produced in the shielding so it is important to keep the resistance of the shield high enough that the eddies remain low. This is a matter of what paint is used and how thick to apply it. MG Chemicals makes a reasonably priced carbon-based spray-on shielding (838AR) which works well for typical IB coils. If an aggressively low resistance is needed (for e.g. PI) then MG makes a nickel-based paint (841AR). Both are available as brush-on paint as well, which is sometimes easier to apply.

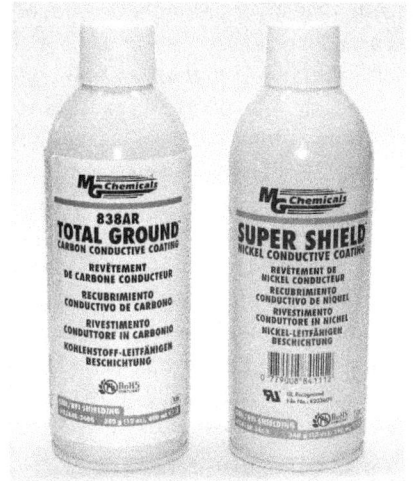

Fig. 8-22: **Shield Paints**

The coil shield must be connected to a circuit node and that depends on the detector design. In a proximity or PI design using a mono coil the shield is connected to the DC side of the coil. This might be ground or a power supply rail. Usually the shield of the coil cable is used for making this connection to the main circuitry (Figure 8-23a). A potential problem in using the DC side of the TX coil is that there is a TX current flowing through it and this can produce ground shield noise. Whenever possible, a preferred solution is to connect the shield to the circuitry using a dedicated connection. That is, the coil cable requires three conductors: TX+, TX−, and Shield (Figure 8-23b).

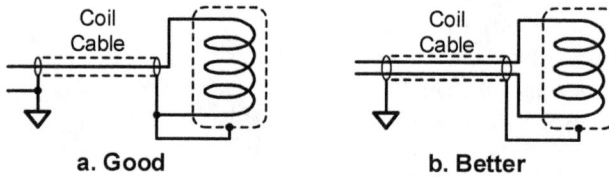

Fig. 8-23: **Mono Coil Shield Connection**

In an IB design the shield is usually connected to the DC side of the RX coil. Unlike the TX coil, the RX coil is connected to a high impedance preamp so the current is negligible. If the coil cable is a dual coax type then the coax shield of the RX side is the shield connection to the main circuitry (Figure 8-24a). A better solution is to use a coil cable with two shielded coax lines (TX and RX), plus an overall shield (Figure 8-24b). In some designs the RX coil might be floating or differential (DOD coil, for example) where neither side is connected to DC. In that case, the coil shield should have a separate connection to the main circuitry that is connected to the RX ground reference (Figure 8-24c).

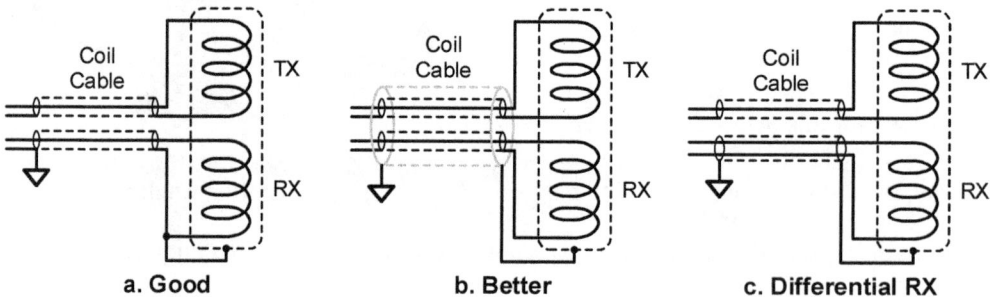

Fig. 8-24: **IB Coil Shield Connection**

8.2.6: Stabilization

If you build a coil that you intend to hunt with then good stability matters. DD coils are especially sensitive to falsing when they are bumped against protrusions like rocks and stubble. Besides impact, overall ruggedness and some level of moisture-resistance should be considered. This section assumes the use of a plastic coil shell.

A common method in industry for producing a stable assembly is to pot the coil(s) in epoxy. This prevents the coils from moving which is especially important with IB designs. A critical characteristic of the epoxy is extremely low cure shrinkage; once the coils are balanced you don't want them to move as the epoxy cures. Often low shrinkage means a longer cure time, but in this case cure time is not important.

Epoxy can be heavy and the coil is one place where weight really matters so it is common to add *microballoons* to the epoxy to lighten it. Microballoons are tiny hollow glass spheres which replace some of the epoxy with air and can be found in hobby shops, Amazon, and, yes, eBay. It is best to mix the microbal-

Fig. 8-25: **Epoxy & Microballoons**

loons into the resin or hardener before mixing the two together. Epoxy can accept quite a large volume of microballoons, perhaps 50% by volume. However, in low-viscosity epoxy, microballoons may tend to float to the surface so it's important to experiment with the mixture to get it where you want it before irreversibly committing to a coil assembly.

WARNING
Microballoons are equivalent to silica dust; take extra caution
to avoid inhalation as it could cause lung problems.

If epoxy is used in a top-only coil shell where the epoxy forms the bottom surface of the coil then normally the final epoxy pour is slightly crowned above the edge of the plastic to produce a rounded edge. This helps it better glide over ground surfaces. With the Nokta coil shown in Figure 8-26 it is easy to see the gray epoxy contrasted with the black housing. The normally black epoxy is gray because it contains microballoons for lighter weight. Notice also that not everything is poured with epoxy; only the coil windings, the center area, and the area where the coil cable enters are epoxy potted. The remaining areas are unpoured to reduce weight. A critical requirement, especially with DD coils, is that the area directly under the coil ears[13] be rigid. Any flexing of the assembly anywhere along the overlap of the

Fig. 8-26: **Nokta Legend Epoxy-Potted Coil**

---

13. Where the lower rod attaches.

---

coils will cause falsing and the coil ears experience a lot of stress. And the larger and heavier the coil, the higher the stress.

When there is a bottom shell the epoxy need not completely fill the top shell and weight can be reduced. If a mono coil is built with top and bottom shells you can often get away with using hot-melt glue to attach the coil to the top shell. Use a high-temperature hot melt for this; a low temperature hot melt can soften in the summer heat (especially if the coil plastic is black) and the coil winding can move or become detached.

In the 1970s and 80s a popular production method for building coils was the use of a foam core with grooves to hold the coil windings. Figures 6-7 and 6-11 show examples of a DD and a concentric coil using foam cores. This method requires a solid disc housing with both top and bottom plastic. A minimal amount of epoxy or hot-melt can be used to hold the windings in the foam and shielding may be applied to either the housing or the foam. If you don't mind the old-fashioned look of a solid-disc coil[14] then foam core construction is an excellent way to build a very lightweight coil.

8.3: Designing the Coil

Now that we have covered all the elements used to make a coil, we now turn to the question of how to design one. If you are building a coil for use with a commercial detector you will normally measure the inductance and resistance of the TX and RX windings on an existing coil and try to replicate those specs in the new coil. Additionally, with any search coil that has separate TX and RX coils (IB or not) it is important to get the phase relationship right. To determine this, you first must establish which pins of the coil connector are TX+, TX−, RX+, and RX−. Often this requires looking at the detector circuitry. Then, drive the TX+ coil pin with a signal generator and observe the signal at the RX+ pin with an oscope. If the coil is induction-balanced, introduce a large piece of ferrite to get a decent RX signal level. The RX signal will either be in-phase with the TX or 180° out-of-phase, at least within a few degrees. Any new coil will need to replicate the phase polarity; if you get it backwards, the coil won't work.

If you are designing a coil and also the detector it goes with, then you get to determine the coil parametrics and phase polarity. For me, the phase polarity is the easy part; I always design coils so that the RX voltage signal is in-phase with the TX voltage drive. It's easy to remember and all circuit designs then also have the same phase polarity. Coil inductances and wire sizes are another matter and often depend on the operation of the detector. A PI mono coil tends to have a fairly low inductance, 300µH is a typical value. But a low power PI pinpointer might use a 1mH coil, and a monstrously pow-

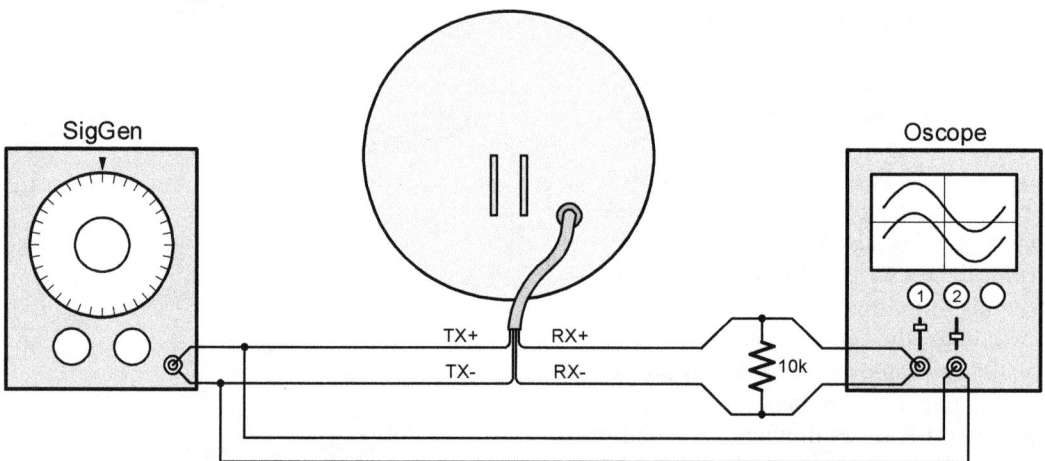

Fig. 8-27: **Coil Phasing Measurement**

14. When hunting in field stubble or loose rock the solid disc coil is usually preferred as a spider coil tends to catch on everything.

erful tow sled for detecting Atocha bars might use a 50μH coil with 10 AWG wire from the local home improvement store.

Likewise, a high frequency VLF nugget detector will typically use low inductance coils and even litz wire to minimize self-eddy losses. A low frequency cache detector can use much higher inductances. With multifrequency detectors the design of the coil must strike a balance between the lowest and highest frequencies of operation.

It is usually the TX coil that demands the most attention in its design. In some detectors the TX coil is part of a free-running oscillator so its inductance can be critical. In a VLF, resistance is important for the TX coil because it determines the amount of phase error between the TX voltage and the transmitted field. The phase error affects the ground balance point which is not a problem if you have adjustable ground balance. But the resistance of copper also has a temperature coefficient and as the coil is moved between sun and shade it can create phase drift in the ground balance point and make for noisy ground conditions. Generally shoot for a maximum 0.05° phase shift over, say, a 5°C shift in temperature. Using a white coil shell instead of black is helpful. See the *Explore* section of this chapter for more details.

For the usual non-inverting VLF preamp with a DC-coupled RX coil the resistance of the RX coil is not important except for thermal noise considerations. Since the resistance and the self-capacitance are not critical then it stands to reason we could add more and more turns to the RX coil until they eventually become an issue. More turns yield a stronger target signal so this seemingly offers us free "transformer gain" in our quest for more depth. However, while more RX turns does produce a stronger target signal it also produces an equally stronger ground signal and equally stronger EMI coupling. For the most part, SNR stays about the same but coil weight goes up. And more RX turns also couples in more TX signal making the nulling operation and null stability more sensitive. There is no free... well, you know the rules.

Capacitance is usually not critical for either coil. For the TX coil you want the SRF to be higher than the TX frequency and that is hard not to do. The TX coil is otherwise resonated with capacitance[15] so the self-capacitance becomes part of this. The RX coil is sometimes resonated, sometimes not, but generally you want the RX SRF to be substantially higher than the TX frequency, especially to minimize phase offset. The absolute offset is easy to adjust for in circuitry but when building multiple coils for a given machine you want all their phase offsets to be similar, and the best way to do that is to minimize them.

With a square wave (or multifrequency) TX drive, the TX coil is not resonated and both the coil resistance and self-capacitance produce non-linearities in the coil current. Whether this is an issue depends on how the RX signal is eventually processed but generally TX coil resistance has a similar effect with multifrequency as with single frequency (phase offset). Likewise, the RX coil resistance and capacitance can further distort the signal. Often the distortion itself is not the issue, but rather the change in distortion with temperature.

When designing a range of different sized coils for a detector both the TX and RX coil inductances are usually held to constant values, and coil resistances are often likewise held constant. On the TX side this maintains a consistent transmit signal in terms of frequency, amplitude, and any undesired parasitic effects. Often the RX side is not as critical but matching the coil parametrics ensures that parasitic effects do not alter, say, the ground balance point.

The remainder of this chapter shows construction methods of several types of coils. Some of the details covered in the earlier coils are not repeated in the later coils. Also, I have tried to present a variety of techniques and many of the methods can be used across different coil types. In short, read the whole chapter (including the *Explore* section) before attempting to build a coil. Useful tidbits are everywhere.

---

15. For single frequency sinusoidal transmitters.

8.4: Mono Coils

A mono coil is the easiest coil to build as it has only a single winding with nothing to balance. Mono coils are used in BFO[16] and PI detector designs. Of these, the PI has higher performance demands so we will start with the simplest of the simple, the BFO.

8.4.1: BFO Coil

The BFO coil has little in the way of critical needs. Most BFOs drive the coil with a free-running oscillator where the coil is resonated with a capacitor to form an LC tank circuit. Therefore, the self-capacitance of the coil is not an issue as it just becomes part of the LC tank[17] and, likewise, the shielding can be simple. But because the self-capacitance is determined by the physical proximity of wires, it is crucial that these wires not move during normal operation. Manufacturers often potted BFO coils in epoxy to ensure rigidity and stability. In this design, a two-piece shell with a foam core will be used.

With any coil design the first place to start is to define what you want. A BFO coil can literally be anything, from a 1/2-inch probe to a 36-inch cache loop. With a big *yawn* we will make the following:

- Diameter = 7 inches (17.8cm)
- L = 1mH nominal
- R_L = 5Ω max
- f_0 > 200kHz

We will use 24 AWG wire and 49 turns should produce a 1mH coil, which will work with the proximity designs in Chapters 9-12. For this design, the coil is wound on a plywood bobbin using bondable magnet wire.

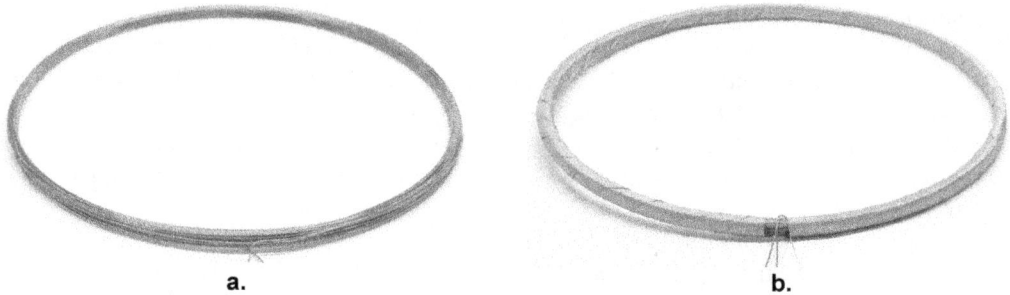

Fig. 8-28: **Coil Winding & Shielding**

Once the coil winding is complete (Figure 8-28a) the next step is to apply the shielding. A very simple method mentioned earlier in the chapter is to wrap the coil with conductive fabric tape. Figure 8-28b shows the shielding applied to the coil winding. Notice that a gap is left in the shield where the pigtails emerge. The gap can be placed anywhere but this is often a convenient location. A drain wire is placed under the last couple of shield wraps and will be soldered to the DC side[18] of the coil winding.

The final step in building a BFO coil is to embed it in a foam core and wire it up in a coil shell. The foam core in this design is 1/2 inch (6mm) thick closed-cell foam board[19] cut to a circle and grooved so the coil can be embedded and secured in place. Use an adhesive that will not dissolve the foam to secure the coil. Silicone sealant works well for this.

The cable selected for this project is RG58A/U coaxial with a stranded center conductor and a 2-pin GX16 style connector is used. A hole is drilled in the shell to accept a Heyco 3237 strain relief

---

16. And practically all other proximity designs. We will lump them all together as "BFO."

17. But, as discussed in the BFO chapter, keeping the coil capacitance within a certain spec is important when designing for interchangeable coils.

18. Either ground or supply, depending on the design of the oscillator.

19. Commonly found as Owens-Corning pink foam board at US building supply centers. If you don't want to buy a 4x8 foot sheet then try an arts-and-craft store, or the usual Amazon.

gland. With the top housing laid up-side-down, adhere the foam coil assembly to the shell, perhaps with silicone sealant. With the connector installed at the other end[20], the cable is inserted through the gland and a hole in the foam core. Solder the cable to the coil and shield as shown in Figure 8-29. Tighten the strain relief to secure the cable. Figure 8-30 shows the complete assembly prior to sealing.

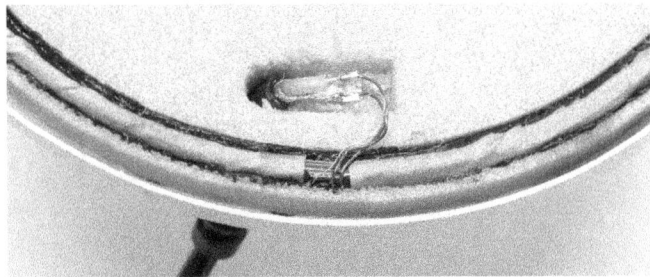

Fig. 8-29: **Connecting the Coil**

The last step is to attach the bottom coil housing and seal the perimeter seam for waterproofness. For coil shells made of black ABS plastic, black ABS cement is available in hardware stores. For other colors or plastics you may have to devise your own cement using plastic shavings (of the same material as the housing) dissolved in MEK[21]. Figure 8-31 shows the plastic shavings, a small copper cup for dissolving them in MEK, and a syringe for applying. Whatever you use, it should be applied in a continuous bead around the seam, making sure that it penetrates the seam. A syringe (10mL or so) makes the job much easier, and a dental syringe (as shown) with a small tip is ideal. Test the syringe first to make sure it is not dissolved by the cement you will be applying; a polyethylene syringe should work. Also, the cement inside the syringe will not harden and gives you more time to complete the sealing process. You might also find it necessary to clamp the shells together during the sealing process as the bottom shell may have a tendency to float out-of-place.

Fig. 8-30: BFO Coil Assembly

WARNING
Use MEK in a well-ventilated area
and wear MEK-resistant gloves.

Fig. 8-31: **Seam Sealing Material**

---

20. It is good practice to install the cable connector on the cable and then check proper connection and continuity before finalizing any more assembly. Measure twice, cut once.

21. In the USA, MEK (methyl ethyl ketone) is listed as hazardous. Many hardware stores only carry something called "MEK substitute" which I have never tried.

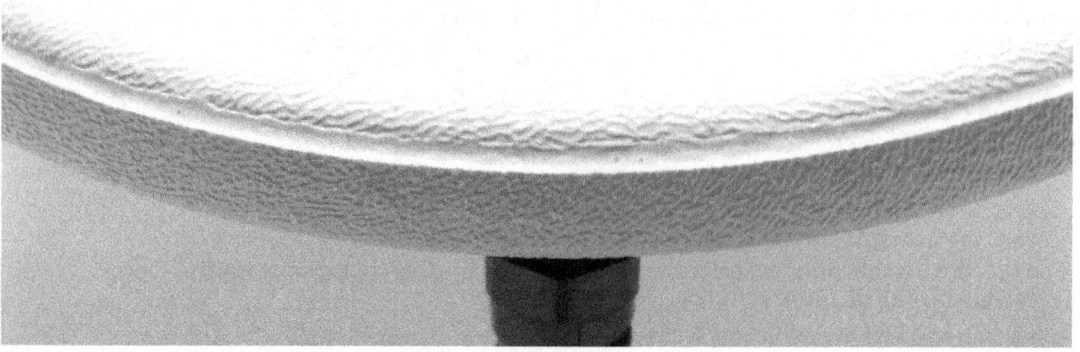

Fig. 8-32: **Seam Seal**

The final measured parameters of the BFO coil are as follows:

- $L = 1005\mu H$
- $R_L = 2.51\Omega$
- $f_0 = 253.5kHz$
- $C_L = 392pF$
- Weight = 10oz|285g

8.4.2: Pulse Induction Coil

The PI coil is similar to a BFO coil in that it is a simple mono coil[22] and requires a shield. How the coil is used is entirely different and requires more attention in its construction. In particular, low self-capacitance is critical in achieving high performance so care is taken to minimize both interwinding capacitance and shield capacitance.

PI coils generally range from $100\mu H$ to several hundred (perhaps 500) μHs. Magnet wire can be used but a thicker insulation reduces interwinding capacitance so a better choice is PVC-insulated stranded wire. Teflon-insulated wire is better yet (it has a lower dielectric constant) but is also pricey and difficult to source. PI circuits tend to run high pulse currents (up to several amps) so wire gauge is a bit on the heavy side — 20-24 AWG. Coil resistance may or may not be important depending on the detector design. For example, Minelab *SD* and *GPX* series require low resistance whereas the White's *TDI* does not.

This coil will be designed for use with the several PI projects later in the book. The target parameters are:

- $L = 300\mu H$ nominal
- $R_L = 2\Omega$ max
- C_L as low as possible
- Diameter = 12 inches (30cm)

This design will use 19 turns of 24 AWG Teflon-jacketed stranded wire. Teflon is difficult to bond so small zip ties are used to secure the winding so it can be removed from the nail board (Figure 8-33). Finally, the winding is tightly wrapped with electrical tape for stability. As the electrical tape is wrapped the zip ties may be cut off.

For this coil we want to minimize the winding-to-shield capacitance which is just a matter of adding more spacing between them. After wrapping the coil with electrical tape, add a layer of plastic spiral wrap before applying the shield tape. If you are trying to hit aggressively low capacitance a second layer of spiral wrap may help. More than that gives a diminishing improvement. Spiral wrap is cheaply avail-

---

22. PI detectors often use an IB DD coil, see the DD coil construction section for that. It is also possible to design a PI to use a separate RX coil that is laid directly on top of the TX coil. That is, it will look like a mono coil but has two windings. The construction methods covered here still apply to this type of coil, except that the cable will need 2 coax conductors.

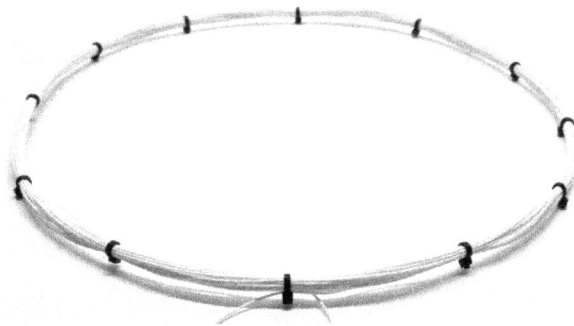

Fig. 8-33: **PI Coil Winding**

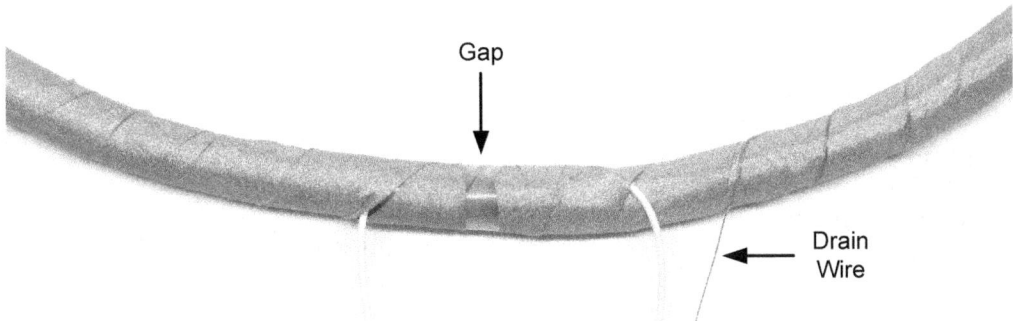

Fig. 8-34: **PI Shield Details**

able for different bundle diameters from the usual eBay/AliExpress sources. In this design layers of 6mm and 10mm wrap were used. The shield tape is wrapped around the spiral wrap with a drain wire inserted near the end. Figure 8-34 shows the details of the shield. Again, note the gap that is left in the shield.

The same RG58 used for the BFO coil could also be used here but it has a capacitance spec of 100pF/m and for a typical cable length of 1.5m would add 150pF to the total coil capacitance. We can do better with the selection of a Vandamme video cable[23] which is 57pF/m. It is a thicker cable and so requires a larger Heyco strain relief.

The connector for this coil is a 5-pin version of the GX16 type used in the BFO coil. This connector has been commonly used many Minelab PI models from the *SD2000* to the *GPX5000* and is also found on some other detectors like the White's *TDI*, Interfacion *QED*, and Algoforce *E1500*. The 5-pin connector can be used with either a mono coil or a DD coil[24]. In the case of a DD coil, the TX coil is connected to pins 4 and 5 and the RX coil to pins 1 and 2.

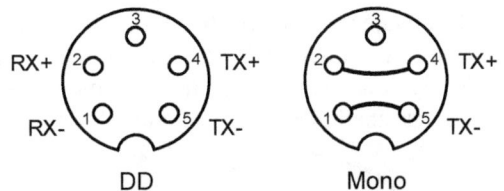

Fig. 8-35: **5-pin Connector Wiring**

For a mono coil, the connector needs two bridge wires added so that the RX pins are also connected to the TX coil. Figure 8-35 shows the connector wiring for both options.

Attach the connector to the cable according to the mono option. Install the Heyco gland in the top shell and feed the cable through the gland and solder it to the coil. The center conductor attaches to one end of the coil winding. The cable shield attaches to the other end of the coil winding and to the coil shield.

---

23. In this case, their Red Series 2-way multicore coax. This cable has two coax cores but only one is needed for a mono coil, the other is unused. Both cores would be needed in a PI DD coil design.

24. For a Minelab-compatible coil (mono or DD), the total TX resistance should be under 0.5Ω. The RX coil may have a higher inductance & resistance; a typical Minelab DD is $L_{RX} = 430\mu H$ and $R_{RX} = 15\Omega$.

The coil assembly is seated into the plastic shell and the cable pulled through the strain relief as needed. The strain relief is then tightened. Although we could pot this coil in epoxy, instead we will use hot-melt[25] to secure it to the top shell and then attach a bottom shell to complete the assembly. This also makes for a light weight coil. The top and bottom shells can be welded together using a cement as described in the BFO coil construction.

The final coil assembly is shown in Figure 8-36. Notice that the wires connected to the cable are tacked down in several places; you don't want wires moving around inside the coil during use. The resulting measured parameters of the coil are:

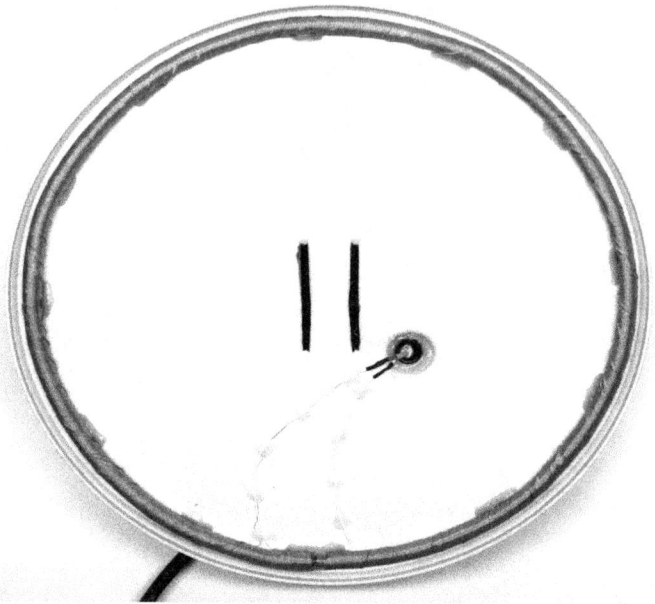

Fig. 8-36: **PI Coil Assembly**

- $L = 297\mu H$
- $R_L = 1.84\Omega$
- SRF = 667kHz
- $C_L = 192pF$
- Weight = 19.7oz|560g

8.5: DD Coils

As we saw in Chapter 6 the DD can be round or elliptical and can have overlaps that vary from straight to bowed. This design will be a 10-inch|25cm round coil with a straight overlap. The first thing to figure out is exactly how much overlap is required to achieve induction balance. Figure 8-37 shows 3 examples of different overlaps that maintain an overall round coil shape. It is necessary to find the exact overlap that achieves induction balance; you cannot simply make it whatever you want, With magnetic FEM software it is possible to simulate the coils and determine the exact geometry; otherwise it is done through trial-and-error. If you are fabricating a spider-style coil shell for a DD design then you will obviously need to determine the exact overlap geometry in order to make a compatible shell. If you use a solid disc shell then there is plenty of room for adjustment.

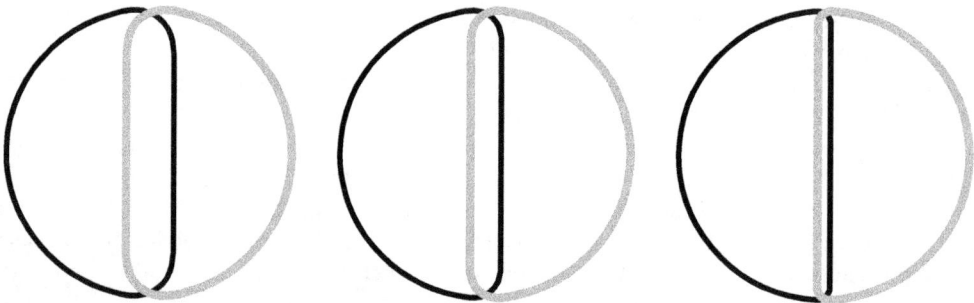

Fig. 8-37: **DD Overlap Examples**

---

25. Use a high-temperature hot melt, or a high-temp silicone adhesive. Adhesion to the plastic shell can be improved by scuffing the surface of the plastic.

Some DD coil designs take no special care in exactly how the two coils overlap. For optimally low capacitive coupling the overlaps should occur at 90° angles. Figure 8-38 shows the same overlap spacing as the middle coil above but with 90° overlaps[26]. This is especially important in high-frequency detectors and wideband (PI and multifrequency) designs.

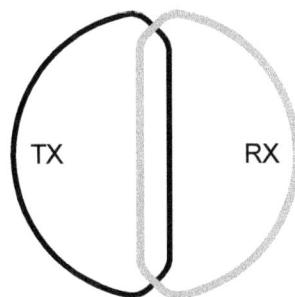

Fig. 8-38: **90° Overlaps**

8.5.1: Windings

This coil will be designed for the TR/VLF projects in Chapters 14-17. The target parameters are:

- d = 10.5 inches (26.7cm) (nominal)
- L_{TX} = 1mH nominal
- R_{TX} = 2.7Ω max
- C_{TX} — not critical
- L_{RX} = 4mH nominal
- R_{RX} = 20Ω max
- C_{RX} — not critical
- f_{TX} = 10kHz
- IB null ≤ −60dB @ 10kHz

Using a combination of experience and guesswork, the coil design is shown in Figure 8-39 with all dimensions labeled. The effective radius of the coil (132 mm) is the result of a specific coil housing. For this design we will use a nail board[27] to wind the coils. The TX and RX coils have identical dimensions so a single nail board will work for winding both coils.

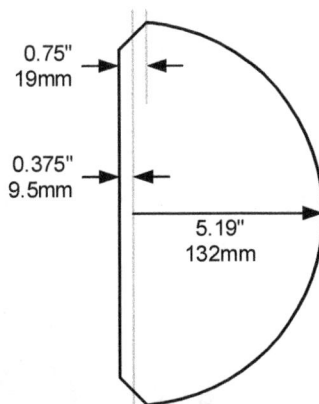

0.75"
19mm

0.375"
9.5mm

5.19"
132mm

Fig. 8-39: **Coil Dimensions**

We now need to estimate the number of turns (N) required and the wire gauge needed. From the dimensions, the circumference of the coil is about 674 mm. This is equivalent to a circular coil with a radius of 107 mm and we can use this in an inductance calculator to get a first estimate of N. Instead of a target inductance of 1mH, we instead use the ratio of the area of the D winding and the area of the equivalent circular coil to scale up the target inductance to 1.052mH[28]. The coil calculator also needs the diameter of the wire and for that we'll use an initial guess. For the TX coil, using 0.511 mm wire (24AWG) results in N = 42. We now use this to estimate the wire resistance which is

R = N × circumference × wire resistivity

For 24 AWG the resistivity is 84.02mΩ/m so the total resistance will be 2.32Ω. This is below the stated goal[29] and does not include the coil cable so 24AWG is probably sufficient. Upon winding the coil, it was found that 42 turns gave 1.013mH and a measured R of 2.40Ω.

The RX inductance of 4mH was chosen as a middle-of-the-road value. The same process as used to figure out the TX coil can be used on the RX coil, which results in a calculated 82 turns of 0.36 mm wire (27 AWG) to produce a 4mH coil with 9.3Ω resistance. An actual coil of 85 turns results in 4.033mH and 10.0Ω.

---

26. Also see Figure 8-26.

27. See Figure 8-9 for the actual nail board used.

28. See section B.7 in Appendix B for an explanation.

29. See 8.11 as to why 2.7Ω was chosen,

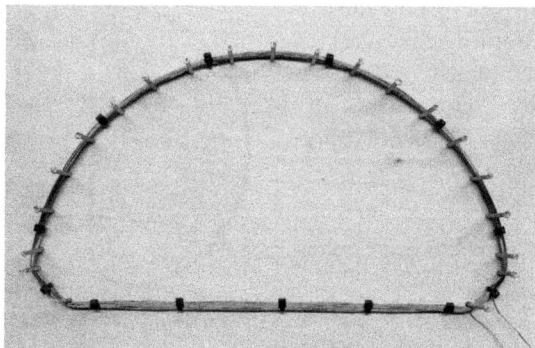

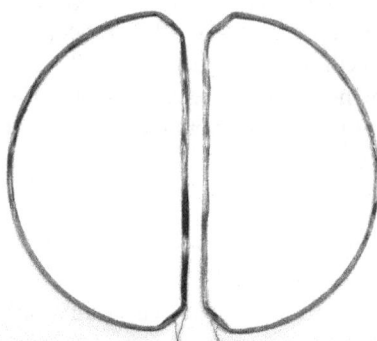

Fig. 8-40: **DD Nail Board and Completed Coil Windings**

8.5.2: Shielding

This design will use a two-piece clam shell with carbon-based shield paint applied to both halves. You don't want paint on the outside of the shells so tape them off before you spray. It is also a good idea to wipe down the interior with isopropyl alcohol to remove any residual solvent or release agent left over from the plastic manufacturing. Also drill a hole for the cable gland if not already provided. For a DD coil the cable entry point is usually on the TX side. Install and secure the gland. Apply 2-3 thin coats of shield paint with dry time in between, ensuring good coverage on both the bottom and side walls. Spraying the shield paint in a thick layer often produces crackle and craze which will diminish the effectiveness of the shield and is also more likely to cause potato chipping.

Fig. 8-41: **Top Shell: (L) Bare (R) Shield Painted**

8.5.3: Cable & Connector

Next comes the cable and connector. For our cable we will use S-video which is commonly available, often in local stores. It is a double coax and while it's not an ideal choice (it has stranded shields instead of braided) it is easy to obtain and will do. The connector is again a GX16 type but now with four pins. Many commercial detectors use a 5-pin variant and that will work just as well.

This is also the time where we need to attach a thin drain wire (30 AWG works well, preferably tinned bus wire) to the shell using the conductive fabric tape mentioned in the BFO coil construction. You can place it anywhere you like but the best place is out of the RX coil interior and definitely out of the overlap area. In Figure 8-41(R) you will see that the drain wire is taped down at the rear of the top housing close to the cable entry. The bottom shell will also need a drain wire.

It is critical at this point that you have planned out both the pin connections and the coil phases. This must be done taking into account both the coil and the detector. If you are winding a coil for an existing detector then the pinout and polarity are already defined. If you are defining your own then the

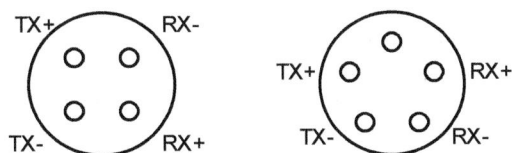

Fig. 8-42: **Example Pin Connections**

main consideration in the connector is to put as much distance as possible between the TX+ and RX+ pins. Figure 8-42 shows examples. Attach the connector to the cable.

8.5.4: Assembly & Nulling

Insert the coil end of the cable through the gland and solder the coils and drain wire to it paying close attention to proper connection. Lay the coils into the top shell; normally the TX coil is laid in first which places the RX coil a little closer to the ground. It is best to space the TX coil slightly away from the shielded shell; scrap pieces of shell plastic work well. Also place spacers between the coils at the overlap points to minimize coupling capacitance, if desired. Finally, thicker spacers (small wood or foam blocks, perhaps) can be added under the RX coil to keep it roughly parallel to the TX coil. In my example coil, the housing was not deep enough to accommodate spacers. One solution is to wind the coils with a "flatter" profile, which absolutely requires a winding bobbin.

At this point you will need to connect the TX coil to a signal generator running at the desired frequency and the RX coil to an oscilloscope. It is best to trigger the oscope with the TX signal from the signal generator. The first task is to check the TX/RX phase polarity. If it is incorrect, reverse the connections to one of the coils. Once the phase is correct, adjust the overlap of the coils to minimize the RX signal level. With care you should be able to get the signal level down to 10 millivolts or so, assuming a TX drive amplitude of 10v. As you adjust the overlap also adjust the position of the coils within the shell to get them centered, straight, and parallel, and preferably not touching any of the walls of the shell.

When you have them where you want them, tack down the TX coil to the shell with hot-melt glue. Preferably apply the glue over the spacers. Let this harden, check the null, and re-adjust the RX coil if needed. Apply hot-melt glue at the overlap points and allow to harden. Hot-melt has a tendency to move slightly as it hardens so you may need to adjust the null yet again. Since the overlaps are now glued you can no longer easily alter the overlap but small nudges to the coils can move them enough to alter the null. Finally, tack the RX spacer blocks to the shell and the RX coil to the blocks. Check the null a final time.

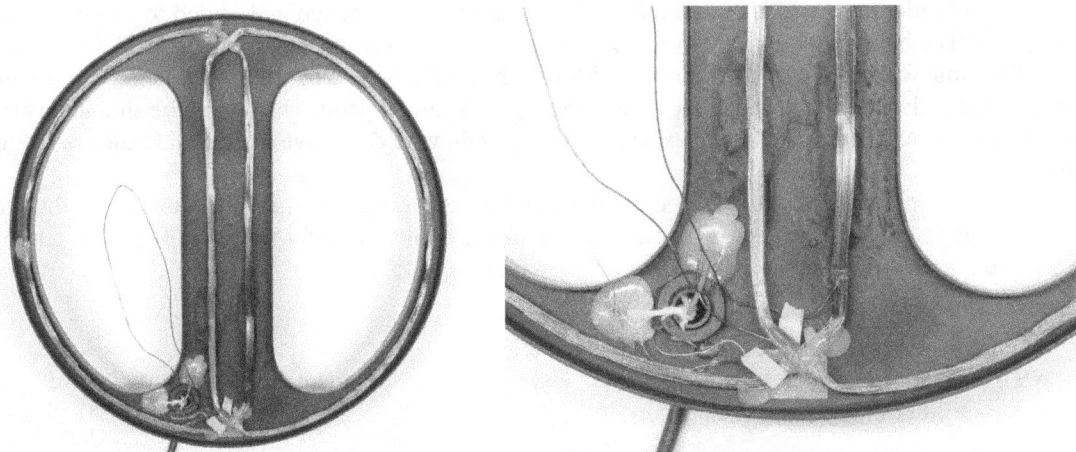

Fig. 8-43: **DD Coil Assembly (L) and Close-Up (R)**

8.5.5: Potting

With the coil windings now tacked down it's time to pour the epoxy. Before doing so an extra short length of drain wire needs to be soldered to the RX– connection. This wire will protrude through the epoxy and attach to the lower shell's shield paint with a piece of conductive tape. Before pouring the epoxy make sure the drain wire for the bottom shield and the tweak wire loop are sticking straight up.

The coil should be level to ensure the epoxy has even coverage. Mix up the desired epoxy and slowly pour it into the shell until both coils are just barely covered. Pouring epoxy into the narrow channels presents a challenge; a large syringe (50mL or so) works well for this. Re-stir the epoxy before every syringe refill as the microballoons tend to float out of the mixture. Allow the epoxy fully cure[30].

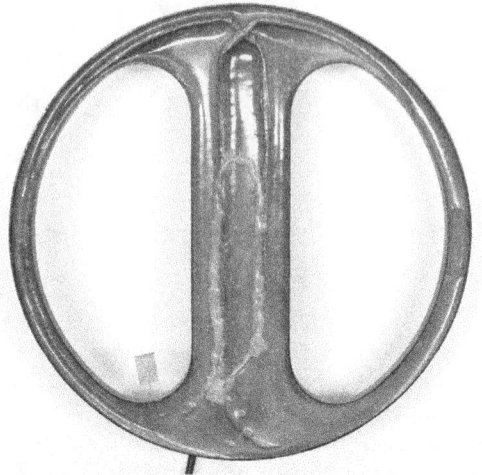

Fig. 8-44: **Potted Coil**

Figure 8-44 shows the potted coil. The epoxy was mixed roughly 60/40 by volume with microballoons to reduce weight. Notice the shield wire with tape applied, ready to attach to the bottom housing.

8.5.6: Tweak Wire Option

The DD coil null is exceptionally sensitive to coil movement. This why you want an epoxy with the least amount of cure shrinkage as possible so that the coils don't shift during the epoxy pour. If you find that movement is unavoidable there is a fix for this. When winding the TX coil leave an extra-long pigtail (10-inch|25cm) on the ground side of the winding. Solder the end of the pigtail to the TX– pin on the cable as usual and leave the resulting loop of wire sticking straight up in the air during the epoxy pour (this can be seen in Figure 8-43). After the epoxy cures lay the loop of wire on the epoxy surface. Move it around, adjusting the position and spread to re-null the coil. If it only seems to make it worse, twist the loop to flip it over and invert its phase, and try again. When the null is tweaked back in, use small pieces of tape to hold down the wire loop and use a high-temperature hot melt glue to completely anchor the wire loop to the epoxy surface. You can see the tweak wire loop in the center of the coil, secured with hot melt glue along its entire length, in Figure 8-44.

8.5.7: Final Null & Sealing

This particular coil design has a bottom shell that must be attached and sealed to make the coil waterproof. The bottom shell has been shield-painted and must now be electrically attached to the top shell. The drain wire that was left protruding through the epoxy should be trimmed as short as possible and attached with a piece of conductive tape to the top shell shield paint. The drain wire should be kept short because it will be unsecured; an excessively long drain wire can move around and cause false signals.

Press the bottom shell onto the top shell. Watch for a null shift due to the presence of the additional shield, see page 145 for this discussion. The last step is to seal the seams for waterproofness using a solvent cement as described in the BFO coil construction on page 151.

8.5.8: Results

The final measured parameters for the DD coil are
- $L_{TX} = 0.93mH$
- $R_{TX} = 2.8\Omega$

---

30. Epoxies often don't fully cure for days. During this time there can be minute movements of the coil windings which can affect the induction balance. Furthermore, epoxy outgasses during curing and you don't want this to happen after the bottom shell has been sealed. **Don't rush this step**, let the coil sit for a week before proceeding.

- $L_{RX} = 3.86\text{mH}$
- $R_{RX} = 10.2\Omega$
- IB null = −70dB @ 10kHz

Notice that the inductances of both coils are somewhat lower than were the raw windings. This possibly happened because the coils were slightly reshaped to fit the housing and to get the overlaps flatter. Also, the coil resistances are slightly higher because they now include the cable resistance. R_{TX} does not quite meet the goal of 2.7Ω and if this is actually a critical requirement then using 23 AWG wire will solve the issue.

The final null for this coil is shown in Figure 8-45. For roughly an 8v p-p TX signal, the residual RX signal is about 2.4mv p-p, which is a −70dB null. The goal was −60dB which is what any well-nulled coil should achieve, so this is better than good. The residual RX signal also appears to be largely in-phase with the TX signal which means the R-null is exceptionally good. This is often the case with DD coils. Finally, the finished weight of the coil is 16 ounces (450g) which is pretty good for a 10.5-inch coil. Less epoxy could have been used and perhaps more microballoons mixed in for an even lighter coil.

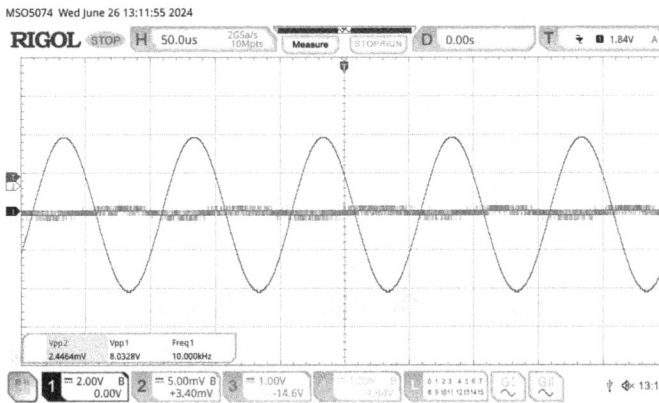

Fig. 8-45: **DD Final Null**

8.5.9: OO and 00 Coils

The DD coil is by far the most popular overlapped coil but the OO (double round) and 00 (double elliptical) also show up in commercial detectors. Building them is done in the same manner as the DD coil. The amount of overlap required for a DD coil depends on exactly how the "D"s are shaped and can also vary with overall size and is therefore usually determined experimentally. With OO coils — and 00

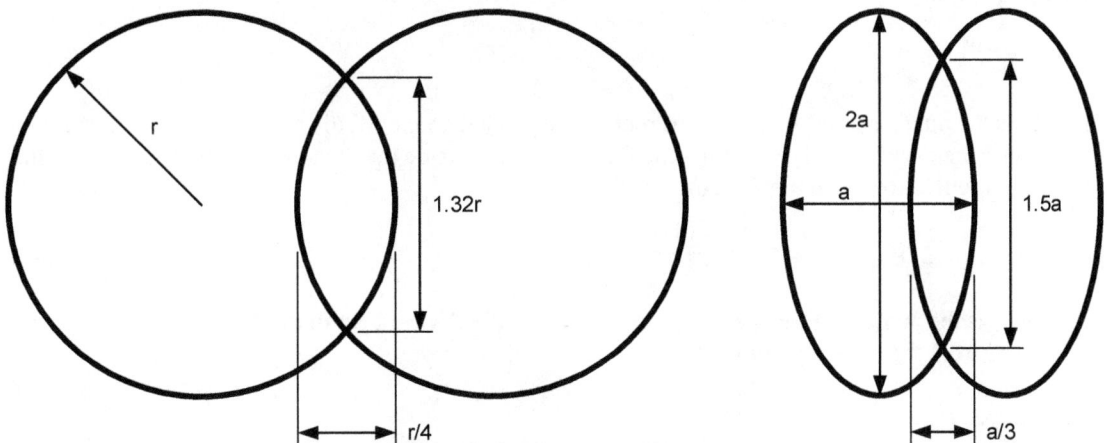

Fig. 8-46: **OO & 00 Standard Overlaps**

coils using specific aspect ratios — the amount of overlap will be consistently the same regardless of size. Figure 8-46 shows the proper overlaps for round OO windings and for 00 windings with a 2:1 aspect ratio.

8.6: Concentric Coils

The concentric coil is a bit more difficult to build than an overlapped coil but with a little care excellent results are possible. The concept of the concentric IB was discussed in Chapter 6; we have a large TX coil, a small RX coil, and a TX bucking coil (feedback, or FB[31]) closely spaced to the RX coil. The bucking coil uses the same wire as the TX coil but is wound in the opposite direction. The negative magnetic field of the bucking coil cancels the positive magnetic field of the TX coil and creates a magnetic null exactly at the RX coil. Figure 8-47 illustrates.

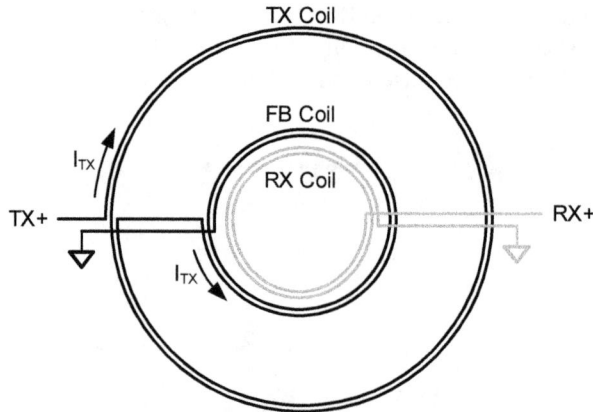

Fig. 8-47: **Concentric Coil Wiring**

8.6.1: Ideal Calculations

In a typical concentric coil the RX coil is one-half the diameter of the TX coil. It does not have to be so; one-third to two-thirds the TX diameter make for a reasonable range. When the RX coil is placed inside the TX coil the TX magnetic flux is strongly coupled and a voltage is induced across the RX coil. For a given TX field strength, the voltage induced on the RX coil depends on the area of the coil and the number of turns in the winding (the area-turns).

The bucking coil uses the same wire as the TX coil but is wound in the opposite direction. What we need at the RX coil is for the total magnetic flux coupled from the bucking coil to exactly cancel the total magnetic flux coupled from the TX coil. We know from Equation 3-3 that the flux density at the center of a coil is

$$B = \frac{\mu N I}{d} \qquad\qquad \text{Eq 8-5}$$

where N is the number of turns, I is the current, and d is the diameter of the coil. This applies to both the TX coil and the bucking coil. Assuming the flux is uniform inside the area of the coil then the total flux coupled to the RX coil from the TX coil is

$$\Phi_{TX} = B_{TX} \cdot N_{RX} \cdot A_{RX} = \frac{\mu N_{TX} I_{TX}}{d_{TX}} \cdot N_{RX} \cdot A_{RX} \qquad\qquad \text{Eq 8-6}$$

where N_{RX} is the number of turns and A_{RX} is the area of the RX coil. Similarly, the total flux coupled to the RX coil from the bucking coil is

---

31. In discussions I tend to use the term "bucking," but for subscripts "FB" is shorter. Occasionally you will see the abbreviation "BX" for bucking.

$$\Phi_{FB} = B_{FB} \cdot N_{RX} \cdot A_{RX} = \frac{\mu N_{FB} I_{FB}}{d_{FB}} \cdot N_{RX} \cdot A_{RX} \qquad \text{Eq 8-7}$$

For the flux of the bucking coil to cancel the flux of the TX coil we simply make them equal:

$$\Phi_{FB} = \Phi_{TX} \quad \text{therefore} \qquad\qquad\qquad\qquad \text{Eq 8-8}$$

$$\frac{\mu N_{FB} I_{FB}}{d_{FB}} \cdot N_{RX} \cdot A_{RX} = \frac{\mu N_{TX} I_{TX}}{d_{TX}} \cdot N_{RX} \cdot A_{RX}$$

$$\Rightarrow N_{FB} = N_{TX} \cdot \frac{d_{FB}}{d_{TX}}$$

Since the coils are wired in series $I_{FB} = I_{TX}$.

Let's say, for example, that the RX coil is half the diameter of the TX coil. With a uniform TX field only 25% of the flux inside the TX coil cuts through the RX coil. The bucking coil has the same diameter as the RX coil so the bucking coil needs half the number of turns as the TX coil for its total flux to equal the 25% TX flux that is seen by the RX coil.

8.6.2: Actual Reality

Everything discussed in the previous section was predicated on a uniformly distributed TX field, which is not the case. The flux density is weakest at the center of the TX coil and strongest at the windings. Figure 8-48 shows the relative measured flux density across the plane of a 10-inch|25cm diameter circular coil. It is also possible to calculate the flux density anywhere inside the coil but it involves some painful math[32]. The flux density in Figure 8-48 is normalized to the center of the coil; in other words, the flux density one inch (25mm) from the windings (−4 and +4) has twice the flux density than at the center of the coil (0).

An RX coil which spans half the TX coil diameter sees less than the expected 25% of the total magnetic flux; it's closer to 17%. Therefore the bucking coil needs fewer turns than previously calculated. Additionally, the coupling of the bucking coil to the RX coil won't be 100% but maybe 90% or so. In other words, this is the part of the design where you have to figure it out through trial-and-error. The easiest solution is to intentionally wind the bucking coil with a few too many turns; using the approximation of a uniform field (Equation 8-8) is a safe over-estimation. Then hook everything up to an oscope and remove bucking coil turns until the balance is "close enough." The next section examines "close enough."

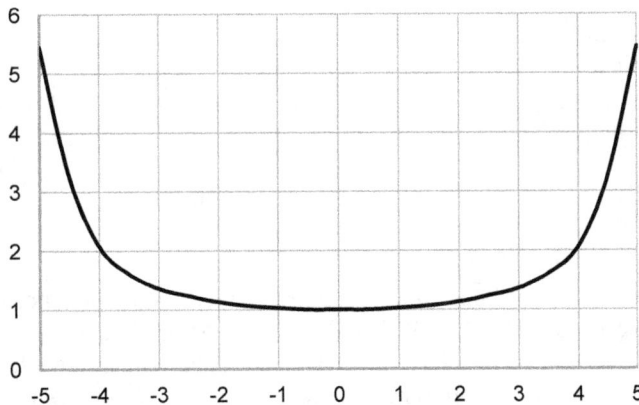

Fig. 8-48: **Actual TX Field Distribution**

---

32. A surface integral of elliptic functions. http://nbviewer.jupyter.org/github/tiggerntatie/emagnet-py/blob/master/index.ipynb has some useful field calculators that do the painful math for you. Also see footnote 1 on page 113.

8.6.3: Tweaking the Null

Since coils are wound with an integer number of turns it is usually impossible to get a good null just by adjusting the number of turns on the bucking coil. You will always end up a little under-nulled or a little over-nulled. The overall null is less sensitive to the turns of the TX coil so you can also add or subtract turns to the TX coil to get a closer null, but it is still likely to end up a little high or a little low.

With the other search coils, any residual imbalance can be removed simply by moving the RX winding around slightly, or maybe just one edge of the RX winding in the case of a DD coil. Residual imbalance is not so easily removed in the concentric coil. In Figure 8-48 we see that the flux density is fairly flat in the middle 40% of the TX coil so moving the RX coil around will have little effect[33]. Even if you can move the RX coil enough to improve the null, doing so also moves the center peak of the target response, making it more difficult to pinpoint a target.

To get a better null we add some extra wire to one lead of either the TX/FB coil combination or the RX coil, and this extra wire is then formed into a small loop and moved around the plane of the concentric assembly (and perhaps deformed) until the null is minimized. In the White's search coil (Figure 8-49), the single strand of wire (taped down) is an extension of the TX bucking coil that is moved around to fine-tune the balance.

This "tweak wire" can come from the signal side of the TX coil, the DC side of the bucking coil, or either side of the RX coil. An advantage of using the TX or FB coil is that it typically provides more tweak sensitivity than does the RX

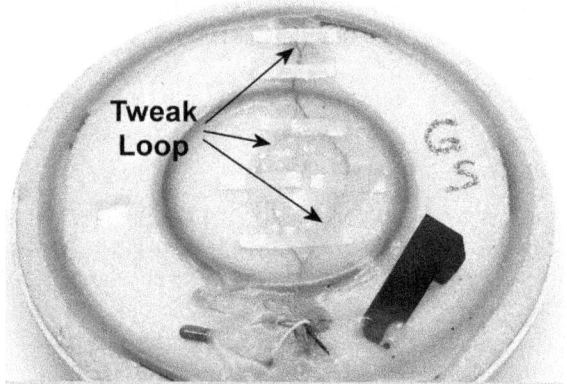

Fig. 8-49: **White's Concentric Details**

coil. Furthermore, the tweak wire is usually taken from the DC side of the FB coil to minimize unwanted capacitive coupling of the TX signal into the RX coil.

8.6.4: Coil Design

Our concentric coil will again be designed for the TR/VLF projects in Chapters 14-17 and closely follow the specs for the DD coil. The target parameters are:

- d_{TX} = 8.75 inches (22.25cm)
- L_{TX} = 1mH nominal
- R_{TX} = 2.7Ω max
- C_{TX} — not critical
- d_{RX} = 4 inches (10cm)
- L_{RX} = 4mH nominal
- R_{RX} = 20Ω max
- C_{RX} — not critical
- L_{FB} — TBD
- R_{FB} — follows R_{TX}
- C_{FB} — not critical
- IB null $\leq$ −60dB @ 10kHz

All of the discussions in the DD coil design regarding resistance and capacitance apply here and so we will use the same wire gauges: TX = 24 AWG and RX = 27 AWG. There is an additional coil — the bucking or "feedback" (FB) coil — which uses the same 24 AWG wire as the TX coil.

Since the TX coil and the FB coil are wired in series the total TX inductance is slightly less[34] than that of the TX coil alone. However, we will target 1mH for the TX coil alone and that comes to 44 turns

---

33. This positional insensitivity makes the concentric more immune to impact falsing.

34. Remember, the FB coil is wired 180° out-of-phase and will decrease the overall TX inductance.

of wire using 24 AWG. This is still close enough for our goal of minimal thermal phase drift. At 4-inches|10cm the RX coil is the usual one-half TX diameter seen in most concentrics. The number of turns comes out to 133 which, for 27 AWG wire, makes a fairly "square" winding that is 1/4-inch tall by 1/4-inch thick.

The housing for this project is a spoked-style (open center) top housing only. Since there is no bottom housing the top housing will be completely filled with epoxy using two epoxy pours, with the bottom shield paint applied in between the epoxy pours. The cable is again a 48-inch (120cm) length of S-video dual coaxial cable and the connector remains a 4-pin GX16 type.

8.6.5: Winding the Coils

The TX coil is stand-alone and may be wound in the usual way[35]. In commercial production the RX and FB coils are most often wound together on a bobbin; the RX coil is wound first, and the bucking coil directly on top of it. Because they are literally touching each other the risk of capacitive coupling is high[36] and it is critical they be wound in a certain way. Figure 8-50 shows a typical schematic for the TX, FB, and RX coil connections. The turn direction of each coil is important. Note that one side of the FB coil is ground, as is one side of the RX coil. In practice one or both may be connected to a power rail but it is "DC" in any

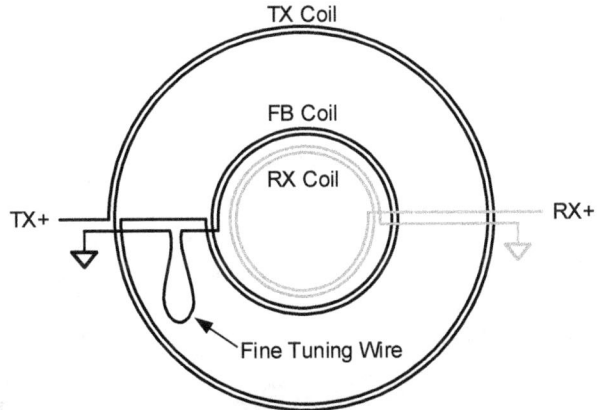

Fig. 8-50: **RX and FB Coil Wiring**

case. The inside winding of the FB coil (closest to the RX coil) should be its DC side and the outside winding of the RX coil (closest to the FB coil) should be its DC side. This places the DC sides of the FB and RX coils in contact and minimizes capacitive coupling by minimizing the signal levels. Even then, it is wise to put a small spacer between the RX and FB windings. Obviously it becomes important to keep track of the winding ends to ensure they all get connected correctly.

8.6.6: Feedback Coil

We know that the TX coil has 44 turns and the RX/TX diameter ratio is 1/2 so, first order, the FB coil requires 22 turns. Because the B-field is non-uniform across the area of the TX coil this estimate is high by as much as 50%. To get the right number of turns, initially overwind the FB coil; let's say 16 turns in this case. Then hook everything up according to Figure 8-50 and remove turns from the FB coil until the RX signal is as low as you can get it. For this design, 13 turns produced a slightly over-nulled RX coil and 12 turns was quite under-nulled, so 13 turns was best and the remaining null will be tweaked in later with the fine-tuning wire loop.

Winding the FB coil on top of the RX coil is best done with a winding bobbin and the use of bondable magnet wire. In this case the coils were wound on a ply-wood bobbin with a 1/4-inch (6mm) coil thickness. After the RX coil was wound and bonded, 8 layers of 1/4-inch

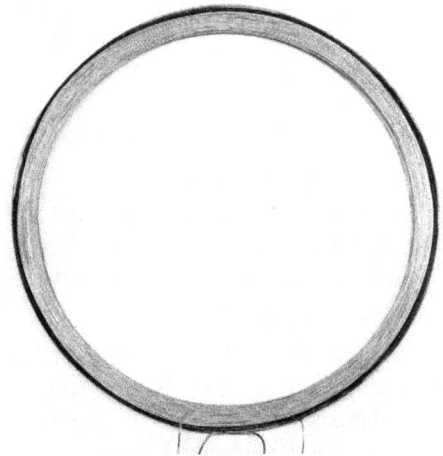

Fig. 8-51: **Finished RX and FB Coils**

---

35. That is, whatever way you usually wind a coil.

36. See the *Explore* section on R-null.

wide tape was wrapped on top of the RX coil as a spacer before winding the FB coil (Figure 8-51). Remember when winding the FB coil to leave an extra-long ground-side pigtail for fine tuning.

8.6.7: Shielding — Part 1

Before placing the coil assembly in the housing we need to apply shielding. The easiest method is to spray the inside of the shell with carbon-conductive shielding paint. As with the DD coil, first install the strain relief gland. Tape a drain wire to the shell using the aforementioned fabric shield tape, preferably near the cable entry point to minimize drain wire length and to keep the drain wire away from the RX coil. The drain wire will be soldered to the coil cable, and we also need a drain wire to protrude through the first epoxy pour to contact the bottom shield. This can all be a single wire, see the assembly details in Figure 8-52. Finally, mask off the shell for spraying and apply 2-3 thin coats of shield paint, allowing dry time in between.

8.6.8: Coil Assembly

With the coils wound they need to be positioned and connected. Assuming the cable and connector have been assembled, insert the cable through the gland and tighten the gland. Solder the coil wires to the proper cable points, as well as the shell drain wire. Recall that the FB-ground pigtail was made extra-long, and this extra length is what we will use to fine-null the coil. Figure 8-52 shows the coils set in the housing and connected; note that the solder connections are hot-melt glued on top of masking tape, not directly to the shield-painted housing which could cause short circuits. Again, pay close attention to the directions of the windings per Figure 8-50. Tack the coils to the housing with hot-melt glue. In best practice, the coils should be slightly spaced away from the shield paint surface.

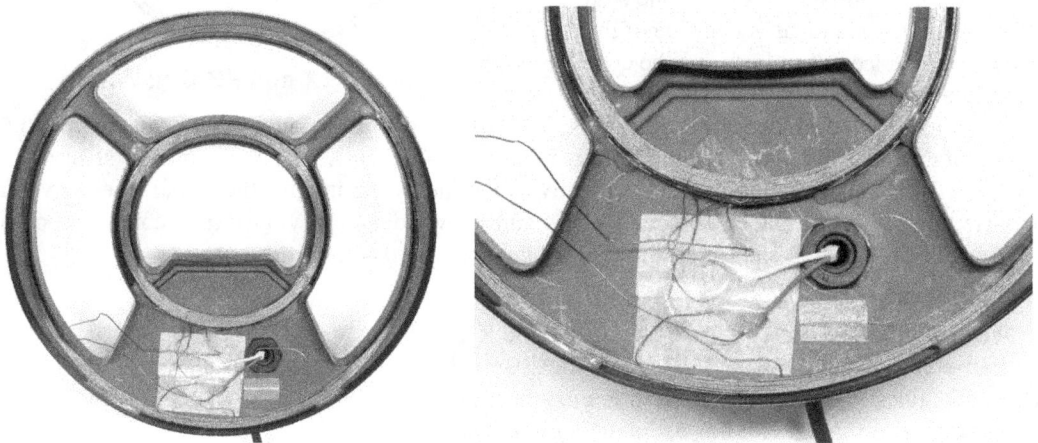

Fig. 8-52: **Concentric Coil Assembly (L) and Close-Up (R)**

Now is a good time to do a test fit to be sure everything will work; once the epoxy is poured it is too late. Connect a signal generator to the TX+FB coil and an oscope to both the TX+FB and RX coils. Short the FB coil and make sure the RX signal is in-phase with the TX drive. Un-short the FB coil and move the fine-tuning wire loop around; There should be a position where you get a nearly perfect null.

8.6.9: Potting — Part 1

Pull both the shield wire and the fine-tuning wire loop straight up into the air. Carefully level the coil housing so the epoxy flows evenly over the coils. The first epoxy pour should be lightened with microballons as much as possible, at least 50% by volume. A large syringe is useful in filling the narrow channels and controlling the amount of epoxy. Re-stir the epoxy mix each time before re-filling the syringe; microballons tend to float out of the mixture. Unlike the DD design where a bottom housing hides a bad epoxy pour, this design needs a little more care. It is true that the second epoxy pour can be used to hide a bad first pour, but a good first pour makes everything else much easier.

Pour enough epoxy to just cover the coils completely. Normally with a coil having two epoxy pours the first pour needs to cure thoroughly so there is no null shifting later on, especially after the fine tuning wire is set and epoxied. This problem is prevalent in overlap (DD) coils which are far more sensitive to minute coil movements, whereas concentric coils are fairly immune. So it is possible in this case to proceed with the second shield and fine nulling after just a day of curing.

8.6.10: Shielding — Part 2

The top-side shielding was applied to the plastic top shell. In this design there is no plastic bottom shell on which to apply shielding so we apply it directly to the surface of the first epoxy pour. First, bend the exposed drain wire down (snip to a reasonable length if needed, 1-inch|25mm should do) and tape it to the epoxy surface using a piece of conductive fabric shield tape. Then mask off the coil and again spray 2-3 thin coats of shield paint over the epoxy surface, allowing dry time in between coats.

8.6.11: Fine Nulling

After the final coat of shield paint has dried, connect the TX coil to a signal generator running at the desired frequency and the RX coil to an oscilloscope. It is best to trigger the oscope with the TX signal from the signal generator. Lay the fine-tuning wire loop on the epoxy surface. Move it around, adjusting the position and spread to minimize the RX signal. If it only seems to make it worse, twist the loop to flip it over and invert its phase, and try again.

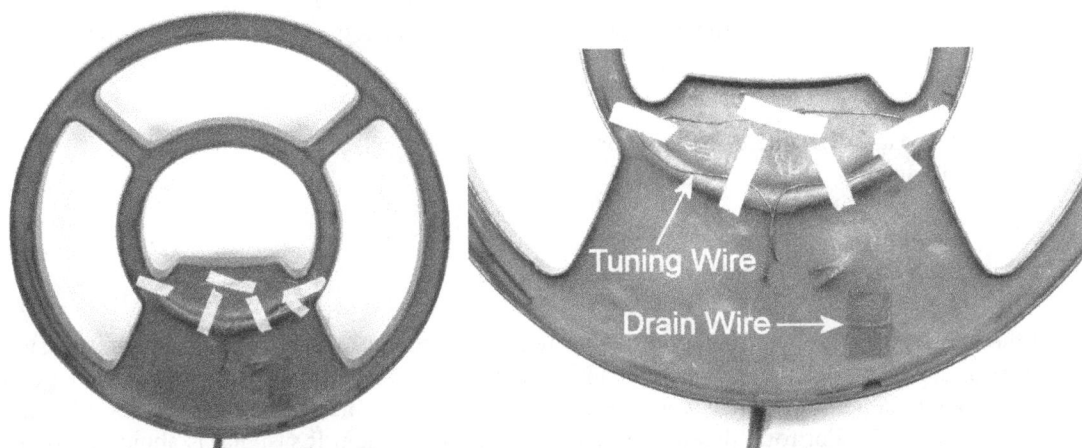

Fig. 8-53: **Concentric Final Null**

The concentric design often has a worse R-null than other coil designs. You are therefore likely to see the RX signal reduce in amplitude to some minimum level, then the phase rolls by roughly 180°, and the amplitude begins increasing but with an opposite phase. That minimum amplitude is the best you will be able to achieve without doing something to compensate for the R-null problem. See the *Explore* section for further discussion. For now, get the best null possible and use small pieces of tape to hold down the wire loop.

8.6.12: Potting — Part 2

Having the loop *exactly* level is more critical for the final epoxy pour. Also ensure that the fine-tuning wire loop is tightly attached to the first epoxy surface. In short, make sure that nothing will protrude through the final epoxy pour.

The final epoxy pour can be lightened with microballoons or not. The only reason not to use microballoons is if you want a nice-looking gloss black finish. Personally, I value light weight over the aesthetics of something I never look at so, for this example, microballoons it is.

Use a syringe to slowly apply the epoxy and re-stir the epoxy mix each time before re-filling the syringe. It is best to apply the epoxy gradually all around the coil until you build it all up to the edge of the plastic. If the epoxy mixture is viscous enough, you should be able to slightly overfill the shell without epoxy running over the edge which will give the epoxy a nice rounded finish and make the coil easier to slide over the ground. If necessary, slow down the process a bit to let the epoxy thicken enough to do this.

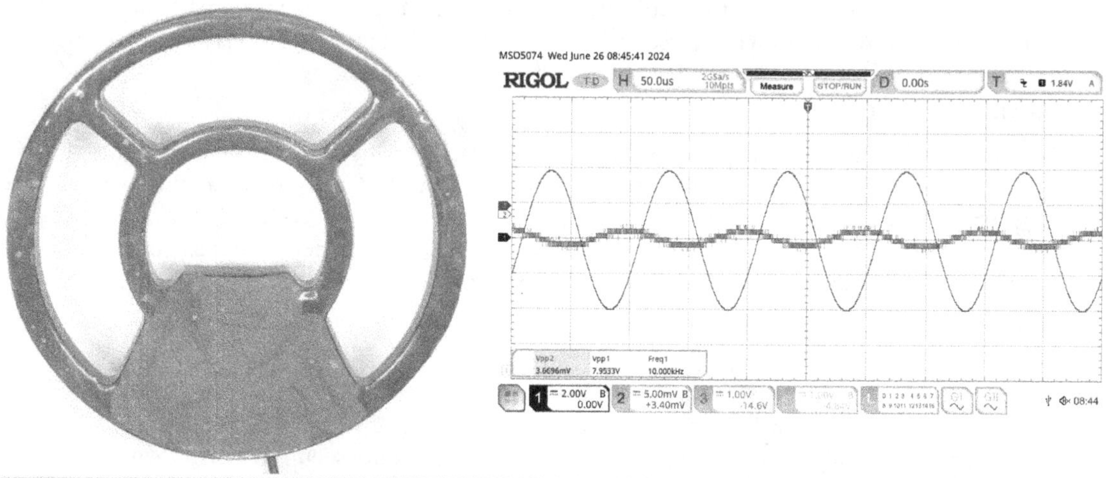

Fig. 8-54: Concentric Final Epoxy Pour & Final Null

Once complete, don't move the coil for day until the epoxy sets. Then, let it fully cure, up to a week.

8.6.13: Results

The final measured parameters for the concentric coil are
- L_{TX} = 1.04mH
- R_{TX}(max) = 3.3Ω
- L_{RX} = 4.04mH
- R_{RX} = 8.0Ω
- IB null = −66.7dB

All of these except R_{TX} are within the initial desired specifications. The raw TX coil had a measured resistance of 2.6Ω but once the bucking coil and cable were added the final TX resistance increased to 3.3Ω so a slightly larger wire gauge would be needed to fix this problem. The final null for this coil is shown in Figure 8-54. For roughly an 8v p-p TX signal, the residual RX signal is about 3.7mv p-p, which is a −66.7dB null. The goal was −60dB which is what any well-nulled coil should achieve, so this is plenty good enough. The residual RX signal has a slightly more than 90° phase shift relative to the TX signal which means the R-null is a limiting factor in getting a better null. This is often the case with concentric coils. Finally, the finished weight of the coil is 16 ounces (450g) which is decent (but not great) for an almost 9-inch coil.

8.7: Coaxial Coil

By now you should have a pretty good feel for how IB coils are built. This section will cover the unique aspects of a coaxial design. The easiest approach in building a coax coil is with a foam carrier in a clamshell housing. This is how Karbowski (Figure 6-22) and C&G did it. The TX and RX coils can all be the same diameter (Karbowski) or the RX coils can be made smaller (C&G).

Home improvement centers sell foam insulation sheets in different thicknesses. A 2-inch|5cm thickness will work, or glue together what you can get to make the thickness you want. Cut the foam to the desired inner diameter[37]. Wind the TX coil in exactly the center of the foam. Wind the RX coils on a form and attach them to the top and bottom surfaces of the foam. The RX coil on the top side of the foam should have an extra-long pigtail that will be used as the "tweak" wire.

Prepare the coil housing & cable. Wire the coils to the cable with the RX coils connected in series but out-of-phase. Plug everything up to test circuitry and an oscope. Monitor the null and adjust the top-side RX tweak loop to fine-null the coil. Assuming a clamshell housing that has been shield painted on

---

37. Every coaxial coil I've ever seen is round, but you can make it any shape you want as long as you have a housing to put it in.

the inside, you may find that the null shifts upon final assembly of the housing. If so, overcompensate the null so that it comes out where you want it in the end.

8.8: Figure-8 Coil

The figure-8 coil consists of a TX coil with two equal RX coil "halves" inside it (Figure 6-15). There are a few ways to make the RX coil:

1. Make a nail board and continuously wind the figure-8 in its final form.
2. Wind a single long rectangular coil and twist it at its center to form a figure-8.
3. Wind two individual RX coils, one for each lobe of the figure-8, and wire them out-of-phase.

The *Bigfoot* coil used method 3 (Figure 6-16). This design can be placed in a clamshell housing (as with the *Bigfoot*) or could be placed in a top shell and epoxy-poured, similar to the Tesoro *Cleansweep*. Nulling can be accomplished by very slight movements of the RX coils, including slight deformations. You can also use a tweak nulling wire.

8.9: Probe Coil

The coil designs covered so far are intended for swinging detectors. Almost universally, pinpointers use a solenoidal coil (Figure 8-5) which is our next topic. Depending on the circuit design, a pinpointer can use a mono coil or an IB coil. A mono coil is found in the Garrett *Pro-Pointer* which is an energy-theft design and the Fisher *F-Pulse* which is a PI design. An IB coil is used in the White's *TRX* which is a VLF design. Probe-style coils can be made for regular detectors as well. For a while, Sunray made a popular probe coil for various detectors such as the Minelab *Sovereign*, and Eric Foster produced probes for some of his PI models.

A major difference between a probe coil and a regular coil is that a probe coil uses a ferrite rod instead of having an air core. The ferrite rod keeps the required number of turns to a reasonable level and helps focus the magnetic field.

The mono probe coil is the easiest to build. It consists merely of magnet wire wrapped around a ferrite rod. The type of ferrite material will affect the results and it is best to try several for comparison. This is especially true for a PI probe coil as some ferrites have enough hysteresis that the resulting coil has poor decay settling. It is generally preferable not to wind the wire directly in contact with the ferrite rod but rather put a spacer between the rod and the windings. Heat shrink tubing works well for this.

The IB version of the probe coil is a bit more complicated. The TX coil is still wound on the ferrite rod but the RX coil is not. Instead, the RX coil is wound with an air core but still placed just ahead of the TX coil. This will produce significant direct coupling in the RX coil that must be balanced out. An easy way to do this is to use an extra-long pigtail from the RX coil and reverse-wind it around the TX coil windings until IB is achieved. This should only require a few turns and, unlike with a planar coil, fractional turns are possible. While it is easy, it is tricky to get it right. Once IB is achieved it is imperative to lock everything in place so it does not move. A coating of epoxy works well.

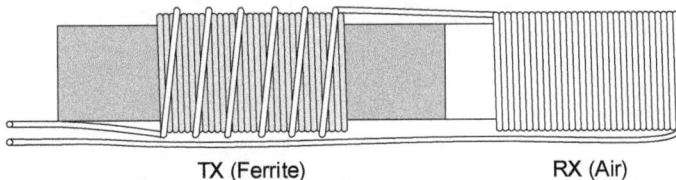

TX (Ferrite)   RX (Air)

Fig. 8-55: **IB Probe Design**

As with other IB coils, an additional tweak might be needed after the epoxy cures. Since a probe coil is solenoidal and not planar, a tweak wire does not work very well. Instead, consider using a small piece of ferrite shielding tape[38] and moving it around the assembly to adjust the null. Tape it in place at the optimal location. If the null shifts over time, the ferrite tape can be repositioned.

---

38. Ferrite shielding tape can be found at the usual DigiKey/Farnell/Mouser supply houses.

8.10: Water Coils

Detectors are often used for water hunting — both wading and diving — where the coil is submerged in either fresh water or salt water. Besides the obvious requirement that the coil housing must be completely waterproof, two other issues should be considered. First is that a waterproof coil connector is a must for diving and should be strongly considered even if you only intend to wade in shallow water, especially in salt surf. It is far too easy for water to splash onto the connector and a GX16 connector (for example) provides no protection from seepage. In salt water this can lead to corrosion not only of the connector pins but salt water can also wick into the cable and corrode the conductors and shield.

With commercial detectors, the waterproof M12 connector is the predominant choice for VLF and multifrequency models. However, most M12 connectors are only rated for 60 volts so they are not appropriate for PI designs. For PI, a larger connector (such as M16[39]) is required; pay particular attention to the pin-to-pin voltage rating.

The other issue is that of buoyancy. In building land-use coils we tend to place a high premium on light weight. This can make water hunting exhausting due to the effort it takes to keep the coil submerged. A seemingly simple solution is to add a weight to the coil or coil rod to counter the buoyancy. However, an air-filled coil (such as the foam-cored BFO coil or the completely hollow PI coil presented in the chapter) will compress under even slight water pressure. This will stress the seam sealing and eventually cause leaks.

The best solution for a water coil is a solid epoxy construction as shown in the concentric example. Furthermore, instead of making the coil as light as possible, mix just enough microballoons into the epoxy so that the coil is neutrally buoyant for shallow water hunting (wading) or slightly negatively buoyant for scuba diving. Getting the epoxy mixture just right may take some experimentation.

Explore

8.11: TX Resistance

The phase between the voltage and current of the TX coil is ideally 90° but the coil resistance changes this slightly. As long as the resistance is stable the resulting phase error can be adjusted out in circuitry. But temperature changes will cause the resistance to change and create an error in the ground balance point. This section derives the math to compute the resulting phase error for a given temperature change.

The phase shift between the drive voltage and the coil current is

$$\phi = \arctan\left(\frac{\omega L}{R}\right) \qquad\qquad \text{Eq 8-9}$$

At two temperatures (T and $T+\Delta T$) we get

$$\phi_T = \arctan\left(\frac{\omega L}{R}\right) \qquad\qquad \text{Eq 8-10}$$

$$\phi_{T+\Delta T} = \arctan\left(\frac{\omega L}{R + \Delta R}\right)$$

We will not consider that L changes with temperature although physical expansion and contraction can produce minute changes in L. Copper has a temperature coefficient (TC) of ~0.4%/°C so the second equation becomes

$$\phi_{T+\Delta T} = \arctan\left(\frac{\omega L}{R + 0.004R \cdot \Delta T}\right) = \arctan\left(\frac{\omega L}{R(1 + 0.004\Delta T)}\right) \qquad \text{Eq 8-11}$$

---

39. A particular favorite of mine was the 16mm connector used on the Tesoro *Sandshark*. It was a unique design manufactured by Methode and was discontinued, with no second source.

Taking the tangent of both sides results in

$$\frac{\omega L}{R} = \tan(\phi_T)$$

$$\frac{\omega L}{R(1 + 0.004\Delta T)} = \tan(\phi_{T+\Delta T})$$

Eq 8-12

Substituting the first equation into the second gives

$$1 + 0.004\Delta T = \frac{\tan(\phi_T)}{\tan(\phi_{T+\Delta T})}$$

Eq 8-13

One of the properties of the tangent function is that as the angle is very close to 90° (as is the case here) and $\Delta\phi$ is small (we assume) the ratio above approaches

$$1 + 0.004\Delta T = \frac{90 - \phi_{T+\Delta T}}{90 - \phi_T}$$

Eq 8-14

And with a little rearranging we get

$$\Delta\phi = \phi_T - \phi_{T+\Delta T} = 0.004\Delta T(90 - \phi_T)$$

Eq 8-15

This equation is a bit weird so how do we make use of it? Let's say we want to design the TX coil so that, over a 5°C shift in temperature, the phase shifts no more than 0.05°. Let's plug those in:

$$0.05° = 0.004 \cdot 5 \cdot (90° - \phi_T)$$

Eq 8-16

We can solve for the resulting initial maximum phase shift:

$$\phi_T = 90° - \frac{0.05°}{0.004 \cdot 5} = 87.5°$$

Eq 8-17

Finally we can say that if our coil has an initial phase shift of 87.5° at some temperature then if the temperature climbs by 5° the phase will be no less than 87.45°. We can plug this initial phase shift into Equation 8-9 and solve for R. We will use the concentric design example which was 1mH and intended to work at 10kHz:

$$87.5 = \arctan\left(\frac{2\pi \cdot 10,000 \cdot 1\mathrm{mH}}{R}\right)$$

Eq 8-18

$$R = 2.74\Omega$$

In the concentric design the TX coil had a 10-inch|25cm diameter with N = 38 for a total linear length of 100 feet/30m. The wire gauge necessary to achieve this resistance is 24 AWG and, in fact, the raw TX coil had a measured resistance of 2.6Ω. But once the bucking coil and cable were added the final TX resistance increased to 3.3Ω so a slightly larger wire gauge would be needed.

8.12: Modified Concentric Design

The normal method of designing a concentric is to place a TX bucking coil near the RX coil. An alternate way is to place an RX cancelation coil near the TX coil. Figure 8-56 shows both methods. We'll call the alternate design the *concentric-modified* (CM) coil to distinguish it from the usual coplanar-concentric (CC) coil.

The CM coil is no more difficult to build than the CC coil. The number of turns required of the cancelation coil is related to the turns of the RX coil and the ratio of the TX and RX diameters:

$$N_{RXB} = N_{RX} \cdot \left(\frac{d_{RX}}{d_{TX}}\right)^2$$

Eq 8-19

With the CC coil you normally calculate the closest whole number of turns for the bucking coil which is

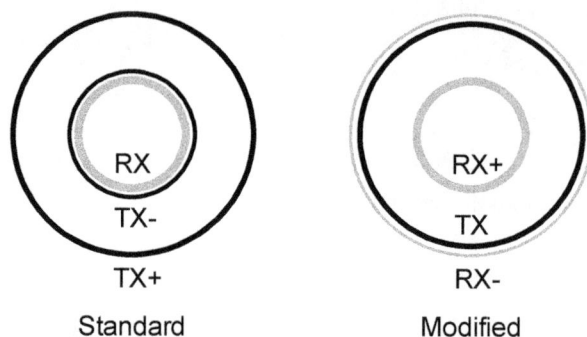

Fig. 8-56: **Standard & Modified Concentric Design**

never exactly right, so you use a tweak winding to dial in the IB null. With the CM coil you also calculate the closest whole number of turns for the cancelation coil which also will never be exactly right. But this is a case where you can then tweak the RX coil windings to dial it in much closer, especially if the RX coil is $d_{TX}/2$ or smaller and has a larger inductance. A small RX coil tends to have quite a few turns of wire so adding or subtracting a few to tweak in the IB null will have a minor effect on inductance and sensitivity. This still won't be perfect so a final tweak winding will still be needed, but it will be a smaller adjustment.

Full construction details are not given here; we are in the *Explore* section where you have to do some of the work yourself. I will also leave it to the reader to determine the advantages and disadvantages of this approach.

8.13: Butterfly Coil

The Minelab *GPZ7000* uses a coil called the DOD (Figure 6-8). This has a center TX coil and two outer RX coils wired out-of-phase. A closely related design is called the *butterfly* coil; see Figure 8-57. This type of coil configuration is commonly used in security walk-through detectors but not in ground search detectors. An almost-identical design is the figure-8 coil (e.g. *Bigfoot*) but its RX coils are oriented north-south instead of east-west.

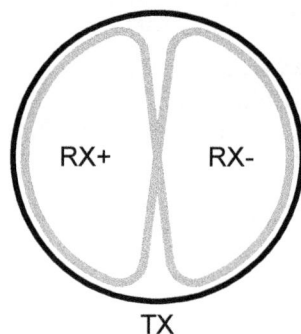

Fig. 8-57: **Butterfly Coil**

The advantages of the butterfly coil are

- Good EMI cancelation
- Natural ground cancelation
- Natural salt cancelation
- No lift-off effect
- First-order coil null

Like the figure-8 there is a center dead zone (Figure 7-24) but in this case the dead zone does not result in potentially missed targets. Both lobes of the RX coil are now being swept across the target so the target produces a positive response directly followed by a negative response with the null in between[40]. The result is that the normal target signal looks like a first derivative signal common in a standard VLF design[41] (Figure 8-58). Because of the unusual response of this coil it cannot be directly applied to a normal VLF detector. Rather, a detector has to be designed (or at least modified) to use this coil which is likely why it is such a rarity. See the *Explore* section of Chapter 18 for more information.

---

40. Or, when the coil is swept in the opposite direction, the response is negative-positive.

41. This is covered in a later chapter, but peak ahead to Figure 16-21 for a comparison.

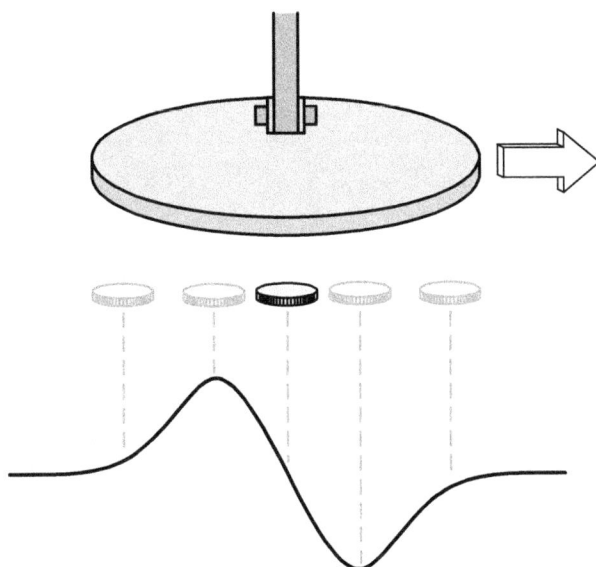

Fig. 8-58: **Butterfly Response**

8.14: X-Null vs R-Null

Nulling a coil involves some kind of mechanical adjustment to get the RX coil in a state of zero magnetic flux linkage. If magnetic flux is the only coupling method between the TX and RX coils and the coils are perfect inductors then any residual coupling to the RX coil will be exactly in-phase or exactly out-of-phase with the TX field. When you first start to null a coil and the RX signal is large it does indeed appear to be exactly in-phase (or exactly out-of-phase). But as you try to zero the null you will likely see it reach a non-zero minimum level, then the phase will roll 180°, and then it will begin to increase. That is, you can never get it completely zeroed. Figure 8-54 shows a concentric null exhibiting this behavior.

The reason for this is that the coupling is not purely in-phase. There will always be a very slight amount of out-of-phase coupling, mostly due capacitive coupling between the TX and RX coils. We call the purely in-phase portion of the null the *X-null* and the quadrature-phase portion the *R-null*. Chapter 16 will explore quadrature demodulation schemes and these terms will make more sense.

Capacitive coupling between the TX and RX coils or between the coils and the shielding are the chief suspects and is why the overlaps in a DD coil should be at 90° and preferably with a small spacer between them. It is also why the DC side of the RX coil should always face the DC side of the bucking coil, and also preferably with a small spacer between them. R-null problems are more prevalent in both concentric designs and in high frequency gold detectors but can be corrected for. With e.g. 30 AWG wire, make a small loop (1-inch|25mm in diameter) and solder the ends together. After minimizing the X-null

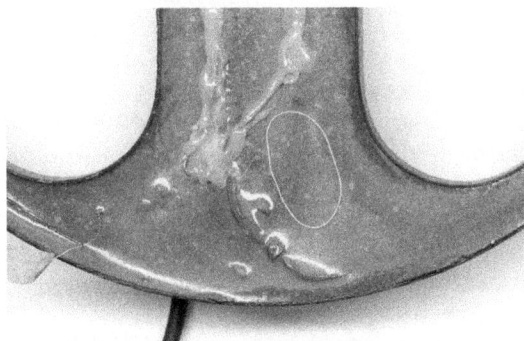

Fig. 8-59: **Shorted Loop Trick**

lay this loop in the coil (DD, concentric, etc., it doesn't matter) and move it around. You will find positions where the R-null is practically zero and the overall null can be (almost) completely zeroed.

8.15: Gold Coils

VLF gold detectors typically run at a high frequency; above 20kHz and (currently) as high as 80kHz. Besides using lower inductances, coils made for gold detectors require a little more care. As mentioned in the last section capacitive coupling between TX and RX will cause nulling difficulties. Care must be taken to minimize coupling and, if needed, a "tweak" R-null loop can be used to compensate. It is always better to avoid the R-null loop if possible as it can become a dynamic target over variably mineralized ground.

At high frequencies skin effect becomes a problem in the TX coil. This is where the coil current creates a magnetic field within the wire itself and that field pushes the current flow out to the surface of the wire. This increases the AC resistance of the wire (see Table 8-1) which imposes a higher voltage phase error. Litz wire is often used in high frequency coils to minimize this problem.

8.16: Multifrequency Coils

The problems with multifrequency coils largely depend on the frequencies used. The White's *V3*, for example, uses 2.5kHz, 7.5kHz, and 22.5kHz. Even 22.5kHz is not aggressively high. However, an imposed design requirement for the *V3* was to use coils that were compatible with the *DFX*. While the TX coil was a reasonable 540μH, the RX coil was ~15mH which was far too high for 22.5kHz. Getting the coil to work became an engineering challenge and a production headache. A lower inductance RX coil and an otherwise normal coil design would have been a much better path.

The Minelab *Equinox*, on the other hand, has a low frequency of 5kHz and a high of 40kHz. This is a more aggressive range and requires designing a coil with the same care as you would do for a gold machine. A coil that is well-nulled at a low frequency will often lose its null at a high frequency due to capacitive coupling and potentially require some "tweak circuitry," perhaps a compensation capacitor judiciously added somewhere or the addition of an R-null wire loop. Done right, the additional cap or loop can get the high frequency null corrected, hopefully without a lot of impact on the low frequency null.

In general, coils that tend to have a problematic R-null (like concentrics) are difficult to null well over a broad range of frequencies. This is likely why you rarely see concentrics made for multifrequency detectors.

8.17: Automated Coil Identification

It is often useful when a detector can automatically determine what coil is attached. For example, different coil sizes produce different depth-versus-signal strength scales, so knowing whether a 5-inch|12.5cm coil or a 12-inch|30cm coil is attached will allow you to present proper depth numbers. Another example is with the case of PI detectors, where a DD coil might be processed differently than a mono coil in order to provide TX-on processing for iron ID. Or a figure-8 coil (*Bigfoot*) which needs true 4-quadrant phase processing.

A popular way to implement coil ID is to place a small PCB in the coil that has a micro on it, whereby the detector communicates with the coil's micro to determine the ID of the coil. The coil micro can return a simple code that the detector micro must recognize, but this creates the possibility that new coils could be developed later that the detector would not recognize without a software update. A better solution is for the coil to return a block of data that describes key attributes of the coil. For example, whether the coil is a concentric, DD, or figure-8. The size of the coil might be useful, but even more so than that is depth scaling information.

Many commercial detector models that have a micro in the coil are also using the micro as a security feature so that aftermarket coils cannot be developed and used with that model detector. In this case, the detector sends the coil a piece of data, the coil micro encrypts the data and sends that back to the detector, which compares it to a locally encrypted data using the same algorithm. If the algorithms don't match, the comparison fails and the coil is illegitimate. This is used to restrict aftermarket coils for those who believe aftermarket coils should be restricted.

8.18: Alternate Shield Contacting

When using a clamshell housing (mated top and bottom shells) in which the inner surfaces are shield-painted there arises the issue of connecting the shield surfaces to the cable shield. In the prior DD design simple drain wires were used. In the 1980s and 90s when clamshell concentrics were the norm White's used a small conductive foam block to press against the bottom shell and contact the shielding. That is, contact was made as soon as the shell halves were pressed together.

Another option that makes contact as the shells are closed is to place some conductive fabric tape around the outer perimeter of the inner shell (the shell half that fits inside the other shell half) and the inner perimeter of the outer shell. It is not necessary to apply tape to the entire perimeter; short pieces at the four cardinal points should be plenty. Be sure the tape does not interfere with sealing. Figure 8-60 illustrates.

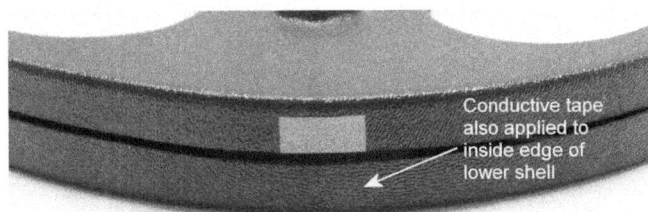

Conductive tape also applied to inside edge of lower shell

Fig. 8-60: **Alternate Shield Contact**

8.19: Conductors in the Coil

In building a coil it is necessary to solder wires together. In low-frequency detectors the additional metal from a solder blob is not a problem. But in gold detectors designed to find sub-grain gold nuggets it can be. We've previously noted that there should be no opportunity for any of the wires inside the coil to move around, otherwise falsing can occur. It would seem that as long as the solder connection is securely anchored the same will hold true.

A solder blob in the coil will generate its own eddy currents and as long as the blob doesn't move relative to the coil the eddy response never changes, and the blob simply produces what is likely a constant R-null error. However, when the coil is swept over a strong enough hot rock, the hot rock momentarily alters the TX field which alters the solder blob's eddy response. The result is that the hot rock can cause the solder blob to respond in a way that sounds exactly like a small nugget.

Therefore, in high-frequency coils the solder connections inside the coil should be kept as small as possible with as little excess solder as possible.

8.20: Aluminum Wire

A primary consideration in designing and building search coils is to make them as light as possible. Some companies make aluminum[42] magnet wire which can seemingly reduce weight. Since the conductance of aluminum is lower than copper, resistance-critical coils (usually the TX) will need to use a larger wire gauge which negates some of the advantage of aluminum's lighter weight. The conductivity of aluminum is 61% that of copper but its weight is 30%, so increasing the wire gauge to equalize resistance still results in a coil winding that is about half the weight of copper.

A major problem with aluminum wire is getting a reliable connection. Aluminum is notoriously difficult because it instantly forms a layer of aluminum oxide which makes soldering impossible. Connecting aluminum magnet wire to, say, the copper leads of the coil cable will probably require special interface connectors, which increases the dead-metal[43] content of the coil. For a number of years White's produced *XLT* coils which used aluminum magnet wire for the TX winding, but they eventually abandoned the practice due to long-term reliability problems.

---

42. "Aluminum" in the USA, "aluminium" just about everywhere else.

43. That is, undesired metal in the coil which could cause problems.

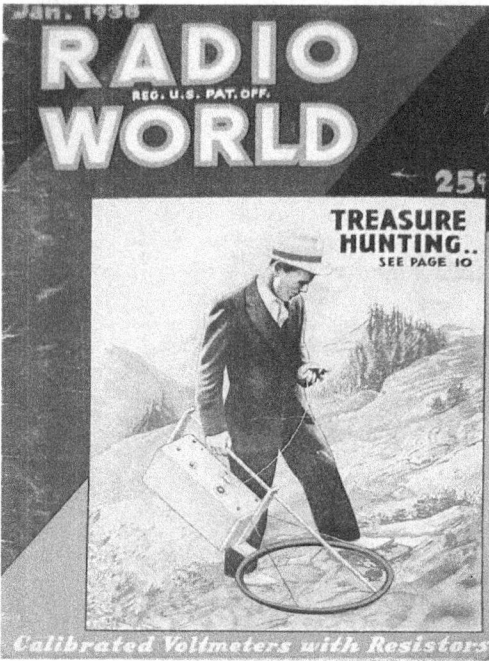

RADIO WORLD

Jan. 1938

REG. U.S. PAT. OFF.

25¢

TREASURE HUNTING..
SEE PAGE 10

Calibrated Voltmeters with Resistors

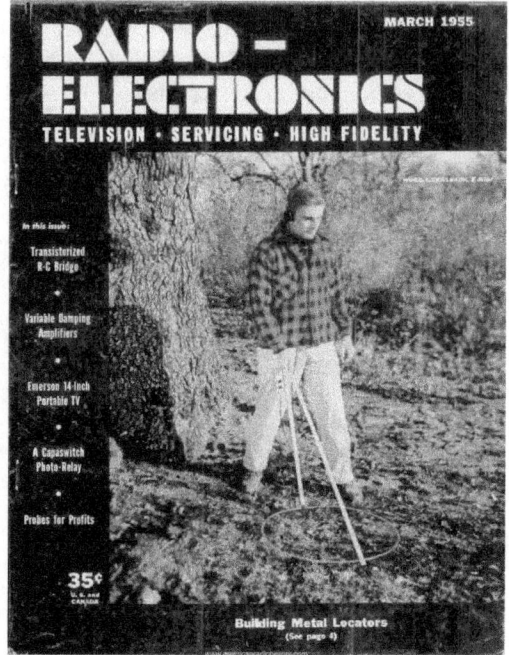

RADIO - ELECTRONICS

MARCH 1955

TELEVISION · SERVICING · HIGH FIDELITY

In this issue:

Transistorized
R-C Bridge
•
Variable Damping
Amplifiers
•
Emerson 14-inch
Portable TV
•
A Capaswitch
Photo-Relay
•
Probes for Profits

35¢
U. S. and
CANADA

Building Metal Locators
(See page 4)

DIRECTORY of SHORT-WAVE NEWSCASTS

POPULAR ELECTRONICS

SEPTEMBER
1962

35
CENTS

BUILD
Reactance Demonstrator
Tuning Fork Oscillator
Use Mobile CB for P.A.
Super-Strong Magnet

METAL LOCATOR
Ultra-Sensitive Design
Build for $25

(See p. 37)

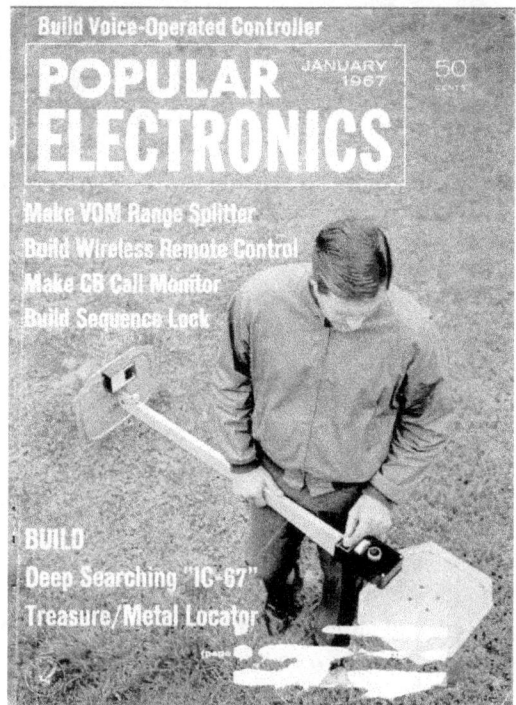

Build Voice-Operated Controller

POPULAR ELECTRONICS

JANUARY
1967

50
CENTS

Make VOM Range Splitter
Build Wireless Remote Control
Make CB Call Monitor
Build Sequence Lock

BUILD
Deep Searching "IC-67"
Treasure/Metal Locator

PART 3
Proximity Methods

Energy Theft

"Everything should be made as simple as possible, but not simpler."

— *Albert Einstein*

The proximity method was not the first metal detection technology nor is it a particularly good technology, and it's certainly way past its prime now. Why we begin these technological investigations with proximity designs has less to do with their status or utility and more to do with simplicity. The circuitry can be sparse, the coil is a mono winding, and the operational theory is relatively easy. These make it a good starting point for building an understanding of more complex and useful technologies.

9.1: Overview

When an air-core search coil is used in a free-running oscillator, the frequency and amplitude of the oscillation will be altered by either magnetic targets (such as ferrite mineralization) or eddy targets. A purely magnetic target such as ferrite increases coil core permeability which increases the coil inductance and decreases the frequency, but also raises the coil Q which increases the amplitude. An eddy target steals energy from the magnetic field which makes the coil inductance appear lower and increases the frequency, and this also de-Qs the coil[1] which decreases the amplitude. Certain iron/steel targets combine both magnetic and eddy effects and the oscillator response can go either way.

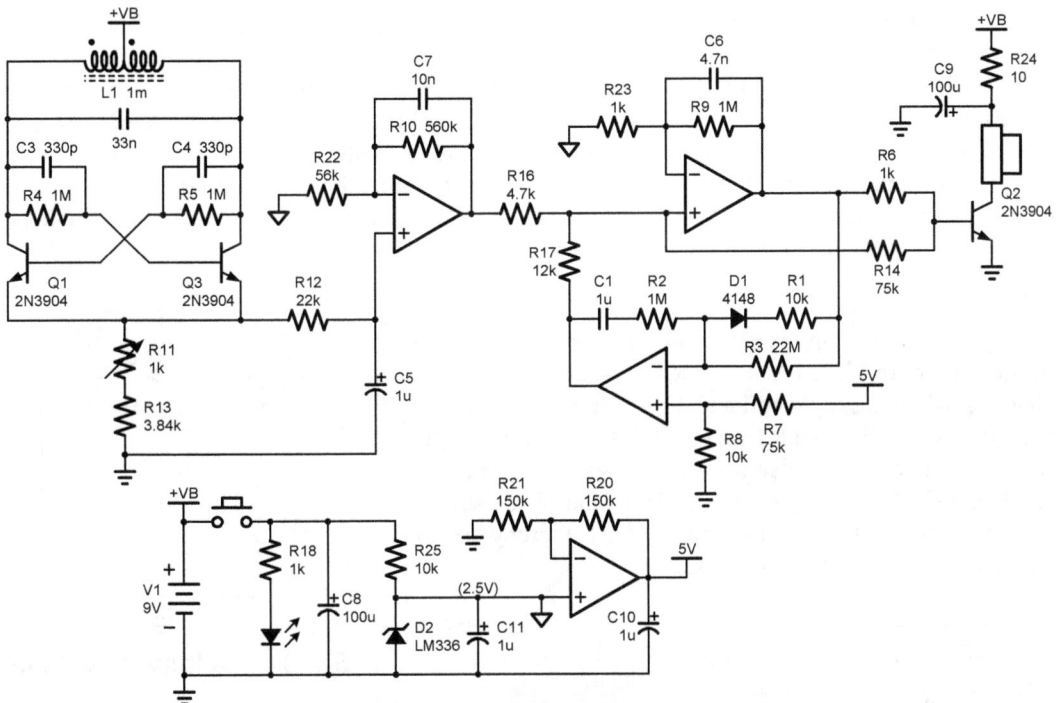

Fig. 9-1: **White's Bullseye Pinpointer**

---

1. An eddy target appears as an additional resistive load to the TX tank circuit.

The *Energy Theft* design focuses on the oscillator's amplitude change and is the simplest of the proximity methods. In the next chapter we will look at frequency change effects via the BFO design. This approach is called energy-theft as that is what happens with eddy targets — they steal energy from the oscillator[2]. An older name for it is *loaded oscillator* or *loaded-loop oscillator*, because the tank circuit is loaded by the target. The proximity effect not only affects the amplitude of the oscillation but also the loading on the transmitter power supply. The energy-theft design is used in modern pinpointers like the Garrett *Pro-Pointer* and the original versions of the White's *Bullseye* and Minelab *Pro-Find*.

9.2: Block Diagram

A block diagram for the energy theft design is shown in Figure 9-2. The oscillator drives a peak detector which monitors the amplitude of oscillation. If the amplitude drops due to eddy targets, a comparator triggers at a set threshold level and turns on an audio response. A rising amplitude caused by a purely magnetic target is typically ignored, otherwise ground mineralization would be detected. Iron targets have both magnetic and eddy responses, so targets which have a dominant magnetic response will be ignored as well.

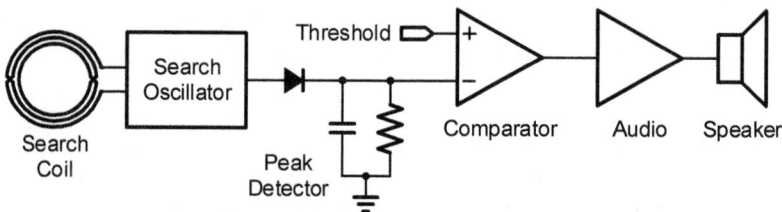

Fig. 9-2: **Loaded Oscillator Block Diagram**

The only requirement for the search oscillator is that it be free-running with its frequency set by the search coil inductance; that is, an LC-tank oscillator. The frequency is not critical so we can choose whatever we want. The response circuit can be as simple as an LED or a DC-driven buzzer or vibration motor. Besides that, the only critical part of the design is dealing with temperature issues in the peak detector to avoid drift.

The block diagram shows the peak detector as a simple diode rectifier with a low-pass filter. This works just fine, with the caveat that the forward voltage drop of the diode has a typical temperature drift of around −2mV/°C. If the comparator threshold is a simple DC voltage — for example, taken from a potentiometer — then temperature changes can cause erratic operation. Two possible solutions are to create a similar temperature coefficient in the threshold voltage, or to design a temperature-stable peak detector. We'll investigate both solutions.

9.3: Oscillator

Many detector designs over the years have used the Colpitts oscillator[3] (Figure 9-3). It's a single-transistor free-running oscillator in which a simple coil and 2 capacitors set the oscillation frequency. R1, R2, and R3 set up a DC bias current in Q1 in the typical manner. The collector of Q1 drives the LC tank circuit. C1 and C2 form a capacitive divider (and phase shift) to provide positive feedback to the emitter of Q1, thereby creating oscillation. The frequency of oscillation is

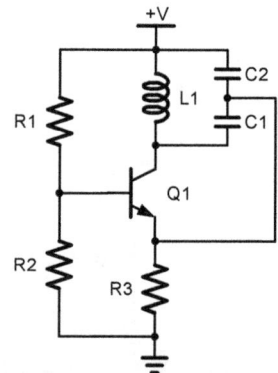

Fig. 9-3: **Colpitts Oscillator**

$$f = \frac{1}{2\pi\sqrt{LC_T}} \qquad \text{Eq 9-1}$$

2. Recall that eddy targets behave like a transformer secondary. This behavior actually happens with all induction metal detector designs but we rarely look at this effect in them.

3. Invented in 1918 by Edwin Colpitts, although he used a vacuum tube.

Inside the Metal Detector

where

$$C_T = \frac{C1 \cdot C2}{C1 + C2}$$ Eq 9-2

That is, the inductance L1 in parallel with the series combination of C1 and C2 forms the resonant circuit.

In the previous editions of this book, the Colpitts design used was slightly different as shown in Figure 9-4. Theoretically it is identical to that of Figure 9-3; in one case C2 is connected to +V and in the other it is connected to ground. Assuming there is zero resistance through whatever battery or power supply is powering the circuit then the two methods are equivalent. But in reality the power supply circuit has resistance and probably capacitance and inductance as well so the two versions may oscillate at slightly different frequencies even with the exact same component values.

For this design we will arbitrarily choose an oscillation frequency of 12kHz[4] and a coil inductance of 1mH. This means the combined capacitance C_T is

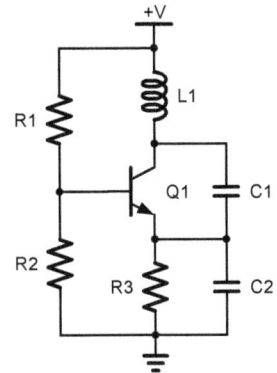

Fig. 9-4: **The Other Colpitts**

$$C_T = \frac{1}{\omega^2 L} = \frac{1}{(2\pi \cdot 12\text{kHz})^2 \cdot 1\text{mH}} = 175.9\text{nF}$$

Normally in a Colpitts oscillator the caps are ratioed unequally, with C2 up to 10 times the value of C1. Choosing C1 = 1μF and C2 = 220nF (both standard values) yields a C_T of 180.3nF. If we manage to wind a search coil of exactly 1mH the search frequency is now

$$f_S = \frac{1}{2\pi\sqrt{1\text{mH} \cdot 180.3\text{nF}}} = 11.85\text{kHz}$$

In engineering, we call that "close enough." Again, the oscillation frequency in this topology is not critical so this will work just fine.

The base resistors (R1 and R2) are chosen to set the base bias voltage at about 1/3 - 1/4 of +V. The emitter resistor (R3) sets the coil's bias current which, along with resistive losses in the tank circuit, determines the peak-to-peak current swing of the coil. We want the current swing to be as high as possible for achieving a maximum B-field but the peak current swing will be limited by the voltage swing of the coil ($v_L = L \cdot di_L/dt$). The maximum voltage swing is approximately twice the supply voltage; that is, the DC value at the collector of Q1 is +V, and that voltage can swing down to ground and up to $2 \times +V$. This means that the peak coil current will be

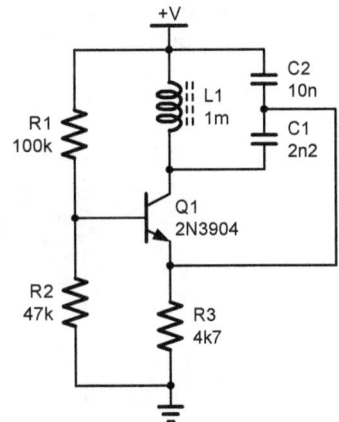

$$I_p = \frac{1}{L}\int V_p \cos(\omega t)\, dt = \frac{V_p}{\omega L}$$ Eq 9-3

Supposing +V = 5V, then I_p = 67mA. Most of the coil current is recycled in the LC tank circuit so that the transistor circuit does not have to provide it. That is to say, the bias current for Q1 will be much lower than the peak coil current and therefore R3 is best arrived at empirically. The final oscillator design is shown in Figure 9-5.

Fig. 9-5: **Oscillator Circuit**

---

4. The same as the Garrett ProPointer.

9.4: Search Coil

Notice that the inductor in Figure 9-5 has double-dashed lines next to it; this indicates the coil is a ferrite-core inductor instead of an air-core inductor. This design is intended to be a pinpointer so the search coil will be a narrow cylindrical winding on a ferrite rod. Using a ferrite-core coil reduces the number of turns (by increasing the core permeability) and better focuses the magnetic field. As an added benefit, the ferrite core also reduces the effect of ground mineralization because the coil already has a high-permeability core.

The ferrite rod is specified to be a Fair-Rite 4078377511 with a diameter of 9.5mm and a length of 50mm. This is just long enough to wind 150 turns of 27 AWG wire in a single layer which yields an inductance of 1mH. A smaller gauge wire will also work, but a larger gauge wire would either require two layers of winding or changing the design for less inductance. Other ferrite rods will also work, though they will require changes to the number of turns to achieve 1mH or, again, wind the coil to a different inductance and adjust the capacitors to whatever frequency is desired. One of the advantages of this design is that almost anything will work.

Figure 9-6 shows the coil diagram and a finished coil. 1/4" (6mm) tape is wrapped around the ends of the winding to hold it in place and a coating of clear cement is applied over the rest of the winding. Other options are to wrap the finished coil completely with tape or dip it in epoxy. Chapter 8 emphasized the need to shield coils but with this coil we can get away without shielding it. The single-layer solenoidal winding has a very deterministic interwinding capacitance that is fairly stable to external influences.

Fig. 9-6: **Search Coil**

9.5: Peak Detector

Figure 9-7 shows the oscillator with a peak detector circuit added to it. Transistor Q2 is used as an emitter-follower which effectively makes it function as a diode, but during conduction its β prevents the Colpitts circuit (Q1 etc.) from being unnecessarily loaded. R4 and C3 form the peak detector's low-pass filter. The peak detector waveform is basically a DC voltage with a very small amount of ripple.

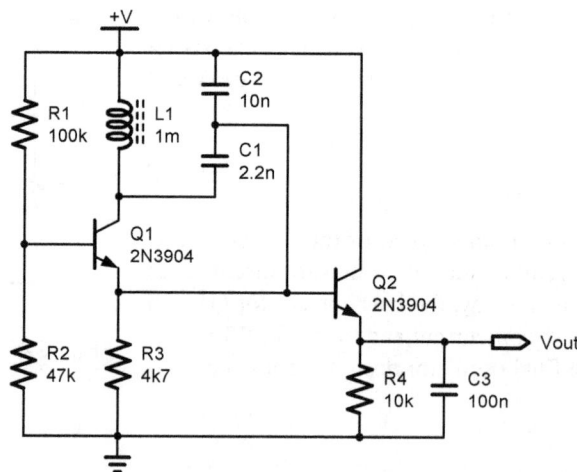

Fig. 9-7: **Oscillator with Peak Detector**

9.6: Threshold

Figure 9-2 shows that the peak detector output is compared to a threshold voltage. The threshold voltage can be produced by a simple potentiometer but for lower temperature drift the threshold circuit should match the peak detector circuit. In Figure 9-8, R5-RV1-R6 establish a variable reference voltage and Q3 is another emitter-follower whose temperature characteristics match that of Q2. R7 biases Q3 and C4 provides some noise filtering.

The nominal voltage level at the peak detector output is about 4V so R5 and R6 should be chosen so the threshold can be adjusted above and below this level. A good range is ±10% to account for residual temperature drift. The threshold level will also vary with battery voltage but so will the peak detector output, though they will not necessarily track.

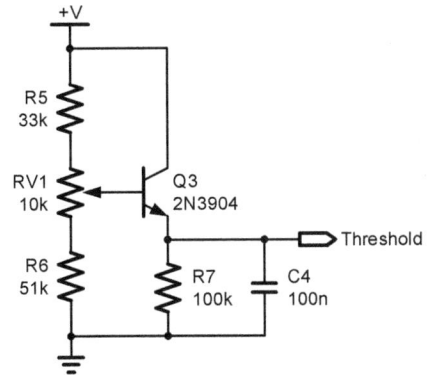

Fig. 9-8: **Threshold Circuit**

Design: Energy-Theft Pinpointer

At this point we have all the elements needed for a final design. The subcircuits as discussed so far are exactly replicated in the complete schematic shown in Figure 9-9. The only new addition is the gain stage which amplifies the difference between the peak detector and threshold circuit outputs by a factor of 1000. When the peak detector output drops below the reference level due to target loading, the output of IC1 drops. This turns on Q4 and drives a response element.

There is no audio oscillator circuit in this design so the response element cannot be an ordinary passive speaker; it must be a self-oscillating buzzer or beeper. Other options are an LED or a vibration motor. The waveform at the emitter of Q2 should be a DC voltage with a slight amount of ripple. In the absence of a target, adjust RV1 until the beeper just begins to sound off, then back off on RV1 for silence. This should be the point of peak sensitivity.

The energy theft design is fairly forgiving of component selection. As mentioned before, if the coil or capacitor values are different the oscillator will probably still work but perhaps at a different frequency or amplitude. Practically any general purpose NPN transistors can be used. Enough component variation, intentional or unintentional, might alter the nominal peak detector voltage. If you build the circuit as shown and it doesn't work, you may need to tweak a resistor or two, especially R5 and/or R6. This can be true for circuits throughout this book, and we will try to point out these "opportunities" as they arise.

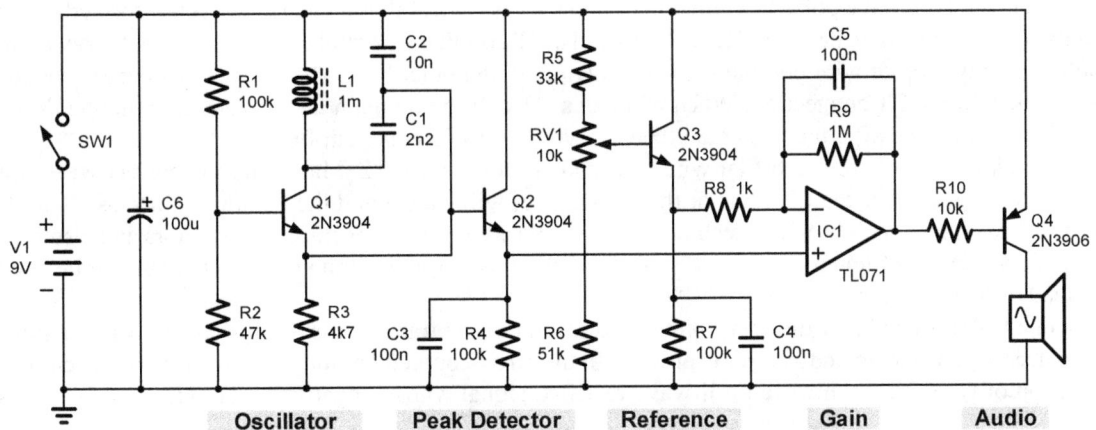

Fig. 9-9: **Energy-Theft Pinpointer Schematic**

Design: VCO Pinpointer

Instead of using a comparator to determine when the amplitude has dropped below a threshold, the peak detector could instead directly drive a voltage-controlled oscillator (VCO) circuit as shown in Figure 9-10. For a nominal peak level the VCO will produce a certain tone; eddy targets can cause the tone to rise and magnetic targets could cause the tone to fall. This means that ground mineralization will affect the tone which may normally be bothersome, but there may be applications where this behavior is desired. This approach makes having a temperature-stable peak detector critical.

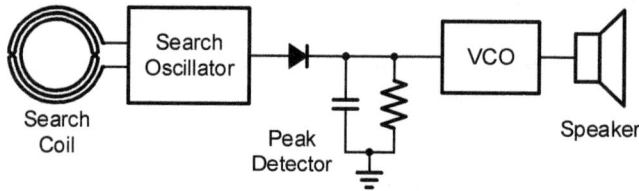

Fig. 9-10: **VCO Version**

This design replaces the comparator and self-beeper with a tone generator for the audio and a true speaker which can produce a rising-pitch response as metal gets closer. This offers the user a perception of distance. Besides adding a VCO, we've also switched gears a little and replaced the Colpitts oscillator with a differential cross-coupled oscillator (Figure 9-11). The purpose of this book is to teach you metal detector circuits and so we'll occasionally toss in something new, whether it's needed or not. If you are a fan of Mr. Colpitts, his oscillator could still be used to drive the new VCO circuit.

The cross-coupled oscillator comes with a new twist: it uses a center-tapped (CT) coil. While CT coils are not all that common in metal detectors they do show up from time-to-time, mostly when a TX coil is driven differentially as done here[5]. The two halves of the inductor are magnetically coupled and behave as a single inductor. If each half-inductor has an inductance of L

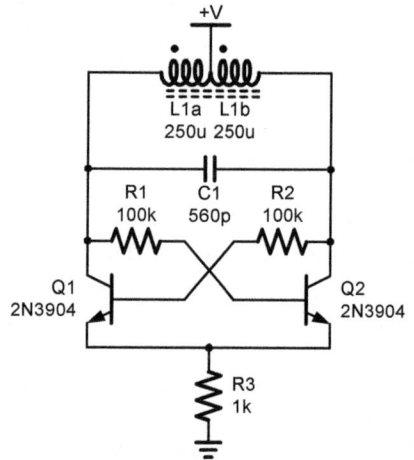

Fig. 9-11: **Cross-Coupled Oscillator**

then the combined inductor is 4L and therefore the frequency of oscillation is $f = 1/(2\pi\sqrt{4LC})$.

The purpose of the CT connection is to provide a power supply path for the coil current. Magnetically coupled coils have polarity in their coupling so it is critical that they are connected correctly. The "dots" on the coils indicate the polarity. You wind a CT coil like a normal coil, but at the half-way point pull a lengthy loop of wire out and then finish winding the other half. The loop can be cut afterwards and tinned for the CT connection. When winding a CT coil, if you maintain a consistent winding direction both before and after the center tap then everything should come out fine.

Another option is to use bifilar wire as is shown in Figure 9-12. Make sure the correct wires are connected together for the CT node or the inductances will cancel and the circuit will not oscillate. If you simulate this circuit in Spice keep in mind the polarity when connecting the inductors and also provide a coupling coefficient of close to 1. For simulations, a regular transformer element can be used instead of two inductors.

In a Colpitts oscillator the waveform at the emitter of the transistor is a full sinusoid whose amplitude decreases when an eddy target is present. In the cross-coupled circuit, the waveform at the emitter junction of Q1/Q2 is an inverted full-wave rectified signal whose negative amplitude also decreases

---

5. The White's Bullseye in Figure 9-1 is an example. The Teknetics T2 is another.

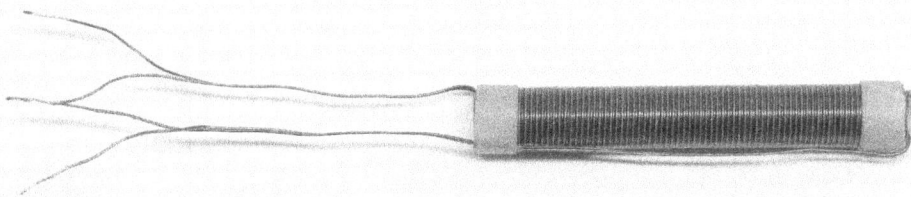
Fig. 9-12: **Center-tapped Coil**

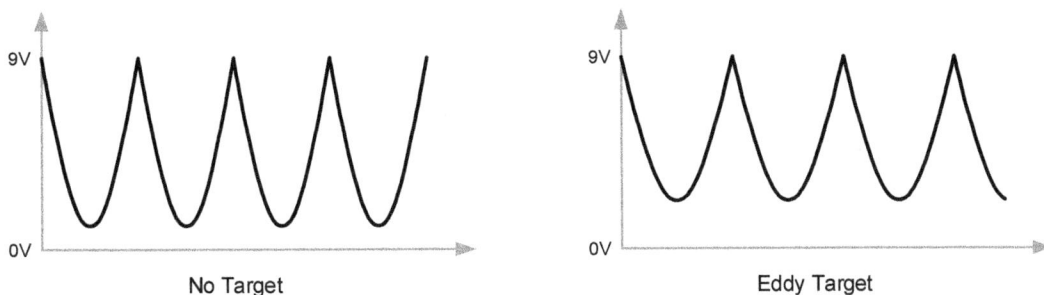

No Target

Eddy Target

Fig. 9-13: **Waveform at R3**

with eddy targets[6] (Figure 9-13). It's possible to bias the oscillator so strongly that the negative amplitude clips on the ground rail and R3 should be chosen to prevent this.

The final schematic in Figure 9-14. If we simply low-pass filtered the voltage at R3 it would be a DC voltage which rises in the presence of an eddy target. However, it would also have a fairly strong negative temperature coefficient. To counter this, Q3 buffers the voltage and cancels some of the temperature drift and Q4 cancels the remainder. Q4 is called a "V_{BE} multiplier" because its collector-emitter voltage is

$$V_{CE} = V_{BE}\left(1 + \frac{R5}{R6}\right)$$

Normally this circuit is used to create a low-impedance voltage offset. In this case, the usual −2mV/°C temperature coefficient of V_{BE} is being multiplied into a higher TC; the schematic shows R5 = R6 so we should expect −4mV/°C. In my particular circuit this worked well. If you find it does not then adjust R5 to zero out the TC.

The voltage at R4,C2 is a reasonably stable DC voltage that rises in the presence of a non-ferrous target. This voltage is applied to an inverting amplifier whose output drives a common-base transistor. The non-inverting input is has a variable voltage applied (via RV1) which creates a threshold control. The output transistor Q5 converts the opamp voltage into a current, and the end result is that the collector current of Q5 is proportional to oscillator loading and this current is used to control the VCO.

The ubiquitous 555 timer[7] makes a dandy VCO. The collector current of Q5 charges C5 and generates a linear ramp. When the linear ramp exceeds the threshold of the 555 timer it triggers the internal reset transistor which briefly shorts C5 to ground (via R12) and the ramp starts over. The effect of this is a sawtooth waveform on C5 whose frequency is proportional to the current provided by Q5. The 555 therefore drives the speaker with a frequency proportional to the detection response. The 555 needs a little help to drive a speaker so transistor Q6 provides a current boost.

---

6. Frequency also decreases, but we don't pay attention to that.

7. We'll see a lot of this chip throughout the book.

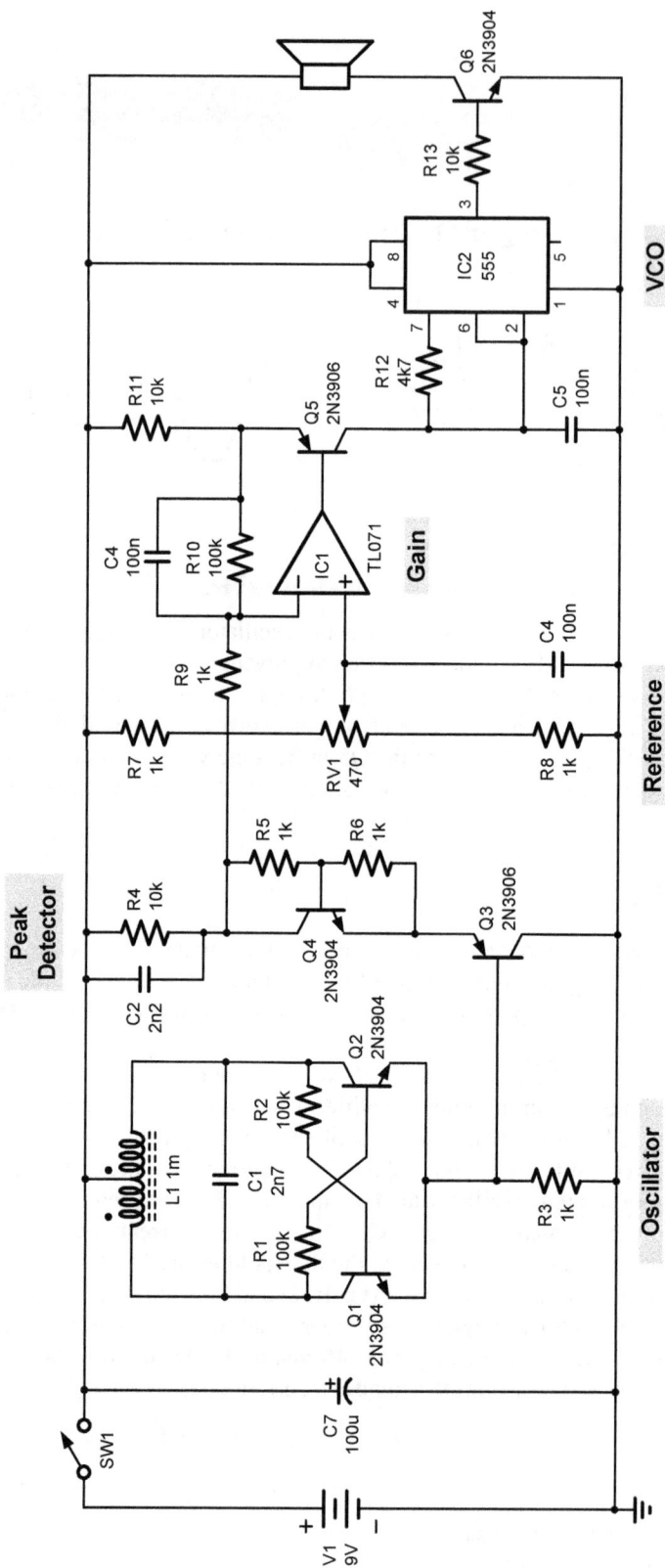

Fig. 9-14: VCO Pinpointer Schematic

Explore

9.7: Improving the Tempco

In the basic energy-theft circuit (Figure 9-9), we stated the purpose of Q3 was to compensate for the temperature drift of Q2, which it does. But there is also Q1 to consider. A transistor V_{BE} typically has a temperature coefficient (tempco) of around $-2mV/°C$ but the overall effect of Q1 is somewhat less than this as seen at the voltage on C3. So the overall circuit will still have a temperature drift, but not nearly as bad as it would be without Q3.

We can add in some additional compensation for Q1's tempco. If the effect of Q1 was 100% at C3, then the obvious solution would be to add a diode-connected transistor in the bias of Q1 as shown in Figure 9-15A. But this over-compensates, so we can partially bypass this transistor with a resistor (Figure 9-15B). Transistor tempco can vary somewhat so this is a case where some trial-and-error is needed, perhaps aided with Spice simulations. A modification of the circuit in Figure 9-9 with a matching transistor and a $15k\Omega$ bypass resistor should do the trick.

In practice a pinpointer is not normally turned on for long lengths of time so temperature stability is not an especially big concern. But if you use this circuit for other purposes it may be, or if you use a Colpitts oscillator in other designs such as a BFO.

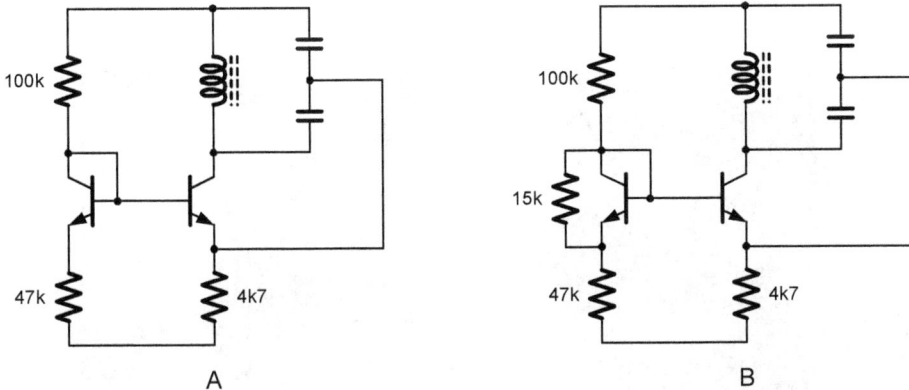

Fig. 9-15: **Extra Temperature Compensation**

9.8: Digital Design

Microcontrollers[8] are the hot item in hobby electronics so we'll include them in our discussions and even present a few micro-based designs to keep the code jocks happy. The Energy Theft method is ideal for converting to a digital approach (Figure 9-16). An analog-to-digital converter (ADC) is used to sample the peak detector and a microcontroller uses the digitized values to determine an audio response. In this way, the user could select either a threshold-based response or a continuous VCO-style response, with or without the effects of magnetic targets.

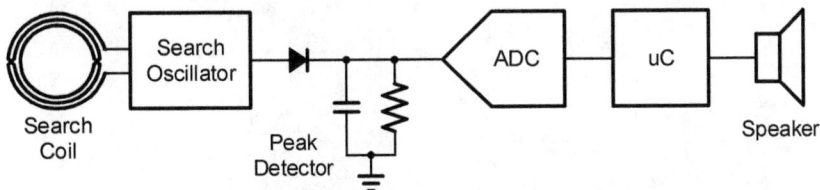

Fig. 9-16: **Digital Version**

---

8. Commonly called a *micro*, or abbreviated uC or μC. A microcontroller differs from microprocessor (uP, also called a *micro*) in that it has on-board ROM and RAM and a host of peripherals like timers, data converters, and display drivers. That is, the uC is more self-contained than the uP, but usually a uP has more raw horsepower.

It is still important to pay attention to the tempco of the peak detector voltage and it may be necessary to sample a dummy reference circuit voltage (the same reference as in Figure 9-9) and subtract it in order to minimize temperature effects. Another option is to add a PNP follower to the emitter of Q2.

Most microcontrollers have an internal ADC with a typical resolution of 10-12 bits which may limit sensitivity. Oversampling and averaging the data can improve the SNR and effectively add more bits of resolution. In general, every factor of 4 adds one bit of resolution. The trade-off is a slower response latency. This issue and more are covered in Chapter 18.

The BFO

"A child of five would understand this. Send someone to fetch a child of five."

— *Groucho Marx*

The beat frequency oscillator (BFO) was the dominant technology in the earliest days of transistorized hobby detectors. BFOs were so easy to design and so cheap to build that new detector companies were constantly appearing and disappearing. Unlike the Energy Theft method shown in the last chapter, literally no one still manufactures BFO models but it is still a useful learning step in understanding metal detector technology.

10.1: Overview

When an air-core search coil is used in a free-running oscillator, the frequency of oscillation will be altered by either magnetic targets (such as ferrite mineralization) or eddy targets. Purely magnetic targets such as ferrite increase coil core permeability which decreases the frequency; eddy targets steal energy from the magnetic field and increase the frequency. Iron targets combine both magnetic and eddy effects and the oscillator response can go either way.

The basic BFO employs two radio frequency[1] (RF) oscillators which are tuned to very nearly the same frequency. The frequency of each oscillator is usually determined by an LC resonant circuit; that is, an inductor and a capacitor. One is called the *search oscillator* and uses the search coil as its inductor; the other is called the *reference oscillator* and uses an conventional inductor. Normally the reference oscillator can be manually adjusted (tuned) to account for component variations and to give the user control over the resulting operation.

The outputs of the two oscillators are fed into a mixer which produces a signal that contains the sum and difference frequency components (plus perhaps other artifacts) of the two input signals. As we will see shortly, the difference frequency is what we are interested in and a low-pass filter at the output of the mixer will remove the other components. Figure 10-1 shows a block diagram of basic BFO design.

As long as the two oscillators are exactly the same frequency the mixer-filter output will have no difference signal, just DC, which produces no audio signal. If the frequency of the search oscillator shifts slightly, then a frequency difference signal will appear at the mixer-filter output. The frequencies

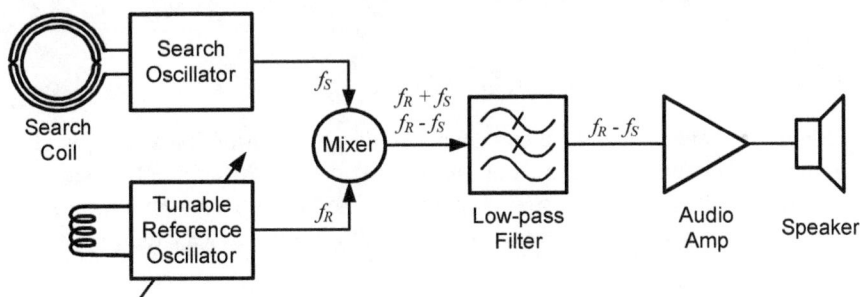

Fig. 10-1: **BFO Block Diagram**

---

1. In this case, RF simply means somewhere above audio frequency.

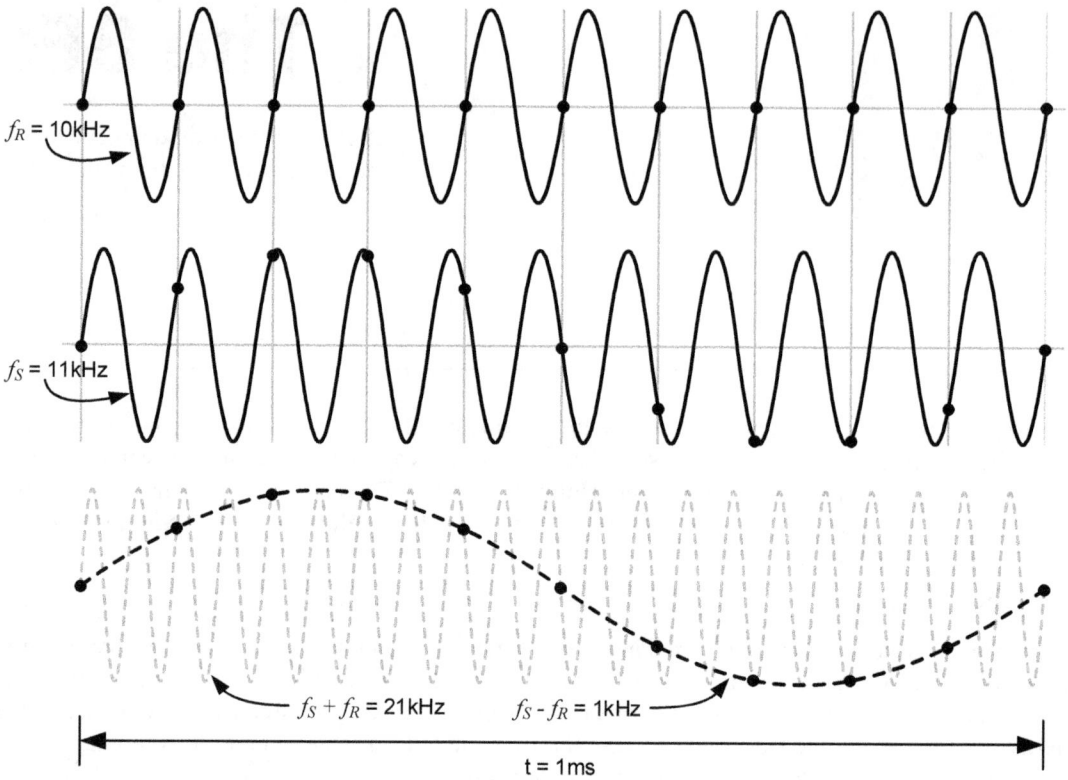

Fig. 10-2: **Mixer signals**

of the two oscillators are normally chosen such that a shift in the search oscillator frequency will produce a mixer output signal that is in the audio frequency range. For example, if the search oscillator is set to 10kHz and the reference oscillator is at 11kHz, then the difference will be a 1kHz signal which can directly drive a speaker. The sum frequency will be 21kHz which is of no use, so the low-pass filter removes this component. In Figure 10-2 these example mixer signals graphically show that the zero-crossing of the reference oscillator signal (f_R) can be mapped to points on the search oscillator signal (f_S), and these points form the low-frequency beat (difference) signal. It so happens that a 21kHz sine wave (shown in gray) also exactly passes through the same points and it is the summation frequency created by the mixer. Our interest is strictly in the difference signal and henceforth we will consider the mixer signal as being only the low frequency component.

The math behind this process is simple and worth a quick look. An ideal mixer performs a multiplication function and if the inputs are sinusoids then

$$v_m = A_R \sin(2\pi f_R) \times A_S \sin(2\pi f_S)$$

Eq 10-1

$$= \frac{A_R A_S}{2}[\cos(2\pi(f_R - f_S)) - \cos(2\pi(f_R + f_S))]$$

where A_R, A_S are the reference and search signal amplitudes. So multiplying sinusoids of two different frequencies produces two new (co)sinusoids at the summation and difference frequencies. Mixers are not ideal, and additional distortion terms (including intermodulation components) usually show up. Most of these will be removed by the low-pass filter and the BFO is largely tolerant of any remaining spurious content.

10.2: Frequency Considerations

There are two frequencies of concern with the BFO: that of the search oscillator and that of the mixer output (the audio frequency). With metal detectors in general, the two primary factors that determine a good search coil frequency are, unfortunately, opposing. Lower frequencies penetrate the soil better than high frequencies which is why modern IB detectors typically use frequencies in the VLF (3-30kHz) range. But higher frequencies generate a better eddy current response, especially in low conductors.

The BFO typically runs at a higher transmit frequency than other detector types because doing so produces a better "difference-frequency" (Δf) sensitivity at the mixer output. Table 1 shows the inductances and operating frequencies of several commercial models. The target effect on the search coil is fairly proportional to frequency so if a target produces Δf = 1Hz at f = 10kHz then it might produce Δf = 10Hz at f = 100kHz. A larger absolute Δf at the mixer output will be easier to hear.

Model	Coil Inductance	Frequency
Garrett Sidewinder	134µH	285kHz
Garrett Master Hunter	143µH	300kHz
D-Tex Deluxe	155µH	360kHz
White's Little Monster	740µH (CT)	100kHz
Bounty Hunter III	39µH	1000kHz
Jetco Treasure Hawk	100µH	500kHz
Relco Pacesetter	309µH	361kHz

Table 10-1: **Commercially used frequencies**

BFO circuits are prone to *frequency locking*. The BFO has two free-running oscillators at or about the same frequency and usually in close proximity on the circuit board. It is common for the oscillators to "pull" each other's frequency through parasitic coupling until they are exactly the same. When the oscillators are locked it takes a stronger target signal to kick them out of lock and get a frequency difference at the mixer.

To minimize the effect of frequency locking the BFO is normally tuned so the oscillators are at very slightly different frequencies and, instead of silence, the audio produces a threshold tone. For example, if the search oscillator runs at 100kHz then the reference oscillator might be tuned for 100.05kHz which produces an audio tone of 50Hz. A target that causes a 10Hz shift would then be heard as either a 40Hz or a 60Hz audio signal, depending on whether it's ferrous or non-ferrous. This will be more sensitive to targets than if the oscillators are set to the exact same frequency.

Besides the absolute frequency shift, we are also interested in the *relative* shift. Let's say that a certain non-ferrous target creates a 0.01% increase in the oscillator frequency while a certain ferrous target creates a 0.01% decrease. Expanding on the prior example, let's add back in the 10kHz oscillator case plus a third possibility:

	BFO1	BFO2	BFO3
Reference Osc	10.05kHz	100.05 kHz	100.005 kHz
Search Osc	10.00kHz	100.00 kHz	100.000 kHz
Nominal mixer freq	50 Hz	50 Hz	5 Hz
Non-ferrous freq	51 Hz	60 Hz	15 Hz
Ferrous freq	49 Hz	40 Hz	-5 Hz
Absolute shift	±1 Hz	±10 Hz	±10 Hz
Relative shift	±2%	±20%	±200%

Table 10-2: **Relative Frequency Shifts**

A 10kHz example (BFO1) that is tuned to a 50Hz threshold results in an absolute frequency shift of only ±1Hz which is a relative shift of ±2%. The 100kHz example (BFO2), also tuned to a 50Hz threshold, has an absolute shift of ±10Hz and a relative shift of ±20%, both of which are 10 times better.

BFO3 has the same base oscillator frequency as BFO2 — 100kHz — but is tuned to a 5Hz threshold instead of 50Hz. This produces the same absolute frequency shift for the non-ferrous target — 10Hz — but the relative shift is now 200%, another factor of 10 improvement. But an odd thing happens with the ferrous target: the 10Hz decrease makes the frequency go *negative*. What is a "negative frequency?" It's the same as a positive frequency, but with a 180° phase shift. That is, the resulting ferrous target frequency will be 5Hz, which is the same as the nominal (no-target) frequency.

At first glance this seems to suggest that the ferrous target won't be detected but the temporal response that occurs as the coil is swept over the target will cause the audio frequency to pass through 0Hz. This means the audio response will start at 5Hz (off-target), decrease in frequency to a null (no response), and then increase to 5Hz again as the coil is centered over the target. As the coil continues its motion, the audio will again decrease and pass through a null before settling back to the nominal 5Hz. This odd behavior is called *frequency rollover* and is similar to frequency aliasing in ADCs. Figure 10-3 illustrates the response for a non-ferrous target without rollover and ferrous target with rollover.

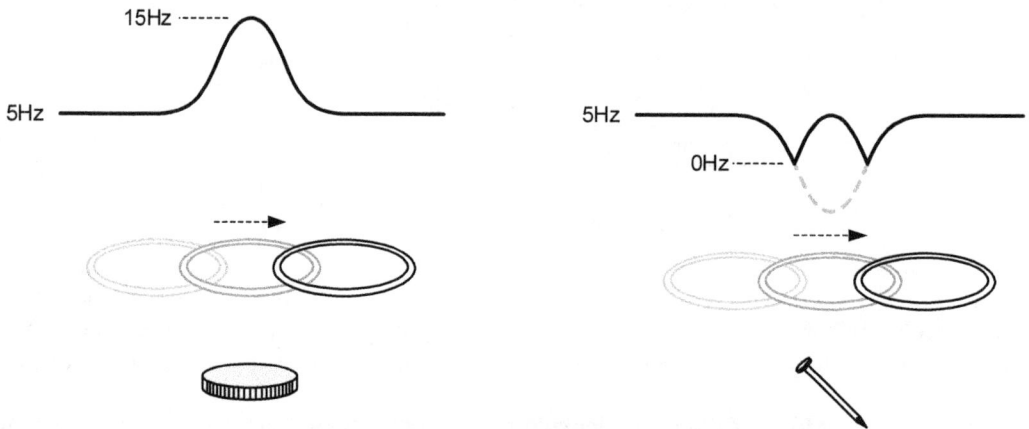

Fig. 10-3: **Sweep Response Without and With Rollover**

An obvious problem with BFO3 is that human hearing doesn't extend down to 5Hz. That is true, yet virtually all BFOs produce audio all the way down to zero Hertz (the null). They do so by using the sinusoidal mixer output to overdrive a gain stage, producing a pulsed audio response which is rich in harmonic energy. Thus a very low frequency will sound like a putt...putt...putt — in BFO-land this is called *motor-boating* and allows us to achieve much larger relative response changes by running the idle frequency closer to zero.

This turns out to be an advantage because human hearing is more sensitive not only to larger frequency (pitch) changes, but also to larger relative changes. Selecting a higher nominal oscillator frequency enables a larger *absolute* frequency change, and tuning the reference oscillator so that the mixer output is closer to null produces a larger *relative* frequency change.

It is, however, important not to tune it too close to DC, for two reasons. First, if the oscillators are tuned too closely to one another then frequency locking is likely. When this happens, it takes a stronger target response to kick them out of lock and the detector loses sensitivity. Secondly, if the nominal audio frequency is too low then very slight changes become imperceptible. For example, a 10% shift from 10Hz to 11Hz is perceptible, but from 1Hz to 1.1Hz is not. A good idle frequency is 5-20Hz.

Figure 10-4 summarizes the audio responses produced by the BFO. There is a null response when the oscillators are running at identical frequencies. Ferrous targets cause the beat frequency to move in one direction while non-ferrous targets move it in the opposite direction, but in both cases the audio increases in frequency. If you have the BFO tuned for a slight idle frequency on the non-ferrous side of null then non-ferrous targets will always increase the frequency while a ferrous target will initially decrease the frequency until it hits the null point and rolls over. If you set the idle tune on the ferrous side of null then ferrous targets will always increase the frequency while a non-ferrous target will ini-

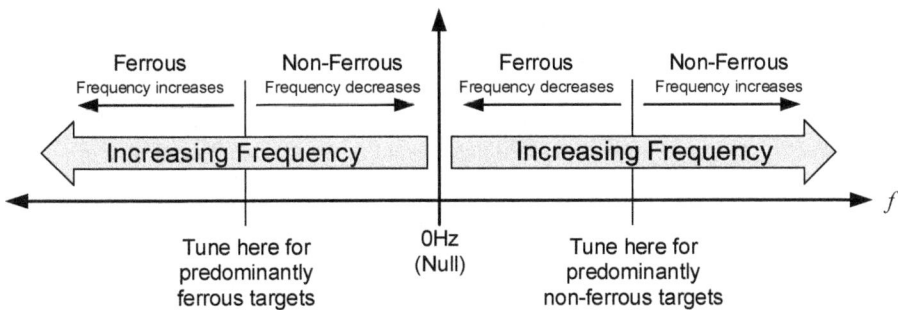

Fig. 10-4: **Summary of Audio Response**

tially decrease the frequency until it hits the null point and rolls over. In this way, the BFO can be used as either a predominantly non-ferrous detector or a predominantly ferrous detector, depending on which side of the null the audio is tuned to.

It should be noted that what a BFO sees as ferrous depends on the frequency. With most BFOs a nail will respond as a ferrous target, but the *Bounty Hunter III*[2] runs at 1MHz (the highest frequency BFO I'm aware of) and sees nails as non-ferrous. That is, the effect of the nail's eddy response (which increases with frequency) overwhelms its magnetic effect (which decreases with frequency). With larger iron targets this can happen at lower frequencies. So while a higher operating frequency invokes a larger absolute frequency change there are trade-offs.

Finally, an object's proximity (depth), size, and type of metal determines how much inductance shift the search coil experiences. Therefore, except for a general ferrous/non-ferrous indication (which may not even be accurate for ferrous) there is no way to determine target type as is often done in discriminating designs. That is, a shallow nickel may give the exact same response as a deep silver dollar.

10.3: Component Sensitivities

The response of the BFO to targets and parasitics depends on sensitivity. Sensitivity is basically what it takes to get the search oscillator frequency to shift by a certain amount. From the preceding discussions we can see that it is desirable to have the oscillator very sensitive to a change in inductance but insensitive to changes in capacitance[3]. The basic L-C oscillator has a frequency that is

$$f_0 = \frac{1}{2\pi\sqrt{LC}}$$
Eq 10-2

where C is the total capacitance, including parasitics. The frequency sensitivity due to *small* changes in L will be

$$\frac{\Delta f}{f_0} \cong \frac{1}{2} \cdot \frac{\Delta L}{L}$$
Eq 10-3

Likewise the frequency sensitivity due to *small* changes in C will be

$$\frac{\Delta f}{f_0} \cong \frac{1}{2} \cdot \frac{\Delta C}{C}$$
Eq 10-4

Targets create the ΔL and in most cases invoke about the same proportional shift regardless of L. That is, whether you choose a 100µH coil or a 1mH coil, a target that causes a 1% shift in one will cause about a 1% shift in the other and in either case the frequency shifts by about 0.5%. Thus for target sensitivity the choice of inductance doesn't make a lot of difference.

---

2. Bounty Hunter began with 3 BFO models that all ran at 1MHz. The name of the company was originally Pacific Northwest Instruments and these first models were called the *Bounty Hunter I*, *II*, and *III*. However, subsequent detectors were all called "Bounty Hunter" with various model names.

3. A proximity detector, on the other hand, usually needs the opposite sensitivities.

ΔC is not influenced by targets except for the case of capacitive ground effect which is largely dealt with by shielding the coil. But C consists of one or more physical capacitors plus the parasitic capacitance of the coil. In general, it is preferable to have the physical capacitor(s) clearly dominate the overall capacitance so that interchangeable coils are easier to make. This suggests that a larger physical C coupled with a smaller search coil L is generally the better choice. Most BFO coils, in fact, have an inductance of 100-150µH (e.g. Table 10-1) which is on the low side compared to other metal detector designs like VLF and PI. But, as we will find out over and over again, there are lots of trade-offs when making design choices.

10.4: Coil Considerations

The super-simple mono coil design is one of the few advantages the BFO has. A BFO can easily be outfitted with loops of almost any size, from a 1-inch probe to a massive loop of 36 inches or more, and is one of the reasons the BFO became known as an all-purpose detector. The smallest coils are sensitive to small near-surface objects like gold nuggets, the mid-sized coils are normally used for coin hunting, and the largest coils were effective on large, deep objects like caches and relics. By the late 1970s Garrett had by far the largest selection of BFO coils.

In a BFO the search coil inductance can be just about any value but the range of coil sizes you might want to build sets a practical lower limit. For example, if the coil inductance is 100µH then a 6"/15cm coil will need roughly 15 turns of wire (99µH). A 24"/61cm coil requires 6.4 turns of wire (100µH). We would have to round this to either 6 turns (89µH) or 7 turns (119µH), either of which is a fairly significant error in the inductance. This translates into an error in the search frequency that will need to be corrected in order to stay within the tuning range of the detector.

Using a 1mH coil reduces this problem. A 6"/15cm coil needs 51 turns (1.008mH) and a 24"/61cm coil needs 21 turns (0.98mH). So a one-turn error at 1mH is less significant than at 100µH. However, a monstrous 36"/91cm coil will get us back to facing the same issue. One solution is to include enough tuning range in the design to accommodate the inductance error. But a wide tuning range can be a challenge so it is better to get the combination of inductance and parasitic capacitance within a tolerance band that can be accommodated by the available tuning range. That is, if the oscillator can be tuned ±3% then you probably want every coil to run within ±1% of some nominal LC combination. This might require putting a "tweak capacitor" in some coils in order to get them within tolerance. For example, for a large coil you would round down to the smaller number of turns (which makes the inductance a little too low) and add whatever capacitance gets the oscillator close to the middle of the tuning range of the detector.

We mentioned shielding in the last section and with BFO detectors in particular a shielded coil is critical. Shielding is covered in more detail in Chapter 8 but the basic idea is to eliminate any variable parasitic capacitance seen by the coil (such as from wet grass) and to also eliminate noise from static electricity (such as from dry dusty soil). The shield attenuates electric field effects but not magnetic field effects, therefore depth is largely unaffected.

Shielding, though, does add to a coil's fixed parasitic capacitance. The shielding capacitance for a large coil versus a small coil will not be linear which, again, can be dealt with through the use of a tweak capacitor inside the coil.

10.5: Drift & Other Annoyances

The discussion on component sensitivities above focused on maximizing target detection and minimizing parasitic effects. Related to this is the problem of drift, which was a notorious drawback of many BFO designs. Since the audio response is directly proportional to the difference between the search and reference oscillator frequencies, drift in either oscillator shows up as a shift in the audio frequency.

Drift is due to component changes over time. In the oscillators, the guilty components are primarily the inductor and capacitor(s) that determine the oscillation frequency. For example, the inductor for the search oscillator is in the search coil and is prone to temperature changes as it is moved from sun to shade, and this problem is made worse if it is in a black plastic shell. This might cause the resistance of the copper wire to change slightly or cause a minute amount of expansion/contraction of the coil winding that shifts the inductance and/or interwinding capacitance. Shifts in the inductance tend to be proportional to the inductance so the value doesn't matter a whole lot. But a lower inductance coil will tend to minimize shifts in the interwinding capacitance.

The physical capacitor can also vary slightly due to temperature changes and is often proportional to the value at hand. For example, if the tuning capacitor is 1nF or 10nF you can get either one in, say, an NP0 or C0G[4] which have a temperature drift of 100ppm/°C so the proportional drift is a wash. And since the classic BFO has two oscillators, we can design them to use the same type and value of capacitor and, if they drift together equally, their drift will (first order) cancel.

Furthermore, oscillators have one or more active components — such as the bipolar transistor in a Colpitts oscillator — which can also be affected by temperature. As with the physical capacitors, if the two oscillators share the same design then the component drift will largely track and cancel. In many other electronic designs this kind of common-mode tracking is enhanced by placing the components physically close together (and perhaps even thermally coupled) so their temperatures track closely, but in the BFO we usually want the oscillators physically isolated to prevent frequency locking.

Another source of drift can be due to changes to the in the oscillator power supply. Again, if the search and reference oscillator run on the same supply and are substantially the same design then these drifts mechanisms should track fairly well.

Component-induced drift tends to be a fairly slow process and requires the user to occasionally retune the reference oscillator. Through the 1970s companies (notably Garrett and White's) marketed "zero-drift" BFO detectors which used a crystal-based oscillator on the reference side to at least eliminate one source of drift. The search oscillator, however, must remain free-running. This seemingly eliminates one half of the problem except that it also eliminates any possibility of common-mode drift cancelation.

Besides drift due to circuit components we also have to deal with environmental changes. Most soils contain some trace amounts of magnetic and conductive minerals including iron oxides and salts. Iron mineralization increases the permeability seen by the search coil and therefore decreases the search frequency. Saline soil can appear to be conductive if there is enough moisture and will increase the search frequency.

The end result is an audio frequency which varies as the search coil is lowered to the ground. This is not really a drift problem but rather a proximity problem. The immediate solution is to retune the detector after the search coil is lowered. Unfortunately, any height variation of the coil as it is being swept can produce false signals. And in areas where ground conditions vary enough occasional retuning will be required for that, as well[5]. There is no easy fix for this and it was a major problem of detectors in general until ground balanced IB designs came along.

10.6: Tuning

In the traditional BFO one of the oscillators must be tunable. LC tank oscillators are most easily varied by changing L and/or C. Finding variable inductors and variable capacitors these days is somewhat difficult, and finding them in a user-adjustable format (that is, using a control knob) is almost impossible. Back in the good ol' days the big-box BFO metal detectors used a large air variable capacitor (Garrett *BFO*, Figure 10-5) for the tuning element. These had a shaft which allowed the use of a knob and were also commonly found on radios and televisions of the day. No one designs circuits that

Fig. 10-5: **Air Variable Capacitor**

---

4. Most people call these "En-Pee-Oh" and "Cee-Oh-Gee" caps, but their proper names are "En-Pee-Zero" and "Cee-Zero-Gee." Not a big deal except, perhaps, when typing it into a search engine.

5. This problem mimics component drift.

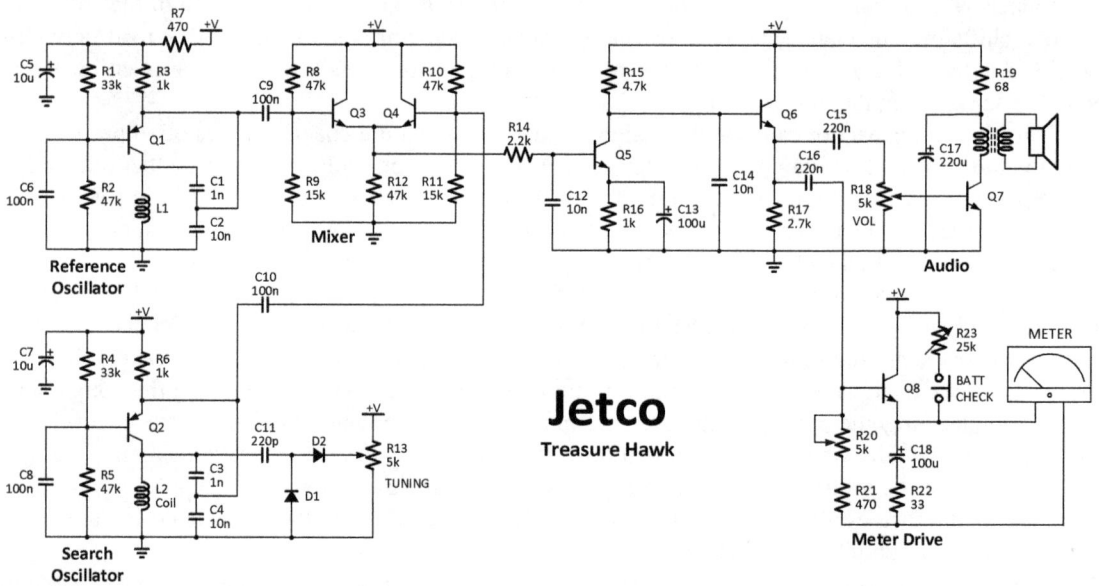

Fig. 10-6: **Jetco Treasure Hawk**

way any more so those kinds of devices are long gone. We need to come up with something more modern.

The common way to provide user adjustments is with a potentiometer so a good solution should try to incorporate a pot. But simply adding resistance into the oscillator doesn't work; it will very slightly vary the frequency but will also severely de-Q the circuit and squash the amplitude.

A better solution is actually found in some of the cheap 70s BFOs (including the Jetco shown in Figure 10-6) which avoided the relatively expensive air variable capacitors and instead used a *varactor diode* as the tuning element. A varactor diode is a diode which is specially doped so that its reverse-bias junction capacitance has a stronger variance with applied voltage.

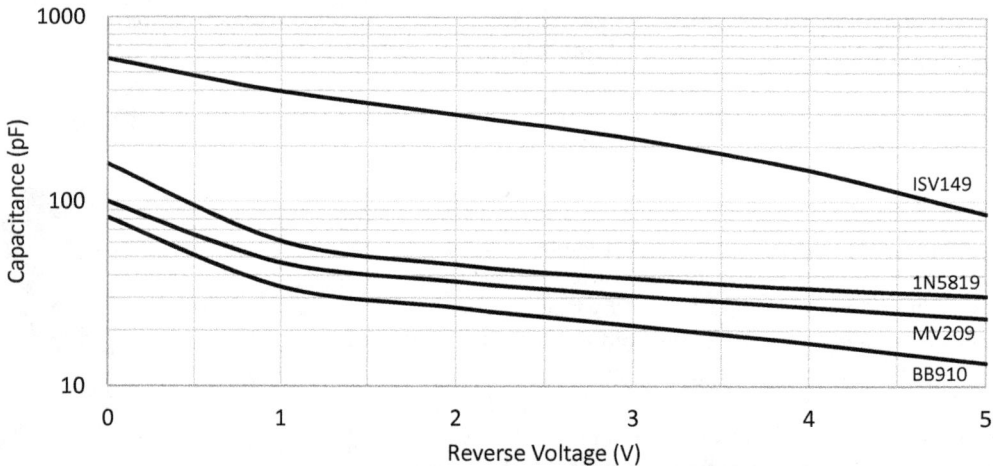

Fig. 10-7: **Depletion Capacitance**

All diodes behave as varactors to some extent, with the depletion capacitance decreasing with higher reverse voltage. Varactors have a more lightly doped junction which results in worse diode characteristics but they are generally not used as diodes so it's a shortcoming that doesn't matter. In any case, even an ordinary diode can behave as a varactor. The graph in Figure 10-7 shows the depletion capacitance vs reverse voltage for three varactor diodes and a 1N5819 Schottky diode.

The varactor requires a DC-coupled variable voltage in order to vary the capacitance and must simultaneously be AC-coupled to the oscillator to have an effect on the LC tank. A potentiometer provides the variable voltage and a wiper series resistor prevents shorting the varactor capacitor when the pot is at its limits. A capacitor couples the varactor to the oscillator. If the coupling cap is large compared to the varactor diode capacitance then the overall capacitance is roughly equal to just the varactor portion. Figure 10-8 shows a complete varactor-based variable capacitor.

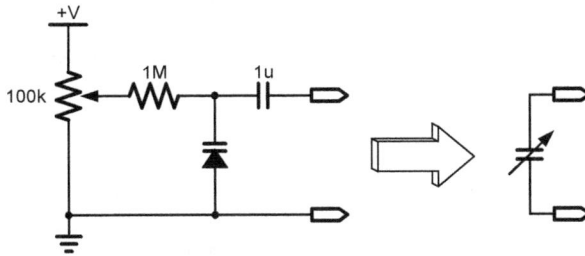

Fig. 10-8: **Varactor Tuning Circuit**

Let's take a look at how the varactor affects frequency. The best varactor in Figure 10-7 (the ISV149) can be varied from 85pF to 600pF with (let's say) a mid-point value of 330pF. Suppose this is part of a (nominal) 100kHz oscillator[6] that uses a 1mH coil. The total required capacitance is

$$C = \frac{1}{(2\pi \cdot 100\text{kHz})^2 \cdot 1\text{mH}} = 2.53\text{nF}$$

This equates to a fixed capacitor of 2.2nF plus the mid-point varactor value of 330pF. If the varactor is then adjusted to its minimum and maximum values then the oscillator frequencies will be

$$f_{min} = \frac{1}{2\pi\sqrt{1\text{mH} \cdot 2.8\text{nF}}} = 95.1\text{kHz}$$

$$f_{max} = \frac{1}{2\pi\sqrt{1\text{mH} \cdot 2.295\text{nF}}} = 105.3\text{kHz}$$

This gives about ±5kHz of tuning range which seems really good. However, in practice we won't get that. The reason is because the voltage across the varactor is not constant but is rather the full AC swing of the oscillator offset by the applied DC voltage. This means that the effective varactor capacitance is an average of a large portion of the curve so instead of a ΔC of 515pF it is likely to be somewhat less. Determining the effective capacitance range is best done experimentally in the intended circuit.

For the most flexibility the tuning circuit should be applied to the reference oscillator. This allows us to optimize the search oscillator for the desired search coil inductance while the reference oscillator is optimized for the tuning circuit. The design circuit will clarify.

10.7: Practical Considerations

Building a couple of oscillators seems simple but there are some pitfalls to watch out for. Capacitors normally have a tolerance of ±10% and sometimes as high as ±20%. There are two ways to account for this. The first is to physically measure the capacitors and select accurate values. However, it's not important that the capacitors are absolutely accurate but that between the search and reference oscillators the capacitors match, assuming the inductors also match. The goal is for the resulting frequencies to match, not that they achieve a specific frequency.

The second method is to use extra "padding" capacitors to tweak a given capacitor closer to its desired value. Capacitors in parallel add so when laying out a PCB throw in a couple of parallel padding caps in each of the oscillators. Again, it's not necessary to increase or decrease a capacitor to an exact

---

6. For now, just a generic LC tank circuit.

value, it's only necessary to alter a capacitor to match the corresponding cap in the other oscillator. As such you can choose to alter the lower of two capacitor values, thereby always padding upwards. There should never be a need to decrease a capacitor.

Another issue that's been mentioned is that of search coil inductance. Machine-wound search coils often have a tolerance of 2-3% and the tolerances of hand-wound search coils are a bit higher. In addition, if you want multiple coil sizes then matching the inductance and interwinding capacitance of a 4 inch coil with that of a 24 inch coil is not very likely. It is possible to compensate for coil variability by placing a portion of the resonant capacitor inside the coil and adjusting its value according to the variation in the coil parametrics. It's typically sufficient to place 10-20% of the resonant capacitor inside the coil, depending on the self-capacitance of the coil relative to the total capacitance.

The reference oscillator uses an off-the-shelf inductor which is always placed inside the control box. To minimize the possibility of coupling from any extraneous magnetic field (a speaker, for example) it is best to use a shielded inductor. If that's not readily available use whatever is, including the option of winding your own[7] which has the added benefit of more control over the final value. Although the reference inductor does not necessarily need to have the same inductance as the search coil it is often done that way. Even then the inductances are unlikely to exactly match so, again, padding capacitors can compensate.

Design: *Classic BFO*

This design will be a (mostly) discrete all-analog BFO set up for 100kHz operation. From the block diagram in Figure 10-1 the design needs a couple of oscillators, a mixer, a low-pass filter, and an audio driver. A classic BFO design that follows this block diagram is found in the Jetco *Treasure Hawk* (Figure 10-6) and this design will be similar to the Jetco.

10.8: Oscillator Design

The *Energy Theft* design in Chapter 9 introduced the Colpitts oscillator (shown again in Figure 10-9[8]) which was popular in BFO designs. With some minor effort it can be made variable as required for tuning the reference oscillator.

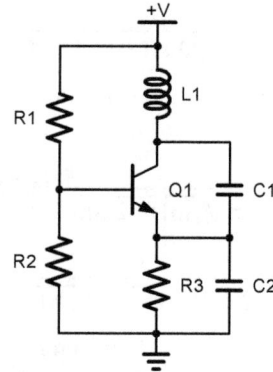

Fig. 10-9: **Colpitts Oscillator**

R1, R2, and R3 set up a DC bias current in Q1 in the typical manner. The collector of Q1 drives the LC tank circuit. C1 and C2 form a capacitive divider (and phase shift) to provide positive feedback to the emitter of Q1, thereby creating oscillation. The frequency of oscillation is

$$f = \frac{1}{2\pi\sqrt{LC_T}}$$

Eq 10-5

where

$$C_T = \frac{C1 \cdot C2}{C1 + C2}$$

Eq 10-6

That is, the inductance L1 in parallel with the series combination of C1 and C2 forms the resonant circuit. Note that the Jetco circuit uses PNP oscillators; it can be done either way.

Both oscillators will be designed around an inductance value of 1mH. This means the combined capacitance C_T is

---

7. See the *Explore* section for details.

8. We will use the alternative version from Figure 9-4 just to be different.

$$C_T = \frac{1}{\omega^2 L} = \frac{1}{(2\pi \cdot 100\text{kHz})^2 \cdot 1\text{mH}} = 2.533\,\text{nF}$$

Starting with the search oscillator we'll first remove 20% of C_T to place inside the search coil for coil parametric matching. This will be 507pF[9], leaving us with 2.026nF for C1 and C2. Normally in a Colpitts oscillator the caps are ratioed unequally, with C2 up to 10 times the value of C1. Choosing C1 = 8.2nF and C2 = 2.7nF (both standard values) which yield a C_T of 2.031nF. If we manage to wind a search coil of exactly 1mH the search frequency is now

$$f_S = \frac{1}{2\pi\sqrt{1\text{mH} \cdot (2.031\,\text{nF} + 507\,\text{pF})}} = 99.90\,\text{kHz}$$

In engineering we call a result that close "a miracle." The search oscillator design is shown in Figure 10-10. As mentioned in the last section, we should be prepared to add padding capacitors to either oscillator; in this case two (C1a and C1b) are placed in parallel with C1-C2. To complete the search oscillator we need to design the search coil, everything else is pretty mundane. We'll get to that soon.

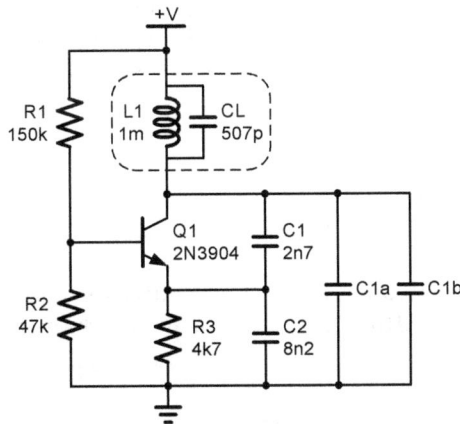

Fig. 10-10: **Search Oscillator**

We now turn our attention to the reference oscillator which includes the varactor tuning circuit of Figure 10-8. The combined circuit is shown in Figure 10-11. As mentioned before the effect of the varactor capacitance is not straightforward because of the large AC voltage imposed on it. The best way to determine its effect is experimentally.

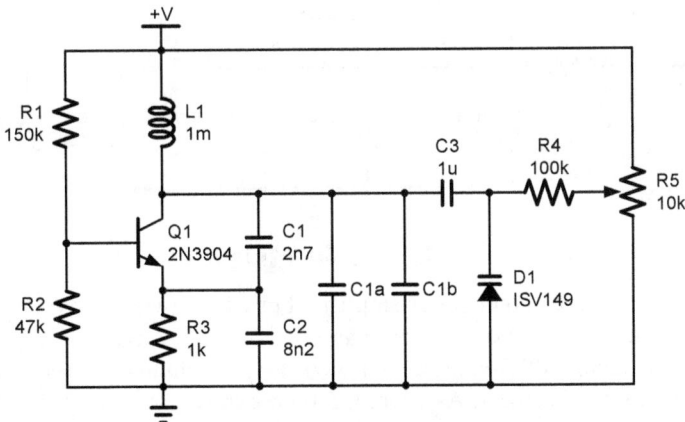

Fig. 10-11: **Reference Oscillator**

---

9. Keep in mind that part of this will be the self-capacitance of the coil.

The reference oscillator starts out with the same main capacitor values as the search oscillator even though it does not include the in-the-coil tweak capacitor. The varactor circuit will add some amount of minimum capacitance so we'll figure it out as we go. Without the varactor tuning circuit the oscillator in Figure 10-11 runs at a frequency of 116.5 kHz[10]. Adding the tuning circuit results in a frequency range of 105.8 - 113.8 kHz. Therefore the effective tuning capacitance range is 90-396 pF. We would like a tuning range of at least ±2kHz and preferably ±3kHz so this is enough. If we needed more tuning range, one solution is to reduce C1 and C2 so that the varactor capacitance has more effect. In reducing C1 and C2 we can either accept a higher operating frequency or we can also increase L1 to maintain 100kHz. Another option is to add an additional parallel varactor to D1. This will almost double the varactor capacitance range.

Whether the search oscillator frequency ends up near the middle of the reference oscillator frequency range depends on how well the search coil matches the reference coil, how well other components match between the oscillators, and on how close the mid-value of the varactor capacitance is to the 10% coil tweak cap. Most likely, some fine tuning will be in order. This can be achieved by measuring the frequencies (with an oscilloscope or a frequency counter) or by monitoring the mixer output and padding one of the oscillators until they match. We'll save this exercise for later.

Notice that the output signals from the oscillators (to be sent to the mixer) are taken from the emitters. The collector signals could also be used (and some BFO designs do this) but they are really too large and are also more sensitive to loading effects.

10.9: Mixer/Filter

Now that we have both oscillators designed they need to drive a mixer/filter stage to produce a difference signal. There are a number of ways to do this; all that is really necessary is to apply the signals to a non-linear junction so even a simple diode could be used. The Jetco circuit feeds the signals to emitter-coupled followers. For this design we will use a diffpair similar to the Jetco but take the output off one of the collectors; see Figure 10-12. In addition to mixing the signals, this also provides a modest amount of gain.

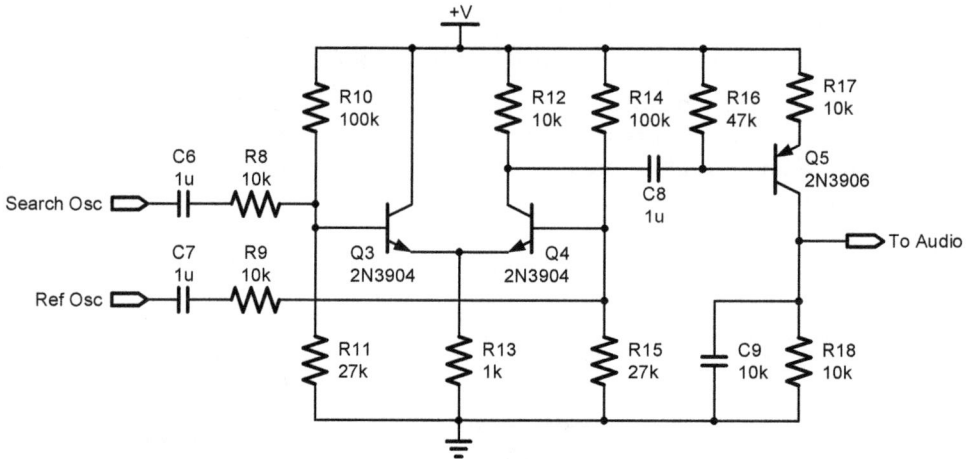

Fig. 10-12: **Mixer/Filter**

The next stage (Q5) is a common emitter amplifier but with a twist. The transistor is not biased in the active region as in the usual fashion so it only amplifies negative dynamic signals coupled through C8. R18-C9 form a low bandwidth load (about 1.6kHz) which shunts the 100kHz carrier frequency while passing the envelope waveform. As such, Q5 both extracts and amplifies the envelope of the mixer signal. Because it takes negative dynamic signals to activate Q5, its output is nominally 0V and

---

10. The numbers presented here are for a particular circuit build. Results will vary due to component tolerances.

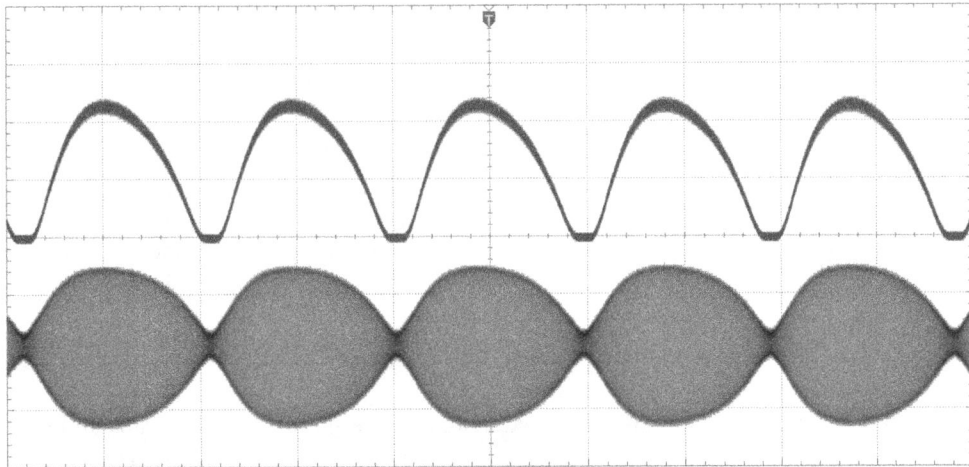

only exhibits positive dynamic signals at its collector. Figure 10-13 shows the mixer waveform[11] (at Q4-collector, lower waveform) along with the filtered signal at Q5's collector (upper waveform). Q6 is an emitter-follower that buffers the signal for driving the audio amplifier.

10.10: Audio

The final piece of the puzzle is the audio stage. The mixer provides a filtered low-frequency signal that may need further amplification. A few transistors and resistors could be used to make a decent audio driver but an easier and better solution is to use the LM386 audio amplifier. This chip has been around for a long time and continues to do the job well and cheaply. The default gain is 20 and the inputs will work all the way to ground so the connection is simple, as shown in Figure 10-14. An emitter-follower stage (Q6) buffers the filter output and uses a potentiometer (RV2) to provide a volume control. An audio taper pot works best here.

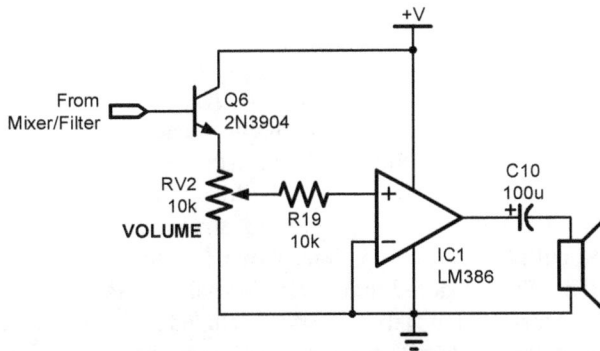

Fig. 10-14: **Audio Stage**

10.11: Power Supply

So far all the circuitry has been shown powered from a "+V" supply rail with no mention as to what that might be. The circuit could be powered directly from a 9V battery but it will slowly drift as the battery discharges. This may be acceptable because the circuit's total power consumption is low and battery discharge will be quite slow, so retuning for this purpose will not happen often.

---

11. The mixer waveform has been offset on the oscope so that it does not overlap the filter waveform; it actually has a positive DC offset. The filter waveform is positioned with no offset, which shows that it bottoms out at 0V.

The majority of power supply drift is due to the applied varactor voltage. In Figure 10-8 the potentiometer is connected to +V and ground so if +V is the battery then the varactor voltage is proportional to battery voltage. A simple way to stabilize the varactor voltage while continuing to directly use the battery is to add a zener diode as in Figure 10-15.

While this eliminates the largest power supply drift mechanism other subcircuits (especially the oscillators) may still vary with supply voltage. It is good practice

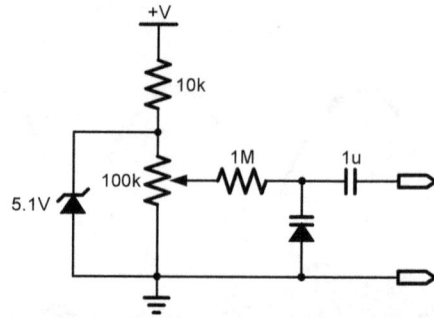

Fig. 10-15: **Voltage-Stabilized Varactor Tuning**

with metal detector circuits to use voltage regulators to provide stabilized power rails and we will do that here. The final design (Figure 10-16) includes an LM317 variable regulator. With a rather high drop-out voltage of 1.7V a 9V battery will be usable down to 6.7V which probably leaves a little bit of unused juice. A lower drop-out voltage (such as 250mV with the LP2951[12]) will allow a more thorough use of the battery, or might allow a higher rail voltage of, say, 6V. The designs in this book will have opportunities for improvement and these opportunities will often be pointed out.

10.12: Search Coil

In the discussion of the search oscillator little mention was made of the search coil other than its inductance value. One reason why we chose a simple value of 1mH for the search coil is to make for an easier selection of the inductor in the reference oscillator. Actually, the reference and search oscillators don't even need to use the same inductance values but doing so results in less explaining and unnecessary side-tracking. In any case, the coil specs for this design are as follows:

- Diameter: 7" (17.8cm)
- Number of turns: 49
- Wire: 24 AWG magnet wire
- Inductance: 1mH (nom)
- Resistance: 5 Ω (max)

Chapter 8 covers the construction of a basic BFO coil in detail. A difference with this coil is that we need to place a fixed capacitor inside the coil. The total coil capacitance value should be 507pF (for this particular design) and this includes the coil's self capacitance and cable capacitance. Therefore the internal cap value cannot be finalized until the coil is built and the SRF is measured. If the coil is to be epoxy-potted then all this must be done, the cap(s) added, and the epoxy potting done last.

As an example, a coil built to the specifications above[13] — including shielding and cable — has a measured SRF of 253.5kHz. The measured inductance is 1.005mH so the overall parasitic capacitance is 392pF. Therefore the physical capacitor that must be included is 115pF. The closest standard value is 100pF but we should instead use two 56pF caps in parallel to get a bit closer, or perhaps 100pF in parallel with 15pF for an exact value. A few percent is close enough.

10.13: Calibration

The two oscillators in the BFO design are likely to be mismatched enough that a zero beat frequency is beyond the tuning range of the varactor. If so, one of the oscillators will need padding capacitors added. To determine this, measure the frequencies of both oscillators by probing the BJT emitters to prevent the probe from loading the oscillator. This can be done with an oscilloscope[14] or a frequency counter. Before measuring the reference oscillator, make sure the tuning pot is set to its center position.

---

12. The LP2951, like every other variable output LDO regulator, is only available in a surface mount package.

13. See 8.4.1

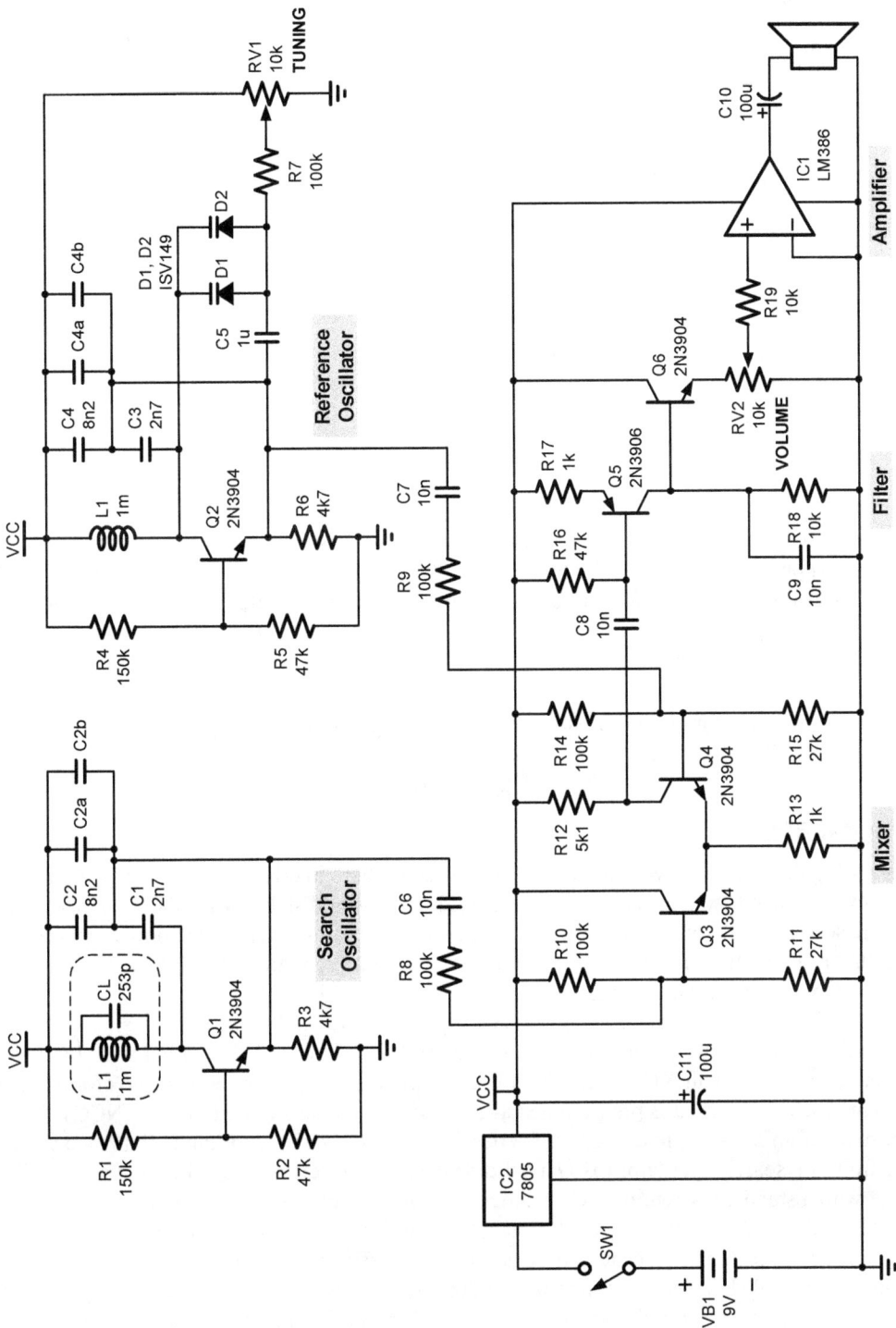

Fig. 10-16: Complete BFO Schematic

14. A digital oscope with frequency measurement will usually work well enough. Use the averaging feature to improve the stability of the measurement. Trying to discern a few percent difference simply by looking at the oscillator waveforms (say, on an analog oscope) will be difficult. In that case, if a frequency counter is not available then the best method is to monitor the mixer/filter output and use trial-and-error to find the required padding cap.

The padding caps are added to the oscillator running at the higher frequency. The value needed can be determined by either trial and error or by calculation. For calculating the padding caps, recall the oscillator frequency is given by

$$f = \frac{1}{2\pi\sqrt{LC}}$$

We can lower this frequency by Δf by adding ΔC:

$$f - \Delta f = \frac{1}{2\pi\sqrt{L(C + \Delta C)}}$$

If we take the ratio of these equations we get:

$$\frac{f - \Delta f}{f} = \sqrt{\frac{C}{C + \Delta C}}$$

Or

$$\Delta C = C\left[\left(\frac{f}{f - \Delta f}\right)^2 - 1\right]$$

f represents the current frequency of the oscillator and $(f\text{-}\Delta f)$ is the frequency we'd like to achieve. As an example, suppose the search oscillator runs at 98kHz and the reference oscillator (with a total C_T of, say, 2.533nF) runs at 100kHz. We'd like to lower the reference oscillator from 100kHz to 98kHz:

$$\Delta C = 2.533\text{nF} \cdot \left[\left(\frac{100\text{kHz}}{98\text{kHz}}\right)^2 - 1\right] = 104.4\text{pF}$$

Therefore adding a 100pF cap to the oscillator will probably get it close enough.

Explore

10.14: Digital BFO

In discussing drift (10.5) it was noted that Garrett and others had replaced the drifty analog reference oscillator with one that uses a crystal. This provides rock-solid stability for the reference side but, of course, the search oscillator is still subject to drift. And the tuning circuit must now be placed in the search oscillator, which may limit flexibility. Still, it's an improvement so let's see what can be done with more modern components.

Figure 10-17 shows an alternate approach. A Colpitts search oscillator creates a sinusoidal waveform at a particular search frequency; let's stick with 100kHz. This is AC-coupled and clamped to a smaller voltage, then fed to an XOR gate which is our digital mixer. The reference waveform is generated from a microcontroller that is programmed as a *numerically-controlled oscillator* (NCO). By using an NCO instead of a fixed-frequency crystal (as per the Garrett circuit; see Figure 10-19) we can move the tuning from the search oscillator (as Garrett did) back to the reference oscillator and even implement tuning with a standard potentiometer. A block diagram is shown in Figure 10-17.

The NCO will need a frequency resolution sufficient for near-zero beat tuning. A simple PWM function cannot be used for this because it does not have sufficient resolution. As an example, if the PWM counter runs at 4MHz then to create a 100kHz reference would require a divide-by-40 and the next higher and lower integer divider settings would result in a minimum tuning step of ~2500 Hz. An NCO, on the other hand, has a frequency tuning step equal to the micro clock frequency divided by the size of the counter:

$$f_{step} = \frac{f_{clk}}{2^N}$$

Eq 10-7

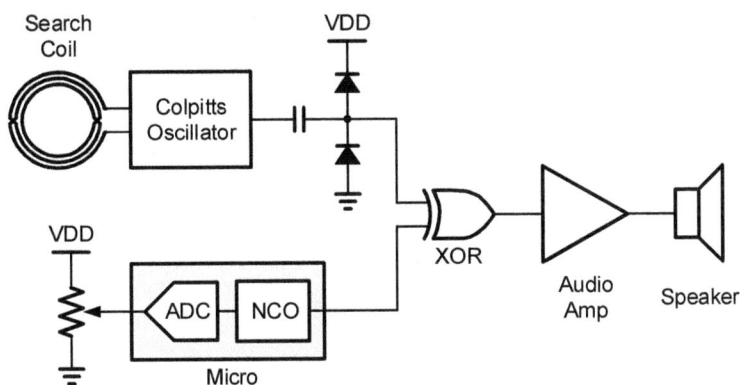

Fig. 10-17: **Digital BFO Block Diagram**

A 4MHz micro with a 16-bit counter yields a frequency step of 61Hz. Better than 2500Hz but still not good enough; we'd like something around 1Hz or so. This would demand at least a 22-bit counter. Many of the lower-end 8-bit micros have only 8-bit and 16-bit counters but often they can be cascaded to form a 24 or 32-bit counter.

In this design Timer0 and Timer1 are cascaded to form a 24-bit timer but only 22 bits are used for the NCO. The NCO step size is therefore $4MHz/2^{22} \cong 1Hz$. The tuning range is controlled by a potentiometer which is read by the micro's ADC; the ADC is 10 bits so the range is $1024 \times 1Hz \cong \pm 500Hz$. We would then just need to set the center tuning frequency of the NCO to be the same frequency as the search oscillator.

An advantage of this approach is that you can easily alter the center frequency of the NCO to match whatever coil you plug in. That is, it is no longer necessary to tweak each coil to be within the tuning range of the reference oscillator as was noted in the standard BFO design. A momentary switch can be used as a "tuning reset;" center the tuning pot, press the switch, and the NCO adjusts itself to be within 1Hz of the search oscillator. For this to work you will also need a way to measure the frequency of the search oscillator which can be accomplished with the same 24-bit cascaded timer temporarily applied to the input-capture feature of the micro.

However, simply counting the length of a single cycle of the TX oscillator results in a large potential error. For example, if the TX runs at 100kHz and the counter runs at 4MHz[15] then a single TX cycle is 40 counts. If the TX is 101kHz the result is still 40 counts. In fact, the resolution of a single-cycle count is, again, approximately 2500Hz as mentioned before. We need far better than that and can achieve it by counting multiple TX cycles. For example, if 10 cycles of 100kHz are counted then that is 400 clock ticks at 4MHz and the resolution is now 250Hz. So by selecting a larger number of cycles per measurement we can achieve any resolution we want, within the counting limit of the counter. That limit is, again, a count of 2^{24} so the best we can do is 419,430 TX cycles which is a resolution of about 6mHz. Besides being way more resolution than is necessary such a long measurement will take 4.2 seconds. A resolution of 1Hz can be achieved by measuring 2500 TX cycles and will only take 25ms.

10.15: Digital Non-BFO

The Digital BFO concept presented above has a feature whereby the frequency of the search oscillator is measured in order to set the reference oscillator. The next obvious step is to dispense with the reference oscillator and just measure the search oscillator frequency on a continuous basis[16]. When a

---

15. 4MHz is used in these discussions because a commonly-used 8-bit PIC micro clocked at a commonly-used frequency of 16MHz will have a peripheral clock of 4MHz. However, some 8-bit PICs can be clocked at 32MHz and better chips (especially 32-bit Cortex micros) can be clocked much higher frequencies and the peripherals can run at the full clock speed. Better micros also include one or more 32-bit counters.

16. Technically this ceases to be a BFO as it does not produce a beat frequency, but we'll stretch the definition a little to say that we are still comparing the search oscillator with the micro's counter clock.

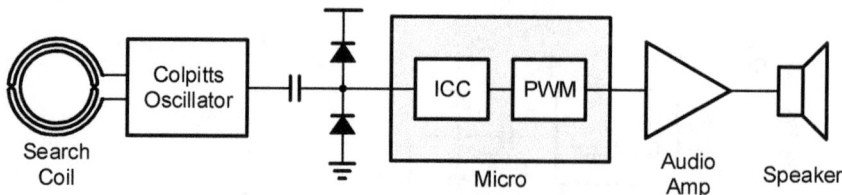

Fig. 10-18: **Digital Non-BFO Block Diagram**

target causes the frequency to increase or decrease, the frequency measurement count increases or decreases and we can use that to control an audio response.

In the Digital BFO the search oscillator frequency was measured with a 24-bit timer created by cascading an 8-bit timer with a 16-bit timer. The recommended resolution in that design was 1Hz because it was being used to set the center frequency of the reference oscillator, and 1Hz was plenty. In this design sensitivity is directly related to the measurement resolution so more is better, up to the point of the target signal-to-noise ratio. The trade-off in higher resolution is that longer measurement cycles create a response lag. For a 1Hz resolution the lag is 25ms which is not noticeable. A latency of 100ms starts to become noticeable so we will limit our measurement to 100ms (10,000 TX cycles) which is a resolution of 0.25Hz.

Note that doubling the TX frequency (to 200kHz) doubles the resolution (to 0.5Hz) for the same total measurement time (100ms). However, a BFO running twice the frequency experiences twice the target Δf so, overall, everything evens out. A higher micro clock can offer a higher resolution for the same response latency or reduce the latency for the same resolution.

10.16: Harmonic Mixing

In the "Digital BFO" section an NCO was used to generate a 100kHz reference signal. This signal is a square wave and is mixed with a squared-up signal from the search oscillator. Square waves are composed of a fundamental frequency plus a bunch of odd-order harmonics. As such, the harmonics are also mixed with each other and thrown into the overall results.

When mixing square waves it is possible to mix a frequency with its harmonic and still produce a baseband output. In the Digital BFO, instead of generating a 100kHz reference with the NCO we could instead generate a 300kHz reference. Even though the search oscillator is running at 100kHz, it has been squared up and therefore presents a decent amount of 300kHz harmonic content to the mixer. Mixing this directly with a 300kHz reference signal will produce a desired baseband signal. The output of the mixer is low-pass filtered and fed to the same audio stage as before. A third-harmonic system should, in theory, achieve 3 times the Δf sensitivity of a fundamental system.

10.17: Stability

An analog BFO design with matched oscillators should have reasonable stability over temperature and power supply changes since the oscillators will (first-order) track each other. In the digital design there is only one analog oscillator so more attention must be given to stability. A regulated power supply is even more important, and a fully temperature-compensated oscillator is also critical. A method of temperature-compensating the Colpitts oscillator was presented in the *Explore* section of the previous chapter.

10.18: Discrimination

By the early 1970s it was clear that TR detectors were superior for most types of hunting and in the mid-70s the introduction of VLF and ground balance signed the death warrant for BFO technology. Garrett maintained a BFO model in their product line until 1981 probably to satisfy the old-timers who still wanted one. The Garrett *Master Hunter BFO* was perhaps the pinnacle of BFO design even as the BFO was gasping its last dying breath; the full schematic is shown in Figure 10-19. It combines frequency shift detection with oscillator loading effects to give a rudimentary method of discrimination. The oscillator load detection method was covered in detail in Chapter 9.

Inside the Metal Detector

Garrett
Master Hunter BFO

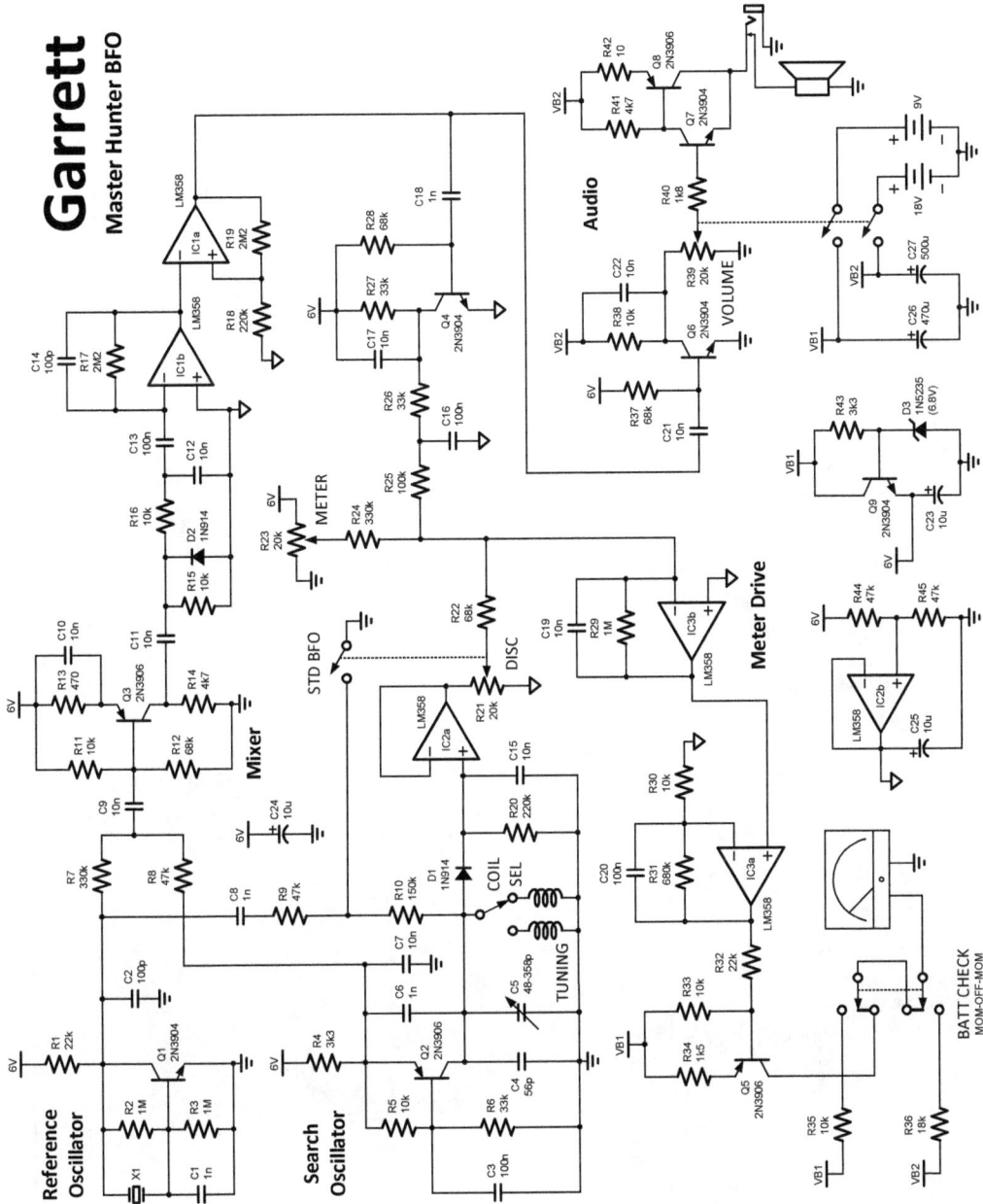

Fig. 10-19: *Garrett Master Hunter BFO Schematic*

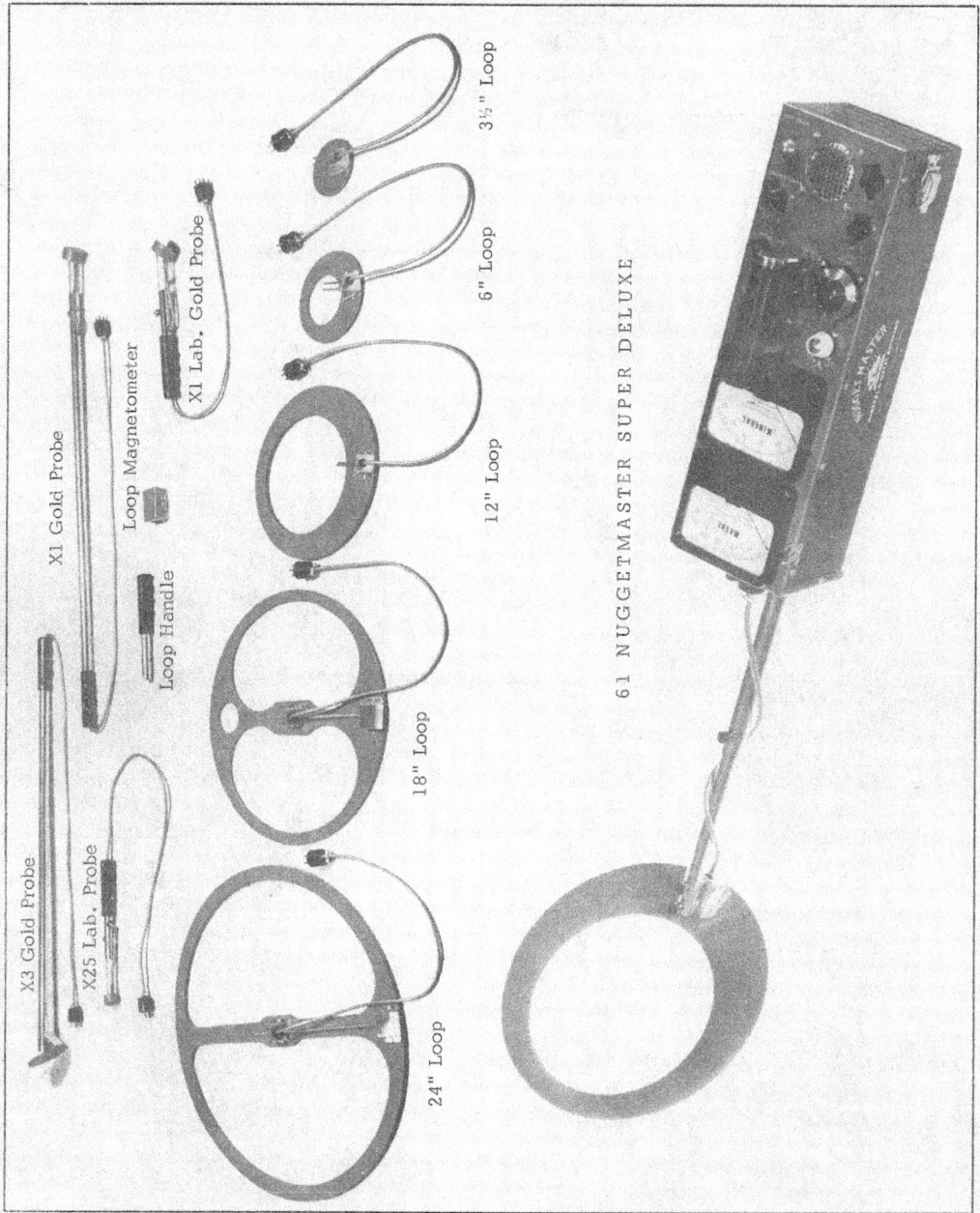

X3 Gold Probe

X1 Gold Probe

Loop Magnetometer

X25 Lab. Probe

Loop Handle

X1 Lab. Gold Probe

3½" Loop

6" Loop

12" Loop

18" Loop

24" Loop

61 NUGGETMASTER SUPER DELUXE

<table>
<tr><td>C
H
A
P
T
E
R</td><td>11</td><td><h1>Off-Resonance</h1></td></tr>
</table>

> "[As a youth, fiddling in my home laboratory] I discovered a formula for the frequency of a resonant circuit which was $2\pi\sqrt{LC}$ where L is the inductance and C the capacitance of the circuit. And there was π, and where was the circle? ... I still don't quite know where that circle is, where that π comes from."

> — *Richard Feynman*

11.1: Overview

All of the proximity methods covered so far use a free-running oscillator where the search coil and a capacitor form a tank circuit that determines the oscillation frequency. When an LC tank is driven with a signal generator via a resistor then its frequency response looks like that in Figure 11-1.

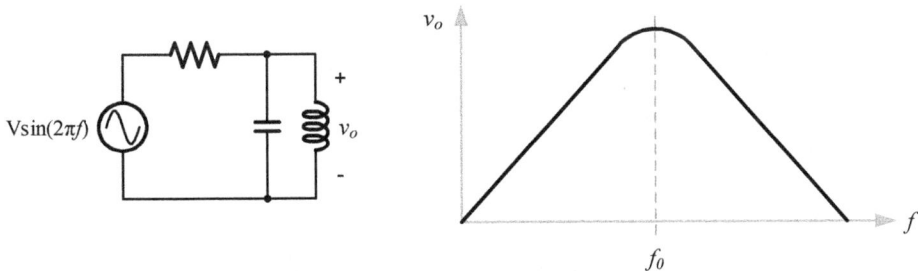

Fig. 11-1: **LC Tank Response**

The peak of the curve is the natural resonant frequency f_0 of the LC tank and is the frequency of the free-running oscillators we have been using (recall Equation 9-1). In the BFO chapter we saw that the coil inductance is affected by nearby targets which causes the frequency to either rise or drop slightly.

Instead of using a free-running oscillator and allowing targets to alter the frequency, we could instead drive the tank circuit at a forced frequency and see what happens when targets are presented. This forms the basis of the *off-resonance* method where, as the name implies, the LC tank is driven at a frequency somewhere lower or higher than its natural resonant frequency. This technique saw success in Allan Hametta's "A.H. Pro" line of detectors in the late 1970s and early 80s[1]. Figure 11-2 shows the basic premise. Except for the drive oscillator, this approach looks remarkably like the energy theft approach of Figure 9-2, yet the results are remarkably different.

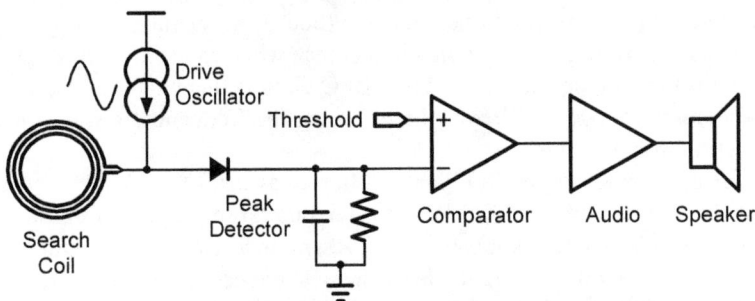

Fig. 11-2: **Off-Resonance Block Diagram**

---

1. Heathkit also produced an off-resonance design in 1978, the Cointrack GD-1190.

11: Off-Resonance   **207**

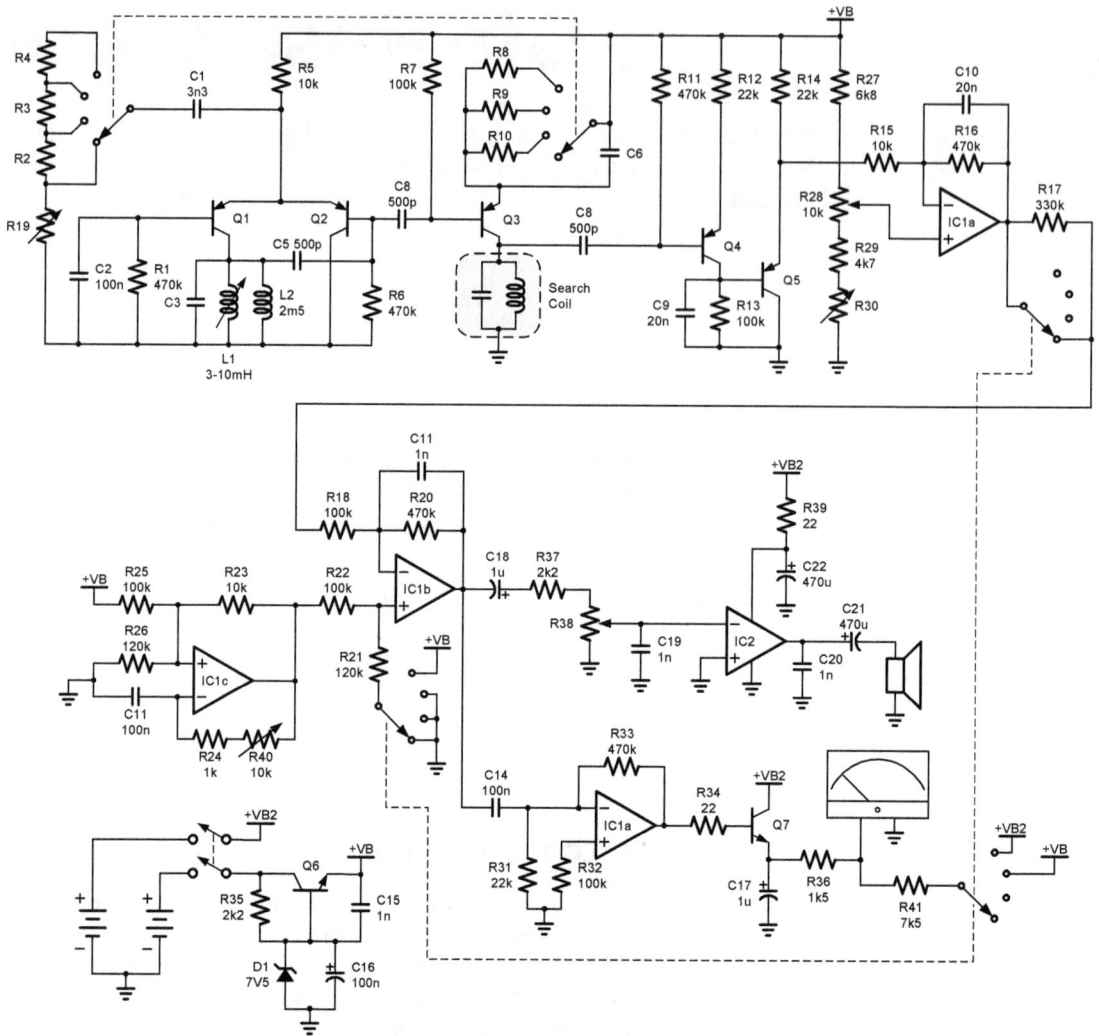

Fig. 11-3: **A.H. Pro Schematic**

11.2: Theory

To see how this works, let's suppose we adjust the drive frequency f_d to be exactly on-resonance. We can then introduce targets and see what the effect is on the tank voltage. For now, we'll use two targets, simply called ferrous (e.g., iron) and non-ferrous (eddy). As explained in the *Energy Theft* and *BFO* chapters, a ferrous target increases the coil's inductance whereas a non-ferrous target steals energy and decreases the inductance. In the energy-theft and BFO circuits this will cause both a frequency shift and a change in oscillation voltage. But here we are forcing the frequency to stay constant so we only get a change in the oscillation voltage.

Consider the case of a ferrous target. In Figure 11-4a we have an LC tank with a resonant frequency of f_0 and driven at that frequency ($f_d = f_0$). When a ferrous target is presented the coil inductance increases and the resonant frequency is lowered to $f_{FE} = f_0 - \Delta f$. However, the tank is still forced to oscillate at f_d so the result is that the oscillation amplitude is reduced by some Δv.

Now consider the case of a non-ferrous target. This will decrease the inductance and raise the resonant frequency of the tank to $f_{eddy} = f_0 + \Delta f$. In Figure 11-4b we see that the effect is the same: the oscillation amplitude is reduced. When the coil is forced to oscillate at its resonant frequency it is the case that all targets — ferrous and non-ferrous — cause a reduction in the oscillation amplitude. After all,

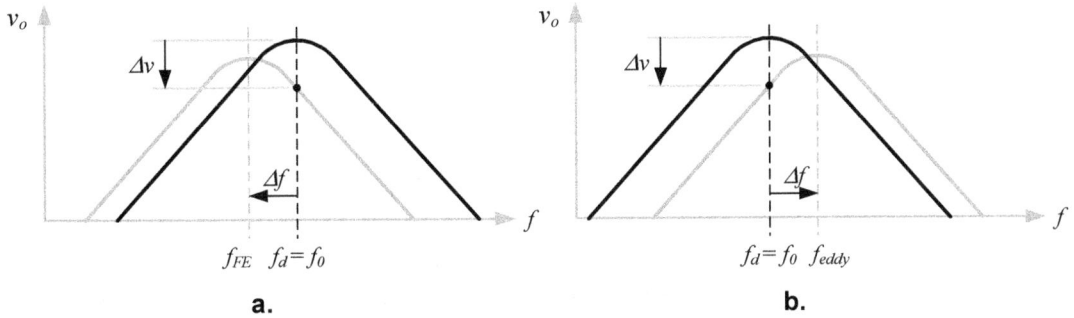

Fig. 11-4: **On-Resonance Target Effects**

when starting at peak resonance there is nowhere to go but down.

Next, suppose that the same tank circuit is driven at a frequency somewhat above f_0. This will force a reduction in the oscillation amplitude (compared to peak resonance) which means that it is possible for a target to either further decrease the amplitude, or to increase it. In Figure 11-5 we see again the LC tank forced to oscillate but at a driven off-resonant frequency $f_d > f_0$. When a ferrous target is presented (Figure 11-5a) the resonant frequency is forced lower and, even though the oscillation amplitude should theoretically increase, the net result seen at f_d is a reduction in amplitude. However, in Figure 11-5b a non-ferrous target raises the resonant frequency and the amplitude at f_d increases.

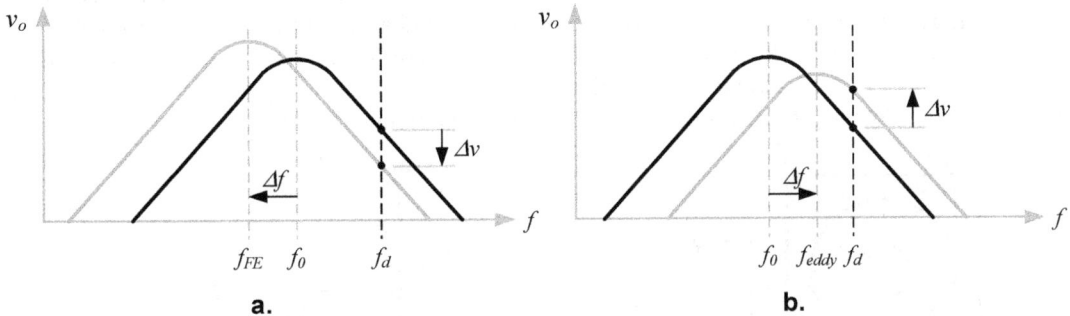

Fig. 11-5: **Off-Resonance Target Effects**

Although not shown, it is also possible to drive the tank on the low side of resonance ($f_d < f_0$) in which case ferrous targets will tend to cause an increase in amplitude and non-ferrous will cause a decrease. However, we will see from the data that the low-side drive effects are not as pronounced or useful.

From this it is obvious that the off-resonance method gives us a way to discriminate between ferrous and non-ferrous. BFO also does that, but off-resonance offers the ability to even discriminate between different eddy targets. What is not shown in Figures 11-4 and 11-5 is the effect targets have on the Q of the coil. A ferrous target increases the inductance of the coil and, everything else equal, also increases the Q of the coil. This slightly "sharpens" the band-pass response since Q = ωL/R. Conversely, eddy targets which steal energy from the system decrease the coil inductance and also de-Q the coil, making the band-pass response broader. The effect depends on how lossy the target is, with the lowest conductors (such as small foil) having high resistivity and high loss. A lossy target can de-Q the coil while having little effect on the resonant frequency.

Figure 11-6 shows the response for a low conductor target which has lowered the Q of the tank (there is still a slight frequency shift, which is not shown). If the drive frequency is at f_{d1}, then the presence of the target will produce a drop in the oscillation amplitude ($-\Delta v$). If the drive frequency is at f_{d2}, then the target will cause no apparent change in the amplitude. If the drive frequency is at f_{d3}, then the amplitude will increase ($+\Delta v$). So depending on the offset of the drive frequency we can reject or accept a given target, and this becomes our discrimination control.

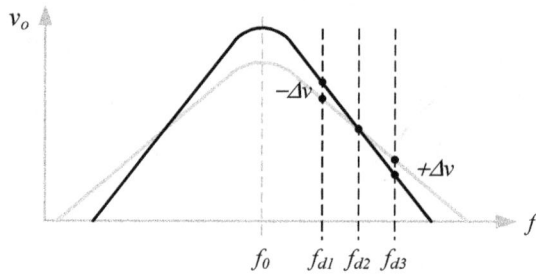

Fig. 11-6: **Low Conductor Effect**

This works up to a point. It is especially effective for lossy low conductors such as thin aluminum. It is less effective for high conductors such as copper and silver coins, but it is rare that we want to reject those targets. Notice that as the drive frequency approaches the crossover point (f_{d2} in Figure 11-6) the response gets weaker. This is similar to TR discrimination, where setting the discrimination to reject a particular conductivity will also progressively weaken conductivities that approach the discrimination point.

11.3: Responses

Let's look at some target responses for a driven coil, both at resonance and off-resonance. The test circuit in Figure 11-7 is used to drive the coil. The change in the oscillation voltage is measured by using a peak detector comprising a rectifying transistor Q1 and an RC filter. For this test, the coil is a solenoidal pinpointer type with a diameter of 12mm, a length of 54mm, and an inductance of 1mH. A tank capacitor (C1) of 63nF resonates the coil at approximately 20.2kHz. Table 11-1 shows the results for the resonant frequency f_0 plus some negative and positive frequency offsets.

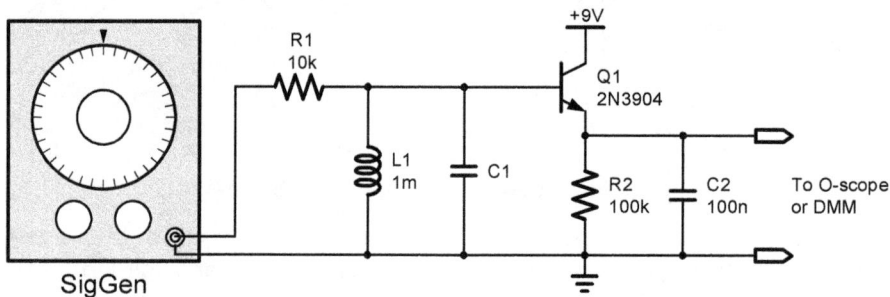

Fig. 11-7: **Test Circuit**

	Amp (Vpp)	Ferrite	16d Nail	Bottle Cap	25mm Sq Foil	US Nickel	Pull Tab	Screw Cap	US Zn Cent	US Quarter
$\Delta f = -750$	3.99	210	25	-60	-50	-170	-150	-420	-180	-275
$\Delta f = -500$	5.43	340	20	-150	-110	-300	-260	-690	-300	-450
$\Delta f = -250$	7.50	370	-35	-400	-260	-555	-170	-1080	-480	-680
$\Delta f = 0$	8.75	-110	-170	-620	-360	-460	-460	-560	-320	-410
$\Delta f = 125$	7.98	-360	-190	-540	-275	-300	-220	-350	15	140
$\Delta f = 250$	7.14	-370	-165	-400	-200	-120	-80	100	150	380
$\Delta f = 500$	5.23	-250	-90	-160	-65	60	60	510	200	435
$\Delta f = 750$	3.88	-160	-50	-75	-25	70	60	420	140	290

Table 11-1: f_0 = 20kHz (Target responses in mV)

Inside the Metal Detector

From this we see that negative frequency offsets will eventually give a positive response on an iron nail. For positive offsets, even the steel bottle cap is consistently rejected, a rarity with any metal detector design. The silver US quarter is quickly accepted with a small positive offset while the zinc cent is barely accepted. A slightly lower offset will certainly begin rejecting the zinc cent. Progressively larger offsets begin accepting more targets. It is notable that in these tests, the small foil target is never accepted. At f_0 = 20kHz, with large offsets above 1500Hz foil simply does not respond at all.

There are additional things to note from the data. The "Amp" column shows that the amplitude of the tank circuit decreases on either side of resonance which contributes to weaker target signals as the drive frequency gets farther from the resonant frequency. Even if the data is corrected for amplitude variation, we will find that a given target response peaks at a certain offset frequency and then decreases for higher offsets. This means that lowering the discrimination level may reduce sensitivity to high-conductive silver coins, a phenomenon not usually seen in TR or VLF detectors.

For positive offsets, ferrite always produces a negative signal. This means that mineralized ground will produce a negative signal that is proportional to the height of the coil above ground. It is possible to compensate for this by tuning the detector with the coil lowered to the ground but variations in sweep height can still cause either falsing or target masking. It is the same problem found in TR detectors.

It is further useful to look at the behavior at other resonant frequencies. Table 11-2 shows the results at 50kHz. As should be expected, a higher tank frequency requires a higher off-resonance to achieve similar discrimination results. Otherwise, the results are similar to the 20kHz data except that there is an overall reduction in target strengths. This suggests that lower frequencies may have an advantage. However, at f_0 = 50kHz and with offsets above 5000Hz foil begins to give a positive response.

	Amp	Ferrite	16d Nail	Bottle Cap	25mm Sq Foil	US Nickel	Pull Tab	Screw Cap	US Zn Cent	US Quarter
$\Delta f = -1000$	8.75	225	-30	-350	-380	-400	-330	-750	-300	-440
$\Delta f = -500$	10.26	150	-60	-440	-480	-425	-350	-800	-290	-430
$\Delta f = 0$	10.58	-20	-75	-400	-430	-280	-220	-470	-140	-190
$\Delta f = 250$	10.50	-75	-80	-350	-400	-200	-150	-280	-60	-60
$\Delta f = 500$	10.34	-130	-80	-300	-340	-120	-80	-90	5	50
$\Delta f = 750$	10.02	-160	-80	-255	-280	-50	-30	75	60	140
$\Delta f = 1000$	9.62	-180	-70	-200	-230	0	20	200	100	200
$\Delta f = 2000$	7.90	-175	-50	-75	-90	110	100	420	150	280

Table 11-2: f_0 = 50kHz

11.4: Transmitter

With off-resonance, the highest sensitivity is achieved when the circuit driving the LC tank has a high output impedance. A simple solution is to use a single transistor current source as shown in Figure 11-8. Either a high-side drive or a low-side drive will work, but the low-side drive is usually easier to implement. Ideally, the waveform that drives the transistor should be a sinusoid but it is much easier to use a square wave and that works just fine. Usually a duty cycle of 25% is

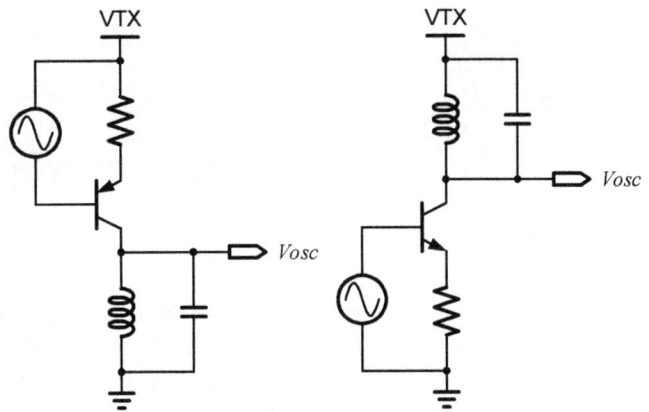

Fig. 11-8: **Simple TX Drive**

sufficient to make things go; a higher duty cycle often produces excessive distortion in the drive current.

The driving waveform for the transistor can be created a number of ways. This will be an all-analog design so we'll use a 555 timer to generate a variable-frequency drive pulse; see Figure 11-9. We will use a resonance of 20kHz so the 555 timer should be designed for an oscillation drive range of 20kHz - 21kHz. Q1 provides the high-impedance drive and D1 allows v_{osc} to swing below the drive transistor without saturating it. Assuming a power supply voltage of +5V, the output signal will therefore be a sine wave with a mid-voltage of +5V. Figure 11-10 shows a measured TX waveform and the driving pulse.

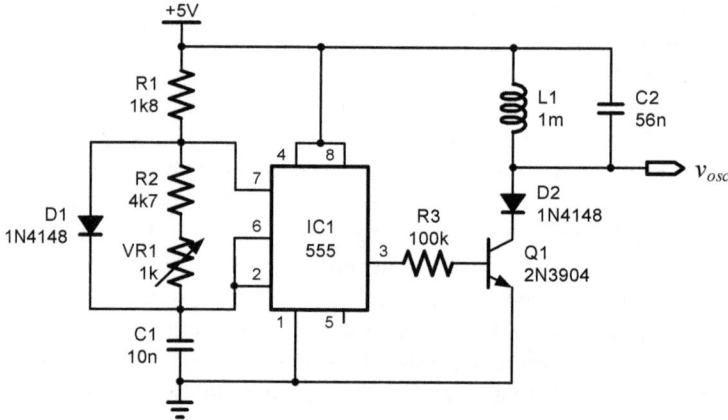

Fig. 11-9: **Complete TX Circuit**

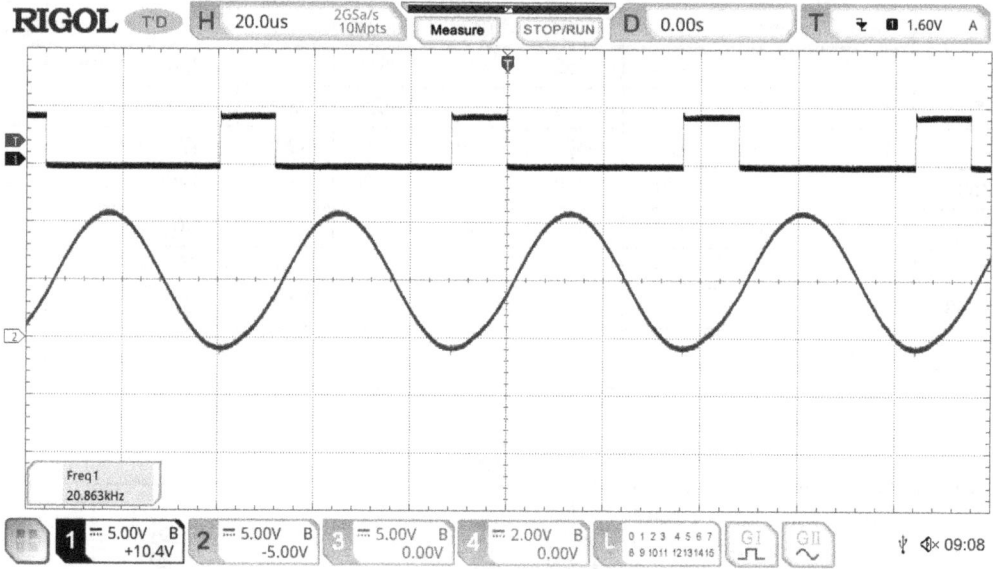

Fig. 11-10: **TX Waveform**

11.5: Peak Detector

The peak detector used in the test circuit of Figure 11-7 is the same as that used in the energy-theft circuit of Chapter 9. However, the output of the energy-theft oscillator is a ground-referred signal and the off-resonance oscillator in Figure 11-9 is referred to +5V. To keep the peak detector output in the range of 0-5V we can either use negative peak detector or we can capacitively couple the oscillator in order to make it ground-referenced. Figure 11-11 shows both options.

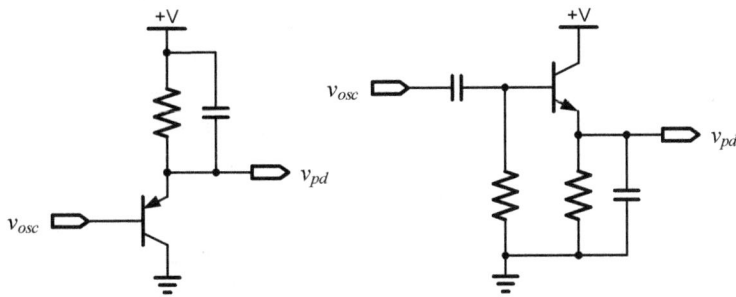

Fig. 11-11: **Peak Detector Options**

Either way will work so let's arbitrarily choose the capacitor-coupled approach. Next, a single transistor peak detector has a temperature drift problem that was explored in the energy-theft design. In this case, we'll deal with it by using a PNP-NPN double-follower circuit, see Figure 11-12.

In the energy-theft design the output of the peak detector is fed to a comparator which determines if it has exceeded a reference threshold voltage. Making the threshold voltage variable effectively creates a method of "tuning" the threshold of the detector. We could do the same thing here. However, it's been noted that when changing the frequency drive for setting the discrimination point, the oscillation amplitude also changes. Therefore the idle peak detector output level will also change. This means the detector will have to be manually retuned every time the discrimination level is changed. The earliest A.H. Pro models featured this annoyance.

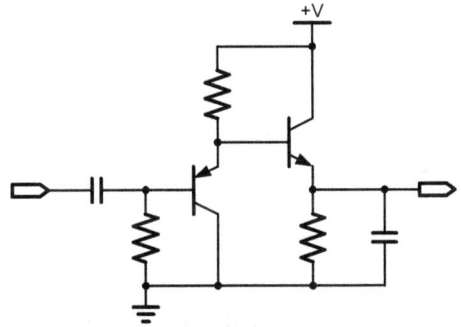

Fig. 11-12: **Double Follower**

The solution is to add an autotune circuit to the peak detector. Figure 11-13 shows the complete peak detector with the autotune circuit. When the switch is closed the integrator forces the voltage on

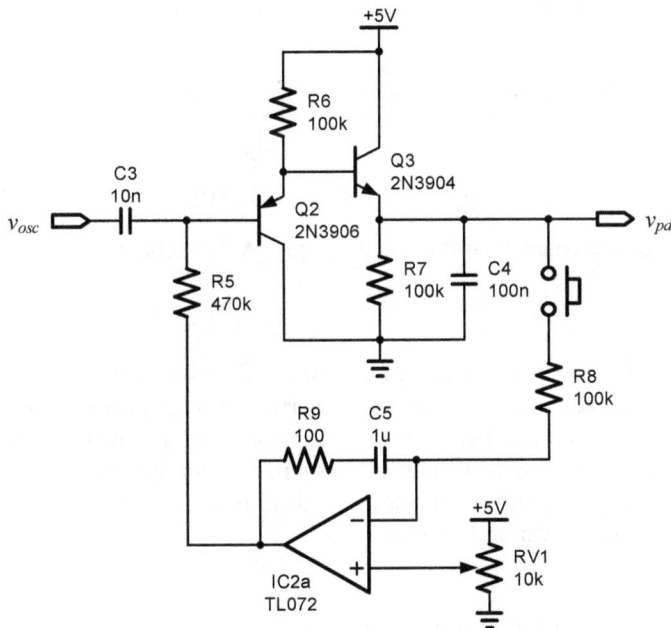

Fig. 11-13: **Autotune Peak Detector**

R5 to whatever level is needed so that the output voltage of the peak detector matches the reference level set by RV1. When the switch is open, the integrator holds that position via the charge on C5[2], assuming the integrator does not drift. It is therefore critical that the integrator opamp has super-low bias currents at the input pins. Use a JFET-input opamp at a minimum, and preferably a CMOS opamp.

After changing the discrimination level, a press of the switch (perhaps a momentary push button on the handle) will return the peak detector output to the prescribed voltage level. If the switch is left closed then the autotune circuit will continuously work to keep the output at a fixed point. In this case, the integration resistor R8 can be set to a higher value to make the retune operation deliberately slow so that the detector functions in a motion mode. It will automatically retune itself when the discrimination is changed, but it will also retune itself if you hover over a target. The retune should not be excessively quick.

Design: Off-Resonance Detector

The key elements of the off-resonance design have been presented so let's go through the remaining circuitry. Because the peak detector has a variable retune circuit that sets the nominal output level, the comparison stage that follows can use a fixed threshold reference voltage instead of the variable reference used in the energy-theft design. That is to say, it is the retune reference that now sets the threshold.

In the energy-theft design an opamp gain stage was used to compare the peak detector signal to the reference signal and this design does the same. Instead of simply toggling a buzzer when a target is detected, this design will present a more nuanced audio. The 20kHz transmit frequency is divided by 32 to produce a 625Hz audio tone that "chops" the output signal from the gain stage. A dual 4-stage binary counter (74HC393) divides the 555 output by 2 (stage 1) and by 16 (stage 2) and also gives us a square wave. Figure 11-14 shows the concept. The result is a fixed-tone audio whose loudness is proportional to target strength.

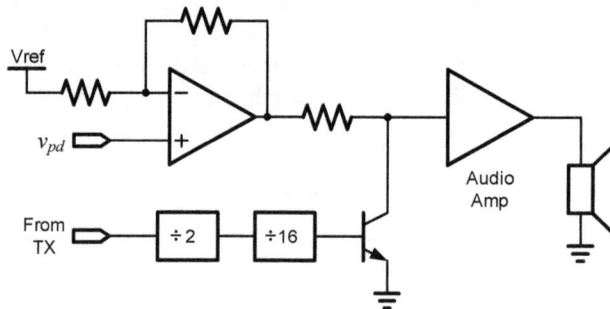

Fig. 11-14: **Audio Circuit**

The remaining circuitry consists of the power supply. A full schematic of an off-resonance design is shown in Figure 11-15.

11.6: Calibration

Because of variations in the search coil inductance and tank capacitor, the nominal resonant frequency may vary and the exact offsets required for particular discrimination points may also vary. It may therefore be necessary to either tweak the tank capacitor (up or down) to get the desired resonant frequency, or to tweak the 555 oscillator adjustment range to get the particular discrimination points correct. The latter is mostly only needed if selectable discrimination points are implemented (as in the A.H. Pro) rather than a linearly variable discrimination.

---

2. Resistor R9 seemingly does nothing, but is often needed for stability. Many opamps are unstable when driving a high-capacitive load, so R9 adds some resistance to the load.

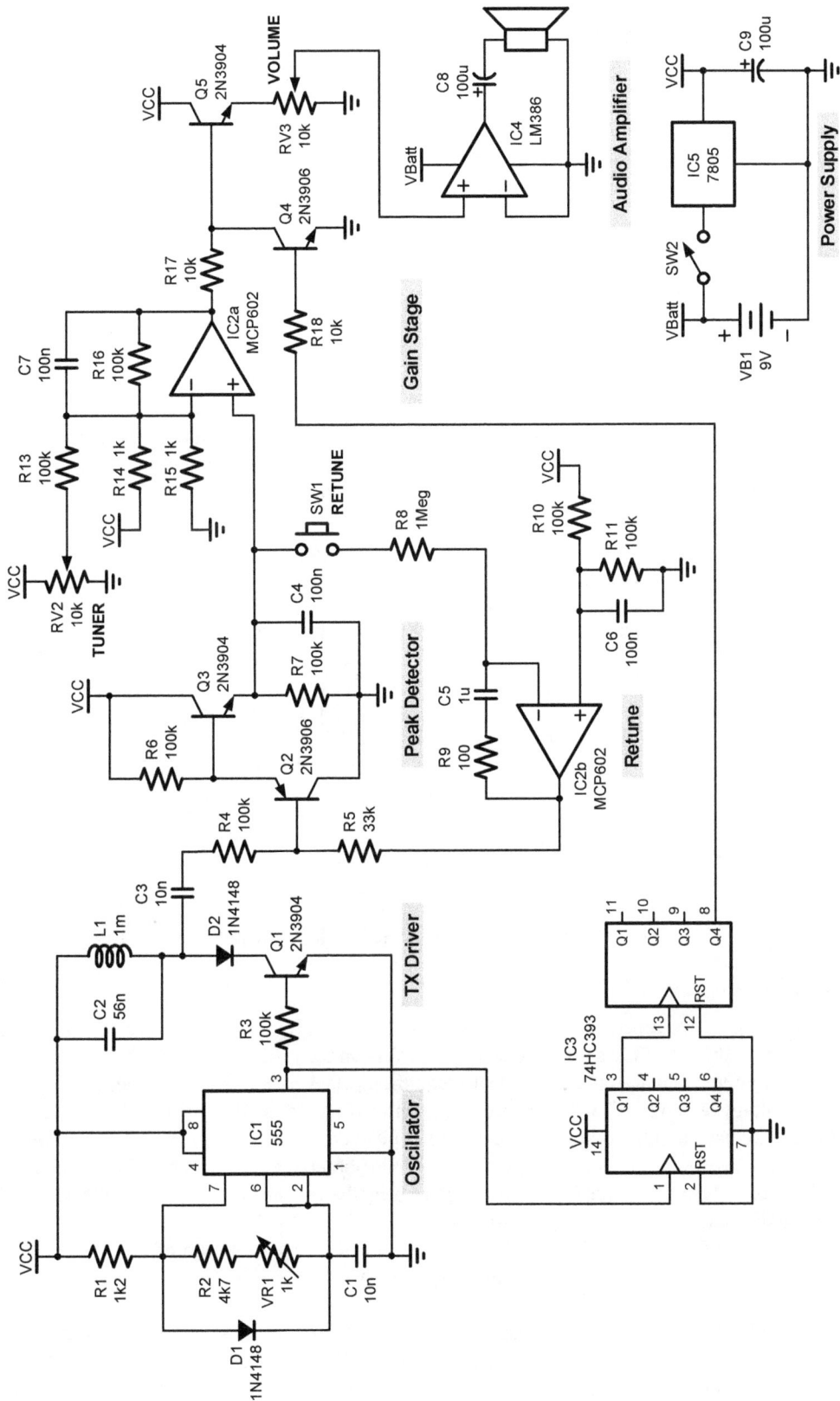

Fig. 11-15: Complete Off-Resonance Schematic

Explore

11.7: TX Amplitude Control

It was noted in the target data that the TX amplitude drops as the frequency offset is increased. This effectively reduces sensitivity when discrimination is reduced to include more targets. A method of maintaining a consistent oscillation amplitude could correct for this. The A.H. Pro design (Figure 11-3) uses a discrete method of adjusting discrimination (switched resistors R2-R4) and, at the same time, also changes the drive strength of the coil oscillator via R8-R10.

Our design circuit uses a linear discrimination adjustment. The autotune circuit produces a signal that corrects the offset of the peak detector for decreasing amplitude. If, instead, the autotune signal could be used to control the TX oscillator amplitude then the nominal output of the peak detector would remain consistent for different frequency offsets because the oscillator amplitude is kept consistent. There are methods of implementing amplitude control in oscillators but they heavily depend on the oscillator topology. As with so many other *Explore* topics, it is "an exercise left to the reader."

Amplitude correction could be under user control via a push button switch, as is done in the main design. It can also be continuous as Figure 11-16 suggests, which means that it becomes a motion detector. While this means you cannot hover over a target, an advantage is that broader mineralization offsets will be tuned out as long as the coil height is maintained. In other words, a crude form of ground tracking. It requires a judicious choice of the integrator RC time constant such that it is not too fast for targets and not too slow for ground.

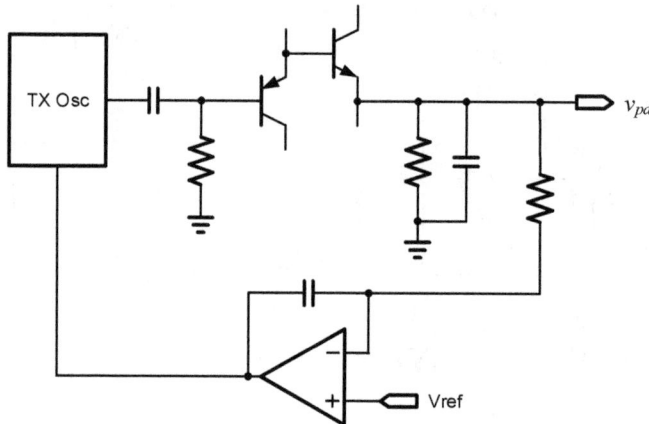

Fig. 11-16: **Automatic Amplitude Control**

11.8: Digital Design

Once again, a microcontroller can replace much of the analog circuitry in Figure 11-15. Instead of using a 555 timer to drive the TX circuit we can use a PWM in the micro. An ADC channel can be used to read a potentiometer that sets the amount of frequency offset. Another ADC channel samples the peak detector voltage, and a third ADC channel reads another pot for setting the threshold level.

A DAC can be used to adjust the offset level of the peak detector in the same way the autotune circuit does in the analog design. If your preferred micro does not have a DAC, then a PWM and capacitor can be used to fake a DAC by adjusting the duty cycle. Finally, another PWM can be used to provide a variable tone audio output. Figure 11-17 illustrates.

As pointed out in Chapter 9's *Explore* section, the typical ADC resolution of 10-12 bits will limit sensitivity so oversampling and averaging may be needed. Another solution is to add a gain stage to the peak detector.

In the all-analog design it may be necessary to calibrate the offset range to the actual resonant frequency, which can vary due to component tolerances. In the digital design, the true resonant frequency can be found at start-up by varying the PWM and finding the highest peak detector reading, assuming

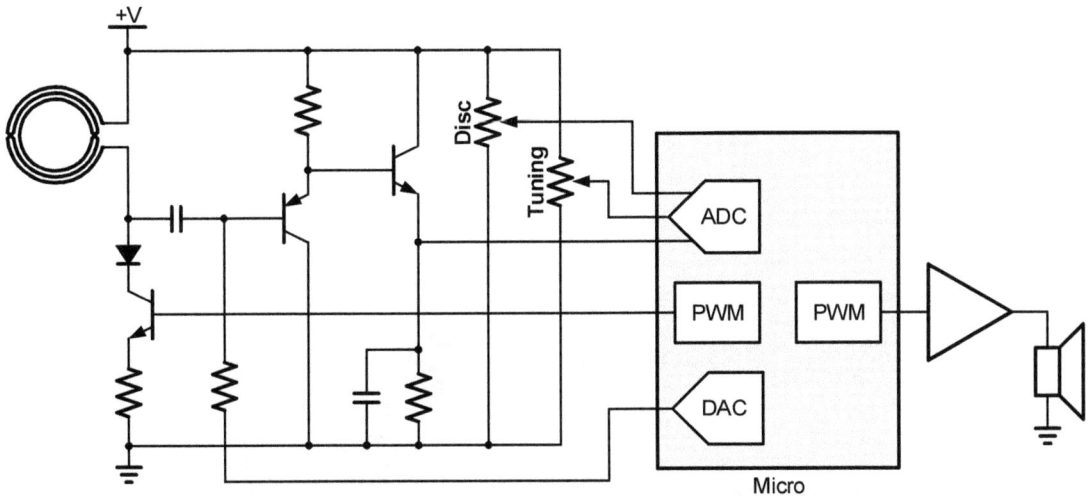

Fig. 11-17: **Digital Off-Resonance Design**

the coil is held in the air. Then the offset range can be calculated and the proper offset applied according to the potentiometer setting.

12 Phase-Locked Loop

"The future ain't what it used to be."

— *Yogi Berra*

Signetics was an early pioneer in designing and manufacturing integrated circuits and notably developed several unique analog chips. One, the NE555 timer, is still very widely used today[1] and has found its way into many metal detector designs. Another Signetics first in 1969 was the integrated phase-locked loop (PLL), the NE565.

Because Signetics produced very uniquely capable chips, their data books[2] contained a lot of application circuits to show how their chips could be used. One such application was a metal detector using the NE565 PLL chip, reproduced in Figure 12-1. Although they included this application circuit for many years, it appears that no manufacturer ever adopted it in a commercially produced metal detector. In fact, I've found no commercial application of a PLL design of any kind, yet it occasionally appears in literature, such as the LM561 variant in the November 1973 issue of Practical Wireless or the LM567 variant in Charles Rakes' book *Building Metal Locators: A Treasure Hunter's Project Book* (1986). Nonetheless, it is an interesting approach and one that — due entirely to the Signetics data book — many people over the years have built and experimented with[3].

Signetics is long gone[4] so the days of their awesome-cool specialty chips are sadly over. The

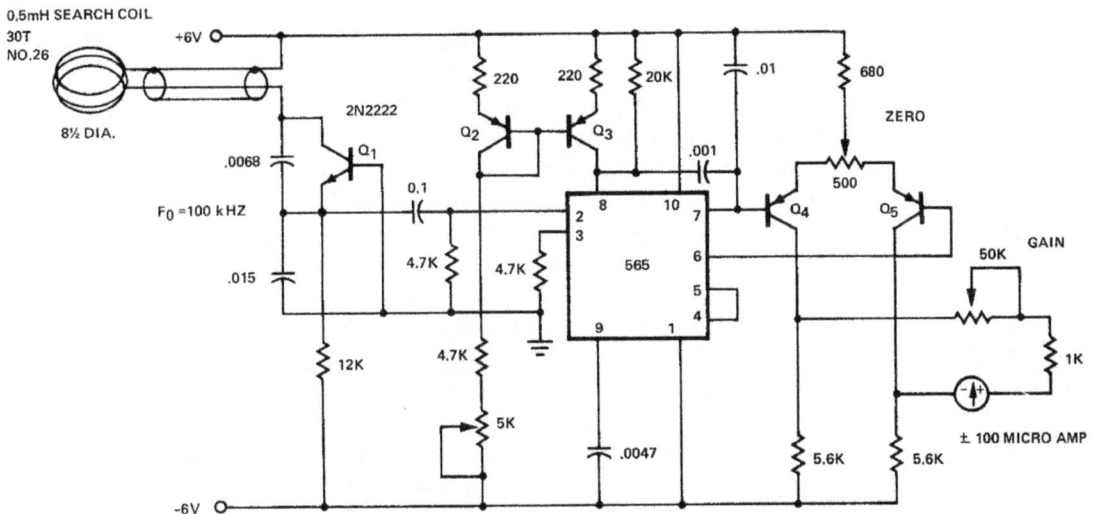

Fig. 12-1: Signetics NE565 Application Circuit (1972)

1. According to Wikipedia, the 555 stands as the most popular IC ever produced.
2. Old Signetics data books are worthwhile acquisitions for a good engineering library.
3. A variant of the Signetics app circuit was published in Electronics World (Sept 1992); several people who have attempted to build this version say it doesn't work.
4. Signetics was bought out by Philips Semiconductor, which was acquired by NXP, which then merged with Freescale (which used to be Motorola), and the whole thing was bought by Qualcomm.

NE565 and other versions of it (like the LM565 from National Semiconductor) are now obsolete and the only reasonable source of chips is eBay. Several years after the introduction of the NE565, RCA developed their own PLL chip — the CA4046 — which, likewise, was produced by numerous companies as part of the CD4000 series[5] and is still available as a catalog item. Our PLL design will use this chip.

12.1: Overview

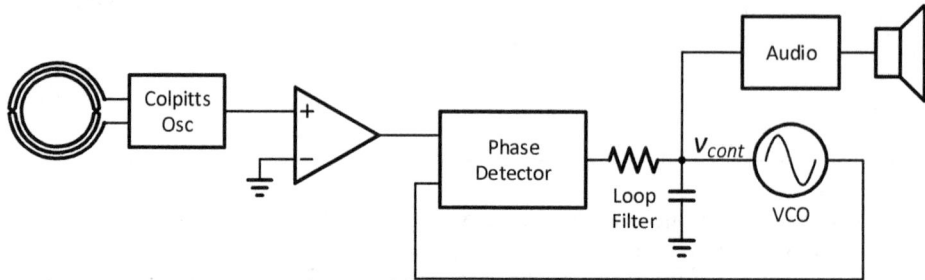

Fig. 12-2: **PLL Detector Block Diagram**

A block diagram of the PLL metal detector is shown in Figure 12-2. Like the BFO, the search coil is part of a free-running oscillator whose frequency will change in response to nearby magnetic or metal targets. The oscillator frequency is squared up with a comparator and fed to the phase detector of the PLL. The other side of the phase detector is connected to the PLL's voltage-controlled oscillator (VCO). When the PLL locks, the VCO will be running at the same frequency as the search coil oscillator.

When a target creates a frequency shift in the search oscillator the PLL will force the VCO frequency to track the shift. It does so by altering the control voltage V_{cont} to the VCO via the RC loop filter. This voltage can then be monitored by a response circuit, from a simple comparator and LED to a 555-based VCO audible response.

12.2: Inside the 4046

Before diving into a complete design it is first necessary to understand the 4046[6] chip. A functional diagram with typical external connections is shown in Figure 12-3. The input signal is first amplified and buffered; the input amplifier is self-biased so the input signal may be AC-coupled. The input signal is then applied to a phase detector along with the VCO output.

There are two phase detectors in the 4046: a standard XOR gate and an edge-controlled memory network. You can use either one but not both. The XOR gate is simpler but requires both inputs to have a 50% duty cycle. It can also lock on to harmonics which sometimes can be beneficial. The edge-controlled detector can take any duty cycle and is better-behaved regarding its lock range.

The output of the phase detector is applied to a low-pass loop filter to provide a control voltage to the VCO. This will increase or decrease the VCO frequency as needed until both the frequency and phase of the VCO signal is synchronized with that of the input signal. It is possible to include a frequency divider between the VCO output and the phase detector which will multiply the VCO frequency (compared to the input frequency) by the division rate.

The nominal VCO frequency is controlled by R1 and C1. R2 provides the ability to offset the lock range of the VCO which is useful in improving the $\Delta v/\Delta f$ which, for a PLL metal detector, means improved sensitivity. 4046 data sheets have formulae and graphs for determining R1, R2, and C1 and the reader is encouraged to experiment with a raw 4046 to see how they interact.

---

5. Much of the CD4000 logic family also shows up in the 7400 logic family. For example, besides the commonly available CD4046 there is also the 74HC4046. In most cases a 74HC4000 chip is functionally identical to its CD4000 brethren. However, there is a slight difference in the case of the 4046: the CD4046 has a zener diode feature on pin 15, whereas the 74HC4046 has a third phase comparator option. In our designs we will not be using pin 15 so either version will work.

6. Again, either the CD4046 or the 74HC4046; we will ignore pin 15 and simply call them the "4046."

Inside the Metal Detector

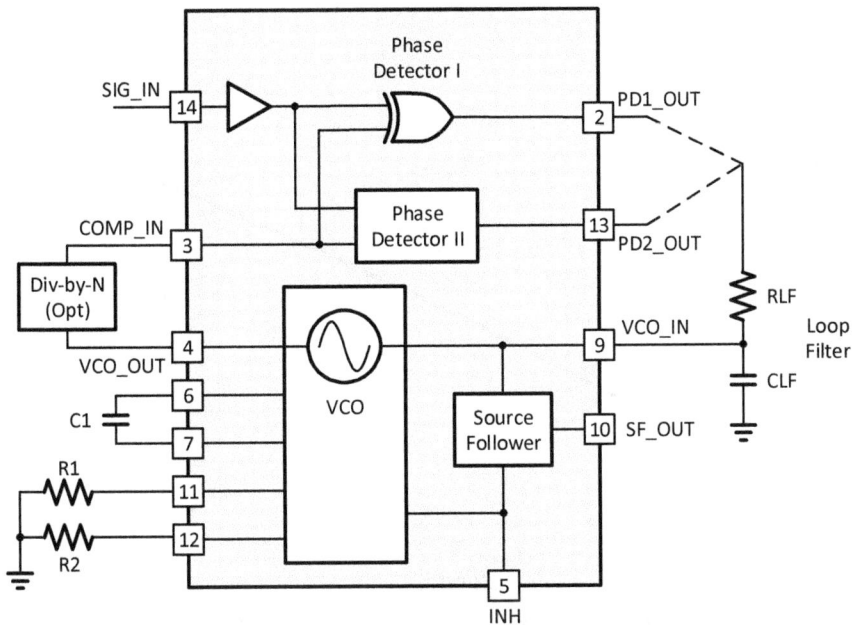

Fig. 12-3: **4046 Functional Diagram**

12.3: Testing the 4046

It is useful to build a simple test circuit for the 4046 to see how it works and what will be necessary to make it into a reasonable metal detector. The first useful thing to do is to see what values of R1 and C1 are needed to create a particular frequency. This can be found in a graph in the data sheet but the graph doesn't offer good resolution so it will only get you in the ballpark.

Let's say we want to power the 4046 from VCC = 5V and we want the VCO to run at a center frequency of 100kHz. A "center frequency" implies that the VCO control voltage (VCO_IN) is at 1/2 VCC, or 2.5V. So connect a 2.5V supply to VCO_IN (a resistor divider will also work). For R1 = 1.8kΩ and C1 = 1nF the VCO runs at about 98.5kHz, close enough for now.

Next, we can sweep VCO_IN and see how the VCO frequency changes with voltage. This is shown in Figure 12-4. The relationship is fairly linear but with slightly less slope at the highest control voltage. What we are interested in is the best $\Delta v/\Delta f$, or the lowest slope possible. The slope at VCO_IN = 2.5V is 10mV/kHz and the slope at VCO_IN = 4V is about 14mV/kHz. This suggests there is a slight advantage in running the VCO with a high nominal control voltage. With R1 and C1 altered for 100kHz at VCO_IN = 4V (R1 = 6.8kΩ, C1 = 1nF) it gets even better; the $\Delta v/\Delta f$ is now 32mV/kHz, a three-fold improvement. It was mentioned earlier that R2 can potentially improve this further; a value of 1.5kΩ gives a final $\Delta v/\Delta f$ of 39mV/kHz.

A final test is to see the PLL actually work. For this, connect a 100kHz square wave (5V logic) to pin 14, short pins 3 and 4, and replace the voltage drive on pin 9 with a loop filter, using pin 13 for the phase detector. The loop filter bandwidth is not critical; an excessively low bandwidth will make the VCO frequency response slow (remember, we'll eventually be sweeping over a metal target) and an excessively high bandwidth will cause frequency instability (jitter). Use 100Hz; RLF = 15k and CLF = 100nF will work. If you probe both the input waveform (pin 14) and the VCO waveform (pin 4) you should see 100kHz square waves that are slightly out-of-phase. If you measure the voltage at VCO_IN (pin 9) it should be a DC waveform somewhere close to 4V.

With a BFO, we often see frequency shifts of only 10Hz, maybe less. For a $\Delta v/\Delta f$ of 39mV/kHz this means a 10Hz shift due to a target produces a loop filter change of only 390μV. This tells us that a decent amount of gain will need to be applied to the control voltage to be able to resolve weak responses.

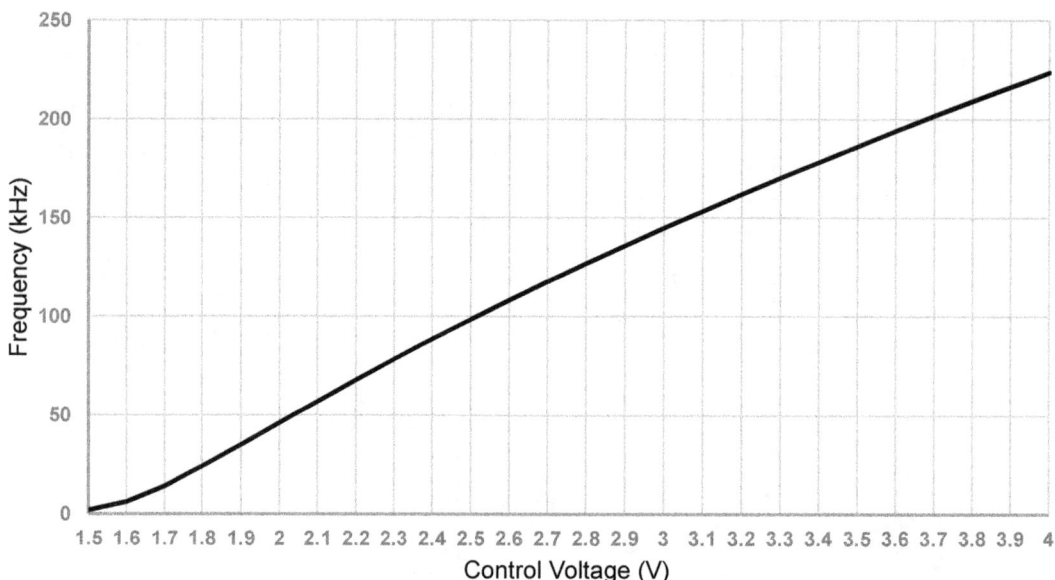

Fig. 12-4: **VCO Frequency vs Control Voltage**

Design: PLL Detector

A complete schematic is shown in Figure 12-5. The search oscillator is the same Colpitts design used in Chapter 9 and the coil is the same design as well. This results in the same operating frequency of 100kHz. The Colpitts' sinusoid is squared up by a common-emitter stage and fed to the phase comparator. The VCO is set up with a nominal oscillation frequency of 100kHz at VCO_IN = 4V. The loop filter is set to 100Hz.

The loop filter control voltage is fed to the non-inverting side of a gain stage. The 4046 also provides a buffered version of the control voltage via a source-follower on pin 10 but this will potentially have a temperature drift. Because the nominal control voltage is around 4V the gain stage must be offset down to a lower voltage, whatever level provides a reasonable threshold audio sound. This is achieved with a variable offset voltage generated by R9, R10, and VR1. VR1 varies the amount of offset and becomes the "tuning" control.

The voltage divider can be reduced to a Thevenin-equivalent circuit of a DC voltage and a series Thevenin resistance, where the series resistance is the denominator of the gain:

$$G = 1 + \frac{R_{11}}{R_{TH}} \qquad \text{where} \quad R_{TH} = R_9 \parallel (R_{10} + R_{V1})$$

From this the gain will vary a bit versus the pot setting, from 100 to 167. At a gain of 100, a 10Hz shift in the oscillator should produce an almost 400mV change at the output of the gain stage.

Finally, a 555 timer generates a 600Hz square wave. The 555 output then drives a signal chopper (Q3) which chops the signal at the input of an LM386 audio amp. The result is a constant 600Hz audio signal whose amplitude increases for non-ferrous targets and decreases for ferrous targets. If the tuner is set for a low audible tone then non-ferrous targets will increase the loudness and ferrous targets will make the audio go quiet.

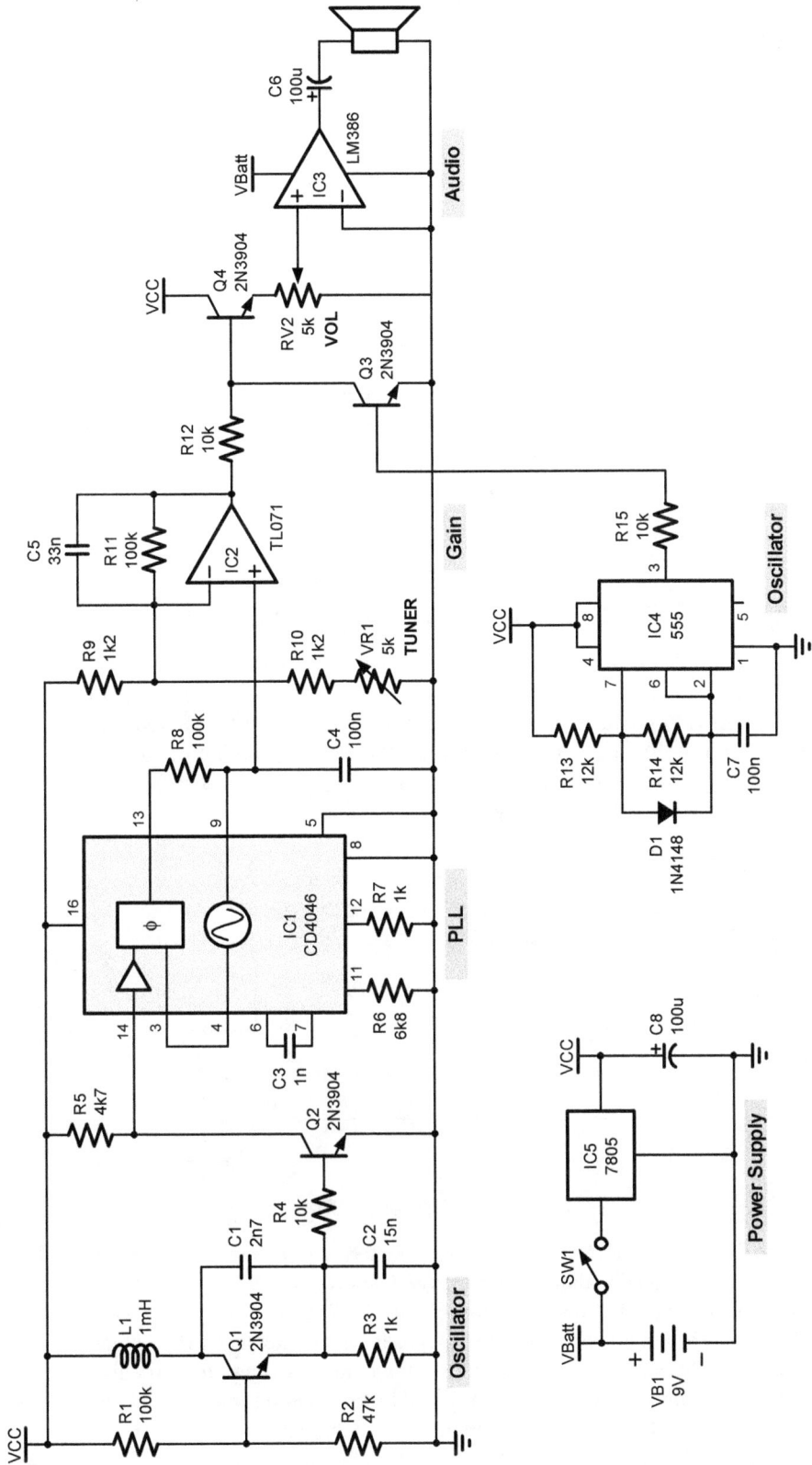

Fig. 12-5: **Complete PLL Schematic**

Explore

12.4: Frequency Scaling

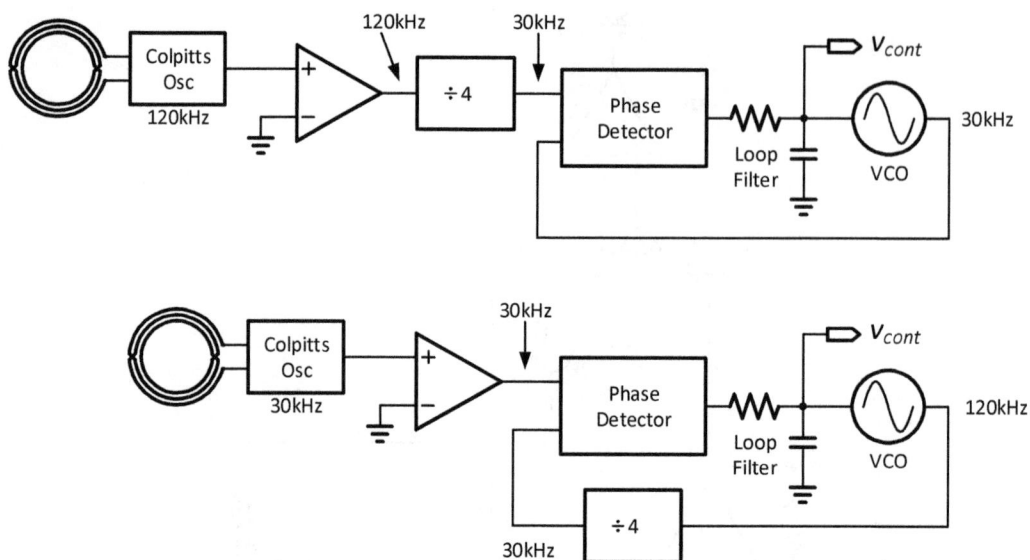

Fig. 12-6: **Frequency Scaling**

In the PLL design the VCO runs at the same frequency as the search oscillator, about 100kHz. It is possible to run the VCO at a lower frequency by dividing down the search oscillator frequency prior to the phase detector. It is also possible to run the search oscillator at a lower frequency by dividing down the VCO frequency prior to the phase detector. Figure 12-6 shows both possibilities using a divide-by-4 as an example. Scaling the VCO to a different frequency (higher or lower) also scales the VCO's Δf response by the same amount and alters the $\Delta v/\Delta f$ loop filter sensitivity.

12.5: Reversed Roles

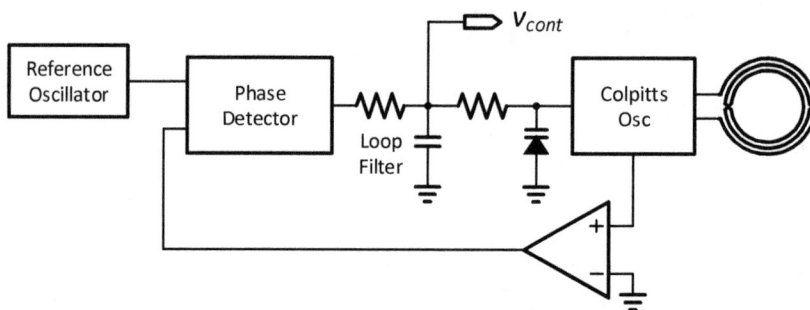

Fig. 12-7: **Design Variant**

A variant to the standard PLL design is to reverse the roles of the oscillators. Instead of forcing the VCO to track the search coil oscillator, place the search coil in the VCO and use a fixed-frequency (perhaps even a crystal) for the reference oscillator. Figure 12-7 illustrates.

Proximity effects that induce changes in the search coil inductance or Q will attempt to change the VCO frequency, but the phase detector will alter the control voltage to maintain the same VCO frequency. That is to say, the search coil frequency will be constant no matter what target is under the coil, but the control voltage will vary and can be monitored as the output of interest.

The drawback to this design is that the VCO must be custom designed. The VCO internal to the 4046 is designed as an RC ramp oscillator and is not suitable to drive the coil. However, we already know how to make an LC-based VCO; recall the voltage-tunable reference oscillator from Figure 10-11. Just remove the potentiometer and drive the varactor with the loop filter output. It may be necessary to apply gain and/or offset to the loop filter voltage and an offset control can be used for tuning the detector.

The 4046 phase detector could still be used or a discrete XOR chip. If using the 4046 for the phase reference you might also use the internal VCO as the reference oscillator, although its open-loop stability is probably not great. Again, a better choice would be something crystal-controlled.

12.6: VLF BFO

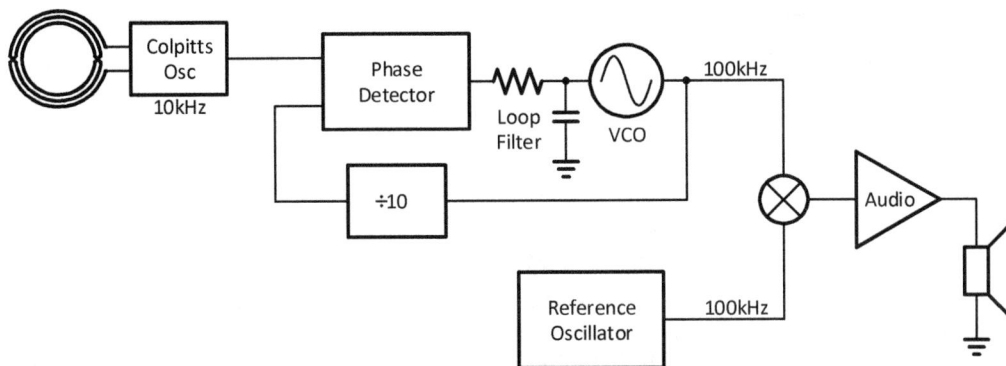

Fig. 12-8: **VLF BFO**

A slightly different application of the PLL is to frequency-multiply a VLF oscillator into the 100-ish kHz range (as in Figure 12-6) and then use the higher frequency in a BFO circuit. Figure 12-8 shows this approach. From Chapter 10 we know that the amount of frequency shift imposed on the oscillator by a target is proportional to frequency. Running the transmit frequency at, say, 10kHz means that the target-induced frequency shift will be small and not a viable option for a normal BFO. Using a PLL to multiply the oscillator by a factor of 10 or 20 means that the target-induced frequency shift will also be multiplied by the same amount. This creates a sufficient mixer shift for the audio.

There are a couple of advantages in this topology. First, the transmitter can now be run at any frequency including the very low VLF range (a few kHz). This may help minimize ground effects. Second, the frequency can be multiplied even higher (say, 250kHz) where it is much easier to make a varactor-tunable reference oscillator. It should be noted that once the transmit oscillator is multiplied to a normal BFO frequency, other methods described in the BFO chapter will also work.

12.7: Multifrequency Induction-Balanced VLF BFO

The reference oscillator in Figure 12-8 can be replaced by a copy of the search oscillator and PLL resulting in a system with two independent search oscillators. Refer to Figure 12-9. Suppose one search oscillator is set to 10kHz and the other is set to 12.5kHz, and the PLL feedback dividers are set so that the PLL-VCO frequencies are both 100kHz. We can then mix these frequencies in the normal BFO manner.

To prevent frequency pulling between the two oscillators it is necessary to use an induction-balanced coil. If a DD coil is used then as the coil is swept across the target the leading coil will respond, followed by the trailing coil. Because the mixer effectively subtracts the two frequencies (Equation 10-1) an increase in one oscillator will produce an increase in the audio tone and an increase in the other oscillator will produce an decrease in the audio tone. This means that as the coil is swept over a non-ferrous target, in one sweep direction you will hear a "wee-woo" audio response and in the other sweep direction you will hear a "woo-wee" response. A ferrous target will produce the opposite responses for the same sweep directions, which ultimately means the only way to tell whether the target is ferrous or non-ferrous is to pay attention to the sweep direction of the coil combined with the audio response.

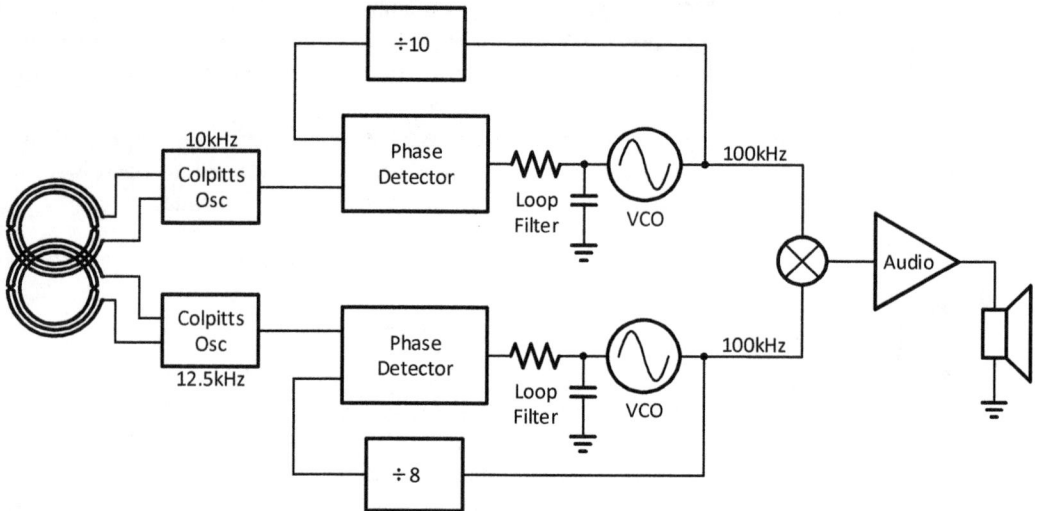

Fig. 12-9: **MF IB VLF BFO**

The search frequencies (10kHz and 12.5kHz in Figure 12-9) should be non-harmonically related to minimize frequency pulling. A wider difference in frequencies would likely improve sensitivity to a broader range of targets. By carefully listening to the strengths of the leading and trailing audio responses it may be possible to distinguish between low and high conductivity targets.

12.8: Digital PLL

Another design possibility is to code the PLL function in a microcontroller. For the basic PLL circuit in Figure 12-2 this doesn't make a lot of sense; you may as well just use the micro to measure the TX frequency directly (as in Figure 10-18) instead of creating a software PLL that alters a DAC output voltage.

However, a digital PLL option that could make sense is to replace the frequency-multiplying PLL in Figure 12-8 with a software PLL in a micro. The TX oscillator can drive a input capture channel and a counter can output a multiplied frequency. A separate counter (or NCO) can provide the tunable reference frequency. Figure 12-10 shows a possible solution; this is an extension of the concept presented in Figure 10-17.

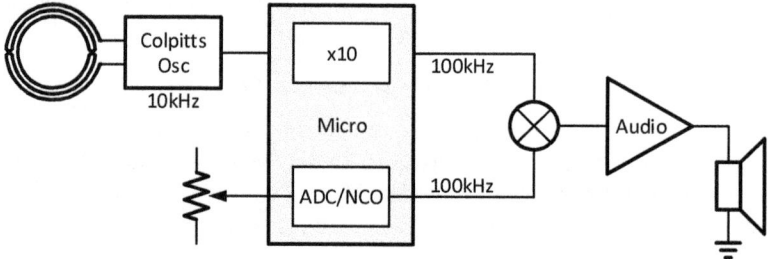

Fig. 12-10: **Micro-based Version**

PART 4
Induction Balance

BOUNTY HUNTER
RED BARON
METAL/MINERAL
DETECTOR
'78 CATALOG
Engineered for TOMORROW'S Treasure Hunter by PNI, Inc. • 1309 W. 21st Street, Tempe

"The Name That Means Treasure!"
Tesoro
DEEP SEARCH
VLF/TR/CPR (PATENT PENDING)
METAL DETECTORS
TESORO ELECTRONICS, INC.

Step into a world of adventure, intrigue and the excitement of anticipation with the worlds first programmable detector . . .
THE MARK I
CoinComputer
TeKnetics MARK I
THE DEEPEST DEPTH . . .
THE SLOWEST SWEEP . . .
PLUS THE AMAZING NEW TARGET SELECT . . .
TeKnetics, Inc.

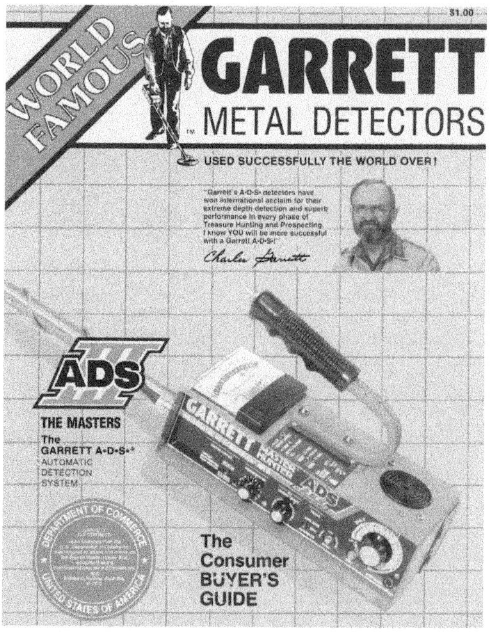
$1.00
WORLD FAMOUS
GARRETT
METAL DETECTORS
USED SUCCESSFULLY THE WORLD OVER!
ADS III
THE MASTERS
The GARRETT A-D-S-*
AUTOMATIC DETECTION SYSTEM
DEPARTMENT OF COMMERCE
UNITED STATES OF AMERICA
The Consumer BUYER'S GUIDE

13

Transmit-Receive

"The most exciting phrase to hear in science, the one that heralds new discoveries, is not Eureka! I found it but That's funny ..."

— Isaac Asimov

The first applied metal detector — in 1881 with Alexander Graham Bell — was an induction balance design. Somewhere along the way, IB detectors became known as transmit-receive (TR) detectors and for a hundred years or so after Bell the basic IB/TR was a mainstay in metal detector design. The predominant detectors of today are usually called VLF designs but they are still the latest in a long evolution of IB. Modern motion-VLF designs have far more capability and complexity than the old basic TR designs, and instead of jumping directly into a complex design we will start with the basics and build our knowledge.

13.1: Overview

The original TR detectors were very primitive in their operation. The need to use vacuum tube circuitry and run the whole thing on batteries limited the complexity of designs. Bell's approach predated even vacuum tubes (and even his batteries were home-made) so he was limited mostly to a clapper device and an earpiece. This required the clapper to run at a frequency that was directly audible, as we saw in the Bell detector project in Chapter 3.

With the availability of vacuum tubes, designers began increasing the transmit frequency well beyond the audible range — even beyond 100kHz — in order to achieve a higher power efficiency. If this were done with a continuous-wave (CW) sinusoid, another circuit would then be needed to create an audible tone. Today we would not think twice about adding a 555 timer, an opamp, or even a small micro to handle such a task but in the 1920s even one more additional tube was a big deal. Designers took a short cut and used a *self-quenching oscillator*[1] instead. The oscillator normally runs at a high frequency[2] but its output is gated at a lower (audible) frequency. The resulting waveform is shown in Figure 13-1. Although the oscillator runs at about 100kHz, the gated pulses have a rate of about 650Hz.

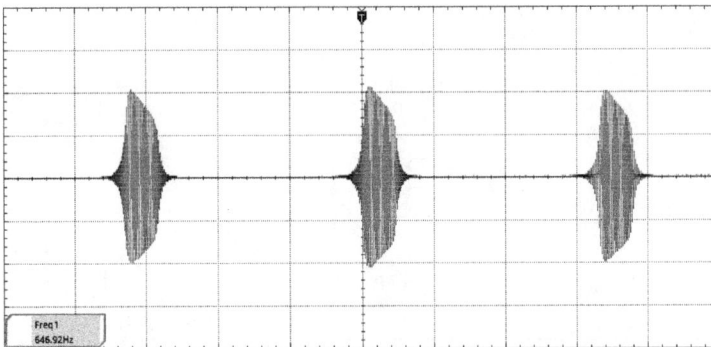

Fig. 13-1: **Quenched Oscillator Waveform**

---

1. Or *squegging* oscillator, where *squegging* is a (very rough) contraction of *self-quenching*.
2. Often called the *carrier frequency*.

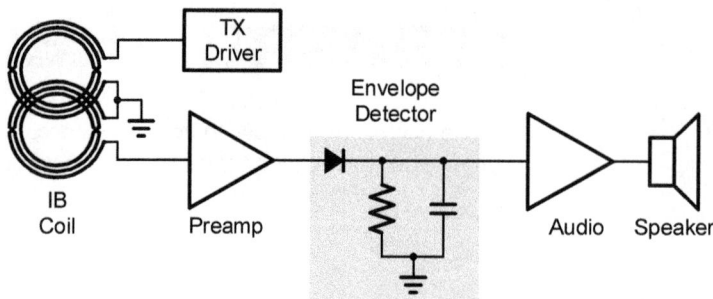

Fig. 13-2: **Block Diagram**

With an induction-balanced coil the RX signal is nominally zero in the absence of a target. A target (whether ferrous or non-ferrous) then couples the same gated oscillation into the RX coil. The signal is amplified and applied to an envelope detector which removes the carrier frequency and leaves only the signal modulation, which is our audible signal. This can be further amplified and sent to a speaker. Figure 13-2 shows a block diagram.

While this approach was favored by vacuum tube designers, it survived well into the era of discrete transistor design, even into the 1960s and 1970s. Detectors from the big companies like Fisher and White's all the way down to kit designs sold by Radio Shack and Heathkit were often quenched oscillator TR designs. Perhaps the last self-quenching designs to be sold were the *Mark 1* and *Mark 2* detectors made by Bill Hays, up through about 2007[3]. The Bounty Hunter *TR-500* shown in Figure 13-3 is almost identical to the Hays *Mark 2* design.

You will notice the *TR-500* has no provision for discrimination. Indeed, for the most part the quenched oscillator approach did not have discrimination. There was some work toward implementing

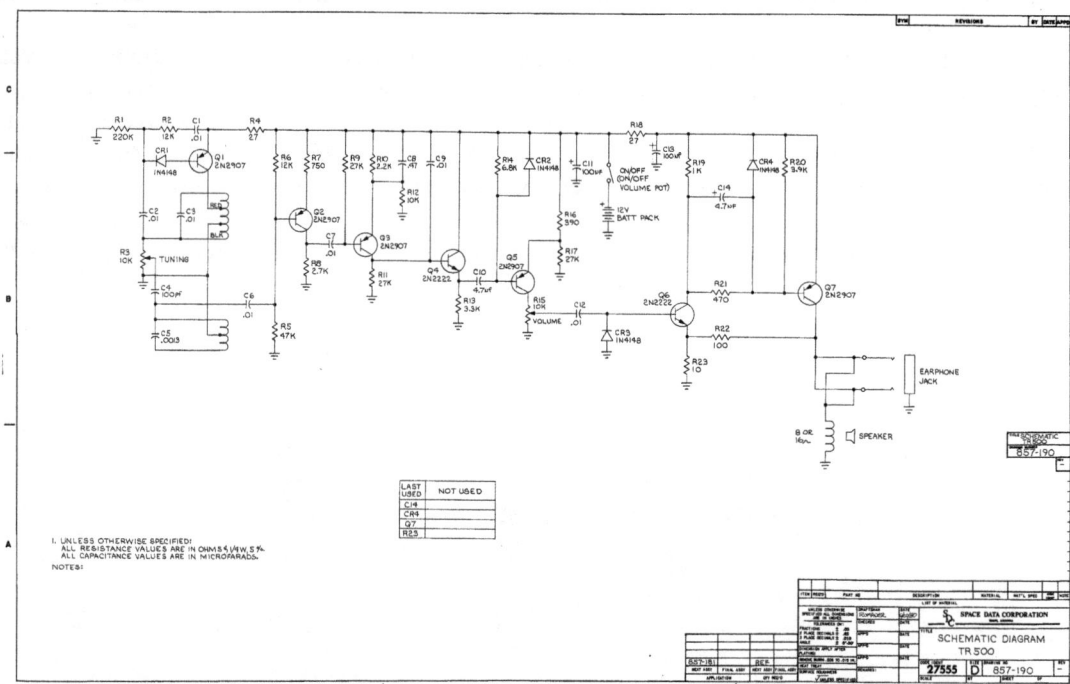

Fig. 13-3: **Bounty Hunter *TR-500***

3. As far as I can tell, the *Mark 1* was introduced in 1962 and the *Mark 2* in 1963. At 45 years or so, they likely hold the record for the longest production run of a metal detector model.

Inside the Metal Detector

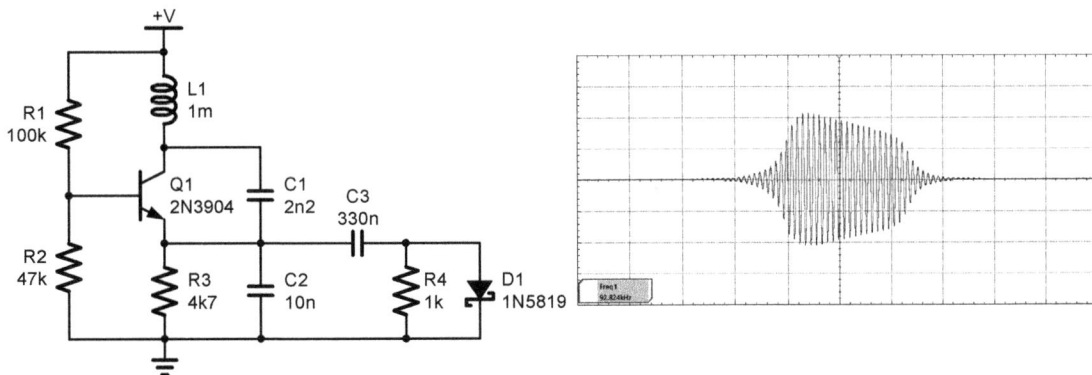

Fig. 13-4: **Transmit Circuit & Waveform**

discrimination (and maybe even released in a few models) but by that time the synchronous sampling method was taking over and the vast majority of TR-Discriminators used this approach (Chapter 14). However, the quenched oscillator TR detector had a strong run from the 1920s until the 1970s.

13.2: Transmit Circuit

If you review the schematics for a variety of TR detectors the first thing you will notice is that the quenched oscillator designs are usually complex and involve multi-tapped coils with multiple feedback paths. The *TR-500* is actually one of the simpler designs. It is possible to make a quenched oscillator that can use a mono coil.

Such a transmitter circuit (Figure 13-4) looks very much like our old friend, the Colpitts oscillator, used in Chapter 9 and beyond. And it is, with the addition of an RC circuit and a diode. These additional components create the self-quenching action as seen in the waveform. The oscillator design is much the same as for the *Classic BFO* design using the same 1mH coil and an oscillation frequency of about 100kHz. The values for C3 and R4 are determined experimentally[4] to give a gate frequency close to 600Hz which was a popular audio frequency for TR detectors. The diode is a 1N5819 Schottky but any Schottky diode should work and, in fact, a 1N4148 diode will work but not quite as well.

C3 and R4 create a pole at 482Hz which is not exactly equal to the gating frequency of 600Hz (per Figure 13-1) but is in the ballpark. Changing C3 and/or R4 will change the gating frequency. As C3 decreases it also begins to affect the carrier frequency. If needed, C1 and C2 can be altered to re-adjust the carrier frequency.

13.3: RX Circuit

Figure 13-5 shows the receive circuitry. In the spirit of nostalgia this initial design will be accomplished with discrete transistors[5]. The RX coil[6] is connected to a common-emitter amplifier. The gain of this stage is $G = R6/r_e$ where $r_e = V_T/I_E$ and $V_T = kT/q = 26mV$ at 27°C. To solve for the gain we need to calculate the bias current. It is

$$I_E = \frac{VB - 0.7}{R6 + \dfrac{R5}{\beta + 1}} = \frac{8.3\,V}{1k + \dfrac{150k}{\beta + 1}}$$

Eq 13-1

The 2N3904 has a β of 100-300 so let's do the calculations for low, nominal, and high β:

---

4. Self-quenching oscillators are difficult both for hand-analysis and simulation.

5. We will not be so nostalgic as to consider vacuum tubes.

6. The requirements of which will be discussed in the next section.

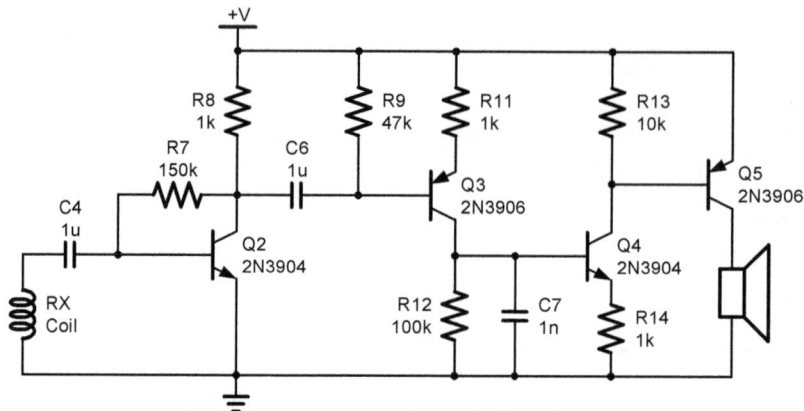

Fig. 13-5: **Receive Circuit**

β	I_E	G	V_C
100	3.34mA	128.5	5.66V
200	4.75mA	182.8	4.25V
300	5.54mA	213.1	3.46V

The reason for doing this short exercise is to show that designing transistor-level circuits requires a little more attention to parametric variation than when dealing with opamps. In this case, a change in transistor β will alter both the gain and the output DC level. The output DC level potentially limits available signal swing; that is, too much or too little offset can clip the signal.

It is not uncommon to substitute parts based on local availability. For example, the 2N3904 is popular in the US but may be less so in other places. Substituting Q2 may require changing R7 depending on the β. A starting point for R7 is $β \cdot R8$ (using the nominal transistor β) but it's best to select R7 to achieve a collector voltage of around 5V (assuming a 9V supply).

The next stage (Q3) is another common emitter amplifier[7]. This transistor is not biased in the active region as in the usual fashion so it only amplifies negative dynamic signals coupled through C6. R12-C7 form a low bandwidth load (about 1.6kHz) which shunts the 100kHz carrier frequency. As such, Q3 both extracts and amplifies the envelope of the RX signal. Because it takes negative dynamic signals to activate Q3, its output is nominally 0V and has only exhibits positive dynamic signals at its collector. Figure 13-6 shows the transmit waveform along with the signal at Q2's collector and at Q3's collector.

Q4 and Q5 form the final gain stages and drive the speaker. With a nominal 0V at the collector of Q3 both Q4 and Q5 will be completely off and there will be no sound. Therefore, it will take some amount of RX signal to activate Q3, Q4, and Q5. For this reason, when substituting any of these transistors there is probably no need to alter any other component values.

Normally we would apply some kind of threshold adjustment to bias everything to have a very slight amount of signal. In designs where the output audio is driven by an independent oscillator the threshold bias is just a DC offset voltage or current. But in this design it is the RX signal itself that needs an offset. This can be done by slightly offsetting the IB coil null so that there is a very small amount of coupling from the TX coil to the RX coil. That is, we don't want a perfect null. This is the method that was used in the old two-box locators (such as the Fisher model shown in Figure 1-5) whereby the tilt of the RX coil would adjust the threshold. This is more difficult to accomplish in a pancake-style coil[8] so we'll consider a different approach.

---

7. An almost identical circuit was used in the *Classic BFO* design, and you may find this explanation evokes a sense of déjà vu.

8. Although this was exactly done on the 1950s Detectron *Model 27*; a knob conveniently located on top of the search coil adjusted the null.

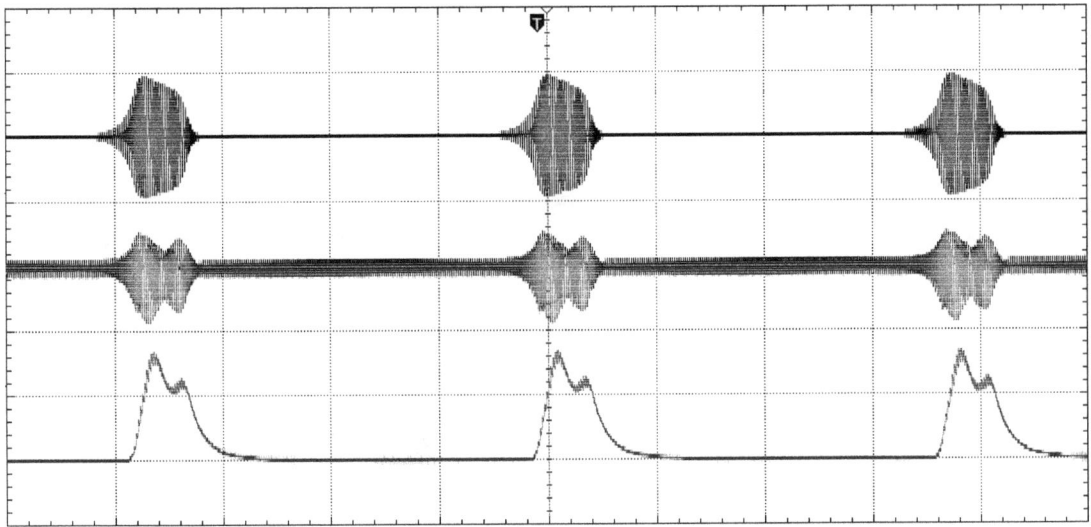

Fig. 13-6: **Transmit; Q2 Collector; Q3 Collector (Target Present)**

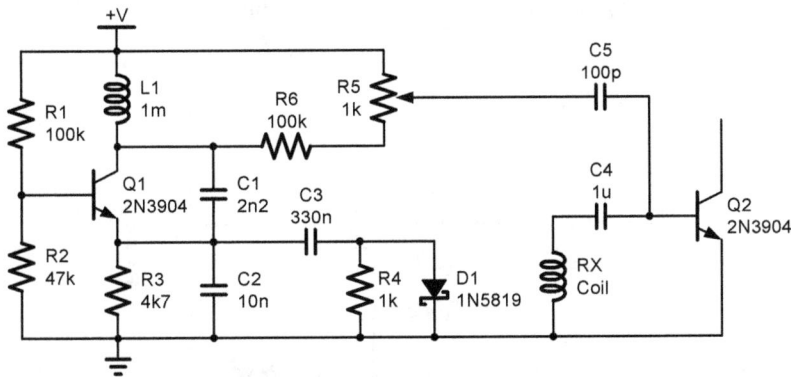

Fig. 13-7: **Threshold Adjust**

Figure 13-7 shows the transmitter and RX input stage with an additional path that couples a small portion of the TX signal into the RX input. This provides an adjustable offset signal for setting a threshold level.

13.4: Search Coil

In this design, the only requirement for the search coil is that it be induction-balanced. Otherwise, just about anything can be made to work. The TX circuit was presented with a 1mH TX coil simply to keep it consistent with the Colpitts design in the BFO chapter. The TX coil can otherwise be anything that can be made to oscillate in that circuit.

The requirement for the RX coil is that it is able to support a 100kHz modulated signal. This means its self-resonant frequency must be higher than the carrier frequency of 100kHz, at least by a factor of two. Except for that, more turns on the RX coil will improve the effective transformer gain and offer better sensitivity. If you want to build a coil and maintain compatibility with future TR and VLF designs then use TX = 1mH and RX = 4mH. Whether it's concentric, DD, or something else doesn't matter.

Since this design is so insensitive to coil parameters another option is to use an IB coil from a commercial detector model. Again, just about anything will work; you will need to adjust C1 and C2 in the TX circuit to get the proper 100kHz oscillation and probably nothing will need to change on the RX side.

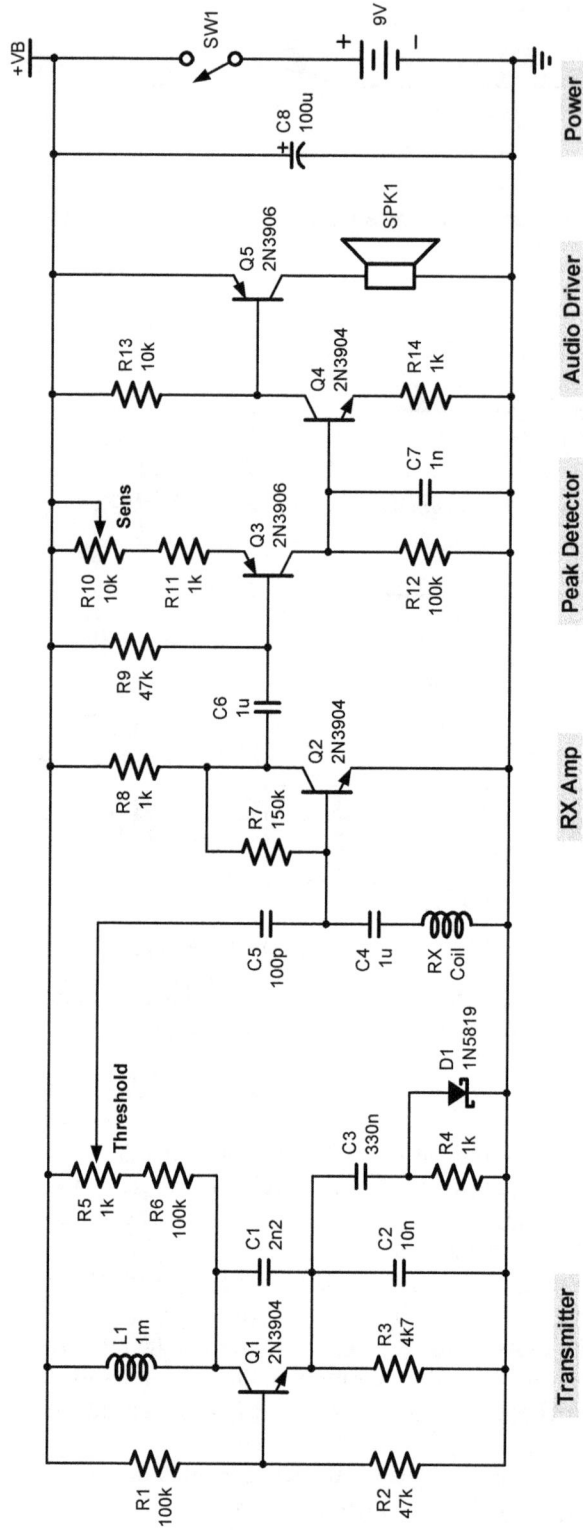

Fig. 13-8: **Basic TR Complete Schematic**

Inside the Metal Detector

Design: *Basic TR*

Figure 13-8 shows the complete TR design. It is identical to everything presented so far with the addition of a battery and power switch, plus a way to adjust sensitivity. This is a very power-frugal circuit; the current consumption from a 9V battery is:

- No target, no threshold   5.5mA
- Threshold   7-8mA
- Full target audio   50mA

For a typical 550mA-hr 9V alkaline battery this circuit will probably run 40 hours or more. Current consumption is dominated by Q2 and could be improved by reducing its bias current to around 1mA. This again underscores why this design was popular for so long: even in the days of vacuum tubes it was a very efficient circuit that did the job.

The TR detector is barely a step up from a BFO. The oscillator still runs at a similar frequency and many of the shortcomings of BFO are present here. Raising and lowering the coil over mineralized ground will evoke a fairly strong response which makes hunting a tedious effort over anything but very mild ground. The main benefit of TR over BFO is the use of an IB coil which offers improved sensitivity.

There is no real discrimination and, unlike the BFO, this circuit will not distinguish the negative frequency shift of ferrous targets, so it is truly an all-metal detector. However, ferrous targets do result in both a negative frequency shift of the carrier and a drop in amplitude in the TX oscillator, effects that one could possibly monitor (see Chapters 9 & 10) to implement ferrous versus non-ferrous discrimination. This is the direction some companies were headed when the introduction of synchronous demodulation nailed this coffin shut.

Design: *Improved TR*

The *Basic TR* is a classic all-transistor design using a quenched oscillator. Let's now consider a TR detector without a modulated transmit signal and with slightly more modern circuitry. In this design the TX circuit will be an improved sinusoidal oscillator as shown in Figure 13-9. This circuit was used in the first two editions of *ITMD* in the *Raptor* IB/VLF project. It is easier to achieve a low-distortion sine wave than with a Colpitts oscillator, a feature that becomes increasingly important in high-performance VLF designs. The frequency of operation is the usual

$$f = \frac{1}{2\pi\sqrt{LC}} \qquad \text{Eq 13-2}$$

A well-nulled induction-balanced coil will normally result in no RX signal in the absence of a target. Because the TX signal is now a sinusoid, when a target is presented to the coil the RX signal will also be a sinusoid of the same frequency. If the sinusoid is in the audio range then all we

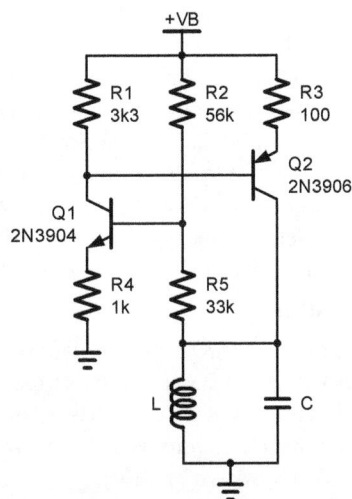

Fig. 13-9: **Sinusoidal Transmitter**

need to do is amplify it and send it to a speaker. This will require a TX frequency in the 500Hz - 2000Hz and Figure 13-11 shows what this design might look like.

While this is certainly possible, most detectors run at frequencies well above this range and for this we need to consider another way to create the audio response. Let's suppose we want this design to run at 10kHz. Not only do we need a separate audio generator but we also need to find a way to detect the RX signal and to trigger the audio response.

Figure 13-10 shows a block diagram of the approach used in the *Improved TR* design. The RX signal is amplified and fed to an envelope detector in the same manner as in Figure 13-2 except that the

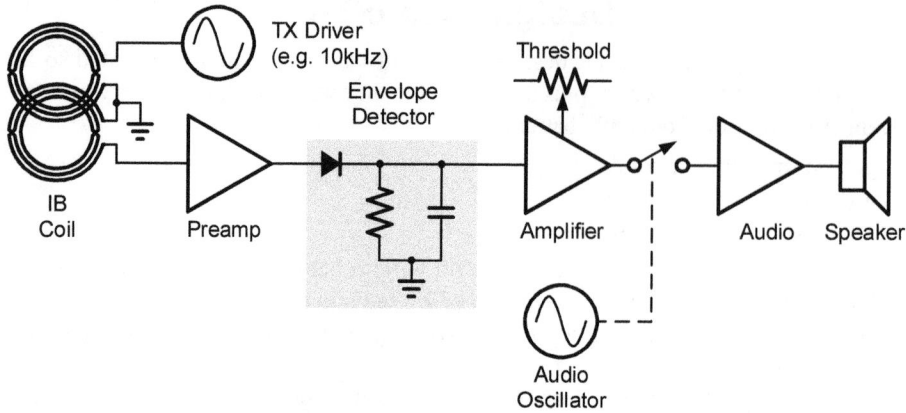

Fig. 13-10: **Improved TR Block Diagram**

RX signal is now an ordinary sinusoid instead of a modulated sinusoid. The envelope detector produces a signal that follows the amplitude of the RX signal so it needs a low frequency response of a few Hertz to perhaps 50Hz for a swept target. This signal is further amplified, an adjustable threshold offset is applied, and a gating switch controlled by an audio frequency clock "chops" it to produce an amplitude-modulated audio signal that drives the final audio stage.

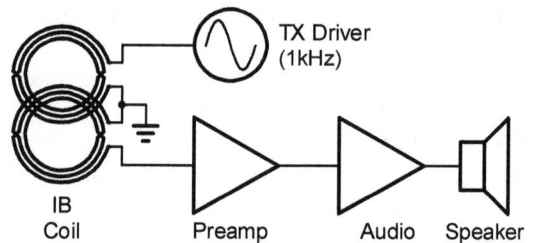

Fig. 13-11: **Direct-Drive TR Detector**

This is all a bit more complicated than the basic TR but has a few advantages. First, we have complete control over the TX frequency and do not need to choose such a high value as in the basic TR. The improved TR can be operated from 1kHz to 100kHz or more. Second, the threshold adjustment is easier to apply. Third, we have more control over the audio frequency and audio quality. Finally, the use of opamps makes the circuit calculations easier.

The complete schematic for the improved TR is shown in Figure 13-12. The power supply regulates the 9V battery down to 5V and also creates a "virtual ground" (2.5V) using a rail splitter. Notice the different ground symbol used for analog ground versus "real" ground[9]. This is a common technique in metal detector design. The audio oscillator is an opamp-based design that generates a 600Hz square wave. This drives Q4 which chops the output of the final gain stage.

Calibration

The *Improved TR* design uses an induction-balanced coil. If the coil is perfectly balanced then the output voltage of IC1a will be virtual ground, as will be the capacitor of the envelope detector. Absent any offsets, it will take a fairly strong target signal to overcome the diode threshold and produce an increase in the envelope detector voltage. If the coil is not perfectly balanced, then the amount of required "overdrive" will vary. To compensate for this, coil trimmer RV1 is added to inject a variable offset in the preamp.

To calibrate the system to the search coil, start RV1 at its maximum value and RV2 (Gain) at its minimum value. Set the Threshold pot (RV3) to its midpoint setting. While monitoring the output of IC1b, adjust RV1 until the output of IC1b just begins to increase from its nominal value. If the output of IC1b is below 2.5V, continue decreasing RV1 until it reaches 2.5V. The circuit is now ready for operation; adjust the Gain pot for stability and the Threshold pot for a faint audio sound.

---

9. The first two editions of *ITMD* used the triangle symbol for regular ground, a holdover from my prior life as a chip designer at Analog Devices. This edition uses the more traditional 3-bar symbol for regular ground (usually the low side of the battery) and the triangle symbol is reserved for "special" grounds, such as in this case.

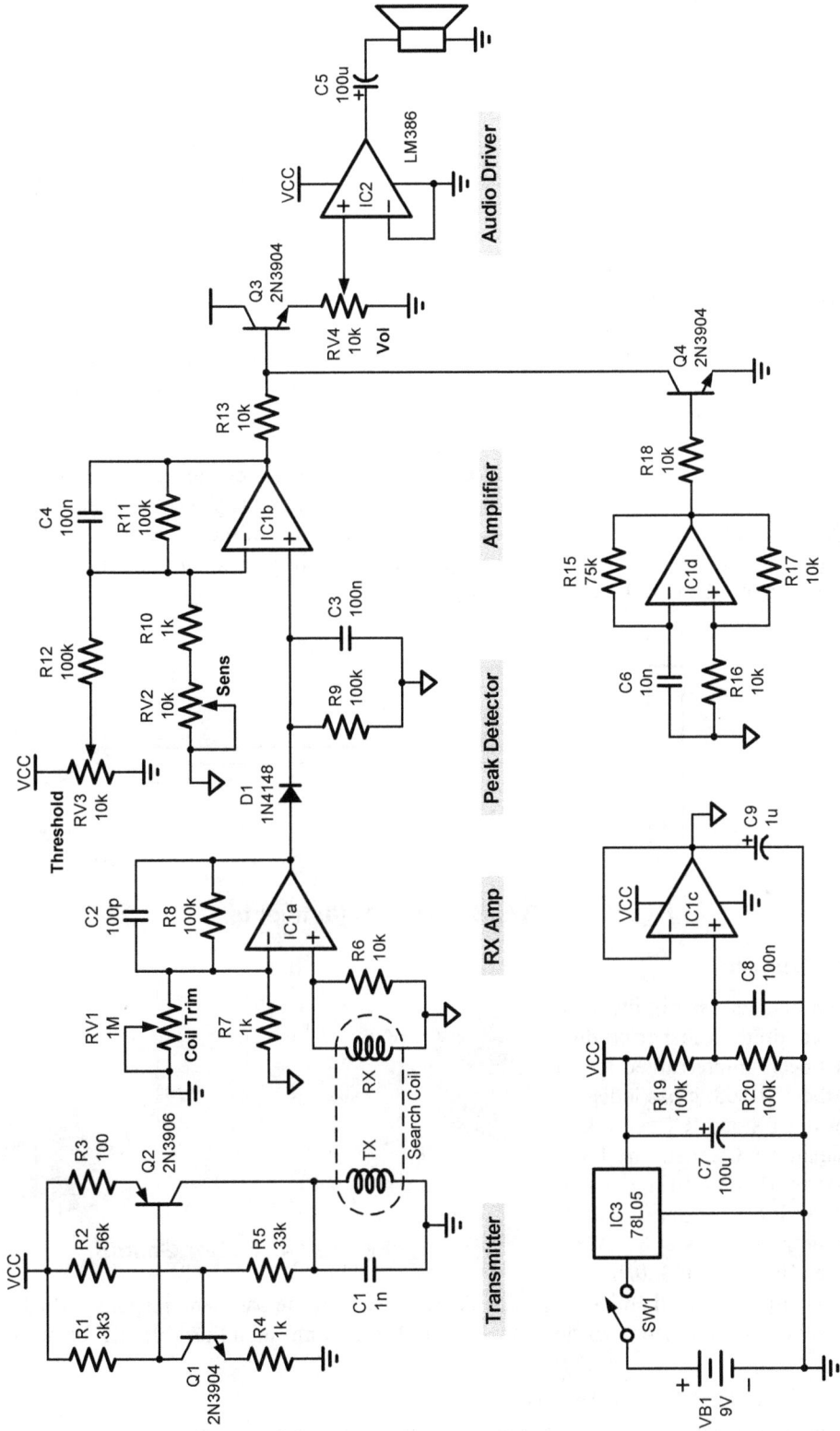

Fig. 13-12: Improved TR Complete Schematic

Although arguably an improvement over the basic TR it still has the problem of ground effect, though selecting a low transmit frequency helps.

Explore

13.5: Two-Box Detectors

Beginning in the 1920s several manufacturers (notably Fisher in the 1930s) produced so-called two-box locators[10]. They consisted of a vertical transmitter box (usually in the rear) and a horizontal receiver box (usually in front) connected by two rails. Even through the 1950s the boxes and the rails were made of wood.

The modulated TR circuit presented in this chapter was ideal for separate boxes. More modern detector designs use synchronous demodulators which require a synchronizing clock signal from the transmitter to the receiver. Therefore the circuits cannot be entirely separated. With the TR circuit the quenched oscillator TX circuit can be placed in one box with its own battery and the RX circuit placed in the other box with a battery. Except for the threshold adjustment that was added in Figure 13-7 there is no physical connection between the circuits. And the two-box locator has an alternate way to adjust the threshold.

It was mentioned before that intentionally unbalancing the coils will create a threshold signal in the receiver. The two-box system uses orthogonal coils whereby the RX coil is placed in alignment with the axis of the TX coil and therefore no lines of flux cut through the face of the RX coil. If the RX coil is then tilted slightly it will begin to see flux linkage and there will be a small amount of TX signal directly coupled into the RX coil. Figure 13-13 illustrates.

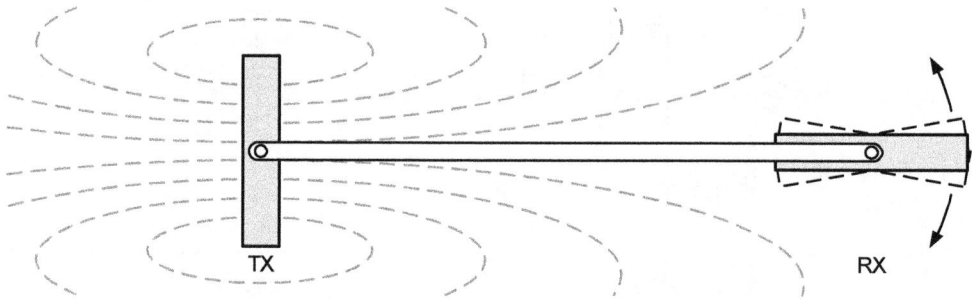

Fig. 13-13: **Two-Box Null Adjustment**

13.6: Fisher *Gemini*

The *Gemini* (and its sister utility locator, the *TW-6*) continues a Fisher product legacy that has been uninterrupted for 85 years. Unlike the TM-808, it has independent boxes for the TX and RX yet it does not use a modulated TX signal. The TX is a sinusoidal oscillator that runs at $81.92kHz$[11]. The RX uses an asynchronous demodulator that runs at 81.6kHz. This produces an IF output of 320Hz.

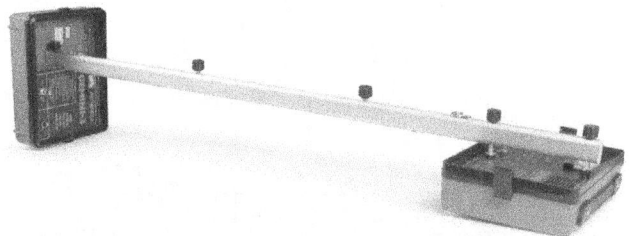

Fig. 13-14: **Fisher *Gemini***

The IF is band-pass filtered and an envelope detector extracts the baseband response which drives the audio circuitry. As with prior two-box designs, the RX coil has an adjustable tilt for nulling the

---

10. More commonly called "RF locators" back then.

11. The reason for this seemingly odd frequency is that utility locators have largely settled on a few particular frequencies, one being 82kHz.

orthogonal coil system or slightly de-nulling it to give a threshold audio. Besides the normal use as a two-box metal detector, the boxes can be removed from the connecting rod[12] and used to trace underground pipes and cables. The transmitter box can be placed over a metal pipe or cable and the TX signal inductively couples to it, and there is also a signal jack whereby the TX signal can be directly connected to the pipe or cable. The RX box can then be used separately to trace the underground signal.

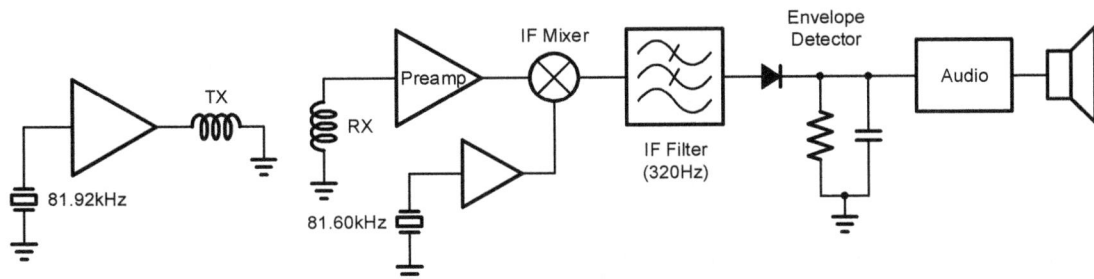

Fig. 13-15: **Gemini Block Diagram**

13.7: White's *TM-808*

Until White's closed their doors they made one of the most popular two-box style detectors in the *TM-808*. The last version of the *TM-808*[13] was basically a stripped-down *Coinmaster 6000* circuit with orthogonal coils. Because it uses synchronous demodulators the TX and RX circuitry and their coils cannot be separated so the circuitry is in a single housing with the coils attached. The RX coil cannot be tilted so the threshold adjustment is implemented electronically.

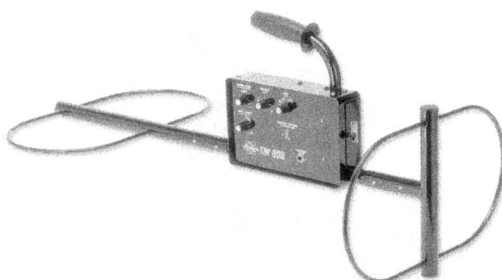

Fig. 13-16: **White's *TM808***

A somewhat unique aspect of this design is the use of solid aluminum rods for the search coils. The rods are bent into oval loops and the ends shorted together. Before they are shorted a toroid coil is placed around the rod which creates a transformer. For the TX side the toroid coil is a primary and the rod loop forms a single-turn secondary. On the RX side the rod loop is a single-turn primary and the toroid coil is the secondary. In Chapter 4 the system of the TX coil, target, and RX coil were described as forming a double transformer system. In the *TM-808* the TX coil and RX coil each add another transformer to the system.

13.8: Other Two-Box Options

Besides the White's *TM808* and Fisher *Gemini*, Nexus has a "two-box" detector model called the *Pathfinder* and XP makes the *XTreme Hunter* which is based on the *Deus II*. Ceia (Italy) has the *DSMD* model but it is intended for the military market. Tinker-Raser (USA) has the *Model 505* acquired from Detectron which is almost identical to the Fisher *Gemini* in both circuitry and mechanics.

Garrett has produced an orthogonal coil add-on accessory for some of their detector models. The *Bloodhound* was made for the older ADS series and the *Treasure Hound* for the CX and GTI series. XP also sells the *XTreme Hunter* coil as an accessory for owners of the *Deus 2*. If you are wanting to build a two-box locator and would like to avoid the daunting task of making the coil system, consider finding a Garrett 'hound coil set and designing around that.

---

12. The early Fisher units used two wooden connecting rails; the Gemini has a single aluminum rail.

13. Jim Karbowski originally designed this concept and sold hand-built units under his 3D Electronics brand. This was a traditional TR discriminator design. Later he sold the design to Discovery Electronics who produced the *TF-900* and contract manufactured the *TM-600* for White's and the *CS-9000* for C-Scope. White's then brought the design in-house as the *TM-800* and later redesigned it as the *TM-808* with a more modern VLF circuit.

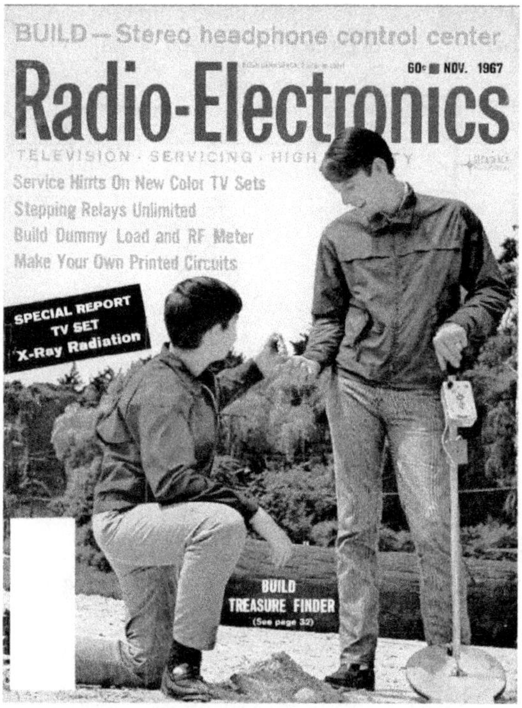

BUILD — Stereo headphone control center

Radio-Electronics
60¢ ■ NOV. 1967
TELEVISION · SERVICING · HIGH FIDELITY

Service Hints On New Color TV Sets
Stepping Relays Unlimited
Build Dummy Load and RF Meter
Make Your Own Printed Circuits

SPECIAL REPORT
TV SET
X-Ray Radiation

BUILD
TREASURE FINDER
(See page 32)

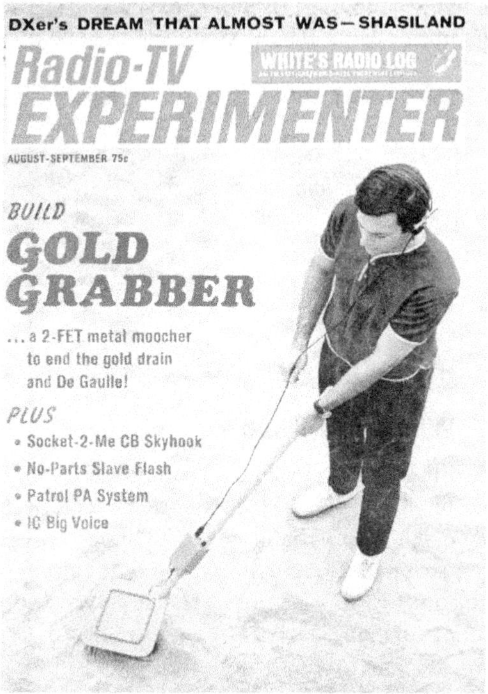

DXer's DREAM THAT ALMOST WAS — SHASILAND

Radio-TV
EXPERIMENTER
WHITE'S RADIO LOG
AUGUST-SEPTEMBER 75¢

BUILD
GOLD
GRABBER

... a 2-FET metal moocher
to end the gold drain
and De Gaulle!

PLUS

• Socket-2-Me CB Skyhook
• No-Parts Slave Flash
• Patrol PA System
• IC Big Voice

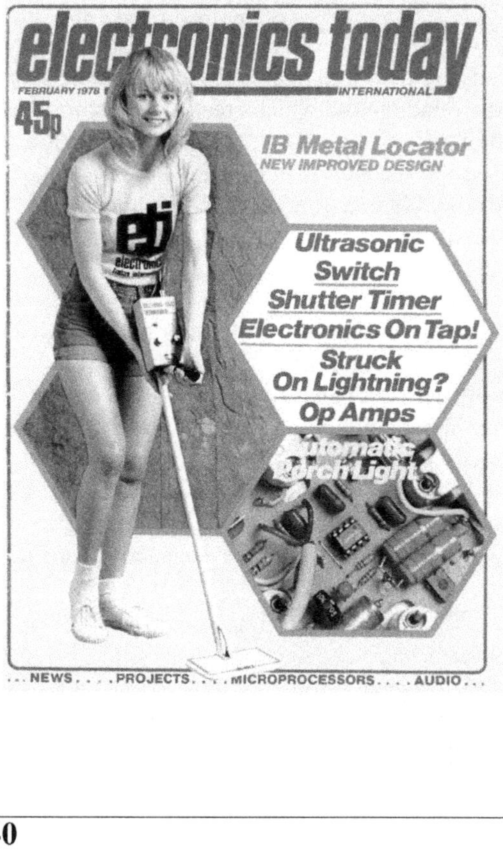

electronics today
INTERNATIONAL
FEBRUARY 1978
45p

IB Metal Locator
NEW IMPROVED DESIGN

Ultrasonic
Switch
Shutter Timer
Electronics On Tap!
Struck
On Lightning?
Op Amps
Automatic
Porch Light

...NEWS... .PROJECTS... .MICROPROCESSORS... .AUDIO...

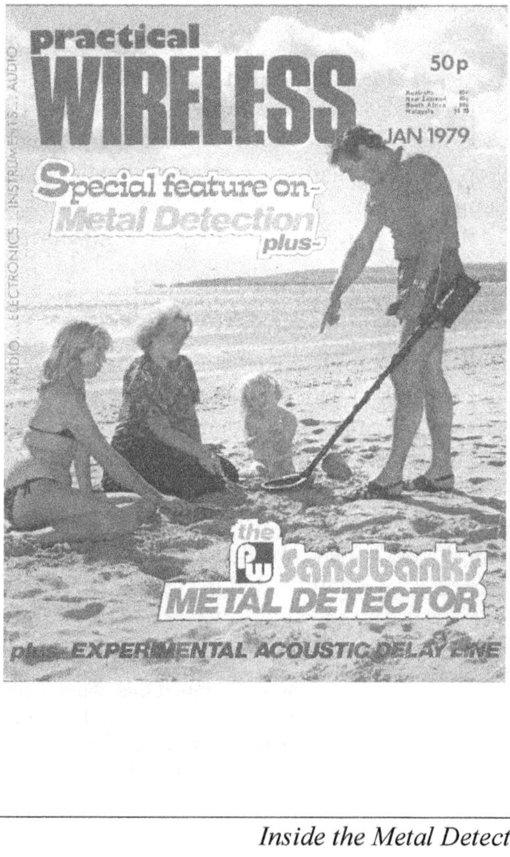

practical
WIRELESS
50p
JAN 1979

Special feature on
Metal Detection
plus

the
Pw Sandbanks
METAL DETECTOR

plus EXPERIMENTAL ACOUSTIC DELAY LINE

CHAPTER 14

TR Discrimination

"…the only 100% reliable discriminator is a shovel."

— *White's Electronics XLT Engineering Report*

In the evolution of induction balance detectors the next progression was the addition of discrimination. The development that enabled discrimination was the synchronous sampling circuit in the receiver. These designs are widely known as *TR-Discriminators* and were popular until VLF-Discriminators took over around 1980. One of the first discriminators was the Technos *Phase Readout Gradiometer* (PRG[1]) in 1974 which, ironically, was perhaps the pinnacle of TR-Discriminators (it had a meter that reported the phase of the target, much like a modern VLF-Discriminator.). Within a year or so most major manufacturers had added TR-Discriminators to their catalogs. A typical example is the Bounty Hunter *550D* shown in Figure 14-1.

The older TR designs used a self-quenching transmit oscillator so that a received target signal could readily generate an audio signal without the need for a separate audio circuit. This originated during the days of vacuum tubes when adding more tube circuitry in a battery-operated design was some-

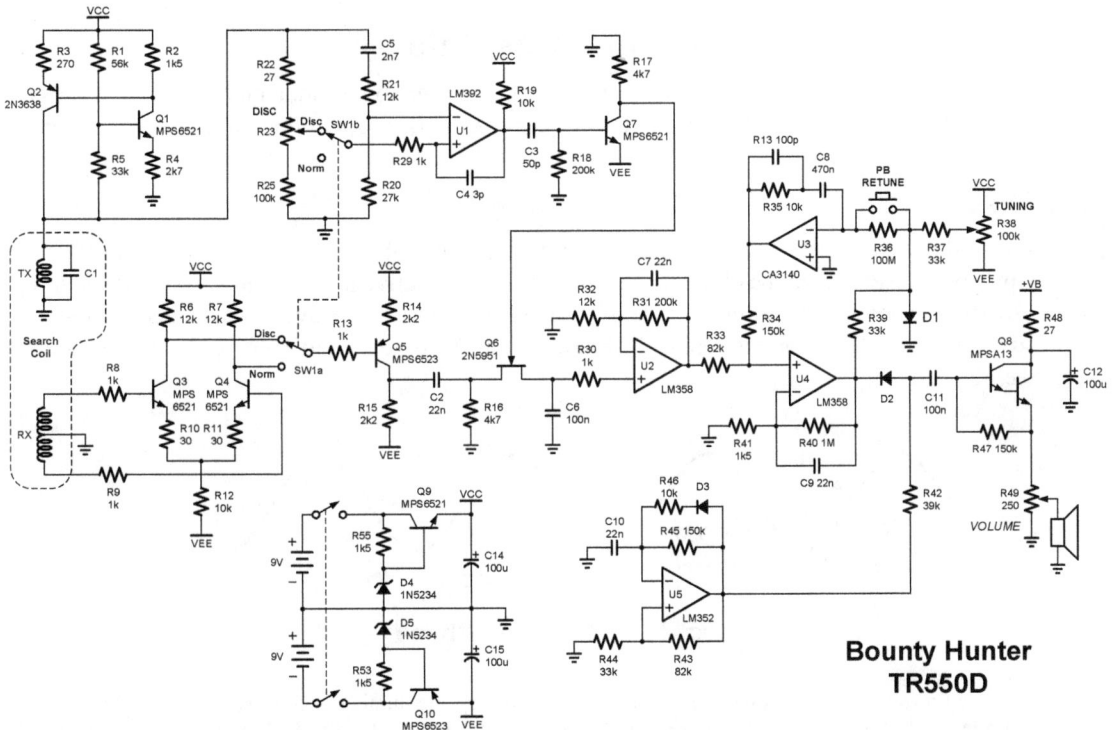

Fig. 14-1: **Bounty Hunter TR-550D**

---

1. The PRG was also among the most expensive on the market at $1800 ($11,330 in 2024 dollars). Supposedly only 50 total were ever made. I'm still looking for one.

thing to avoid. When the TR-Disc came along it was during the semiconductor revolution. Not only were discrete transistors rapidly replacing tubes but integrated circuits were also being introduced[2]. As such, even the earliest TR-Disc designs used a pure sinusoidal transmit signal and created the audio signal separately. In fact, the phase discrimination method used in the TR-Disc requires an unmodulated (continuous sinusoidal) transmit signal. Like the TR detector, many TR-Discriminators ran at a fairly high TX frequency, often around 100kHz. The combination of a sinusoidal transmitter and a synchronous sampler produces a response that depends on the phase of the target's eddy response.

Figure 14-2 shows a block diagram of the TR-Disc design. The voltage drive of the TX oscillator is used to derive a clock signal for a demodulator that follows the RX preamp. Because the demodulator is clocked at exactly the same frequency as the TX signal it is called *synchronous demodulation*. The key to achieving discrimination is to apply a variable phase shift to the demodulator clock. The relationship between the demod clock phase and the level of discrimination will be the focus of this chapter.

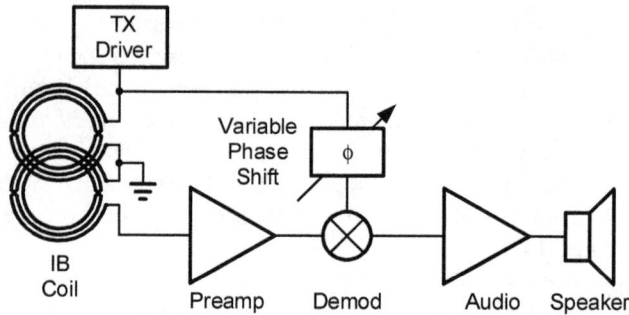

Fig. 14-2: **TR-Disc Block Diagram**

IB Fundamentals

Before diving in to a TR-Disc design we need to first cover some fundamentals concerning synchronously-sampled induction-balanced systems in general. This information applies to the remainder of the chapters in Part 4 of the book.

14.1: Phase

A sinusoid has three properties: frequency, amplitude, and phase. Frequency and amplitude are absolute properties[3]. Phase can be either an absolute term or a relative term. As an absolute term, it represents a particular point on a complete cycle of the sinusoid. For a sine wave, 0° is the zero-crossing in the positive direction. The absolute phase increases, from 0° to 360°, and 360° is also 0° of the next cycle. See Figure 14-3.

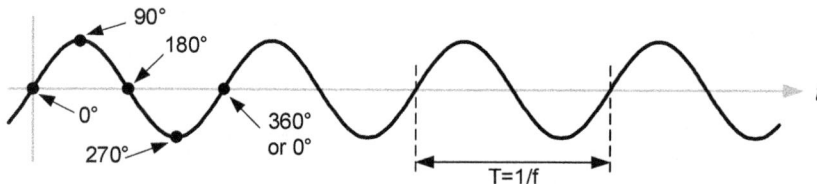

Fig. 14-3: **Absolute Phase**

---

2. During those times detector manufacturers would sometimes tout the number of transistors in their advertising. For example, in 1976 Bounty Hunter advertised that their *Professional* model had "2 Field-effect transistors, 3 Silicone [sic] transistors, and 1 22-transistor integrated circuit."

3. Although amplitude can be described as a relative term, such as "-10dB," which means 10dB below some other amplitude. But generally we say a waveform has such-and-such an amplitude, and such-and-such a frequency, but we don't say it has such-and-such a phase unless we are comparing it to another waveform.

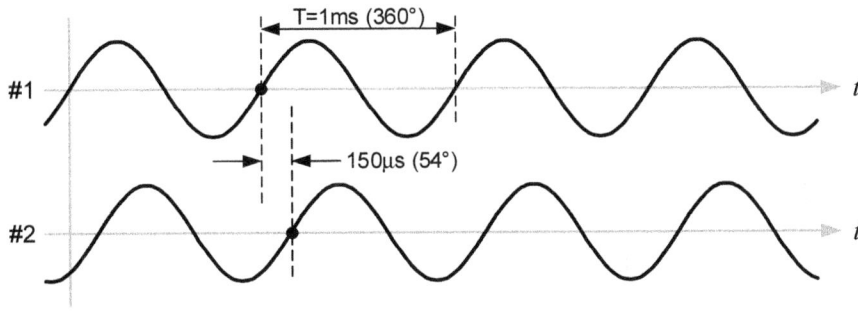

Fig. 14-4: **Relative Phase**

Relative phase refers to the phase difference, or phase shift, between two sinusoids of the same frequency. The phase shift can be measured by comparing the zero crossings or the peaks of the two waveforms. As an example (Figure 14-4), a 1kHz sine wave has a measured zero-crossing delay 150μs compared to a reference 1kHz sine wave. The phase shift is $150\mu s/1ms \cdot 360° = 54°$.

In the example, sinusoid #2 is said to *lag* sinusoid #1 by 54°. Equally, sinusoid #1 is said to *lead* sinusoid #2 by 54°. Either sinusoid can be considered to be the *reference* waveform. Phase can be measured on an oscope using this method, or with a phase meter such as the HP3575A.

14.2: Reference Signal

Modern IB metal detectors look at relative phases, in particular the phase shift that occurs between the transmitted signal and the received target signal. Ideally, it's the same phase shift that occurs between the transmitted magnetic field and the target response magnetic field. As such, construction of the IB coil matters because it contains two coils which are part of a double transformer system. Chapter 8 covers IB coil construction including the recommended phase relation between the TX and RX coils (Figure 8-27).

Using the transmitted magnetic field as the reference signal requires access to that signal. Fortunately the transmitted magnetic field has exactly the same waveform as the TX coil current. Figure 14-5 shows an idealized IB system with signals and phases. The transmitted magnetic field is assigned the reference phase angle of 0°, as we've been consistently using throughout the book. The TX current is also in-phase with the TX magnetic field while the TX coil voltage leads by 90°; that is, it has a relative phase angle of −90°.

Whatever field distortion is created by ideal (purely reactive) ground is also in-phase with the TX field so it has a relative return phase of 0°. A magnetic target will exhibit a reverse magnetic field phase-shifted by some amount φ° (limited to a range of 0° to 90°), whereas an eddy target has a reverse magnetic field phase-shifted by an amount φ + 90° (with a range of 90° to 180°). Back at the receiver, the target field induces a voltage in the RX coil but with an additional leading 90° phase shift due to the

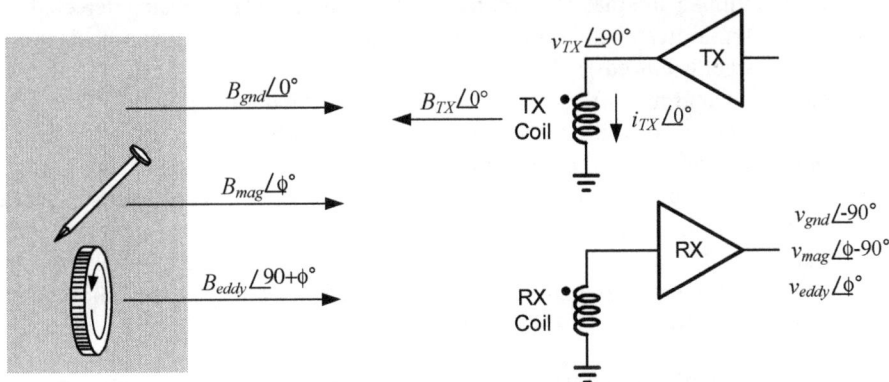

Fig. 14-5: **Idealized IB Operation (Referenced to i_{TX})**

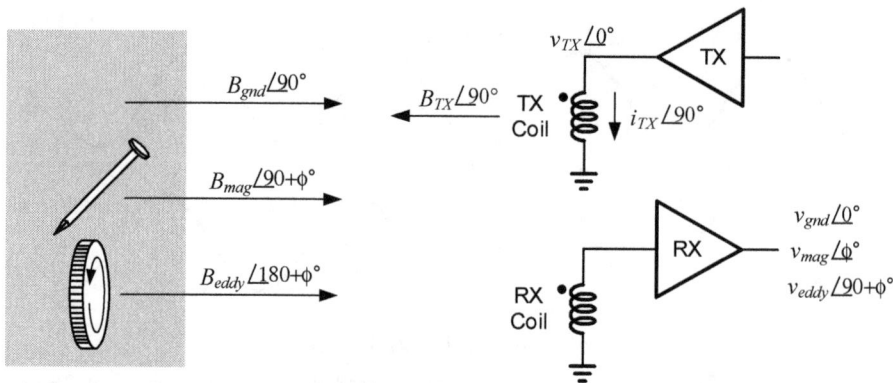

Fig. 14-6: **Typical IB Operation (Referenced to v_{TX})**

induction derivative ($\varepsilon = -N \cdot d\Phi/dt$). All explained in Chapter 4. Note that the RX coil is connected with the same phase polarity as the TX coil (dot-conventions), as explained in Chapter 8.

Assuming the RX amplifier imposes no additional phase shift, the end result is that the receiver voltage is "in-phase" with the transmitter *voltage*. That is, if ground imposes a phase shift of 0° in the magnetic field, we can also see that the receive *voltage* has no phase shift when compared to the transmit *voltage*; they are both −90° in the diagram. Figure 14-6 relabels this for clarity now using the *transmit voltage* as our 0° reference; this adds 90° to all signals.

After all this trouble, why did we not just start the book out by defining the transmit coil voltage to be our phase reference point? It's because the core of metal detecting physics lies in the magnetic excitation and magnetic field responses. How we create the magnetic excitation (the reference) or process the responses (the results) can vary and can incur errors, but the magnetic physics does not vary. For example, we'll see in a later chapter how the resistance of the TX coil produces a phase error in the transmit voltage compared to the TX current and magnetic field. However, we have easy access to the TX voltage but not the TX current so we will use that as our phase reference.

In the above discussion it was assumed that the preamp does not add to the phase shifts. In most modern designs the preamp is a non-inverting configuration which normally has no phase shift, but it is common in single-frequency designs (the focus of this section of the book) to configure the preamp as a band-pass filter to suppress wideband noise (mostly from EMI) from the RX coil. So it is not safe to say that the preamp has no phase shift and we will see later how to deal with that.

14.3: Target Signals

For a transmitted magnetic field of a given frequency and amplitude, a target will respond with the same frequency[4] but with a different amplitude and phase. A typical IB design uses the transmit signal as the phase reference against which the received signal is compared. From this, the phase angle of the target signal can be determined and that often indicates what kind of target is being detected. The amplitude of the received signal indicates the relative strength of the target, which can be used to calculate the depth at which the target is buried.

To help understand synchronous IB systems we will standardize on a few select targets: ferrite, a 16d common nail, a small piece of foil[5], a US nickel, and a US quarter. The plots in Figure 14-7a show the transmit current waveform (i_{TX}), plus the magnetic field responses (B_{TGT}) of these targets, all following the conventions of Figure 14-5. Figure 14-7b shows the same target responses but following the conventions of Figure 14-6 where the reference is v_{TX} and the responses are voltages as seen on the RX coil or at the output of an ideal preamp. It is important to remember that the eddy phase responses are dependent on the transmit frequency (recall Equation 4-9) and for this chapter (and subsequent chapters on IB designs) we will standardize on 10 kHz[6].

---

4. Ignoring distortion effects, which can occur with magnetic (ferrous) targets.
5. 1" or 25mm square, household variety.

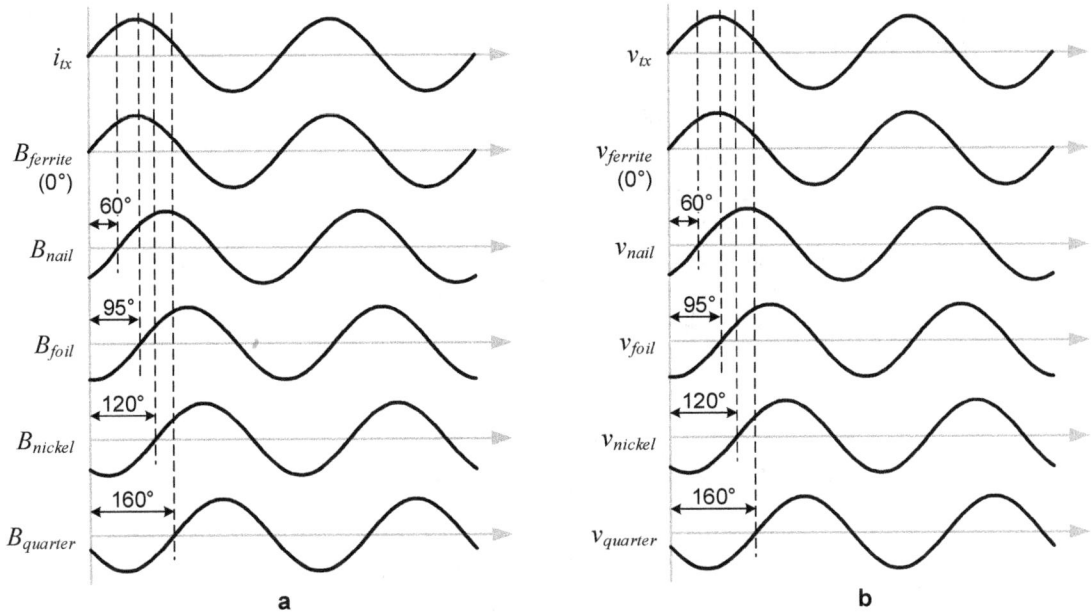

Fig. 14-7: **Target Phase Shifts**

Because the overall range of signal responses spans from 0° to 180° it is convenient to plot various target responses on a phase diagram. Phase diagrams in trigonometry (and also polar plots in circuits) are formatted with 0° on the right side (the 3 o'clock position) and increasing counterclockwise to 180° on the left side (the 9 o'clock position). In this book all metal detector phase diagrams have 0° on the left side and phase increases clockwise to 180° on the right side[7]. The reason for using this seemingly backward format is because in real metal detectors with target ID meters that is exactly the way the meters are formatted: magnetic responses are on the left — starting with ferrite/0° at the far left) — and non-ferrous on the right — ending with large silver coins. Figure 14-8 shows the very first target ID meter from the Teknetics 8500[8] which helped establish this standard.

Figure 14-9 shows a phase diagram for the targets in Figure 14-7. Each target arrow represents a "vector" which has a phase angle and a magnitude. For now all the magnitudes are the same but real responses will have vector lengths that depend on the strength of the

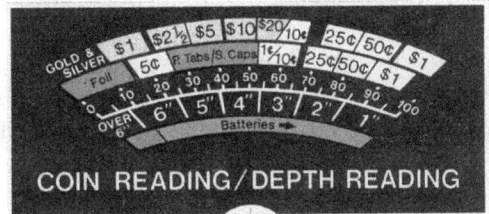

Fig. 14-8: **Teknetics 8500 TID Meter**

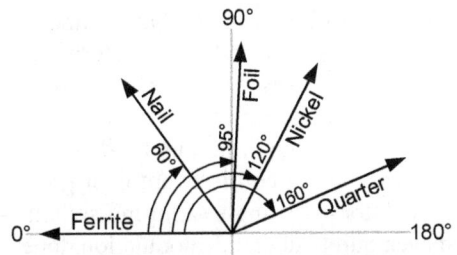

Fig. 14-9: **Phase Diagram**

6. 10kHz is the geometric mean of the band practically all modern IB detectors fall into, 1kHz - 100kHz, and is close to the geometric mean of the real VLF band, 3-30kHz. It is also a good all-purpose frequency and has been commonly used by manufacturers.

7. Many others use the same format, but not all. In patents you will occasionally find 0° on the right, and sometimes find the whole phase plot rotated 90° so that 0° is at the top and 180° is at the bottom.

8. The meters on first Teknetics TID models only went down to foil on the far left. Later detectors expanded this to include part of the ferrous range, and eventually all the way down to ferrite. It should be noted that before the Teknetics 8500 several detectors (Compass *Coin Magnum* TR and various BFO designs from Garrett and Bounty Hunter) had meters that indicated "Trash/Treasure" or "Bad/Good," but still with ferrous to the left and non-ferrous to the right.

RX signal. This will be covered in more detail as we proceed. Notice that the target vectors are spread across the 180° space rather evenly, which is why these particular targets are chosen.

Demodulation

A *demodulator* is a circuit that extracts the modulated information from a waveform. In the case of a metal detector the receive signal has been amplitude- and phase-modulated by a target. The process of demodulation also performs a frequency translation to baseband. The BFO mixer circuit in Chapter 10 behaves in a similar manner and could be used in the TR-Disc. In fact, the Bounty Hunter *Red Baron* used analog mixers (LM1496) for demodulation. But the vast majority of designs use a simple sample-and-hold[9] circuit as shown in Figure 14-10. The sample-and-hold is also easier to visualize.

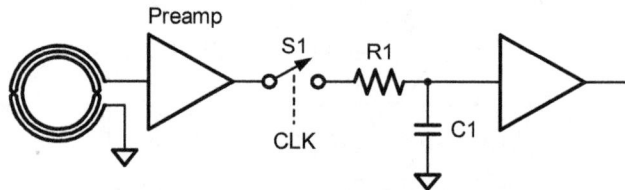

Fig. 14-10: **Sample-and-Hold**

14.4: Synchronous Demodulation

Demodulation is a form of mixing whereby a signal is multiplied by another waveform. This act performs a frequency translation of the input signal; see Equation 10-1 for the case of mixing two sinusoids. If the input signal is 100kHz and the mixer clock is 99kHz then the mixer output will be the difference, or 1kHz[10]. When the input signal and the mixer clock are exactly the same frequency (say, 100kHz) we call this *synchronous demodulation*. The output signal is essentially a DC value, assuming sinusoidal signals. We call this the *baseband* because it doesn't get mixed down any further[11], and it's where we perform the final signal processing.

Synchronous demodulation has been the basis of almost all metal detector designs since the early 1970s when it replaced the use of modulated TX oscillators as seen in the last chapter. Having the transmitter and receiver co-located on the same circuit board makes this quite easy, and the low operating frequency of metal detectors (sub-100kHz) eliminates most of the problems that RF systems face when trying to mix directly to baseband without an IF stage. Finally, the bandwidth of the baseband signal is almost always under 50Hz — determined by how fast the user can sweep the coil over a target — which, again, makes synchronous demodulation an easy solution.

Besides translating the input signal to baseband, a demodulator also extracts the modulation components of the signal. In a metal detector that is usually the amplitude and phase of the input signal which vary as the coil is swept over ground and targets. This chapter will start out with the simplest demodulator — a single sampling switch — and future chapters will look at more complex schemes and even purely digital demodulation done inside a microcontroller.

14.5: Sampling

Consider a sinusoidal signal applied to a switch as shown in Figure 14-11, where the switch is clocked exactly in-phase with the sinusoid. When the switch is closed the signal is passed through and when the switch is open the output is zero. The result is that only the positive half of the sinusoid appears at the output. We'll call this "0° sampling."

---

9. The sample-and-hold amplifier (SHA) circuit is commonly used at the front-end of an ADC to capture a moving signal and hold it statically while the ADC converts it to a digital value.

10. There is also a summation term that is usually removed with a low-pass filter.

11. Many radio systems mix down to an *intermediate frequency* (IF) before finally mixing down to baseband.

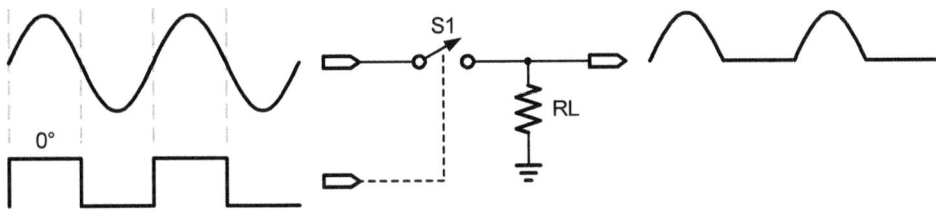

Fig. 14-11: **0° Sampling**

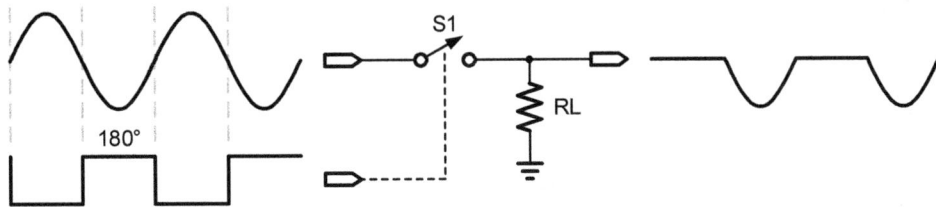

Fig. 14-12: **180° Sampling**

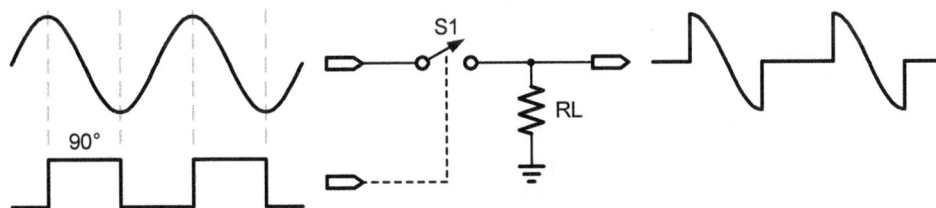

Fig. 14-13: **90° Sampling**

Now let's shift the sampling clock by 180° and sample exactly out-of-phase with the sinusoid (Figure 14-12). The result is that only the negative half of the sinusoid appears at the output. We'll call this "180° sampling."

Finally, let's look at the case exactly in-between, where the sampling clock is shifted 90° from the sinusoid (Figure 14-13). During the time the switch is turned on the sinusoid transitions from the positive peak to the negative. This is "90° sampling." From these three examples it is easy to visualize the output waveform for any clock phase shift.

14.6: Averaging

Suppose now that instead of a simple load resistor the output of the sampling switch is connected to an averaging circuit. For 0° sampling the average would be some positive voltage, for 180° sampling it would be a negative voltage of the same magnitude, and for 90° sampling the average would be zero. It is apparent that regardless of the phase setting the average can never be higher than the 0° case nor lower than the 180° case. We can plot the output average versus the clock phase with the resulting curve shown in Figure 14-14. This

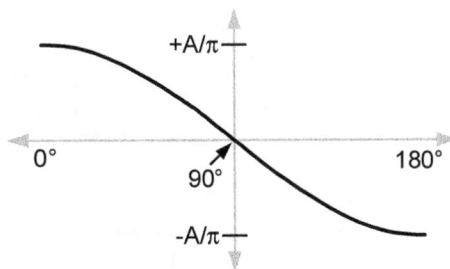

Fig. 14-14: **Sampling Average vs Phase**

curve represents the effective gain of the demodulator with a peak gain equal to the amplitude of the input sinusoid divided by π. To see how to reach this conclusion visit the *Explore* section[12].

12. *Explore* derives the average assuming there are no return-to-zero dead zones every half cycle. Also see Footnote 16.

The averaging circuit used in most metal detectors is an RC filter as shown in Figure 14-10. When the switch is closed the input signal is applied to the RC circuit and effectively averaged, assuming the RC time constant is much slower than the signal. When the switch is open, the input is isolated and the capacitor holds its voltage. The demodulator is typically connected to a high-impedance buffer or amplifier to minimize hold-mode droop and the whole circuit is often referred to as a *sample-and-hold amplifier*, or SHA. This particular circuit is called a *half-wave demodulator* because it averages only one-half of the period of the input signal. In early TR-Disc designs the switch was usually a simple JFET transistor; modern designs use an analog switch such as a 4066 or 4053.

14.7: Target Responses

Let's apply the five targets shown in Figure 14-7 to the demod and see what happens. The clock signal is derived directly from the TX voltage waveform and we'll start with 180° sampling. We're going to call this the "90° discrimination" setting and will explain why shortly. In Figure 14-15 the results on the left are without averaging and what the output waveforms will look like if we pretend the capacitor is removed. The results on the right (with averaging) represent reality where the RC converts the waveform to a simple DC average. The 'no capacitor' waveforms help illustrate what is happening as either the target phase or the clock phase varies.

Ferrite produces no phase shift so with the capacitor temporarily disconnected the output is a negative half-cycle voltage waveform. With the capacitor added to the circuit, the RC filter averages this to a fairly strong negative voltage.

The nail is phase-shifted by 60° so a large portion of its negative signal is passed through, plus a small portion of its positive signal. The overall averaged signal is negative, though not as strongly negative as the ferrite. The foil target has slightly more positive signal than negative signal so its average will be slightly above zero. For the nickel, a large portion of its positive signal is passed through plus a small portion of its negative signal, with the overall averaged signal being positive. The quarter has an average that is even more positive than the nickel.

In this example foil is very close to having an average value of zero volts because it has a phase response just 5° past 90°. A target with an exactly 90° phase shift will have an exactly zero volt average. All non-ferrous targets above 90° will have positive demod voltages and all ferrous targets below

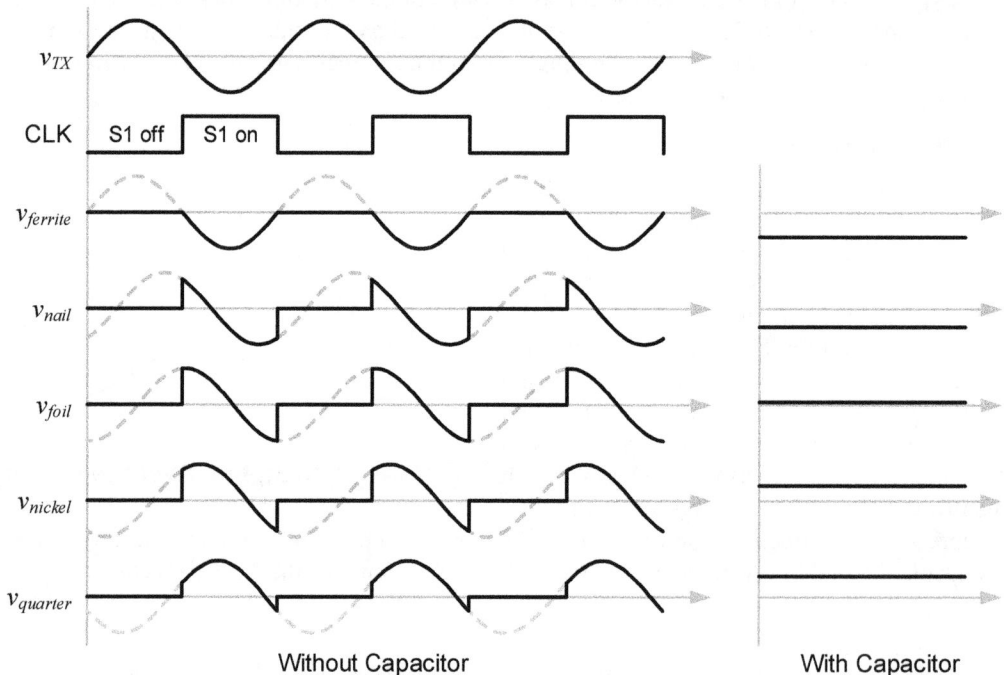

Fig. 14-15: **Demod Waveforms: Disc = 90°**

90° will be negative. We therefore say that this example has a 90° *discrimination threshold*[13] even though the demod clock phase is 180°. Magnetic targets approach a high of 90° but never quite reach it, and eddy targets approach a low of 90° but also never quite reach it, so technically 90° is a "no-man's land." But our small piece of thin foil is very close, as is salt water. If the output of the demod is applied to an audio circuit which responds only to positive-going signals then all ferrous targets will be rejected and non-ferrous targets will be detected. Discrimination is achieved.

14.8: Variable Discrimination

Suppose we now delay the clock signal by an additional 15°, to 195°. This is a discrimination setting of 105°. Figure 14-16 shows the new results, and we can now see that foil has a negative voltage and a nickel has a smaller positive voltage. Everything just below the nickel will be rejected and everything else will be accepted. Continuing to increase the clock delay gives us a variable discrimination control. We could increase it until the quarter is rejected, or decrease it to accept nails and other ferrous metals.

If we wanted a full range of discrimination control from ferrite all the way to Atocha bars then the variable delay would need to have a range of 90° to 270° in order to achieve a discrimination threshold of 0° to 180°. Depending on the locale or the type of hunting the discrimination range is usually abbreviated. For example, in US coin hunting it is rarely desirable to discriminate above a Lincoln/Wheat cent (roughly 150°) because that is where all the high-value silver coins reside. But eliminating the pesky zinc-Lincoln cent[14] (135°) can be a benefit in some places, even if it means losing nickels. On the ferrous side, relic hunters might want to pick up iron, but iron rarely falls below 40° or so. Therefore a minimum of 40° would capture most iron and a maximum of 150° would prevent losing any silver, for a total range of 110°. Other locales may require a different range.

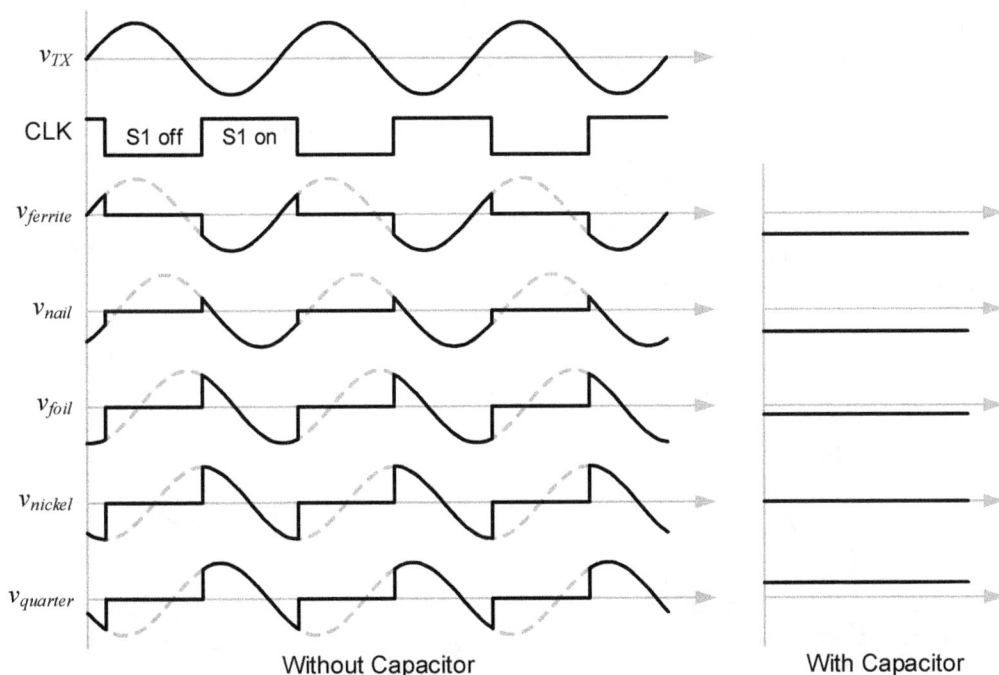

Fig. 14-16: **Demod Waveforms: Disc = 105° (CLK = 195°)**

---

13. We'll also refer to this as the *discrimination setting* or *discrimination phase*. Just keep in mind that the required demod clock phase is 90° greater than the desired discrimination phase.

14. Affectionately known as Zincolns, abbreviated Zn¢.

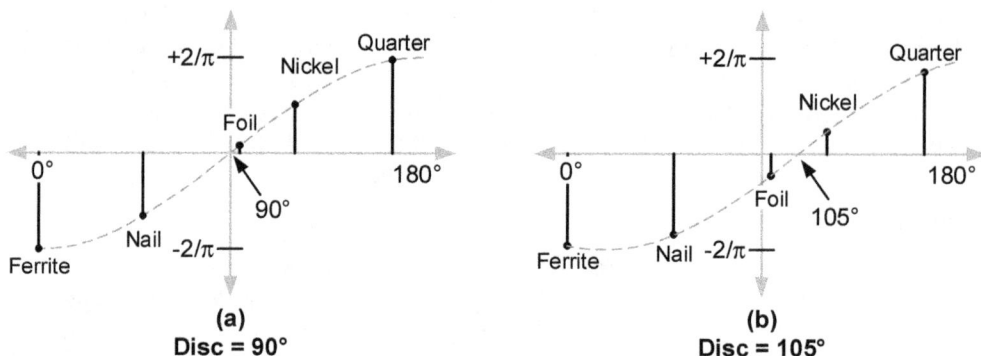

Fig. 14-17: **Signal Attenuation**

Discrimination done this way comes with a steep price. As targets approach the discrimination cut-off point their averaged voltage response is progressively weakened. We can take the sampling response curve[15] in Figure 14-14 and add targets at their respective phase responses to see this effect. As an example, in Figure 14-15 the discrimination cutoff is 90°, and foil comes in at about 95°. But its averaged value will be just slightly above 0 volts, whereas a quarter will have a strong response. Figure 14-17a shows the attenuation curve that results from the demodulator action for the same 90° cutoff. The x-axis indicates target phase and the y-axis is demodulator conversion gain. The dashed line represents the attenuation curve with its zero-crossing equal to the phase setting — 90° in the (a) example. The vertical vectors indicate various targets at their phase on the x-axis, and the direction and height of the vector indicate polarity and signal strength, assuming all have the same signal strength at the demod input.

Signals that are exactly in-phase with the demod clock (like Atocha bars) or exactly out-of-phase (like ferrite) will have a maximum amplitude equal to the demodulator conversion gain[16] of $2/\pi$, or about 0.636. For other signals the response amplitude follows a sinusoidal envelope where the zero-crossing is the discrimination cutoff. Note that in Figure 14-17a the signal from a US nickel is weakened by roughly half when the discriminator is set to accept all non-ferrous targets.

The discrimination curve can be slid to the left (lower disc phase) or to the right (higher disc phase) and the target vectors will change accordingly. For example, Figure 14-17b shows the attenuation curve for a discrimination cutoff of 105°, the same as Figure 14-16. Increasing the discrimination threshold to reject foil further weakens the responses of nickels and even affects quarters.

All of the target plots have been for a TX frequency of 10kHz. As we know from Chapter 4 the eddy current phase shift depends on frequency. Figure 14-18 shows the normalized (same amplitude) responses for foil, nickel, and quarter at 10kHz and 100kHz. At 100kHz the phase responses get pushed

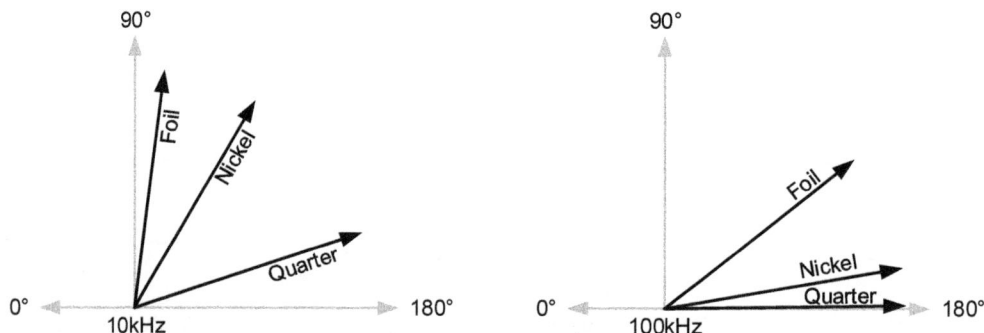

Fig. 14-18: **10kHz vs 100kHz**

15. The curve is inverted here to make the ferrous side negative and the non-ferrous side positive.

16. This assumes a normalized input amplitude of 1. Note that in Figure 14-14 the conversion gain was $1/\pi$ instead of $2/\pi$. That's because in an actual SHA demodulator the cap voltage is held during the switch "off" time instead of returning to zero.

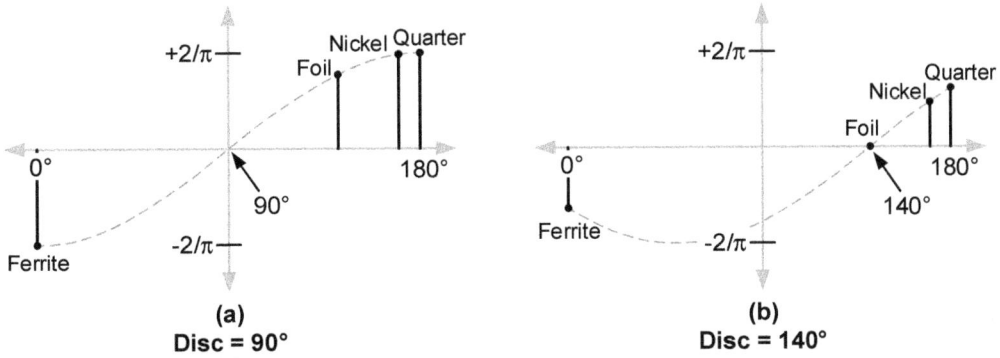

Fig. 14-19: **Signal Attenuation @ 100kHz**

closer to the 180° axis. Ferrite ground will show little phase shift and other ferrous targets are not considered because their behavior versus frequency is uncertain and our concern here is how discrimination affects non-ferrous targets.

For the TR-Disc this will improve the responses of low conductors when discrimination is applied. Figure 14-19a shows again the case where 90° discrimination is applied to the 100kHz. Comparing this to the 10kHz case in Figure 14-17a we see that foil especially has a much stronger response and the nickel is also improved. However, now it will take far more demodulator phase shift to discriminate out the foil. Figure 14-19b shows what happens when the disc phase is exactly adjusted to the foil phase; foil disappears but, again, the nickel is greatly attenuated and now the quarter is as well. This is why the old TR discriminators had such a bad reputation for depth loss in Disc mode and an especially infamous weakness for US nickels; experienced hunters would try to stay in all-metal mode to avoid this issue.

Running at a higher frequency does have a benefit. Because it takes a larger phase shift to eliminate, say, foil the ferrite (ground) signal also sees more attenuation. Since the ground response is fairly continuous this will help with falsing and masking. Other magnetic targets like nails[17] may see an increase in their response but they generally remain below 90° and therefore the response is still negative.

14.9: Polar Plots

Before commencing with a design, it is useful to show an alternate way of plotting responses. Figure 14-15 and Figure 14-16 show time-domain plots which are tedious and difficult to read. Figure 14-17a-b are Cartesian phase responses which are a little easier to interpret.

It is also possible to plot the target vectors in a polar plot as shown in Figure 14-20. For the polar plot the demodulator attenuation envelope is now a double circle[18]. Target responses are plotted as phase vectors from 0° on the far left to 180° on the far right. The y-axis is at 90° and, in this case, the

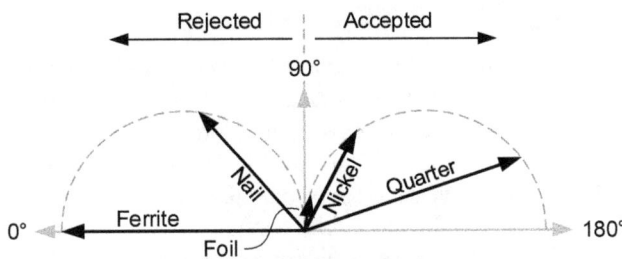

Fig. 14-20: **Polar Plot**

17. Purely magnetic targets should see little phase shift but practical ferrous targets can have either a viscous component or an eddy component (or both) that causes their phase shift to move with frequency, perhaps even into the "non-ferrous" quadrant.

14: TR Discrimination
251

Disc = 75°; stronger non-ferrous targets

(a)

Disc = 90°; foil is accepted but weak

(b)

Rejected 105°
Accepted

Rejected 125°
Accepted

Disc = 105°; foil is rejected, nickel weaker but accepted

(c)

Disc = 125°; nickel is rejected, quarter is progressively weaker

(d)

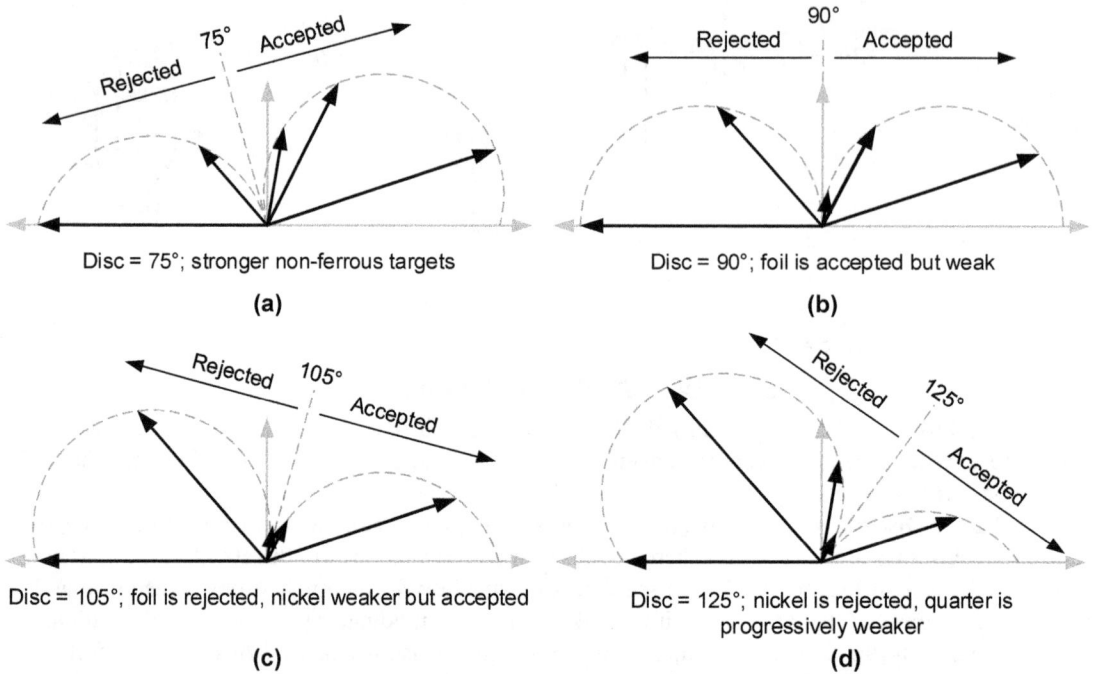

Fig. 14-21: **Polar Plots**

discrimination phase is also at 90°. Signal polarity is determined by which side of the disc phase the vector falls: vectors to the left of the disc phase are negative, those to the right are positive.

Several different discrimination phase settings are shown in Figure 14-21 with the resulting target responses. Changing the disc phase only rotates the attenuation circles and the accept/reject division line; the phase angles of the target vectors do not change but their amplitudes do. Compare Figure 14-21b with Figure 14-15 and Figure 14-17a as different ways of representing the results of a 90° disc phase. Likewise Figure 14-16, Figure 14-17b, and Figure 14-21c all represent the same 105° disc phase.

Admittedly this presentation of the effects of disc phase takes a bit of getting used to but in subsequent chapters[19] polar vector plots will be the standard way we view target responses. As the disc phase progresses from 75° to 125° notice how the quarter signal gets weaker. Also notice that when the disc phase is 90° — the supposed "accept all non-ferrous targets" position — foil targets are virtually suppressed. This might be a Good Thing if you are wanting to ignore foil, but it might be a Bad Thing if you are wanting to detect small jewelry and thin necklaces. However, when the disc phase is backed down into the iron range (Figure 14-21a) the responses of low conductors like foil and nickels are greatly improved. Even the quarter is a little stronger, though Atocha bars would be a little weaker.

14.10: Demodulator Clock

The eventual goal for the TR-Disc design is an adjustable discrimination phase of 70° - 140° which will discriminate from the high side of the ferrous range (in order to boost low conductor responses) up to just beyond the US zinc cent. But for a given discrimination phase the actual clock phase required is 90° greater, resulting in a clock phase range of 160° to 230°. If it were that simple we could use the conceptual circuit in Figure 14-22a. However, the RX signal also experiences some amount of phase shift from the RX coil to the demod input. The amount depends on the front-end design but the additional phase shift must be accounted for in the phase delay for the demod.

18. Figure 14-20 obviously shows two half-circles. We are generally only interested in the two quadrants with a positive Y-axis but we could also process the negative Y-axis quadrants which would result in an attenuation envelope of two complete circles.

19. And in patents and other publications.

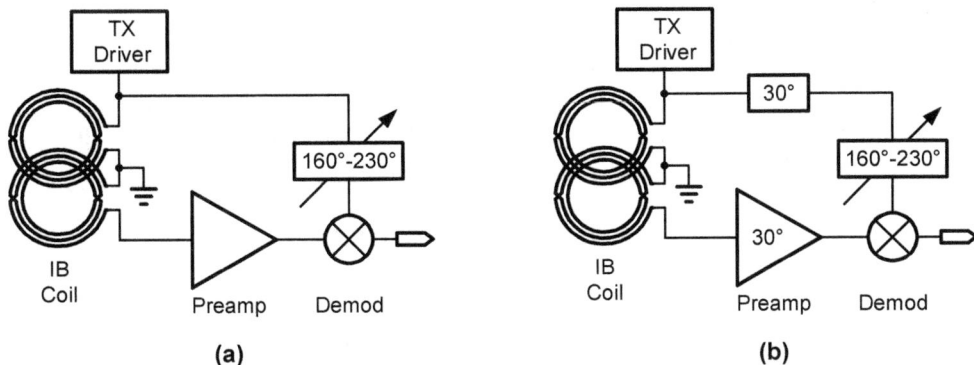

(a)   (b)

Fig. 14-22: **Adjustable Phase Delay**

In Figure 14-22b the preamp is shown to have a 30° phase shift. The demod clock therefore needs an additional 30° phase shift to compensate for the preamp. This can be combined with the desired variable range to give a total phase shift of 190° to 260°. Of this, 180° can be achieved through a logic inversion so in reality we need a variable phase shift circuit with a range of 10° to 80°.

The TX driver voltage is a sinusoid so the phase shift circuit can be implemented many ways. Figure 14-23 shows a possible solution using an "all-pass filter" which can produce a phase shift of 0° to 180° with no attenuation effects[20]. The overall phase shift of the all-pass filter stage is given by[21]

$$\theta = -2 \cdot \arctan(\omega\tau) \hspace{4cm} \text{Eq 14-1}$$

where ω is the angular TX frequency ($2\pi \cdot 10\text{kHz}$ for this design) and $\tau = R \times C1$.

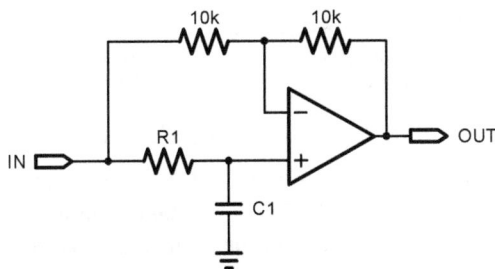

Fig. 14-23: **All-Pass Filter**

Figure 14-24 shows the complete phase delay circuit. In this case potentiometer RV4 allows changing phase and a minimum phase shift is set by R31. For the values shown the minimum delay is 10.4° and the maximum delay is 80.4° which meets our goal. Component value tolerances (especially the

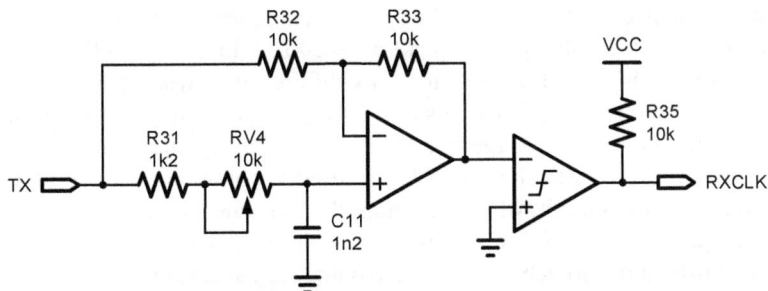

Fig. 14-24: **Phase Delay Circuit**

---

20. Well-documented in electronics texts and fairly easy to analyze.

21. The negative sign signifies that it is a phase *delay*.

capacitor and the pot) can result in phase shift errors so it is not unusual to insert an additional trimmer potentiometer to adjust out these errors. The alternative is to hand-select critical components.

A final comparator stage in Figure 14-24 converts the sinusoidal output of the all-pass filter into a rectangular clock waveform for driving the demod switch. Note that the comparator polarity gives us the final 180° inversion mentioned earlier. The resulting RXCLK has a phase of 190° to 260° compared to the TX voltage waveform.

Remaining Circuitry

For the purposes of this chapter the demodulation scheme is the most important concept to learn. The rest of the design can be recycled pieces from previous designs. The TX circuit can be any sinusoidal oscillator such as a Colpitts. In this design we will continue with the TX circuit from the *Improved TR* design of Chapter 14. The preamp is a normal moderate-gain non-inverting opamp. The circuitry following the demod does need a little discussion.

The demodulator output is a near-DC voltage that has a nominal zero value and goes negative for rejected targets and positive for accepted targets. From that it sounds like we can just apply the demod output to a comparator and be done with it. The problem is that a ground signal pushes the demod output negative and ground tends to be a persistent signal that would end up masking out desired positive target signals. Figure 14-25 illustrates.

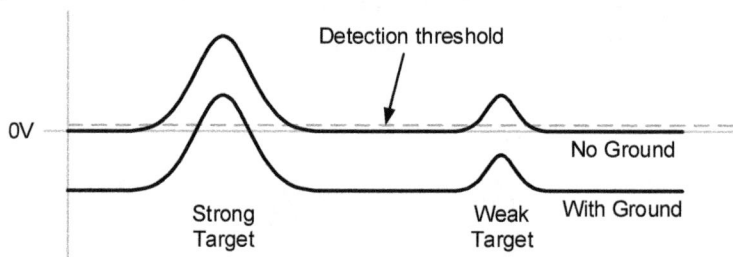

Fig. 14-25: **Ground Masking**

If ground mineralization is reasonably consistent and the user sweeps the coil at a consistent height above the ground then the ground signal will be fairly constant as shown in Figure 14-25. Targets, however, have a fairly fast transient response. The solution then is to add a retune circuit.

14.11: Retune Circuit

In the heyday of the TR-Disc a majority of designs had a manual retune circuit which statically held a tuning level. Many implementations had a push-button retune switch. Example retune circuits can be seen in the Bounty Hunter *TR-550D* schematic at the beginning of the chapter, and also in the design presented in the *Off-Resonance* chapter.

In this design the retune circuit will be applied to the demodulator amplifier as shown in Figure 14-26. With the push button closed, the integrator is in a feedback loop around the amplifier. Ground is applied to the non-inverting input of the integrator; any difference between *Vout* and ground produces a current through R13 which charges C4. The resulting voltage change at *Vc* provides an input offset to the amplifier. Since both the retune voltage and the signal of interest are applied to the non-inverting input of the amplifier (IC2a) resistors R8 and R11 are used to combine the two voltages. R8 and R11 form what is called a "seesaw" circuit, which becomes obvious when the circuit is analyzed.

In practice, the user adjusts the DISC control which alters the DC offset of the demod voltage on C2, then presses the push button switch to re-zero the output of the amplifier. Let's suppose this is done with the loop raised in the air. When the push button is released the voltage at *Vc* is held on the capacitor (assuming the opamp has no leakage current) and the audio remains stable at the desired level set by a threshold adjust in the next stage.

Now the loop is lowered to the ground and mineralization forces the demod voltage to go negative. This forces *Vout* to also go negative resulting in the audio going silent which can mask target responses.

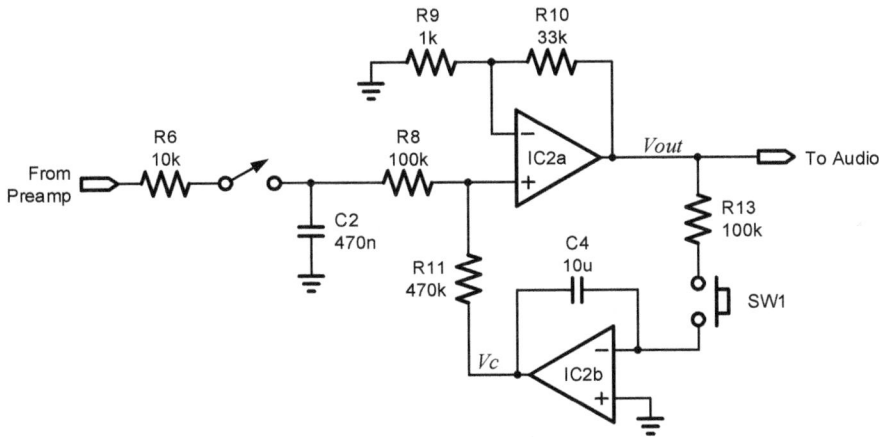

Fig. 14-26: **Retune Circuit**

A quick press of the push button will restore *Vout* to 0V by forcing *Vc* to a new offset level to compensate for the ground signal. This allows for quick retuning in variable ground conditions while permitting zero-motion target responses. If the button is held closed then the retune circuit operates continuously which creates a motion mode of detection.

It should be noted that, when the switch is open, the ability to maintain a constant *Vc* depends on selecting an opamp with exceptionally low input bias currents, such as one with FET inputs. A poor opamp choice will result in a constantly drifting threshold. Pressing the button will still return it to the proper tuning point but you will be pressing the button often.

14.12: Audio

BFO designs have an audio response that is pitch-based; that is, the audio pitch rises with an increase in target strength. The majority of TR-Disc designs had audio that was a constant pitch and varied in loudness. We'll start with that approach; see Figure 14-27.

The output of the demod amp (Figure 14-26) feeds the input v_{in} to the audio gain stage. This signal is a DC voltage with a nominal (no target) value of about zero volts, and a non-discriminated target signal goes positive. Amplifier IC3a is non-inverting so that the output is a positive signal when a target is present. Potentiometer RV2 injects an offset current into the feedback which creates a user-adjustable threshold offset at the output of the amplifier. Normally this is adjusted to create a slight positive offset.

The output of the amplifier drives the common-emitter transistor Q5 via diode D1. D1 compensates for the offset and temperature drift of Q5, and R19 provides a DC bias current for D1. A clock signal is applied to the base of Q4; when the clock is high Q4 pulls the base of Q5 to ground which shuts off Q5 regardless of the target signal level. This gating signal chops the voltage applied to the LM386 and produces a fixed pitch tone, with a loudness that rises in response to a target.

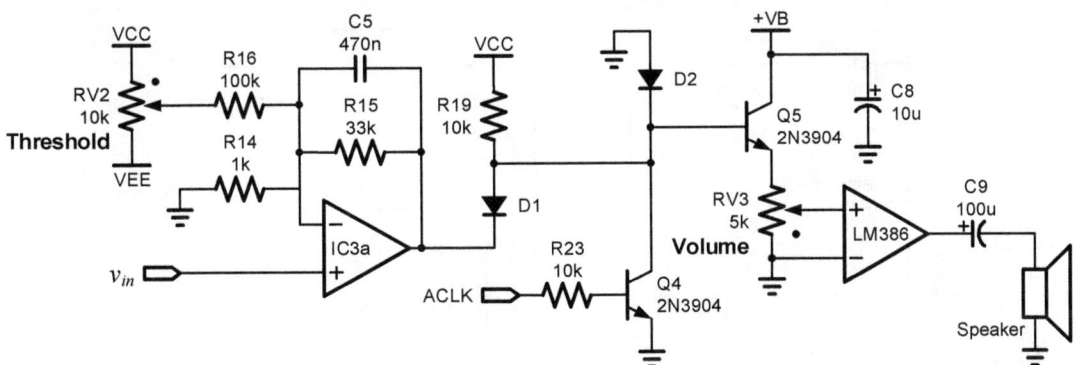

Fig. 14-27: **Audio Circuitry**

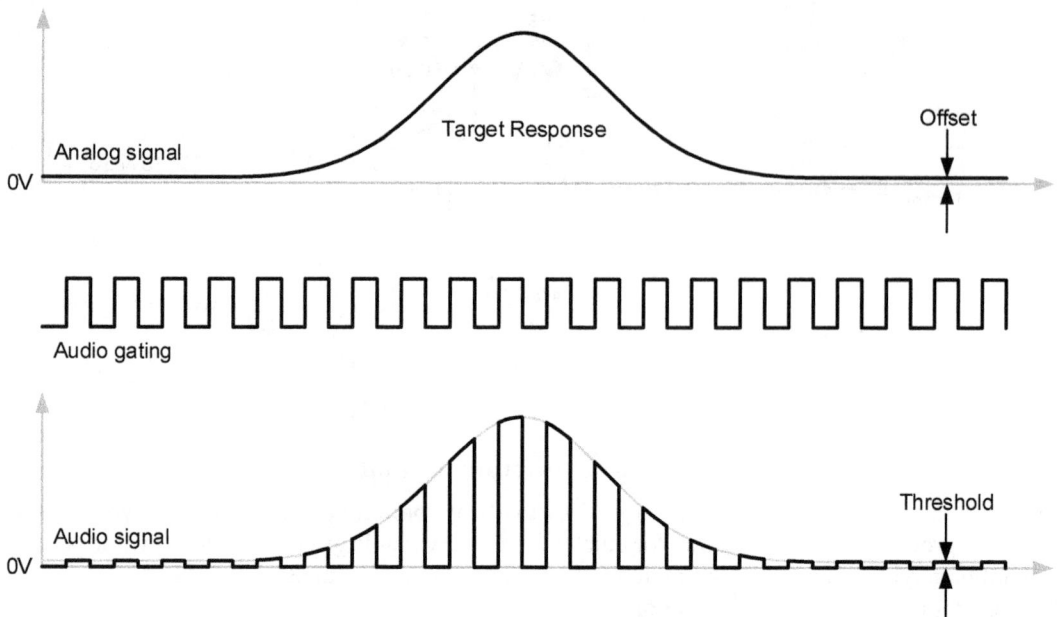

Fig. 14-28: **Audio Response**

Figure 14-28 illustrates the waveforms described above. A typical target response has a "bell" curve seen in the top waveform. The middle waveform is the audio gating clock which can be any desired audio frequency; for this design it is set to around 600Hz. The bottom waveform shows the result of "chopping" the analog signal. The target response causes a rise in the loudness of the audio signal. Notice the small DC offset in the signal which is set by RV2 and creates a slight no-target sound. This is our audio threshold.

In Figure 14-27 the audio gating clock (ACLK) can be generated in a number of ways. TR-Disc detectors predated microcontrollers so typically the audio clock came from a 555 timer or an opamp relaxation oscillator. In the final design we will use the latter.

14.13: Power Supply

The final portion of the design is the power supply. As with prior projects we will use a 9V battery as the power source. The battery will directly drive the transmitter, clock phase, and audio circuits. The remaining analog circuitry will be powered from a +5V linear regulator, plus a charge pump and −5V linear regulator. Figure 14-29 shows the complete power circuitry.

Some people may find offense at running a 5V linear regulator from a 9V battery as it wastes about 30% of the battery power in the voltage drop across the regulator. A better solution would be to use a

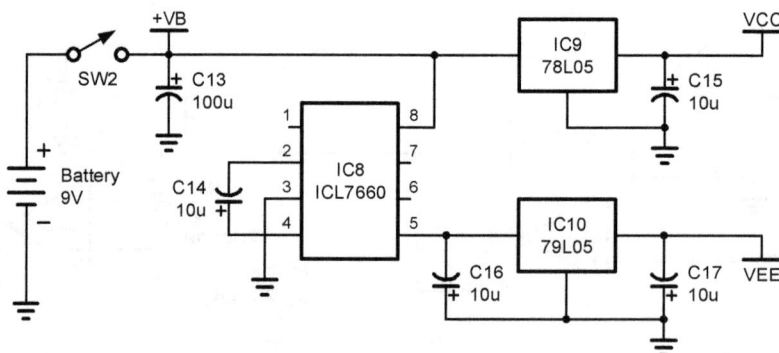

Fig. 14-29: **Power Circuitry**

switching regulator as many commercial designs do but care must be taken to minimize the coupling of switching noise into the analog circuitry. Another option is to use a higher voltage linear regulator, such as 6V or 7V[22]. This does not reduce overall power but does increase headroom in the analog receiver chain, practically for free. A similar argument can be made for the negative regulator.

Design: *TR-Disc*

This project will combine what has been presented so far into a basic TR-Discriminator. Figures 14-30 and 14-31 show the complete schematic for the *TR-Disc* design. You may notice there are gaps in the numbering of components — for example, there is no R4, R5, or R7. The next chapter will further develop this circuit with additional capabilities and uses the same numbering for common components.

The preamp is a straightforward non-inverting opamp with a gain of 101. The RX coil is lightly loaded with a 10kΩ resistor to prevent ringing due to its self-capacitance. The 100kΩ∥100pF feedback impedance provides for low-pass filtering with a cutoff frequency of 15.9kHz. This creates a 32° phase shift at 10kHz, close to what we used in the earlier hypothetical example (Figure 14-22b).

The demodulator SHA consists of a JFET sampling switch with R6 and C2, plus non-inverting opamp IC2a. IC2a provides a fairly high impedance load for the demodulator cap, limited by the retune balancing resistors R8 and R11. The other half of the opamp, IC2b, is used for the retune circuit. The retune's integrator/hold circuit requires a low input bias current to avoid capacitor droop which would show up as tuning drift. The JFET-input TL072 is used for this although it is marginal for this function. With a maximum input bias current of 200pA, IC2b can experience a droop rate of up to 20μV/s. Increasing the value of C4 will help but so will a better opamp. See the *Explore* section for more tips on the retune circuit. Finally, notice a resistor has been added in series with C4; this is for opamp stability as some opamps become unstable when driving a highly capacitive load.

The final gain stage (IC3a) includes a variable current injection circuit for adjusting the audio threshold level. In the heyday of TR-Disc detectors this was usually called the Tuning control. The gain stage feeds a simple transistor buffer, volume potentiometer, and LM386 with audio chopping provided by Q4 and oscillator IC3b. The oscillator[23] is set up to run at approximately 600Hz. A rising voltage at the output of IC3a will cause an increase in audio loudness, as desired.

The TX oscillator is recycled from the *Improved TR* design which, in turn, was lifted from the *TR-550D*. Because the TX voltage swings from roughly +VB to −VB it needs to be reduced before feeding the all-pass filter phase shifter. C10 AC-couples the signal and R29-R30 reduces the amplitude.

IC5b is the same phase-shifter implementation discussed earlier in the chapter except that a buffer stage has been added to isolate the effect of the variable DISC potentiometer from the transmitter signal divider[24]. The phase shifter drives an LM393 comparator which converts the sinusoidal phase shift signal into a logic signal for the JFET demod switch. Recall that a negative V_{GS} on the JFET turns it off while $V_{GS} = 0V$ turns it on so the load resistor (R35) for the open-collector comparator is connected to ground instead of VCC.

Finally, let's take a look at the overall signal operation. Earlier in the chapter the demodulator (and its timing) was designed so that a non-discriminated target would cause a rising voltage at the demodulator. This action is preserved by the non-inverting IC2a and also by the non-inverting IC3a. For this design, a valid target produces positive signals all the way to the audio stage.

However, it is possible that, for example, IC3a could be designed as an inverting amplifier, which would produce a *falling* voltage at the audio stage for valid targets. There are several ways to deal with this:

---

22. 6V is available in the 78L06. 7V can be realized using a variable linear regulator such as the LM317. However, these old regulators also have a high drop-out voltage which then limits how low the battery can discharge and still be useful. If you want to run the battery down to 7V, then the 78L05 may be the best you can do. A better choice is to use a modern low-dropout regulator.

23. Another opamp circuit that is well-documented in other places.

24. Technically, this isn't much of an issues in this design but it will be an issue in the design presented in the next chapter.

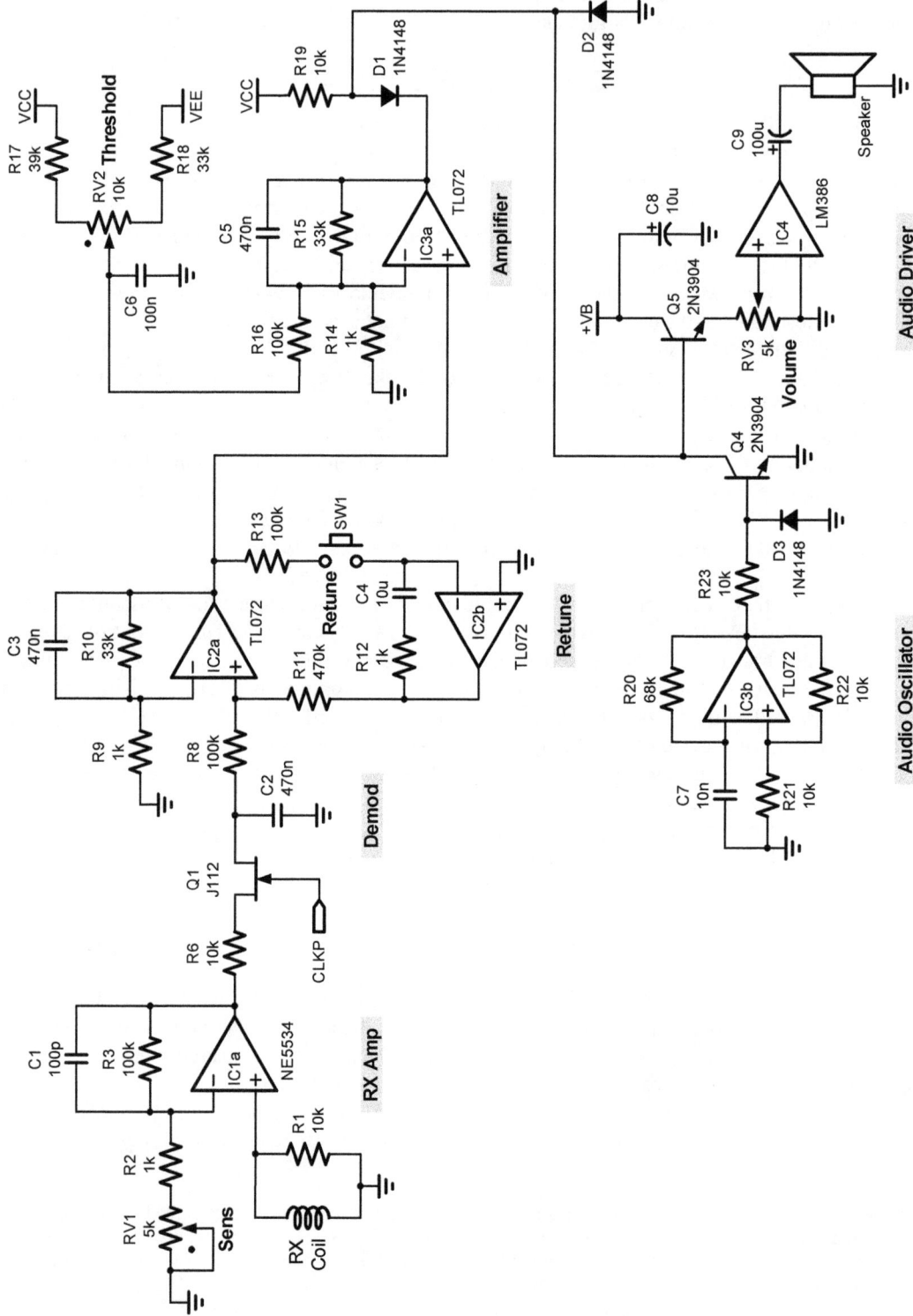

Fig. 14-30: **TR-Disc — Main Circuitry**

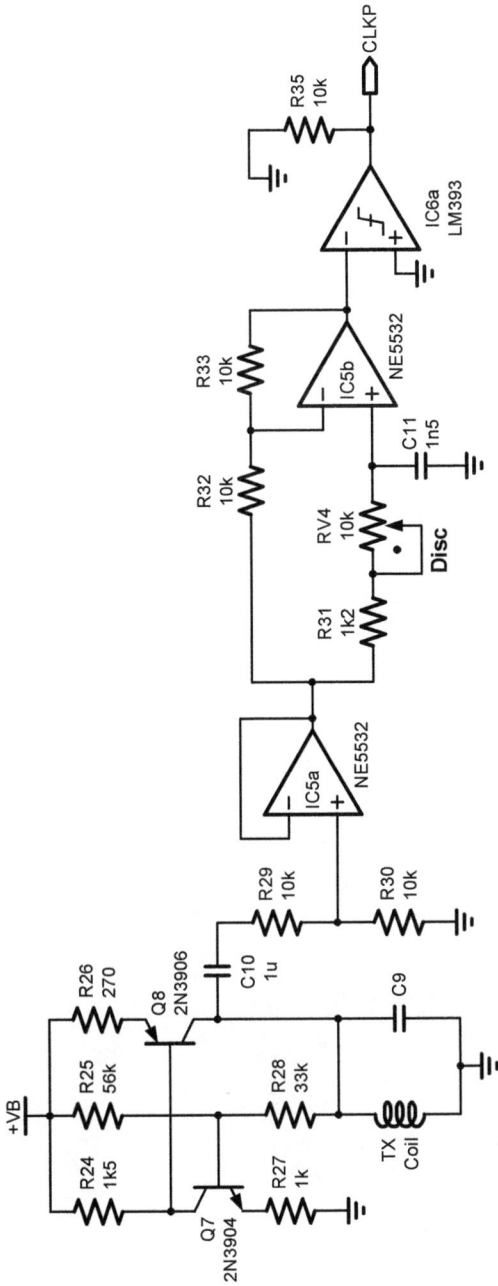

Phase Reference

Transmitter

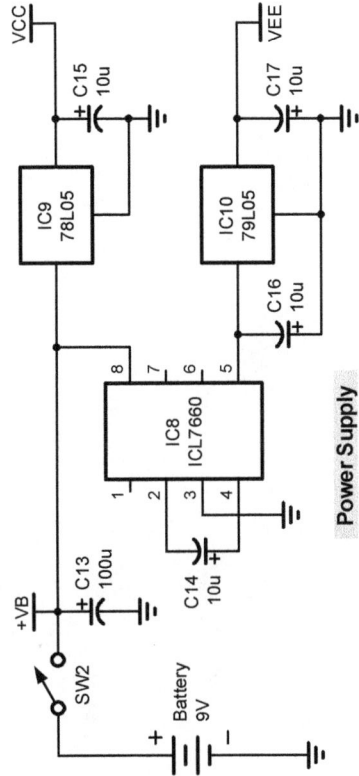

Power Supply

Fig. 14-31: TR-Disc — TX/Phase & Power Supply Circuitry

- Design an audio stage whereby a falling voltage at the output of IC3a will cause an increase in loudness.
- Invert the demod timing polarity by reversing the polarity of IC6a. Then the demod output will decrease for valid targets, but IC3a will increase.
- Flip the RX coil (or the TX coil) over to invert the signal polarity.

It is important to pay attention to the signal processing "big picture" and understand how to deal with such issues. Sometimes the easiest way to fix a problem is not the best. For example, flipping the RX coil over to invert the RX signal polarity is certainly a simple solution but if you are sharing a coil design amongst multiple projects (as we do in this book) then that may cause a problem elsewhere. Besides, it is my general philosophy to always design coils so that, for a ferrite target, the RX voltage is in-phase with the TX voltage. This is easy to remember and all coils will come out the same.

Explore

14.14: Band-pass Filtering — Part 1

It is common in single frequency metal detector designs to use a band-pass preamp. The *TR-Disc* design has a low-pass preamp; adding a single capacitor to the feedforward leg makes it a band-pass. Since the preamp is a non-inverting topology, the low frequency gain bottoms out at 0dB but for the circuit shown in Figure 14-32 that is still a 20dB reduction. This will help suppress low frequency EMI like AC mains. For the component values shown, the low-pass pole is set to 33.9kHz and the high-pass pole is 3.39kHz.

Figure 14-33 shows the frequency and phase response for both the low-pass and band-pass preamps using the NE5534. The low-pass phase delay[25] at 10kHz is a little more than 20° (actual measured was 28.8°) while the band-pass phase delay is closer to 0° (measured was -2.4°). In a band-pass preamp, when the high-pass pole and the low-pass pole are selected to be geometrically equal distances from the operating frequency then the phase shift will be 0°, assuming an ideal opamp. As an example, for a 10kHz operating frequency we could select 10kHz/3 (3.33kHz) and $10\,\mathrm{kHz} \times 3$ (30kHz). A non-ideal opamp has an internal high-frequency pole that creates a phase shift of its own, with higher-speed opamps creating less phase error. A 0° phase shift may seem to be another advantage of the band-pass preamp over the low-pass preamp but, in reality, it doesn't make much difference; we can easily compensate for whatever preamp phase shift we get.

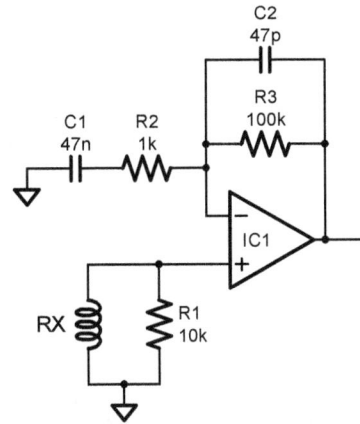

Fig. 14-32: **Band-pass Preamp**

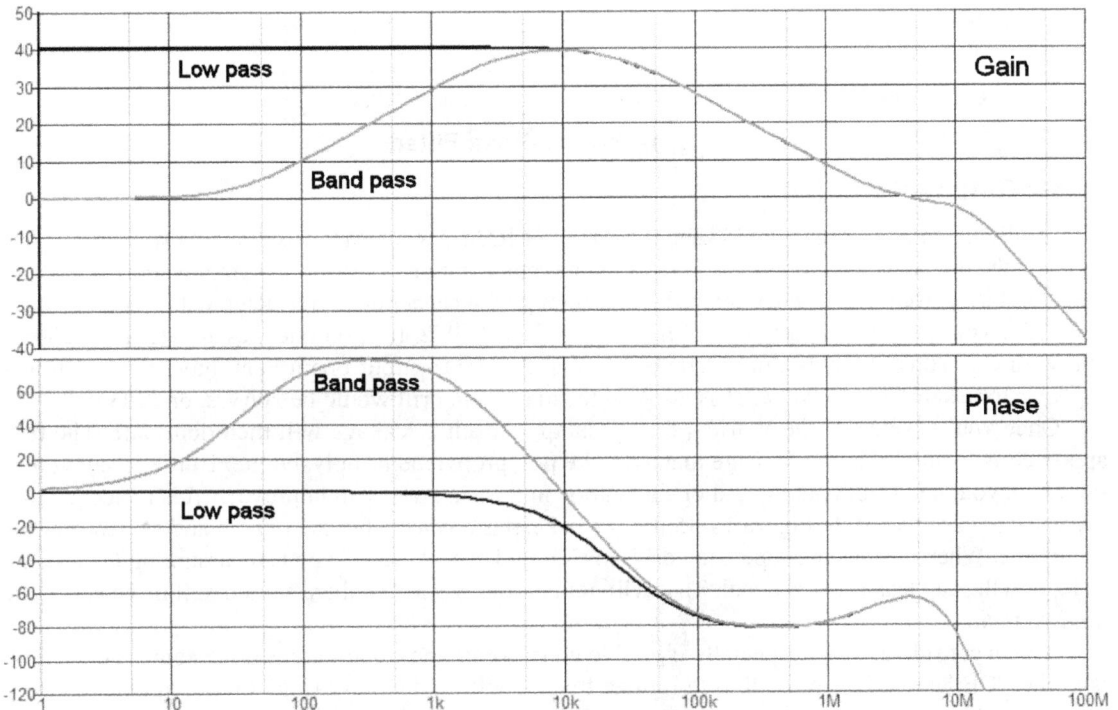

Fig. 14-33: **Band-pass Preamp Response**

---

25. Remembering that a negative phase shift is a phase delay.

14.15: All-Pass Filter Tuning Range

In the project design a tuning range of 10° to 80° was achieved with an all-pass filter, and it was stated that the all-pass filter was capable of a phase shift range of 0° to 180°. However, the high phase shifts are approached asymptotically resulting in a non-linear adjustment range. For example, Figure 14-34 shows the phase response (black curve) using a 1nF capacitor with a 100k linear potentiometer at 10kHz. The maximum phase shift is approaching 165° at a declining rate. Changing the capacitor to 10nF gives an extended look at the non-linear behavior at even higher phase shifts (gray curve). For decently linear operation this circuit should be limited to around a 0° to 90° adjustment range. If some amount of non-linearity is acceptable then an adjustment range up to 120° (and perhaps a littler higher) may work.

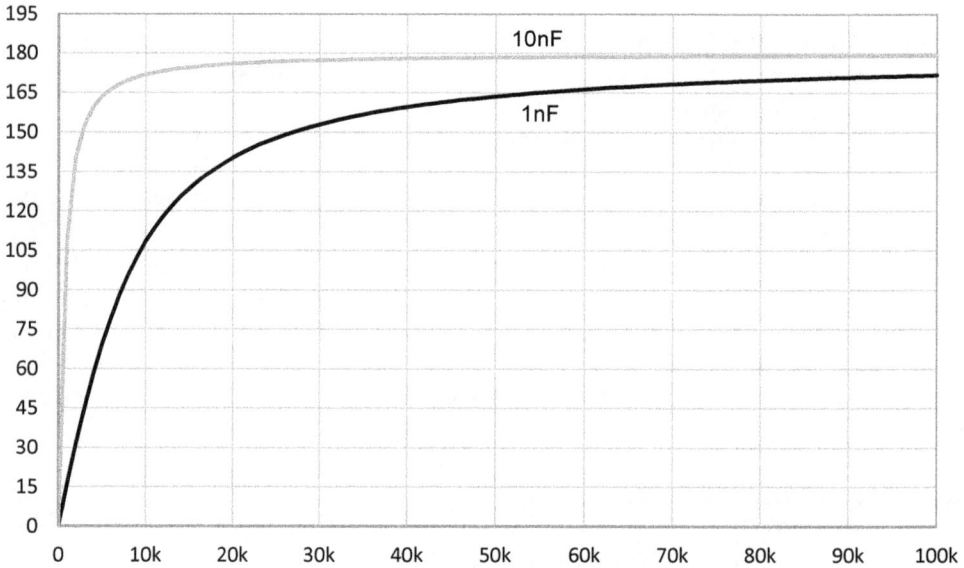

Fig. 14-34: **All-Pass Filter**

14.16: Retune Revisited

Ideally the retune circuit should keep the output of IC2a at a constant offset level. Over time, it will drift and the user will occasionally need to press the retune button. There are two major sources of drift: the input bias current (I_b) of the opamp flowing through the capacitor, and leakage within the capacitor itself. The voltage drift due to I_b is $dv/dt = I_b/C$. The TL072, for example, has an $I_b(max) = 200pA$ and with $C = 10\mu F$ the maximum drift is $20\mu V/s$. A CMOS-input opamp can have a much lower $I_b(max)$; for example, the LMP7721 is 20fA so the maximum drift would be 20nV/s, or $72\mu V/hr$.

Once you've selected the World's Best Opamp, capacitor leakage will then dominate. The best approach is to choose a low-leakage material like polypropylene or polystyrene. But let's say you're limited in your choice of opamps and/or capacitors. In this case a way to improve drift is to add another capacitor to cancel drift. In Figure 14-35 the second cap is connected to the (+) pin and switched at the same time. When the switches open, the drift on C4b will cancel the drift on C4a, assuming I_{b+} is equal to I_{b-} and the cap leakages are equal. That will not be exactly true but there should still be considerable improvement.

A final consideration — especially if you've gone to extremes to minimize other sources of drift — is in the PCB layout. It is critical to minimize trace lengths, add guard rings to the input nodes, and apply a conformal coating to these nodes to prevent moisture-induced leakage. Clean everything with isopropyl alcohol, being sure to remove all flux before applying the conformal coating. Even clean and coat the capacitors as they are subject to surface plastic leakage.

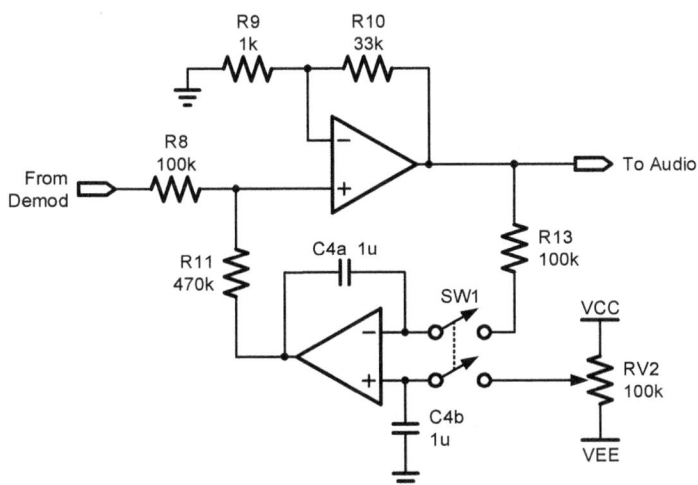

Fig. 14-35: **Improved Retune Circuit**

14.17: Autotune Circuit

The autotune solution is a simpler method to reduce ground effect but did not become popular until VLF motion discriminators. It consists of a high-pass filter following the demodulator as shown in Figure 14-36. The tau of the high-pass filter is selected so that the slow ground signal is largely suppressed and faster target signals pass through. Following the high-pass filter is a threshold stage. Signals above the threshold are heard and signals below the threshold are not.

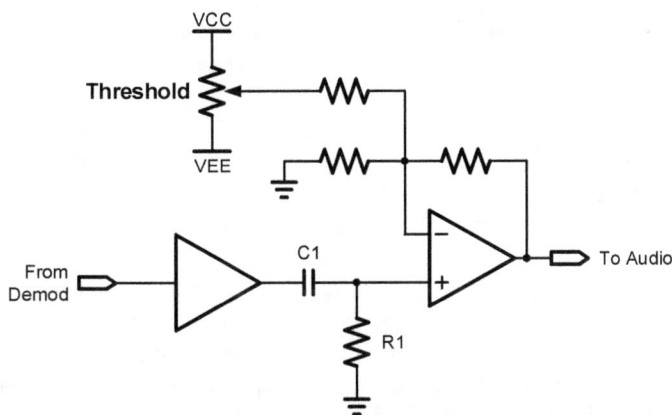

Fig. 14-36: **Autotune Circuit**

The autotune method relies on a ground signal that does not vary much and does so slowly. This requires level ground, consistent mineralization, and a level coil sweep. While this situation may seem improbable it is not uncommon for typical coin hunting venues like parks, schools, and yards. Autotune also requires coil motion for target detection; if the coil is held steady over a target, the target will get tuned out.

The retune circuit in Figure 14-26 can also function as a slow autotune if R13 is increased in value. Some detectors — notably Tesoro models — did just this and had a slow autotune mode (retune switch closed, high value for R13) and zero-motion mode (retune switch open) with a fast manual retune by switching in a low value for R13.

14.18: Demodulators — Part 1

The demodulator is a frequency-translation circuit; the input and output frequencies are not the same. This has two important consequences. First, the demod gain isn't strictly a gain; it depends on the signal, both waveform and frequency. Instead of gain, the demod is characterized by *conversion gain*. Second, because of frequency translation, demods cannot be simulated in Spice using frequency-domain analysis. They can only be simulated in the time-domain, making it difficult to see how they respond over frequency. There are simulators (such as *SpectreRF* from Cadence) that include a method called *periodic steady-state* analysis which can produce a frequency-domain response for demodulators and oscillators[26], but such software is stratospherically expensive.

The sample-and-hold demodulator has been used in detectors since the mid-70's. George Payne's 1977 patent (US4030026, filed in '74) kicked off sampling receivers at White's and was subsequently adopted by everyone else, possibly because Payne worked for everyone else[27] at some time or another.

The half-wave (HW) demod produces an average of the sampled input signal. If the input is $A \cdot \sin(\omega t)$ and is sampled for the first half-period then the average is:

$$v_o = \frac{2}{T} \cdot \int_0^{T/2} |A \cdot \sin(\omega t)| \, dt$$

or

$$v_o = \frac{2A}{T} \cdot \int_0^{T/2} \sin(\omega t) \, dt$$

which, for $T = 1/f$, evaluates to

$$v_o = \frac{2A}{T} \cdot \left[-\frac{1}{\omega} \cos(\omega t) \Big|_0^{T/2} \right] = -A \frac{2f}{2\pi f} \cdot [\cos \pi - \cos 0] = \frac{2}{\pi} A$$

Thus the HW demod has a conversion gain of $2/\pi$, or about 0.636. During the "hold" half-cycle the voltage is simply held on the capacitor and has no effect on the averaging. Note that the SHA conversion gain does not depend on the time constant (tau) of the SHA. The tau affects other things like the amount of ripple seen riding on the output average and, perhaps more importantly, the response speed.

The conversion gain analysis was for a purely reactive (0°) signal, which gives a maximum output voltage. Any other sampling phase will result in a smaller output voltage, but the conversion gain is still considered to be $2/\pi$. For a purely resistive signal (90°), it's visually obvious from Figure 14-15 that the output voltage *should* be zero. In general this depends on the demod design but the SHA demod is well-behaved in this respect.

Since we use the synchronous demod to extract target phase information, it is useful to plot the response of the demod vs input phase. Sweeping the phase[28] from 0° to 180° is sufficient to get a complete picture as shown in Figure 14-37. The input signal is still normalized to 1v peak so we see right off that the peak output at 0° is 0.636, our expected conversion gain. The quadrature output can be seen by zooming into the 90° value and, accounting for an additional 1ms delay due to the tau, the (averaged) output is exactly zero. At 180° the output is −0.636. If you think this curve looks familiar, see Figure 14-17.

While it is not possible to run a classical AC analysis on a demod using Spice, it is possible to fake the AC response with a time-domain analysis. Suppose the demod is clocked at, say, 10kHz. If the input signal is slowly swept from f_1 to f_2 then the output of the demod will be a frequency response. The out-

---

26. In a previous life I used *Spectre RF* to aid in the design of a low phase noise 2.4GHz VCO for a GSM receiver.

27. Besides White's, Payne worked for Bounty Hunter and Compass, and co-founded Teknetics and Discovery. Interestingly, the original Bounty Hunter Red Baron (Payne design, ~1977) used the LM1496 double-balanced mixer (Gilbert cell design) for the demods. Most likely the reason for this was to avoid his '026 patent with White's in which he used a simple JFET for the sampling switches.

28. In Figure 14-37 *time* is the swept parameter (from 0 to 1 second) but phase is set up to be proportional to time. How this is done will be explained in a few more paragraphs.

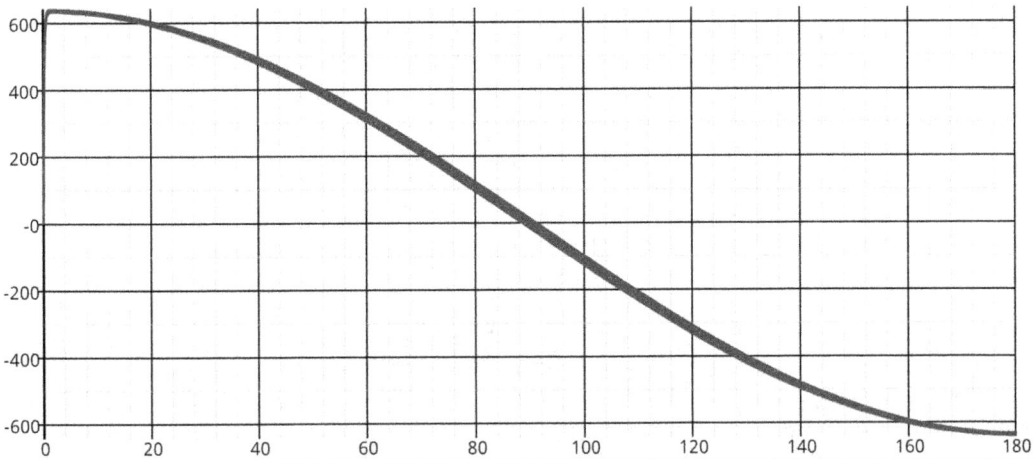
Fig. 14-37: **Phase Response**

put signal will have both positive and negative excursions as it should, but in Spice we can simply take the absolute value to get a more palatable-looking FFT-ish response plot. Figure 14-38 shows such a plot for an input swept from 0.1Hz (essentially DC) to 20kHz with the demod clocked at 10kHz.

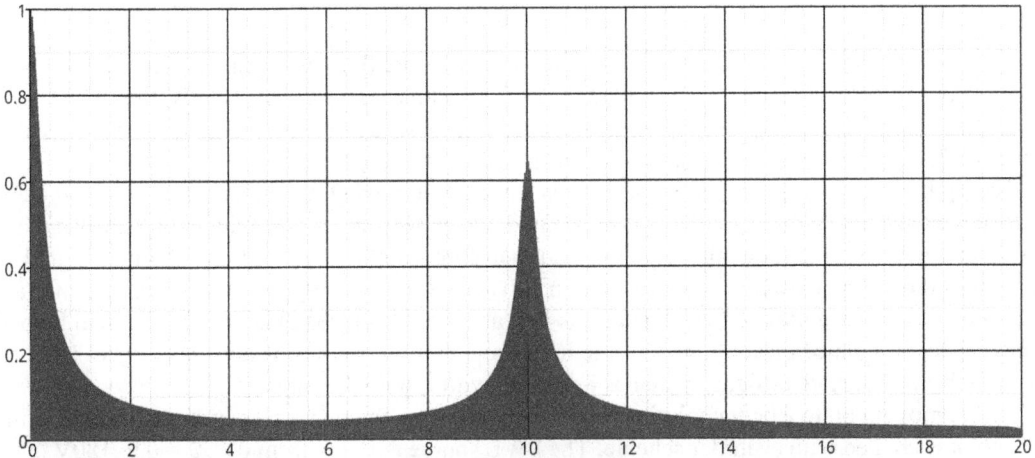
Fig. 14-38: **Half-Wave Demod Frequency Response**

Let's look at this a bit. First, this kind of simulation can take a while to run. The frequency needs to be swept rather slowly and for each demod clock period you will want perhaps 100 sim points. For this sim the input was swept linearly from 0.1Hz to 20kHz over a 10 second span[29] and the maximum transient step time was 1μs. This results in a 10 million point simulation which can take several minutes on a moderately fast computer.

Second, the sim requires the use of "arbitrary sources" which are not a standard part of Spice but are found in simulators with XSpice enhancements. I use a program called Easy-Spice[30] which has this, but the popular (and free!) LTSpice does as well. The sim schematic is shown in Figure 14-39 and deserves its own discussion.

Switch S1 along with R1 and C1 form the sample-and-hold circuit to be tested. S1 is clocked with a 10kHz square wave (V2). The input signal is created by a combination of a piece-wise linear (PWL)

---

29. This is a time domain sim and normally the x-axis represents time. In the case of Figure 14-38 the x-axis should be 0 to 10 seconds with the peak at 5 seconds, but it's been fudged by re-labeling the x-axis to represent frequency in kHz.

30. Easy-Spice was sold as an add-on to Easy-PC, but is no longer available. It was an offshoot of SiMetrix Spice.

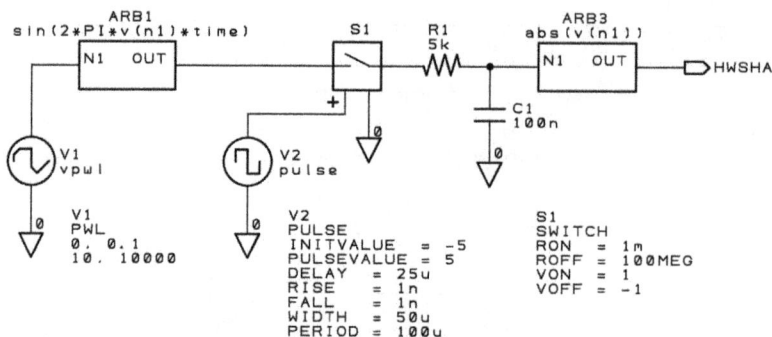

Fig. 14-39: **Half-Wave Spice Schematic**

source (V1) and an arbitrary source (ARB1). The PWL source creates a linear ramp from 0.1V at t=0 to 10,000V at t=10s. This "voltage" is applied to the arbitrary source which has a transfer function defined as $v_{out} = \sin(2 \cdot \pi \cdot v_{in} \cdot t)$. That is, the PWL voltage is actually used as the frequency variable in the arbitrary waveform function with time t being the simulation time, which is swept from 0 to 10s. This would seemingly create a 1v-peak sine wave with a frequency that is swept from 0.1Hz at the beginning of the simulation to 10kHz at the end of the simulation. However, the math works out so that

$$f_1 = v_1$$
$$f_2 = 2v_2 + v_1$$

where f_1, f_2 are the starting and ending frequencies and v_1, v_2 are the starting and ending voltages. When $v_2 \gg v_1$ the ending frequency is approximately $2v_2$ or, in this case, 20kHz. Finally, the output voltage of the sample/hold is applied to another arbitrary source (ARB2) which outputs the absolute value of the input. The result is the plot we saw in Figure 14-38.

Getting back to that plot, we see that there is a peak output when the input signal frequency equals the demod clock frequency. And that is how metal detector signals are demodulated: the demod clocks are synchronous with the transmit frequency and therefore synchronous with the received signal. This peak equals 0.636, the expected conversion gain[31]. There is an even higher peak in the response at DC where the conversion gain is 1, which suggests that any signal (or noise) at or near DC will be passed through. That is a major disadvantage of the half-wave demod because it does not cancel DC offsets and, in high-performance designs, demands a preamp with low flicker noise[32].

Before moving on to a better solution let's briefly return to the phase response plot in Figure 14-37. Such a plot is created with a similar scheme. The PWL source is swept from 0V @ t=0 to 180V @ t=1s; the arbitrary source has a transfer function defined as $v_{out} = \sin(2 \cdot \pi \cdot 10000 \cdot t + v_{in})$ where v_{in} now represents a phase variable; time is swept from 0 to 1s; and the absolute value function is omitted.

The next logical demod topology to consider is a simple enhancement to the half-wave demod whereby the entire signal is averaged, not just half of it. The *full-wave demod* is shown conceptually in Figure 14-40, and Figure 14-41 shows the time-domain waveforms for the usual targets when the demod clock phase is 180°. Compared with the half-wave responses in Figure 14-15 the full-wave demod produces demodulation throughout the entire TX cycle, not just half of it. Although the conversion gain of the full-wave demod is identical to that of the half-wave (0.636) the full-wave has significant benefits.

The simulation schematic is shown in Figure 14-42. The result of the simulation is shown in Figure 14-43 with the half-wave demod result also included (light gray). While the two responses are virtually

---

31. Actually, careful inspection shows the peak is 0.619. That is because the input signal at exactly 10kHz was not exactly at 0°. This can be corrected by tweaking the phases of the demod clocks a little, but is generally not worth the effort.

32. Most opamps have an "input noise" spec v_n which represents wideband thermal noise, often spec'd at 1kHz. Flicker noise is a near-DC increase in the noise level due to other causes. Many opamps do not list a flicker noise spec but it is often seen in the data sheet frequency domain plot of v_n.

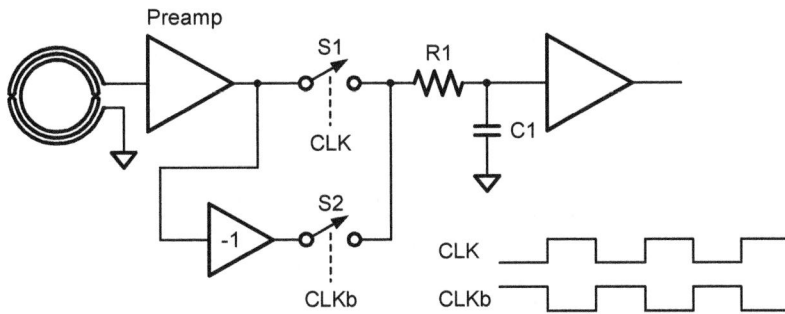

Fig. 14-40: **Full-Wave Demodulator**

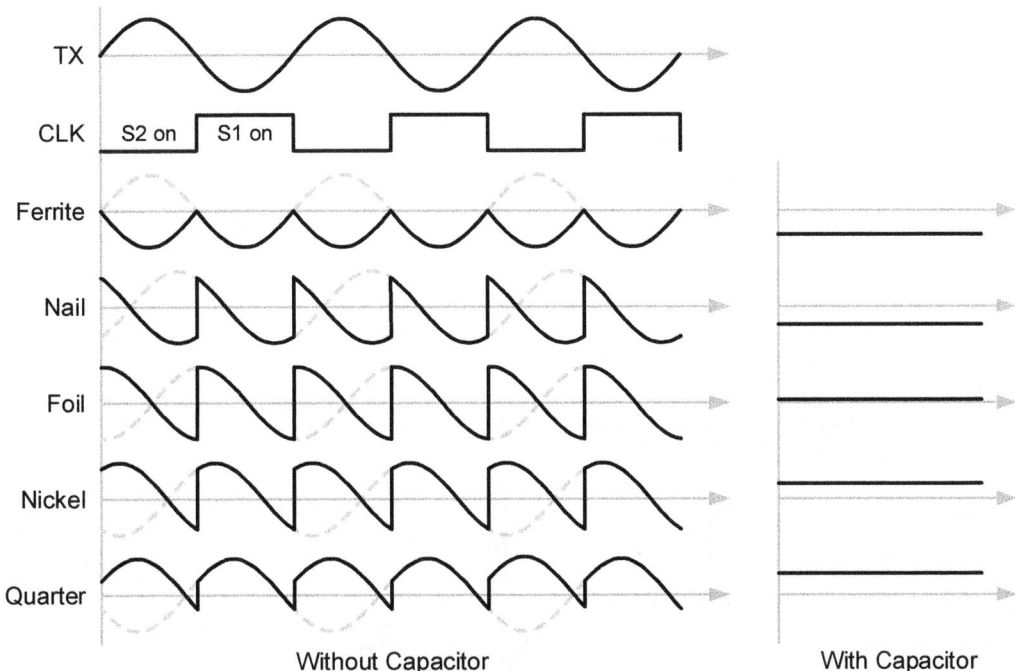

Fig. 14-41: **Full-Wave Waveforms**

identical in the region of the operational frequency (10kHz) there is an immediately obvious benefit in low-frequency suppression. Because the full-wave demod adds an inverted version of the input signal every other sample, a purely DC signal will exactly cancel. Low frequency signals will likewise be heavily attenuated, including Earth field effect, and which provides significant relief to the flicker noise demands on the preamp. This is why half-wave demods are rarely seen in modern designs. There also appears to be a slight improvement in frequencies above 10kHz but it is due to the half-wave response having more ripple; remember, this is a transient simulation[33].

In narrowband detector designs (i.e., single frequency) the input signal to the demod is usually band-pass filtered to attenuate noise outside of the operational frequency. Not so with wideband detector designs (multifrequency and PI) and since noise can also be wideband it is useful to consider the frequency response beyond twice the operational frequency. To do this we simply redefine the PWL source that sets the effective frequency limits to, say, 100Hz to 100kHz[34], still maintaining a 10kHz demod clock. The resultant frequency response plot is shown in Figure 14-44[35].

---

33. Ripple is also the reason why DC is not exactly zero.

34. By sweeping the PWL source from 100V to 50,000V.

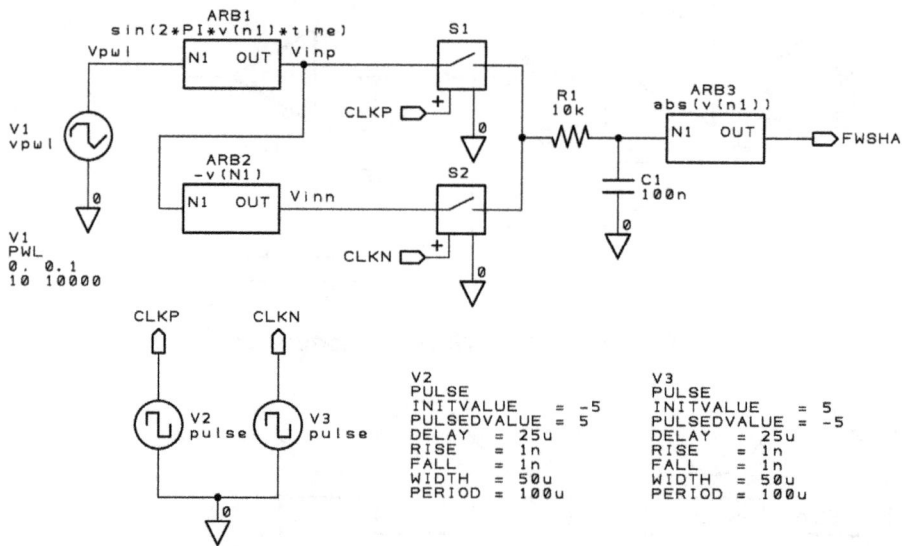

Fig. 14-42: **Full-Wave Spice Schematic**

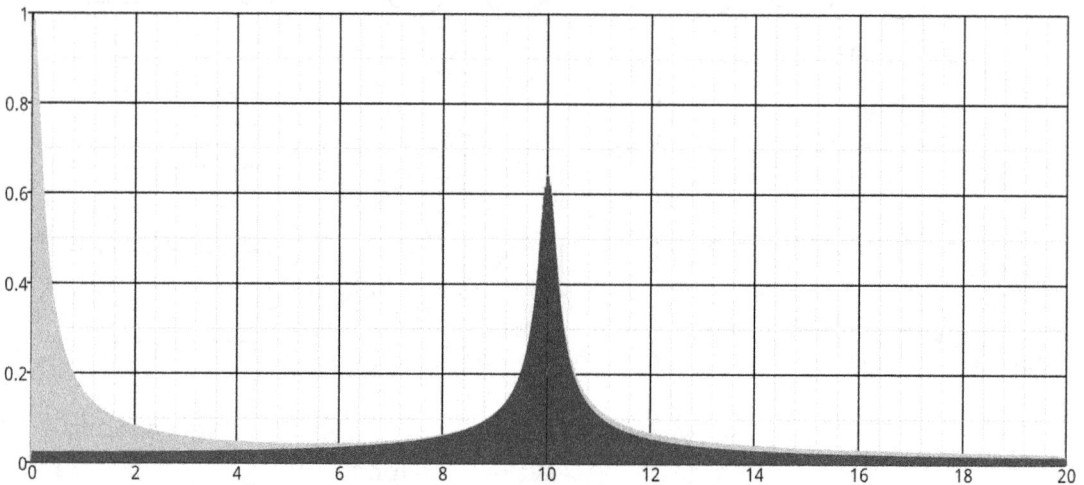

Fig. 14-43: **Full-Wave Demod Frequency Response**

The synchronous 10kHz response has the expected conversion gain of 0.636. There are additional harmonic responses at 30kHz, 50kHz, 70kHz, and 90kHz. As with DC, even-order harmonic responses cancel. This shows that odd harmonics of the synchronous frequency are passed through, but at a decreasing rate. The $2/\pi$ conversion gain derived for the 10kHz synchronous signal doesn't apply to the harmonics and the rate of decrease exactly follows the $\sin(x)/x$ response expected of a sampled system. For fclk = 10kHz the fundamental and harmonic conversion gains are:

---

35. Again, this is a time-domain simulation but the x-axis has been re-labeled to represent frequency (in kHz) to make it more readable.

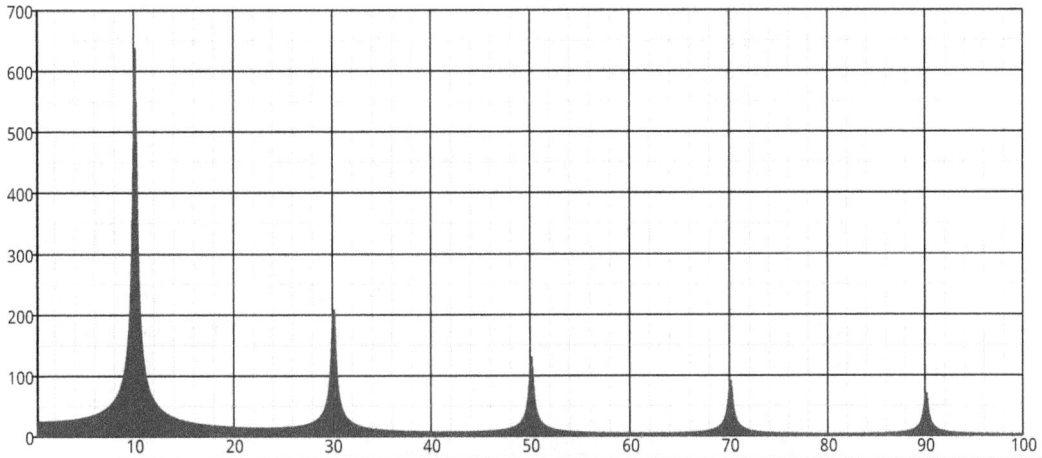

Fig. 14-44: **Full-Wave Demod Extended Frequency Response**

fin	G
10kHz	0.636
30kHz	0.217
50kHz	0.128
70kHz	0.091
90kHz	0.072

At these harmonics wideband noise will alias directly into the demod output so knowing the conversion gains can help in calculating overall noise.

The astute reader will have noticed an inconsistency between the Spice schematics for the half-wave and full-wave demods: the half-wave demod has a 5kΩ resistor while the full-wave demod uses a 10kΩ resistor. This would seemingly give the full-wave version a time constant that is twice that of the half-wave. But they actually have the same time constant because the half-wave demod is only active half the time; the other half of the time the resistor has no effect on the output. That is to say, the resistor in the half-wave demod functions at twice its actual value.

This is demonstrated in Figure 14-45 which shows the step response for both demods. The demod taus are 1ms and both settle to the 0.636 conversion gain in about 5 taus. Note that the half-wave (light gray) has a higher ripple, another drawback of that topology. The only advantage of the half-wave demod is that it is simpler; only one switch, one clock, and no inverting signal amp.

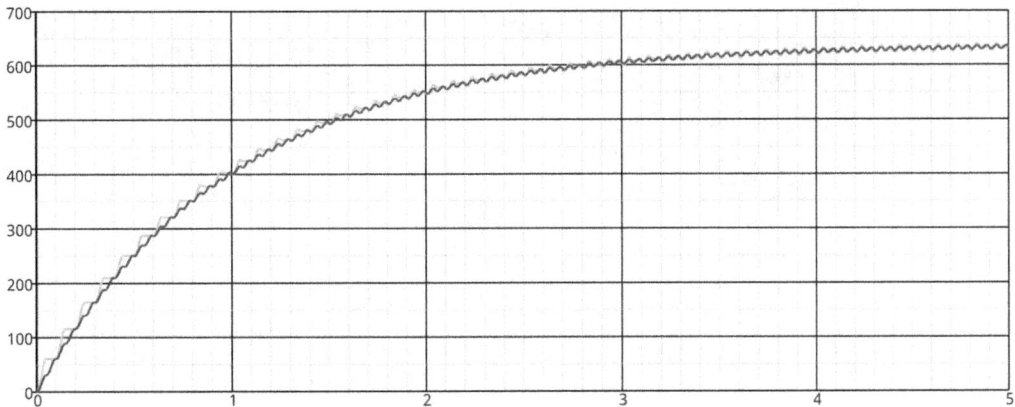

Fig. 14-45: **Step Response: Half Wave vs Full Wave**

While on the topic of settling, let's wrap this up with another look at response speed. Figure 14-45 shows the response for a 1ms tau, but is this a good choice? The answer to that depends on how fast the user is sweeping the coil, plus the size and depth of the target, and even the size of the coil. The actual target signal may end up only a few 10's of milliseconds wide so a 1ms tau seems plenty fast. However, users are demanding faster target separation to deal with trashy ground and while DSP techniques are a big part of achieving fast separation, DSP cannot fix slow demodulators.

It turns out that ripple and response speed are directly related; if you want a wicked-fast response you will get wicked ripple. Three examples are shown in Figure 14-46: tau = 0.1ms, 1ms, and 10ms (Vin is still 1v peak). Note that the sim is now 50ms, long enough for the 10ms tau to fully settle. The fastest response has a ripple 650mv p-p, the middle response is 65mv p-p, and the slowest is 6.5mv p-p. Thus the ripple is proportional to speed. The amount of ripple can affect the stability of target responses, especially when tonal or visual target ID is implemented.

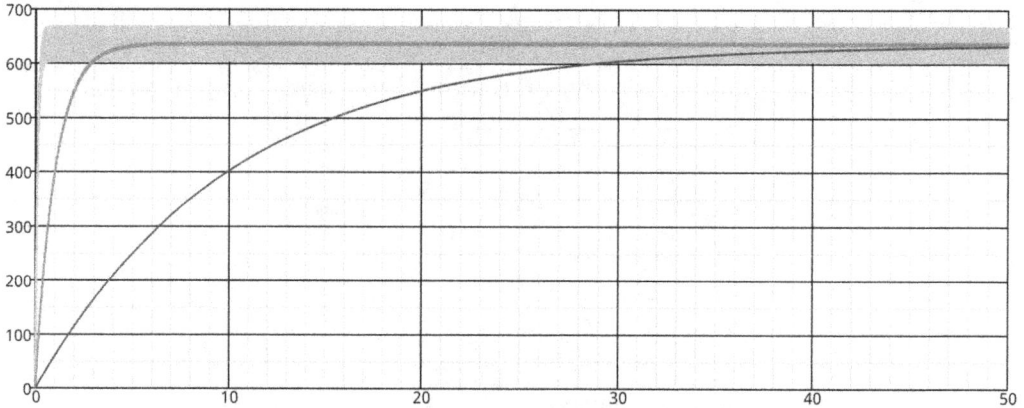

Fig. 14-46: **Step Response: Tau = 0.1ms, 1ms, 10ms**

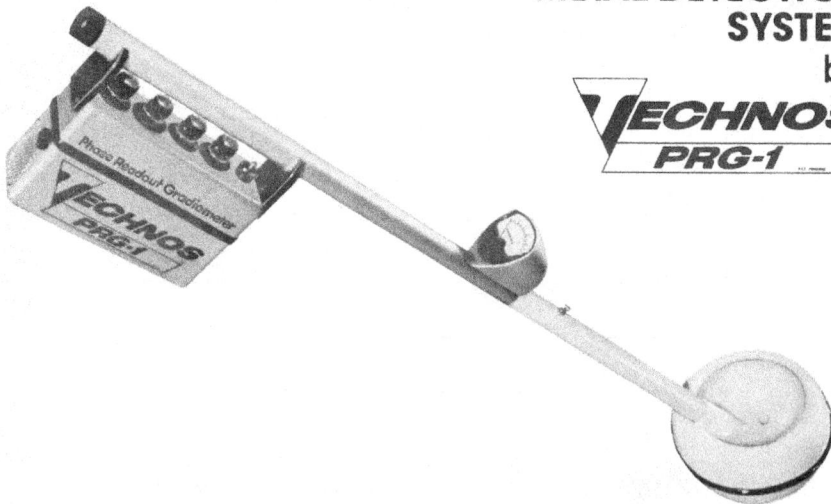

The **Phase Readout Gradiometer**

METAL DETECTION SYSTEM

by

ᐯECHNOS PRG-1

Inside the Metal Detector

15 VLF Ground Balance

"After all, the struggle between 'It's good enough'
and 'I can do better', is part of the engineer's DNA."

— *Bill Schweber* (Editor of Planet Analog)

VLF stands for *Very Low Frequency* and is formally defined as the radio spectrum between 3kHz and 30kHz. However, many so-called VLF detectors fall outside this frequency range so for metal detectors it's a loosely used term. Although VLF and TR designs are both induction balance, in Chapter 2 we decided that VLF designs incorporate ground balance whereas TR designs do not. This chapter explores the methods of achieving ground balance.

It is possible that detectors with ground balance actually predated detectors with discrimination. I've heard that there were military mine detectors as early as the 1940s that were VLF designs with ground balance but I don't have a solid reference to confirm that[1]. In the hobby market, the first VLF-GB[2] designs appeared at about the same time as the first TR-Disc models; the White's *Coinmaster V* (1.75kHz) was a notable early example[3] (Figure 15-1).

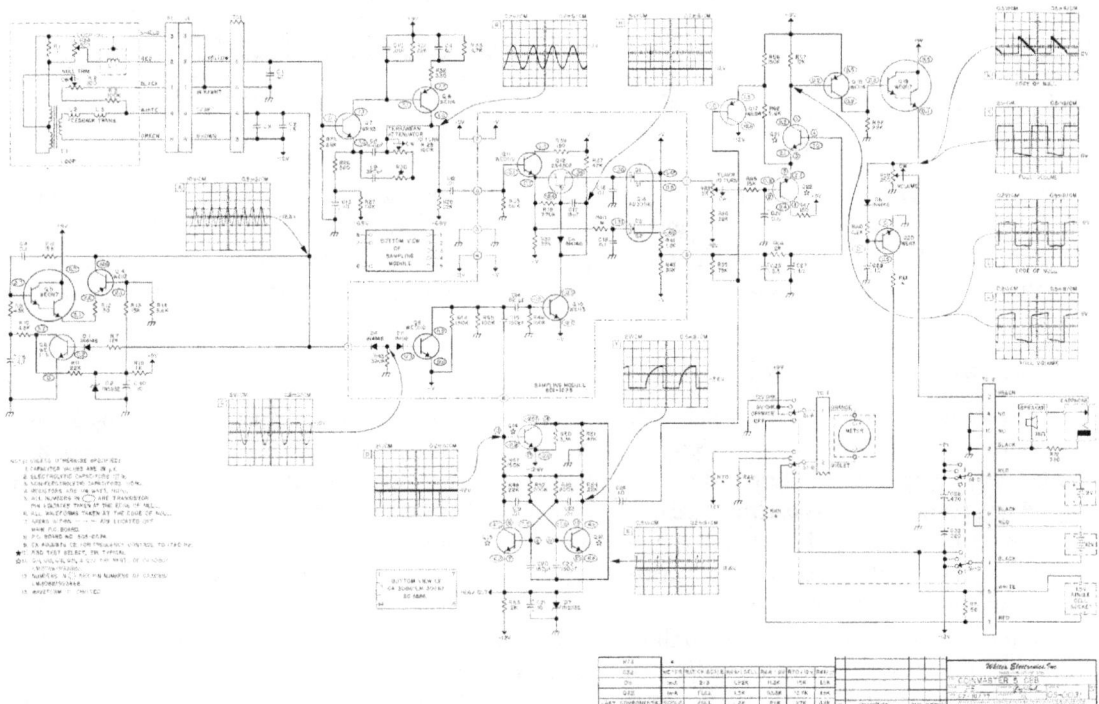

Fig. 15-1: **White's** *Coinmaster V*

---

1. It's hard to imagine ground balance without synchronous demodulation which WW2 detectors did not use.

2. GB is short for *Ground Balance* and will be used extensively throughout the rest of the book.

3. White's coined the popular term *Ground Exclusion Balance*, or GEB, for their ground-balanced VLF design.

Ground Response

15.1: Loss Angle

Before exploring how to achieve ground balance we must first determine how ground behaves. When magnetic and metal targets are interrogated with a sinusoidal magnetic field their responses will exhibit a phase delay, as covered in Chapter 4. Purely magnetic targets, including the responses of soils and rocks, are limited to a phase delay between 0° and 90°.

Chapter 4 briefly explored ground behavior by substituting ferrite for ground. It was shown that ferrite has a magnetic response with a phase angle of approximately 0° which is considered an ideal lossless response. The phase response of a purely magnetic substance is called the *loss angle*. Most real soils and rocks have a loss angle close to 0° and on up to 5-6°, with some really weird soils being as high as 15°. Ideally the loss angle does not vary with frequency but practically it will to some extent, with the extent depending largely on the amount of salts and viscous mineralization in the soil.

15.2: Frequency Response

Table 4-1 shows the frequency response for the loss angle of a typical ferrite plus a volcanic hot rock which serves as an example of what severe ground might look like; these are repeated in Table 15-1. Besides the fact that different soils have different loss angles, the loss angle of the soil even in a particular location can vary somewhat over distance. Usually this is a minor problem with the variation being gradual. In a worst-case scenario it is possible to swing the coil over a back-filled trench where the mineralization has a very sudden change in loss angle and can create a strong ground response[4]. With experience this can be recognized and the coil swept parallel to the trench line to minimize the problem.

	1kHz	2kHz	5kHz	10kHz	20kHz	50kHz	100kHz
Ferrite (#33)	0°	0°	0°	0°	0°	0°	0.2°
Volcanic rock	3.7°	4.1°	5.0°	5.9°	6.6°	7.8°	8.9°

Table 15-1: **Soil Loss Angle vs Frequency**

Regardless of the soil or the frequency a ground balance circuit needs to work for loss angles in the range[5] of 0° to 15°. The method of doing this is exactly the same as was used in the TR-Disc circuit where clocking the demodulator at a particular phase would create a null response for a particular target. In the case of ground balance, we will select a demod phase that creates a null response for the ground.

15.3: Advantages of Low Frequency

The earliest VLF-GB designs ran at the very low end of the VLF range. For example, the White's *Coinmaster V* ran at 1.75kHz[6], the Compass *Relic Magnums* used 4kHz, and Garrett chose 5kHz for their *Groundhog* models. Modern GB designs often run at much higher frequencies, even above 30kHz for gold nugget detectors. The difference between then and now is that those early designs had less gain so a lower frequency helped produce stronger target responses — especially with silver coins — as will be demonstrated in the next section. Lower frequencies also penetrate mineralization more easily, resulting in better detection depth.

Even today the deepest detectors for high conductive silver coins often operate at the low end of the VLF range. For example, the Minelab BBS & FBS models were renowned for hitting deep silver. They were multifrequency designs but the predominant frequency was 3.125kHz. In fact, many modern multifrequency detectors include single frequency modes that reach down to 3-4kHz specifically for hunting deep silver.

---

4. I once discovered an especially bad one on the Oregon State University campus.

5. For most places 15° is probably excessive, but we'll have a design that should work anywhere.

6. Actually below the VLF range.

Demodulation

The TR discriminator uses a single synchronous demodulator. If the demodulator is clocked 180° out-of-phase with the transmit signal then the resulting demod output is negative for ferrous and positive for non-ferrous with a (roughly) null response for tiny foil[7]; refer to Figure 14-15. It so happens that ground mineralization is at the extreme end of the ferrous range so highly mineralized ground will produce a strongly negative signal at the demod output. We saw that this can be mitigated by using a retune circuit but requires careful sweep techniques and is still prone to be noisy.

If, instead, the demod is clocked at a phase of 90° with respect to the transmit signal then the ground signal will be sampled with equal negative and positive portions and the output of the demod will be zero. This will be true even if the coil is raised and lowered, making the detector response substantially immune to ground mineralization. We call this effect *ground balance* or *ground cancel*. It is important to understand the detector still receives the ground signal and the preamp gain must be low enough that a reasonably strong ground signal does not create an overload condition. But after the demodulator the ground portion of the signal is removed.

Figure 15-2 shows the demod timing[8] that produces a null signal on ferrite. Although the clock is delayed 90° from the TX voltage the null occurs for a target (ferrite) which has a 0° response shift, therefore we call this a 0° *ground balance point*. This exactly follows the convention used in the TR-Discrimination chapter whereby the discrimination point is the phase of the reject/accept null, which is always 90° from the actual clock phase delay. It is worthwhile to compare Figure 15-2 with the TR-Disc timings in Figure 14-15. All targets now have a positive demod output signal regardless of whether they are ferrous or non-ferrous, with foil having the strongest result since it has a roughly 90° phase shift. This was an early "feature" of ground-balancing detectors; in GB mode there was no ability to discriminate, it was truly an "all-metal" mode. Many detectors had a mode switch to select between VLF-GB

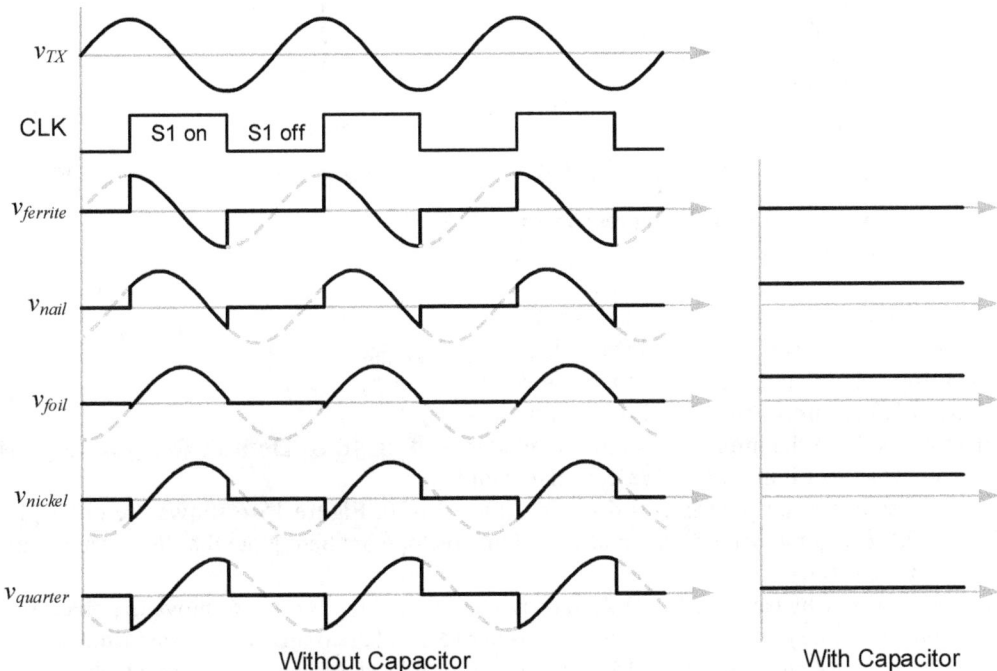

Without Capacitor   With Capacitor

Fig. 15-2: **Ground Balance Signals**

---

7. Also recall that this was referred to as "180° sampling" and resulted in a "90° discrimination threshold."

8. We will continue to use 10kHz for the transmit frequency. Keep in mind that a different TX frequency will result in different target phases. We will also continue using voltage waveforms at the RX coil.

and TR-Disc; you could do one or the other, but not both at the same time. It was common to hunt in GB mode and then switch to Disc mode to check a target.

In Figure 15-2 the quarter is somewhat weak because it is approaching a relative phase of 180°, and an Atocha bar would be almost invisible. As we did for the TR-Disc, we can create a response plot for the VLF-GB that includes the attenuation effect of the demodulator. Figure 15-3 shows how the quarter response is heavily attenuated compared to a nickel. Notice that the demodulator attenuation envelope now has two nulls, one at 0° and one at 180° (compare with

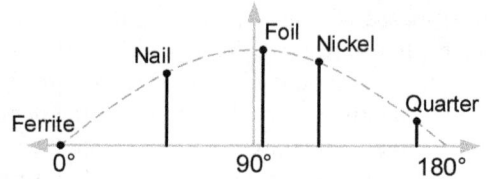

Fig. 15-3: **Demod Response (10kHz)**

Figure 14-17a). The 0° null completely suppresses the ferrite (ground) response.

So far we have continued to use 10kHz as our preferred frequency. With the TR-Discriminator we noted that a higher frequency was somewhat effective in helping certain targets overcome the attenuation effect of the demodulator. With the demodulator now set for ground balancing a lower frequency will do likewise. Figure 15-4 shows the normalized (same amplitude) responses for foil, nickel, and quarter at 10kHz and 1kHz. At 1kHz the response phases get pushed closer to the 90° axis which, for a demodulator set for ground balance, is closer to the peak of the response envelope.

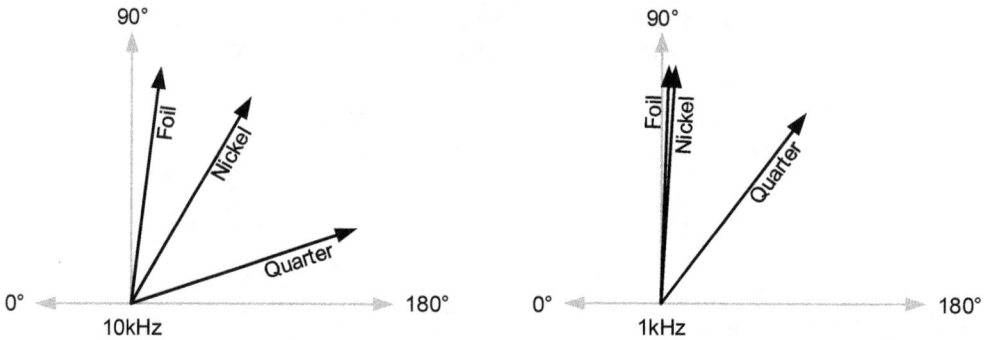

Fig. 15-4: **10kHz vs 1kHz Responses**

Figure 15-5 shows the 1kHz target responses with the demodulator attenuation envelope applied. As we've noted before, the phase of ferrite (ground) changes very little with frequency so the ground signal will still be nulled. Ferrous targets will vary depending on their alloy, shape, and size. What we are really interested in is the improvement in high conductors like silver coins. Note the improvement in the quarter from Figure 15-3 to Figure 15-5. We can also plot

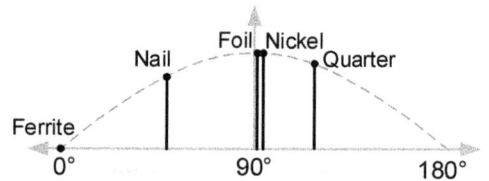

Fig. 15-5: **Demod Response (1kHz)**

these responses in a polar plot as was done in Figure 14-20. Figure 15-6 shows the polar plots for 10kHz and 1kHz with the demodulation attenuation envelope applied. Note that the circular envelope has a null at 0° and 180°.

A major reason why early VLF-GB designs ran at only a few kHz was to move the phase response of high conductors away from the 180° null. Another reason has to do with the raw (undemodulated) target response. The signals in Figure 15-4 are all drawn with the same amplitude which is somewhat misleading. At lower VLF frequencies skin effect makes thin targets like foil have a weak response while thick high conductors like silver coins have a (relatively) much stronger response. Using a low VLF frequency means low conductors which are naturally weak suffer little additional demodulator attenuation, and high conductors that do suffer demodulator attenuation are naturally stronger to begin with. Chapter 16 will cover this issue in more detail.

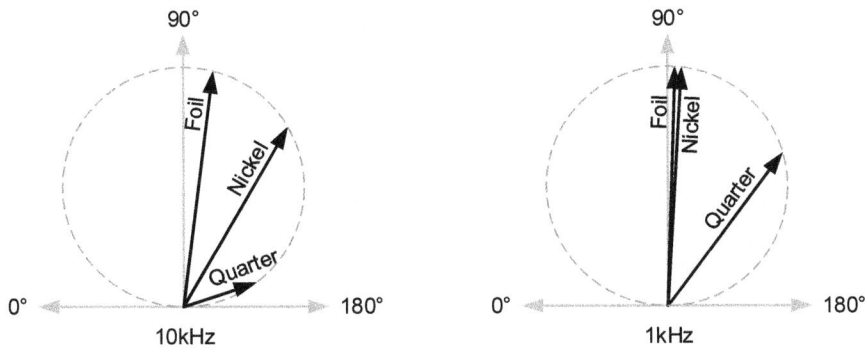

Fig. 15-6: **10kHz and 1kHz Polar Plots**

15.4: Block Diagram

The preceding section is a continuation of the discussion in the TR-Disc chapter on demodulator response versus demod clock phase. It is now obvious that VLF-GB is achieved in exactly the same way as TR-Disc, just using a different clock phase. For clarity the block diagram of the VLF-GB system is shown in Figure 15-7 and looks remarkably identical to the TR-Disc diagram in Figure 14-2. The only difference is the magnitude and range of phase shift applied to the demodulator.

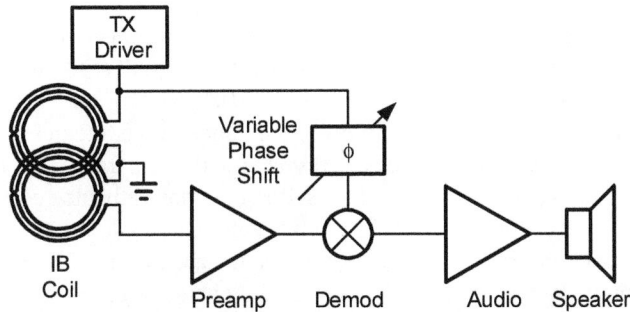

Fig. 15-7: **VLF-GB Block Diagram**

15.5: Demodulator Clock

The loss angle of ground mineralization has some amount of variability so simply fixing the demod clock at 90° will not completely eliminate most ground. Instead, the demod phase is made adjustable so the user can vary it according to ground conditions. The *VLF-GB* design in this chapter will largely recycle the circuitry of the *TR-Disc* design but with a slightly different demodulator clock phase shift. Setting the phase for target discrimination is a pretty inexact exercise, especially in an all-analog detector. With ground balance, a small error in the phase setting[9] can result in an audible ground signal so we'll pay a little closer attention to the details and add a method for trimming the low side.

This design will target a loss angle range of 0° to 15° for the ground balance. As with the *TR-Disc* design it is necessary to compensate for the phase shift of the preamp which is 30°. Therefore the GB null range is 30° to 45° which requires a clock delay of 120° to 135°. This would normally require a fixed delay of 120° and a variable delay of 0-15° but we can do the same trick as we did in the *TR-Disc* design and subtract 180° to reduce the magnitude of the fixed delay. Doing so results in a required delay of −60° to −45° (a negative phase delay is actually a phase advance) with the 180° rotation implemented in a reversal of the comparator polarity.

Figure 15-8 shows how this is done. A phase advance is achieved with a series capacitor and shunt resistor. The shunt resistor is shown as a variable element to vary the phase shift but in reality it would

---

9. Many of the old analog VLF detectors used a 10-turn potentiometer for the ground control.

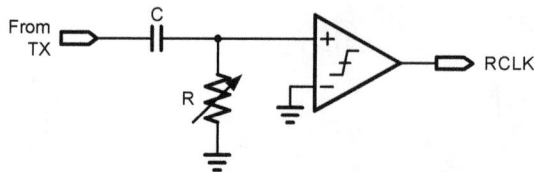

Fig. 15-8: **Basic Phase Shift Design**

be implemented as a fixed resistor for the fixed shift portion and a potentiometer for the variable shift portion. The phase shift imposed by the RC network is

$$\theta = \arctan\left(\frac{1}{\omega \tau}\right)$$

$$= \arctan\left(\frac{1}{2\pi f \cdot RC}\right)$$

Eq 15-1

or

$$RC = \frac{1}{2\pi f \cdot \tan(\theta)}$$

Eq 15-2

From this we can plug in a minimum and maximum θ and calculate the range R needs to be for a given capacitor value.

Let's suppose we implement this scheme and calculate the component values that exactly compensate a 30° preamp delay. Now suppose the preamp phase is actually 28° measured instead of the expected 30° due to the preamp feedback capacitor being 10% low. If our ground has a loss angle of less than 2° then it will fall below the minimum ground balance range, meaning we will not be able to null it. In other words, it appears to be a negative ground signal compared to the range we've designed for. The same situation can happen if the capacitor used in the phase advance circuit comes in high and produces more phase shift than expected. That is, too little phase delay in the preamp or too much phase advance in the demod clock can produce a situation where the ground will not balance.

In a one-off home-brew detector this can be corrected by hand-selecting capacitors. In a production environment this is a tedious solution so instead the phase shift is intentionally designed for capacitor (and resistor[10]) tolerance spread. This would normally create a lot of variability in the ground balance adjustment so a method is included to trim the low end of the range to a consistent starting point, usually with ferrite as the reference. The resistor in Figure 15-8 can be implemented as a fixed resistor, plus a potentiometer for ground balance control, plus a trimpot to adjust the starting point.

To exemplify, let's flesh out our design and include capacitor tolerances of ±20%. The nominal preamp delay of 30° can now have a range of 24° to 36°. We assume the minimum of 24° in calculating the fixed phase shift but include the possibility of an additional 12° in the trimpot adjustment. Additionally, the cap used in the phase shift can be ±20% so we start with +20% and, again, use the trimpot to compensate. The required fixed phase advance for a minimum GB setting is now 66° and the RC needed for this is

$$RC = \frac{1}{2\pi f \cdot \tan(\theta)}$$

$$RC = \frac{1}{2\pi \cdot 10000 \cdot \tan(66)}$$

$$= 7.09 \mu s$$

---

10. Modern surface-mount resistors are normally ±1% and are usually ignored. 1% or 2% capacitors are available but expensive, and low-cost caps are typically ±20%, though ±10% are often very reasonable.

Inside the Metal Detector

This is the combined RC needed regardless of the tolerance of C. Let's arbitrarily choose R = 1kΩ and C = 7.09nF and say that this value of C represents a situation where it is 20% above nominal. This means the nominal value is 5.91nF and a 20% low value which is 4.72nF. Therefore, to maintain a phase advance of 66° we have the following:

C	R
4.72nF	1.5kΩ
5.91nF	1.2kΩ
7.09nF	1.0kΩ

From this we can see that a delay resistor that is a 1kΩ fixed resistor in series with a 500Ω trimpot will work for a capacitor that is 5.91nF ±20%. This only accounts for the variation in the delay circuit cap, not the preamp cap. We assumed the minimum preamp delay of 24° so now let's consider its maximum delay of 36°. The resultant phase advanced needs to be 54° so RC is now

$$RC = \frac{1}{2\pi \cdot 10000 \cdot \tan(54)} = 11.56\mu s$$

The capacitor value does not change so the required resistor values are now

C	R
4.72nF	2.45kΩ
5.91nF	1.96kΩ
7.09nF	1.63kΩ

The only one that matters is the low cap value for which the total resistor value needs to be 2.45kΩ. At the other end of the spread (low preamp delay, high delay cap) the resistor needs to be 1kΩ. This requires a 1kΩ fixed resistor and a 1.45kΩ trimpot.

That does it for the circuit variations but we have not yet addressed the necessary phase variation to balance ground. Again, we want to be able to adjust for a loss angle of 0-15° and the above calculations assumed a loss angle of 0° in order to find the starting point for the phase advance circuit. Let's now add a ground adjustment of 15° and see what happens. The worst-case condition occurs at the other end of the scale: maximum preamp delay (36°) and minimum phase shift cap value (4.72nF). The total phase advance we want is now 39°:

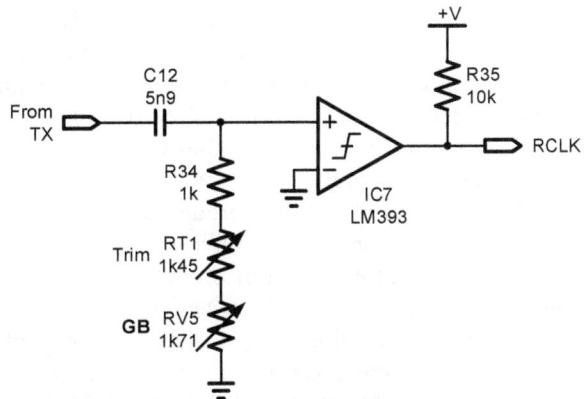

Fig. 15-9: **Complete Phase Shift Design**

$$RC = \frac{1}{2\pi \cdot 10000 \cdot \tan(39)} = 19.65\mu s$$

The resultant total resistor value is 4.16kΩ which is 1.71kΩ higher than the case where the ground adjust is at 0°. Our circuit now looks like that in Figure 15-9, with components labeled as they will be in the final design.

The values in Figure 15-9 are mostly non-standard in electronics. The cap can be made up of 2 or more parallel standard values and the pots could be increased to 2kΩ, but even that's not very common. An alternative is to re-scale the values to get them more in line with available components. If instead C12 = 1nF, then RT1 = 5.91kΩ, RV5 = 8.55kΩ, and R34 = 10.11kΩ. Except for the cap these are also not common but we can increase the pots to 10kΩ which are very common. It's perfectly fine for the pots to have a bit more adjustment range than needed. Since a 10kΩ trimpot is 70% more than needed we could actually reduce the

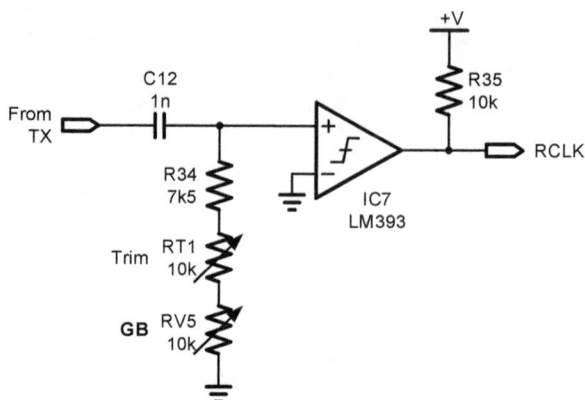

Fig. 15-10: **Optimized Phase Shift Design**

fixed resistor to 6kΩ. However, to allow for 20% trimpot value variation we'll use 7.5kΩ. The final phase shift circuit is shown in Figure 15-10.

Calibration is done by bobbing a ferrite target at the search coil and adjusting the trimpot until the R channel has a minimal (hopefully zero) response. Before doing this it is necessary to first set the GB pot to an appropriate ferrite position. That would seemingly be at its minimum value (0Ω) but instead it should be set slightly above minimum, perhaps 5-10% higher. Once calibrated. this will allow you to adjust the GB a little below the ferrite loss angle. This is important because the loss angles of ferrites vary somewhat and the ferrite you choose to calibrate with may not represent a near-zero loss angle.

If this whole process seems confusing, well, it is. This is typical when designing for component variation and often there can be multiple requirements[11] that must be dealt with simultaneously. It is important to understand these issues, especially if you are designing circuitry for mass production where it must work for all possible process corners.

15.6: Hybrid VLF/TR-Disc

The VLF detector was introduced in the early-mid 1970s at about the same time as the TR-Discriminator. When comparing the basic VLF design to the TR-Disc design it is clear there is little difference in the circuitry. Both use a single demodulator with the primary difference being the phase range for the demodulator clock. In the case of the VLF, the demod clock is phase-adjusted to eliminate ground. In the TR-Disc, it is phase-adjusted to discriminate varying levels of metal conductivities.

With a single demodulator it is not possible to accomplish both at the same time but it is possible to switch the demodulator between ground elimination mode and discrimination mode. It wasn't long before companies began offering a hybrid VLF/TR-Disc detector and these quickly became the dominant type until the VLF motion discriminator was invented. Examples were the Bounty Hunter *840*, Garrett *Groundhog*, and Compass *Magnum 320*.

A hybrid VLF/TR-Disc design can be implemented by designing the demod clock phase shift circuit to cover an extended phase range that will allow it to either ground balance or discriminate. This would require a variable phase shift in excess of 140° which is difficult. Another method is to include both individual phase shift circuits for the demod clock and then add a method to select which one drives the demod. See Figure 15-11. The advantage to this approach is that the unit can be ground balanced in VLF mode and the discrimination point set in TR-Disc mode; the user can then normally hunt mineral-free in VLF mode, quickly switch to TR-Disc mode to check a target, and then switch back again to VLF mode all without needing to adjust any phase control. This is, in fact, how practically all VLF/TR-Disc models were designed and many had a thumb push button or toggle switch for quickly changing the mode.

---

11. Another issue not covered here is temperature variation.

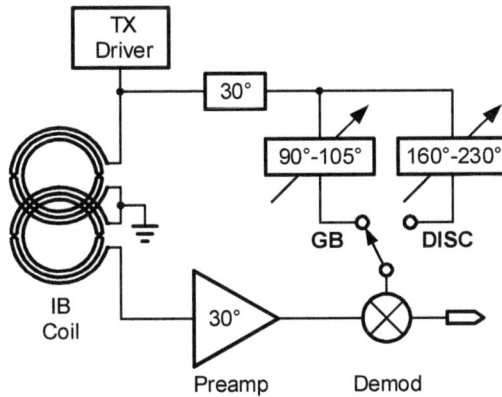

Fig. 15-11: **VLF/TR-Disc Demod Clock Selection**

Design: *VLF/TR-Discriminator*

The *VLF/TR-Disc* design is an enhancement of the *TR-Disc* design and, as mentioned in the previous chapter, inherits the same part numbering for easy comparison. The biggest change is the addition of a second demod clock phase circuit for the VLF-GB option and the ability to switch between it and the original TR-discrimination demod clock. Buffer IC5a presents a low-impedance drive to both clock phase circuits so they do not interact with each other.

Additionally, this design now uses a full-wave demodulator instead of the half-wave demodulator used in the *TR-Disc* design. As seen in Chapter 14's *Explore* the full-wave demod suppresses low-frequency components which improves Earth field rejection and reduces both power-line noise and 1/f (flicker) noise of the preamp.

Notice that the Ground Balance pot is placed in such a way that one side is connected to ground[12]. For example, we could have swapped the locations of RV1 and R2. Since both pots are typically panel-mounted user controls it is good practice to keep minimal signal levels on their connections to minimize the chances of noise pick-up or signal contamination.

---

12. As is the sensitivity pot, even in the *TR-Disc* design, for the same reason.

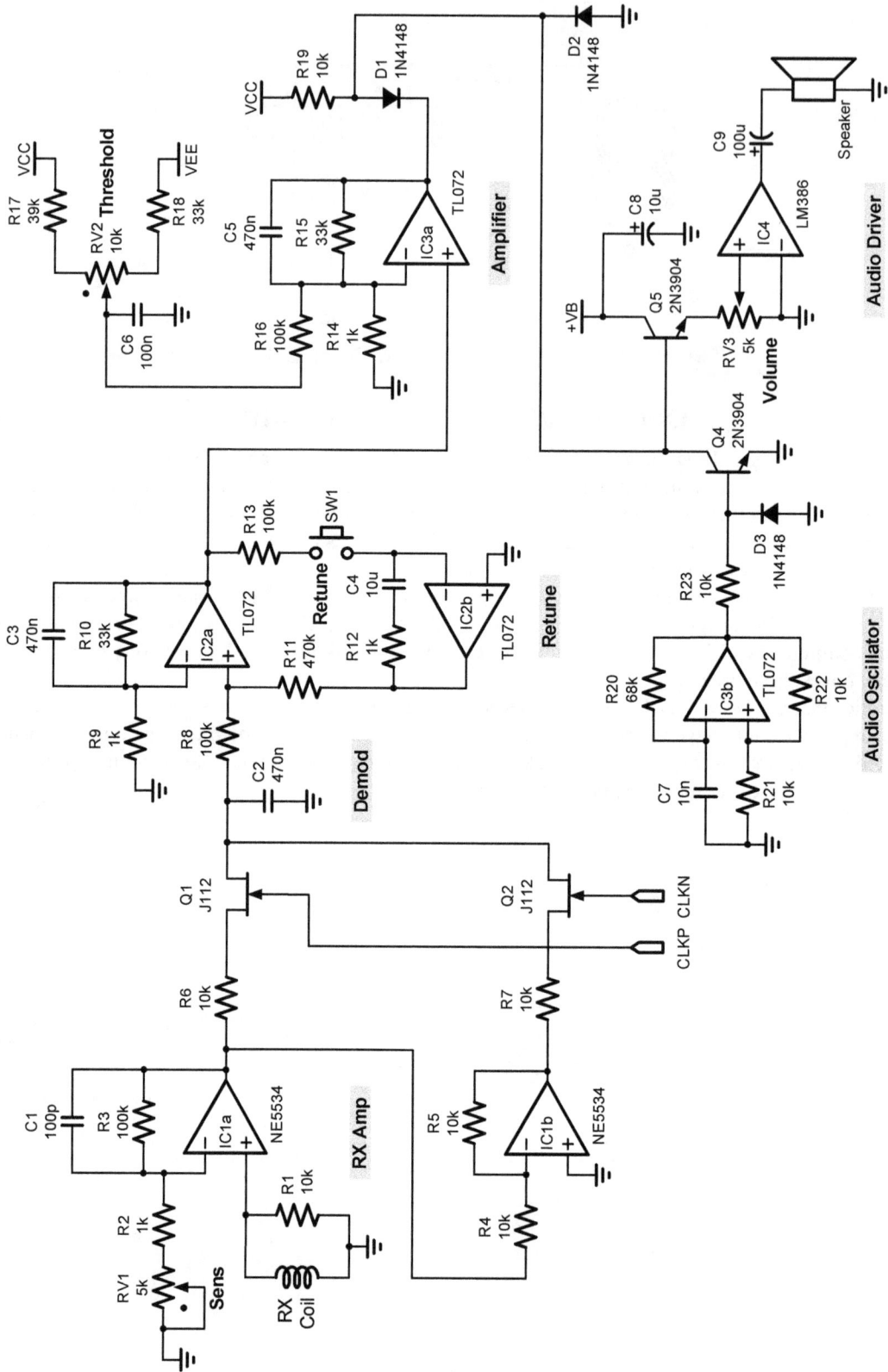

Fig. 15-12: VLF-TR — Main Circuitry

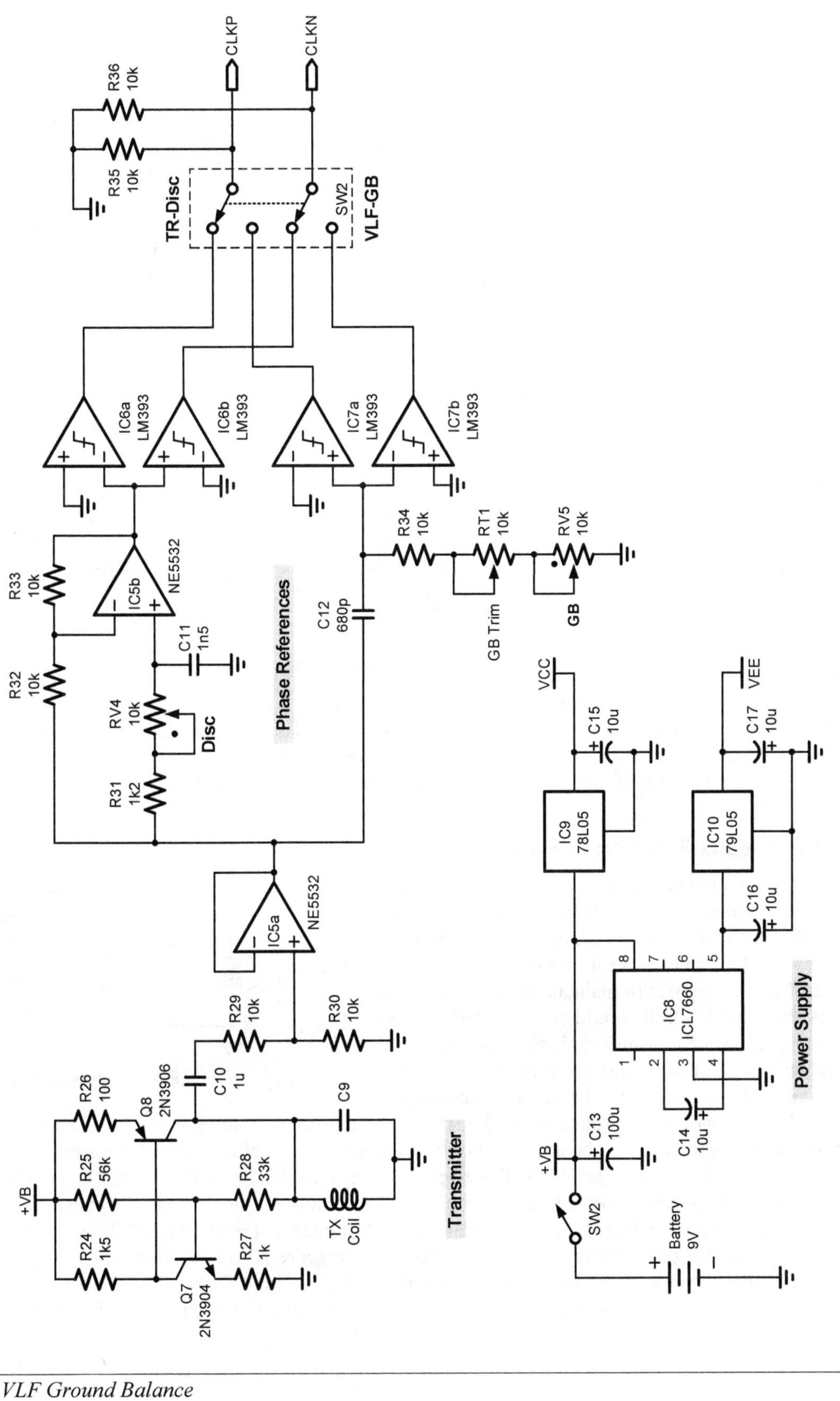

Fig. 15-13: **VLF-TR — TX/Phase & Power Supply Circuitry**

Explore

15.7: Parallel Detection System

The *TR-Disc* design clocks the demodulator with a variable phase that creates a null point somewhere between iron and moderately high conductors. The *VLF-GB* design clocks the demodulator with a variable phase that creates a null point for ground. The example design in this chapter combines the two techniques in a switchable VLF-GB/TR-Disc design, but they cannot be run at the same time. As a precursor to the next chapter we will briefly consider a simultaneous system.

Running both methods simultaneously requires two complete analog demodulator channels, one for GB and one for Disc. The GB channel will respond positively for any metal but not for ground; it can be used as an all-metal trigger signal. The Disc channel can then determine whether the target is "bad" or "good" and either suppress the audio or not.

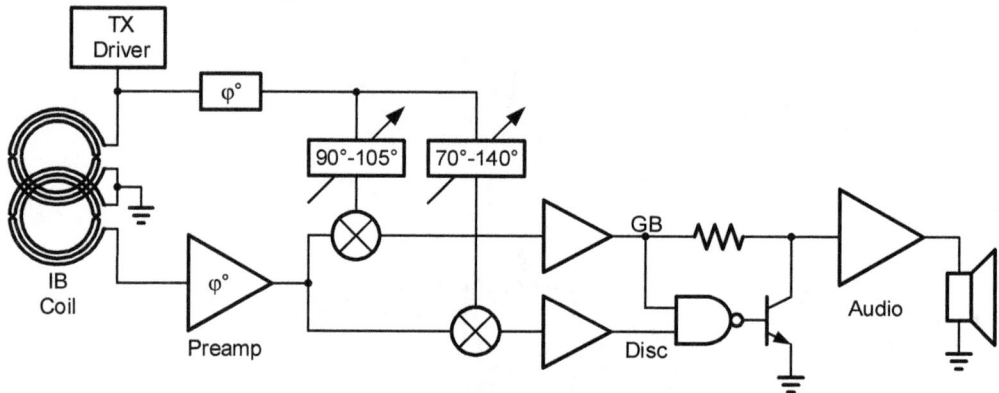

Fig. 15-14: **Simultaneous GB/Disc System**

A major problem with this approach is that the Disc channel will still be affected by ground as it was for the TR-Disc, whereby even variations in the height of the coil above the ground can cause noise or depth reduction. We'll see in the next two chapters how this is dealt with in a modern VLF-Disc design, but for the purposes of this chapter we'll leave it as a potential better-than-no-discrimination-at-all solution.

15.8: Band-pass Filtering — Part 2

In some commercial designs the RX coil is connected to a non-inverting opamp with only a load resistor, and sometimes without the load resistor. Other designs place a parallel capacitor across the RX coil to make it into a band-pass filter for the purpose of reducing EMI. To maintain the highest possible target sensitivity the RX coil should be resonated to exactly the same frequency as the transmitter. However, doing so results in a steep phase slope at the frequency of interest.

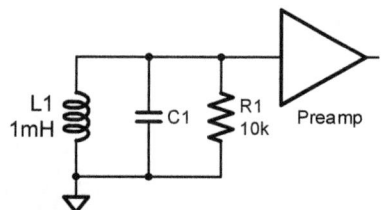

Fig. 15-15: **Band-pass RX Coil**

Figure 15-15 shows a 1mH RX coil resonated with capacitor C1 and additionally loaded (damped) by R1. If the TX frequency is our favorite 10kHz then the cap value for 10kHz resonance must be 253.3nF. For this, the resulting amplitude and phase response at the preamp input is shown in Figure 15-16. The amplitude response is massively boosted at 10kHz and greatly suppressed elsewhere, something we would surely like to see. But the phase response is almost vertical at 10kHz, meaning that any phase noise on the transmitter will produce a wildly varying phase on the RX coil. Additionally, the phase modulation due to a target response will also vary wildly.

In the example above the resonant system formed by L1, C1, and R1 has a Q of 159 at 10kHz. What has been ignored is the series resistance of the RX coil which is usually significant and will

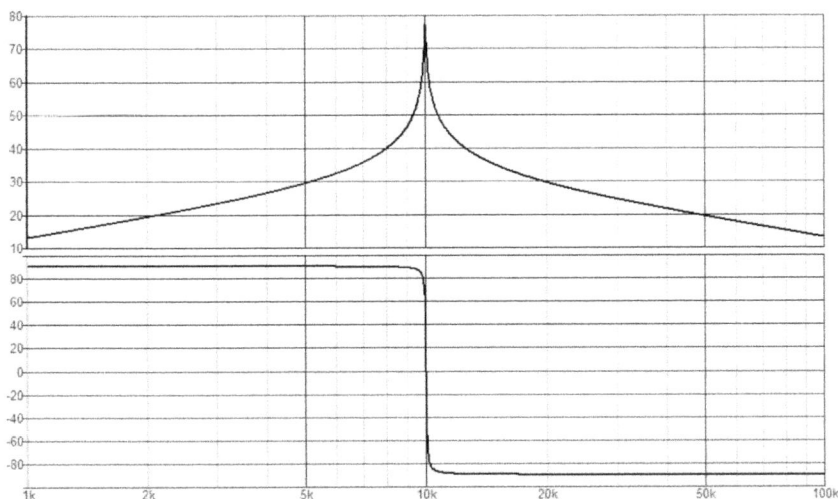

Fig. 15-16: **Ideal Resonated RX Coil**

reduce the Q considerably. Let's say the RX coil has a diameter of 4" (10cm) and is wound with 30AWG wire resulting in a series resistance of 7.5Ω. If R1 is removed then the Q of the resonant system is now 8.4 and can never be higher; adding R1 back in will only reduce it further, though not by much (8.0).

The new amplitude and phase responses are now depicted in Figure 15-17 (black curves). While the RX coil's series resistance de-Qs the circuit and somewhat tames the steepness of the phase transition, it is not enough. What most designers do is offset the RX resonant frequency away from the transmitter frequency where the phase transition is less steep. Suppose instead that the RX coil is tuned to 20kHz by setting C1 to 63.3nF, resulting in the gray curves in Figure 15-17. The phase slope is now significantly less at 10kHz and may be low enough so that any TX phase noise (as well as target phase modulation) will cause only a slight error in phase measurements. At the same time, there is still a fairly significant suppression of noise, over 45dB if we extrapolate from the operating frequency to where 50/60Hz resides. A drawback is that any noise at 20kHz is boosted but this can be mitigated with a low-pass filter in the preamp.

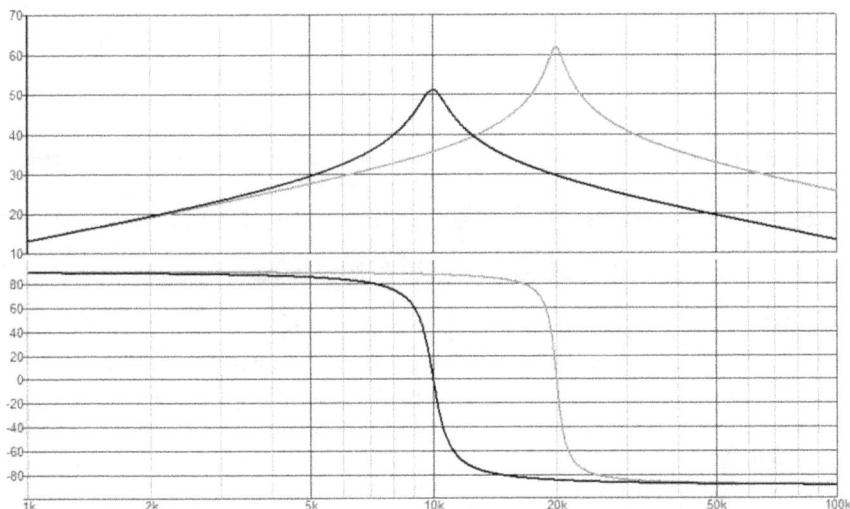

Fig. 15-17: **Typical Resonated RX Coil**

15.9: Demodulators — Part 2

All of the demodulators presented so far have been of the sample-and-hold variety, either half-wave for explanatory discussions or full-wave for actual designs. Both were analyzed in the *Explore* section of Chapter 14. The sample-and-hold demods have so far been shown with a single-pole RC circuit. In commercial metal detector circuits it is common to see the demod RC split into a double RC:

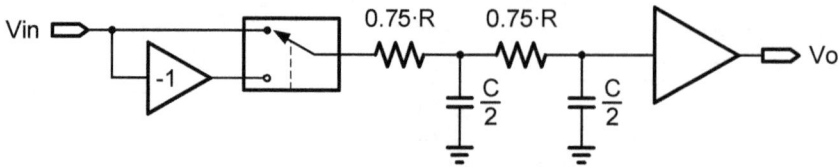

Fig. 15-18: **2-pole SHA Demodulator**

This forms a 2-pole LPF which ideally has twice the roll-off slope of the single-pole LPF. An unbuffered RC-RC achieves the same −3dB frequency (overall tau) as a single pole RC when $C_{2p} = 0.5 \cdot C_{1p}$ and $R_{2p} = 0.75 \cdot R_{1p}$. Figure 15-19 shows the Bode plots for three filters — a 1-pole (dark gray), 2-pole (medium gray), and 3-pole (light gray) — all with a −3dB cut-off of 100Hz. The values used are:

	R	C
1-pole	5300Ω	300nF
2-pole	3975Ω	150nF
3-pole	3092Ω	100nF

It should be noted that the base values (5300Ω, 300nF) were selected for no particular reason and other values could be chosen to achieve the same results.

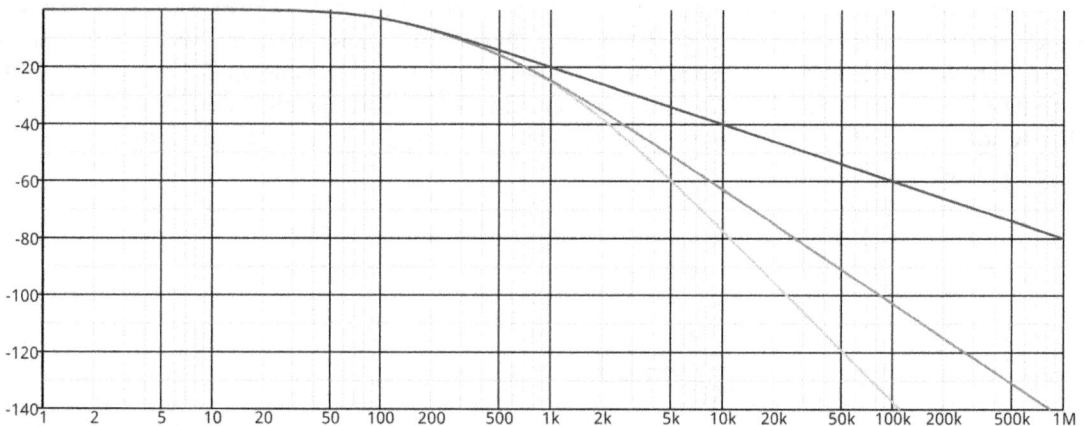

Fig. 15-19: **1-pole, 2-pole, & 3-pole Responses**

Splitting the RC into multiple poles does not affect the conversion gain; it is still exactly $2/\pi$. And since this is being applied to a demodulator the Bode plot alone is insufficient to say more poles are better; it needs to be looked at using a time-domain frequency response as was done in Chapter 14. Figure 15-20 compares the 1-pole (light gray) and 2-pole (dark gray) responses and shows the sampled noise performance is also improved. For the same equivalent tau, the low-frequency attenuation is over 10x better with closer-in frequencies attenuated to a lesser amount. Ripple is also significantly reduced. It otherwise behaves almost identically to a 1-pole SHA in every other respect: conversion gain, step response, and phase response. Internal noise generation will be slightly better because the NBW is a little tighter ($1.22 \times BW$ vs $1.57 \times BW$).

Inside the Metal Detector

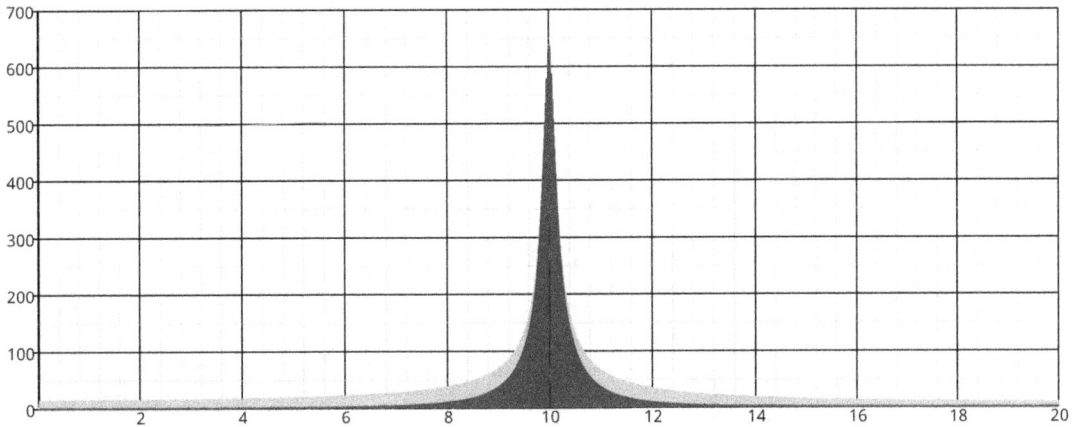

Fig. 15-20: **1-pole vs 2-pole SHA Responses**

If a 2-pole filter is so much better, then what about 3-poles? Or 4? Yes, more poles continue to reduce conversion noise (and internal noise) and have no effect on conversion gain. Simulations show that the 3-pole SHA begins having some oddities in the step response that might be undesirable. But when using a SHA demod there is no reason not to use at least a 2-pole RC.

The 2-pole response has both poles aligned using the above calculations for R & C. It is possible to achieve the same results using non-identical capacitor values but there is an inherent advantage in using the same cap value for both poles: it results in a slight statistical improvement in overall channel-to-channel cap matching[13]. That is, if you buy capacitors with a 6-sigma tolerance of ±5%, placing 2 caps in parallel yields a 6-sigma tolerance of $5\%/\sqrt{2} = \pm 3.5\%$. In general, N parallel components improves matching by $\sqrt{N}$, a trick commonly used in IC design. This can be applied across any number of poles; for example, Figure 15-21 shows a 2-pole SHA implemented with 6 capacitors which (assuming 5% inherent matching) yields an overall 2% accuracy. Hand-matching a few capacitors is not a big deal for home-brew designs so this trick is mostly advantageous in a production environment.

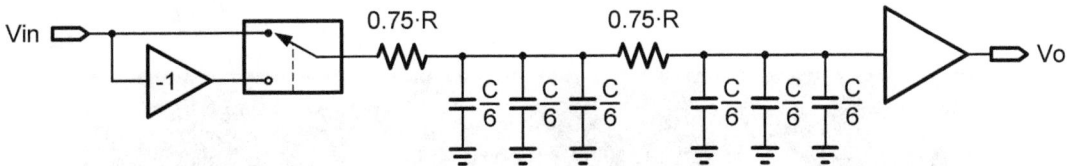

Fig. 15-21: **Improved Cap Matching**

15.10: Mode + Retune

When switching between VLF-GB and TR-Disc it will often be necessary to also press the retune button to restore a proper threshold. In the Garrett ADS models of the early 1980s these actions were combined into a SPDT momentary switch at the end of the handle for on-the-fly mode changes and simultaneous retuning. Conceptually this can be done as in Figure 15-22.

SW4 is a momentary-off-momentary style switch that can be conveniently located for easy access. A J-K flip-flop[14] changes state depending on the selection and controls two sections of a CD4053 triple SPDT analog switch. These directly replace SW2 in the VGB-1 schematic. At the same time, the NOR gate energizes the remaining CD4053 switch (regardless of selection) which replaces SW1 and invokes the retune function.

---

13. Channel-to-channel cap matching becomes important in a multichannel design, such as a motion VLF detector with X & R channels or a PI design with target & ground channels.

14. You can also use a flip-flop's Set/Reset function to avoid using the clock pin.

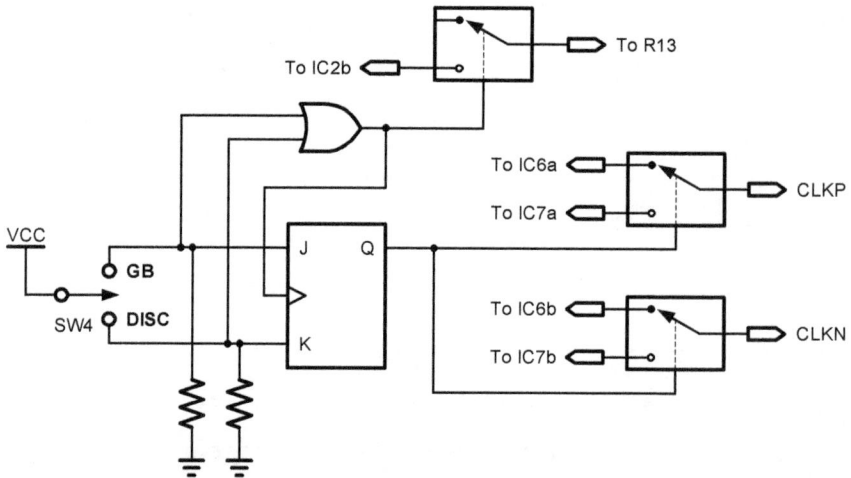

Fig. 15-22: **ADS-Style Mode/Retune Select**

A quick press of SW4 to either mode selects that mode and does a retune. Pressing and holding SW4 (to either mode) allows you to adjust the tuning threshold.

16

Motion VLF

"Each problem that I solved became a rule which
served afterwards to solve other problems."

— Rene Descartes

A single demodulator is used in a TR discriminator to sample the receive signal with a clock that is nominally in-phase with the transmitter. This produces a positive signal for non-ferrous targets, a negative signal for ferrous targets, and mineralized ground is a negative signal that, if strong enough, can overwhelm target responses (see Chapter 14). The same single demodulator is used in a VLF ground balancing design but samples the receive signal with a clock that is 90° phase-shifted from the transmitter. This produces a null signal for ground mineralization and positive signals for all metal targets, both ferrous and non-ferrous (see Chapter 15).

For a brief time metal detectors were sold with a mode switch that could select between all-metal VLF and TR-Disc[1] but could not do both at the same time. In 1977 George Payne changed that with the Bounty Hunter *Red Baron*[2], a detector that could both eliminate ground and discriminate at the same time. This break-through in metal detector design was called *VLF motion discrimination* because the technique required the coil to be in motion in order to detect targets. This chapter will explore how this is achieved.

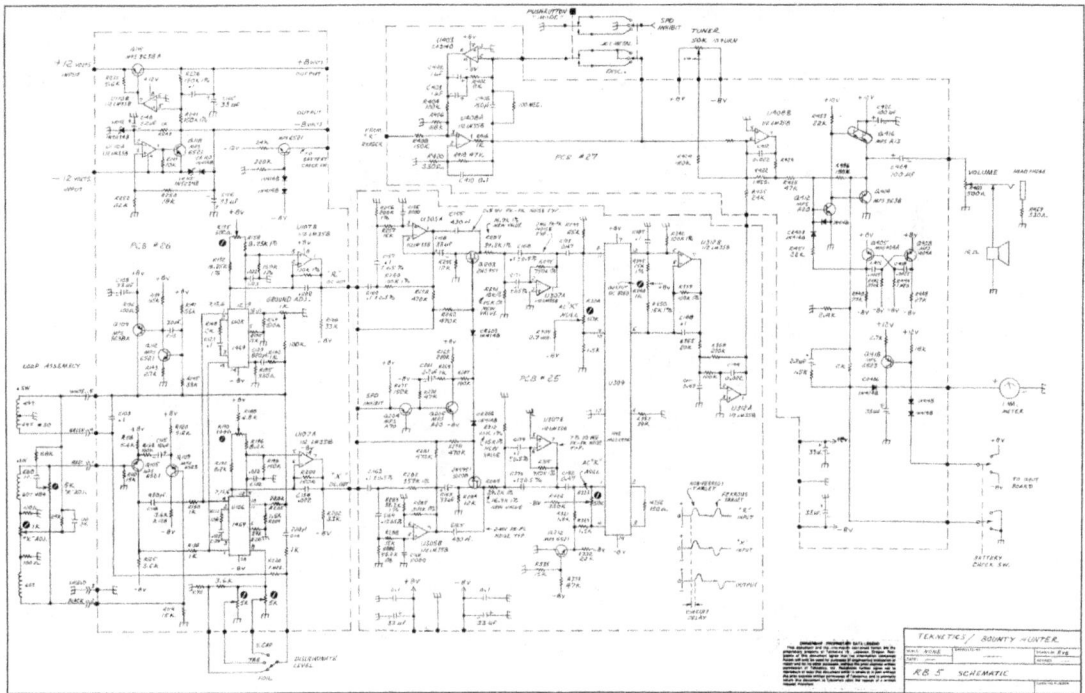

Fig. 16-1: **Bounty Hunter RB5**

1. See the *VLF-TR Disc* design in Chapter 15.

2. The *Red Baron* — or *RB7* — had two siblings, the *RB5* and *RB3*. Figure 16-1 shows the *RB5*.

16.1: Quadrature Demodulation

Since a single demodulator can produce two desired effects when using two different clock phases, an obvious[3] design step is to include both of the demodulators and run them in parallel using both clock phases. One demod must be ground balanced. The other demod would normally have an adjustable phase to implement variable discrimination but, as we will soon see, it is not necessary to make this demod phase adjustable; we can simply set it 90° from the ground balanced demod phase. This will produce two signal responses: a full-time ground-balanced signal that has a positive response to all metal targets, and another signal which is negative on ferrous and positive on non-ferrous targets. Since one demodulator is clocked 90° from the other demodulator, this process is called *quadrature demodulation*[4]. Figure 16-2 shows a block diagram of a typical quadrature demodulation scheme.

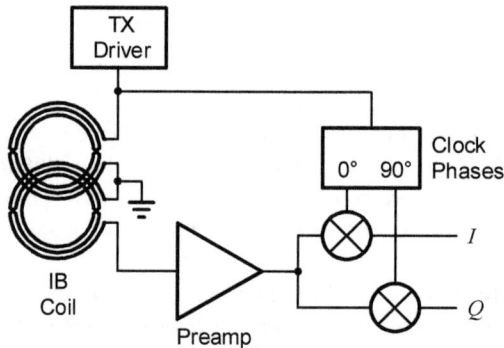

Fig. 16-2: **Quadrature Demodulation**

Recalling Chapter 4, a system which transmits a sinusoidal magnetic field will produce a target response which is also sinusoidal but with an altered amplitude and phase. When clocked with a synchronous reference (the transmit signal) the outputs of the quadrature demodulator are quasi-DC signals which can be used to determine the amplitude and relative phase of the input signal. The signal which results from a 0° clock phase is called the *in-phase* component (signal I in Figure 16-2) and the signal which results from a 90° clock phase is called the *quadrature-phase* component (signal Q in Figure 16-2). These are often abbreviated as the I and Q signals.

To see how this works let's consider the demodulators to be simple half-wave samplers as we did in Chapters 14 and 15. There are now two demodulators, one for I and one for Q:

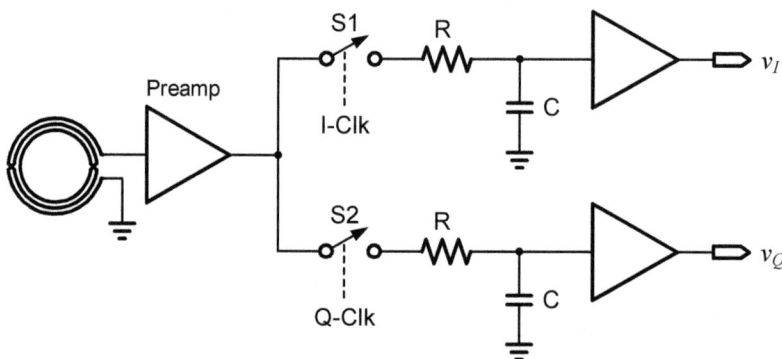

Fig. 16-3: **Half-Wave I/Q Demodulators**

---

3. Everything is obvious in the rear-view mirror.

4. A technique commonly used in radio systems, including cell phones. Also note that we are still using synchronous demodulators, so we can also call this *synchronous quadrature demodulation*.

We can do the same kind of waveform analysis as before, first imagining the caps are removed to see the demodulation waveforms, and then with the caps added in to see the resultant averaged DC levels. Consider that the transmit waveform is a sinusoid, and assume it to be defined as the 0° phase[5]. Also consider the "target" to be ground mineralization, which is assumed to produce a 0° phase shift in the receive signal. The resulting I & Q signals are:

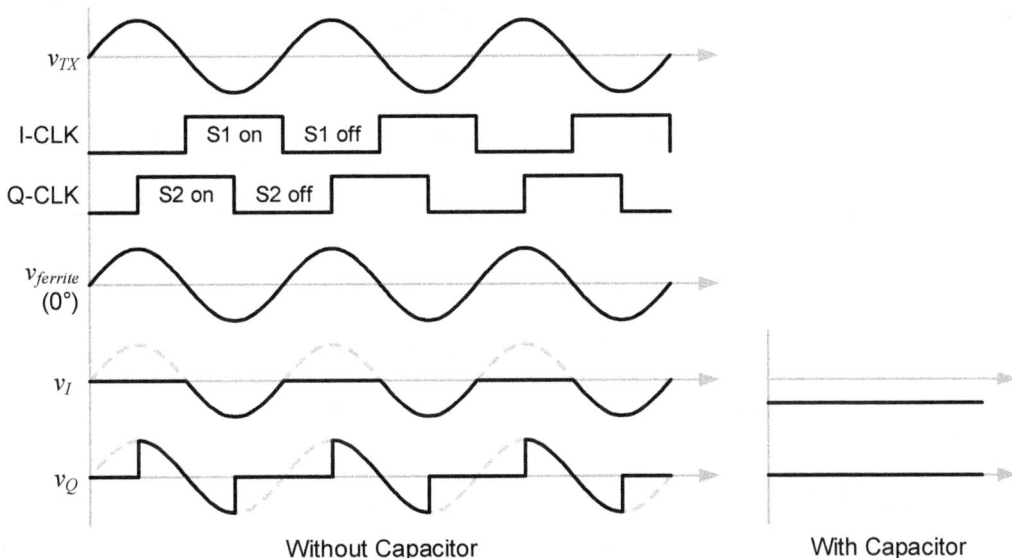

Fig. 16-4: **I & Q Signals for Ferrite**

Again, this is a repeat of what was presented in Chapters 14 and 15. The I-CLK and v_I signals were seen in the TR-Discriminator, and the Q-CLK and v_Q signals were seen in the VLF-GB discussion. With the quadrature demodulator we now get both signals, and we see that mineralization has a strongly negative I signal and a zero-average Q signal. Note that these signals continue to be presented as half-wave signals even though, in practice, we prefer to use full-wave demodulators. It is just easier to visualize the concept using half-wave.

Let's look at some more targets. The next four sets of plots show the I/Q responses for a nail, foil, US nickel, and US quarter. The TX waveform (in gray) is in included along with the target signal.

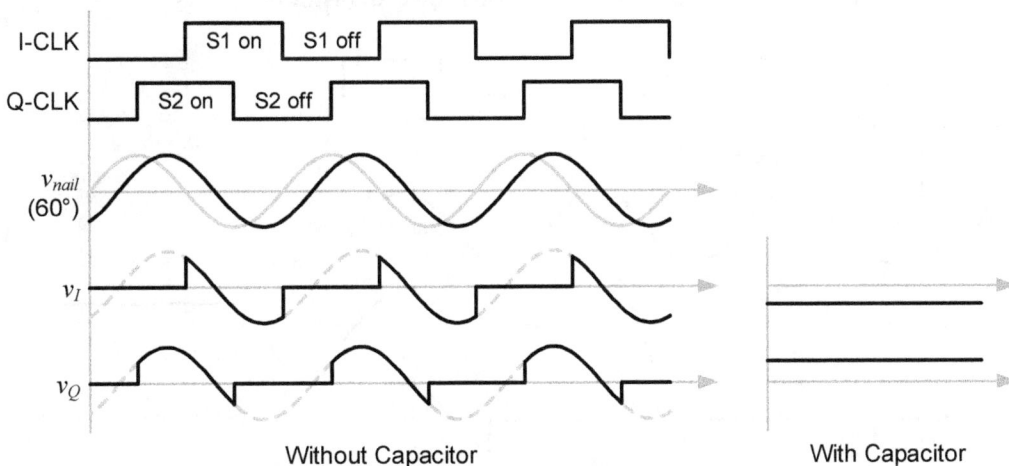

Fig. 16-5: **I & Q Signals for a Nail**

---

5. Note that we are using the TX voltage (not current) and RX voltages as we did in Chapters 14 & 15.

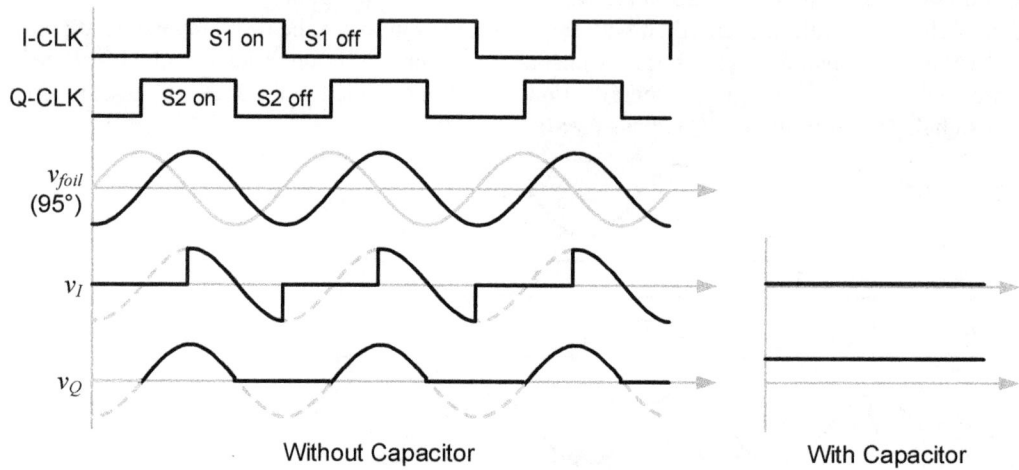

Fig. 16-6: **I & Q Signals for Foil**

I-CLK

S1 on | S1 off

Q-CLK

S2 on | S2 off

v_{foil}
(95°)

v_I

v_Q

Without Capacitor

With Capacitor

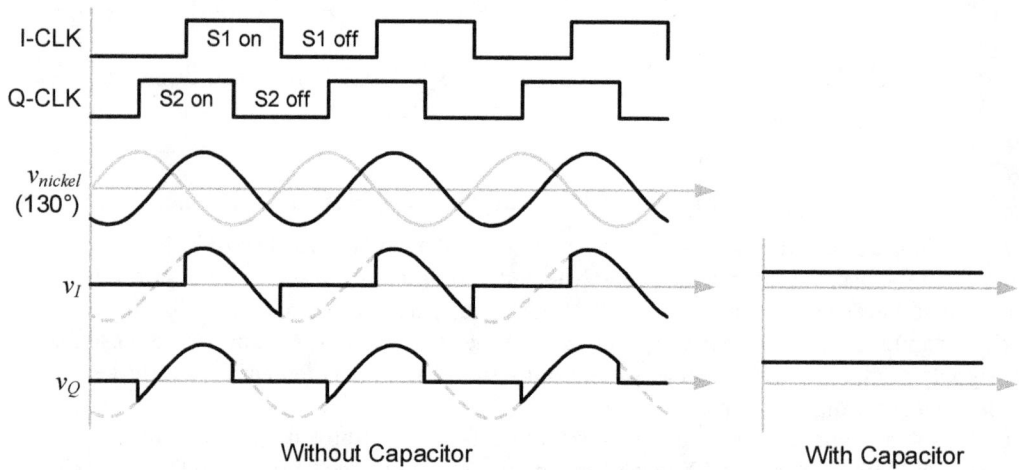

Fig. 16-7: **I & Q Signals for a US Nickel**

I-CLK

S1 on | S1 off

Q-CLK

S2 on | S2 off

v_{nickel}
(130°)

v_I

v_Q

Without Capacitor

With Capacitor

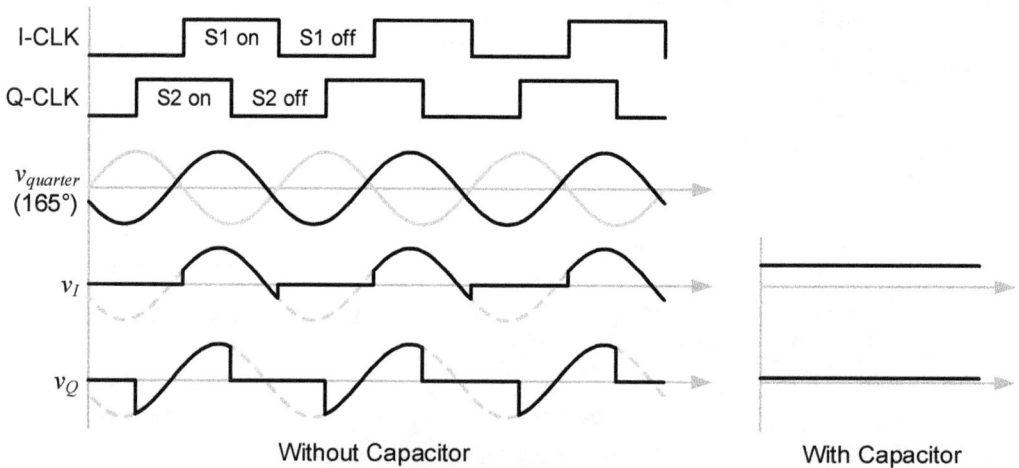

Fig. 16-8: **I & Q Signals for a US Quarter**

I-CLK

S1 on | S1 off

Q-CLK

S2 on | S2 off

$v_{quarter}$
(165°)

v_I

v_Q

Without Capacitor

With Capacitor

16.2: Target Vectors

In each of the above cases, when the capacitors are included the demod output pairs are DC voltages with some positive or negative value (or perhaps zero). These voltage output pairs can be plotted on a Cartesian graph to see how they relate to one another. In Figure 16-9a the I-Q demod voltages are plotted as points in space, with each point located by its DC values V_I and V_Q as shown in the previous few plots. In Figure 16-9b they are drawn as vectors, each with (0,0) as the vector origin.

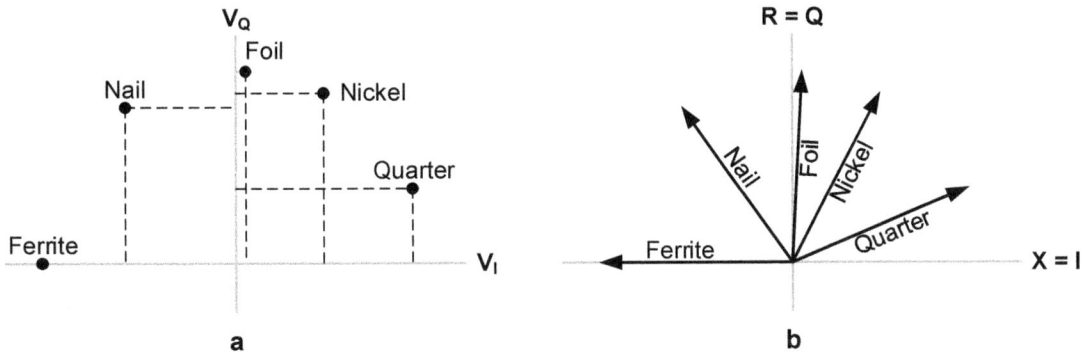

Fig. 16-9: **Plot of I & Q Signals**

If we use the negative I-axis as the 0° reference[6] then each vector represents the phase response of each item. This is called a *target vector plot*. We can see that ferrite and magnetic/ferrous targets will fall in the 0°-to-90° quadrant while eddy/non-ferrous targets will fall in the 90°-to-180° quadrant. Quadrature demodulation therefore produces voltages that represent the target phase responses that were described in Chapter 4. From this we could discriminate targets based on their phase response. This is why it is not necessary to vary a demod clock phase in order to discriminate as we had to do in the *TR-Disc* design. All the information we need for discrimination is contained in the quadrature voltages.

While the terms "in-phase" and "quadrature" are completely correct for this kind of demodulation and are commonly used in radio design, they are rarely used with metal detectors[7]. Instead, the in-phase signal is commonly called the *reactive* (X) signal and the quadrature signal is called the *resistive* (R) signal. By "reactive" we mean the portion of the signal that changes only as long as the driving signal is changing — think of it as the part of the signal that *reacts* (in real time) to the transmit signal. By "resistive" we mean the portion of the signal that can linger after the driving signal is completely removed. In non-ferrous targets that would be the eddy decay, and in ferrous targets it would be the collapse of the induced magnetism[8]. From here on, the X/R designators[9] will be used instead of I/Q.

From Figure 16-9 we can see that ferrite is purely reactive and small foil is almost purely resistive. Large silver coins tend toward the positive X-axis which means they are highly reactive but with a near-180° phase shift. A superconductor[10] will have no resistive component and will lie exactly on the X-axis, opposite to ferrite. In all cases the magnitude of the response vector represents the strength of the received target signal. With no target the demodulator voltages are at (0,0), assuming the coil has a perfect induction balance null. As a target gets into range the vector grows from (0,0) and gets longer as the target gets closer, but ideally remains at the same vector phase. Figure 16-10 illustrates a US nickel response at three hypothetical depths.

---

6. As mentioned before in the book, this is opposite the usual convention in trigonometry but conveniently places our target responses exactly as they appear on a typical VDI display.

7. Although Minelab did use the term "Multi-IQ" in their *Equinox* & *Vanquish* models.

8. The time it takes to go from a point on the B-H curve back to zero. In hot rocks, it is a magnetic viscosity decay.

9. It is fortunate that the reactive (X) axis corresponds to the traditional X-axis. It is also a deliberate choice. You may see cases (in patents, for example) where X-R phase plots are rotated 90° so that the reactive (X) axis is vertical; there are even cases where they are rotated and then mirrored. The signals are all doing the same thing, they are just drawn weirdly. Keep this in mind if you read and study other sources.

10. Or our practical "close enough" example, the Atocha bar.

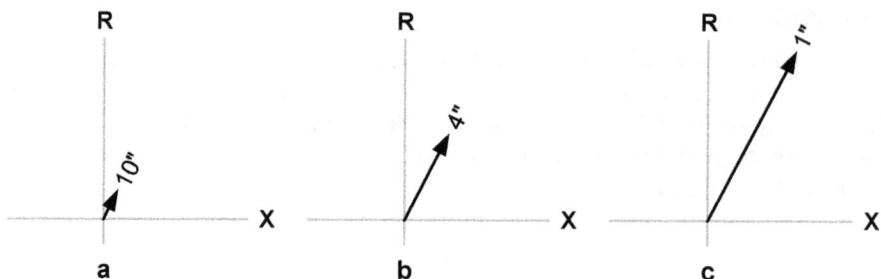

Fig. 16-10: **US Nickel Response vs Depth**

16.3: Demod Clocks

In the chapters on TR-Discrimination and VLF-GB there was only one demodulator and therefore one demod clock. It was generated from the TX voltage through a simple phase delay, with the delay either setting the discrimination level or the ground balance point. A VLF-Disc design uses quadrature demods that are clocked 90° apart, with one demod (the resistive) clocked in a manner that results in a nulled ground signal. As was seen in the VLF-GB chapter, ideal ground should have a 0° phase response but real ground has a slightly elevated phase and can vary from place-to-place. The VLF-GB therefore required some range of phase adjustment for the demod clock to deal with variable ground. The quadrature VLF has the same requirement as shown in the block diagram in Figure 16-11.

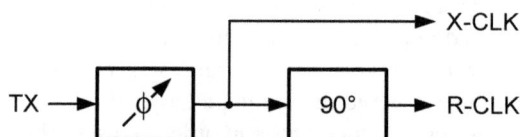

Fig. 16-11: **Demod Clock Block Diagram**

The variable delay can be lifted directly from the *VLF-TR Disc* design. The 90° quadrature shift can be produced by a simple RC-CR phase shift circuit as shown in Figure 16-12. The values for R_2 and C_2 are such that

$$f_{TX} = \frac{1}{2\pi R_2 C_2} = \frac{1}{2\pi\tau}$$

Eq 16-1

where f_{TX} is the transmit frequency. Every 1% error in the RC tau produces a quadrature phase error of about 0.6° so we should strive for no more than a few percent error. The comparators provide logic-level drives for demod switches.

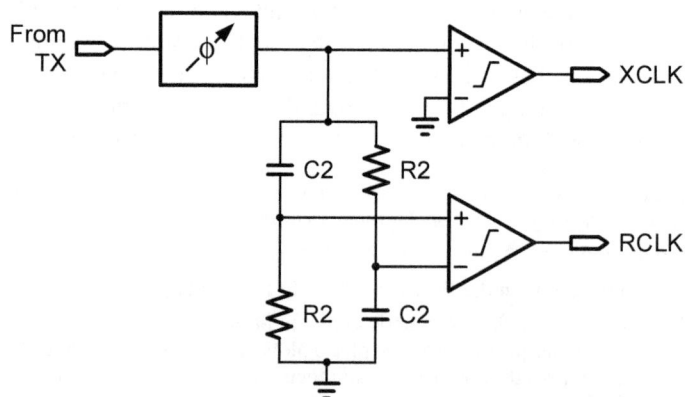

Fig. 16-12: **Example Demod Clock Circuit**

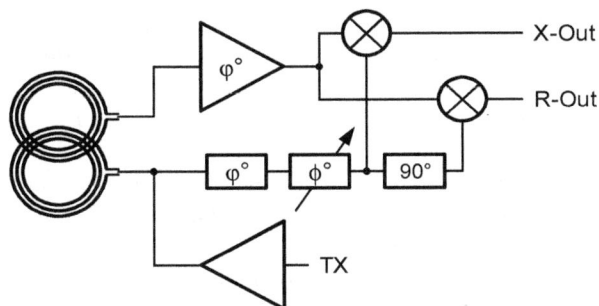

Fig. 16-13: **RX Path Delay**

Besides compensating for variable ground, the initial delay block must also compensate for any delays in the RX path up to the demodulators. Figure 16-13 shows a more complete situation whereby the delay from the TX signal to the XCLK must also account for RX signal delays from the RX coil to the demodulators in order to produce a phase shift that can achieve ground balance. This was covered in detail in the *VLF-TR Disc* design and will not be repeated here.

16.4: Ground + Target Responses

The output signals from the demodulators represent static responses. That is, if a nickel is presented 4" from the coil and held stationary then the demod outputs will move from (0,0) to some (X,R) voltage levels (per the vector response in Figure 16-10b) and stay there until the nickel is removed. The same thing will happen with ground; when the coil is lowered to mineralized ground there will be a large static ferrite-type response which remains until the coil is raised.

Let's suppose the coil is lowered to the ground and the resulting demod voltages form a vector response similar to the "Ferrite" vector of Figure 16-9b. As long as the ground is homogeneous and the coil is maintained at a constant height above the ground, this ground response will not vary in phase or amplitude. Now suppose the coil is swept over a nickel. The resulting signal from the nickel will add to the ground signal both before the demods and after. Since the various demod output signals can be represented as vectors, it is a simple matter to do a vector addition to determine the overall ground+nickel response. Figure 16-14 shows what this looks like for the three nickel depths illustrated in Figure 16-10. Vector **G** is the constant ground signal. Vector **T** is the target (nickel) response transposed from the origin (Figure 16-10) to the new offset created by the ground signal. The overall resultant signal (ground + target) is the vector labeled **G+T**.

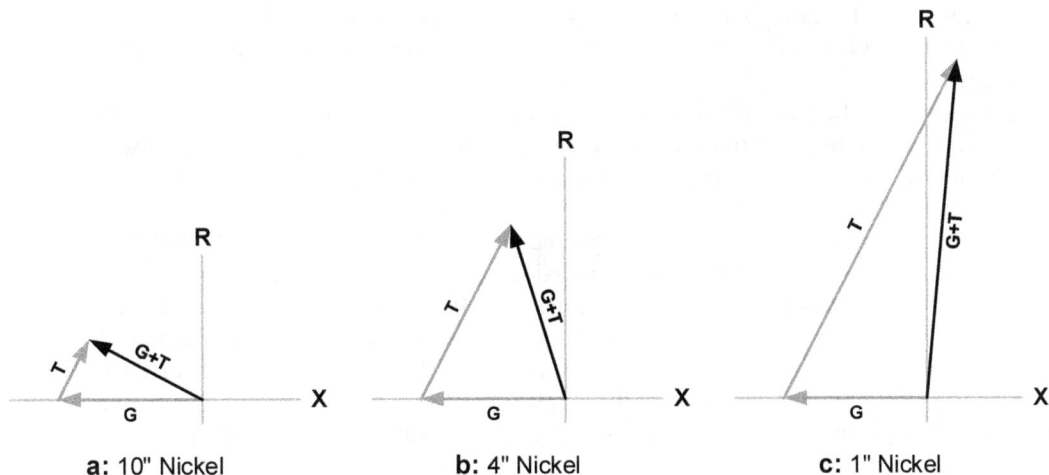

a: 10" Nickel   **b:** 4" Nickel   **c:** 1" Nickel

Fig. 16-14: **Ground+Nickel Response vs Depth**

A nickel 10" deep results in an overall vector (Figure 16-14a) that is not far removed from the ground vector, perhaps encroaching into the typical iron target range. A stronger response from a 4" deep nickel produces an overall vector (Figure 16-14b) that is solidly in the iron range. A shallow 1" deep nickel has a strong enough response so that the overall vector (Figure 16-14c) finally ends up in the non-ferrous quadrant, though looking much like a foil target. A stronger ground signal would pull even a shallow nickel into the ferrous quadrant.

All of this shows that for static signals the composite response (ground + target) is extremely dependent on the strength of the ground and the type (phase) and depth (strength) of the target. To make matters worse the ground is not really homogeneous and, even if it was, users cannot maintain a perfectly constant coil height above the ground while swinging. Therefore, even multiple passes over a given target can produce different composite phase responses. This is why a TR-Disc design performs so poorly in mineralized ground.

Modern metal detectors do not suffer from this behavior so there is obviously a solution. In order to get the composite responses in Figure 16-14 to behave like the ideal target-only responses in Figure 16-10 we need a way to force the ground response vector to have a zero amplitude so it affects the target response vector as little as possible. This can be accomplished using a high-pass filter.

16.5: High-Pass Filter

Figure 16-15 shows a typical target response combined with the ground response with the coil in a normal sweep motion, as seen at the outputs of the demodulators[11]. The target has a fast response when the coil passes over it whereas the ground is a slowly varying signal. For a typical 1m/s swing speed the ground signal might be around 1Hz while the target signal bandwidth is closer to 20-30Hz. Both typically are proportional to the swing speed of the coil so as a user swings faster the effective frequency of the ground signal increases but so does the target signal.

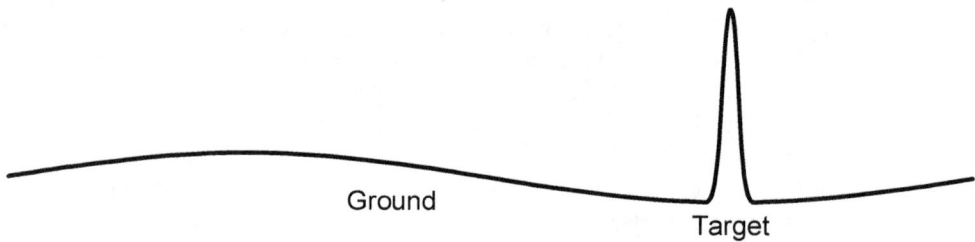

Fig. 16-15: **Time-domain Response of Ground and Target**

It is this difference in signal bandwidths that allow us to suppress the ground signal with a high-pass filter. Figure 16-16 shows a high-pass response with a target signal T_1 on the edge of the high-pass plateau at 20Hz and a ground signal G_1 at 1Hz. The total suppression of ground is 25dB, or $G_1 = 0.056 \times T_1$.

Suppose the user begins swinging the coil twice as fast. G_2 is now 2Hz and T_2 is 40Hz; the frequency ratio remains the same but the suppression of ground is now only 19dB because T_2 is on the high-pass plateau. There is a 6dB loss in ground suppression, meaning that the target-to-ground signal ratio is now halved.

A simple solution to this is to place both ground and target signals on the roll-off slope as in Figure 16-17. Now regardless of sweep speed the suppression of ground is consistent. A drawback, however, is that the gain of the target signal varies with sweep speed. If the user slows down the sweep speed, both the ground and target signals will drop in amplitude. If the sweep speed becomes too slow then the target signal will be severely attenuated to the point that it will no longer exceed the threshold level and will not be detected, or it will get buried in the system noise. This suggests there is a minimum practical sweep speed for a given filter design or, conversely, the filter should be designed for a desired range of

---

11. When the ground is highly mineralized or the target is deep, the amplitude of the ground signal can be much higher than the amplitude of the target signal.

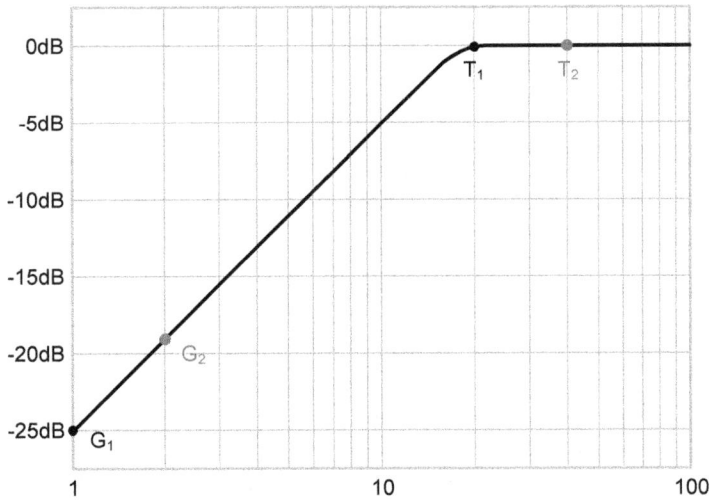

Fig. 16-16: **High-Pass Filter: Slow vs Fast Sweep**

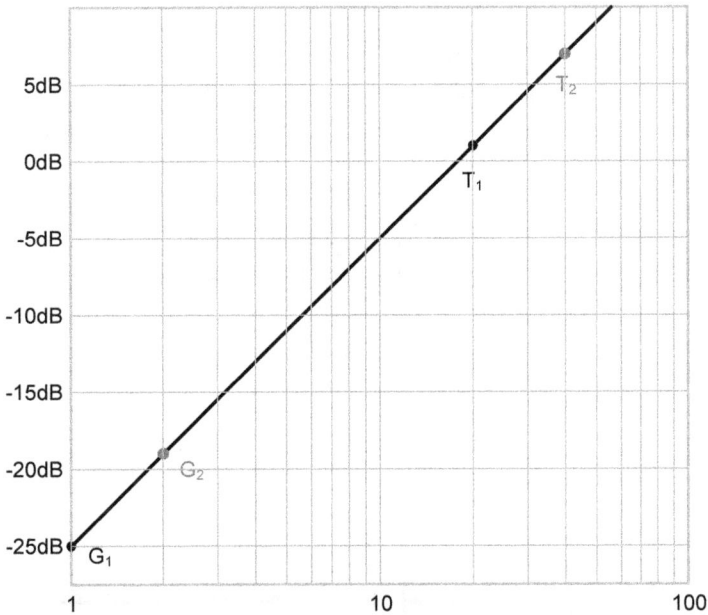

Fig. 16-17: **Differentiator: Slow vs Fast Sweep**

sweep speeds. Because this filter is only effective when the coil is in motion we call it a *motion filter* and discriminating VLF detectors using this technique are called *motion discriminators*.

Although this is still effectively a high-pass filter (there will eventually be a plateau) it is being used as a *differentiator*. In VLF detectors the motion filters are commonly called *derivatives*. Most modern VLF detectors have two motion filters in series, a first derivative stage and a second derivative stage. Figure 16-18 illustrates. Two stages offer better ground rejection than one, and we will see in the next section that the two derivatives also produce convenient signal timing.

Each demod channel shows three signals: a raw signal (X, R), a first derivative (X', R'), and a second derivative (X", R") [12]. A typical VLF detector operating in a normal discrimination mode will use the second derivative signal which offers good ground rejection. But the first derivative can also be use-

---

12. In mathematics, X' is called "X prime" and X" is called "X double-prime."

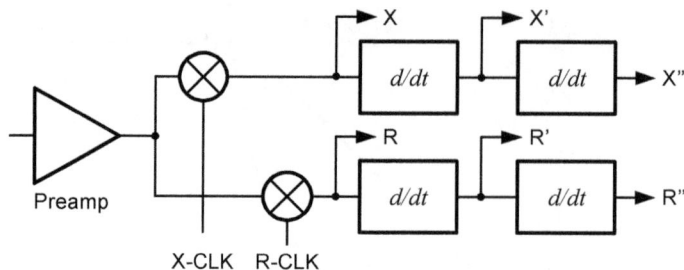

Fig. 16-18: **Block Diagram of Motion Filters**

ful as we will see at some point. Even the raw signal with no ground rejection at all has value in a static pinpoint mode.

Although practically all modern VLF detectors use a 2-filter system the very first motion discriminators[13] had four series high-pass filter stages[14] after each demodulator. It was only later[15] that this was simplified to two filters, and at that point the earlier designs became known as *4-filter* detectors and the newer designs as *2-filter* detectors. The last 4-filter production model was probably the White's *XL Pro* in 2006, the last of a long line of analog VLF-Disc designs that began with the *6000/D* in 1978[16]. See the *Explore* section for more coverage of 4-filter designs.

16.6: Time-Domain Filter Responses

In suppressing the ground signal, the derivative function of the high-pass filters alters the time-domain waveforms. If the target signal were to fall far enough inside the passband (point T2 in Figure 16-16) then the signal in Figure 16-15 would look like this:

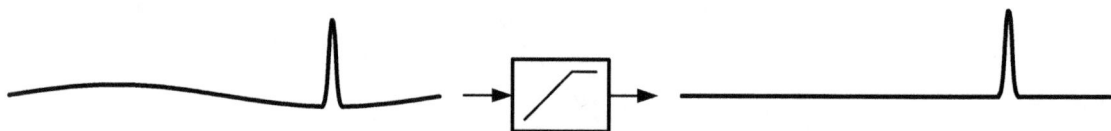

Fig. 16-19: **High-Pass Response**

The ground response is suppressed while the target response is relatively unchanged. However, if the filter functions as a derivative for both target and ground (per Figure 16-17) then the target response appears as in Figure 16-20. In this case the target response still comes through relatively strong but as a derivative of the input signal.

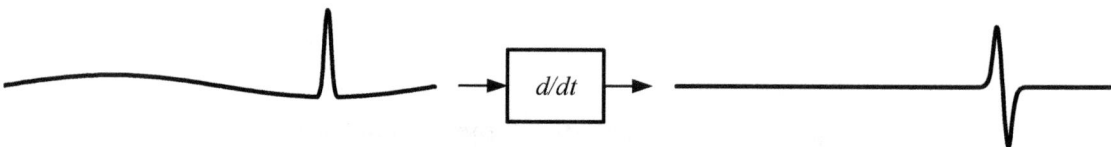

Fig. 16-20: **Derivative Response**

As mentioned previously modern VLF designs employ two derivative stages. Figure 16-21 shows the target responses at the outputs of the demod (X), the first derivative stage (X'), and the second derivative stage (X"). Although labeled as 'X' channel signals, the 'R' channel behaves the same way.

There are a few things to note here. First, the three signals occur as the coil is swept over the target and are time-aligned. Therefore, the peak of the X signal aligns with the zero-crossing of the X' signal

---

13. Such as the Bounty Hunter *Red Baron* and White's *6000D*.

14. Or two second-order filter stages.

15. With the Fisher *1260X* in 1982.

16. The *6000/D* was my first serious detector.

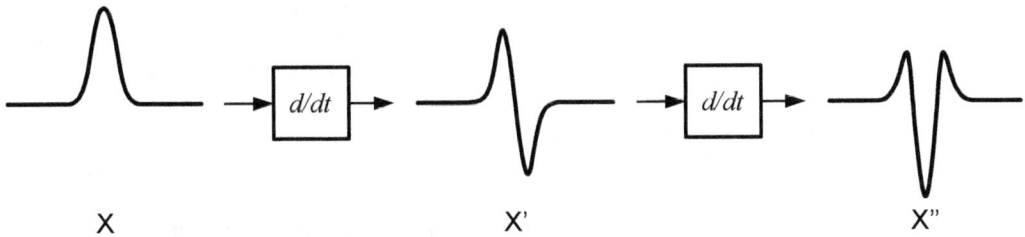

Fig. 16-21: **Raw, 1st Derivative, 2nd Derivative**

and also with the negative peak of the X" signal[17]. The zero-crossing of the X' signal can be useful for producing a timing marker for signal processing.

Second, the X signal and the X" signal both have their largest peaks directly over the target, whereas the X' signal has its zero crossing directly over the target. This is fortuitous since the X" signal is typically used for creating an audible target response and so the peak audio signal will tend to be aligned with the real-time position of the coil over the target, except for signal processing latency.

Third, the purpose of the filters is to remove the (relatively) slowly varying ground signal. We'll soon rediscover that the R channel is phase adjusted to completely null the ground so it would seem that this channel does not need the filter stages. However, we will also see in the next chapter that both channels are used for discrimination and it is important that the signal delays through the channels match. Therefore, not only do both channels require the same applied filter stages, but the filter delays must be well-matched which requires matched components. Typically it is the capacitors that need the most attention and are often hand-matched to 1%.

Finally, the signals in Figure 16-21 are drawn in a mathematically ideal manner. Real filters are not quite so ideal. There will be a slight time delay through each filter and the symmetry of X' and X" will not be so perfect. If the filtering is done in DSP then the waveforms will be more ideal in shape (though quantized) but there will still be a processing time delay for each filter. Figure 16-22 shows example analog responses on the left (the 2nd derivative is inverted) and digital responses on the right. Note the propagation time delays and imperfect symmetry, especially in the 2nd derivative curves.

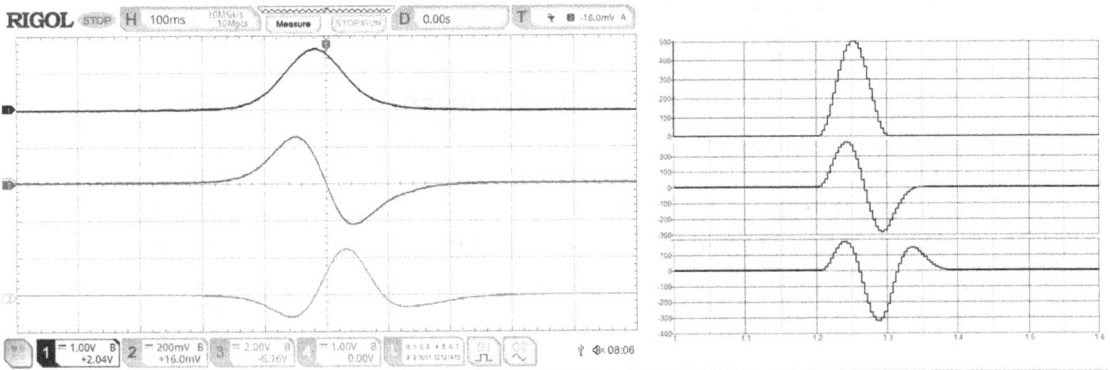

Fig. 16-22: **Real Filter Responses**

16.7: Ground Balance

While high-pass filtering suppresses the relatively slow ground response it does not eliminate it. For the purpose of being able to positively signal the presence of a metal target in mineralized ground we need one of the two channels to be phase-adjusted so that the response to typical ferrite is zero. We saw how to do that in Chapter 15 and the technique is the same here.

---

17. The X" signal is often drawn as inverted so that the peak is positive.

Because ferrite has a phase of nearly 0° its response lies primarily along the negative X-axis which makes the R channel nearly devoid of any ground signal, assuming the phase shift in the preamp is compensated for. Once the demodulator clock phases have been adjusted for ground balance the R channel will be completely free of any ground signal and any signal seen on the R channel is assumed to be a target. The R channel is therefore used as a *target channel* and goes positive for both ferrous and non-ferrous targets. Meanwhile, the X channel sees a significant amount of ground which is suppressed through filtering.

The target channel is used as an indicator of the presence of a metal target. Both channels can then be used to determine the phase of the target and potentially discriminate out unwanted targets. The advantage of this scheme is that the target channel does not rely on filtering to remove the ground signal. This results in an overall quieter target channel and ground suppression is independent of sweep speed.

16.8: Demod Clocking

Recall from Figure 16-12 that the XCLK is derived directly from the TX signal and that the RCLK is delayed by a fixed 90° from the XCLK. Let's assume there is no phase shift in the RX preamp and that ground has an ideal 0° phase shift. The resulting situation is illustrated in Figure 16-23a with the variable phase set to zero and the resulting X-R clock phases exactly aligned at 0° and 90°.

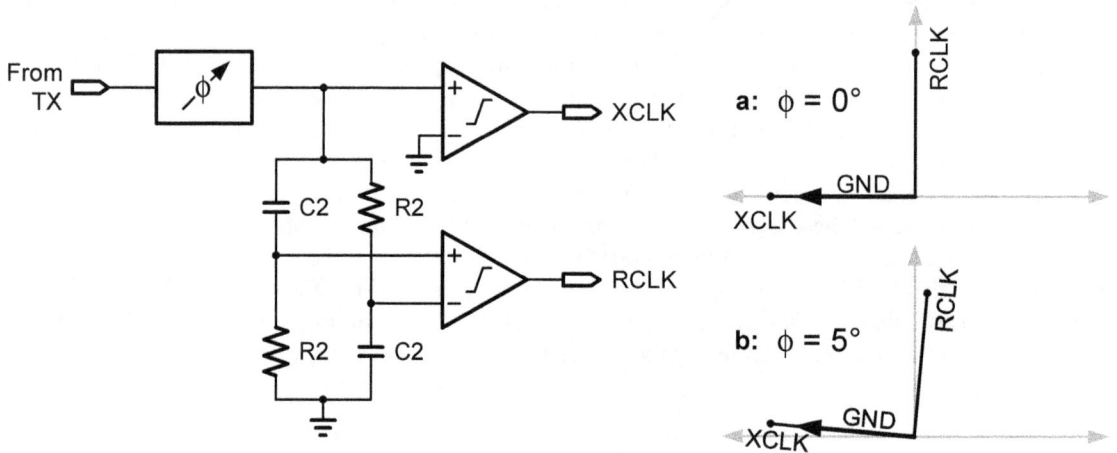

Fig. 16-23: **Ideal Ground**

Now consider ground that has a 5° loss angle. In order to achieve ground balance the RCLK must be phase shifted by an additional 5° per Figure 16-23b. This action rotates both clock phases which effectively rotates the entire coordinate system. However, the input signals to the demodulators — in particular the target response phase shifts — are still the same so this rotation will impact target discrimination.

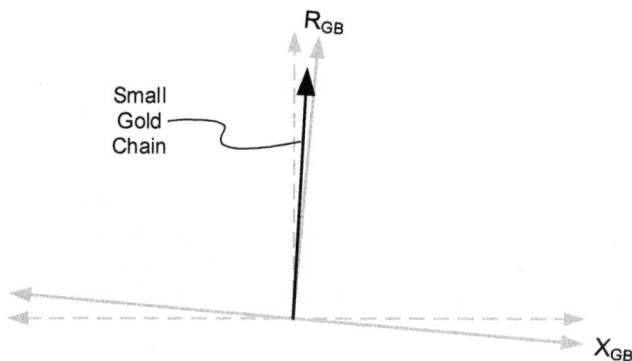

Fig. 16-24: **Target Error Due to Rotated Coordinates**

Figure 16-24 shows the 5° rotated coordinate system (X_{GB}, R_{GB}) along with the "true" coordinate system (dashed lines). Suppose that the coil sweeps over a fine gold chain which has a 93° phase shift; that is, 3° to the *right* of the unrotated vertical axis. In the rotated coordinate system it appears to be 2° to the *left* of the vertical axis and will now produce a negative X response. In other words, rotating the coordinate system has now made the gold chain appear to be a ferrous target. This is why with some detector models it is important that you do not discriminate the last 5-10° of the ferrous region if you want to get extremely low-tau targets like jewelry or gold nuggets, especially in lossy ground.

There is a partial solution to this and a more complete solution. The partial solution is to simply change the clocking scheme so that phase shift needed to ground-null the R channel does not affect the X channel; see Figure 16-25. In this case only the R channel will rotate as the ground balance is adjusted, The X channel stays fixed so that, at a minimum, its ability to correctly produce a negative voltage for ferrous targets and a positive voltage for non-ferrous targets is preserved and low-tau targets (like the gold chain example above) will continue to identify as non-ferrous. However, because one axis is phase-adjusted and the other is not, strict quadrature (90°) is not preserved and the demod voltages will no longer accurately represent the true target phase. It's an improvement, depending on your point-of-view.

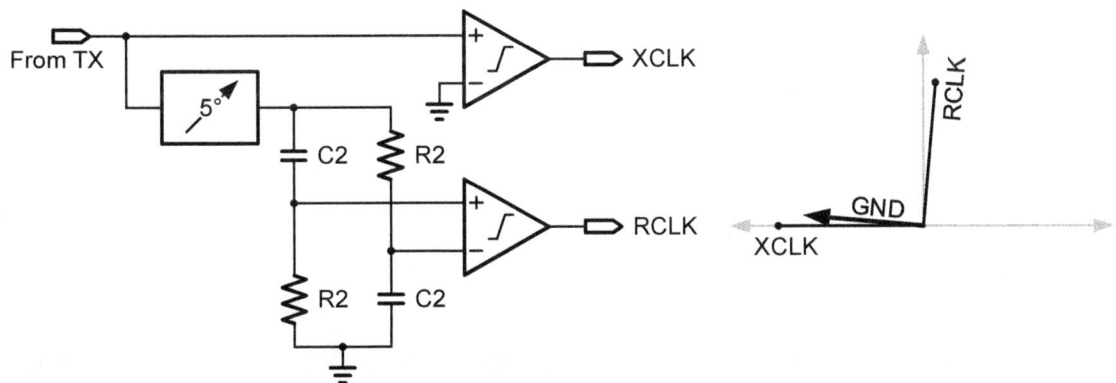

Fig. 16-25: **Ground Balance by R Rotation Only**

16.9: The G Signal

A more complete solution is to use a pair of channels with perfect quadrature timing for discrimination and a separate ground-balanced signal for use as a target indicator. This new signal can be derived from the X and R signals with some simple math. Since lossy ground (again, we'll use a 5° loss angle) produces a large negative X signal and a small positive R signal then the simple solution is to add a slight amount of the X signal to the R signal such that the resultant signal is zero for ground only. Any other demodulator outputs, whether from ferrous or non-ferrous targets, will always produce a positive voltage for this new signal. As such, it can be used as a ground-free target indicator that does not alter the coordinate system. We call this the *targeting signal* and therefore this new channel is the *target channel*.

We will call this signal the "G" signal (since it is ground-balanced), with the general equation being

$$G = kX + R \hspace{4cm} \text{Eq 16-2}$$

where G = 0 at balance. Since R = A · sin(ϕ) and X = A · cos(ϕ) then the required k at balance is

$$k = \tan(\phi) \hspace{4cm} \text{Eq 16-3}$$

If you paid attention then you'll notice Equation 16-3 is missing a negative sign. That's because our polar coordinate system for detectors is backwards from normal trigonometry. As an example, if the ground loss angle is 5° then we have

$$k = \tan(5°) = 0.087$$

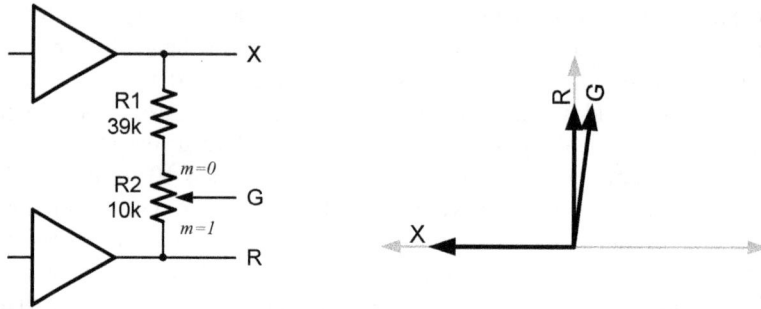

Fig. 16-26: **Creating a G-Signal**

Since the ground phase can vary, a potentiometer is used to select the correct ratio of X and R and the pot becomes the ground balance control (Figure 16-26). A fixed resistor (R1) can be used to offset the adjustment range. The equation now becomes

$$G = X \cdot \frac{(1-m)R_2}{R_1 + R_2} + R \cdot \frac{R_1 + mR_2}{R_1 + R_2} = \left(\frac{R_2}{R_1 + R_2}\right)\left((1-m)X + \left(m + \frac{R_1}{R_2}\right)R\right)$$

Eq 16-4

where m is the relative position of the potentiometer ($0 \le m \le 1$). The minimum angle is $0°$ ($G = R$) when $m = 1$ and the maximum angle ($m = 0$) is when

$$G = \left(\frac{R_2}{R_1 + R_2}\right)\left(X + \frac{R_1}{R_2} \cdot R\right) = 0$$

or

$$\frac{R}{X} = \frac{R_2}{R_1} = \tan(\phi) \Rightarrow \phi = \arctan\left(\frac{R_2}{R_1}\right)$$

Again omitting a negative sign for the backward coordinate system. For the values in Figure 16-26 the range is $0°$ to roughly $14°$ which should be sufficient for just about any ground.

The G signal can be created from the raw demod outputs, the first derivative outputs, or the second derivative outputs. If it is created with the raw demod signals then the G signal can be used for either motion discrimination or non-motion pinpoint[18]. It is also easier to set the ground balance using a static mode. Be aware, however, that a static G signal from the demod outputs is subject to drift from DC offsets or from a poorly nulled coil. This can be cured with high-pass filtering or a retune circuit.

16.10: Salt Balance

The purpose of ground balance is to create a signal that is nulled to the ground mineralization. Normally, the loss angle of the ground is in the range of 0-$15°$. Wet salt sand has a phase angle that is at the lowest end of the non-ferrous region, or slightly higher than $90°$. Some single frequency detectors have been designed with a ground balance that is adjustable all the way to salt, allowing them to be usable in salt conditions. Figure 16-27 illustrates[19].

With normal ground balance, a G channel can be created by adding a small amount of the (negative) X channel with the R channel. With salt ground, the R signal is now large and the X signal is small but positive; the best approach to create a salt mode G channel is to subtract a small amount of R from X.

Figure 16-27 also includes two target vectors: a Small Gold Ring (SGR) and a Big Silver Coin (BSC). In normal ground, the Big Silver Coin has a fairly weak G component but in salt ground it has a

---

18. This is why you will occasionally see a detector where the GB control works in disc mode but not in pinpoint or all-metal mode. A few of the later Tesoro models have this shortcoming.

19. The phase angle of wet salt sand varies with salinity. The illustration suggests a $5°$ range and that's on the high side.

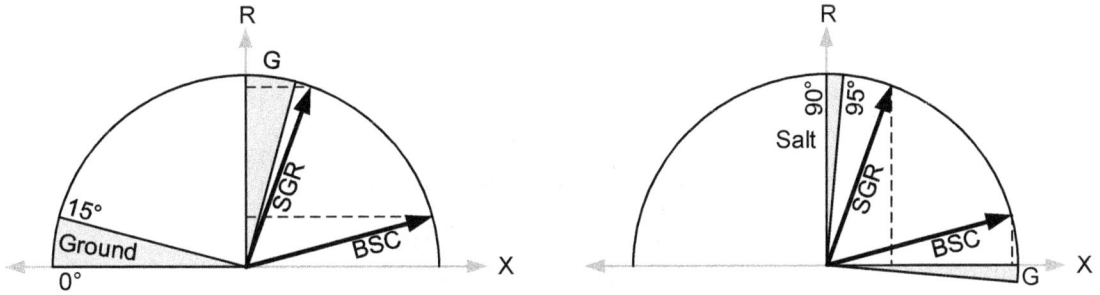

Fig. 16-27: **Normal vs Salt Ground Balance**

significantly stronger G component. Recall that the G component is our all-metal targeting signal. The Small Gold Ring is opposite; it has a strong G component in normal ground but a significantly weaker G component in salt ground. In fact, it is possible to have a small low conductive target that falls right on top of the salt phase, in which case it will look exactly like a ground signal and be nulled.

Single frequency salt balance tends to work best at white sand beaches. Where beaches have a significant amount of magnetic black sand mixed in, the ground signal becomes a composite of a ferrite signal and a salt signal, meaning the overall ground phase can be anywhere between 0° and about 95°. In such cases a single frequency detector with automatic ground tracking to salt might work, but preferably a multifrequency detector should be used.

16.11: Frequency Considerations

In Chapter 4 it was shown that a metal target in an incident magnetic field produces its own reverse magnetic field via eddy currents that are generated by induction. Due to skin effect, both the strength and phase of the reverse magnetic field increase as the frequency of the incident field increases. We can plot this vectorially as shown in Figure 16-28. Four specific vectors of a given target are shown for four frequencies. Different targets of different sizes, thicknesses, and conductivities will have different specific amplitude and phase responses but they all follow the same semicircular frequency response.

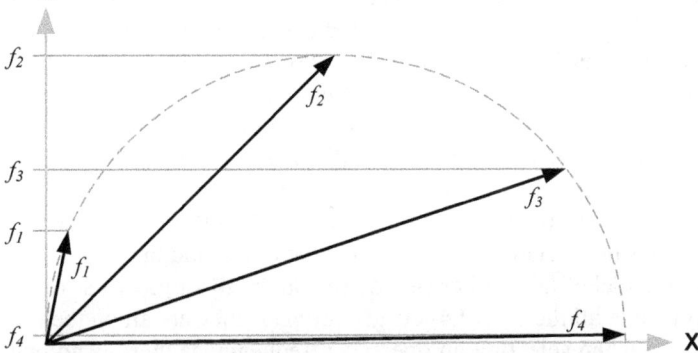

Fig. 16-28: **Target Response vs Frequency**

Since the highest frequency produces the strongest target response, it seems logical that a VLF design should operate at a high frequency. However, the all-metal target response is taken from the R channel[20] and the response at f_4 has an exceptionally weak R response. Of the vectors shown, f_2 has the strongest R response and would represent the optimum frequency for that particular target.

In general, the strongest R response always occurs at a target phase angle of 45° from the R-axis, or 135° overall. From this we can calculate optimum frequencies for particular coins based on their tau since we know[21] that $\phi = \arctan(\omega\tau)$. For $\phi = 135°$ we have

---

20. Or the G channel, which is pretty close in phase to the R channel.

21. See Section 4.12.

$$f_{opt} = \frac{1}{2\pi\tau}$$

Using the data from Table 4-7, the optimum frequencies of common US coins are:

	Tau	f_{opt}
Cent	66.1µs	2.4kHz
Nickel	10.2µs	15.6kHz
Dime	80.8µs	1.97kHz
Quarter	146.5µs	1.09kHz
Half	184.8µs	860Hz
Dollar	310.1µs	510Hz

Table 16-1: **Optimum R-Response Coin Frequencies**

The frequencies above have been referred by some as the "resonant" frequency and by others as the "-3dB" frequency. They are not resonant frequencies, and they can be considered −3dB only in the sense that they occur at a phase shift of 45°. They are simply the frequency at which the R response is maximum. If we used a different method of creating a target signal then the optimum frequency might well be somewhere else.

This explains the well-known correlation between the detector operating frequency and sensitivity to different targets. Low frequency detectors would be a better choice for hunting deep silver, whereas high frequency detectors are better suited to low-tau targets like small gold nuggets. We can plot the responses of the above US coins at 10kHz (Figure 16-29) and find that while a nickel projects a good, strong response the larger silver coins do not. However, this plot is misleading in that it doesn't show that physically larger targets (like a US silver dollar) produce a stronger response than, say, a US dime. Also, given the same size, a high conductive target (like a silver coin) has a stronger response than a low conductive coin, like a US nickel.

To account for these factors each target should have its own frequency response curve, with larger, stronger targets having a larger curve. When this is included, a comparison of the target vectors yields a better balance as seen in Figure 16-30. The US silver dollar response is almost twice as strong as the half dollar and its vector is truncated so the other coin vectors and their response circles can still be clearly seen. The R values are compressed and difficult to see so they are expanded to the left of the plot. In the case of 10kHz, the nickel response is still has the strongest R response but now the silver dollar is stronger than the other coins, as we tend to expect.

For many years White's ran their detectors at 6.6kHz. The relative responses (Figure 16-31) show that a nickel R response is now relatively weaker than the dollar and the half dollar. Further reductions in frequency will see the nickel fall further (it's on the downhill slope of its frequency semicircle, for decreasing frequency) while all the other targets get stronger (they are all on the uphill slope).

This is the trade-off when selecting an operating frequency as there is no one best frequency for almost any application. Although silver coins tend to favor a low frequency, gold coins have a much wider range, as does jewelry. In relic hunting, targets will likewise be all over the spectrum, possibly including iron. In nugget hunting, nugget taus can range from 1µs (f_{opt} = 159kHz) to 100µs or higher (f_{opt} = 1.59kHz). This is partly why multifrequency detectors have become popular; the different simultaneous operating frequencies can produce a wider range of optimal responses.

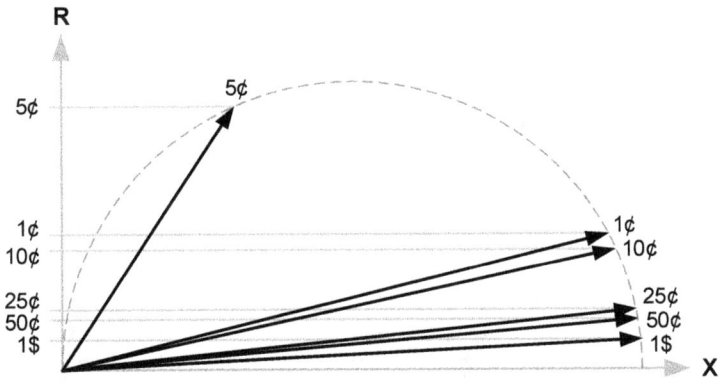

Fig. 16-29: **Not-Quite Relative Responses at 10kHz**

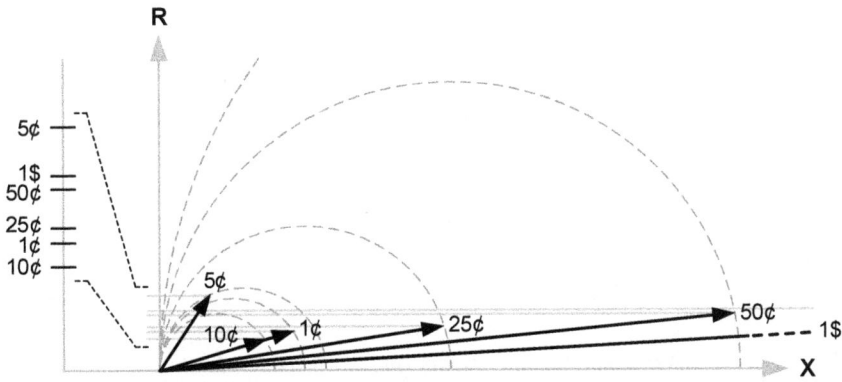

Fig. 16-30: **Relative Responses at 10kHz**

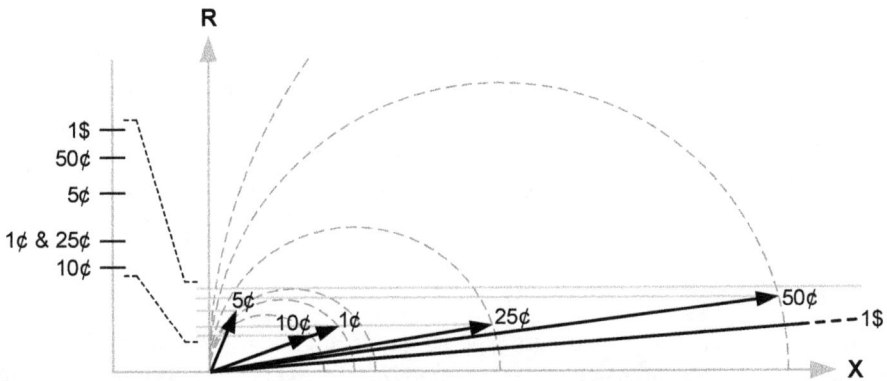

Fig. 16-31: **Relative Responses at 6.6kHz**

Design: *Motion VLF*

The basics of motion VLF have been covered so let's use this to create a working design. As designs get more complicated it is useful to draw a block diagram to see how everything fits together:

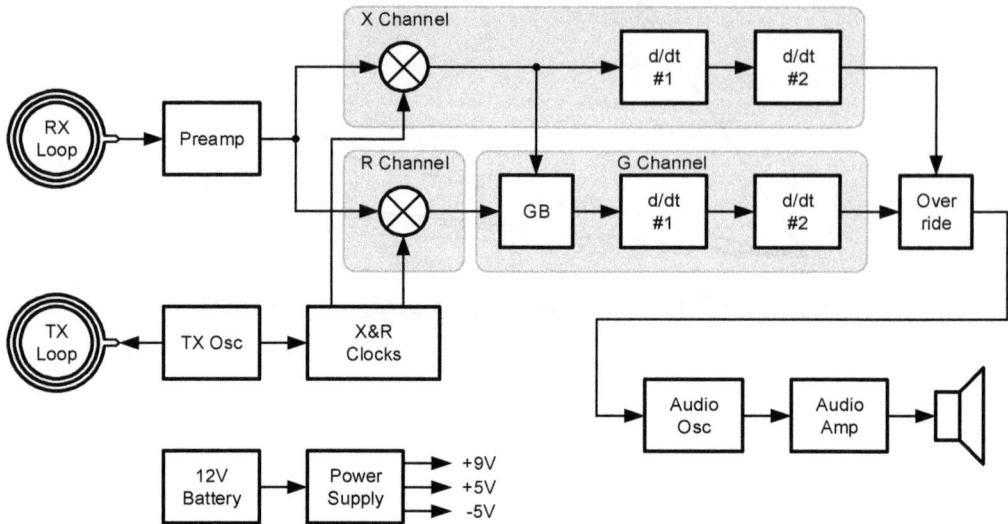

Fig. 16-32: **Motion-VLF Block Diagram**

The G channel will be derived from the raw demodulator outputs. The second derivative G signal provides the target signal for the audio, and we'll use the second derivative X signal as a simple ferrous/non-ferrous override that enables or blanks the audio. This gives us a crude form of discrimination.

Figures 16-34–16-37 show the full schematics for the *Motion-VLF* design. A discussion of each section follows.

16.12: Transmitter

The transmitter (still at 10kHz) and phase reference circuitry for the X & R demods is mostly pulled from Chapters 14 and 15 where it is covered in detail. Notice that the demod clock phases are adjustable for calibration purposes but otherwise do not have the broad adjustment ranges seen in the previous chapters. That's because, in this design, ground balance is not achieved by rotating the demod clocks but rather by using the post-demod summation method presented previously.

The amount of calibration range needs to be sufficient to account for the preamp phase variation and other circuit tolerances; usually 10° is plenty. This will be further discussed in the Calibration section.

16.13: Preamp

The preamp is a non-inverting band-pass type with a fixed gain of 100. A capacitor is added to the feedforward leg of the opamp and its pole is set to 5.9kHz, whereas the feedback pole is set to 15.9kHz. Since they are fairly symmetrical about the 10kHz operating frequency, the overall preamp phase shift should be close to 0°. The preamp also includes an inversion stage (IC2b) with a gain of −1 for driving the full-wave demodulators.

16.14: Demodulators

Whereas the *TR-Disc* and *VLF-GB* designs used JFET sampling switches, this design will move to the 4053-style analog switches commonly found[22] in more modern designs. We also continue with the full-wave demod but now use a 2-pole SHA to improve noise performance. The SHA time constant is

---

22. Some designs instead use the 4016 or 4066 switch, which are 4xSPST instead of 3xSPDT.

6.3ms which should provide for a quick response. Following each SHA is a non-inverting gain stage with a gain of 22 and a bandwidth of 16Hz.

16.15: Motion Filters

The next step in the design is the motion filters. These are realized using two cascaded band-pass filters. The "frequency" of the target response depends on the size and depth of the target, the size of the coil, and the user sweep speed. For a typical coin several inches deep, a 10"/25cm search coil, and a nominal sweep of 1 m/s the frequency of the target response is about 15Hz. Therefore the low pole is set to 34Hz to give us a little room for faster sweep speeds and smaller coils. The high pole (feedback RC) is used to suppress high frequency noise and should be set at or just above the low pole. We will use 34Hz here as well. Besides filtering, the differentiators provide some much needed gain. Figure 16-33 shows a Spice simulation of a single differentiator stage.

Fig. 16-33: **Differentiator Simulation**

The X channel and G channel filters are identical, and that should be taken as literally as possible. It is important that the X and G channel filters have closely matched delays so corresponding resistors and capacitors (such as R26,R28 and C13,C15) should match to, say, 1%. You can easily buy 1% resistors but the capacitors may need to be hand-matched. If using dual packaged opamps (like the TL072) then it is best to pair the first opamps in both filters together, and pair the last opamps together.

16.16: Audio Circuitry

The audio stage follows the design in the last chapter with a few changes. The G2 signal is applied to a potentiometer for adjusting the overall sensitivity and further amplified by IC8b, which includes a provision for injecting a variable offset current for a threshold adjustment. The tone generator is now derived from the TX frequency. The 10kHz XCLK signal is fed to a 74393 binary counter chip and divided by 16 to create a 625Hz audio tone. This tone then chops the target signal via Q5. The final audio amp is the familiar LM386, preceded by a volume control.

The Override block in Figure 16-32 allows the X channel signal to temporarily mute the audio when a ferrous target is present. Recall that the X channel is negative for ferrous and positive for non-ferrous, and we can use this to override the audio chopper and mute the tone fed to the LM386. This is done using the open-collector output of IC9a.

16.17: Power Supplies

This design uses a 12V battery pack. A 9V linear regulator powers the transmitter and audio and also feeds a charge pump inverter chip (ICL7660). The output of the charge pump is −9V and that is

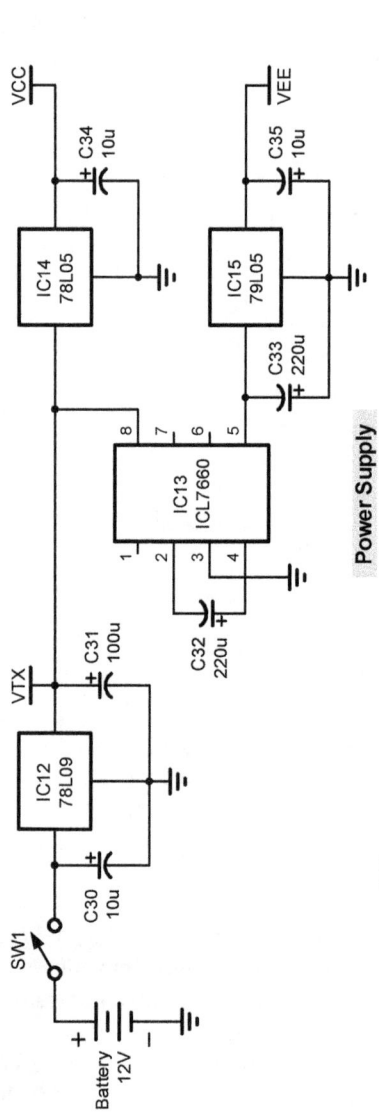

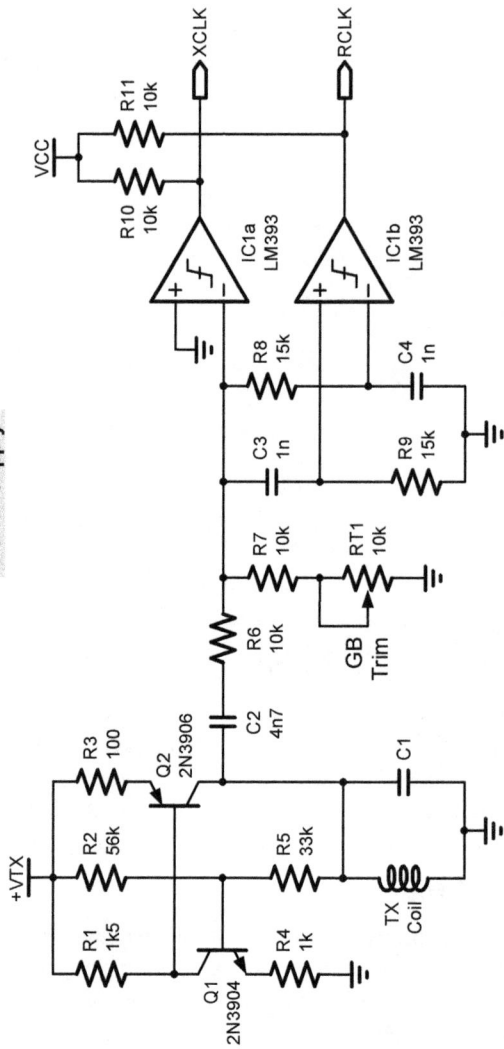

Power Supply

Transmitter

Phase References

Fig. 16-34: **Motion-VLF — TX & Power Supply**

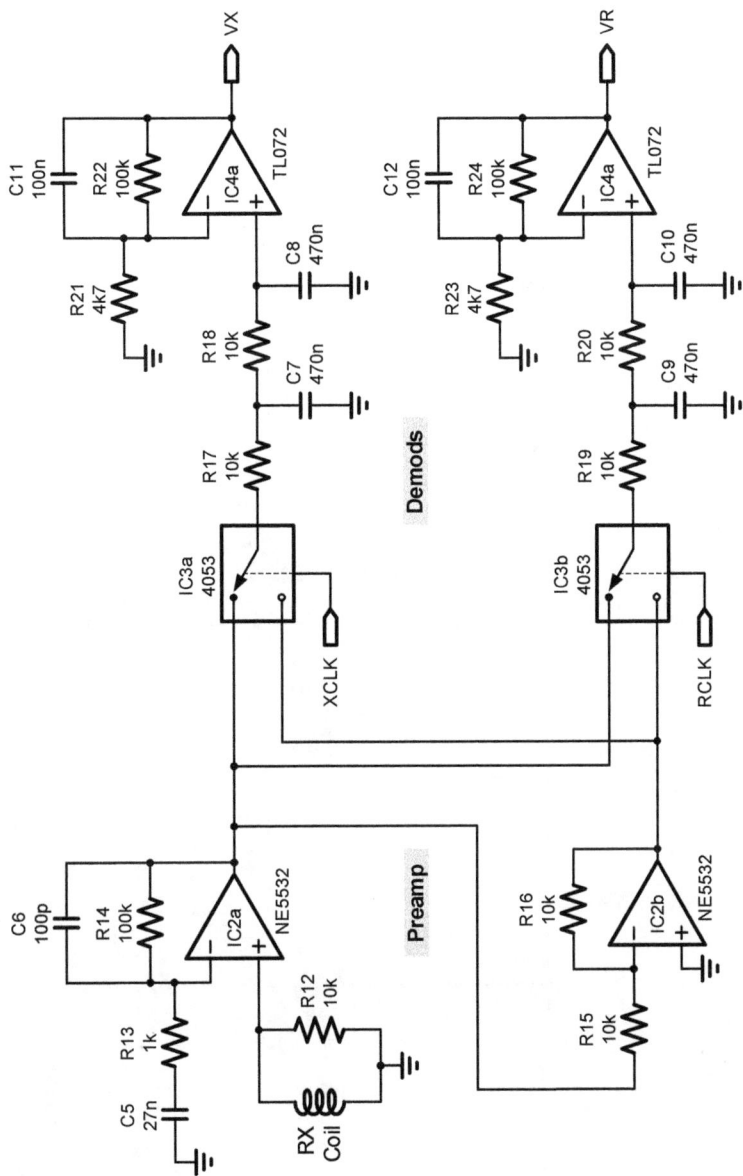

Fig. 16-35: **Motion-VLF — Preamp & Demods**

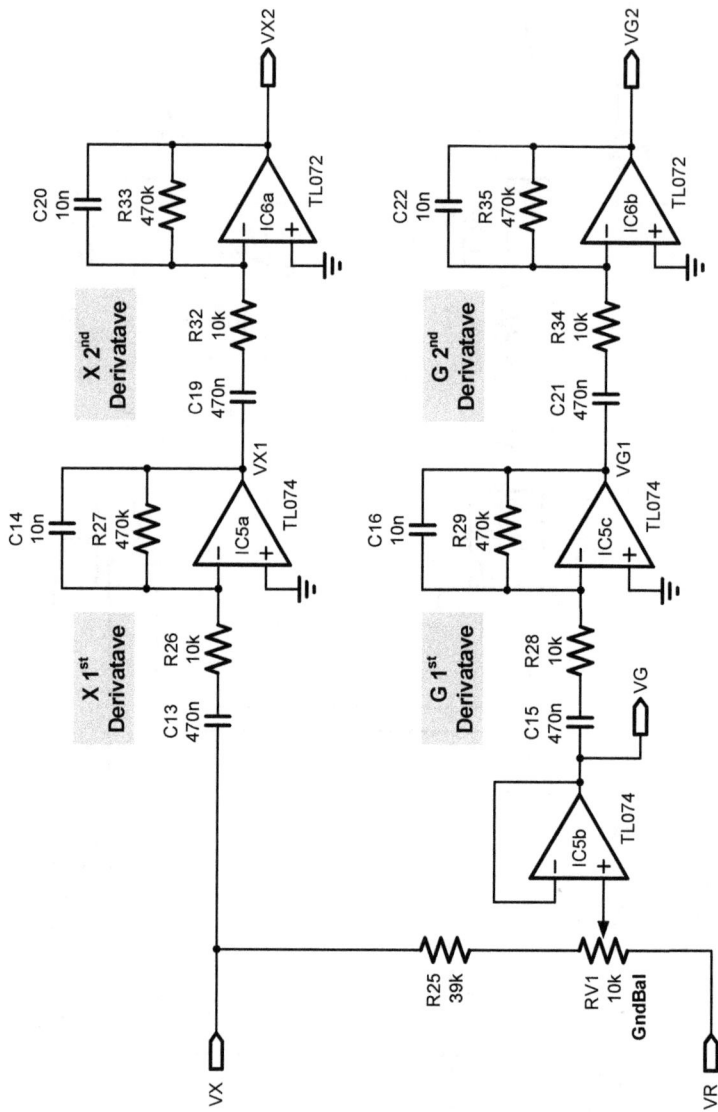

Fig. 16-36: **Motion-VLF — Filters**

Inside the Metal Detector

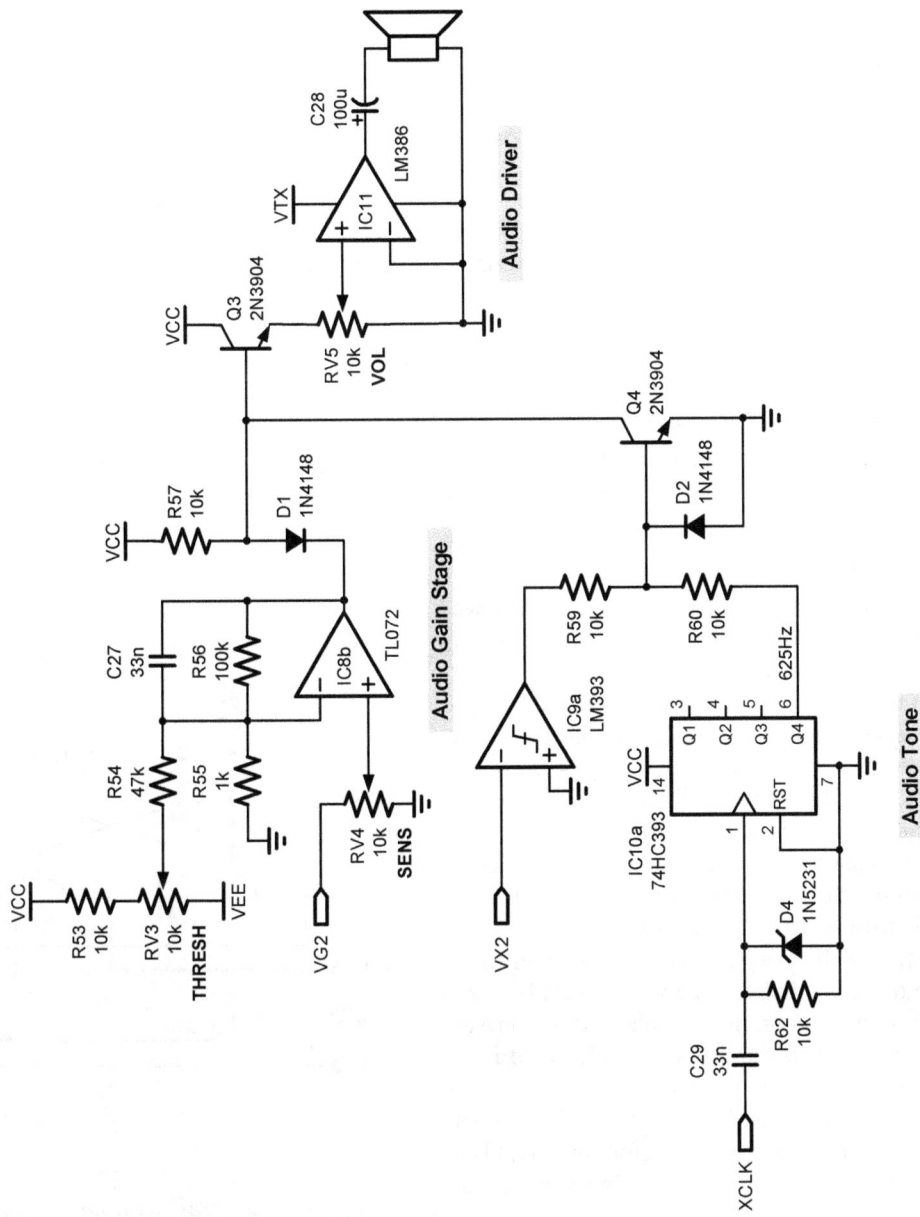

Fig. 16-37: **Motion-VLF — Audio**

regulated to −5V. Another regulator creates the +5V supply directly from the battery. The ±5V supplies power most of the analog circuitry, with battery ground used for analog ground throughout.

16.18: Calibration

The only calibration needed is to set the default ground balance point:

1. Set potentiometer RV1 to about the 10 o'clock position.
2. Press and hold the retune button.
3. Adjust RT1 until the audio is nulled while bobbing a ferrite.

Calibrating with RV1 at 10 o'clock offers more adjustment range on the positive side than on the negative side. But ferrite should be pretty close to the negative limit so there is little need for setting the balance point lower, plus it offers some margin on the low side in case the ferrite used for calibration has a higher-than-expected loss angle.

In normal operation RV1 is adjusted for ground balance. The retune button needs to be held during ground balance, and you will listen for the leading response (too much GB) and the trailing response (not enough GB) while bobbing the coil over ground. When there is little or no response, then the GB setting is optimal.

Explore

16.19: Demodulators — Part 3

The *Explore* sections of Chapters 14 & 15 covered the basic SHA-type demodulation circuit that is popular in metal detectors. There are other demodulator topologies that have been used over the years and they will be (briefly) presented here.

16.19.1: Gilbert Cell

Demodulators implemented with analog switches (such as the 4066 or 4053) are, by far, the dominant method in metal detector designs. An alternative that appeared in the Bounty Hunter *Red Baron* and is popular in RF designs is the use of the double-balanced mixer circuit, also known as the *Gilbert cell*. Figure 16-38 shows a typical Gilbert cell and is similar to the LM1496, the IC used in the *Red Baron*.

By using a square wave as the clock it can function much in the same way as the analog switch. But the switching transistors could also be set up with a broader range of linearity (by adding degeneration resistors) and, for example, a sine wave could be used as a clock signal. This produces a narrowband mixer which can reduce noise and is especially useful in multifrequency designs.

Why the *Red Baron* designs used the LM1496 (it does not appear to have been used before or since) is a mystery. Likely it was to avoid infringing patents that used analog switches.

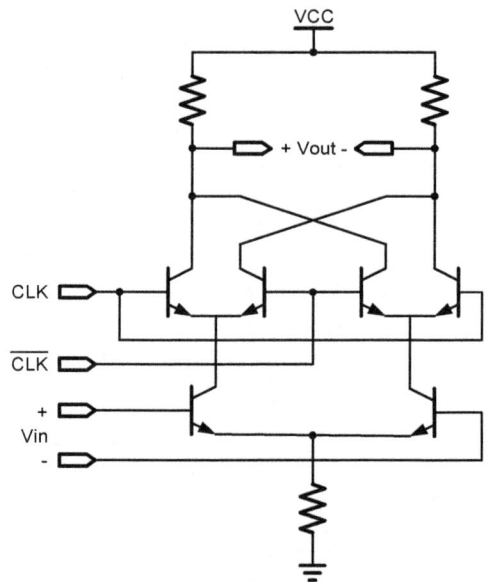

Fig. 16-38: **Gilbert Cell Mixer**

16.19.2: Sampling Integrator

Found largely in pulse induction designs, the sampling integrator is shown in Figure 16-39. The input resistor R1 performs a voltage-to-current conversion and, while the switch is closed, the current charges the integrator cap C1. The integrator is not DC-stable so a large feedback resistor R2 provides stability and also slowly drains the integrated signal from the cap to prevent overload. The sampling

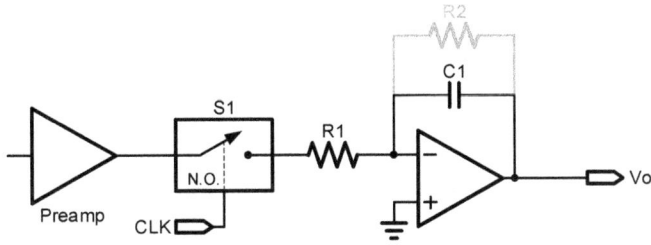

Fig. 16-39: **Sampling Integrator**

integrator demod performs much the same as the SHA demod so why it's popular in PI designs is a bit of a mystery. Possibly an early designer chose it and it stuck.

16.19.3: Differential Demod

The full-wave SHA used in the *Motion VLF* design requires a normal-polarity preamp signal plus an inverted version of the preamp signal. An alternative is to use a differential SHA/integrator which eliminates the need for the inverted preamp signal. Figure 16-40 illustrates. The drawback with this approach is that it doubles the number of capacitors which must be matched. It also prevents the use of a 2-pole RC as presented in the *Explore* section of Chapter 15.

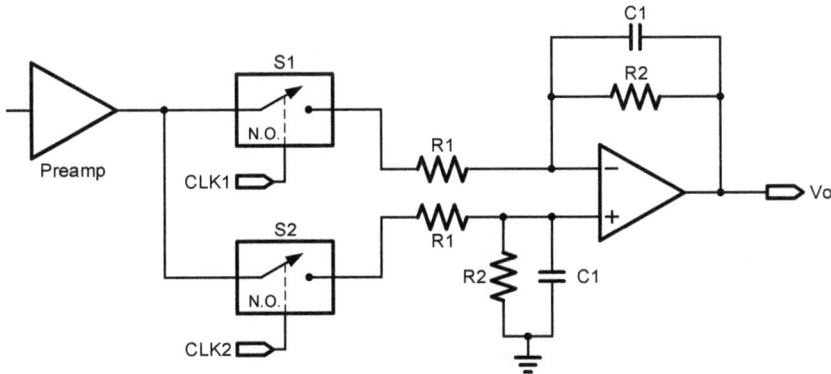

Fig. 16-40: **Differential SHA**

16.19.4: White's *6000* Demod

Another interesting demod topology was used in the White's *6000* series detectors. Like the previous method it dispenses with the need for an inverted preamp waveform. Operation is akin to a switched capacitor circuit.

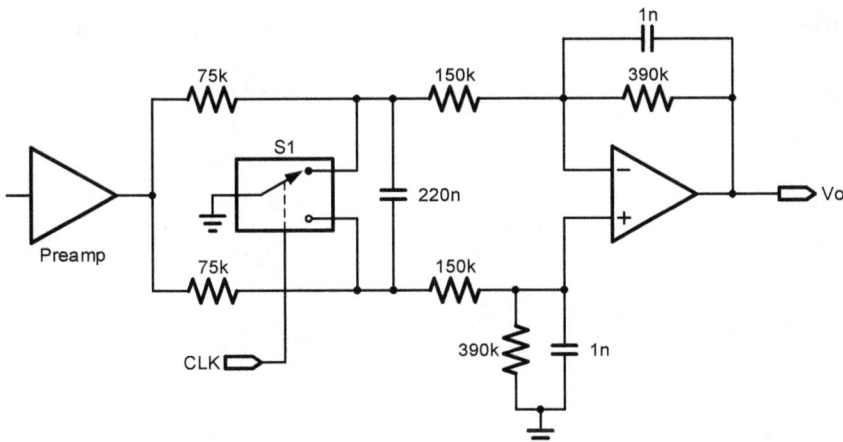

Fig. 16-41: **White's 6000 Demod**

16.19.5: Feedback SHA

In an early TR model White's used a clever method of creating a full-wave demodulated signal from a half-wave demod; that is, a full-wave demod using only one clock. An equivalent diagram is shown in Figure 16-42. Getting a good balance between the two half-wave responses requires a precise gain and feedback ratio.

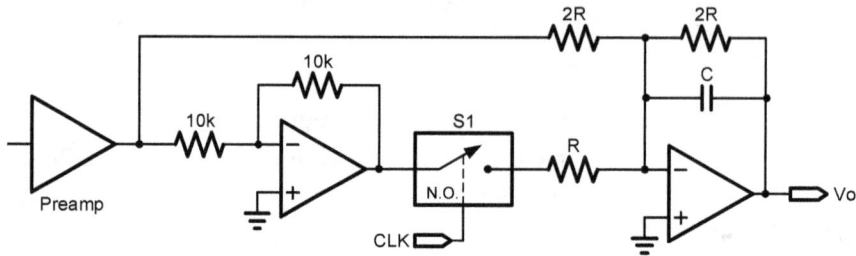

Fig. 16-42: **Half-Wave with Feedback**

16.19.6: Switched-Cap Demod

Another topology that does not require an inverted preamp signal is the switched-cap demod used in many of the Fisher and Teknetics models designed by David Johnson. A minor drawback is that it is not fully balanced. In one cycle the preamp charges C1, in the other cycle it charges C2 but with C1 in series which produces some asymmetry. This is minimized through the ratio of C1 and C2 and for moderate performance detectors is sufficient.

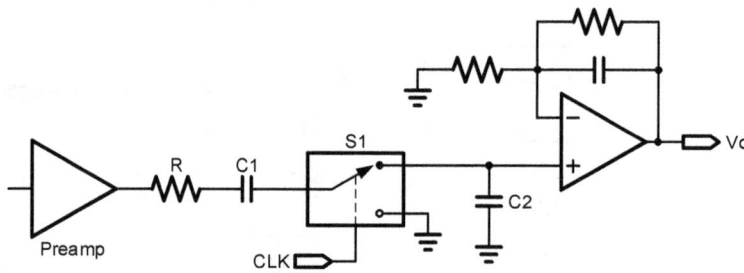

Fig. 16-43: **Switched-Cap Demod**

16.19.7: Improved Switched-Cap Demod

Figure 16-44 shows an improvement I made to the standard switched-cap demod shown previously. In this case, the switched capacitor C1 is applied to an inverting integrator instead of a non-inverting buffer. C1 therefore sees ground in one clock phase and virtual ground in the other clock phase, resulting in symmetrical operation with a single demod clock and no preamp inverter stage.

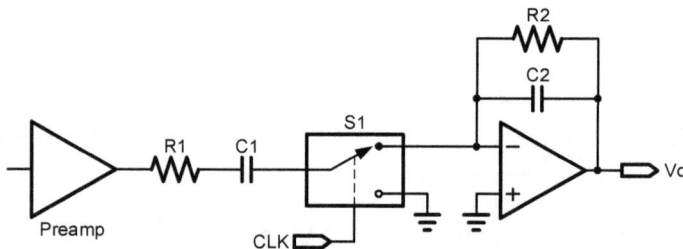

Fig. 16-44: **Improved Switched-Cap Demod**

16.19.8: Tayloe Demod

Tayloe demodulation has little to do with how the demodulator is designed and instead describes an alternate timing scheme. In the majority of VLF detector designs the demod clocks are run at a 50%

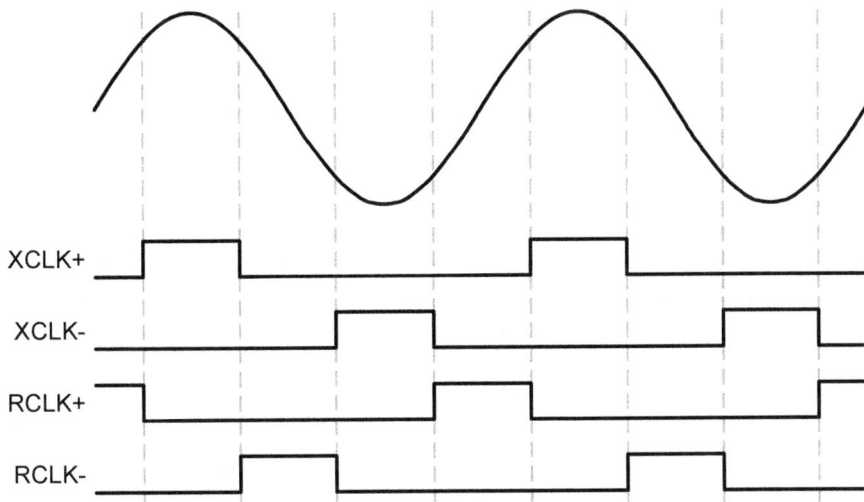

Fig. 16-45: **Tayloe Timing**

duty cycle and the X & R clocks are skewed by 90°, such as the I_{CLK} and Q_{CLK} in Figure 16-4. Tayloe demodulation narrows those pulses to a 25% duty cycle as seen in Figure 16-45.

The results are equivalent to the normal 50% timing but may offer some minor benefits in phase sensitivity or SNR. For example, notice that the X clocks only sample the peaks of the signal which are low-slew, and that the R clocks only sample the high-slew portions of the signals. This means that, for small phase variations, the X channel may have less sensitivity and the R channel may have more sensitivity. This might prove beneficial in some designs.

16.20: Band-pass Filtering — Part 3

In Part 2 we saw that the addition of a parallel capacitor to the RX coil is often used to bandwidth limit EMI noise. It also has an effect on the preamp's opamp noise. The opamp input current noise is converted to a voltage noise by the RX coil's impedance. Suppose the RX coil is parallel-resonated at the same frequency as the TX, and suppose there is also a parallel 10kΩ damping resistor. At resonance, the parallel LC components will have a very high impedance so the opamp input pin will see only the 10k resistor. This means the input current noise is multiplied by the resistor value and converted to a voltage noise. If, for example, the i_n spec is $1pA/\sqrt{Hz}$ then the resulting voltage noise will be $10nV/\sqrt{Hz}$. This may be higher than the opamp's voltage noise spec e_n.

Fortunately, parallel resonance results in a fairly narrow bandwidth and outside of resonance the impedance seen by the opamp input is low. So the additional integrated noise from this configuration may be minimal, but it's something to watch for.

It is also possible to series-resonate the RX coil and this is often seen in designs from David Johnson. In this case, at exactly the resonance frequency the series L and C become a low impedance (regardless of the load resistor) and the opamp's input noise current is shunted and not a point of con-

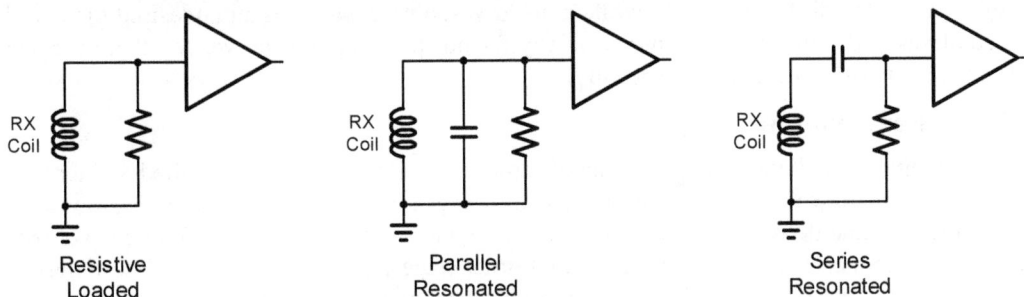

Fig. 16-46: **RX Coil Options**

cern. However, away from resonance the impedance seen by the noise current approaches the load resistor and can be an issue. Also, with a series resonated RX coil the Q of the coil is considerably reduced by the load resistor, although it is also largely under the control of the designer. This can be a benefit in that the phase slope is greatly reduced and may allow an RX resonance closer to the TX resonance.

16.21: G Channel Demod

Earlier in the chapter we considered a way to create a ground-free "G" signal by combining a small amount of X signal with the R signal. Some metal detector designs create a G signal by adding a completely separate demodulator channel that is phase-rotated to be nulled to ground. This means there are now three demod channels: X and R for the usual target signals, and G which is ground balanced.

Figure 16-47 shows the resulting block diagram and timing using the earlier example of a 5° loss angle for ground. It is now the G channel which is completely free of ground and, since both ferrous and non-ferrous targets create a positive response, is used to indicate the presence of a target. If the G channel indicates a target then the X and R channels can be used to identify or discriminate the target.

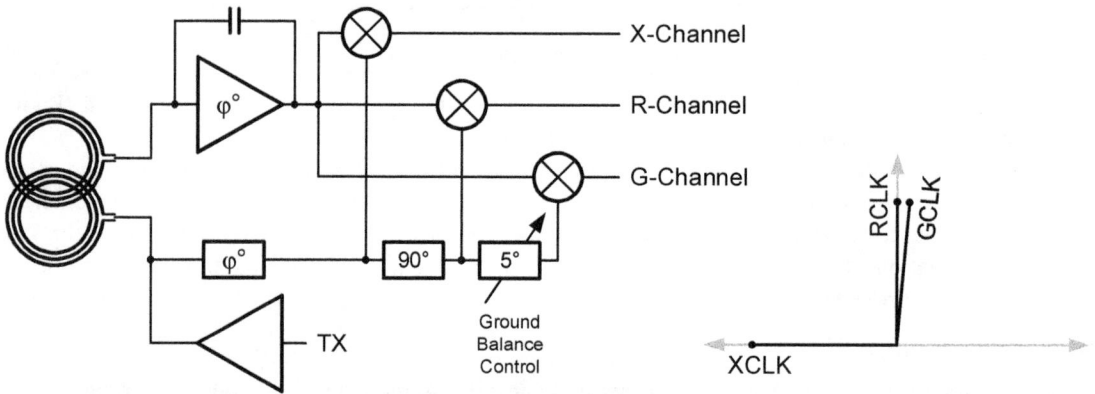

Fig. 16-47: **G Channel**

16.22: I (X) Signal Polarity

This chapter introduces the concept of quadrature demodulation whereby the "in-phase" (I, later called X) signal is derived by sampling the signal with an I-clock that is actually 180° out-of-phase with the TX voltage. See Figure 16-4. This produces a ferrite I signal with negative polarity. Technically, this is wrong. In quadrature demodulation the I-clock should be in-phase with the TX and therefore ferrite should produce a positive I signal, and sometimes you will see this convention used in patents and other publications.

If this convention had been followed then ferrous targets would have a positive X response and non-ferrous targets would have a negative X response. The reason for inverting the I-clock is to produce the opposite: negative for ferrous and positive for non-ferrous. The demodulation can be done either way and the subsequent circuitry designed as needed, but it's easier to think of a positive-going X signal as being a desirable non-ferrous target while a negative-going X signal is an undesired piece of iron. This also aligns with some VDI schemes whereby ferrous targets are negative, and there are no VDI schemes where non-ferrous targets are negative.

16.23: 4-Filter Designs

It was mentioned that the first generation of motion discriminators were so-called 4-filter designs. This means that there were four sequential derivative stages. Figure 16-48 shows the response waveforms for the raw, 2nd derivative, and 4th derivative signals as measured from a White's *XL Pro*. The derivatives are produced by two sequential second-order stages, therefore the first and third derivatives are not directly measurable. In a 4-filter design, each successive signal is the derivative of the previous

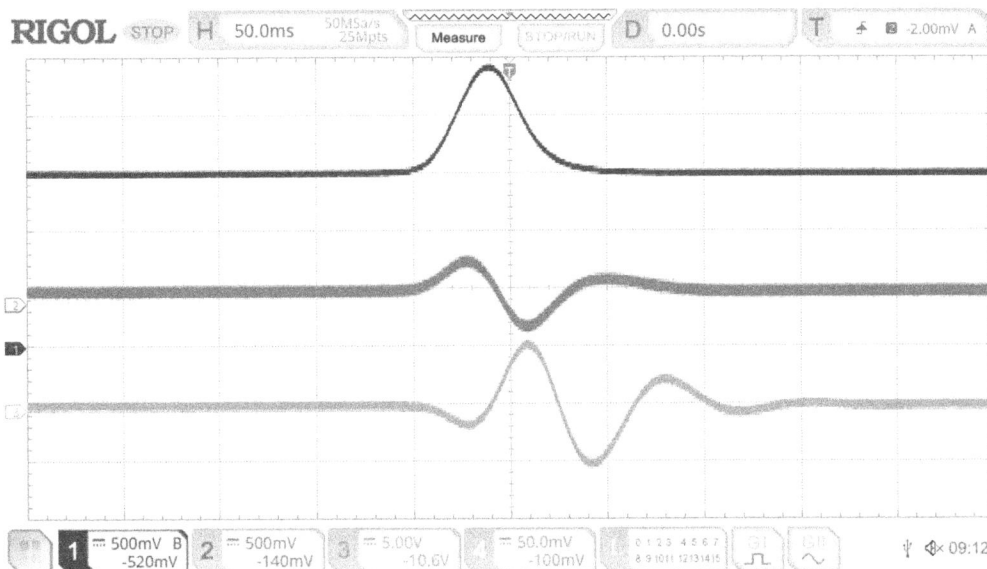

Fig. 16-48: 4-Filter Response Waveforms (Raw, 2nd, 4th)

signal, meaning that it is a measure of the rate-of-change. As such, each successive signal has a lower amplitude than the previous signal, and more severely so when the input signal is slow to start with. In the earliest 4-filter designs (starting with the Bounty Hunter *Red Baron*) the coil had to be whipped with a fast motion in order to keep the output of the fourth filter at a strong enough level to get decent depth. A normal sweep rate (by today's standard) would severely reduce depth. Later 4-filter designs improved on this, but the 2-filter designs pretty much eliminated the problem.

The reason motion VLFs began as 4-filter designs was because the engineers at the time felt that it was the only way to sufficiently suppress the ground signal. But using four filters has another drawback: the final signal tends to ring with a long tail which can create extended target responses. In some cases circuitry (like a clamp in the Compass Vari-Filter design) was added in an attempt to suppress this and other bad behavior. Again, the move to the 2-filter topology created much better behaved signals.

16.24: Demod Switch Selection

The design in this chapter uses a "4053"-style SPDT analog switch for the demodulators. This and the "4066"-style SPST analog switch are commonly found in commercial designs, both VLF and PI. Both types of switches can be found in the CD4000 and 7400 families of logic chips. It is important to choose the appropriate chip for the design, which depends both on the voltage swing of the analog signal and the high/low logic levels of the clock.

The CD4053, for example, is rated up to 20V (±10V) for the analog supply[23] so it will work for the *Motion VLF* design, which is ±5V. At this supply voltage the required high logic level for the clock is 3.5V which will also work since our clock is also rail-to-rail (10V). However, if you were to drive the demod clocks from a 3V micro the high logic level might be insufficient.

In the 7400 family there are numerous "technologies" to choose from, but for practical purposes we will only consider the 74HC and 74HCT families. If you wanted to drive the demods from a 3V micro then the 74HCT4053 has TTL-compatible clock inputs which work well with 3V logic regardless of the analog supplies. However, the HCT family is sometimes limited to 5.5V max[24]. The 74HC4053 is rated to a higher analog voltage (10V) but, like the CD4053, requires a 3.5V high logic level on the clock.

Later in the book, the PI designs use mostly 4066-type switches and the same considerations apply. The 4053 chip has connections for VCC and VEE (analog supplies) plus a "VSS" connection for the

---

23. This can depend on the manufacturer.

24. Ditto on the above footnote.

clock ground. That is, the clock logic has a reference independent of the analog supplies. The 4066 only has VCC and GND, and the clock logic is referred to GND. In many designs, GND is actually analog VEE so depending on the clock source this may or may not work. If you want complete control over the clock reference (like with the 4053), then instead of the 4066 use a 4316, such as the 74HCT4316 which has pins for VCC, VEE, and GND, plus a master enable pin.

VLF Discrimination

"I'm not in favor of any discrimination of any form."

— *(US Senator) Rand Paul*

In the previous chapter a motion VLF design was presented that had rudimentary discrimination. The X signal was used to blank the audio when a target is ferrous. This chapter will expand on that concept to implement a variable discrimination that can span the entire range of target phase responses.

17.1: Demod Rotation

Once we have motion-filtered signals we need to process them for potential target information and to generate responses. It may seem obvious from the plot in Figure 16-9a that we can simply digitize the X and R signals and write some code that determines target phase from the voltage levels. That is largely true and represents the approach used by practically all modern designs, but it eliminates the fun of doing it in analog circuitry which is what engineers had to do in the days before micros[1].

In the VLF-GB chapter we saw that a single demodulator can be ground balanced and used to indicate any detected metal, whether ferrous or non-ferrous; we call that the R channel. And in the TR-Disc chapter we saw that a single demodulator can be phase-adjusted to distinguish between accepted and rejected targets. When the disc phase is set to 90° that corresponds to the X channel as defined in the last chapter. It stands to reason that we could implement discrimination in a VLF-Disc design exactly as it is done in a TR-Disc design: by adjusting the phase of the X channel[2].

Figure 17-1 illustrates what this looks like vectorially. Again, using an example ground phase of 5°, the balanced R axis is labeled R_{GB} and results in a ground balanced R signal. The X axis (labeled X_{Disc}) can be rotated by adjusting the phase of the X demod clock to create a variable discrimination. Since the discrimination phase is always 90° from the clock phase (see Chapter 14) it is useful to add an imaginary R_{Disc} axis to visually see where the discrimination break point will be. In Figure 17-1 the X_{Disc}

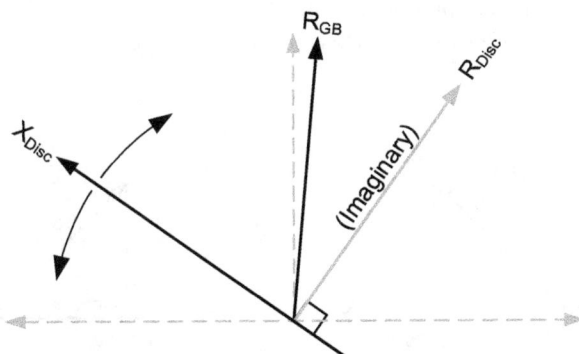

Fig. 17-1: **Discrimination via X Phase Adjust**

1. The circuits in this book begin with 1881 technology but then skip an entire 50 years of vacuum tube design. However, no one but brain-damaged audiophiles still design vacuum tube circuits, yet there are lots of fine people who appreciate an all-analog solution to a problem, or don't know how to code a micro.

2. This is the method used in the *Raptor* project in *ITMD-1|2*.

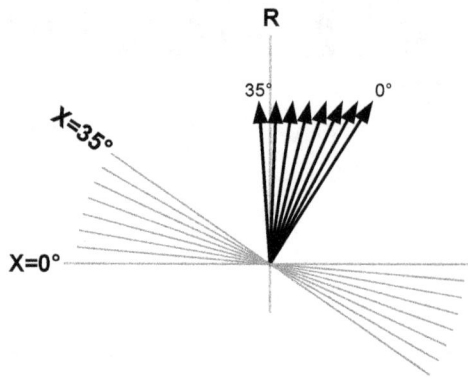

Fig. 17-2: **Discrimination by X-Phase Rotation**

phase is 35° which places the discrimination phase (R_{Disc}) at 125°. This is sufficient to discriminate foil and nickels.

Many early VLF-Disc designs implemented discrimination in exactly this manner. Recall, though, that the TR-Disc method had a major drawback in that targets progressively closer to the discrimination point were progressively weakened, resulting in a poor reputation for lackluster performance in Disc mode. This effect was not quite as bad in VLF-Disc models which used a fixed (ground-balanced) R demodulator as the all-metal detection signal. Figure 17-2 shows how rotating the X channel from 0° to 35° eventually moves the target phase (a US nickel) to a negative X value[3]. Because the R channel is fixed for ground balance (in this example at 0°) the target's R response remains constant and the overall vector amplitude only reduces slightly. However, it is easy to imagine that a high conductor with a normally large X-value and small R-value could have far more attenuation as discrimination is increased. In fact, adjusting the X phase by 35° to reject the nickel reduces the sensitivity to a US quarter by 25%.

17.2: Ratiometric Discrimination

A better way of setting the discrimination came about in the early 1980s. Suppose we have a VLF-Disc with matched X & R channels, with the X channel clocked at 0° and the R channel clocked at 90°. The outputs of the two channels are voltages that represent the target vector. For example, if the X channel is at −1V and the R channel is at 0V then that would be a 0° signal which is ideal ground. If X = 0.707V and R = 0.707V then that would be a target at 135° which might be a pull tab. Figure 17-3a plots these two examples on a polar plot. Notice that both vectors have an overall amplitude of 1V. This is a good way to normalize the vectors for discussion and we'll see later that it doesn't matter what their

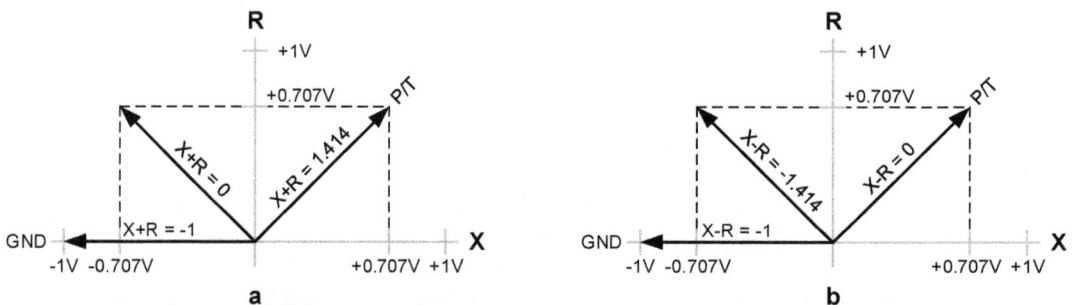

Fig. 17-3: **Discrimination Using X+R and X-R**

3. Note that as the X-axis rotates clockwise the target vector rotates counter-clockwise. This is a confusing concept best illustrated with animated graphics, which this book does not support. See instead *The Hogwarts Guide to Metal Detector Design*.

amplitudes are[4]. And even though the X & R components are real voltages, the vectors are really a mathematical construct and for this we can drop the "volts."

Suppose we add the two channel voltages and call our new signal the "D" signal: D = X+R. For our ferrite target the sum is D = −1. For the pull tab example the sum is D = 1.414. Thus ferrite has a negative sum and the pull tab has a positive sum. So where does the sum cross zero? X+R = 0 for a 45° target where X = −0.707 and R = +0.707. A 45° target is somewhere in the iron region. So if we used X+R for a discrimination threshold then we would reject a few iron targets and accept everything else.

What if we subtract the channels so that D = X−R? Now the zero crossing occurs when X = +0.707 and R = +0.707, which is right where our hypothetical pull tab is located (Figure 17-3b). So with X−R as the discrimination threshold everything below the pull tab is rejected and everything above is accepted.

Now suppose D = X+kR, where k is between −1 and 1. We've seen the two extremes: 45° when k = +1 and 135° when k = −1. When k = 0 then the X channel alone determines the discrimination point and the zero crossing occurs at 90°. What this means is that for a fully variable k we can adjust the summation break point anywhere between 45° and 135° and use this to control discrimination. What if we want more than 135° on the high side, for example to knock out those pesky Zincolns at 155°? Or less than 45° on the low side to accept more of the iron range? Simply expand or reduce the range of k on the minimum or maximum side as needed. k can be calculated as follows:

$$D = X + kR = 0$$

$$X = -kR$$

$$\frac{X}{R} = -k$$

And, trigonometrically[5],

$$\frac{R}{X} = -\tan\theta$$

So

$$k = \frac{1}{\tan\theta} \qquad\qquad\qquad \text{Eq 17-1}$$

For example, to achieve a range maximum of 155° the required k = −2.14. And for a range minimum of, say, 30° the required k = +1.73. So an adjustable k from +1.73 to −2.14 will yield a discrimination range from 30° to 155°.

Now that we know the math we can implement a practical solution. In the early all-analog designs this was done by adding a potentiometer between the X and R channels to create an adjustable trigger threshold for discrimination. If this all seems eerily familiar, it's because we did the same thing for creating the G signal in the last chapter; see Figure 16-26. So how do we add a potentiometer between X and R to create a G signal, and also add a potentiometer between X and R to create a D signal? In the case of the G signal, we needed G = kX + R = 0, where k is usually small, perhaps no more than 0.2. In the case of discrimination we need D = X ± kR, and k can be 2 or more.

Figure 17-4 shows how this is done with a simple potentiometer and an offset resistor (compare with Figure 16-26). For the values shown,

$$D = \frac{1.5 - m}{1.5} \cdot X - \frac{m}{1.5} \cdot R \qquad 0 \le m \le 1 \qquad\qquad \text{Eq 17-2}$$

with the derivation identical to that of Equation 16-4 and m representing the relative position of the pot. Notice that we need an inverted R channel, and the resulting discrimination range is only 90° up to 153°, which is just below foil and up through most typical aluminum trash. Assuming 10kHz operation,

---

4. Amplitude is an indication of target depth, which we're not concerned with right now.

5. The unexpected negative sign is (again) because our polar coordinate system for detectors is backwards from normal trigonometry.

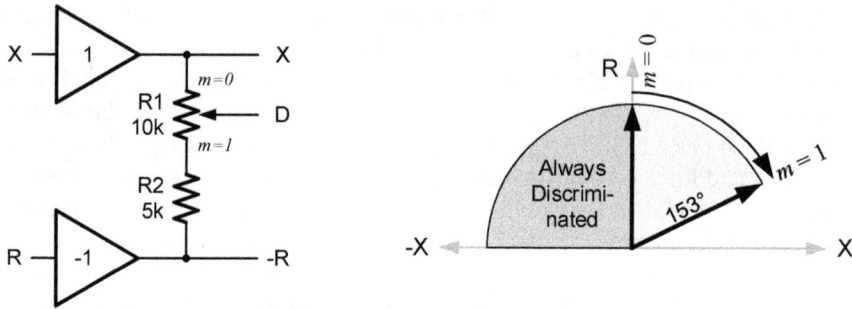

Fig. 17-4: **Simple Discrimination Control**

as usual. R2 can be changed to adjust the upper limit of the discrimination; if R2 = 0 then discrimination can extend all the way to 180°. (But why?)

A simple pot and offset resistor cannot lower the discrimination break point below 90°. This is evident in many analog detectors where the disc mode at its lowest setting cuts out everything in the ferrous range. If we want more discrimination range at the low end then we need more circuitry. Figure 17-5 shows an example of how this can be done. An additional gain of 2 is applied to the R channel and, with an inversion stage, a potentiometer can select between +2R and −2R. The subtraction opamp performs the operation 2R−X so the overall output function is $v_o = X \pm kR$. In this case the total range is ±4 which produces a discrimination range of 14° to 166°. If a different range is desired then the R channel gains can be adjusted or fixed series resistors can be added to either side of the potentiometer. The final design offers such an example.

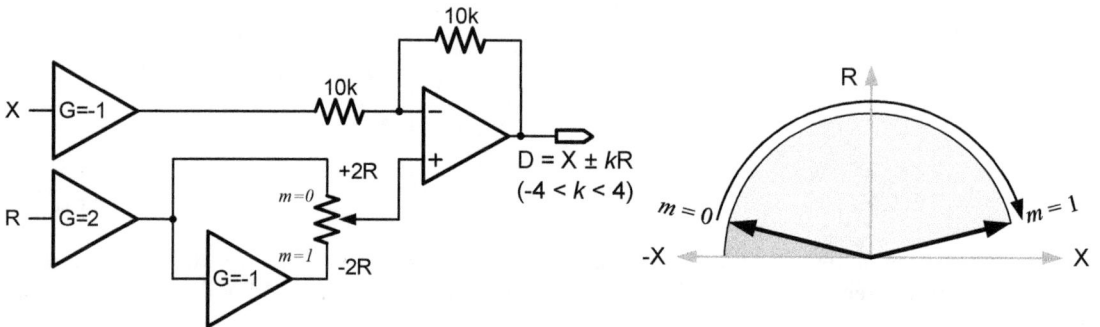

Fig. 17-5: **Expanded Discrimination Circuit**

Like the G signal, the D signal can be derived from the raw signal, the first derivative, or the second derivative. The raw signal is a poor choice because the X channel will potentially have a large and variable ground signal which will corrupt the D signal. The first derivative stage removes a lot of the X channel ground signal, and the second derivative removes even more. For this reason, the discrimination signal is usually taken from the second derivative stage.

17.3: Applying Discrimination

Now that we know how to create a discrimination signal we need to apply it somehow. In the *TR-Disc* design, the demod output drives the audio but can be phase-adjusted so that a negative response is produced for a target that is discriminated out, and no audio is generated. In the VLF-Disc case, we might use the ground-free G channel but it has a fixed phase (at least relative to targets) and produces a positive signal no matter the discrimination level setting, so it will produce a beep for all targets.

However, there is a new signal being generated that is negative for discriminated targets and positive for non-discriminated targets. This signal could directly drive the audio but it has a familiar problem that was seen in the TR-Disc chapter. As targets approach the discrimination break point the amplitude of this signal drops off as shown in the typical plot[6] of Figure 17-6. Therefore, using this sig-

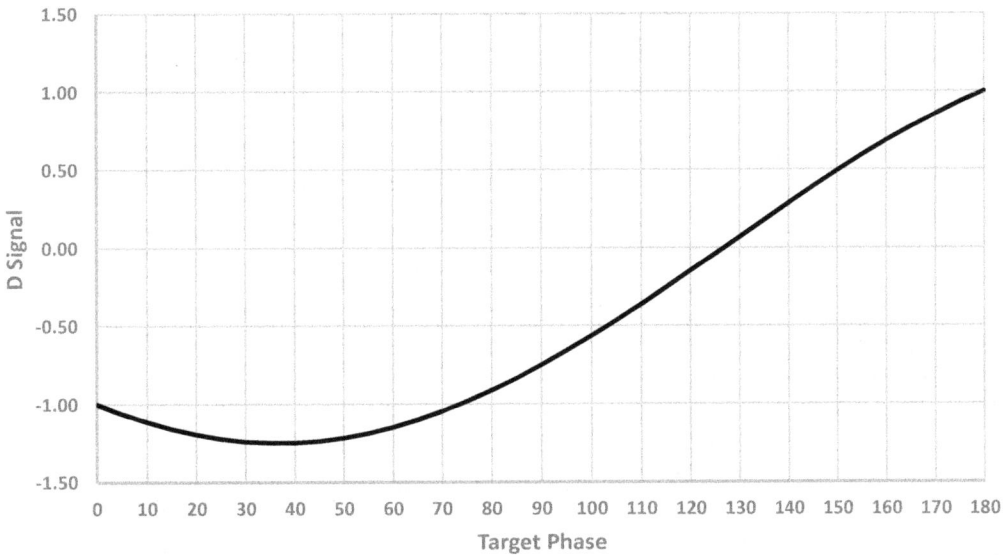

Fig. 17-6: **Discrimination Signal Response**

nal to directly drive the audio results in progressively weaker responses for targets that are progressively closer to the break point.

One way around this problem is to apply the D signal to a comparator to get a strong response no matter the target's proximity to the discrimination break point. Many detector designs do just that, and the result is a strong beep on any non-discriminated target, no matter the size or depth of the target, up to the limit of detection.

Another approach is to use the G signal for the audio and use the polarity of the D signal to gate the audio response as shown in Figure 17-7. A tone generator creates a fixed audio tone that chops the output of the G channel's all-metal signal. However, a comparator from the D signal can override the tone signal (assuming the comparators have open-collector outputs) and disable the chopper switch when the detected target is discriminated. This approach is also commonly found in commercial designs. The advantage is that the strength of the audio response of a target will be proportional to the strength of the received signal. That is, a deep target will sound deep, and a shallow target will sound shallow.

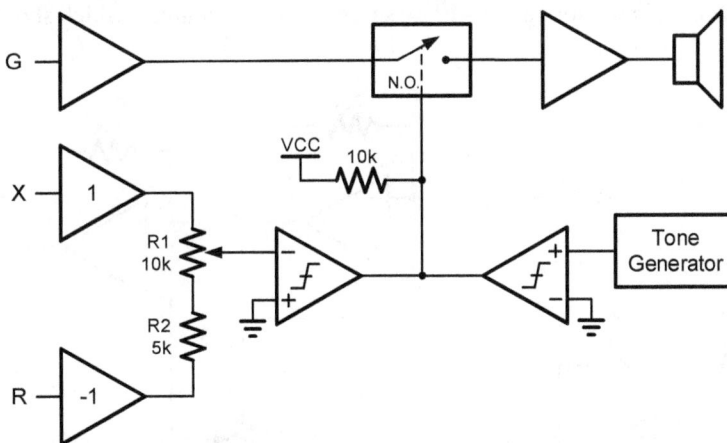

Fig. 17-7: **Discrimination Gating Circuit**

6. Figure 17-6 is specifically for the circuit in Figure 17-4 with the pot at mid position ($m = 0.5$). Note the zero crossing is at 116°, which is just below a US nickel. Other curves can be created for other disc circuits and at other pot settings, but they will all look similar to this one,

Design: *VLF Disc*

This design will build upon the *Motion-VLF* design presented in the last chapter. We will continue with an all-analog design; digital design considerations will be presented in the next chapter. It is again useful to draw a block diagram of the design to see how everything fits together; see Figure 17-8. The G channel will be derived from the raw demodulator outputs, and the Disc signal from the second derivative outputs. Notice that the G channel also includes two differentiator stages; the reason for this is so that the time latency for the G signal matches that of the D signal.

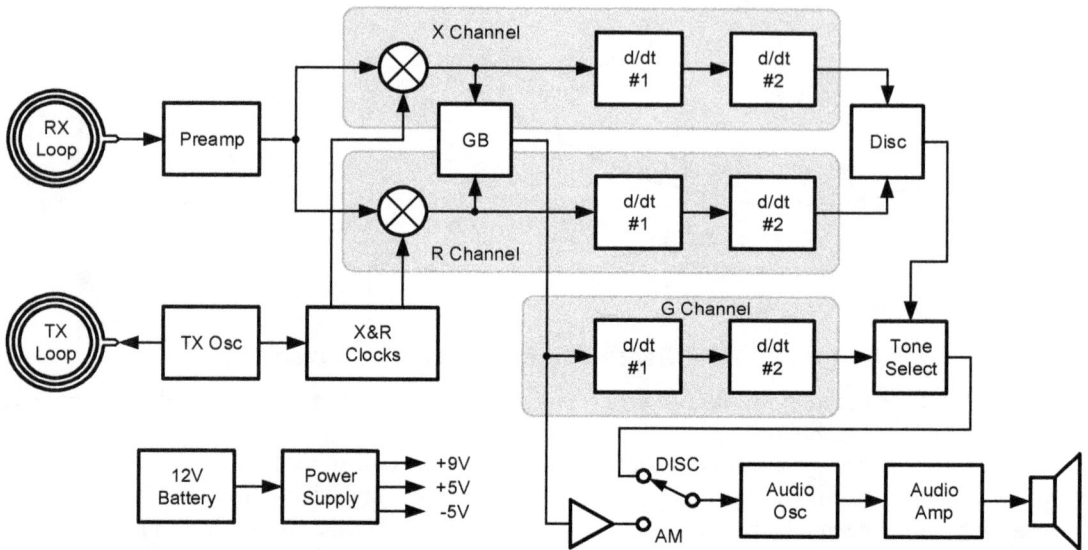

Fig. 17-8: **VLF-Disc Block Diagram**

A new provision that is added is the ability to switch between Disc and All Metal (AM) modes. The AM mode should preferably be a static non-motion mode for pinpointing so it uses the raw G signal which requires additional gain.

17.4: Discrimination Circuitry

The discrimination circuitry is conceptually the same as in Figure 17-5 and is shown in its entirety in Figure 17-9. The discrimination range will be set to 40° as a minimum which should pass most iron,

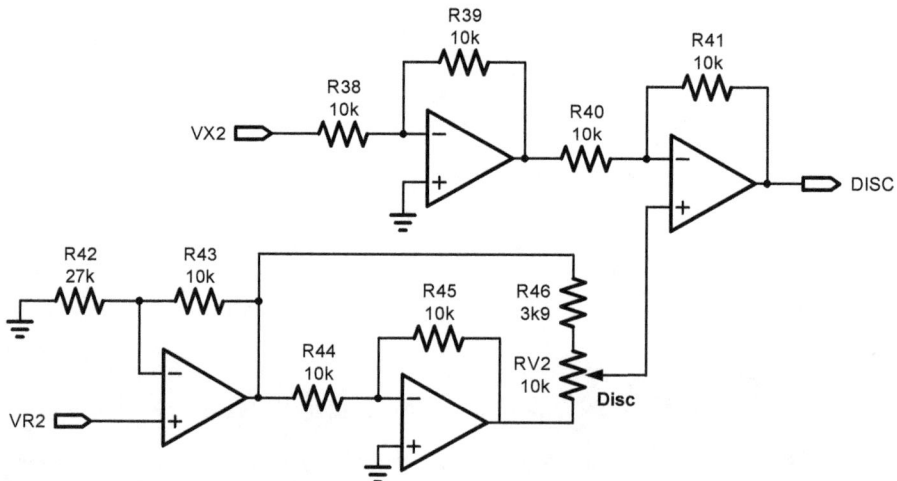

Fig. 17-9: **Discrimination Circuit**

and 160° as a maximum which goes right up to high conductive coins. The output of this circuit is a positive voltage for accepted targets and a negative voltage for rejected targets.

17.5: All-Metal and Mode Selection

Discrimination mode uses the second derivative motion signals to ensure fairly ground-free signals in the discrimination circuitry. This design adds a selectable all-metal mode which also can be a motion mode, but for pinpointing a non-motion operation is much preferred. So for the all-metal mode we will use the static G signal.

When pinpointing a target it is important that the all-metal mode have no less sensitivity than the discrimination mode. That is, you don't want to detect a target in discrimination mode, switch to all-metal mode to pinpoint, and have the target disappear. Since all-metal uses the raw G signal and bypasses the derivative gain stages, we need to add some more gain to the all-metal path. The problem is that any offset in the G signal can quickly cause the all-metal circuitry to saturate. To solve this, we turn again to the retune circuit that's been used a few times before. Figure 17-10 shows the all-metal gain stage and retune circuit.

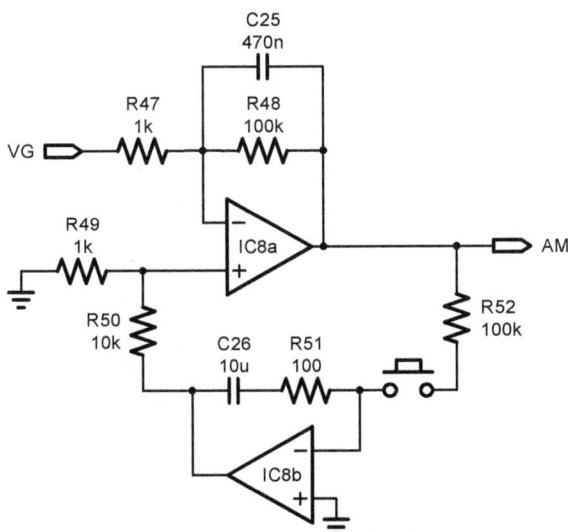

Fig. 17-10: **All-Metal Gain Stage**

Notice that the gain stage uses an inverting opamp (IC8a) which might seem incorrect. We want the all-metal signal to have the same polarity as the second derivative G signal so that the audio stage is driven in the same manner. The raw G signal passes through two inverting derivative stages to produce the second derivative G signal (G'') which suggests G'' has the same polarity as G, and therefore the all-metal gain stage should be non-inverting. However, recall that in Figure 16-21 the second derivative has a center lobe with a peak opposite in polarity from the raw signal. This means the second derivative has an extra polarity inversion, and therefore the all-metal path must also be inverted to match.

17.6: Audio Circuitry

The remainder of the design consists of the power supplies and audio. The power supplies will be lifted from prior designs and are included in Figure 17-11. The audio stage is an enhancement of the design in Figure 17-7 and is shown in Figure 17-14. Instead of disabling (nulling) the audio on discriminated targets, a low tone is produced. For the tone generator, the 10kHz XCLK signal is fed to a 74393 binary counter chip and divided by 16 to create a 625 Hz audio tone for non-discriminated targets (same as the *Motion VLF* design), and divided again by 4 to create a 78 Hz audio tone for discriminated targets. The selected tone then chops the target signal via the collector-coupled transistors Q4 and Q5. The final audio amp is the familiar LM386, preceded by a volume control.

Power Supply

Phase References

Transmitter

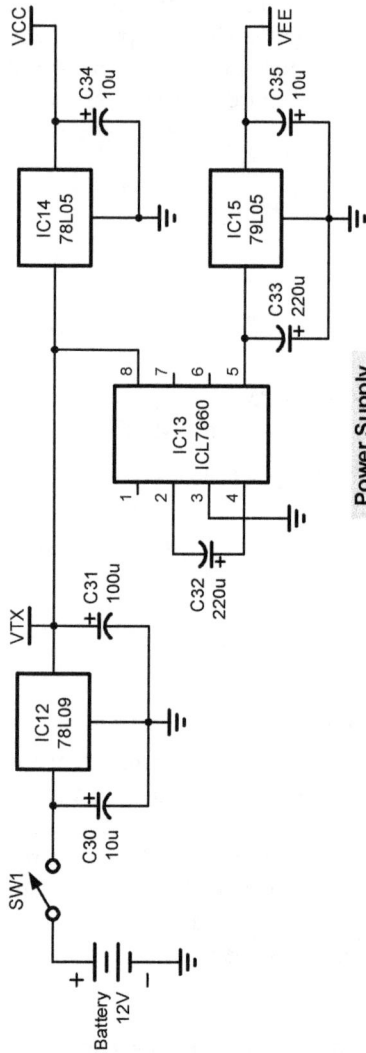

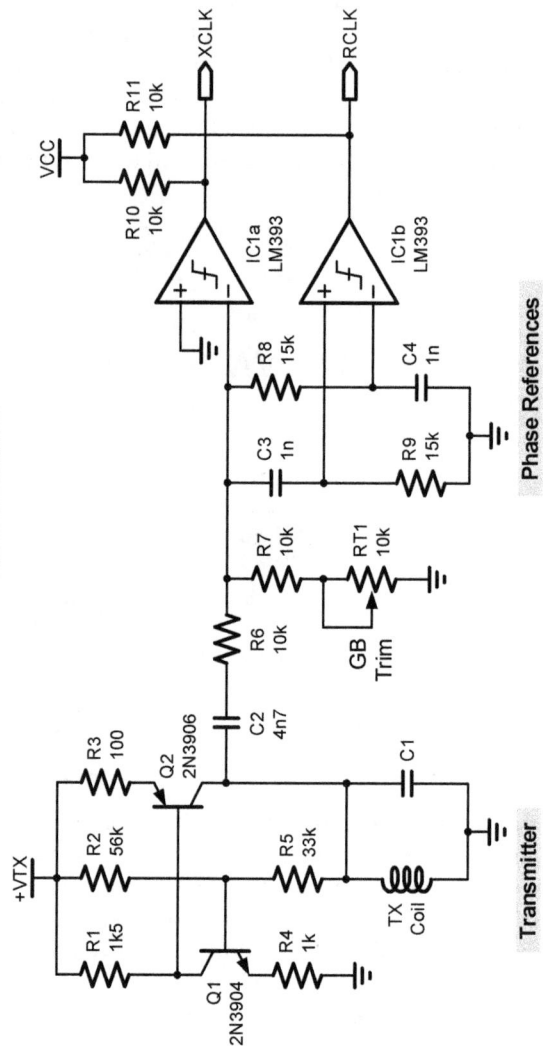

Fig. 17-11: **VLF-Disc — TX & Power Supply**

Inside the Metal Detector

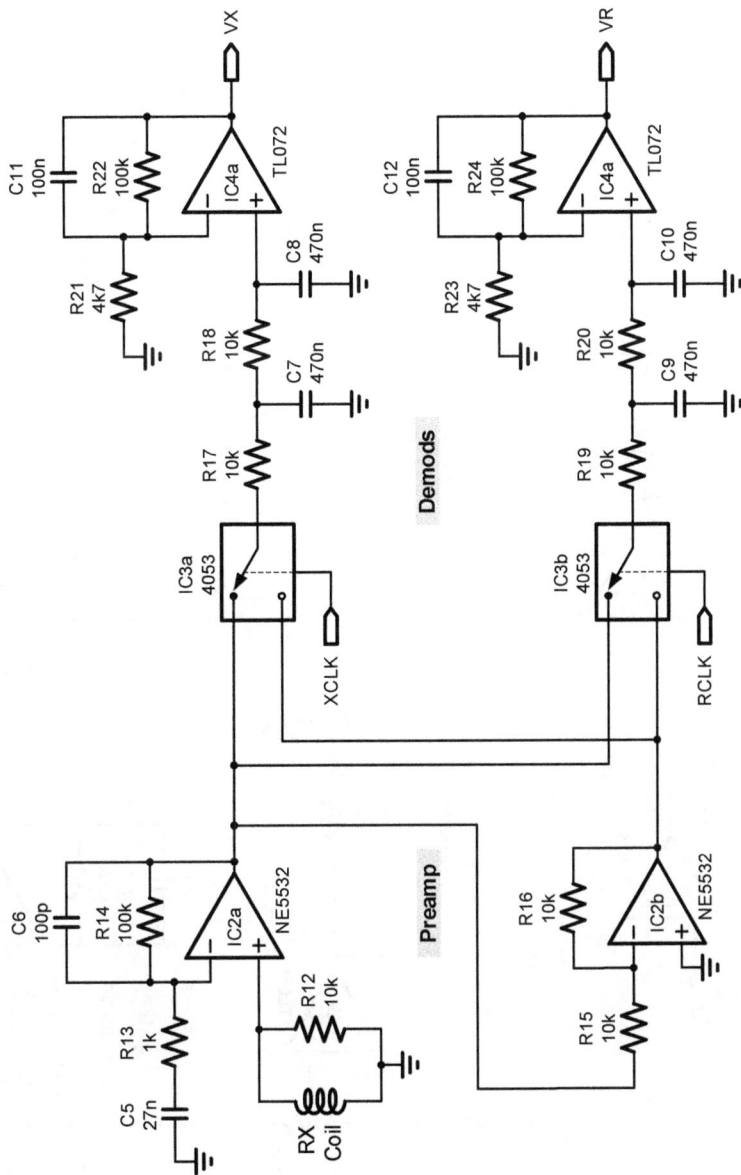

Fig. 17-12: **VLF-Disc — Preamp & Demods**

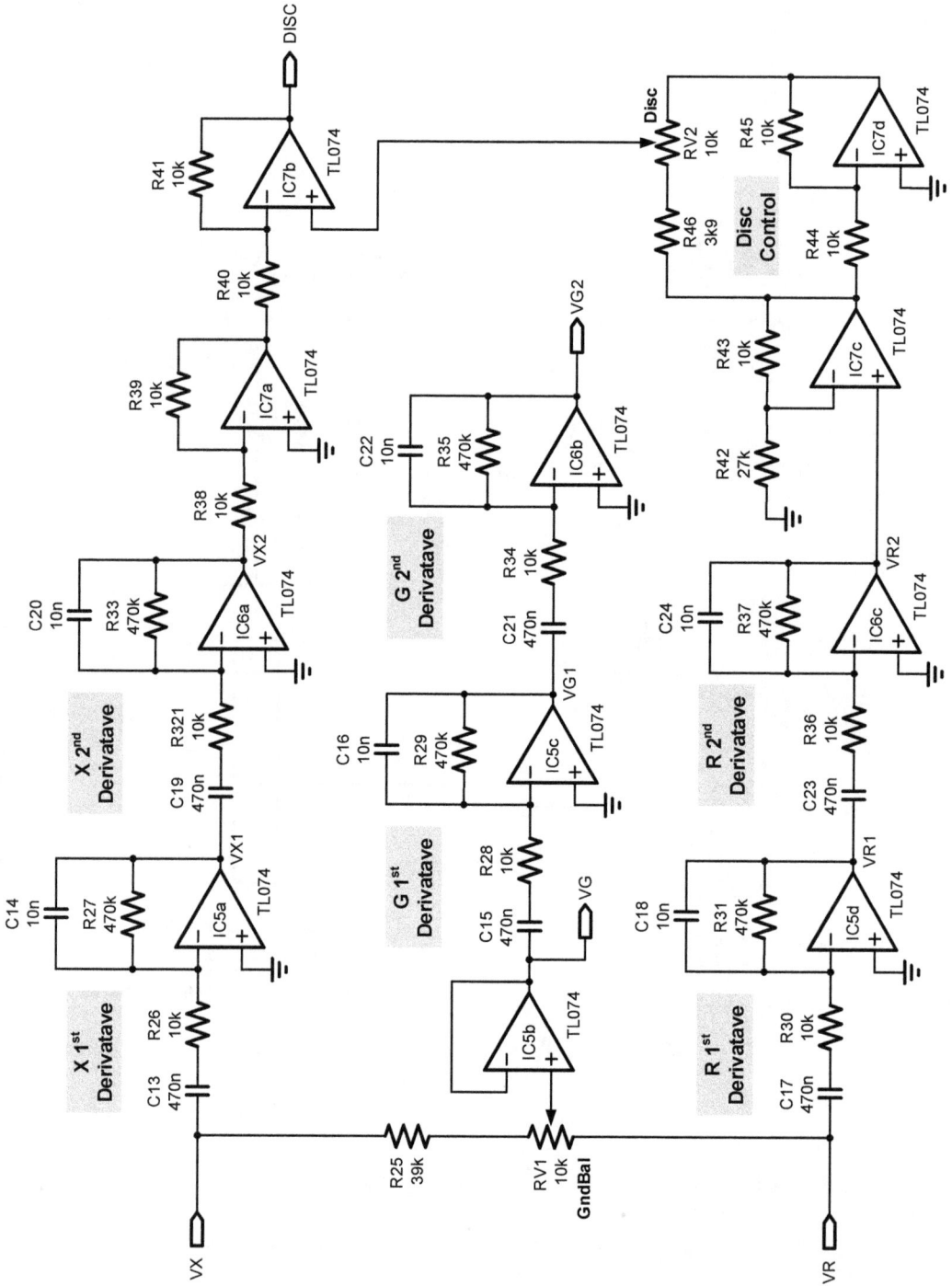

Fig. 17-13: **VLF-Disc — Filters & Disc**

Inside the Metal Detector

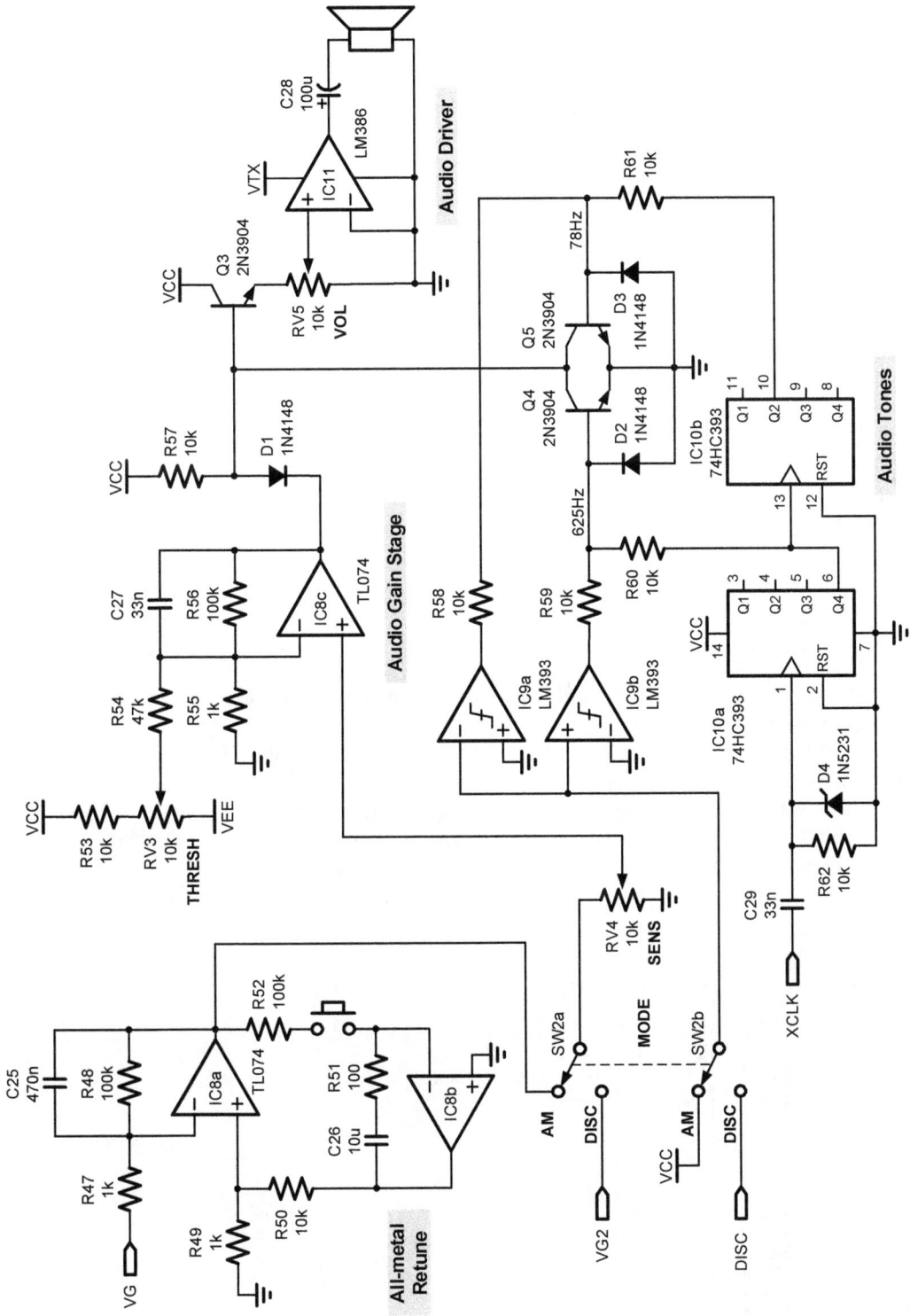

Fig. 17-14: VLF-Disc — Power & Audio

In Disc mode the target signal is taken from the second derivative G channel (VG2) whereas in AM mode the target signal comes from the raw non-motion G channel via an additional gain stage. In either case, the signal is applied to a potentiometer for adjusting the overall sensitivity and further amplified by IC8c, which includes a provision for injecting a variable offset current for a threshold adjustment.

17.7: Complete Schematic and Calibration

Figures 17-11 through 17-14 show the complete schematic for the *VLF-Disc* design. The only calibration needed is to set the default ground balance point:

1. Set the detector in AM mode.
2. Set potentiometer RV1 to about the 10 o'clock position.
3. Press and hold the retune button.
4. Adjust RT1 until the audio is nulled while bobbing a ferrite.

Calibrating with RV1 at 10 o'clock offers more adjustment range on the positive side than on the negative side. But ferrite should be pretty close to the negative limit so there is little need for setting the balance point lower, plus it offers some margin on the low side in case the ferrite used for calibration has a higher-than-expected loss angle.

In normal operation RV1 is adjusted for ground balance. This can be done in either AM or Disc mode. In AM mode the retune button needs to be held during ground balance, and you will listen for the leading response (too much GB) and the trailing response (not enough GB) while bobbing the coil over ground. When there is little or no response, then the GB setting is optimal. In Disc mode it is much the same, except that you will hear a leading low tone (too much GB) or a leading high tone (not enough GB).

Explore

17.8: Multi-Tone Audio

In modern micro-based designs, implementing multi-tone audio is almost trivial. But multi-tone audio began in the days of all-analog detectors. This chapter's *VLF-Disc* design already shows how to implement two-tone audio using analog methods, where the polarity of the Disc signal selects the tone. If more tones are desired, then we need more than just the one Disc signal.

Recall from Figure 17-4 that the Disc (D) signal can be extracted using a simple resistive divider from the X and R signals. We can extract more such signals by replacing the potentiometer with a string of resistors (Figure 17-15). Each resistor tap can be designed to produce a zero crossing at some particular phase response, thereby creating "zones" between successive zero crossings where different tones can be enabled.

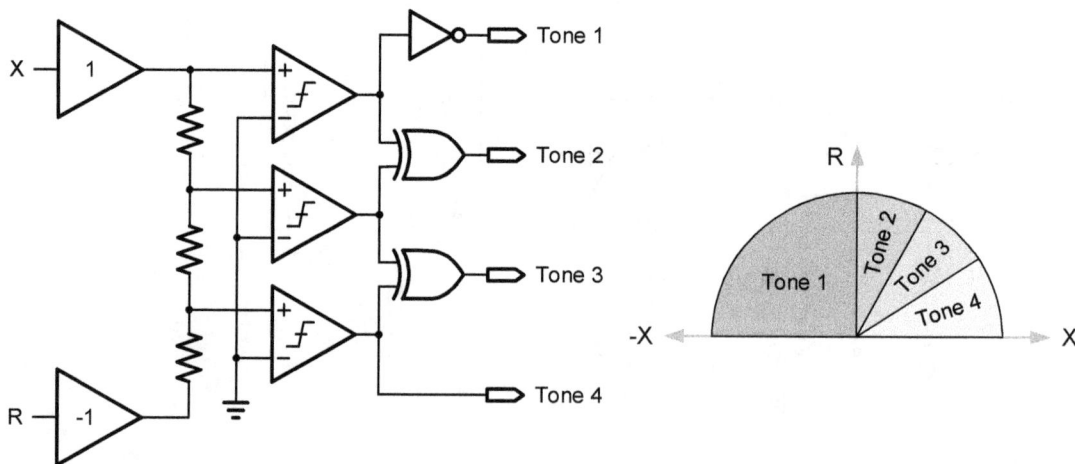

Fig. 17-15: **Multi-Tone Audio**

The simplest way to do this is to use a comparator at each tap to generate a "thermometer code" set of digital outputs, and then XOR each adjacent output to convert the thermometer code to a single active "elevator code." These are terms commonly used in flash ADC design, and the method shown in Figure 17-15 closely resembles a flash ADC[7]. The single active code is then used to select which tone to send to the audio.

In the example shown, four zones are created which can select four tones. This can be more finely divided to more zones if desired, but the analog circuitry required quickly makes a digital solution using a micro far more appealing. Finally, this "Tone ID" resistor divider can be run in parallel with the discrimination selection circuit of Figure 17-4. The discrimination signal could then override the tone selection by either blanking out the audio or generating a unique rejection tone.

17.9: Channel Gains

The discussions on motion filters have not addressed how much gain the channels should have. Both the X and R demods have the same gains, and both the X, R, and G channel differentiation stages were likewise designed with the same gains. Let's look at what those gains should be. This discussion will only consider the X and R channels, and assumes the R channel is ground balanced. If a design has a G channel, then substitute G for R below.

In general, we want the gains of the X and R channels to be as high as possible while avoiding signal saturation of the stages under certain conditions. Since the R channel is ground balanced, that condition is whatever near-target overload you're willing to tolerate. Let's say you want to barely reach R channel overload on a US quarter at a depth of 2 inches, so you set the overall R channel gain for this condition. A quarter that is less than 2 inches deep will cause overload, but you can lift the coil a bit to eliminate the overload. A US silver dollar, however, might still overload at several inches. So will other large targets but, again, just lift the coil to get a good reading.

The dominant X channel signal will likely be due to ground mineralization so we need to make sure the X channel does not overload in the strongest ground mineralization you might encounter. Obviously we could also lift the coil in the presence of ground overload. But while temporarily lifting the coil for a shallow target is not unreasonable, constantly keeping the coil a few inches off the ground to avoid overload is unreasonable. Since the ground signal is fairly static, it is the X channel demod gain we need to mostly worry about and it should be designed assuming the coil is in contact with the worst likely ground. After the demod, the derivative stages remove the static ground signal.

It is not necessary that the overall gains of the X and R channels be equal. In fact, because the R channel is devoid of a static ground signal, the gain of the R channel demod can be set higher than the X channel demod. It is common to see the R channel gain 4 to 8 times higher than the X channel. Since the R channel is also the target channel, a higher gain improves depth without the risk of ground overload. Gain variations are usually applied in the demod amplifier stages. The differentiation stages that follow are usually matched because of the importance of having well-matched signal delays. The filter stages are not only identical, but the Rs and Cs are often matched to 1% or so to ensure equal delays.

Finally, when the gains of the X and R channels are not equal, the math behind the discrimination circuitry (Equation 17-2) also changes. That is, a non-ferrous target at 135° (pulltab, maybe) will no longer have equal X and R demod voltages; the R voltage will be higher because of the higher gain. While the discrimination circuit design is still basically the same, calculating the resistances will need to account for the gain differences. The explanation in Section 17.2 should be sufficient to figure this out.

17.10: First Derivative Designs

Most VLF designs use two stages of differentiation. Some designs, notably those used in gold prospecting, may only use a single differentiator. These detectors are typically run in all-metal mode with no discrimination. Therefore, having the second differentiator to produce a target-centered discrimination response is unnecessary.

---

7. See patent US6172504, literally using the term "flash phase analysis."

The usual approach to a VLF prospecting design is to run it at a very high frequency (40-80kHz) for better sensitivity to small gold, and to use the rectified first derivative signal to drive a VCO audio response. The rectified first derivative produces a slower rising edge as the target is approached, and then abruptly dives to the zero crossing where it is truncated. See Figure 17-16. The resulting sound is commonly known as a "zip-zip" response.

Fig. 17-16: **Zip-Zip Response**

17.11: Non-Quadrature Sampling

So far this chapter and the last have focused on quadrature sampling, where the X channel is purely reactive and the R channel is purely resistive and are set 90° apart, with an additional G channel that is phase-rotated for ground balance (Figure 16-26). Occasionally you will run across designs where the X and R channels are not 90° apart. Suppose that the R channel is at 90° but the X channel is rotated by 45° as in Figure 17-17. What we saw in Figure 17-2 is that a clockwise rotation of the X channel causes a counter-clockwise rotation of non-ferrous target vectors, perhaps enough to move them to the left of the R axis. Once this happens, that non-ferrous target will have a negative X response, exactly like a ferrous target. Figure 17-17 shows how a US nickel projects a positive R value but a negative X value. It can be confusing to look at; just remember that a vector's projection lines always intersect the axes at 90°.

At first glance this seems to be something we'd rather not have. However, it's possible to use this effect to simplify a design. In Figure 16-26 we see that by adding X and R signals it is possible to discriminate a first-quadrant signal. In Figure 17-4, subtraction can be used to discriminate second quadrant signals. But it takes a complex circuit (Figure 17-5) to discriminate across both the first and second quadrants.

By rotating the X-axis we can effectively move second quadrant signals into the first quadrant and then use the simple additive method to discriminate across both ferrous and lower conductive non-ferrous targets. This method is used in some Tesoro designs.

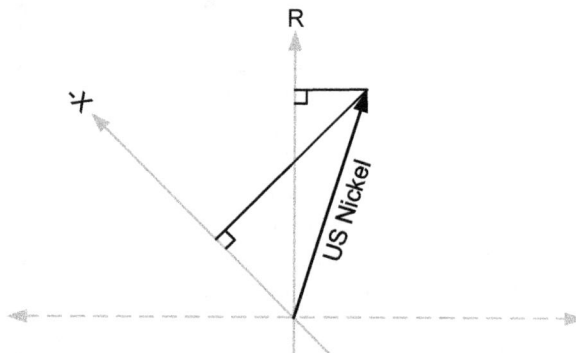

Fig. 17-17: **Non-Quadrature Sampling**

18

Digital Techniques

"People who are really serious about software should make their own hardware."

— Alan Kay

The previous chapters in this section have progressively developed the induction balance from (almost) the earliest TR method up to a more modern VLF design with ground balance and motion discrimination. All of these efforts have been realized in analog circuitry, yet virtually all contemporary commercial detectors are designed around a microprocessor[1] with most of the analysis done in the digital domain. This chapter will explore those techniques. Unfortunately, a detailed and thorough treatment of these methods would easily fill another book so this chapter will be at a higher level.

Digital Design

From the perspective of circuitry a digital metal detector is easier to design than an analog type. However, the complexities of what needs to be done (filtering, ground balance, discrimination, etc) are moved into microprocessor code. Many people are not programmers and this presents what appears to be an unclimbable wall. This book isn't thick enough to teach the basics of microcontrollers; there are many other books that do this task well. But if you're code-shy, perhaps this chapter will motivate you to grab some climbing gear and get over the wall.

The VLF-Disc project from the last chapter provides an excellent starting point for making a digital detector. All digital designs require an analog-to-digital converter (ADC) to convert analog signals into digital signals, a process called *digitizing*. There are two broad approaches to this shown in Figure 18-1.

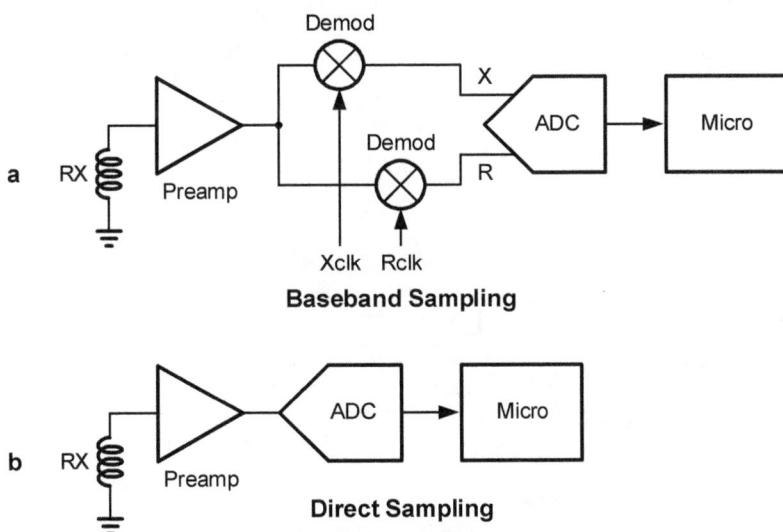

Fig. 18-1: **Baseband Sampling vs Direct Sampling**

---

1. In detectors we're almost always talking about a microcontroller but we'll be a little loose with terminology. In general, we'll just call it a *micro*. See also the footnote on page 185.

The first is *baseband sampling*, whereby the ADC is placed after the demodulators and digitizes the X and R baseband signals. The first digital detectors[2] used baseband sampling and it continues to be a popular method. The second approach is *direct sampling* whereby the ADC is placed after the preamp and digitizes the raw undemodulated signal. Demodulation is then done digitally. Some of the latest models[3] are using direct sampling. However, the ADC must oversample the raw signal and is preceded with less gain which places extreme demands on the ADC: it must be fast, and needs high precision.

18.1: Baseband Sampling Architecture

In baseband sampling, the first decision is where to sample the baseband signal. Modern motion discriminators are known as 2-filter designs because they use a double high-pass filter (or derivative) on the baseband signals to attenuate the slow ground signal. The filters can be implemented in analog circuitry or in digital code; it's possible to design the first filter in analog and the second in code, or both in analog, or both in code. See Figure 18-2. All three methods have been used in various commercial designs. The benefit in using analog filters is that it offers the chance to amplify the signals and correctly offset them for the ADC. But the purpose of this chapter is to cover digital techniques so we will assume both filter stages are in code as in Figure 18-2c.

A simple baseband analog front end (AFE) schematic is shown in Figure 18-3. The preamp and full-wave demods are borrowed from the *VLF-Disc* design in the prior chapter. The TX driver is an

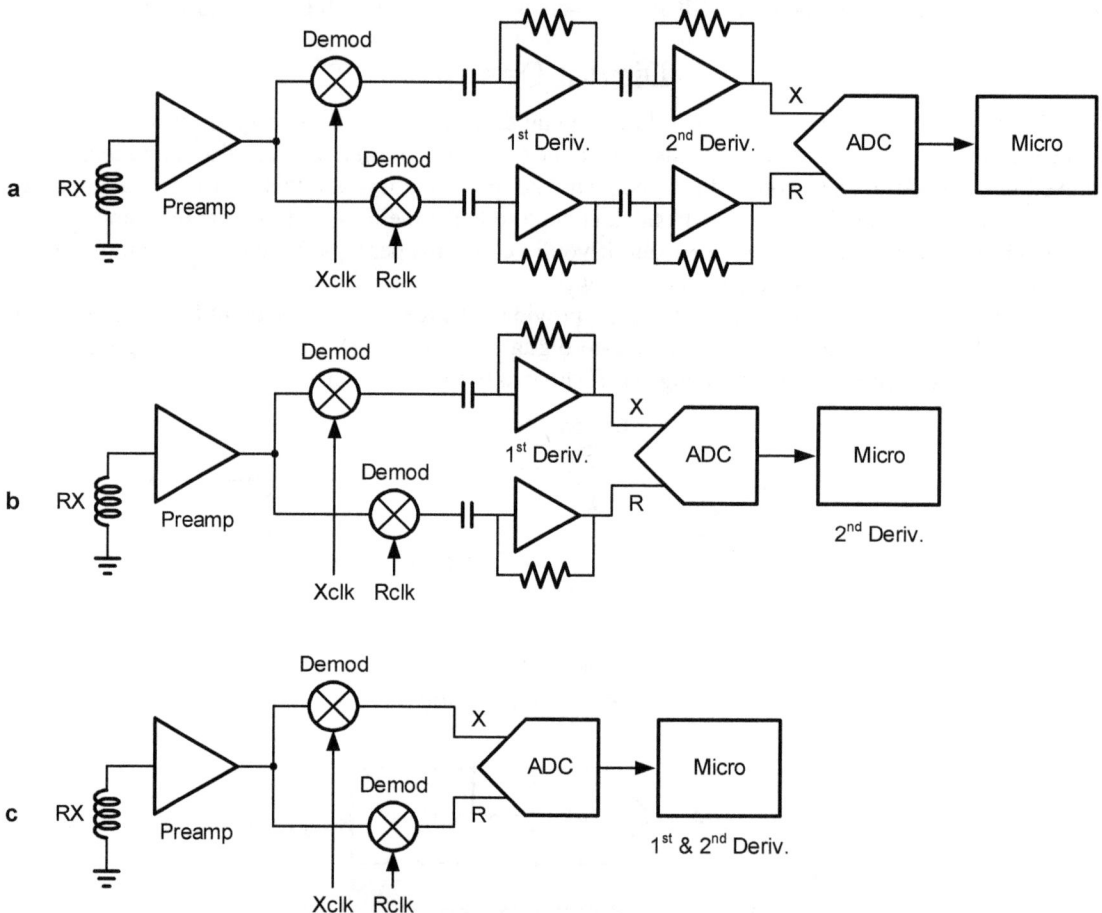

Fig. 18-2: **Baseband Sampling Options**

2. The White's *Eagle* was the first in 1987.

3. The Minelab *X-Terra* series was first in 2005; many of the latest models like the Minelab *Equinox*, XP *Deus 1 & 2*, Garrett *Ace-Apex*, and Nokta *Legend* are direct sampling.

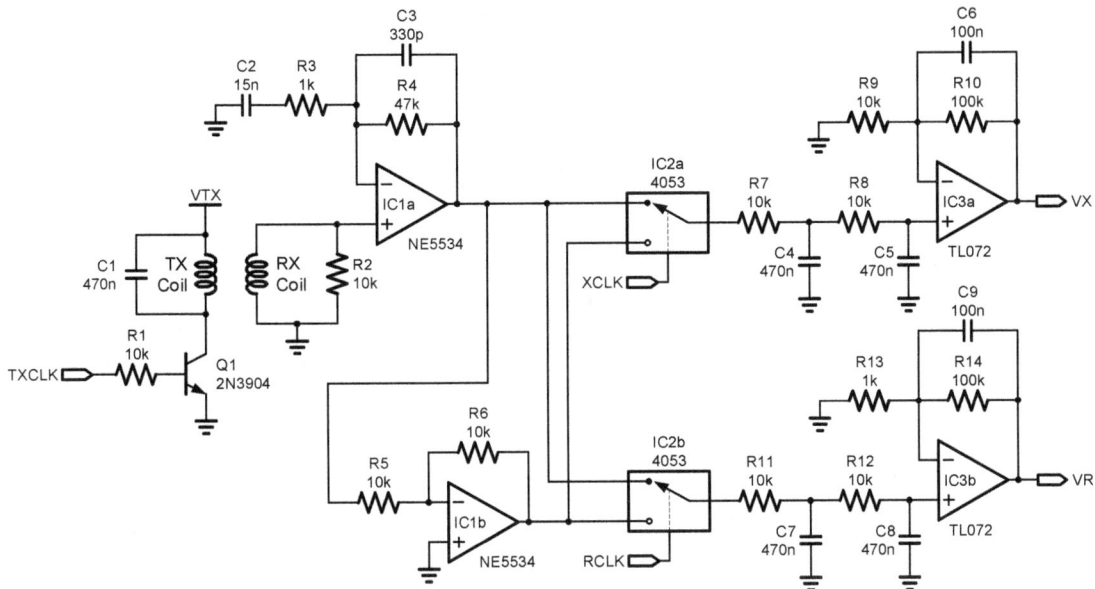

Fig. 18-3: **AFE Schematic**

NPN transistor clocked from the micro. There is no need for the discrete demod clock circuitry as shown in Figure 16-12; instead, the demod clocks can be generated by the micro. Add an ADC, a micro, power supplies, and an audio driver to Figure 18-3 and you have a complete modern VLF metal detector.

18.1.1: Choosing the ADC

The signals VX and VR are digitized by an (as yet-to-be-determined) ADC. The two primary considerations for the ADC are the number of bits and the sample rate. Number of bits is dictated by the desired SNR which will depend on the SNR of the AFE. Therefore it is important to determine the SNR of the AFE, maximize it for highest performance, and then select an ADC that does not degrade it.

In most detector designs the AFE SNR is dominated by the preamp, in particular the voltage noise of IC1a and the thermal noise of R3. The total noise at the output of IC1 is

$$v_n = \sqrt{4kTR_3 \cdot \left(\frac{R_4}{R_3}\right)^2 + 4kTR_4 + v_{nl}^2 \cdot \left(1 + \frac{R_4}{R_3}\right)^2} \qquad \text{Eq 18-1}$$

where v_{nl} is the input-referred voltage noise of IC1a. This comes to $256.1 nV/\sqrt{Hz}$ for the values shown. This noise is amplified further by the conversion gain of the demodulator (G=0.636), and the gain of the post-demod amplifier (G=101 for the R channel). Therefore, the total noise at VR is $16.5 \mu V/\sqrt{Hz}$. The noise bandwidth at this point is 25Hz so the integrated noise is 82.3μvrms.

The maximum signal level of VR is 5vpp (1.77vrms) so the maximum SNR at this point is 86.6dB. VR was used for calculating the SNR because it is the signal that triggers the all-metal response. Note that the final gain of the X channel is less than that of the R channel. The reason for this is that the X channel signal is where almost all of the ground signal resides, therefore it usually has a lower gain to avoid ground overload. The R channel, which is largely devoid of ground signal, can be run at a higher gain without the concern of ground overload. This is fortunate as the R channel signal determines the target sensitivity of the detector.

If the AFE has an output SNR of 86.6dB then this is also the SNR at the input of the ADC, and the ADC will degrade this SNR due to its own internal noise. In choosing an ADC we preferably don't want to degrade the SNR by more than, say, 1dB. This requires that the ADC be at least 6dB better than the SNR of the input signal. In other words, for this design we need an ADC with an SNR of 92.6dB, or about 15 effective number of bits (ENOBs). In the world of ADCs the ENOB rating is usually one or two bits less than the actual number of bits, meaning we will want to use at least a 16-bit ADC in order to achieve 15 ENOBs.

With baseband sampling the ADC must digitize two signals: VX and VR. Preferably this is done using a 2-channel ADC with simultaneous sampling so that the two signals are time-aligned. In this case, the ADC can be clocked at a certain sample rate and both digitized signals transferred to the micro simultaneously. If, for example, the ADC has 16 bits of resolution then both results can be transferred in a 32-bit word.

The signals can also be multiplexed into a single-channel ADC and time-aligned in software. In this case the ADC must run twice as fast but each transfer contains only one result. A multiplexer is also needed to alternately select the input signals. Because the X and R data samples do not represent the same points on the curve they must be aligned in software. This can be done, for example, by interpolating a new R value between every pair of R samples, where the interpolated R values are now time-aligned with the X samples. Figure 18-4 shows both of the described methods.

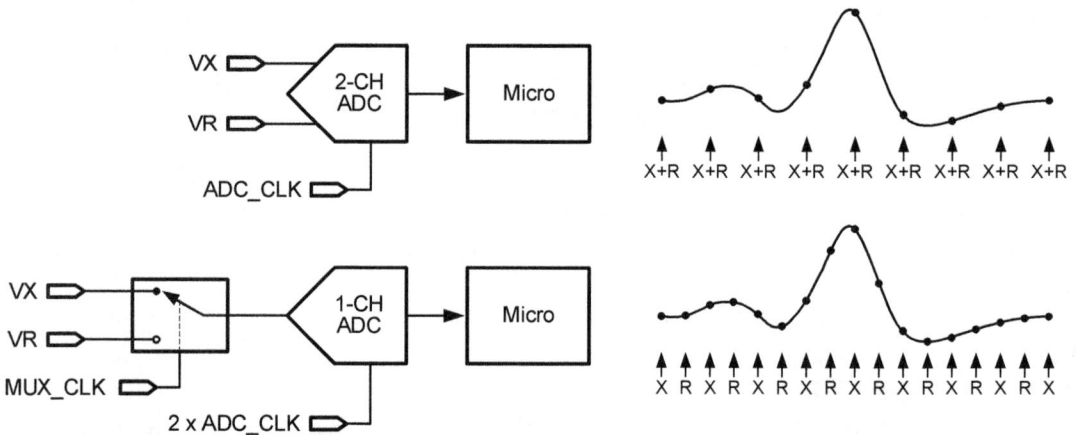

Fig. 18-4: **Two-Channel vs Multiplexed Sampling**

18.1.2: Sample Rate

In a baseband system using analog demodulators, the ADC is sampling a relatively slow signal whose bandwidth is pretty consistent regardless of what TX frequency is being used. Therefore, the required ADC sample rate has nothing to do with the TX frequency. Rather, the sample rate of the ADC is selected to match the microprocessor's input data rate, accounting for any desired oversampling. This will be covered when we consider the processing loop rate.

18.1.3: Oversampling and Dithering

It is also possible to improve SNR by *oversampling* the signals and averaging. Every doubling of the sample rate provides a 3dB improvement in SNR. A 6dB improvement is about one extra bit of resolution and this requires oversampling by a factor of four. Therefore, if a 12-bit ADC oversamples the signal by a factor of 256 then it can effectively produce 16 bits of resolution. Most micros include an internal SAR ADC with 12 bits being a popular resolution and a few going as high as 16 bits. Development-wise, an internal ADC is much easier to implement than an external ADC so it is usually a better first step in learning digital design.

Getting an SNR boost from oversampling requires that the signal be noisier than the quantization noise of the ADC. If it is not, then this can be overcome by the intentional addition of noise, a technique known as *dithering*. Dithering can be added as a random noise to the analog signal and used with oversampling to improve SNR. There is also a technique known as subtractive dithering, in which a digital-driven non-random offset is added to the analog signal and is then digitally subtracted later. This technique is usually paired with oversampling.

Oversampling and dithering are mostly useful in cajoling a few more bits out of an inadequate ADC, especially when using an ADC internal to a micro. Oversampling also places the X and R sample points closer together in time when using a multiplexed single-channel ADC and may eliminate the need to time-align data with interpolation. It is worth noting that some metal detector designs have used these methods. Oversampling and dithering are well documented elsewhere[4] and will not be covered in further detail here.

18.2: The Micro

The micro potentially does a lot, so there are a lot more considerations in choosing one:

- Processing loop rate
- Total processing required in each loop
- Flash requirements for code
- RAM requirements for data storage
- EEPROM for storing user settings
- Timer requirements for TX, demods, and audio
- Internal ADC requirements for pots and/or battery check (or, perhaps, signal sampling)
- Display drive requirements
- SPI or I2C for things like digital pots or an external ADC
- GPIOs for UI inputs (e.g., keypad) or control outputs
- Other desired features like USB

In a very simple system — say, a PI detector in which you only need to generate a few clock signals — you may only need a handful of timer outputs and a few GPIOs, in which case the micro can be small and simple[5]. A comprehensive VLF (and especially multifrequency) design that has a display, keypad, and lots of user options may require a large, fast processor with lots of flash and RAM.

It is wise to initially "overspec" the micro at the beginning of a new design. That is, pick one that has more pins, more memory, more speed, and more peripherals than you think you'll need. At the end of the design, scale back to what is really needed if cost is an issue.

18.2.1: Processing Loop Rate

We mentioned previously that the required ADC sample rate is determined by the *processing loop rate*. The processing loop rate is the rate at which new ADC data is sent to and processed by the micro. A digital metal detector must operate nearly in real-time, with a data rate that is fast enough to get a number of samples of a deep coin-sized target at a typical coil speed of 1m/s. The higher the data rate the better, but at some point the micro will run out of available time to process the data. A fairly typical loop rate in digital designs is 5ms. This means that each meter of ground coverage (at 1m/s) will produce 200 data samples, or 1 sample every 5mm. For imperialists, that's 5 samples per inch. For a medium depth coin target, that might result in 10 samples over the target response curve.

A processing loop rate of 5ms is 200Hz, and this implies that the ADC should have a sample rate of 200Hz, assuming no oversampling. Every 5ms the ADC produces new data and a data ready (DR) signal is used to trigger an interrupt service routine (ISR) in the micro. The data is transferred (typically using an SPI port for an external ADC), the micro processes the data, and then waits for the next interrupt trigger (Figure 18-5). If the ADC is oversampling then the data rate may be much higher than the desired loop rate. In the example of a

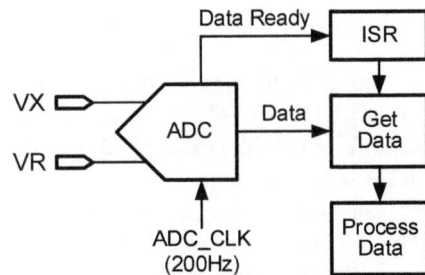

Fig. 18-5: **(Very) Simplified Processing Loop**

12-bit ADC oversampling by 256X the data rate will be 51.2kHz. Many microcontrollers with an inter-

---

4. Especially look at app notes and white papers provided by the chip companies who make ADCs.

5. An 8-pin PIC12F1840 is used in the *Pulse-2 and Pulse-3* designs.

nal ADC have a built-in oversampling mode so that the ADC internally samples (say) 256 times and then delivers the accumulated data at the end of the oversampling period, meaning that data can still arrive every 5ms. If this feature is not available, then the processing loop will need to interrupt on every sample and simply add the new data to an accumulator; a count flag can be used to tell the processing loop when to proceed with other data processing.

If the loop rate is 5ms, then that is how much processing time is available before the next sample. The amount of processing that can be accomplished depends on the clock rate of the micro. For example, if the micro runs at 20MHz and can execute one instruction per four clock cycles[6], then in 5ms we can execute 25,000 instructions. Whether this is enough depends on how much processing is required. Besides sampling the signal, we need to apply filtering, double derivatives, more filtering, discrimination, and audio generation. Optionally there may be ground tracking, audio tones, and perhaps a wireless audio system. If there is a display, then that needs to be updated during the loop, with a graphics display requiring more processing than a segmented display. We'll see in a minute there is a way to cheat the system, but in an aggressive design it is wise to choose a micro with more processing power than anticipated, then scale back to a lower clock rate at the end of the design (or not).

18.2.2: Processing Loop Control

Broadly, there are two ways to control the processing loop. Both methods usually start with a trigger from the ADC[7], most often a data ready pin. The first is to create a round-robin task list that is manually controlled by a task counter. Although the data arrives every 5ms, it may not be necessary to complete all the tasks within that 5ms. For example, updating the audio every 10ms is perfectly acceptable, and the display can be updated at an even slower rate. Of course, basic signal processing must be done every time new data arrives. Figure 18-6 shows a diagram of what the internal processing structure of a round-robin task list might look like.

It is important to pay attention to the execution time of each task. If a task overruns its allotted 5ms then it may create an interrupt error. It is useful to monitor a GPIO test pin with an oscope, whereby the pin is set high at the start of a process and set low at the end of the process. If the ADC data ready pin is also monitored then it is easy to see the time relationship.

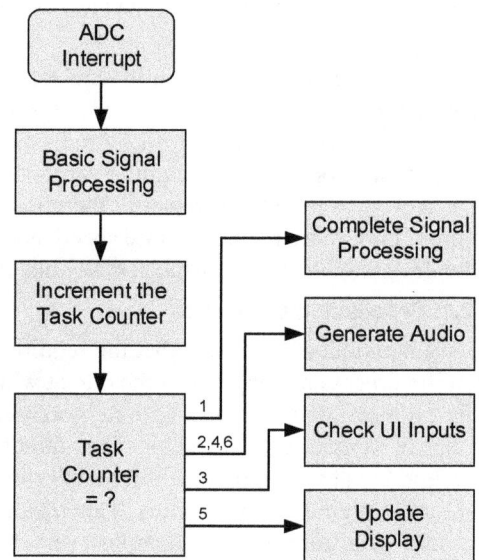

Fig. 18-6: **Round-Robin Diagram**

A more modern method is to use a *real-time operating system* (RTOS) to handle various tasks based on priorities. An RTOS is a piece of software that creates and controls multiple tasks using time slicing methods, much like the typical personal computer does. These tasks run autonomously but can communicate with one another using events and sending data through queues. The RTOS does incur some overhead (both clock ticks and memory) so it is best left to a decent 32-bit micro[8].

Figure 18-7 shows a simple RTOS system with several tasks. A task is initiated by the RTOS and then runs as an endless loop, with time given to it according to its priority. The most important aspect of a metal detector is the speed and quality of the audio response, so the audio task might get the highest priority. Of course, it is necessary for the DSP task to complete its signal processing but even this task is

---

6. Typical of low-end PIC micros.

7. This assumes that the TX and demod clocks are generated by autonomous timing generators, widely available on modern micros. If you are still bit-banging the clock signals, it's time to upgrade your micro.

8. Which should be your choice anyway.

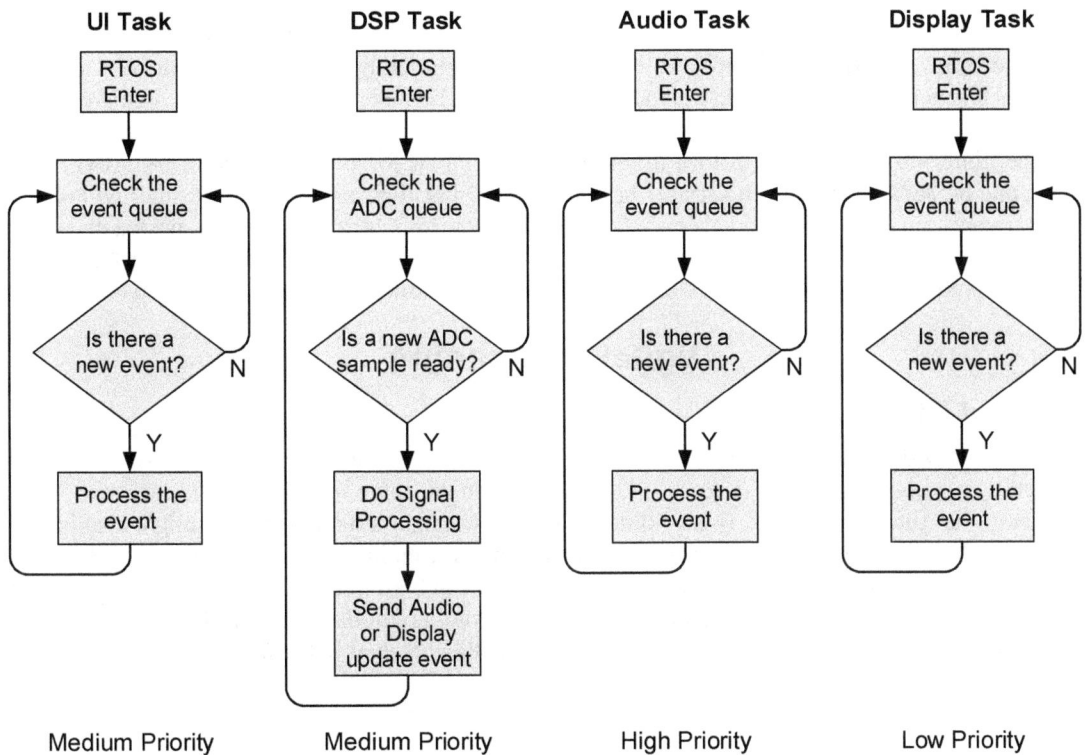

UI Task	DSP Task	Audio Task	Display Task

UI Task / **DSP Task** / **Audio Task** / **Display Task**

RTOS Enter → Check the event queue → Is there a new event? (N loops back) / Y → Process the event

RTOS Enter → Check the ADC queue → Is a new ADC sample ready? (N loops back) / Y → Do Signal Processing → Send Audio or Display update event

RTOS Enter → Check the event queue → Is there a new event? (N loops back) / Y → Process the event

RTOS Enter → Check the event queue → Is there a new event? (N loops back) / Y → Process the event

Medium Priority   Medium Priority   High Priority   Low Priority

Fig. 18-7: **RTOS Diagram**

usually not the highest, as long as it is completed before the next ADC sample. A slightly sluggish display is no big deal as its results are usually held static for maybe a second, so it can have a low priority. However, a sluggish user interface (keypad, pots, etc) will be annoying so this task should not be low priority.

A task that does not finish before the next input data will continue at its next time slice. It is, however, important that the core DSP that must execute on every data input (the input signal filtering, derivatives, and other basics) be complete within the 5ms window, with plenty of time to spare for other tasks to get some execution time. Both approaches (round-robin and RTOS) require you to keep things between the ditches, but the RTOS method is more self-driving.

18.3: Timer Requirements

One of the first programming tasks to achieve in a digital design is to get the TX and demod clocks running. Different micros have different ways of doing this but in any case you want the timing to be an autonomous function; that is, they run in the background and do not require any attention in the processing loop. Bit-banging clock signals should not even be considered except in the crudest and simplest designs[9].

Some micros have multi-channel timers where one timer-counter can control, say, four outputs. These are mostly intended for motor control but work extremely well in metal detectors where the TX and demod clocks are synchronous and can be run from a common timer base. Some micros have single timers which can be triggered from a master source to create synchronous clocks. In either case, the timers are set up with initialization code and started, and potentially never messed with again for the duration of the detector's use.

---

9. For example, the PIC12F1840 used in the *Pulse-2* and *Pulse-3* design does not support autonomous timers but has other attributes such as size (8 pins) and cost ($1.50). It also fits the "crudest and simplest" criterion.

A very useful function to include is a means for slightly altering the overall frequency of the timing in order to minimize EMI noise. Many digital models now include a central frequency and, say, ±10 frequency "offsets" that are either chosen by the user or automatically scanned for the lowest apparent noise. For this, a method for interactively altering the timers is needed, but this action is only executed as the user is making the offset selection and afterwards the timer runs autonomously again.

Other timers to consider are one for the ADC clock, a PWM for audio generation (two if you want stereo), and perhaps a PWM for a variable brightness LCD backlight. A system tick timer will also be needed if you want to track timed events[10]. It is important to think through the entire design and figure out your timer needs before selecting the micro. The same applies to the requirements for ADC channels, capture/compare inputs, GPIOs, DMA channels, and so forth.

Digital Processing

18.4: Input Filter

In an analog design the first thing we do after demodulation is a first derivative, usually directly followed by a second derivative (Figure 16-18). We can do the same in DSP but often we start with a low-pass digital filter. If the ADC is providing data at an oversampled rate then this might also include decimation. A digital filter with a fast roll-off and a notch at the mains frequency (50Hz or 60Hz) is difficult in analog circuitry but relatively easy in DSP. Filters can be implemented as either finite impulse response (FIR) or infinite impulse response (IIR) and design methods can easily fill another book[11]. Figure 18-8 shows a 17-sample raised-cosine FIR filter with a notch at both 50Hz and 60Hz.

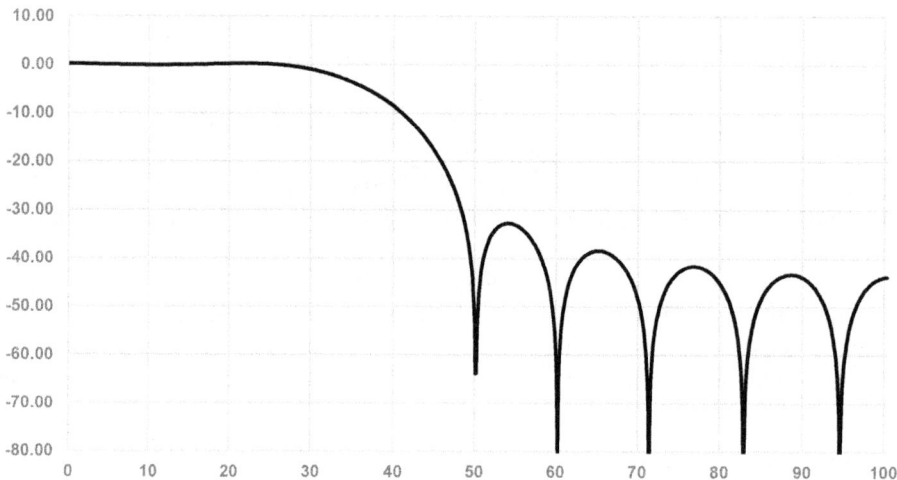

Fig. 18-8: **17-Sample FIR Filter**

Digital filters (whether FIR or IIR) require a history of the signal to work with. For example, the 17-element FIR filter requires the last 17 samples; this is done by feeding the input data into a 17-element array, where the newest data replaces the oldest data (first in/first out, or FIFO). If the loop rate is 5ms then a 17-element array represents $(17 \times 5ms)/2$ ms = 42.5ms of latency which is pushing the upper limit of what we would like. It is possible to split the filter into two shorter filters in series. For example, two 10-element FIR filters in series would have a latency of $(10 \times 5ms)/2 + 5ms = 30ms$ because they are operating largely in parallel, staggered only by a single sample.

If we are using a single input ADC and interpolation is needed, it is possible to combine it with the input filtering by using zero-stuffing. The data arrays fed to the digital filter would now look like:

---

10. Such as: press-and-hold a button for 2 seconds to activate such-and-such.

11. For example, *Digital Filters* by R.W. Hamming. However, there are lots of filter software tools available to assist with designs, many free. The 17-tap filter above was generated by Iowa Hills FIR Filter Designer, V7.0.

X_0	0	X_1	0	X_2	0	X_3	0	X_4	0	X_5	0	X_6	0	X_7	0	...
0	R_0	0	R_1	0	R_2	0	R_3	0	R_4	0	R_5	0	R_6	0	R_7	...

With a FIR filter, it is not necessary to actually create the zero-stuffed arrays. Rather, the filter code can be written so that every other coefficient-data multiplication is simply omitted. Note that in the arrays above, the X array starts with X_0 and the R array starts with 0. In the next cycle, the arrays will still look the same: the X array starts with (a new) X_0 and the R array starts with 0. This effectively decimates the output data by a factor of two.

18.5: Signal Arrays

We've already seen that an input FIR filter requires a short history of the raw signal saved in an array. There are, or course, two arrays because we have X and R signals, and shortly we'll add a third array for the G signal. Beyond the filtered raw signals, there will be first and second derivatives of the X, R, and G channels and possibly other signals. It is often useful to maintain a short history of each of these for signal processing. The required depth of these arrays depends on what kind of processing you want to do and the type of arrays might depend on the quality of the processor or the amount of RAM available. Typically 32-bit integers are more than sufficient (192dB of dynamic range) whereas 16-bits (96dB) might be barely sufficient. It is important to plan for the total data array space needed before selecting a micro, and is another reason to initially overspec it.

18.6: G Channel

We saw in Chapter 16 that a ground-free signal can be mathematically created from the X and R signals (see Figure 16-26) and can be derived from the raw X,R signals, the first derivatives, or the second derivatives. The raw signals will be assumed here, and the G signal can also be created in firmware:

```
G = R + k*X;
```

In a digital detector with manual ground balance k might be derived from reading the setting of a ground balance pot, where the minimum and maximum pot settings represent some predetermined minimum and maximum values of k, easily scalable in software.

Pots are usually read by the micro's internal ADC which means that the range and resolution of the ground balance will be limited by the resolution of the ADC. For example, a 10 bit ADC has, at most, 1024 discrete values. If you want a total of 15° of ground balance range then the resolution will be about 0.015°. This is a very reasonable resolution. Fewer bits will give a coarser resolution and may require reducing the GB range.

For a 15° GB range the value of k will be between 0 and 0.25. Depending on the micro chosen the calculation of G might be done with a floating point calculation or 16- or 32-bit integer math.

18.7: Derivatives

Following the input filter are the first and second derivative calculations. In DSP, a derivative can be calculated as the difference between the current data sample and the previous data sample, divided by the time spacing:

$$\frac{dV}{dt} = \frac{V_n - V_{n-1}}{t_n - t_{n-1}}$$

Eq 18-2

The time spacing is always constant, equal to the processing loop rate (5ms, for example). Since everything at this point is digital data, the denominator is just a constant scalar and can be omitted. The first and second derivative of the X signal would then be:

```
dX1[n] = X [n] - X [n-1];    // First derivative
dX2[n] = X1[n] - X1[n-1];    // Second derivative
```

where X is the X channel filtered ADC data, X1 is the first derivative, and X2 is the second derivative. The R channel and G channel derivatives are done identically.

Depending on the loop rate, the above technique may produce low-amplitude noisy results if the data is spaced too close together in time. That is, if the amplitude difference between X[n] and X[n+1] is small even during the target response then the difference calculation will be closer to zero. A solution to this is to space the difference samples farther apart. Instead of subtracting adjacent samples, we can subtract samples with some amount of spacing between them. For example, subtracting the [n-7] sample from the current sample would be:

```
dX1[n] = X[n] - X[n-7];
```

This leaves six samples in between that, for this instance in time, are not used. However, a further improvement is to include them in the calculation for an averaging effect:

```
dX1[n] = X[n] + X[n-1] + X[n-2] + X[n-3] - X[n-4] - X[n-5] - X[n-6] - X[n-7];
```

That is, we can add the first k sequential buffer elements, and then subtract from that the next k sequential buffer elements to produce a first derivative.

The second derivative can be calculated in the same manner from the first derivative output data. It can also be calculated directly from the raw data by subtracting the first $k/2$ elements, adding the next k elements, then subtracting a final $k/2$ elements:

```
dX2[n] = -X[n] - X[n-1] + X[n-2] + X[n-3] + X[n-4] + X[n-5] - X[n-6] - X[n-7];
```

An additional advantage of this method is that it eliminates the latency between the first and second derivatives (see Figure 16-22) and they are now exactly time-aligned.

18.8: Target Calculations

There are two primary target calculations we need to make: the magnitude and phase. The phase is what determines target VDI and the tone to generate. The magnitude is used to estimate depth and (often) determines the audio volume. Mathematically, the calculations are simple:

$$\text{Magnitude} = \sqrt{X^2 + R^2} \qquad\qquad \text{Eq 18-3}$$

$$\text{Phase} = \arctan\left(\frac{R}{X}\right) \qquad\qquad \text{Eq 18-4}$$

In a higher end micro with good floating point support and a hardware multiplier these calculations are not too taxing. If you are using a lesser micro then some cheating might be in order. For the magnitude, an alternative is to simply add the absolute X and R signals:

$$\text{Magnitude} = |X| + |R| \qquad\qquad \text{Eq 18-5}$$

This gives perfect results at 0° (ferrite), 90° (small foil), and 180° (Atocha bar) but 41% too much magnitude at 45° (small iron) and 135° (aluminum screw cap[12]).

It would seem that a 41% error in the signal is a lot, but depth is not linear with signal strength. If we assume the round-trip signal is proportional to $1/d^6$ then a 41% signal strength error produces a depth error of

$$\Delta\text{Depth} = \sqrt[6]{1.41} = 1.06 \qquad\qquad \text{Eq 18-6}$$

That is, simply adding the vectors produces a maximum apparent depth error of 6%. This is an error of 1cm for a target 16cm deep, not too bad. The assumption that the signal is proportional to $1/d^6$ is true beyond twice the coil diameter (see Figure 3-9); less than that, it is no better than $1/d^5$ which produces a maximum error of 7% (1cm in 14cm).

A better approximation that still avoids multiplication is to add the magnitude of the larger of X and R with half the magnitude of the other[13]. The code for that looks like:

---

12. These target IDs are for 10kHz; other frequencies might indicate different targets at these phase angles.

13. This is called the "αMax+βMin" method. Different values of α and β produce different error curves, and α = 1 and β = 0.5 avoids multiplication. See *Understanding Digital Signal Processing* by Richard G. Lyons.

```
if (X < 0) X = -X;    // Take the absolute value
if (R < 0) R = -R;    // Take the absolute value
if (X > R)
  Mag = X + (R >> 1);
else
  Mag = R + (X >> 1);
```

Figure 18-9 shows a normalized plot of both short-cuts: Equation 18-5 is the black curve and the coded approximation is the gray curve. The coded approximation has a peak magnitude error of 12% which translates into a depth error of 2%.

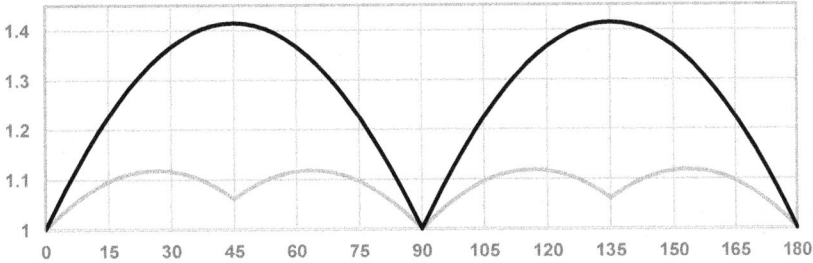

Fig. 18-9: **Magnitude Calculations**

The phase calculation can similarly be kludged. Figure 18-10a shows a vector plot with an example target vector and its X & R components, normalized to 1. Additionally, the unit semicircle (dashed gray) is replace by straight-line approximations. In the ferrous region the approximation is

$$\phi_{0-90} = 90 \cdot \frac{R}{R-X}$$

while in the non-ferrous region it is

$$\phi_{0-90} = 180 - 90 \cdot \frac{R}{R+X}$$

For the example vector of 120°, the approximation is

$$\phi = 180 - 90 \cdot \frac{0.87}{0.87 + 0.5} = 122.8°$$

which is an error of 2.8°. Figure 18-10a shows that the example vector of 120° crosses at about 123°. The error across the full 180° is also shown in the graph to the right (Figure 18-10b). These equations are exactly correct at 0°, 45°, 90°, 135°, and 180° and the maximum error elsewhere is ±4° which is quite tolerable. In most designs the actual target IDs and discrimination settings can be altered to compensate for phase calculation errors.

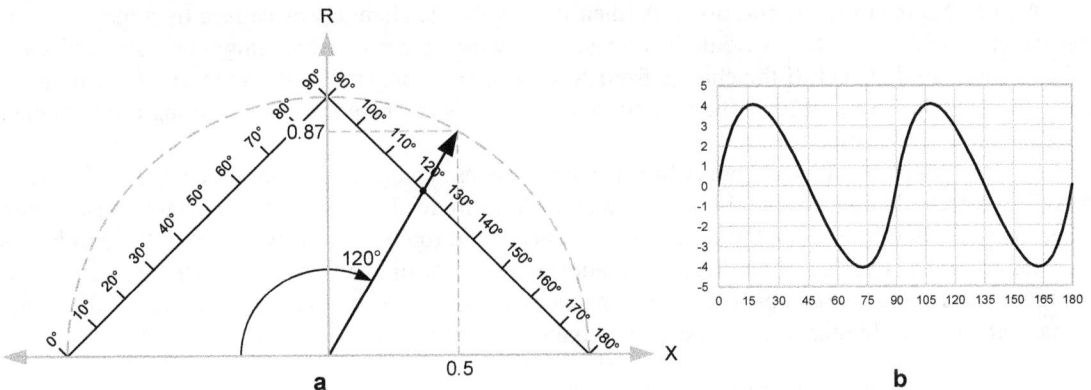

Fig. 18-10: **Phase**

This method still requires a multiplication and a division but may be less taxing than an arctangent calculation. Another method is to consider is a look-up table, which replaces mathematical operations with higher memory requirements. Yet another method is to use a successive approximation algorithm, commonly known as a CORDIC method, which uses only addition and bit-shifting, and a small look-up table. The CORDIC method can be extended to achieve any desired precision, it just takes more time.

18.9: Ground Tracking

Modern digital designs often employ automatic ground tracking (AGT) instead of, or in addition to, a manual ground balance method. AGT consists of the algorithm for computing the tracking and a method of applying the tracking correction. There are several approaches, starting with a purely hardware design whereby discrete circuitry is used to assess ground changes, and corrections are applied to the XCLK and RCLK phases[14]. This is the Hard Way to do it.

This chapter is all about digital techniques so, at a minimum, let's assume that software is used to evaluate the state of the ground. This is normally done by monitoring the amplitude and phase of the raw X and R signals. A brief history of these characteristics is held in a FIFO buffer and changes in the ground phase are used to (somehow) adjust the ground balance.

In a motion VLF detector, high-pass filters are used to remove the ground signal for the purpose of target detection. For the purpose of ground tracking, we'd like to have just the ground signal with the target signal removed. The presence of a target signal can cause the tracking algorithm to track on the target signal and cause the target to disappear. This is especially prone to happen on a weak target signal in the presence of a strong ground signal.

We can illustrate this problem using the vector plot in Figure 18-11. Suppose that the sampled signal is represented by vector A and, because it has had a steady phase for a while, we assume it is a ground-only signal. The signal then moves to vector B and, at a glance, it appears as if the loss angle of the ground has increased somewhat. That may indeed be the case, but it may also be the case that the ground hasn't changed at all and the new vector is the result of a target signal (vector T) getting added to the original ground signal.

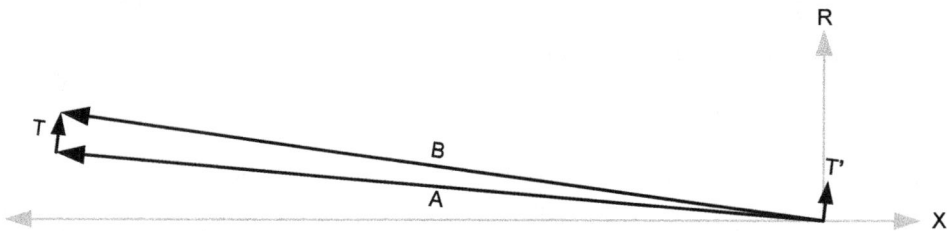

Fig. 18-11: **Ground + Target Vector Plots**

The only way to know for sure is to look at the time response of the change. If the signal changes from A to B and then rapidly returns to A, then it is likely the change was caused by a target. In the absence of ground, such a target would be represented by the vector T′, which might be a small piece of foil or a small gold nugget. If the change from A to B persists longer than the span of a typical target, then it is likely due to a change in the ground loss angle. This is where low-pass filtering can help minimize the rapid target response.

In Figure 18-11 the change from A to B is purely a phase change; the amplitude remains the same. In AGT we are only concerned with phase change. Amplitude changes with a constant phase means either the strength of the ground mineralization has changed (but not the loss angle) or the coil height over the ground has varied, either due to an uneven sweep or uneven ground. A change only in the amplitude means that the ratio of X to R remains the same and therefore the math that creates the G signal has not changed. In other words, the ground balance point remains the same.

---

14. The first ground-tracking models were analog and discrete logic designs, produced by White's.

Since for AGT we are concerned with phase changes, a common strategy is to place limits on how much phase change results in a tracking change, and also the speed at which tracking can occur. Both limits are intended to minimize unintentional tracking on targets. Suppose the current ground has a loss angle of 5°; we might impose an upper track limit of 8° and a positive track speed limit of 1°/100ms. Usually the lower limit and the negative track speed are not as critical because an apparent drop in the ground loss angle is unlikely to be caused by a target, as it would require the target to fall in the third or fourth quadrant. However, if you use, say, the first derivative signal for AGT then a target's bounce-back signal can fall into the third and fourth quadrants.

Depending on the ground conditions, the track and speed limits may need to vary. For example, in mundane ground that does not change much, the phase limits and track speed can be fairly tight. In highly variable ground, the same tight limits might result in poor AGT and that can produce excessive ground noise. There are many possibilities in designing the AGT algorithm.

The output of the AGT algorithm must somehow control the ground balance of the detector itself. In Chapter 15, basic VLF ground balance was achieved in hardware by altering the phase of the demod clock. In Chapter 16, a G channel was created in hardware by a simple resistive divider between the X and R channels. Our AGT algorithm could alter either of these through the use of, say, a digital potentiometer. But in this chapter we're looking for all-software solutions.

The creation of the G channel in Chapter 16 was done per a simple mathematical equation (Equation 16-2). There is no need to do this in hardware, we can more easily do it in software. All we need are the digitized X and R demod signals and everything can be derived from those. The G channel is created in software per Equation 16-2 and the k term can be adjusted on-the-fly by the AGT algorithm.

It seems prudent in a baseband digital design to have the X and R demod signals properly phase-aligned to some standard. This is usually done in the same manner as in Chapter 16, where k is set to some particular starting value and the R signal is phase-rotated (by adjusting its demod clock) to null out a ferrite target. The X demod is clocked at 90° from the R demod. In this manner all units will operate in the same way, with the same AGT limits.

It's possible to let the hardware X and R signal phases fall where they may and do all the phase rotations in software. In other words, we should not need to compensate for the preamp phase shift, we can do that in software during a calibration routine. All of this is true, but there is a caveat. The ground signal can be strong in amplitude and if it is not reasonably nulled *at the demodulator* then there is the risk of signal overload on the R channel. The best practice is to make sure that the R channel is reasonably nulled to a reference point like ferrite, then it will have a minimal signal for most ground conditions. This channel can then be run with a higher gain which improves overall sensitivity since this is also the target channel.

18.10: Discrimination

In analog detectors discrimination is usually a linear control that sets a discrimination threshold. Everything below the threshold is rejected while everything above is accepted. Some analog models also have a notch feature whereby a small region can be selectively removed. This was mostly an attempt in the US market to remove pull tabs while still detecting US nickels.

In digital detectors, there is practically no limitation on how targets can be discriminated. Logically, we can reject or accept any target phase even down to each degree of phase angle. That is, we could reject a target with a 30° phase response while accepting target phases of 29° and 31°. While some digital designs retained the linear discrimination threshold control (example: White's *MXT*), most have adopted a method whereby the user can select multiple discrimination regions using the display and keypad. This is called the *discrimination mask* and will be covered in more detail later.

In early analog detectors, discrimination was implemented by rotating a demodulator phase to reject targets below the demodulator's zero threshold. A side effect was that nearby accepted targets were attenuated (Figure 14-17) and so depth was noticeably compromised. Discrimination in digital detectors relies on the calculated phase response and simple logical comparisons to the discrimination mask to determine, on the fly, whether a particular target response generates an audio alert or not. Target response strengths are always maintained. However, the abrupt nature of the discrimination cut-off gives rise to sputtering audio sounds when passing over a target that produces sampled responses on both sides of the cut-off, especially when audio is produced throughout the target response. This has always been a source of annoyance by users who transitioned from all-analog detectors.

18.11: Misc. Processing

18.11.1: Overload Detection

If a target signal overloads any part of the analog path it becomes invalid for the purpose of magnitude and phase calculations. It is useful to alert the user of an overload condition so they know any audio or visual indication is likely bogus.

In a baseband sampling design it is more likely that one of the demodulators will overload rather than the preamp. Comparators can be placed on the demod outputs to perform window detection so that (for example) a signal greater than $0.9 \times VCC$ or less than $0.9 \times VEE$ will produce a high signal that flags the micro of an overload condition. If there is no concern that the preamp might overload, then overload detection can also be done purely in software by setting alert limit thresholds on the incoming ADC data.

18.11.2: Pinpointing

In a normal search mode the second derivative signal is analyzed because it is relatively free of a ground signal. But the coil must be kept in motion. Detectors usually have a pinpoint mode that, at a minimum, disables discrimination processing so that it processes the all-metal signal. That signal should be ground-free so it will be taken from the G channel and can be the raw, first derivative, or second derivative signal. Preferably, pinpoint mode should be non-motion so the raw G signal should be used.

Although this signal is ground balanced, there is no high-pass filtering for circuit offsets so as soon as the pinpoint mode is triggered the current G signal should be held as a starting reference point for normalizing subsequent new G signals. Grabbing an initial level for a reference point also allows for a technique called *ratcheting* where you can move closer to the target, release and re-engage the pinpoint mode, and this will reset the audio response back to a no-target threshold condition. This makes it easier to zero in on the exact target location, especially for stronger shallow targets.

18.11.3: Coil Identification

A recent trend in commercial detectors is to place a small PCB in the coil that includes a micro. When the detector turns on, it queries the micro in the coil for an ID code. The code might simply be a way to identify the coil type (say, a 6" concentric versus a 12" DD), or it might be used as a security check to prevent the use of aftermarket coils, or both.

Identification of the coil is a useful thing to do. At a minimum, the size and type of coil determines how strong a particular target responds so that, for the purpose of estimating target depth, you might use different scaling factors for different coils. Or perhaps use different motion filter coefficients for concentric versus DD. Or perhaps a specialty coil is made that only runs properly in a special mode, so the coil ID is used to limit operation to that mode.

18.11.4: Coil Accelerometer

Once you have a micro in the coil for coil ID, it is a simple matter to add a 3-axis accelerometer inside the coil. Knowing whether the coil is in motion, or is static, or has reached the end of a swing, or is bobbing up-and-down all can be used to automate tasks and alter DSP. Later sections will mention some instances where an accelerometer is beneficial.

User Interface

All-analog metal detectors tend to have switches and potentiometers for controls and an analog meter for the visual response[15]. The very first digital detector, the White's *Eagle*, went immediately to a keypad and LCD[16], while other digital (and semi-digital) models continued to use switches and pots.

Today, the user interface (UI) of practically all detector designs consists of a keypad, LCD, and audio. The keypad and LCD provide a way for the user to change operational parameters, and the LCD and audio provide real-time ground and target information. Some detectors (especially pinpointers) include a vibration motor for tactile feedback and many detectors have wireless audio. Security walk-through metal detectors often include an Ethernet or wireless interface for the ability to remotely control the operational parameters or monitor traffic.

18.12: Inputs

18.12.1: Keypad

The de facto UI input method for modern detectors is a multi-button keypad. The easiest and most flexible way of reading a keypad is to simply connect each button to a GPIO pin on the micro. If there are more buttons than available GPIOs then it may be necessary to read the keypad as a row-column scanned matrix or using an ADC channel. See Figure 18-12. With dedicated GPIOs a 16-button keypad would require 16 GPIOs, a 4x4 matrix connection only requires 8 GPIOs, and an ADC connection could do the job with only one or two pins.

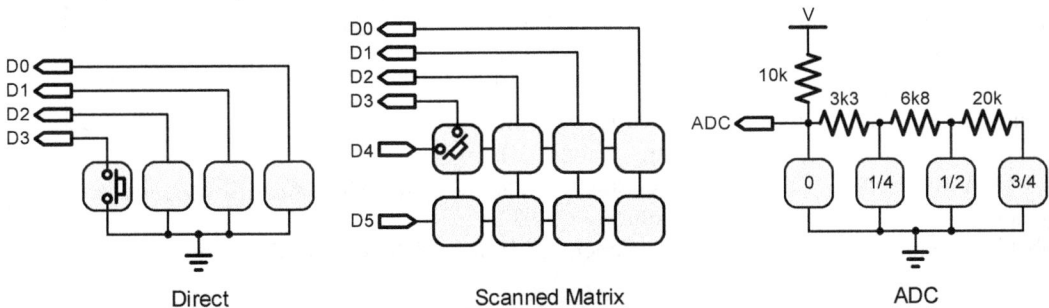

Fig. 18-12: **Keypad Connections**

A direct connection offers more flexibility in that it has the ability to read multiple key presses at the same time. This can be useful for creating "short-cut" keystrokes, like pressing a MENU button at the same time as a MODE button to reset all parameters to factory defaults. This ability is limited in a matrix connection and only possible with an ADC connection when the combination buttons are on different ADC channels[17]. The scanned matrix requires driving one row at a time (D4 & D5 in the example) and reading the columns (D0-D3), which means that it needs to be set up as a recurring background process (RTOS), or something you manually do every so often (round robin).

Besides reading simple button presses, you may also want to check for a button release or a button that is pressed-and-held for a certain amount of time. For example, it is common to press the POWER button to turn on the detector, and to press-and-hold the POWER button for, say, two seconds to turn it off. A good way to check both presses and releases is to use GPIOs that have interrupt-on-change functionality. The interrupt handler can then place the button event into an event queue that is checked by the processing loop. Reading a hold event can be done using a system tick (SysTick) timer and noting the time during the button-press interrupt, and then at the button release interrupt calculating the hold time. A key-hold event can also be placed in the event queue and processed later in the processing loop.

The system tick timer just mentioned is a very useful program element. It is created by running a simple timer in the background and reading its value when needed; times can then be compared for computing delays. A 16-bit timer running at 32MHz can achieve a maximum delay of 65536/32MHz =

---

15. Good examples include the Compass *Challenger* & *Scanner* series; the Garrett *ADS Master Hunter* & *Groundhog* series; and the White's *5000/6000* series. The pinnacle of analog design was probably the White's *XL Pro*, the last of the *6000* series, and had automatic ground tracking, target ID, and depth readout. The only thing it lacked was tone ID, and some hackers have created a solution for this.

16. The *Eagle* wasn't the first model to use an LCD, that honor goes to the Teknetics *9000*. However, the *9000* LCD was mostly a direct replacement for the analog meter of the *8500* and the detector was otherwise an analog design.

17. In the ADC method shown in Figure 18-12, when multiple keys are pressed only the key closest to the ADC will register. An R-2R type circuit might offer more flexibility.

2ms so a prescaler divider is needed to get this to something reasonable. A prescale divide-by-32768 results in a maximum delay of 67.1 seconds with a resolution of 1ms. A 1ms resolution is also a good number for creating simple delay functions. For example, a function called Delay_ms(10) would create a 10ms delay in code by looking at the SysTick counter and then looping until the SysTick counter has increased by 10. In general, you should avoid polling delays except for very short intervals as they just waste processor bandwidth. Also, a SysTick resolution of 1ms means that a delay can have up to 1ms of error; the error can be reduced by increasing the resolution (say, to 100μs) but that will also reduce the maximum delay (to 6.7 seconds, in this case). If you need both fine resolution and a large maximum delay, a 32-bit timer may be necessary.

18.12.2: Knobs

Even though most contemporary designs have only a keypad, it is easy to argue for the inclusion of one or two knobs for very common adjustments. This can be a potentiometer or perhaps a rotary encoder. A potentiometer can be read by an ADC channel of the micro by connecting the outer pot legs to VDD and ground, where VDD is the full-scale voltage of the ADC (usually the micro's supply voltage, 3V or 5V). A 10-bit ADC results in 1024 distinct positions, and in a typical 300° pot rotation this is a resolution of 0.3°,

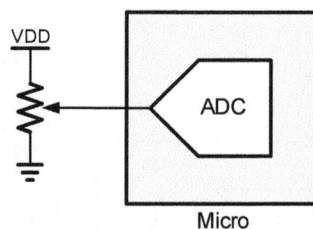

Fig. 18-13: **Potentiometer Connection**

probably way more than enough for most functions. If desired, this can easily be reduced by right-shifting the ADC value to the desired resolution.

A rotary encoder can make some menu adjustments much quicker. For example, suppose you have a manual ground balance option that can balance all the way to salt, so you may want a 95° balance range with a resolution of 0.1°. This comes out to 950 points of adjustment. Instead of pressing a keypad button 100s of times[18], it would be much quicker to rotate an encoder knob. An encoder can be read by either using two capture/compare channels or with two interrupt-on-change GPIOs. Some micros have special encoder input pins specifically for gray-code rotary encoders. For extremely large adjustment ranges (like 950 points) it is useful to look at the speed of rotation and jump by 5 or 10 points when the encoder is turned rapidly.

18.12.3: Touchscreen

So far, no commercial hobby metal detector with an LCD has included touchscreen capability. In an era of touchscreen-everything this seems odd. The most likely reason is that metal detectors are used in dirty/muddy/sandy conditions and this can cause abrasion on the screen. However, there is a good argument for including a touchscreen in addition to a traditional keypad. If the detector has a lot of settings, the touchscreen can provide a much faster way of initially setting up the detector, before getting in the dirt. Once in the dirt, further changes (usually minor) can be done via the keypad.

18.13: Audio

Metal detectors have always tended to use simplistic audio methods, usually generating one tone at a time. This makes the design of the audio driver fairly easy, with only an amplitude and frequency required. For micro-generated audio, a simple solution is use a DAC to create the amplitude and chop it with a switch driven with a PWM. If the micro has an internal DAC with sufficient resolution, then it may be possible to enable/disable its output using a PWM; in other words, both amplitude and frequency are produced on a single pin of the micro. See Figure 18-14 for both approaches.

---

18. Instead of making the user press a button hundreds of times, implement a hold-and-repeat function whereby, as long as the key is held down, a repeated key-press event is generated. Better yet, add a delay timer so that, on initial key press, the repeat rate is, say, 2 events per second but after holding the key down for 3 seconds, the repeat rate jumps to 10 events per second. All of this requires writing an event handler and needs a SysTick timer.

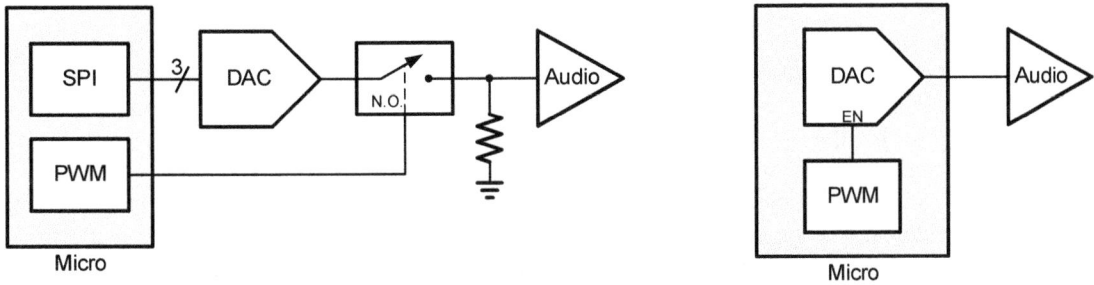

Fig. 18-14: **Audio Drivers**

What resolutions are required of the DAC and PWM? An 8-bit DAC will produce 256 volume levels. At the lowest levels (0, 1, 2, ...) each bit increase will produce an audibly noticeable step in the volume. At the highest levels (... 253, 254, 255) each bit increase adds little additional volume so the transitions are not noticeable. So the audio will sound a little "digital" at very low volume levels and smooth at higher volume levels[19]. A 10-bit DAC will improve this, and a 12-bit DAC will probably have a smooth response over the entire volume range.

The PWM is a little different in that it will have a defined minimum and maximum frequency. Let's say the minimum is 50Hz and the maximum is 1600Hz and the micro clock is, again, 32MHz. This requires a minimum division of 32M/1600 = 20,000 and a maximum division of 32M/50 = 640,000. This suggests a 32-bit timer is needed, but we can use a clock prescaler to reduce these numbers. If we prescale-divide by 1024 then the timer clock is now 31.25kHz; a divide-by-20 gives us 1562Hz and a divide-by-625 gives us 50Hz. This requires a 16-bit timer.

A problem with this is that the frequency resolution at the high end is poor. The five highest frequencies attainable are:

N	f	Δf
20	1562	—
21	1488	74
22	1420	68
23	1359	61
24	1302	57

Table 18-1: **16b PWM Audio Resolution**

If individual tones are assigned to VDI numbers or VDI regions this is not a big problem, it just limits the exact tones that can be used. But if a VCO pinpoint mode is implemented then as the tone climbs higher you will begin hearing the frequency steps; that is, the VCO audio response will not sound smooth. If the decision is to use a 16-bit timer, then the solution to this is to use a prescaler value that will maximize the resolution at the high end. This occurs when the timer clock is $65535 \times 50\text{Hz} = 3.277\text{MHz}$. This requires a prescaler value of just 10 for a micro clock of 32MHz and results in a worst-case Δf of 1Hz.

The use of an 8-bit timer is possible by using a prescaler value of 2510 but this further decreases the frequency resolution. Again, this is not much of a problem for producing discrete tones for VDI regions but will degrade the smoothness of a VCO pinpoint audio.

Finally, a PWM alone can conceptually provide both amplitude and frequency. The frequency is, of course, the frequency of the PWM and volume can be set by the duty cycle of the pulse, with 50% being the loudest and 1% (or 99%) being quiet. The problem is that adjusting the duty cycle alters the harmonics and produces an inconsistent tonal quality. In other words, it doesn't sound as clean as a consistent 50% duty cycle.

---

19. This is noticeable in some Tesoro designs.

18: Digital Techniques

18.13.1: Pinpoint Audio

Most detectors, whether VLF or PI, operate in a motion mode where the coil must be in continuous motion in order to produce a target response. A "pinpoint" mode is often included, whereby the user can switch to a non-motion mode for zeroing in on the location of the target. Because the pinpoint mode is non-motion, it must use the raw (non-derivative) demodulator signals. This means the X channel signal may have significant ground content so a VDI calculation will be prone to error. However, the pinpoint mode is really just for pinpointing, and for this all we need is the R channel (or G channel) signal which should be phase adjusted to be free of ground.

In the early days of pinpoint mode, the audio was a fixed tone with a loudness that was proportional to signal strength. At some point, VCO pinpoint was introduced whereby both the loudness and the tone were proportional to signal strength. In an all-analog design, a pinpoint VCO can be created with a simple 555 timer as is done in the *PI-GB* design (Chapter 22). In a digital design, we prefer to use a PWM in the micro and it, combined with a DAC, can produce the variable loudness and tone needed. See the audio hardware discussion (page 346) for how this can be done and for considerations in PWM resolution on VCO audio.

18.13.2: Tone ID

The Teknetics *8000*[20] introduced audio tone ID[21] in 1986. Until this innovation, all targets produced the same fixed-frequency audio response so that a pull tab sounded the same as a large silver coin. Tone ID assigns different audio tones to different conductance ranges. For example, the following table shows an 8-tone audio system one might use for US coin hunting:

Iron	Foil	Nickel	Pulltab	ScrewCap	Zinc ¢	1¢/10¢	Silver
100	300	400	500	600	700	800	900

Table 18-2: **Example Tone Frequencies (Hz)**

Tone ID can be as coarse or as fine as you like. A 2-tone system (as was used in the *VLF-Disc* design) might assign a low tone to ferrous targets and a high tone to non-ferrous targets. It is common to see 3 tones, 4 tones[22], and on upwards of 8 tones or more. Many detectors have multiple tone ID options the user can choose from. The ultimate tone system is found in the White's *V3*, where the user can assign tones as finely as each of the 191 VDIs, and can program the frequency of each tone.

It is an easy matter to select an audio tone based on the target phase/VDI, but as the detector is swept over the target the phase is not necessarily constant. The question then becomes, what specific point in the target response do we use for selecting the tone? It stands to reason that the target phase will be most accurate when the signal is strongest so the most obvious choice is the peak of the response. This is done on the ground-free second derivative signal, and the target peak of the second derivative corresponds to the zero crossing of the first derivative. Therefore, the process is:

1. Monitor the R channel second derivative until a target detection threshold is exceeded.
2. Then monitor the R channel first derivative until there is a negative-going zero crossing.
3. Then use the second derivative X and R signals to calculate the target phase.
4. Then select the audio tone based on the phase and create a beep.

A next question might be, how long of a beep do we produce? If you make the beep too long then there is a risk that an adjacent target will be masked. If it's too short then it could get lost in other audio chatter. A decent range to shoot for is 250-500ms. If an adjacent target is seen before the end of the primary target's audio beep, then the beep can be truncated to allow for the adjacent target to respond. In Figure 18-15[23] there are four targets that break the detection threshold to create an audio response. Each

---

20. Specifically the *8000 B*-series, I believe.

21. The acronym TID is sometimes used for tone ID, but it is more commonly used for Target ID. We'll use TID for the latter and spell out "tone ID."

22. See Chapter 17's *Explore* section for an all-analog example.

23. For simplicity the responses are drawn as "raw" signals, not second derivatives.

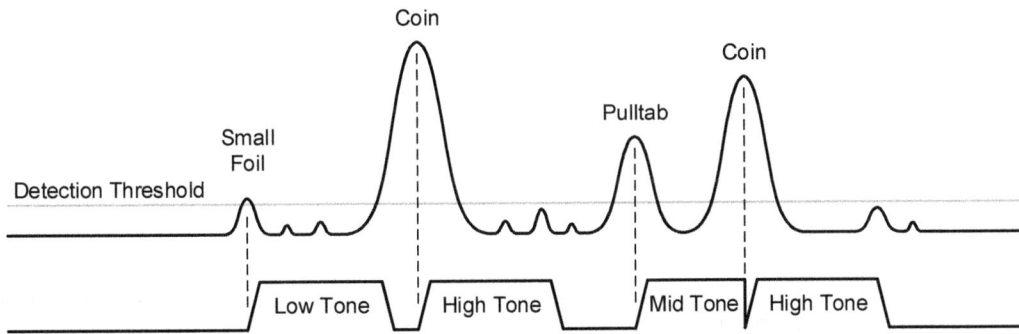

Fig. 18-15: **Peak Tone Response**

response tone is generated starting at the peak of the response and lasting for the prescribed duration. Note that the pull tab response gets truncated because of a nearby coin. A drawback with this approach is that the audio tones do not begin until the peak of the signal which creates a latency between the location of the target and the audio response.

Another approach is to make the audio response track the phase response. In this case, as soon as the target detection threshold is exceeded the audio tone begins to track the target's phase response in real time. Although it eliminates the latency problem, this approach can produce a very busy audio which some people may not like. However, it can also produce more subtle target information to those who can tolerate the barrage of information. For example, the phase response of a bottle cap varies considerably as the coil is swept over, yet the peak phase is often similar to a high-conductive coin. A simple peak-phase audio response will identify the bottle cap as a coin, while phase-tracking audio will tell you otherwise[24].

There are a myriad of other ways to produce audio responses based on first and second derivative signals. Many gold prospecting detectors based their audio responses on a rectified first derivative signal using a simple VCO-style response, with perhaps a low tone for probable iron targets. Because a rectified first derivative signal starts out slowly, reaches a peak, and then rapidly plummets to zero and abruptly stops (Figure 17-16), the resulting audio has a "zip" sound to it.

18.13.3: Mixed Mode Audio

Nautilus, a minor detector manufacturer that specialized in designs for relic hunting, introduced an innovative feature called *mixed mode* audio. With most detectors, you can hear an all-metal signal in, say, the pinpoint mode and a discrimination response (perhaps with tone ID) in discrimination mode. With mixed mode, you hear both signals at the same time. A further enhancement for headphones, *stereo mixed mode*, places the all-metal signal in one ear and the discrimination signal in the other ear.

The result can be a rather busy presentation of sounds, especially if the discrimination audio is the continuous phase-tracking type described previously. But the ability to hear non-discriminated signals is a huge advantage in being able to separate closely adjacent targets where one is rejected and the other is not — for example, a silver coin next to a nail. With just discrimination audio, you are likely to hear a chopped-up response from the silver coin due the nail. With mixed mode, you might still hear the chopped-up response, but you will also hear the all-metal response that clearly shows two adjacent targets, which should steer your sweeping technique to better separate them.

A key requirement for stereo mixed mode is the inclusion of two audio channels. Most detectors include a built-in speaker but so far no model has included two speakers for stereo audio. Practically, detector enclosures tend to be too small for stereo speakers to be effective and the act of swinging the detector continuously shifts the position of the speakers relative to the ears, further reducing effectiveness. Instead, stereo audio should focus on headphone usage, and the audio blended when the internal speaker is used.

---

24. Discrimination "between the ears."

18.13.4: Depth Modulation

A final audio technique to consider is that of *depth modulation*. This is where deeper targets produce a fainter audio signal so that you can tell by the audio response about how deep a target might be. It seems logical that the audio loudness will naturally be inversely proportional to depth but in early analog designs there was often a lot of gain in the audio stage that produced about the same loudness regardless of depth, up to the fringe of detectability. Some modern digital designs offer the selection between modulated and unmodulated audio.

18.13.5: Headphones

Practically all metal detectors include a headphone jack. For decades, the 1/4" jack was a standard but newer, smaller enclosures have forced designs to use the smaller 3.5mm jack. In the case of waterproof detectors, most headphone jacks use a water-tight M12 connector. A bonus with an M12 connector is that extra pins can be used for charging a permanent internal lithium battery or for performing a USB firmware update of the micro, both of which are increasingly popular features.

18.13.6: Wireless

In about the last decade, wireless headphones have become popular because they eliminate the long dangling cord that easily gets caught on brush and is otherwise an annoyance. A key element with wireless headphones is *latency*, the time it takes to encode, wirelessly transmit, and decode the audio packets. High latency can delay the audio response enough that the response is heard after the coil has passed the target[25]. A maximum latency to strive for is 50ms.

Wireless audio can be achieved using a custom solution whereby a defined structure is filled with the current audio requirement (frequency and amplitude) and sent as packets of data. The headphones receive the data and generate the audio tone as specified. For stereo, an 8-byte structure might look like:

```
struct {
    unsigned int LeftFreq;
    unsigned int LeftAmp;
    unsigned int RightFreq;
    unsigned int RightAmp;
}
```

A drawback with this method is that the headphones are then specific to the detector. That is, whatever packet structure is used by the detector must be understood by the headphones. This is the approach used in the Garrett Z-Lynk system.

Another approach to wireless audio is to use an industry standard like *Bluetooth* (BT). In this case, you would generate the actual audio signal in the detector and send it to the BT transmitter. The transmitter encodes the audio for RF transmission; the BT receiver in the headphones decodes the RF back to the original audio and plays it to the speakers. For low latency, it is important to select a proper BT protocol. Normal BT audio has a latency of 100ms or so which is noticeably slow. *AptX Low Latency* (AptX-LL) uses a special BT coding/decoding (CODEC) method that reduces latency to about 30ms which is not noticeable. AptX-LL is a proprietary method owned by Qualcomm but there are inexpensive BT modules available[26] that support it. Off-the-shelf AptX-LL headphones are also widely available, as are transmitter dongles that simply plug into the headphone jack. Another option is the newer Bluetooth *LE Audio* standard that was released as part of the BT 5.2 standard. There are, again, inexpensive modules[27] that support BT LE Audio, but off-the-shelf headphones are still scarce[28].

---

25. Some really cheap off-brand detectors have this problem even when using the speaker.

26. For example, Feasycom.

27. For example, Fanstel.

28. Just because a chip or a module or headphones support BT 5.2 or higher does not mean it supports LE Audio. It must specifically state that LE Audio is supported, or that it includes the LC3 CODEC. Otherwise, assume it does not and that latency will be a disappointing 100ms or more.

Besides audio, wireless can also be used to control the detector. For a handheld device this doesn't seem too important, but for a security walk-through it can make setting up multiple units much easier. In this case, the security of the wireless system is important so that hackers can't meddle with settings.

18.13.7: Tactile Feedback

Some detectors (notably pinpointers) have a vibration motor to give a tactile response. This can be in addition to, or instead of, the audio response. In a few cases (again, mostly pinpointers) the tactile response is the only response; there is no audio option. Tactile feedback can be a critical requirement for those with a hearing impairment.

Tactile response began with early underwater metal detectors in the 1970s. For example, the White's *Goldmaster Amphibian* included a Scuba hood with a built-in "bone phone" which, placed against the user's temple, would vibrate and create the sensation of an auditory response. Today, after-market bone phones are available that plug into the headphone jack and are worn like traditional headphones.

Modern built-in implementations use a vibration motor with an off-center weight to create a physical vibration, the same as in a common cell phone. In mainstream detectors, the Nokta *Simplex* and *Legend* models include a vibration motor in the handle.

18.14: Display

18.14.1: Segmented and Graphics LCDs

Most modern detectors use a segmented monochrome LCD; some use a monochrome graphics display; and fewer yet use a color graphics display. In all cases, outdoor readability is critical. For a segmented or monochrome graphics display, FSTN is probably the best choice. For color, a purely transmissive display is only readable with a backlight, and a maximum backlight is needed in bright sunlight. This requires a lot of power. A transflective display[29] takes advantage of reflected sunlight, making it sunlight-readable with no backlight at all. The drawback is that a transflective display looks a little more "washed out" than a transmissive display.

Micros are easily found with built-in display capabilities, either for segmented displays or graphics. Segmented displays are usually designed with multiple backplanes; for example, a display with 4 backplanes, each with up to 40 segments, can have up to 160 segments. Micros with built-in LCD support will directly interface with this example using 44 pins (4 backplane pins and 40 frontplane pins). Therefore, micros that drive segmented LCDs tend to have a high pin count. You can also use an LCD driver chip which connects to the micro using an SPI port. This allows the use of a smaller micro, but the LCD driver chip will have a high pin count. Cost-wise, the difference between a larger micro with built-in LCD support and a smaller micro and separate LCD driver is usually a wash. However, the separate LCD driver chip can often make the PCB layout much easier, with the ability to place the driver chip next to the LCD connector while the micro is elsewhere on the PCB. In either case, you will need to write all the code that manipulates the segments and keeps track of their status.

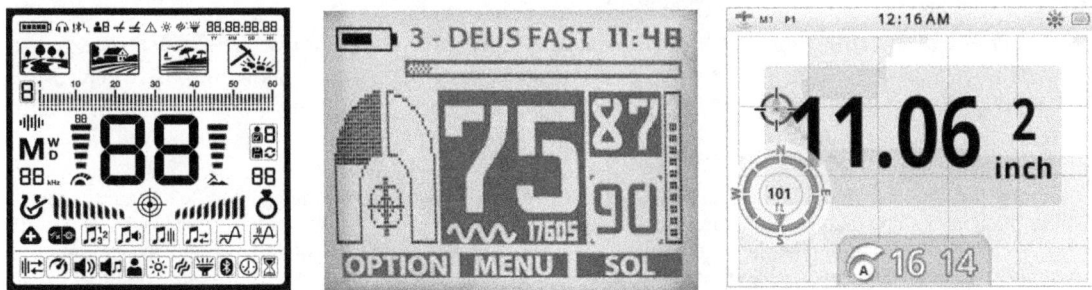

Fig. 18-16: **Segmented LCD, Monochrome Graphics, and Color Graphics Displays**

29. Used on the White's *V3* and Minelab *CTX3030*.

A graphics LCD requires far more computing horsepower because graphical elements must be "drawn." They are also memory-intensive because there can be a lot of pixels. Even a low-end monochrome display of 128x64 (as used in the White's *XLT* & *DFX*) has 8192 pixels, far above the example 160-segment display above. In a monochrome display each pixel only needs as little as 1 bit of memory, so 8192 pixels requires 1024 bytes. A gray-scale display might have 8 bits per pixel, or 8192 bytes.

A color display is even more demanding, and a 320x240 screen with 16 bits of color depth (as used in the White's *V3* and Minelab *CTX3030*) requires 320x240x2k = 154kBytes of memory. This is static RAM, not flash memory, and usually requires external memory chips. Graphics processing is also computationally intensive and often demands a micro that has built-in graphics processing capabilities.

For experimental and developmental purposes, *Nextion* offers an excellent color graphics option with their self-contained displays. A Nextion display is a module that includes the memory and processing in the module. Screen elements are designed in PC-based software and loaded into the module's memory. A micro then communicates with the display via SPI and sends simple commands to place the pre-built elements on the screen. The result is a full color high-resolution display with an interface that is not much more complex than a segmented LCD.

Another option is an alphanumeric display. These have shown up in a few production metal detectors like the Tesoro *Cortés* and White's *MXT*. They are easy to program and support an extended character set, and include the ability to create a limited number of custom characters.

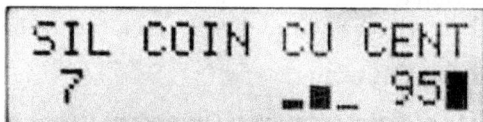

Fig. 18-17: **Tesoro *Cortés* Display**

18.14.2: TID and VDI

Two fundamental pieces of information included in most metal detector displays is a probable target identification and a probable depth. In many cases, "probable" should be "possible" or even "perhaps maybe, we're not sure." It is a very inexact process.

The target identification (target ID, or TID) is a representation of the target phase and is usually reported on a scale that varies amongst the manufacturers; a popular range[30] is 0 to 100. The early analog TID meters were marked with regions indicating various common coins and trash targets. At the time, the focus was on the US market so the "common coins" were all US coins. When a target was swept over, the needle would jump to the appropriate TID and was held in that position for about one second. In most cases, the meter would also indicate the target's depth when the detector was switched to pinpoint mode. Depth was only accurate for predictable targets, namely US coins.

When the White's *Eagle* was introduced in 1987 it included a numerical target ID called the *Visual Display Indicator* (VDI). For that model, the VDI was a number from 0 to 95 where 0 represented iron and 95 represented a very high conductor[31]. Some coins (again, US) were associated with specific VDI numbers; for example, a US nickel was "20" and a US quarter was "83". Earlier detectors with analog meters were also usually labeled with a range from 0 to 100 and also marked with US coinage ranges, so the VDI number was more evolutionary than revolutionary[32]. Figure 18-18 shows the Teknetics *8500* analog meter and the White's *Eagle* display for comparison.

A few years after the *Eagle*, White's introduced the *Spectrum* model where the VDI range was extended to −95 on the low end. The middle of the scale, 0, became the division point between ferrous (−95 to 0) and non-ferrous (0 to +95). This same scale was used throughout the remainder of White's corporate life.

---

30. Historically, the very first TID detectors (Teknetics and White's) used analog meters, and these analog meters were often labeled by their manufacturers as having a default range of 0-100. Most likely, this is why many TID ranges are 0-100. Additionally, the pointer in an analog meter moves left-to-right and having a maximum pointer swing indicate silver is likely why ferrite (0°) ended up on the far left and silver (~180°) on the far right, exactly opposite of the usual mathematical convention. And all this is why I depict vector phase responses from left to right.

31. Such as an Atocha bar.

32. The *Eagle* itself, however, was revolutionary, as it was the first detector to make use of a micro.

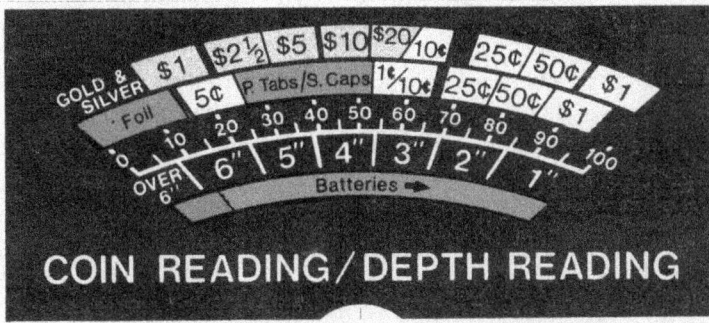

Fig. 18-18: **Teknetics *8500* Meter and White's *Eagle* Display**

Meanwhile, other companies adopted the VDI numerical readout but chose their own ranges, with many of them using the range of 0-99 or 0-100, which mimics a generic analog meter. In all cases, the low end of the scale represents ferrous targets such as iron and the high end of the scale represents non-ferrous targets of increasing conductivity. There is a particular point on the scale that delineates ferrous and non-ferrous[33], called the *ferrous break point* or *iron break point*. This can also vary amongst manufacturers, and even amongst models from a single manufacturer, so different models will often have different VDI numbers for the exact same target. See Figure 18-19.

◀——— Ferrous ———+——— Non-Ferrous ———▶

0°	90°	180°	Raw Phase
-95	0	95	White's
0	15	100	Fisher F75
0	40	100	Teknetics T2
0	35	100	Garrett AT-Pro
0	45	100	Garrett Ace Apex
0	10	60	Nokta Legend
-9	0	40	Minelab Equinox 800
-19	0	99	Minelab Equinox 900
-6.4	0	99	XP Deus & Deus 2

Fig. 18-19: **VDI Ranges for Various Detectors**

One thing to notice is that most TID scales are not symmetrical about the iron break point, White's being the exception[34]. All others have a compressed ferrous range. The reason for this is that there is less need for a high-resolution iron range. Most users simply don't care about the exact nature of the iron targets, and often they will discriminate out the entire iron region.

All of these TID scales roughly map to the phase of the target response, although the mapping may be very non-linear. The target phase response is almost always taken from the second derivative signals where any ground signal has been attenuated. Usually the TID is based on the phase response at the

---

33. We've already seen in Chapter 17 that this is somewhat not true; ferrous targets with substantial eddy responses often end up on the non-ferrous side.

34. Even on White's models the ferrous range is visually compressed so that it gets less meter space than the non-ferrous range. See the V3 display in Figure 18-21.

18: Digital Techniques

peak of the target signal where (we assume) accuracy will be greatest. However, with noise (especially on deep targets) it may be necessary to average several samples around the peak to get reasonably consistent results on each sweep over the target.

We also know from Section 16.11 that a target's phase depends on the TX frequency, and as frequency increases eddy targets tend to move toward higher phase responses. The TID scale can be "normalized" for a particular frequency so that, for example, a selectable frequency detector like the Minelab *X-Terra Pro* will produce a consistent TID no matter which operating frequency is chosen. There is an advantage in maintaining a raw (un-normalized) TID scale versus frequency. A low frequency (say, 2kHz) will expand the TID resolution of high conductors while a high frequency (say, 50kHz) will expand the TID resolution of low conductors.

18.14.3: Spectral Meter

White's scored another first with the *Spectrum* (1991) when it introduced the spectral meter[35]. Instead of displaying a VDI number or simply giving a single metered result, the spectral meter can produce results across the entire VDI range. A well-behaved target like a silver coin would typically still produce a single metered response, but iron targets (especially) would produce a "smeared" response over multiple spectral points. Initially, the *Spectrum* used a 4x20 character display that limited the resolution of the spectral meter, both in terms of VDI resolution and amplitude resolution. In 1994, White's released the *Spectrum XLT* with a graphics display which improved both the VDI and amplitude resolution.

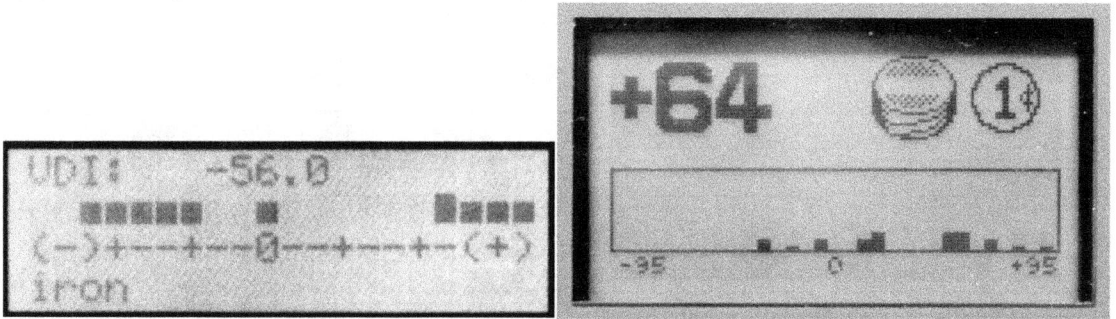

Fig. 18-20: **SignaGraph Meters — *Spectrum* and *XLT***

Over the years, practically all detector manufacturers have adopted some kind of spectral meter, from just a few coarse categories (iron, foil, nickel, pull tab, cent, quarter, dollar) to the finely graduated "SpectraGraph" meter of the White's *V3*. The *V3* used a triple spectral meter to display the spectral responses of each frequency individually, in color no less.

A slightly different approach was pioneered by Minelab in 2003 with the *Explorer* model and called *SmartFind*. Instead of using the dimensions of amplitude (Y-axis) versus VDI number (x-axis),

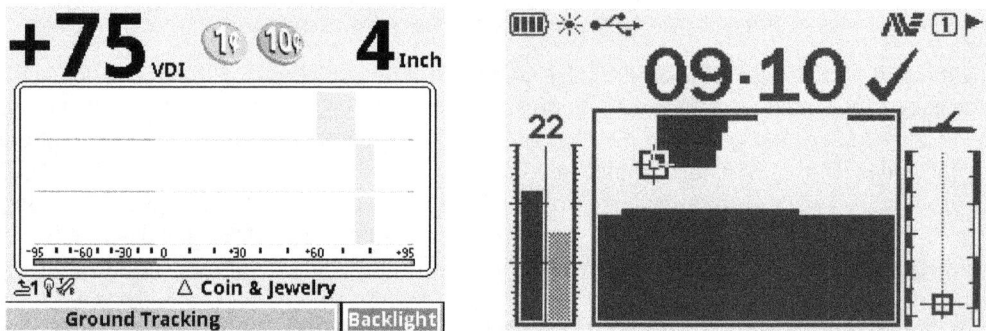

Fig. 18-21: **White's *V3* SpectraGraph and Minelab *E-Trac* SmartFind**

35. White's called it the *SignaGraph*.

they used conductivity versus "ferrousity." Conductivity is largely the same as a non-ferrous VDI, but ferrousity attempts to measure whether a target is ferrous in nature. As we know from Chapters 4, 5, and 24, iron targets can exhibit both a magnetic and an eddy response, often looking more nonferrous than ferrous, with the steel bottle cap being a notable example. However, the magnetic and eddy responses often vary differently with frequency (and the *Explorer* was a multifrequency design) so by looking at two or more frequencies it is possible to determine a target's measure of "ferrousness" (see Figure 24-19 as an example). Even with a single frequency design, the magnetic and eddy responses vary during the sweep of the coil (Figure 5-3) and this can also be used to determine ferrousness if the target is not deep.

Like a simple TID, a spectral meter represents the target's phase response that has been scaled in some way. Unlike a TID, the spectral response usually takes in a lot more samples of the swept response and "bins" those samples according to magnitude and phase. Let's say we sweep over a shallow silver coin and the sample readings are as shown in Figure 18-22a. Each sample has a very consistent phase so the resulting spectral response will be tight. In Figure 18-22b a deep silver target might have readings more divergent, resulting in a spectral response that is broader but still largely signifying a silver coin. Finally, a steel bottle cap (Figure 18-22c) might have some readings that venture into the ferrous region and some that land in the non-ferrous region, producing a very "smeared" response. This approach gives you far more information over the entire swept response than a simple VDI number and is easier to mentally process than an equivalent real-time audio response.

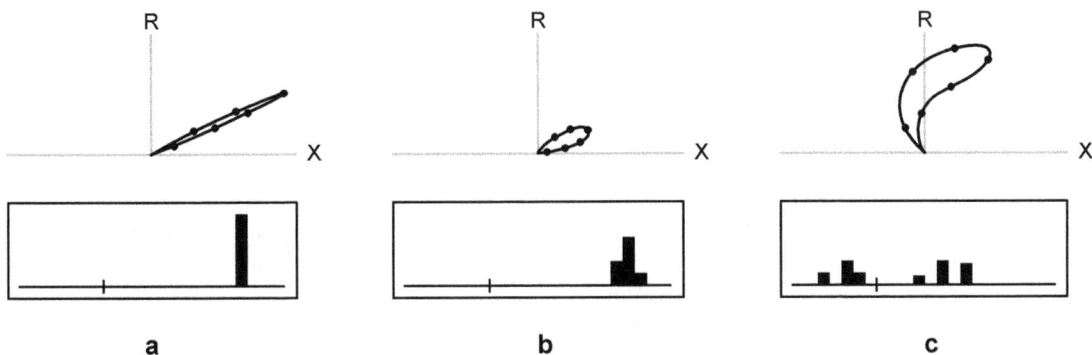

Fig. 18-22: **Spectral Responses**

18.14.4: Category Meter

A category display as seen in the Teknetics *Omega 8500* (Figure 18-23) is the simplest kind of spectral meter to implement. Once the target's phase response is calculated — either from a peak response or an average of a few readings — a simple look-up algorithm can determine which category to select. Many detectors with a category display also show a VDI number, so the selected category

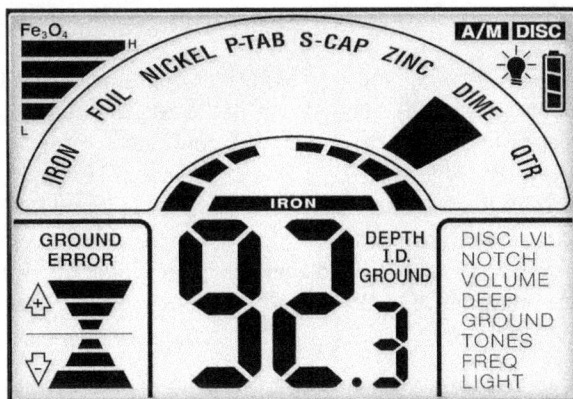

Fig. 18-23: **Teknetics *Omega 8500* Display**

gives a rough idea of the target and the VDI number may offer a finer resolution. Also in detectors with category displays, the audio ID tones are often differentiated by category. That is, if the display has eight segments, then the audio will have eight distinct tones.

18.14.5: Discrimination

Digital discrimination was briefly covered in section 18.10. Most modern detectors allow the user to create a discrimination mask with as fine a resolution as the spectral meter allows. In Figure 18-23, the *Omega* display has spectral meter with eight categories and any of those categories can be rejected. The shorter bar below each of the eight categories indicates accept/reject, and the image shows that only the "P-TAB" category is currently rejected. But we have no finer control than that. Software will define, for example, just what range of phase responses are included when the "P-TAB" category is rejected. That might include something we want, like a gold ring.

If we want finer control over discrimination, then we need a display with a finer resolution spectral meter. In Figure 18-16 the Nokta Legend display (far left) shows a spectral meter with 60 segments. If these were distributed evenly then each segment would represent $3°$ of target phase. But the iron region is compressed to 10 segments for $9°$ of ferrous resolution, and the non-ferrous side has 50 segments which is closer to having $2°$ of resolution.

Some detectors have even finer control. The White's *V3* discrimination is programmable by VDI number with 191 settings, equally divided between ferrous and non-ferrous. The Minelab *E-Trac's* two-dimensional SmartFind has a conductivity resolution of 50 and a ferrous resolution of 35 for a total of 1750 selectable discrimination points. Whether finer control is better is debatable. A target's phase response is usually clean and tight in shallow air tests, but phase spreads with depth and ground mineralization will also alter phase responses. So care must be taken when setting up a fine-resolution discrimination mask to give plenty of margin for phase spreading.

18.14.6: Sweep Averaging

Audio responses are instantaneous but visual responses need not be. A visual TID (whether a simple VDI number, a few categories, or a complete spectral graph) is normally held on the screen for 1 second or more to allow the user time to look at the screen after hearing an audio alert. The screen result can be held for even longer; if the user repeatedly sweeps the target then new results can be averaged with older results to perhaps produce a more accurate TID.

The White's SignaGraph, in fact, includes a setting called *Fade Rate* which controls how long a result remains valid for averaging. Once a result is older than the fade rate, it is deleted from memory. Depending on the length of the fade rate, the maximum expected sweep rate, and the size of the TID method, sweep averaging can consume a modest amount of memory. For example, consider the Minelab *Explorer* with a 2D spectrum of 1750 bins. Assuming that each bin requires 1 byte for a relative signal strength, a single result requires 1750 bytes of data. Now suppose the fade rate is 3 seconds and a user can sweep the target 5 times per second; the required data storage in now 26,250 bytes. Not enormous, but it needs to be considered when choosing the processor.

18.14.7: Confidence Meter

Some detectors include a *confidence meter*. This might be in the form of a segmented bar graph meter, a one-digit number, or a percentile. The idea is that a target whose phase response is strongly consistent through the sweep (Figure 18-22a) gets a high confidence rating, and a target whose phase response varies through the sweep (Figure 18-22b) gets a low rating. However, a target whose response varies significantly through the ferrous and non-ferrous regions (Figure 18-22c) might get an "iron" identification with a high confidence rating.

Deeper targets have a weaker response and their phase will usually vary more, even from one sweep to the next. Therefore, confidence levels tend to drop as targets get deeper.

18.14.8: Depth Readout

While target ID is usually shown in motion mode using the second derivative signals, depth is usually shown in pinpoint mode and is based on the raw signals. For the purpose of audio, the X signal is not needed in pinpoint mode as the R signal (or G signal) alone can represent the relative target strength. However, a shallower dime and a deeper silver dollar can both have the same G signal strength so this approach won't work if we want to create an actual depth estimate.

A way to solve this is to also keep track of the raw X and R signals during pinpoint mode and use them to calculate the target phase, realizing that ground variations can cause errors in the X signal. The target phase is used to identify the probable target type which is then used to scale the depth response so that a shallow dime and deep silver dollar exhibit proper depth readings. Because the raw X signal includes ground, when switching to pinpoint mode grab an initial X reading and use it to normalize all subsequent X readings. For this to work, the user should enter pinpoint mode with the coil to the ground and just off the target, then move the coil over the target while keeping the coil a consistent height above the ground. This is obviously prone to error.

If you are already calculating the target phase in order to scale the depth, then it is possible to maintain a VDI readout during pinpoint mode even when using static signals, knowing that variations in coil height above the ground can alter the target's apparent phase.

Similarly, in motion mode we are already calculating the target phase for the purpose of TID, and it is little extra work to also calculate a depth estimate. However, the motion filters alter the signal strength based on sweep speed so there is no reference point with which to determine depth, unless the coil includes an accelerometer for determining sweep speed and correcting for the motion filters. Even without an accelerometer, we still have access to the raw signals and, again, if we can reasonably normalize the X signal using ground data just prior to hitting the target then a decent depth estimate is possible in motion mode.

18.14.9: Ground Meter

A final display element that is found on a few detector models is the ground meter, used to indicate the severity of the ground mineralization. Like any target, ground has two components: magnitude and phase. Some detectors show a simple bar graph which is likely correlated to the strength (magnitude) of the ground. Some models have a two or three digit readout which is correlated to the phase of the ground. Rarely, a detector includes both. The Fisher *F5* (Figure 18-24) has a 2-digit ground phase readout, plus a 4-bar strength meter.

Fig. 18-24: **Fisher *F5* Ground Meter**

Although most end-users have little interest in the ground information[36], it can be useful in development, especially in determining how ground conditions affect target ID and depth estimates. Ideally, you want both elements — ground phase and strength — and they can be calculated from the raw X and R signals. When it's determined that the signals include a target, that data should be thrown out. That is, the ground meter should be locked while actively swinging over a target. This is the same situation as with automatic ground tracking so if you have implemented AGT in your design then you have everything you need for a ground meter.

Commercial detectors often display a ground phase number that seems to have no correlation to either loss angle or the VDI scale being used. An "ideal" ferrite ground might be 0, or 90, or −90, and increase or decrease from there. My personal preference is to display the true loss angle starting at $0°$ for ideal ferrite and increasing from there. Relative magnitude[37] can be indicated with one or two digits, or a simple bar graph as with the *F5*.

---

36. Surprisingly, some do, and often compare ground meter results with other users on forums. The ground meter results can also be useful when a user complains about performance and it turns out they are hunting in severe ground. In this respect, the meter is also useful in determining when to back off the sensitivity setting.

37. Absolute magnitude will vary based on coil size, coil height above the ground, and TX field strength. It would be possible to calculate absolute susceptibility if the coil has an ID chip, with coil height above the ground being the only variable.

18.14.10: Modes and Menus

Depending on the complexity of the detector, the display may include a mode indicator and a menu for common settings. In Figure 18-23, the *Omega* display has icons in the upper-right corner to indicate Disc or A/M mode. Some detectors are multi-mode with several modes for particular environments. For example, the Nokta *Legend* (Figure 18-16, far left) has the following modes represented by the four large icons:

- Park (Basic coin mode)
- Field (Relic mode)
- Beach
- Prospecting

In a multifrequency detector (such as the *Legend*) the mode often determines which mix of frequencies are transmitted. The mode might also select other internal attributes, such as:

- Whether salt cancel is activated (usually beach mode only)
- Filter settings (extra deep for Field, extra fast for Prospecting)
- The scale of the VDIs
- Target icon set (on a graphics display)
- Audio mode or number of tones

Most detectors allow one or more settings to be changed; some basic settings are Sensitivity, Volume, and Discrimination. For example, those settings and a few more can be found in the lower-right section of the *Omega* display (Figure 18-23). The *Omega* has up/down keys for scrolling though the settings and the user can press the MENU key to edit a particular setting, with the VDI digits used to display a numerical value that can then be altered by the user.

The *Legend* uses icons instead of words for the settings and they are found in the two rows at the bottom of the screen. Icons can produce a more universal appeal that English words cannot; however, a large number of icons (19 just at the bottom of the *Legend* screen) may be confusing, especially to a new user. The Minelab *Equinox* also uses a set of icons for user settings, where the addition of an underscore gives the icon a secondary meaning which may be even less obvious than the initial meaning of the icon.

Setting		Advanced Setting	
🔊	Noise Cancel		
⌇	Ground Balance		
🔊	Volume Adjust	🔊	Tone Volume
🔊	Threshold Level	🔊	Threshold Pitch
🔊	Target Tone	🔊	Tone Pitch
✕	Accept/Reject	✕	Tone Break
⬒	Recovery Speed	⬒	Iron Bias

Fig. 18-25: **Minelab Equinox Settings**

Issues such as these result from the limitations of a segmented display which can only show fixed icons or text. It becomes a careful balance between providing the user all the useful features they want and requiring them to carry the manual with them in the field to figure it all out. A graphics display offers far more flexibility in that it can display whatever you want, even using the same screen space for different purposes. Furthermore, when using text you can provide translations in multiple languages[38].

18.14.11: Other Display Elements

Detector displays will usually include a few more elements than those discussed so far. At the least, most include a battery level indicator, usually a bar graph meter in the shape of a battery cell. Another common element is an overload indicator, sometimes with the word "OVERLOAD," or the letters "OL" using the VDI digits, or a custom icon. Most detector displays also include a backlight and it's useful to include something on the screen to indicate the backlight is on because, in bright sunlight, it may not be apparent, and an unneeded backlight just wastes battery power.

A selectable frequency or multifrequency detector might include an indicator for which frequency(s) have been selected. Often this is mode-dependent so a separate frequency indicator may be

---

38. For example, the Minelab *CTX3030* supports English, Spanish, Portuguese, French, German, Italian, Polish, Russian, and Turkish.

redundant. Most of the newest models include wireless headphone support so there is usually an icon to indicate wireless connection.

18.15: Saving Settings

A good design will save the current mode and any setting changes before power down so the user is presented with the same operation on the next power-on. It is common for many of the settings to be separately saved in the different modes, so that (for example) a Park mode and a Field mode can have different saved discrimination settings. When the user changes modes, the unique settings for the new mode are loaded from memory, while any changes to the previous mode are saved to memory. There may also be certain settings that are global to all modes, such as the wireless connection, the backlight level, or perhaps a master audio volume.

Saving settings requires EEPROM. Many micros include some amount of EEPROM but often it's not much, and some micros have no EEPROM at all. In those cases you will need an external EEPROM chip which will serially communicate with the micro, either with SPI or I2C. It is also possible to create pseudo EEPROM using flash memory but this can be problematic as constant changes to flash memory can cause it to eventually "wear out."

Explore

18.16: Direct Sampling

Figure 18-1 shows a block diagram for a direct sampling scheme. With direct sampling, the signal is presented to the micro as an undemodulated raw (or "RF") signal. Most direct sampling detectors use an audio CODEC chip[39] for the ADC, and a common resolution and sample rate are 24 bits and 192kSps.

Normally, the Nyquist criterion demands that we sample at least twice the maximum frequency but, for quadrature demodulation, that needs to increase to at least four times the maximum frequency. If the sample rate is 192kSps, then the maximum frequency we can sample is 48kHz and, with some room for filtering, even less than that. One thing you might notice about the newer multifrequency detectors is that many of them max out at around 40kHz. This is why. There are some newer audio CODECs that sample up to 768kSps which could support TX frequencies up to 100kHz and with more oversampling.

18.17: Digital Demodulation

With direct sampling, after reading in the raw ADC data and perhaps doing some filtering, the first order of business is to demodulate the RF signal into the necessary X and R baseband signals. This is done using a digital demodulator. In an analog full-wave demodulator we use a simple switch and square wave clock to effectively multiply the signal by $+1, -1$. The same can be done digitally by multiplying the incoming data by $+1, -1$. Figure 18-26 shows what this might look like if the preamp signal is oversampled by a factor of 16.

Multiplying by $+1, -1$ results in a "wideband" demodulation whereby higher-order harmonics and noise can be aliased back down to baseband, thereby degrading the SNR (see Figure 14-44). A commonly used approach in radio system design is to multiply the RF signal by a sinusoid, which results in "narrowband" demodulation that rejects the excess noise and harmonics[40]. In software, this can be accomplished by multiplying the incoming signal by numbers that represent a sinusoid. If the sample rate is exactly four times the TX frequency then the multipliers are $[0, +1, 0, -1]$ for the X channel and $[+1, 0, -1, 0]$ for the R channel. That is to say, you are effectively multiplying by $+1, -1$. In order to take advantage of narrowband demodulation the signal must be oversampled by some amount.

---

39. Such as the PCM1862 from Texas Instruments.

40. The technique described has been used since at least the 1920s in superheterodyne receivers, and since about 1990 in IF sampling and direct sampling software-define radio (SDR) designs. Most likely your cell phone uses direct sampling with narrowband software demodulation.

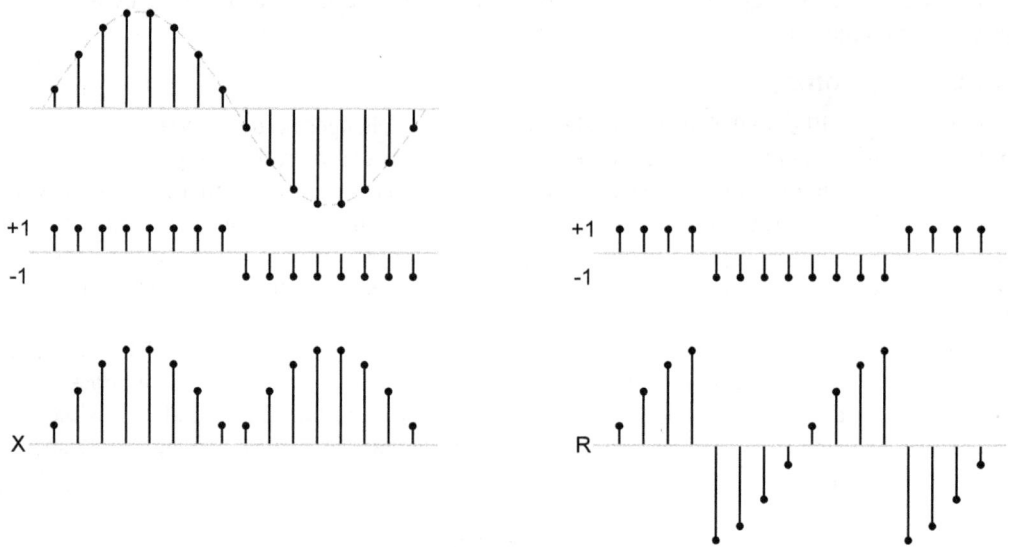

Fig. 18-26: **Digital Demodulation**

It should be noted that Minelab received a patent on this scheme (US7579839) despite its widespread use in radio systems, so avoid implementing anything that might infringe until the patent expires (29 June 2025).

18.18: Hodograph

In plots like Figure 16-10 we often show a target response as a vector on an X-R graph. A metal detector with a graphics screen can actually generate such a plot, called a *hodograph*. The White's *V3* has an optional analysis screen for generating a target hodograph, see Figure 18-27. The screen plot on the left shows a well-behaved silver coin[41], the plot on the right shows an ill-behaved steel bottle cap that produces almost the same VDI.

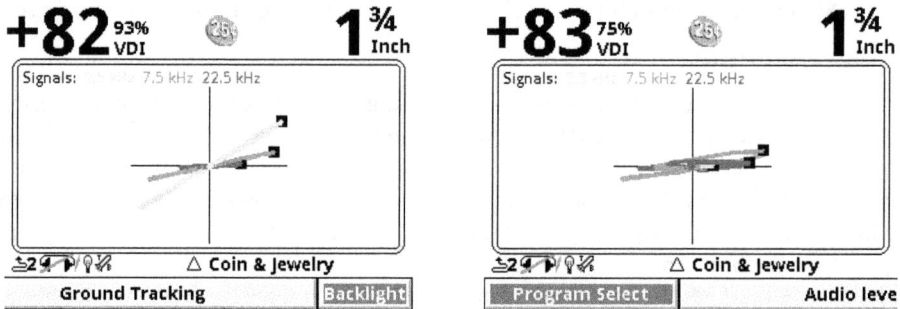

Fig. 18-27: **White's *V3* Hodograph**

Generating a live hodograph is best accomplished with a search coil that includes an accelerometer as it is important to determine the start and stop points in the sweep. The *V3* did it without an accelerometer but it could be tedious to get good, repeatable results.

18.19: Static Mode Improvements

In static mode, the X channel reading is prone to error because of variations due to ground mineralization. In pinpoint mode where the coil is limited to a small area of ground, X channel variations are mostly due to an inability to keep the coil a consistent height above the ground.

---

41. There are 3 frequencies, therefore 3 vectors.

In Chapter 4 the loading effects of targets on the TX circuit were considered. It may be possible to monitor the TX circuit in such a way that the ground strength can be determined independently of the X channel signal, thereby providing a way to correct X channel errors.

18.20: Special Coil Considerations

Mono, concentric and DD coils tend to produce well-behaved Gaussian-like target responses. Some coils — notably those with a normal and an inverted RX coil — can produce bidirectional responses. An example is the figure-8 (Bigfoot) coil where the forward half produces a positive response and the rear half produces a negative response. Some detectors (notably White's *XLT*) are designed to process all four quadrants of the target response instead of the usual two quadrants. This is quite easy to accomplish in software by creating both a positive and a negative target threshold and monitoring the raw R channel signal. A negative signal (quadrants 3 & 4) can then be rotated and treated like a positive signal.

Another example is that of the butterfly coil, such as the DOD coil used for the Minelab *GPZ7000* or the coils typically found in security walk-throughs. In this case, a normal raw target response looks like the first derivative response of a concentric coil. Depending on sweep direction, the leading lobe may be positive or negative; this can be handled in the same way as the Bigfoot coil. Otherwise, an accelerometer in the coil can indicate the direction of coil swing.

In cases of special coils, you will either need a way to manually select special processing or an automated way to detect the type of coil. We've already covered the concept of embedding a micro in the coil for automated coil ID and this is a good example of its utility.

18.21: The Ultimate Programmable Detector

Moving the signal processing from analog circuitry to software opens up a nearly infinite world of possibilities. It is impossible to describe in detail even a subset of what can be done without doubling the size of this book. In commercial designs, one detector stands as a monument of just how far you can go in software[42]: the White's *V3i*.

The *V3* (the model just prior to the *V3i*) was the first hobby detector with a color display and it made good use of it. It presented the VDI spectrum for the 3 frequencies separately in user-definable colors, had a triple-bar pinpoint meter, and a hodograph display. An innovative programmable "live control bar" gave quick access to commonly used settings and Windows-like menu system provided access to everything else. Tones were fully programmable down to the individual VDI and it had optional stereo mixed mode audio and wireless headphones. The *V3i* ("i" for International) added multiple languages and a full on-board context-sensitive help system.

This does not even begin to touch on all the features of the *V3i*. If your goal is to design and build a software-defined metal detector (SDMD) with advanced features then it is worth the time and money to obtain a *V3i* and see what it does.

---

42. Some will say too far.

PERCENTAGE OF SENSITIVITY LOSS
White's vs Brand X

NOTE:
While some manufacturerers claim a loss of sensitivity on coins when adjusted to reject foil of 15%, a loss of 25% when adjusted to reject bottle caps and a loss of 50% when adjusted to reject pull tabs, it should be clear from the chart below that White's has only a 3% loss on most coins when adjusted to reject bottle caps (13% loss on nickles) and only a 20% loss on most coins when adjusted to reject pull tabs.

We do not recommend the operator to adjust out pull tabs because he will then also lose nickles and rings.

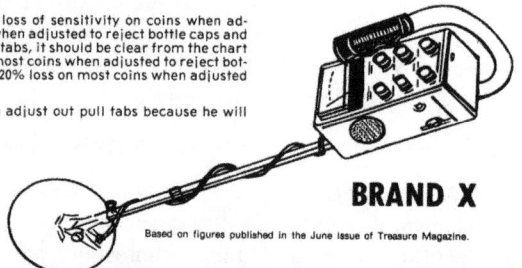

WHITE'S

Based on figures published in the June issue of Treasure Magazine.

BRAND X

SET TO TUNE OUT

PERCENT LOSS ON THESE COINS:	Gum Wrapper	Foil	Bottle Cap	Pull Tab
Quarter	0	0	3%	15%
Penny	0	0	3%	20%
Dime	0	0	3%	20%
Nickel	0	5%	13%	100%

SET TO TUNE OUT

PERCENT LOSS ON THESE COINS:	Gum Wrapper	Foil	Bottle Cap	Pull Tab
Quarter	NO DATA AVAILABLE	15%	25%	50%
Penny		15%	25%	50%
Dime		15%	25%	50%
Nickel		15%	25%	100%

Inside the Metal Detector

PART 5
Pulse Induction

Pulse Induction

"And now I see with eye serene,
the very pulse of the machine."

— William Wordsworth

19.1: Overview

So far all of the designs we have considered are of the continuous wave type, specifically using a sinusoidal transmit current. The current waveform does not need to be sinusoidal nor even continuous; target eddy currents are generated by a changing magnetic field (dB/dt) so any changing magnetic field will do, thus any changing coil current will also do. The faster the magnetic field changes, the greater will be the induced eddy currents:

$$i(t) = \frac{\varepsilon(t)}{Z} = -\frac{dB}{dt} \cdot \frac{1}{Z}$$

Eq 19-1

It stands to reason that a fast-slewing signal, such as a square wave or a pulse, will produce a stronger target response than, say, a sinusoid. A square wave current through a coil is a difficult challenge so, as we have done throughout the book, we will start out with a more basic approach and work our way up.

Pulse induction (PI) has been around since about the 1950s (see Chapter 1). In its most basic form a large current (often more than 1 amp) is generated in the transmit coil, then abruptly turned off. The turn-on is a somewhat slow exponential so the abrupt turn-off is mostly what kicks the target. Because the transmit pulse is a discontinuous signal, the receiver requires different signal processing than with a sinusoid. This chapter and the next few will focus on a progression of PI designs.

Transmitter

Suppose we connect a coil directly to a battery with a switch as in Figure 19-1. When the switch is closed the coil will be energized with a current which produces a magnetic field. When the switch is opened the current instantly drops to zero and the magnetic field collapses. Both the creation and collapse of the magnetic field can produce eddy currents in targets but usually the collapsing field has a higher slew rate.

When the switch is closed, the battery is connected directly across the coil. An ideal battery, an ideal switch, and an ideal coil all have zero resistance, so in the ideal world this would produce an infinitely rising current through the circuit:

$$\frac{di}{dt} = \frac{V_B}{L}$$

Eq 19-2

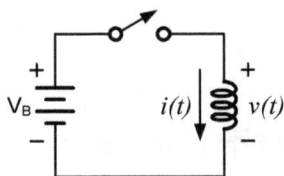

Fig. 19-1: **Ideal Switched Coil**

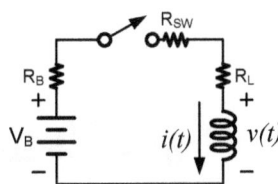

Fig. 19-2: **Reality**

That is, the current rises with a constant slope, meaning that it is a linear ramp to infinity as long as the switch is closed.

However, real batteries, real switches, and real coils have some amount of resistance (Figure 19-2) and the total series resistance changes the current from a ramp to a rising exponential

$$i(t) = \frac{V_B}{R_{eq}} \cdot (1 - e^{-t/\tau})$$ Eq 19-3

where R_{eq} is the equivalent series resistance ($R_B + R_{SW} + R_L$) and τ is the time constant of the RL circuit:

$$\tau = \frac{L}{R_{eq}}$$ Eq 19-4

As in the ideal case, at the instance the switch is closed ($t = 0$) the initial slope of the current (di/dt) is V_B/L. The current then exponentially builds up to a peak value of V_B/R_{eq} and remains there until the switch is opened. A rising exponential takes 3τ to reach 95% and 5τ to reach 99%; as an example, if the coil is 300µH and R_{eq} is 2Ω then $\tau = 150$µs, so it would take 750µs to reach a 99% plateau. In reality, allowing the coil current to fully charge consumes a lot of battery power so the switch is usually opened long before the plateau, perhaps at 1τ or less. The peak current can still be as high as a few amps.

While the series resistance alters the turn-on characteristic, it (ideally) does not alter the turn-off response. When the switch is opened, the current through the coil (ideally) abruptly cuts off regardless of the series resistance and falls to zero in zero time, creating an incredibly fast dB/dt. The overall resulting current (ideally) looks like this:

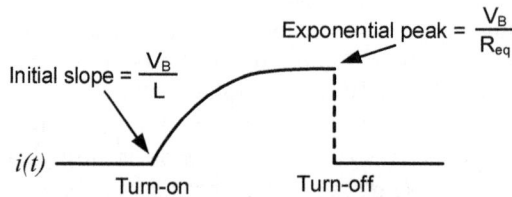

Fig. 19-3: **Exponential Current Waveform**

You'll notice a few "ideallys" in the last paragraph from which you might infer things don't really happen this way. When the switch is opened the current *from the battery* is abruptly cut off. But there is energy stored in the coil in the form of the magnetic field that cannot dissipate instantly. Instead, the magnetic field collapses and induces an EMF in the coil which is stored in the distributed parasitic capacitances[1]. The EMF then creates another current in the coil which generates another magnetic field and the process repeats, dumping the energy alternately between a coil magnetic field ($E = \frac{1}{2}LI^2$) and a capacitive voltage ($E = \frac{1}{2}CV^2$). The resulting coil current waveform looks like that in Figure 19-4.

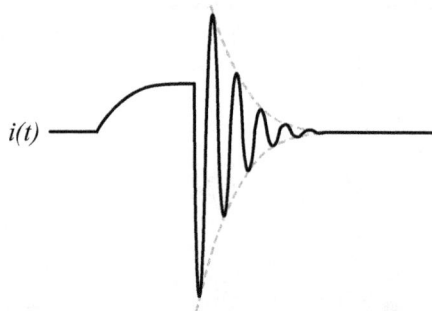

Fig. 19-4: **Undamped Current Waveform (turn-off)**

1. Right now just the coil interwinding capacitance; later we'll consider other parasitic capacitances.

Resistive losses in the coil dissipate the energy and cause the oscillations to exponentially decay to zero. The frequency of the oscillation is

$$f = \frac{1}{2\pi\sqrt{LC_L}} \times \sqrt{1 - \frac{R_L^2 C_L}{L}}$$

Eq 19-5

where L is the inductance of the coil, R_L is the parasitic series resistance[2] of the coil, and C_L is its lumped parasitic capacitance. Equation 19-5 represents the special case of a parallel LC circuit where the L also has a series parasitic R. Except in unusual cases of high capacitance or high resistance the frequency can be approximated by the very familiar:

$$f = \frac{1}{2\pi\sqrt{LC_L}}$$

Eq 19-6

This is the classic equation for the resonant frequency of an LC tank circuit. Every real coil has distributed parasitic capacitance and will exhibit a self-resonant frequency (refer back to Chapter 8, specifically Figure 8-3). The oscillation has an envelope (dashed lines in Figure 19-4) with an exponential decay given by

$$e^{-(R_L/L)\cdot t}$$

Eq 19-7

which has a tau $\tau = L/R_L$. A few people have investigated the use of this ringing response as a means for detecting metal and even discriminating. But for basic PI designs it is considered "undesirable."

What we want is the ideal abrupt turn-off, which is not possible because the parasitic capacitance creates an under-damped response. We can tame this by adding a *damping resistor*[3] to the circuit to convert the stored magnetic energy into heat. Currently our non-ideal coil can be modeled per Figure 19-5 where R_L is the series resistance of the wire (same as before) and C_L is the interwinding capacitance. Since the capacitance and resistance are distributed throughout the coil it is not practical to add more series resistance to the coil. Instead, we add a resistor in parallel with the coil to get the current (and voltage) to zero as fast as possible:

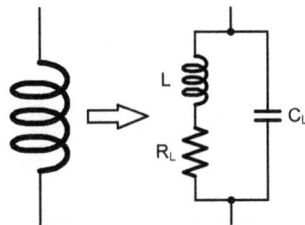

Fig. 19-5: **Coil Lumped Model**

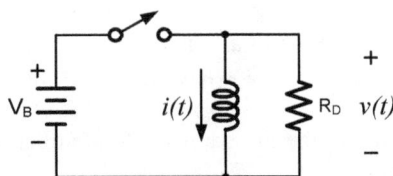

Fig. 19-6: **Damping Resistor Added**

If we ignore the small resistance of the coil then the result is a parallel RLC circuit. The fastest possible settling is achieved with critical damping, which occurs when the resistance is

$$R_D = \frac{1}{2}\sqrt{\frac{L}{C_L}}$$

Eq 19-8

Now our coil current should look like Figure 19-7. Instead of the abrupt turn-off we were hoping for, critical damping gives another exponential response albeit one with a much faster time constant than the

---

2. No longer including the resistance of the battery & switch.

3. Some people refer to this as a "dumping resistor."

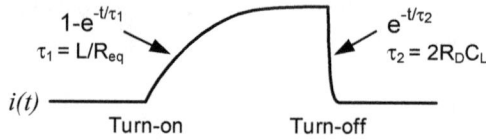

Fig. 19-7: **Damped Coil Current**

turn-on exponential. The turn off tau[4] is given by

$$\tau = 2R_D \cdot C_L = \sqrt{LC_L} \qquad\qquad \text{Eq 19-9}$$

This illustrates that for any given coil inductance we might choose, a lower parasitic capacitance will result in a faster turn-off decay.

19.2: Flyback

Although the current through the coil is what we are most interested in from the perspective of creating a target response, the voltage across the coil is also important because voltage is ultimately what we will be able to look at on an oscilloscope and also what we will process in the receiver electronics. We will again consider the turn-on and turn-off regions separately.

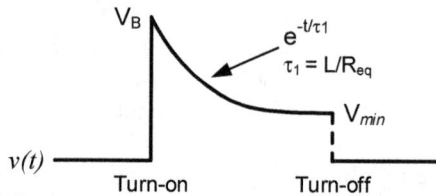

Fig. 19-8: **Voltage Waveform (turn-on)**

At turn-on, the coil opposes a sudden change in current and initially behaves like an open circuit so the voltage across the coil starts out at V_B. As the coil current exponentially rises from zero the coil voltage exponentially falls (with the same tau). When the coil current reaches its exponential maximum the coil voltage is at a minimum value determined by the resistances in the circuit:

$$V_{min} = V_B \cdot \frac{R_L}{R_B + R_{SW} + R_L}$$

The behavior of the coil voltage at turn-on isn't all that important because in a basic PI the turn-on portion of the coil current isn't considered, but it is useful to understand what is going on if you look at it with an oscope.

The coil voltage at turn-off is where the fun happens. Remembering that

$$v(t) = L \cdot \frac{di}{dt}$$

the near-instantaneous turn-off of the current results in a very large voltage spike. "Very large" is typically hundreds of volts, enough that if you touch the right place in the circuit you will feel the fun[5]. As with the current, an undamped coil will produce a voltage that rings and a proper damping resistor will produce a critically damped voltage as in Figure 19-9.

---

4. The turn-off tau looks different than what you would expect from an RL or RC circuit; it is the tau for a critically damped RLC circuit. See Appendix C for a complete derivation.

5. Seriously, though, PI transmitters can produce high currents and high voltages, so be careful when poking around the circuitry. Even if the risk of serious shock isn't very high, the risk of damaging test equipment could be.

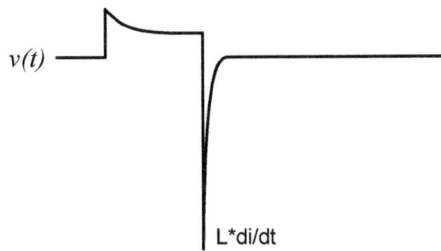

Fig. 19-9: **Damped Voltage Waveform (turn-off)**

The high-voltage spike is called the *flyback voltage* and is shown in Figure 19-9 somewhat exaggerated in its duration but de-exaggerated in its amplitude. Ideally the flyback *voltage decay* follows the turn-off *current decay* as depicted in Figure 19-7. The flyback voltage is illustrated with a vertical turn-off slew followed by an exponential decay. When viewed on a large scale the flyback does not actually begin with a vertical slew but overall has the form[6] of $t \cdot e^{-t/\tau}$. Figure 19-10 shows the expanded flyback pulse for an IRF9640-based TX circuit.

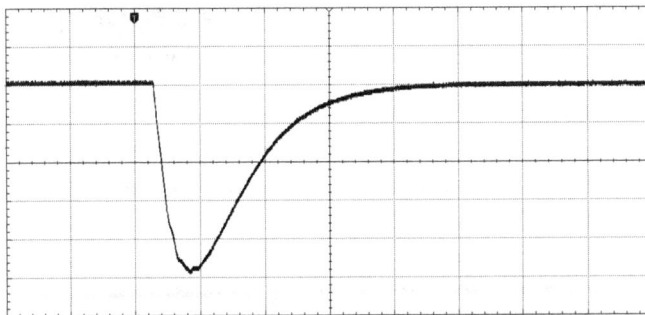

Fig. 19-10: **Turn-off Region Detail (100v/div, 5μs/div)**

Because of the extremely fast slew rates involved during turn-off, exactly what you see on an oscilloscope[7] will depend heavily on the bandwidth of the oscope. The important thing to remember is there are two different time constants: one for turn-on and one for turn-off. During turn-on the charging current follows an exponential curve set by the inductance and (total) series resistance. During turn-off the flyback voltage follows a more complex exponential response determined by the inductance, parasitic capacitance, and parallel damping resistor.

19.3: An Example

Before we proceed, let's tie all this together with an exercise. Suppose we want to design a PI detector with a 300μH coil. By measuring the self-resonance (as demonstrated in Chapter 8) we determine the nominal self-capacitance to be 130pF. We also measure the nominal series resistance to be 2.4Ω. From Equation 19-4 the turn-on tau is

$$\tau_{on} = \frac{300\mu H}{2.4\Omega} = 125\mu s$$

Assuming the resistance of the battery and switch are negligible, this means that it will take roughly 625μs (5τ) for the current to rise to its peak (99+%) value.

---

6. See Appendix C for all the gory details.

7. Be careful when directly probing a flyback voltage. Pay attention to the voltage rating of the probe, a flyback can be several hundred volts. A 100x probe is available specifically for probing high voltages. Mine is rated for 2500V.

Using Equation 19-6 the undamped coil should ring at

$$f = \frac{1}{2\pi\sqrt{300\mu H \cdot 130pF}} = 805.9kHz$$

with the same exponential decay time constant as Equation 19-7. Here is what we should see on an oscilloscope[8]:

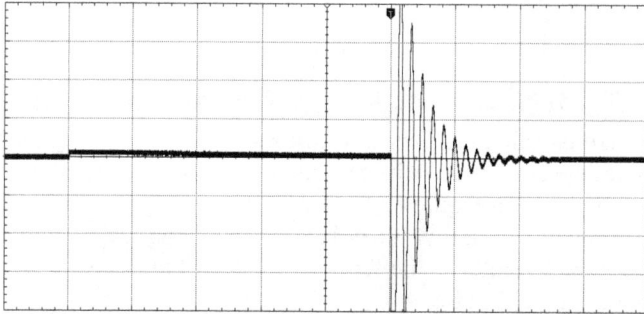

Fig. 19-11: **Undamped Coil Current & Voltage**

To achieve critical damping we add a parallel damping resistor with the value

$$R_D = \frac{1}{2}\sqrt{\frac{300\mu H}{130pF}} = 759.6\Omega$$

The closest 5% standard value is 750Ω. The resulting waveform now looks like this:

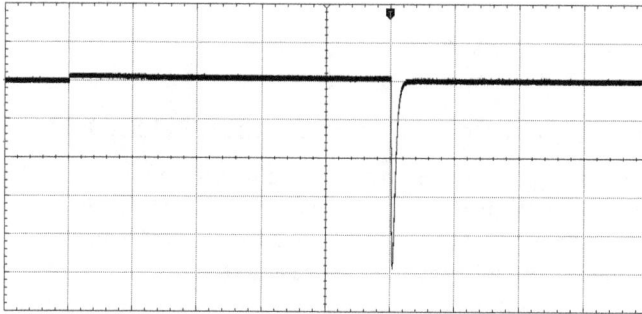

Fig. 19-12: **Damped Coil Current & Voltage**

and the turn-off tau is

$$\tau_{off} = 2 \cdot 750\Omega \cdot 130pF = 195ns$$

If 5τ continues to be the standard for settling, then this suggests that our coil will be 99% turned off in about 1μs.

A few points should be made:

• Equation 19-4 is valid for a series RL circuit, which is not quite what we have because of the coil's parasitic capacitance and the damping resistor. But those elements have practically no effect during turn-on so we can safely ignore them.

• Equation 19-8 is valid for a parallel RLC circuit, which is not quite what we have because of the coil's parasitic series resistance. But the resistance of the wire is almost always small enough in a typical resistively damped PI transmitter that we can safely ignore it.

---

8. Measuring coil current is a bit tricky; using a series resistor alters the current, especially in a PI circuit. In this case, the current was measured with a Tektronix P6302 inductive current probe.

- The calculated turn-on time constant assumes the battery & switch have no resistance. They do, but often the coil dominates. MOSFET switches can be less than $100m\Omega$ and the battery resistance is mitigated with an amply-sized low-ESR tank capacitor next to the TX circuit.
- The turn-on time of $625\mu s$ assumes we want to achieve a 99% settled current. This is almost never done as it consumes a lot of battery power. The trade-offs between turn-on time and other factors will be explored later.
- The turn-off time of $1\mu s$ assumes we want to achieve a 99% settled voltage. However, we'll see later that 5τ is not nearly enough on the decay side of the signal.
- The damping resistor is right at the front end of the detector and will therefore contribute, at a minimum, thermal noise. Carbon and carbon film resistors additionally contribute flicker noise and even though the coil shunts most low-frequency noise a high-quality metal film[9] resistor will help minimize overall noise.
- The job of the damping resistor is to dissipate energy and, depending on the transmitter design, this can make the resistor quite hot. Be sure to use a resistor of sufficient wattage. (We'll cover that in the next chapter's design.)

19.4: Target Responses

What is the purpose of getting a nicely damped flyback decay? Recall that when the coil is turned off, the magnetic field collapses in a fast transient dB/dt; this magnetic transient induces an EMF in a metal target which produces eddy currents. From the sudden dB/dt the eddy currents increase to some value and then decay back to zero with a time constant determined by the electrical characteristics of the target. The counter-magnetic field of the eddy currents also follows this decay and, as long as the time constant of the target is slower than that of the coil, this counter-magnetic field alters the flyback voltage decay as shown in Figure 19-13. In other words, in the absence of a target, the flyback decay is determined by only the coil and damping resistor. But when a target is present, the flyback decay is altered by the counter-magnetic field which decays with the tau of the target. So we can monitor the flyback decay and determine the presence of a target and possibly even discriminate somewhat based on the shape of the altered response.

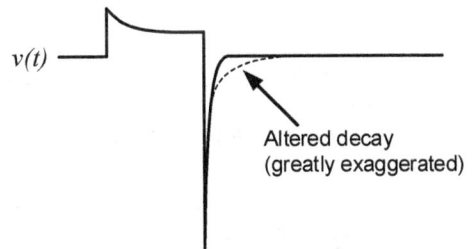

Fig. 19-13: **Target Response**

The altered decay shown in Figure 19-13 is greatly exaggerated; in reality, the decay shift caused by targets will be in the microvolts to millivolts range. Keep in mind that this small perturbation is riding on top of a waveform that peaks at perhaps a few hundred volts. Fortunately, we can design the coil so that its flyback response is faster than a typical target decay, which allows us to look for the target perturbations in a region where the coil flyback is very close to its final settled (zero) value.

19.5: Practical Transmit Switches

In Figure 19-6 the coil switch is depicted as a simple ideal switch. In a real design a solid-state device is used. Because the target's eddy current response depends on a fast-changing magnetic field (Equation 19-1), the time it takes the coil to switch off is important. So the switch must be able to handle a high (albeit short-duration) current and have fast switching, especially turning off. Typically, the switch is a single transistor, either a BJT or a MOSFET.

In Figure 19-6 a *high-side* switch was used to drive the coil; that is, the switch is on the high-voltage side of the coil. Such a switch could be implemented either with a PNP or a PMOS transistor as shown in Figure 19-14. A *low-side* switch could also be used to drive the coil, in which case an NPN or NMOS transistor may be used as shown in Figure 19-15. Regardless of the device used, it will add parasitic capacitance to the coil's interwinding capacitance which increases the overall capacitance that

---

9. Metal oxide film resistors do not share the low noise property of metal film resistors.

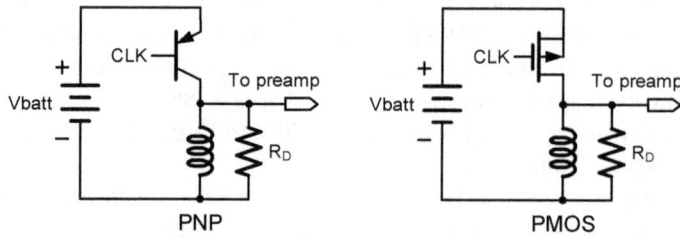

Fig. 19-14: **High-Side Coil Switches**

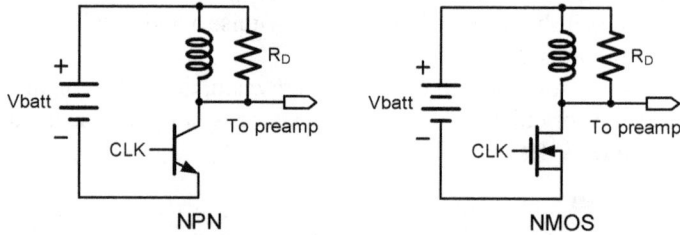

Fig. 19-15: **Low-Side Coil Switches**

determines the damping resistor, per Equation 19-8. For the fastest possible critically damped response, all sources of parasitic capacitance should be minimized and the switch can be a significant contributor. During the turn-on time it will also add series resistance to the coil and battery, but the coil normally dominates so rarely is this a critical factor.

Most modern PI designs have migrated toward using an NMOS switch. While bipolar transistors tend to have a lower parasitic collector capacitance, they can still have a slower turn-off time due to the base transit charge (usually expressed via the C_π model parameter) that must be removed to achieve turn-off. Furthermore, MOS devices have generally replaced BJTs in power applications (especially motor control) so there is a much wider selection of breakdown voltage, peak current, on-resistance, and capacitance than with BJTs. In the choice between NMOS and PMOS, for the same process technology a PMOS device must be roughly 2.5 times larger than an NMOS device to achieve the same channel resistance. This means that NMOS devices generally have lower capacitances and higher breakdown voltages, plus they have a broader selection and cost less than P-channel devices.

Receiver

PI signal processing is similar to that of VLF. A preamp amplifies the RX signal and a synchronous demodulator is used to extract a small dynamic target signal from a much larger static signal. In a VLF, there is a reactive and a resistive channel and the demods are typically run in a continuous-time manner. In a PI, there is no reactive channel because the RX takes place when there is no active TX signal. There is only a resistive (eddy decay) target signal so the simplest PI needs only one demodulator. The demod typically samples in a discrete-time manner and does so as early in the signal decay as possible. Some post-demod amplification, perhaps an SAT circuit, and an audio generator are all that are needed to finish out the design. Figure 19-16 shows the basic stages.

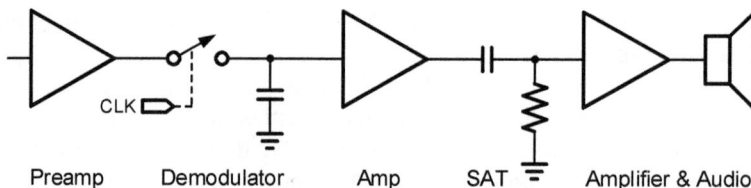

Fig. 19-16: **Basic PI Receiver**

19.6: Preamp

The transmit circuit creates an impulse waveform and the next step is to look at the resulting decay of the waveform for those small target perturbations. Because the target signal deltas are so small, an amplifier stage is needed before any other kind of processing is attempted. As with VLF designs, this amplifier is commonly called the preamp.

A basic PI design uses the same coil for both transmit and receive. The preamp must therefore be connected directly to the coil, yet the coil can produce a flyback of several hundred volts. The preamp can be an ordinary opamp but needs protection from the flyback voltage. Two popular solutions in PI designs are to either clamp the signal at the input to the preamp or to place an isolation gate between the coil and the preamp. The clamp is the simpler solution found in many PI designs:

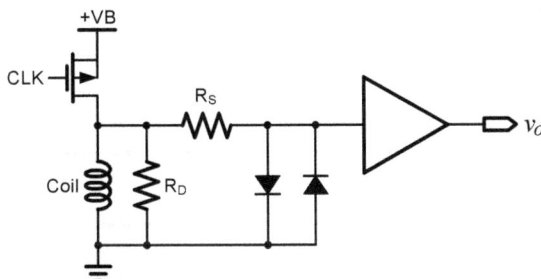

Fig. 19-17: **Preamp Clamping**

R_S plus the parallel diodes form a crude clamping circuit that limits the input voltage to the opamp to roughly ±0.7V, thus clipping both the high-voltage flyback as well as the turn-on charging voltage. A side-effect of the clamp is that, when the diodes are clamping, R_S is effectively in parallel with R_D and the damping time constant is altered. Depending on the design of the preamp, this effect can continue even after the diodes stop conducting. Like R_D, R_S should be a quality metal film resistor with a proper power rating. A typical value for R_S is 1kΩ; a higher value will decrease the diode clamp's transient current resulting in lower power and faster recovery; a lower value will decrease thermal noise. The diodes are commonly 1N4148 types.

Clamping allows the use of an ordinary opamp for the preamp stage. A consideration is whether to use an inverting or non-inverting amplifier topology; either will work with minor trade-offs between the two approaches. Typical gains for the preamp are 100-1000 with higher gains often split between two gain stages. Although the diode clamps prevent the high-voltage flyback from frying the opamp[10], the clipped levels (±700mV) will still overdrive the opamp output, even for low gains. Therefore, the opamp used in the preamp stage must not only have a fairly high gain-bandwidth product (10MHz or more) but also needs to have fast overdrive recovery.

Figure 19-18 shows the output waveform for a non-inverting preamp. The TX drive and the flyback are both clipped to the power rails of the opamp. An inverting preamp will have the same waveform but upside-down.

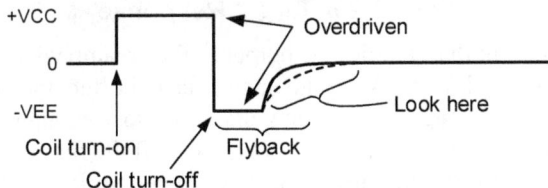

Fig. 19-18: **Preamp Output**

10. And without the diode clamps, this *will* happen.

19.7: Sampling

With the flyback voltage limited and the rest of the decay amplified, we can now look for subtle changes in the decay due to metal targets. This is still problematic because the output signal of the preamp is decaying by a few volts and we want to find a much smaller change on top of this. If we were to simply inspect the voltage decay in real-time it is very likely that noise (both external EMI and internal circuit noise) will mask a weak target signal.

A method which is commonly used in commercial detectors is to sample a short portion of the decay curve, average multiple samples, and then look for variations in the average. The sampled portion of the decay will have some nominal voltage level depending on where the sample is taken along the decay — and a target will cause the level to change. Figure 19-19 shows the waveforms for this process.

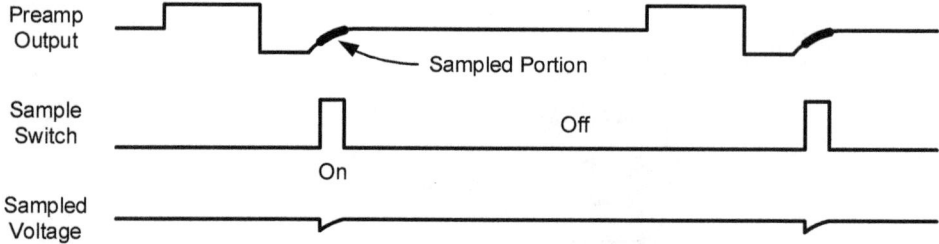

Fig. 19-19: **Sampling Waveforms**

The preamp output swing is limited to the supply rails of the opamp (±V). The sample pulse "looks" at a small portion of the preamp output somewhere between the point at which the opamp comes out of overdrive and the point where the decay has settled to zero. Earlier sampling (closer to the overdrive point) increases target sensitivity but also produces a higher DC offset which can then overdrive the sampling circuit. Most PI designs strive for the earliest possible sampling point[11] but usually offer a variable *sample delay* control to accommodate different coils and different ground conditions.

Figure 19-20 shows what happens when the decay is disturbed by a target (dashed line). The falling edge of the sample switch determines the value of the sampled voltage (assuming a sample-and-hold style operation), and in the presence of a sufficient target it will be consistently different than the nominal (no-target) value.

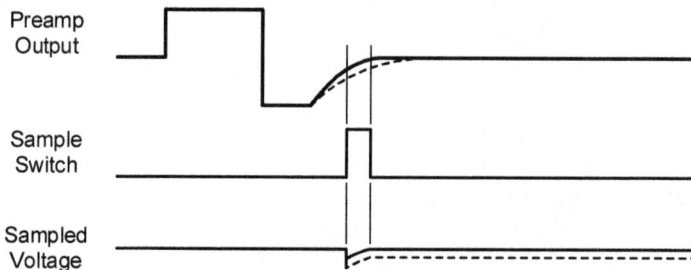

Fig. 19-20: **Target Response**

Figure 19-21 shows a circuit that samples the output of the preamp with a simple switch and capacitor, the same as was done in IB designs. A second opamp stage buffers the sampled voltage on C1 and can provide additional gain if the range of sampled voltages is close enough to zero. Switch SW2 is the sample switch and is controlled by a clock signal that is timed off the falling edge of the TX coil switch. With the switch closed, the capacitor voltage tracks the preamp voltage. When the switch opens, the voltage is held on the capacitor, and a level detector can be used to determine if the voltage has increased or decreased due to a target.

---

11. Which is why fast coil and transmitter designs are desirable; we'll look at this in later chapters.

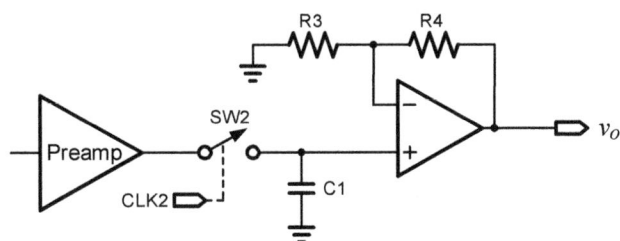

Fig. 19-21: **Sampling Circuit**

The ideal[12] sample-and-hold style sampling circuit responds on a sample-by-sample basis, so it is extremely fast to respond to target signals. But random noise — either from outside the detector, or from the detector's circuitry — can also cause momentary jumps in the sampled voltage. Usually, the pulse rate of a PI detector is at least a few hundred pulses-per-second so as the coil is swept over a target quite a few of the samples have an elevated level due to the target. A noise event, on the other hand, is a random and relatively quick transient that will probably affect only one sample or a very few samples. So to improve the ability to distinguish random noise fluctuations from a real target, a running average of the sampled decay is taken (Figure 19-22). Correlated target samples are reinforced over many samples while random noise gets averaged out.

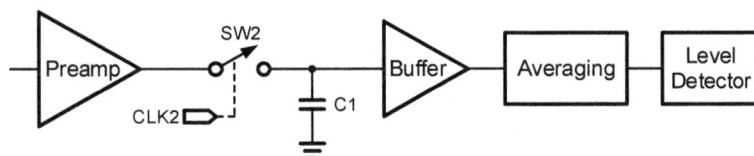

Fig. 19-22: **Averaging**

A simple method of averaging the sampled signal is to add a series resistor to the switch. This resistor, along with capacitor C1, will form a low-pass filter which averages each new sample with the voltage already held on the capacitor. This is the same technique as some of the demodulators used in VLF circuits. Another method is to replace the non-inverting buffer with an inverting opamp and add a capacitor as part of the feedback network. Then we get sampling, averaging, and gain in one circuit. This is known as a *sampling integrator* and is shown Figure 19-23.

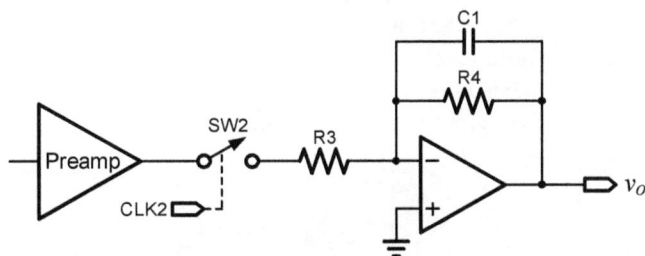

Fig. 19-23: **Sampling Integrator**

During the sample time R3 converts the preamp voltage into a current and the current is integrated by C1. During the hold time the voltage on C1 is largely held, with R4 maintaining long-term DC stability. The time constant formed by R4-C1 determines both how much cumulative signal is integrated and how fast the target response fades away. If the time constant is too low, the integrated signal will be weaker and detection depth will suffer. If the time constant is too high, the target response will be sluggish making pinpointing difficult. A good range for a normal swinging detector is 25-100ms. A detector in which the coil is not normally swung — such as a large-coil deep-seeker, a boat-towed "fish," or a security walk-through — may have a much higher time constant.

---

12. Zero switch resistance.

The output of the integrator is nominally a DC signal that rises or falls[13] when metal is detected. The gain of the sampling integrator is, at first glance, −R4/R3 but this is the DC gain of the opamp assuming the switch is always closed. The gain of the complete sampling stage is

$$G = -\frac{R4}{R3} \cdot \frac{t_s}{T}$$

Eq 19-10

where t_s is the sample width and T is the pulse period; that is, t_s/T is the sampling duty cycle. For typical values R3=1k, R4=100k-1M, t_s = 10μs, and T=1ms the overall gain is in the range of 1-10. Therefore the output voltage swing of the sampling integrator is usually small and needs further amplification to drive an audio stage.

19.8: Self-Adjusting Threshold (SAT)

There is an additional complication: the integrator output voltage can also have a DC offset which depends on where along the decay the preamp signal is sampled. If the sample occurs when the decay is very close to ground (t_2), the integrator offset will be small (Vos$_2$ in Figure 19-24). However, an aggressively early sample time (t_1) means that the preamp signal is sampled closer to the overdrive point which can result in a large offset voltage at the output of the integrator (Vos$_1$). This means that a follow-on gain stage will easily be overloaded by the offset.

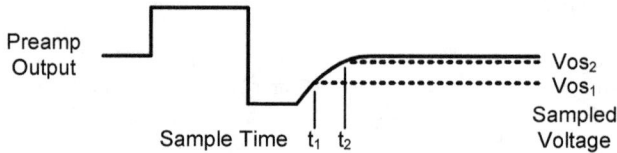

Fig. 19-24: **Integrator Offset vs Sample Delay**

Most PI designs attempt to sample with as short of a delay as possible to maximize sensitivity so the offset can be significant. To make matters worse, different coils can produce different decay rates and make the integrator offset somewhat unpredictable. It is possible to partially address this by injecting an adjustable offset into the preamp but, while this is helpful in maximizing the dynamic range of the sampling integrator, it usually does not address the minor variations in integrator offset which become major variations at the output of the final gain stage.

A solution to this problem is to AC-couple the integrator output signal into the final gain stage; see Figure 19-25. Besides eliminating the offset problem, the AC-coupling also provides a stable signal reference from which a target threshold level can be established. Because of this important function it is commonly called the *Self-Adjusting Threshold*[14] (SAT) circuit and is the reason many PI detectors require some amount of coil motion to produce a target signal.

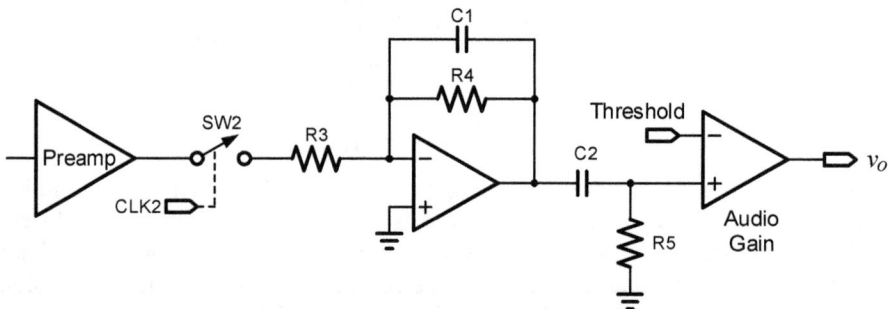

Fig. 19-25: **SAT & Final Gain Stage**

13. Polarity depends on the design of the front end.

14. The term 'SAT' was coined by White's Electronics in the 1990s, along with the term VSAT (variable SAT). I have found no evidence that White's trademarked either term, and both have reached ubiquitous usage amongst detectorists.

Inside the Metal Detector

19.9: Gain & Threshold

After the SAT circuit, the final gain stage amplifies the signal and a threshold level is applied. The gain and threshold are often implemented with user-variable controls; sometimes the SAT is also user-adjustable (so-called Variable SAT, or VSAT). Like the integrator output, the output of the audio gain stage is a DC-like signal which has a nominal level (the threshold) and rises and falls in response to metal. The difference is that the nominal DC level is under the control of a threshold pot and the target signal amplitude is maximized. The only thing lacking is a method to signal the presence of metal, such as an audio response.

19.10: Audio

Standard PI designs are fundamentally all-metal detectors; they do not have target ID or multi-tone responses. The reason for this is we only have a resistive signal to work with; a PI samples the signal after the transmitter has shut down so there is no reactive signal component, unlike with a continuous-time VLF. The most basic units have only a single all-metal response. In previous designs this was done either as a pitch response (BFO, Chapter 10) or a loudness response (TR, Chapter 19). Either one will work here, and while most PI designs use the BFO-like pitch response, our next design will use a simpler loudness response audio circuit as shown in Figure 19-26. The pitch response audio will be considered in a future chapter.

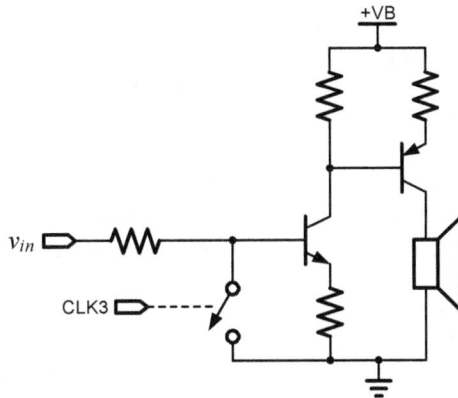

Fig. 19-26: **Audio Circuit**

Complete (Conceptual) Circuit

Throwing everything into one big diagram, Figure 19-27 shows what we have put together so far. Two elements missing are the power supply discussed earlier and something that will generate the needed clock signals. The clock generator has not been discussed at all and can be accomplished a number of ways. It is specific to the actual implementation of the final circuitry so it will wait until the next chapter.

The PI concepts discussed thus far are fairly simple yet are the bases for many commercial detectors such as the Tesoro *Sandshark* and the White's *Surfmaster PI*. The details may vary, such as whether discrete logic or a microcontroller is used to generate timing pulses, or whether a PMOS or an NPN is used to switch the coil current[15], but this basic approach has been widely used for many years. The next chapter will present a practical working design.

---

15. The Sandshark uses a PNP switch and a micro; the Surf PI uses a PMOS switch and discrete logic; the White's TDI uses an NMOS switch and has (in different revisions) used both discrete logic and a micro.

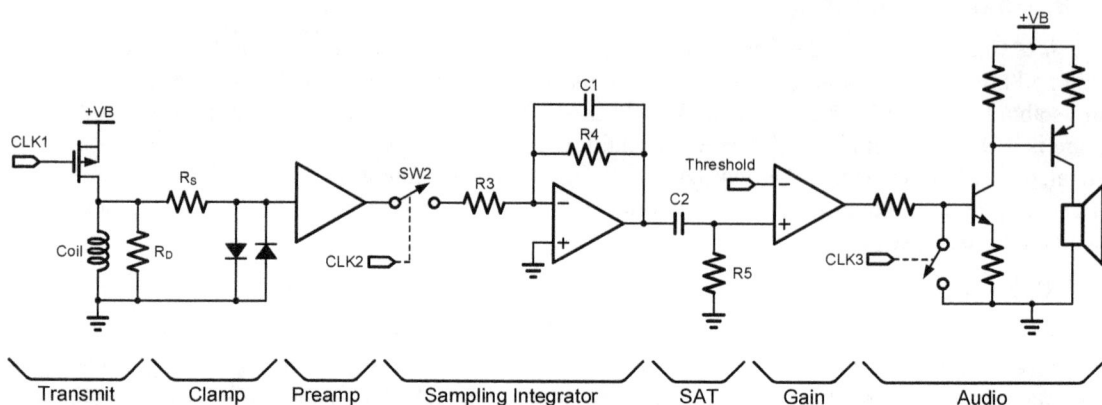

Fig. 19-27: **Complete Conceptual Circuit**

Explore

19.11: Anti-Interference Coils

Building and testing a PI prototype often begins with bench tests, usually done inside the home or workspace. The inside environment is littered with EMI including AC mains, WiFi signals, cell phones, electronic (LED/CFL) lighting, and even the test equipment used for development. Since PI is wideband all this EMI can make testing a challenge.

A way to mitigate the EMI is to build a special anti-interference coil. There are several ways to do this and most require that you plan ahead and include separate TX and RX coil connections. Two such coil types were introduced in Chapter 6: the figure-8 coil (Figure 6-15) and the coaxial coil (Figure 6-21). Other styles are the *top hat* coil (Figure 19-28a) and the *mono-coaxial* coil (Figure 19-28b).

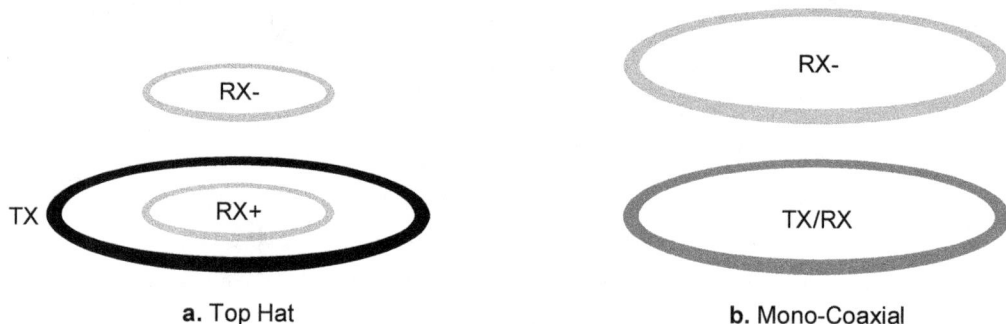

a. Top Hat

b. Mono-Coaxial

Fig. 19-28: **Anti-Interference Coils**

The figure-8 and standard coaxial have already been described. To use them as anti-interference PI coils you will want to design the TX coil with the same specs as a mono coil (say, 300µH for the book designs). The RX coils can also have the same specs as the TX coil, or you can give them more turns if desired. More turns usually improve sensitivity but can also slow down the decay settling and possibly cause ringing. For bench testing it is best not to get aggressive with the RX turns.

Most often individual damping resistors are required across each RX coil to prevent ringing. A single damping resistor across the RX coil pair probably will not work as their magnetic coupling allows energy to slosh back and forth between the coils, rendering the damping resistor ineffective. In VLF designs the figure-8 RX coil can be made by winding one big coil and twisting it into a figure-8. In order to use separate damping resistors, for PI designs it is better to wind each RX half as a separate coil. It is also possible to forego the separate TX coil and just use a figure-8 coil in place of the usual mono coil.

The top hat coil and mono-coaxial coil are both just variants of a standard coaxial coil. The top hat coil has a large TX coil with a smaller coplanar RX coil, plus a coaxial-spaced anti-RX coil[16]. When a target is swept by the coplanar TX/RX coil combination it is mainly detected by the coplanar RX coil, while the coaxial-spaced coil is much farther away and receives very little target signal. Usually the RX coils are one-half the diameter of the TX coil giving the overall look of a top hat. However, they can be made smaller or larger than this, even up to the diameter of the TX coil since there are no induction balance requirements. Once the RX coils equal the diameter of the TX coil then the Top Hat coil can be simplified to a mono-coaxial coil.

The mono-coaxial coil is not really "mono" but for target detection effectively behaves like one. There is a main TX/RX mono coil as in the usual PI design, plus a matching coaxially-spaced RX-only coil. The TX energizes just one coil but the RX sees both coils, requiring a slightly different front-end configuration than was used for the other coils.

Figure 19-29 shows the required front-end changes needed to use these coils.

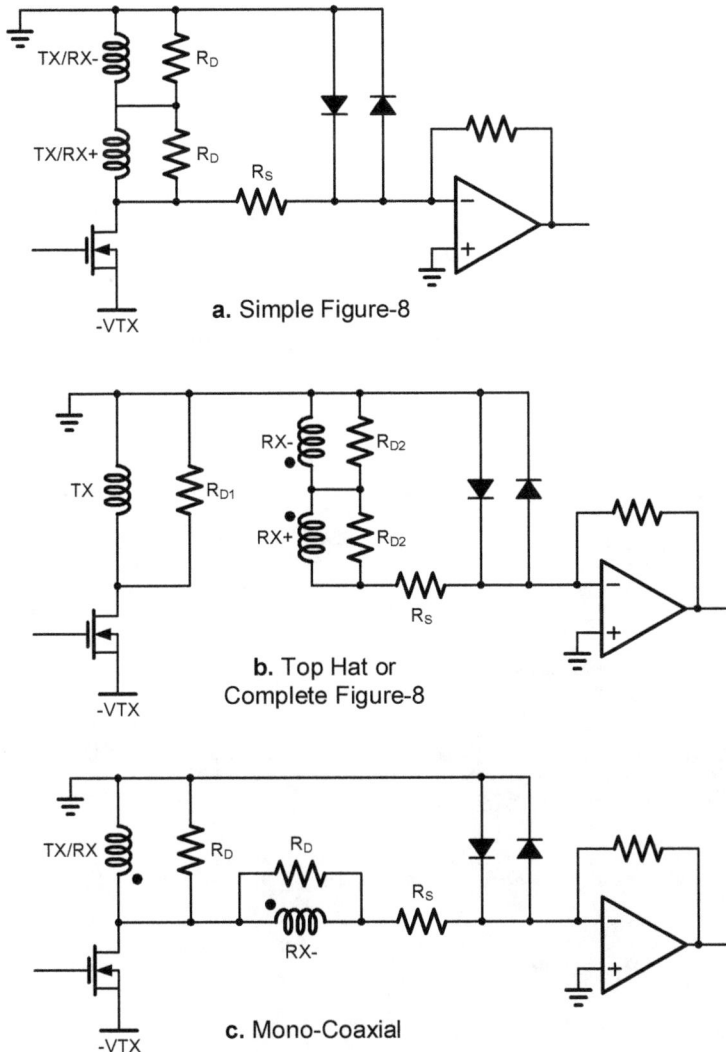

Fig. 19-29: **Front-End Connections for Anti-Interference Coils**

16. That is, connected out-of-phase with the main (coplanar) RX coil.

ELSEC

$1695⁰⁰ PLUS FREIGHT

Most designs of metal detector are not suitable for underwater operation. The ELSEC approach to the problem has been well proved. The system functions by transmitting a high power magnetic pulse from a coil and the analysis of the resulting eddy–currents in surrounding objects. In the type 700 instrument both transmitting and receiving is accomplished by the same coil. The associated electronics uses the most up-to-date solid state integrated circuits giving accurate readings under extremes of mechanical shock and temperature.

The instrument has been constructed for easy use by an aqualung diver whose air tanks etc., would be out of range of the detector. We believe the instrument will become invaluable to the underwater archaeologist, salvage operator, treasure hunter and pipe-line technologist.

UNDERWATER PIPE LINES AND CABLES. There has for a long time been a requirement for an instrument not only capable of finding pipes and cables but also for measuring **depth of bury** of pipes. This may be done by first calibrating the instrument on an exposed pipe either ashore or at sea.

INDICATION is normally given on an internal meter seen through a transparent window. This meter is fitted with illumination to enable the diver to take readings even in muddy water. An output socket is available for a diver-worn earphone, the presence of a tone indicating the presence of metal.

DETECTION DISTANCES. The maximum detection distance possible with the apparatus will depend on the size, shape and electrical conductivity of the object or objects. The higher the conductivity, the greater the detection distance. Iron is a special case and has the largest detection distance due to its magnetic properties. Some examples may be given:

Basic PI Design

"I check my pulse, and if I can find it, I know I've got a chance."

— *Paul Newman*

Now that the concepts of pulse induction has been covered, it's time to step through the process of designing a complete metal detector. This design will start out very basic but, as was done with VLF, will be expanded on in subsequent designs.

Transmitter

20.1: The Coil

In a basic mono coil PI, the same coil performs both transmit and receive functions. This appears at first glance to be an advantage, but in truth it results in conflicting trade-offs. In the previous chapter it was determined that a fast coil is needed, at least fast enough to outrun the target response. The maximum speed of the coil's response is largely determined by its inductance and interwinding capacitance, and the resulting damping resistor required to prevent ringing. Equation 19-9 shows that, for a critically damped coil, both the inductance and capacitance should be minimized. This suggests very few turns, perhaps even a single-turn coil. But the transmit field strength is determined by the coil current multiplied by the number of turns (the *ampere-turns*), and this suggests using lots of turns. Except that, for a given drive voltage, the coil current is inversely proportional to inductance, and because inductance is proportional to the *square* of the turns (N), the end result is that field strength is *inversely proportional* to N. So now we're back to a single-turn coil. Except that coil current, and therefore power consumption, will go through the roof.

These are the conflicting trade-offs faced in coil design, and we have yet to consider the receive side of the equation. There have been PI designs that use a single-turn transmit coil, notably custom units designed by Eric Foster for Mel Fisher in the 1970s, but he had the luxury of running his design off a deep-cycle marine battery. And while fewer turns produce more transmit oomph, the opposite is true on the receiver side. The induced target signal on the receiver coil is proportional to N, so for a stronger receive signal we want more turns.

So what do we do? We compromise. We design the coil with enough turns to get a good receive signal and decent power consumption, yet with low enough inductance and capacitance to maintain good speed. Typically, mono PI coils range from 100µH - 500µH; we will use 300µH for a nominal coil inductance for our projects, which is a value widely used by commercial PI detectors[1] and offers an option for building the circuits in this book without having to build a coil from scratch.

Our standard PI coil will use a diameter of 10" (25cm). Using a coil calculator, the number of turns should be 20, assuming 24 AWG wire. Typically, VLF metal detector coils are made from enameled magnet wire with a bond coating that holds the windings together. For PI detectors, magnet wire may result in excessive interwinding capacitance so often wire with a heavier insulation is used, preferably Teflon due to its lower permittivity. Table 20-1 shows the results of three different wire types.

---

1. In particular, Minelab SD and GP/GPX detectors, plus White's TDI, use 300uH coils, and several aftermarket companies like Coiltek, Nuggetfinder, and Detech make a wide variety of compatible coils.

The coils in Table 20-1 all use 24 AWG wire. For the same number of turns the enameled wire produces a slightly higher inductance because the thinner insulation creates a tighter bundle of wires. Heavier insulation spreads out the bundle and reduces the inductance. To compensate, two more turns of wire were added to the PVC and Teflon coils. Notice that common PVC-insulated wire has less than half the capacitance of magnet wire and that Teflon insulation offers little additional improvement[2]. For the project in this chapter, any of the wire types will work. See 8.4.2 for details on building a PI coil.

Wire Type	N	Inductance	Self-Capacitance
Enameled Magnet	20	300µH	159pF
PVC Stranded	22	301µH	67pF
Teflon Stranded	22	312µH	63pF

Table 20-1: **Coil Self-Capacitance vs Wire Types**

20.2: Transmit Driver

In Chapter 19 the TX circuit was presented mostly as an idealized switch but there was brief mention that the switch is, in practice, usually a single transistor. In the spirit of maintaining continuity with the conceptual circuit of Figure 19-1 we will proceed with a high-side PMOS coil switch as shown in Figure 20-1. An NMOS switch solution will be considered in the next chapter. The PMOS transistor can be an IRF9640 which has a breakdown of 200V. If available, a PMOS device with a higher breakdown is preferred. Q1-R2 converts a low-voltage clock into a 9V/0V voltage swing. Depending on the clock circuit design it may or may not be needed, but our clock circuit will have a +5V/−5V output swing.

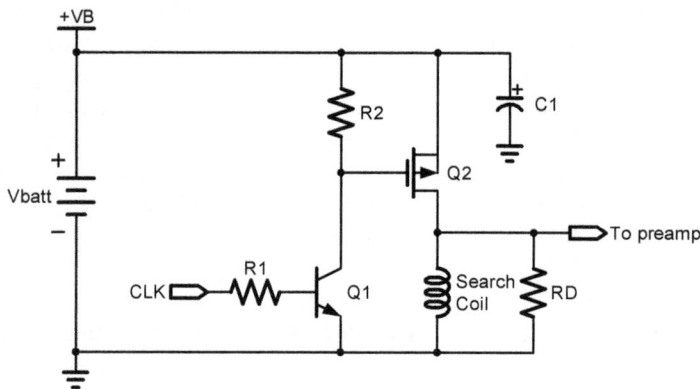

Fig. 20-1: **Transmit Circuit**

An additional element in Figure 20-1 is capacitor C1. Recall that there are three parasitic resistances we are concerned with: the coil, the switch, and the battery. The coil and switch resistance are under control of the designer, but the battery resistance may not be. To avoid performance degradation due to marginal batteries a large electrolytic capacitor is added to the +VB side of the switch. In this way the capacitor, not the battery, supplies the high transient current during turn-on, and is then recharged by the battery during the off-time.

The capacitor value is determined by how much coil current is needed during turn-on, for how long, and what voltage ripple is produced on the capacitor. As an example, if the coil current is 2 amps for 100µs and the desired capacitor voltage change is 100mV, then the capacitor value should be:

$$C = I \cdot \frac{dt}{dV} = 2A \cdot \frac{100\mu s}{100mV} = 2000\mu F$$

Thus a 2000µF cap will supply 2 amps of current for 100µs with a voltage droop of only 100mV.

---

2. Although the Teflon insulation has a lower permittivity, it is also not as thick as the PVC insulation.

Very often a small (1-10 ohms or so) resistor is placed between the battery and the capacitor/coil switch to ensure that the capacitor, not the battery, supplies the current. The advantage of using a capacitor will be lost if it has a high ESR; this is a place where using a "junk box" capacitor can degrade the design. Further improvement in ESR can be achieved by using multiple capacitors in parallel instead of a single big capacitor.

20.3: Damping

Resistor R_D is the damping resistor. Its value can be calculated from Equation 19-8 but that requires knowledge of the total parasitic capacitance at the search coil. That capacitance can be measured, but often we just empirically find the proper damping resistor through trial and error. The damping resistor can end up dissipating a fair amount of power so it is important to select one with a sufficient power rating. We can calculate the power requirements for the two separate operating regions and add them. During the turn-on time the power is V_B^2/R_D. Because this power is only dissipated during turn-on, we can convert it to a time-averaged power by multiplying by the duty cycle. For our design, the TX pulse is 100μs and the overall pulse frequency is 650Hz so the duty cycle is 6.5%. Thus this part of the power is $0.065V_B^2/R_D$. As an example, if V_B = 9V and R_D = 680Ω this comes to 7.7mW.

During the turn-off time the resistor is converting the energy stored in the inductor into heat. The energy in the inductor at the moment of turn-off is $0.5L \cdot I_{peak}^2$ where I_{peak} is the peak current achieved at turn-off. The power dissipated is energy/time where time is the overall period, 1.5ms in our case. Supposing that the peak current is 2A, the peak energy in the inductor is $0.5 \cdot 300μH \cdot 2^2$ which equals 600μJoules. This happens every period so the power is then 600μJ/1500μs = 400mW. Compared to 7.7mW, it is apparent that the power dissipated during flyback damping is dominant.

Good practice suggests choosing a resistor with additional margin, perhaps a 1 watt resistor. The damping resistor can be split into multiple resistors, such as 4 series resistors of 180Ω or 4 parallel resistors of 2.7kΩ, and the power will get divided amongst them, 100mW in this example. It is preferable, especially with a single high-powered resistor, to mount it slightly elevated above the PC board for better airflow, assuming the use of a through-hole resistor. As mentioned in the last chapter, it is better to use a metal film type for the damping resistor to minimize noise.

Receiver

20.4: Preamp

In using an opamp for the preamp, a consideration is whether to use an inverting or non-inverting topology. Either will work and there are, again, trade-offs between the two approaches. For this design a non-inverting opamp will be used as in Figure 20-2.

The gain of the amplifier is 1+R5/R4. R4 is usually chosen to minimize thermal noise contribution; this will be considered later, and for now 1kΩ will suffice. R5 is chosen for gain; some PI designs go as high as 1MΩ for a gain of 1000, some use a lower value to speed up the settling time of the preamp. In this unaggressive design, we'll use 100kΩ for decent gain and good settling.

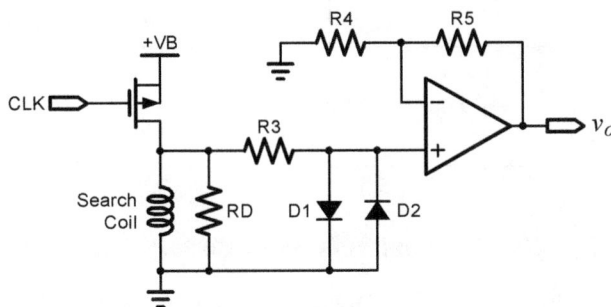

Fig. 20-2: **Preamp**

The clamping provided by R3 and D1-D2 deserves some scrutiny. If the flyback voltage peaks at 200V and R3 = 1kΩ then the peak current through both the R3 and D2 will be 200mA. The peak power seen by R3 is therefore 40 watts but this is only for a brief instance during the flyback, perhaps 5μs out of a pulse rate period of, say, 1500μs. So the time-averaged power is actually around 130mW. This resistor might get warm so it is best to derate it to, perhaps, 1/2 watt or 1 watt. As with the damping resistor, it is best to use a metal film resistor to minimize noise.

20.5: Power Supply

There is still a minor problem with the circuit in Figure 20-2. Most PI designs run the coil directly off the battery because of the high transient currents involved. When there is only a single battery, as in this design, the coil will decay to one or the other battery rails, depending on the type of switch used. Our circuit uses a PMOS switch so the coil decays to the negative battery rail, or ground. Redrawing for clarity and connecting the opamp power supply to the battery gives us Figure 20-3. With the coil switch open, the coil voltage settles to ground so the inputs and the output of the opamp will need to be fully functional all the way to ground. Most opamps don't work well all the way to their power rails and we don't want the design limited to the few that do.

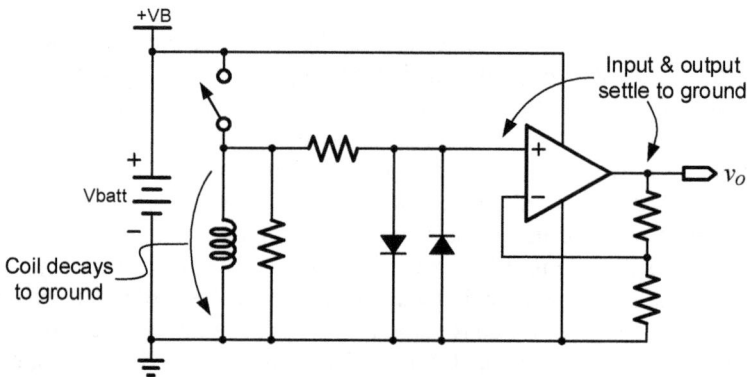

Fig. 20-3: **Clarified Preamp**

There are several ways to solve this problem. One is to simply add another battery for a –VX supply, a common solution in metal detector designs of the 1960s and 70s:

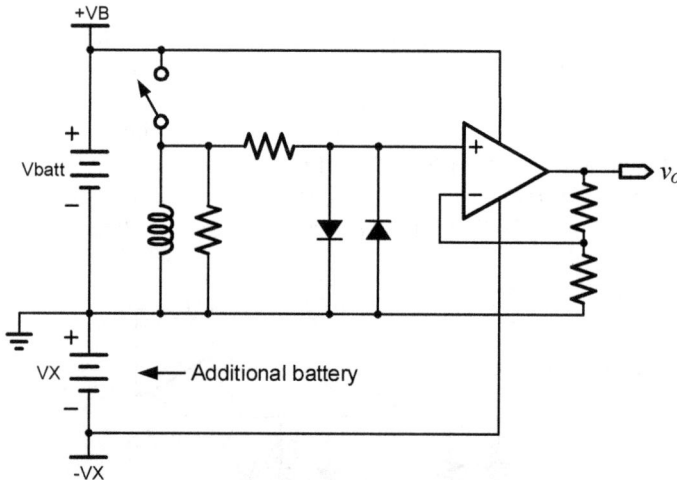

Fig. 20-4: **Undesirable Extra Battery Solution**

A second solution was used in the 1st & 2nd editions of *ITMD*, by capacitively coupling the coil signal to the preamp as in Figure 20-5. The opamp can now be powered from the battery although a new "analog ground" voltage is required. This is easily accomplished using a rail splitter. This solution creates some odd long-term droop in the opamp output signal but works just fine for a simple single-sample PI design. However, we'll want to take this concept further than the previous editions and for that purpose the droop will be a problem.

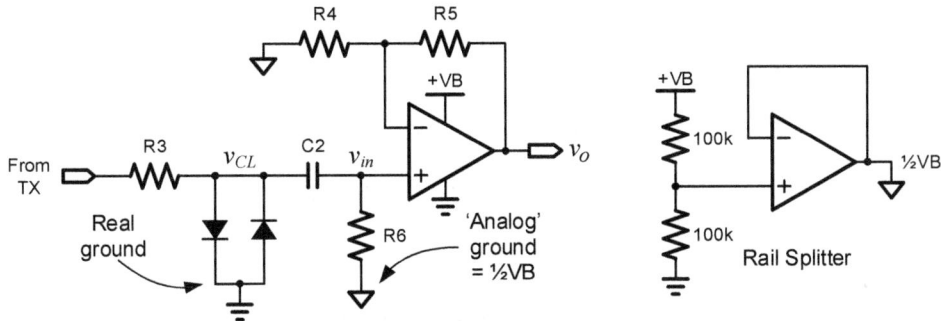

Fig. 20-5: **Undesirable Capacitive Coupling Solution**

An elegant solution found in most modern detector designs is to use a voltage inverter to replace the extra battery shown in Figure 20-4. A voltage inverter provides a −V supply from a +V supply (or vice versa) using a capacitive charge pump. A popular integrated chip for this is the "7660" such as the ICL7660. However, many variants are limited to 10 volts[3] which will work for this design (9V) but not for later (12V) designs. An inverter can also be easily designed using a discrete devices. Both options are shown in Figure 20-6; we'll use the 7660 this time around and move to the discrete-device option in future designs.

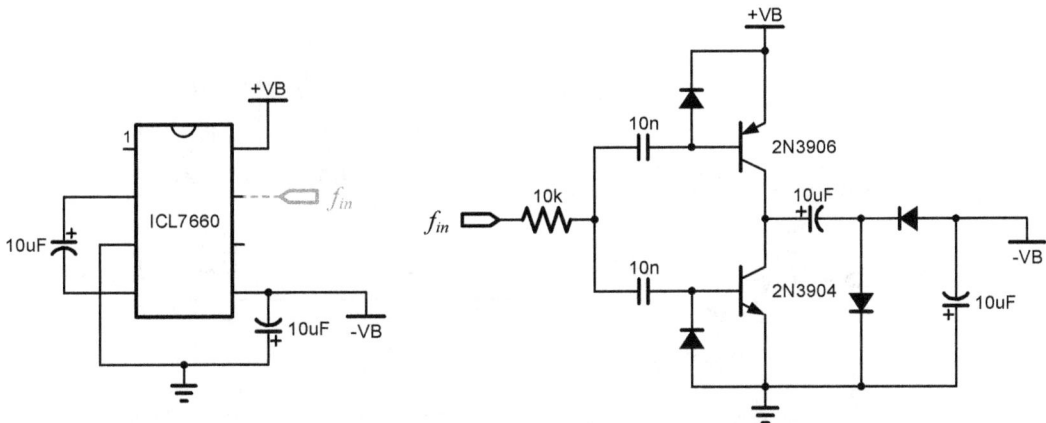

Fig. 20-6: **Voltage Inverters**

Figure 20-7 shows the complete front-end schematic. The (non-inverting) gain is set to 100 and the input signal is clamped by R3 and D1 & D2, which are common small-signal diodes (e.g. 1N4148). Additional capacitor C2 provides high-frequency noise filtering. A larger C2 results in better noise performance but the time constant $R5 \times C2$ should be lower than the minimum desired target time constant, by a factor of 2 or 3. That is, if you want to detect a 10µs target, choose C2 so that its time constant (with R5) is no more than 5µs.

---

3. The ICL7660A is specifically rated for 12V and the ICL7662 is rated for 20V.

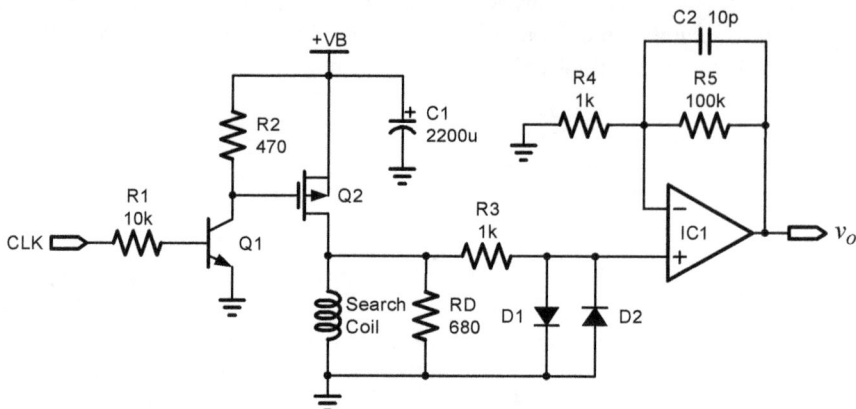

Fig. 20-7: **Complete TX & Preamp**

20.6: Sampling & Gain

The sampling integrator stage (Figure 20-8) uses the same basic design as shown in Figure 19-23. Sampling switch Q3 is a depletion-mode JFET transistor which is "off" when the gate voltage is equal to the source (or drain[4]) voltage, but "on" when the gate is pulled to a lower voltage. Compared to Figure 19-23 the switch and resistor R7 are transposed; this places the JFET source at virtual ground so that its turn-on threshold is constant regardless of the input signal. The functionality is otherwise identical.

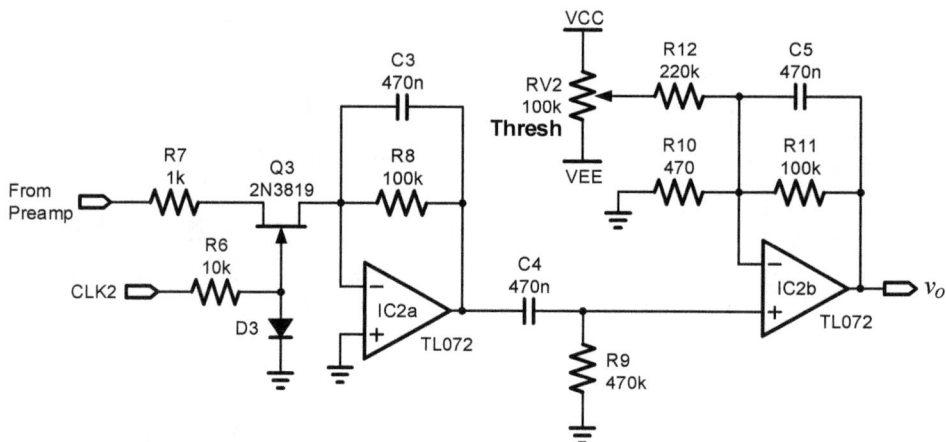

Fig. 20-8: **Sampling Integrator & Gain Stage**

The integrator stage is designed with an integration time constant of $470\mu s$ $(R7 \times C3)$ and a decay time constant of $47ms$ $(R8 \times C3)$. With a sampling rate of 650Hz (or T=1.5ms; more on this later), these time constants will provide sufficient averaging while still having a reasonably fast target response.

The output of the integrator stage feeds an audio gain stage via SAT components C4 & R9 which form a high-pass filter for removing the DC offset. Keep in mind that if you hold the coil steady over a target, the integrator stage will continue to produce a static target signal at the output of IC2a but the SAT circuit will eventually tune it out and the audio will return to a set threshold level[5]. The speed of the SAT circuit is set by its time constant $(R9 \times C4)$, 220ms in our case, so "eventually" is actually quite fast, about 1/2 second.

---

4. Unlike MOSFETs, drain and source are usually interchangeable on discrete JFETs.

5. This is why this piece of circuit is commonly called *Self-Adjusting Threshold*.

The SAT signal is applied to a final amplifier stage (IC2b). Potentiometer RV2 is the threshold adjustment which can inject a small positive or negative current into the feedback path to decrease or increase the output offset of the opamp voltage which allows for adjusting a nominal audio threshold level. The values shown should produce an offset adjustment roughly from -100mV to $+1$V. The gain is set by R11/R10 (approximately 200); this is usually the point where one of the resistors is a potentiometer to create a sensitivity adjustment and later we'll make R10 variable. C5 provides some additional low-pass filtering to help reduce noise, set to the same time constant (47ms) as the integrator stage.

20.7: Audio

Before looking at the audio driver let's review the signal waveforms, levels, and polarities. The PMOS TX switch produces a negative-polarity flyback and targets create a negative deflection in each TX decay curve (see Figure 19-13). The preamp is non-inverting and maintains the negative polarities with clipped peaks of ±5V and a decay level of 0V (Figure 20-9a). The integrator is an inverting stage so a decreasing signal at the preamp is sampled and will create a rising signal at the integrator output (Figure 20-9b). Although the preamp signal variations are seen on a pulse-by-pulse basis, the integrator signal (and subsequent signals) has a much slower response; that is, waveform (a) has a time scale of perhaps 1ms, while (b)-(d) have time scales of 100s of milliseconds. The Gaussian-looking integrator signal is applied to the SAT stage and the output, being a mathematical derivative[6], has both a positive lobe and a negative lobe with a nominal (no-target) level of 0V (Figure 20-9c). Finally, the audio gain is non-inverting so its output (Figure 20-9d) has the same target waveform as the SAT output, but the Threshold adjustment adds a nominal positive offset.

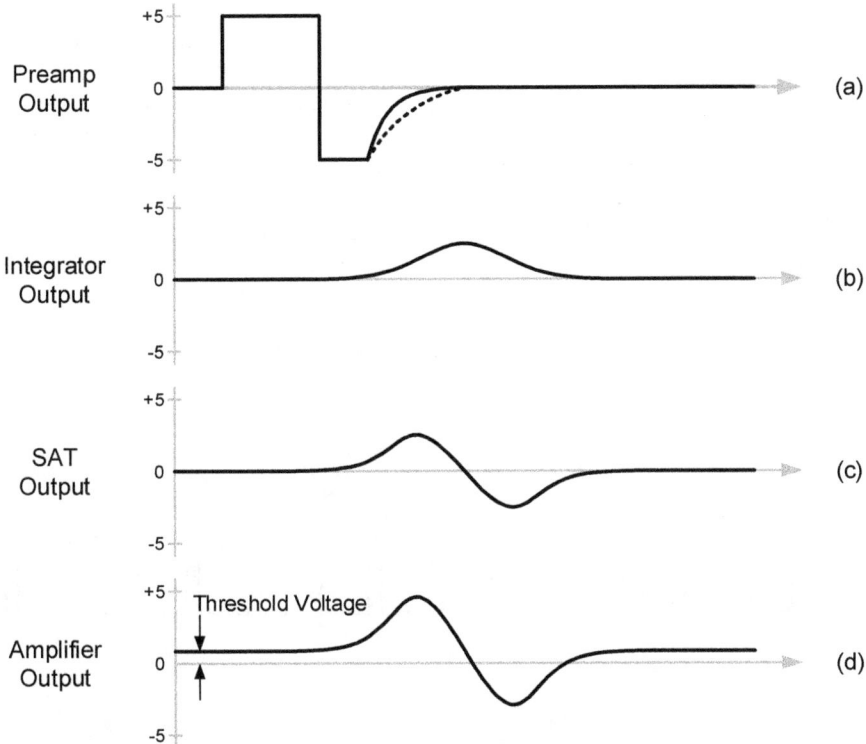

Fig. 20-9: **Signals**

6. Previous VLF designs had basically the same differentiator stages; they were used as high-pass filters to eliminate the slowly varying ground signals but allow fast target signals through. In PI, the SAT stage is also used as a high-pass filter, but in this case to eliminate the initial (and unknown) integrator offset, plus slowly varying circuit drifts (mostly opamp offsets).

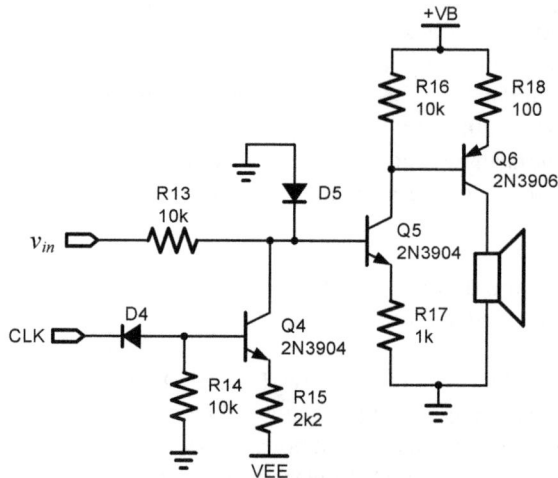

Fig. 20-10: **Audio Circuit**

The final part of the signal path is the audio stage shown in Figure 20-10. The output of the audio gain stage (IC2b) is a DC voltage with a nominal (no target) value of about zero volts, ignoring the threshold bias. When a target is present, the opamp output rises in voltage. Q5 and Q6 are common-emitter amplifiers so that the rising signal from IC2b increases the voltage (and thus the current) applied to the speaker. A +5V/−5V clock signal is applied to the circuit; D4 and R14 convert this to a 0V/−5V clock signal at the base of Q4. When the clock is high Q4 pulls the base of Q5 to about −0.7V (clamped by D5) which shuts off both Q5 & Q6, regardless of the target signal level. This gating signal chops the voltage applied to the audio stage and produces a fixed pitch tone, and the loudness will rise in response to a target. Since the audio gating clock is the same as the TX clock, the audio frequency will be the same as the TX frequency, or 650Hz in our case.

Figure 20-11 shows the analog signal (a) from IC2b, the gating clock (b) applied to Q4, and the resultant signal (c) at the speaker. Note that the threshold offset creates a slight no-target sound, and the bounce-back signal due to the SAT stage results in a portion of dead audio right after the peak target response.

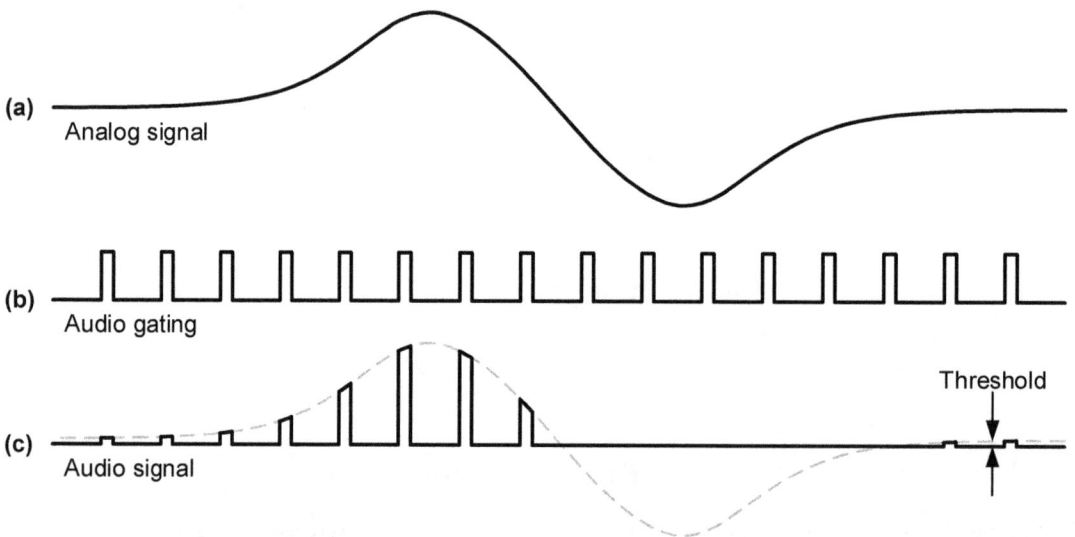

Fig. 20-11: **Audio Response**

Clocking

The final piece of our circuit is the clocking. We need three clock signals: one to drive the coil switch, one to drive the sampling switch, and one to drive the audio chopper. The first order of business is to determine the frequency we want to use and generate the master clock. As already mentioned in the description of the integrator, we will rather arbitrarily choose 650Hz. A higher frequency allows the integrator to time-average more samples but, everything else being equal, drains the batteries more quickly. A lower frequency saves power, but produces a lower signal-to-noise ratio in the integrator. Common PI frequencies run between a 100Hz or so up to many kHz[7], though we need to keep in mind that in this design the master clock is also the frequency of the audio signal. 650Hz will do fine.

As with most subcircuits, there are hundreds of ways to produce a clock signal. A popular circuit with experimenters is the "555" timer configured as an oscillator (Figure 20-12). It is easy, cheap, and flexible, able to generate a wide range of frequencies and duty cycles. However, its single output pin will satisfy only one of our clock signal needs. We'll call it MCLK for the master clock but it's exactly what is needed for the TX clock and will also be used for the audio chopper.

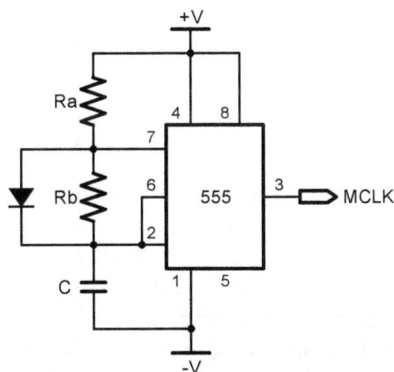

Fig. 20-12: **555 Oscillator**

With the 555 oscillator in Figure 20-12, a diode has been added to the R-C components that set the frequency and duty cycle. The charge cycle is now controlled only by Ra and the discharge cycle only by Rb which gives us more control over the resulting waveform, including the ability to output a narrow positive pulse which we happen to need for the TX clock. This neat little trick is well-documented in 555 applications sources. The frequency and pulse width are given by

$$f = 1/(0.695(Ra+Rb)C)$$ Eq 20-1

and

$$PW = 0.695 \cdot Ra \cdot C$$ Eq 20-2

The clock required for the sampling switch runs at the same frequency as that for the coil switch. It is delayed by some amount and usually has a different pulse width (see Figure 19-19), so we need a circuit that produces a delayed-and-shortened pulse. Once again, there are many ways to accomplish this; one is known as a "monostable multivibrator," which is a fancy way of saying it's a single-shot pulse circuit. The 7400-series logic family includes a chip with 2 complete monostables, the 74221[8]. We can use the first monostable to set the pulse delay and the second monostable to set the pulse width.

Figure 20-13 shows a 74221 configured in a way to give the proper output delay and pulse width. The delay is set by the first monostable at

---

7. Low pulse rates are often used in large-coil deep-seeking PI detectors where, say, a 1-meter coil is pulsed with a very high current and has a slower response than that of a coil with a lower peak current. High pulse rates are typically used in gold detectors designed with extremely fast-settling coils & front-ends.

8. We will use the 74HC221 which has CMOS (rail-to-rail) logic and can tolerate up to a 15V supply. The 74C221 will also work, but the 74LS221 (or any other TTL variant) will not work.

$$t_{delay} = k \cdot \mathrm{Ra} \cdot \mathrm{Ca} \qquad\qquad\qquad \text{Eq 20-3}$$

and the pulse width is set by the second monostable to be

$$t_{width} = k \cdot \mathrm{Rb} \cdot \mathrm{Cb} \qquad\qquad\qquad \text{Eq 20-4}$$

where k depends on the supply voltage and is slightly non-linear. The 74221 datasheet offers a graph of delay versus R and C but you can also find the proper values empirically.

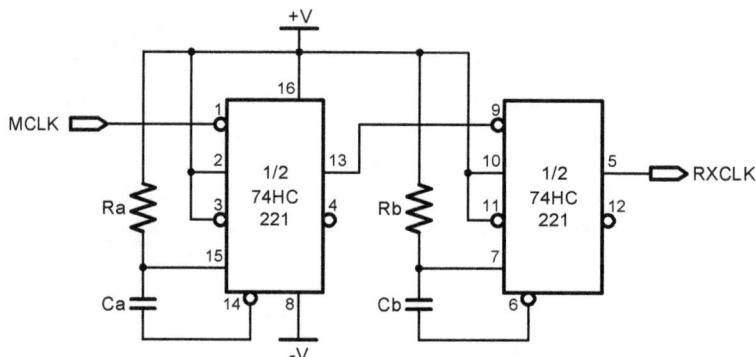

Fig. 20-13: **Monostable Clock Circuit**

Finally, the clock for chopping the audio signal can be any frequency. However, speakers are typically low-impedance and generate large current spikes when they are driven with pulses. These current spikes will usually translate into glitches on the power supply voltages that run the opamps, and these glitches can end up as spurious noise in the signal path. To minimize this problem, the speaker frequency should be the same as the sampling frequency so that glitches correlate in the integrator and become a long-term DC offset. This is another reason why 650Hz was chosen for the master clock frequency. Instead of generating another clock we will simply use the transmit switch clock (MCLK) to chop the audio signal, and 650Hz is a nice frequency for the audio.

Design: *Basic PI (Pulse-1)*

Let's put all the pieces together and make a PI detector. Figure 20-14 shows the complete main circuit and Figure 20-15 shows the power supply and timing circuitry. You should find that the current consumption is 90-100mA at a full 9 volts and drops in rough proportion as the battery voltage drops. Performance also decreases slightly with lower voltage, and as it approaches 6.5 volts it becomes unacceptably poor.

This design is intended to run on a single 9V transistor radio type but should run just fine up to 12V, limited by the ICL7660A. It should be noted that a typical alkaline 9V battery has 2-3Ω of internal resistance when it is new and resistance increases as the battery discharges. This is, again, why a good low-ESR capacitor (C1) at the transmitter is important. The use of a 6-cell AA battery pack will have lower resistance and more capacity and makes a better choice for PI.

The 555 timer (IC3) creates a master clock that drives both the transmitter and the audio chopper. It is set to a frequency of

$$f = \frac{1}{0.695(R19 + R20)C6} = 620\mathrm{Hz}$$

and a pulse width of

$$t_{on} = 0.695 \cdot R19 \cdot C6 = 83.4\mu\mathrm{s}$$

The 74221 monostable multivibrator (IC4) then creates a delayed and shortened pulse for the sampling switch. The delay is

$$t_{delay} = 0.2 \cdot (RV3 + R21) \cdot C7$$

which is variable between 13.6μs and 33.6μs. The sample pulse width is fixed at

$$t_{sample} = 0.2 \cdot R22 \cdot C8 = 11.2\mu s$$

Actual timing numbers will vary somewhat depending on component tolerances.

The remainder of the circuitry closely follows what has been presented throughout the chapter. The preamp gain is 100 and the integrator decay time constant is about 50ms which offers a zippy response. Increasing R8 or C3 will slow the response but increase depth. The SAT tau is 220ms and can be slower or faster by altering R9. The final gain stage has about a 10:1 gain adjustment range.

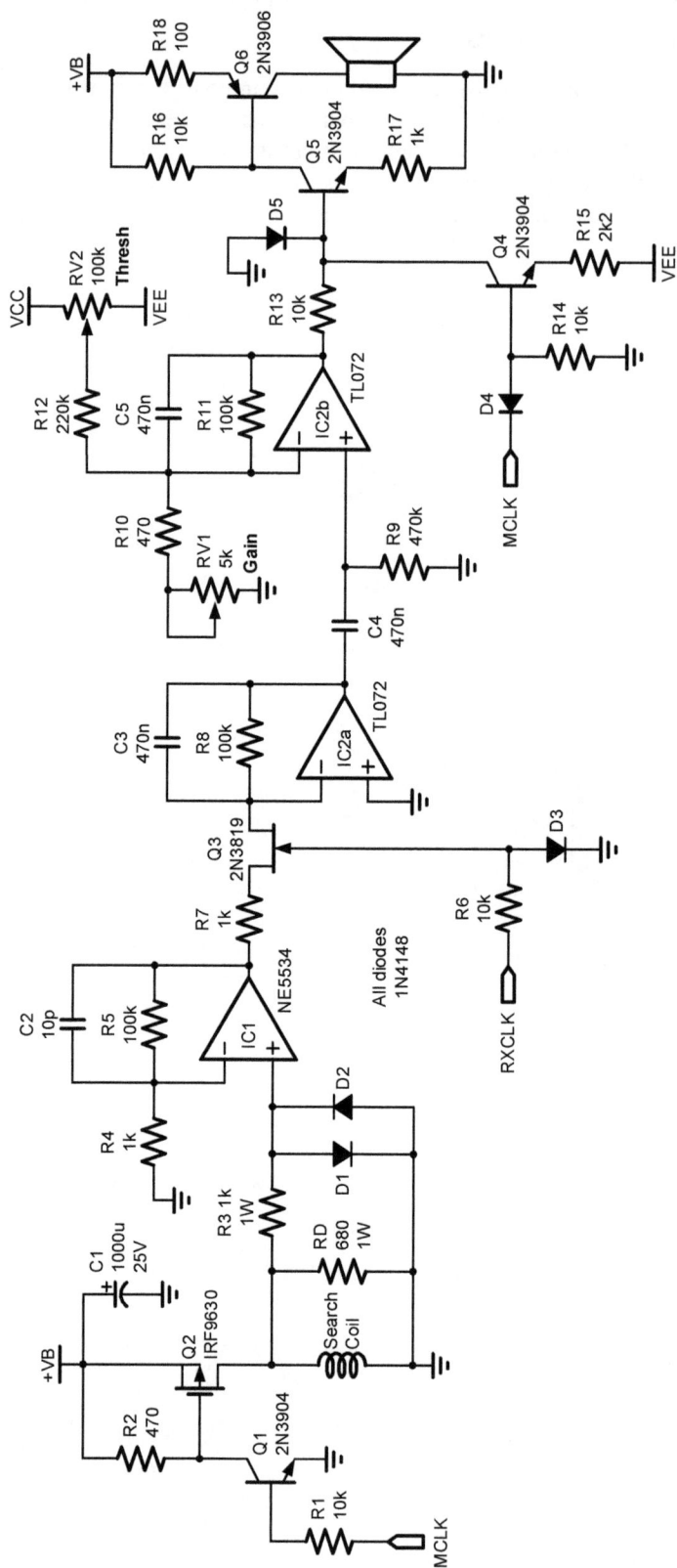

Fig. 20-14: **Pulse-1 — Main Circuit**

Inside the Metal Detector

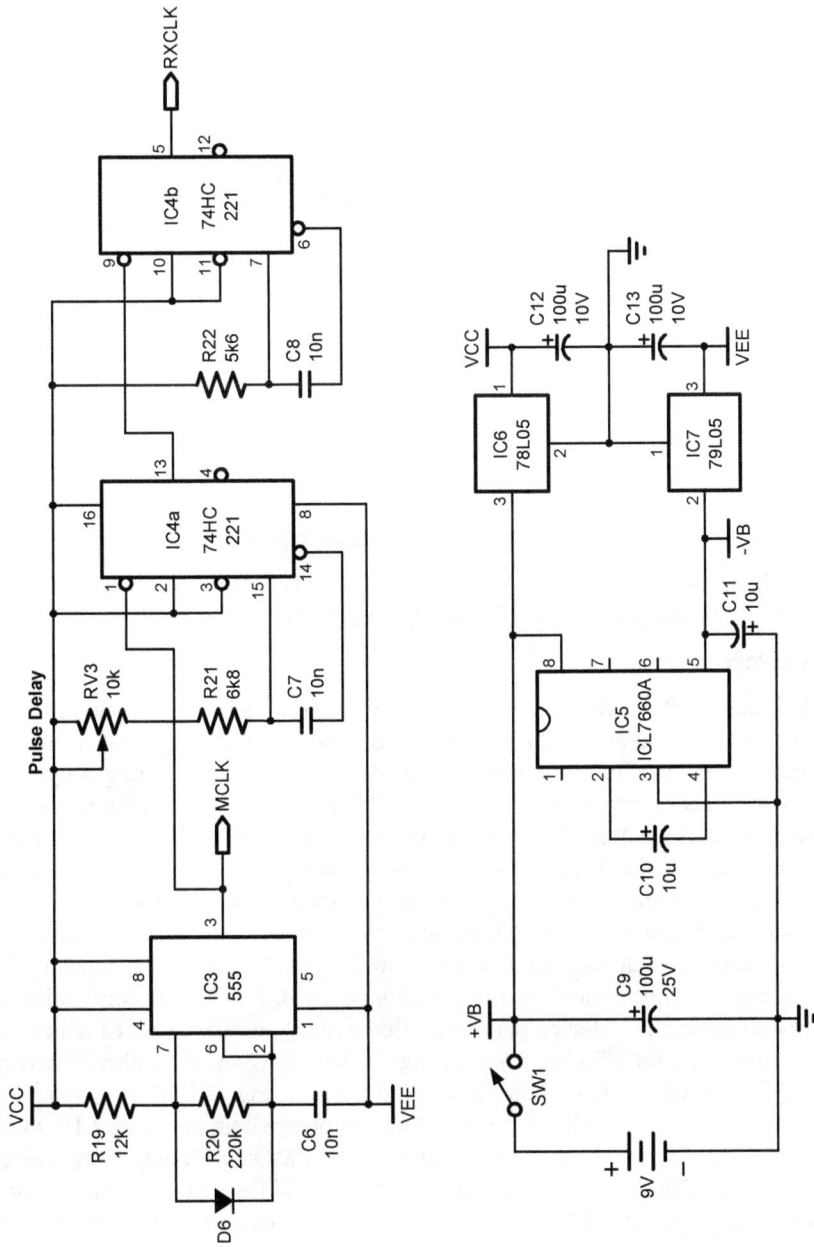

Fig. 20-15: **Pulse-1 — Power Supply & Timing**

Explore

20.8: Coil

The obvious first thing to play with is the coil. If you wound a coil from magnet wire, it probably has a fairly high self-capacitance. Try insulated wire — Teflon if you can get it, PVC otherwise. Carefully compare the flyback decays between the magnet wire and insulated wire coils. Also consider less or more inductance; in addition to the 300µH coil suggested for *Pulse-1*, try 200µH and 400µH. How is settling time and target sensitivity affected?

20.9: Adjustable Damping

Coils with different SRFs and/or different inductances will require different damping resistors. The value of the damping resistor can be calculated (Equation 19-8) but usually it still requires trial-and-error to achieve critical damping. While it's not a good idea to use a potentiometer for the total damping resistor, it is acceptable to partially use a pot as a tweaking element. Suppose you know the damping resistor is close to 680Ω but want to "dial it in" a bit closer; the following circuit results in an adjustment range of 597-736Ω.

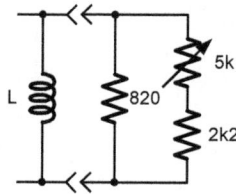

Fig. 20-16: **Adjustable Damping Resistor**

High-voltage flybacks can eventually damage the pot so it is best to use it for development and then replace it with a fixed resistor (or, more likely, multiple combined resistors) for long-term usage.

20.10: Coil Switch

The choice of the IRF9640 for this project is largely predicated on easy availability. It's been around for a while and is cheap and popular. But with a breakdown of only 200V it is a bit underrated for a PI transmitter and will avalanche with all but the slowest coils. Avalanching supposedly causes no long term harm (many, many commercial PIs allow the TX FET to avalanche) but it does cause the FET to heat up and so may require a heat sink, and avalanche recovery slows down the flyback response.

A better choice is a FET that is rated to handle the peak flyback. This can be a bit difficult in the land of PMOS devices[9], but there are some reasonably-priced solutions up to 400V and more pricey ones above that. Besides the breakdown voltage, pay attention to the continuous current rating and the channel resistance. Because PI designs have a fairly low duty cycle, a peak TX current of, say, 2A may result in a time-averaged "continuous" current of 250mA. Still, it is best to limit selections to 1A or more. Channel resistance can be selected per the desired overall on-resistance. Channel resistance usually goes hand-in-hand with the FET's current rating — that is, if you want ultra-low resistance, then the FET will likely have a very high current rating. And the price tag will follow.

Finally, we don't have to use a PMOS device at all. As pointed out in Chapter 19, the device could also be a PNP. Here, selection will be even more limited than PMOS with only a few devices to choose from in the 400V region. The drive circuitry will need to be modified because a high-power PNP typically has a poor beta, perhaps only 20. Figure 20-17 provides an example whereby Q2 is added to provide a beta boost.

---

9. And is a key reason that the next chapter changes the TX switch to an NMOS device.

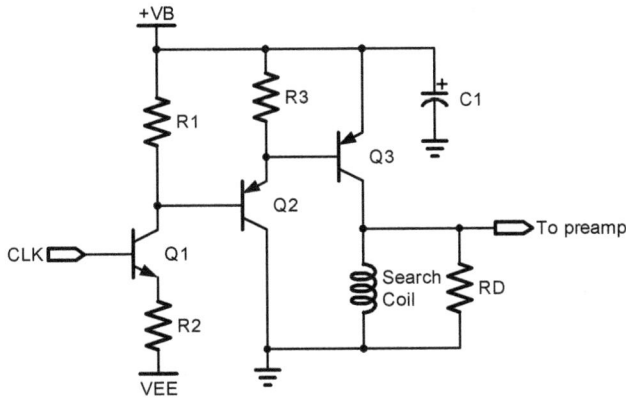

Fig. 20-17: **PNP Drive Circuit**

20.11: Preamp

The NE5534 is an old workhorse opamp, with the operative word being "old." There are many newer opamps with better specs. A good choice for the PI preamp is the MAX410. Perhaps one of the best choices is the LME49990 which is now obsolete and available only from eBay[10]. Key specs to pay attention to are:

- Gain-bandwidth (GBW) — PI front-ends need to be fast, and 10MHz is probably a minimum.
- Noise — *Pulse-1* has a 1k input resistor for the clamps which generates $4nV/\sqrt{Hz}$. The opamp should preferably be no worse, and perhaps a little lower.
- Overdrive recovery — Even clamping the flyback to a single diode drop overdrives the preamp, and the time it takes the opamp to come out of the overdrive condition can be a significant part of the sample delay budget. Overdrive (or overvoltage) recovery is often not specified, or done poorly, so it may require actual testing in situ.

The preamp can also be configured in an inverting fashion (Figure 20-18). For the inverting opamp, the clamping resistor R3 also serves as the feedforward resistor.

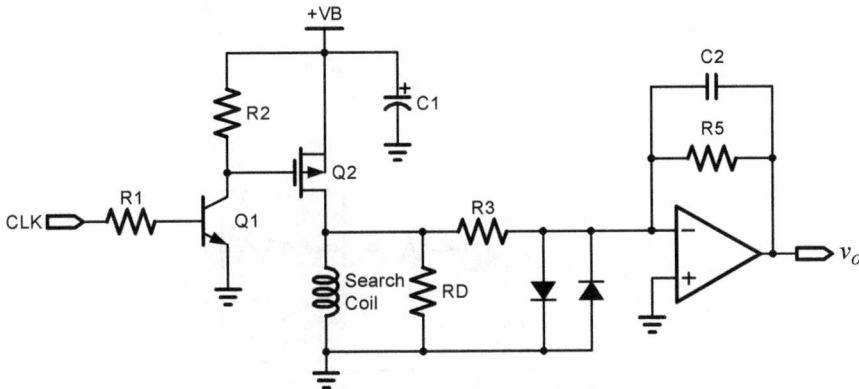

Fig. 20-18: **Inverting Preamp**

20.12: Demod

The *Pulse-1* demod has an attack time constant of 470µs ($R7 \times C3$) and a decay time constant of 47ms ($R8 \times C3$). The attack time constant combined with the sample pulse width determines how much charge a given input voltage transfers to the integration cap. That is to say, it sets the effective

---

10. Beware of fake chips.

gain of the demodulator. You can decrease R7 to increase the gain, but likely you will then need to decrease the gain of the following stage.

The decay time constant determines how long the target signal "lingers" in the integrator. If you increase R8 the response will become broader and slower; decreasing R8 creates a "zippier" response. While it is not possible to completely remove R8 (doing so will make the integrator unstable and it will eventually "rail out") it is possible to modify the circuit so that R8 is only active during the sample time. In Figure 20-19, R8 is connected to the far side of the sample switch so that during the hold time there is no droop on C3. The decay time constant is now $R8 \times C3 \times DutyCycle$, so if the sample pulse width is 10μs and the frequency is 1kHz then the duty cycle is 1% and the decay is now 100 times longer while maintaining the same effective gain.

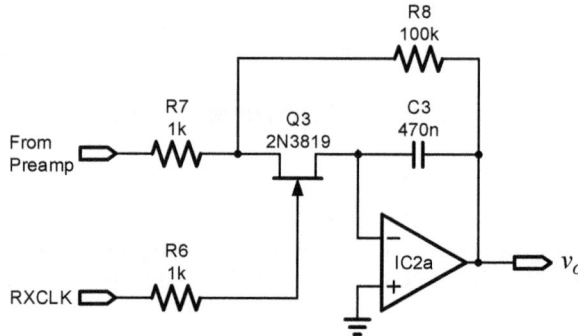

Fig. 20-19: **Leakless Sampling Integrator**

20.13: VSAT

The SAT circuit provides an automatic return to a stable quiescent operating point. The speed at which it does so is established by its time constant $R9 \times C4$. A slower retune speed can sometimes offer slightly more depth but may require a slower sweep speed. A faster retune speed can help in variable ground conditions but may reduce depth. Some PI detectors make this function variable by replacing R9 with a potentiometer; say, 1 MΩ. Another fixed resistor is usually added in series to set a minimum retune speed, perhaps 100k.

IC2 can also be upgraded, but a FET-input opamp is required to avoid an induced offset with R9 due to input bias current. If a CMOS-input opamp is used here, a SPST switch added in series with R9 can be used to switch out R9 and effect a zero-motion mode. The switch will need to be occasionally shorted to account for SAT drift, but would be useful as a short-term pinpoint mode.

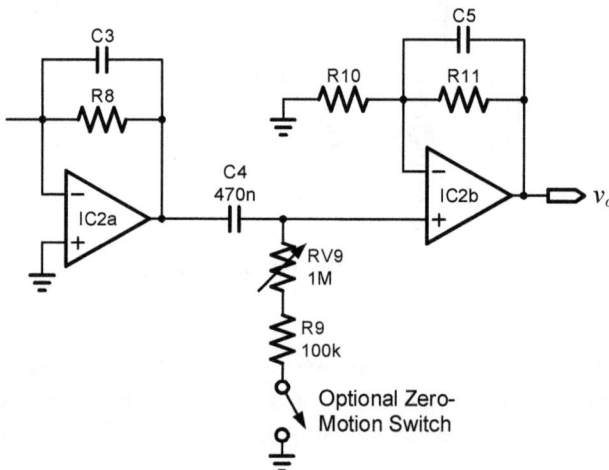

Fig. 20-20: **Variable SAT & Zero-Motion Options**

20.14: Clock

R19 and R20 can be replaced with potentiometers (include series fixed resistors to establish minimum levels) to effect a variable pulse rate and pulse width. Replacing R22 with a pot allows you to vary the sample pulse width for the JFET. This has the effect of altering the integration time, which can trade off selectivity for sensitivity.

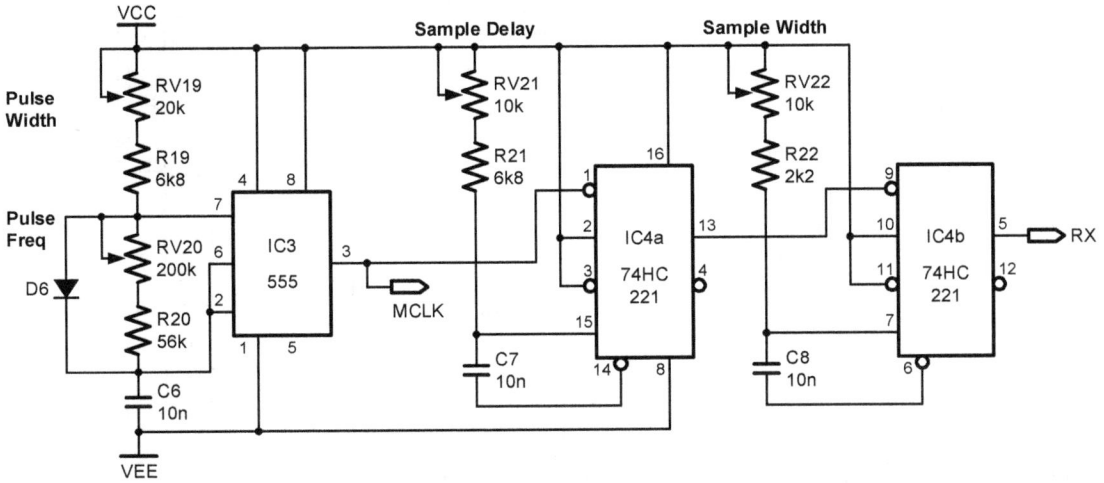

Fig. 20-21: **Variable Clocking**

For the values shown, the following (approximate) ranges are provided:

- Pulse rate   520Hz - 2kHz
- Pulse width   48μs - 190μs
- Sample delay   12.5μs - 28μs
- Sample width   5μs - 25μs

Be mindful that a higher pulse rate and wider pulse width will cause the damping resistor (RD) and input clamp resistor (R3) to dissipate more power and they can run significantly hotter.

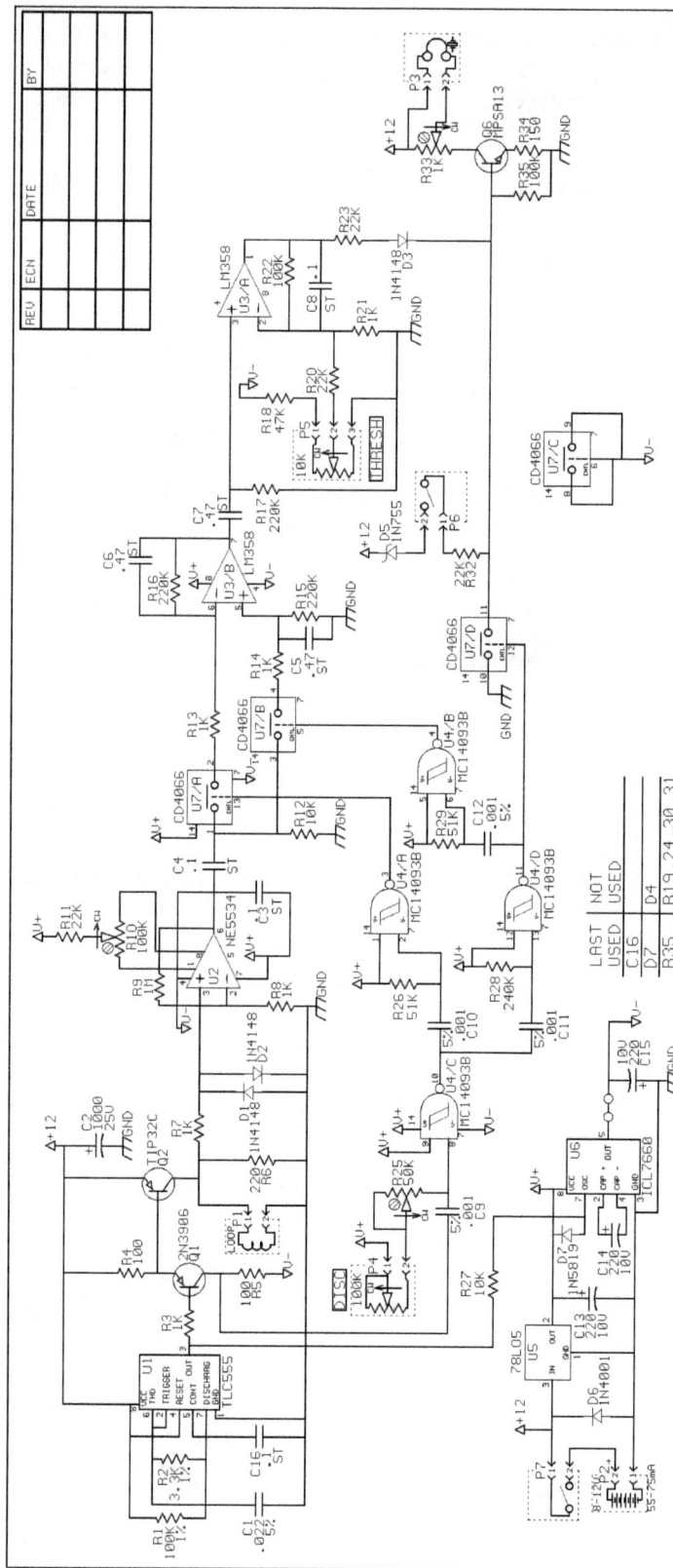

NOTES: UNLESS OTHERWISE SPECIFIED

1. RESISTOR VALUES ARE IN OHMS.
2. RESISTORS ARE 1/4 WATT; ± 5%.
3. CAPACITOR VALUES ARE IN MICROFARADS.
4. CAPACITORS ≥ 1.0 MFD = ALUMINUM ELECTROLYTICS; +80%, -20%.
5. CAPACITORS < 1.0 MFD = POLYESTER FILM; ± 10%.
6. NOMINAL CAPACITOR RANGE OF .001 - .039 = POLYESTER FILM.
7. NOMINAL CAPACITOR RANGE OF .047 - 1.0 = STACKED POLYESTER.
8. ST.=STACKED POLYESTER (PANASONIC "V" SERIES): ± 5%.
9. MONO=MONOLITHIC CERAMIC: ± 5%.
10. AREAS INSIDE DASHED LINES ARE NOT LOCATED ON MAIN PCB.
11. PCB 505-0188 (DWG 15-0387).

LAST	NOT			
USED	USED			
C16	D7	D4	R19,24,30,31	U7/C
D7	R35	U7	Q3,4,5	P3
Q6				S2

(C) White's Electronics Inc. '92

SWEET HOME, OREGON U.S.A.

This print is the property of, and embodies propriety design of White's Electronics, Inc., no part of this design may be used in any way without the written consent of White's Electronics, Inc., Sweet Home, OR 97386.

TOLERANCES UNLESS NOTED			
DECIMAL .XXX ±.005" .XX ±.010" ANGLE ±1/2°			
DO NOT SCALE DRAWING	DRAWN BY JWP	DATE 04-17-92	APPROVED BY
SCALE FULL	CHECKED BY SGE		
TITLE		PART NUMBER 801-1186	
Surfmaster P.I. Schematic		DRAWING NUMBER 15-0393	REV. 0	SHEET 1/1

Improved PI Design

"The biggest room in the world is the room for improvement."

— Helmut Schmidt

The previous two chapters covered PI fundamentals and the resulting project design is extremely simple, not even rising to the level of a most basic commercial detector like the White's Surfmaster PI. In this chapter we will begin to consider more advanced elements of PI design and pay a little more attention to the details that win performance.

Modernized Platform

Pulse-1 in the previous chapter is an all-analog design, with timing created by a 555 timer and logic chips, plus a PMOS-based transmitter and a fairly sub-optimal analog receiver. The goal is — as it was for the VLF designs — to create a platform design upon which we can easily expand with new ideas. So let's modernize *Pulse-1* to something we can work with.

21.1: Transmitter

The first item is to replace the PMOS switch with an NMOS switch. While a PMOS device works just fine, most designs use NMOS because of better availability and better performance. We started with a PMOS simply because a high-side switch is easier to visualize and easier to explain the resulting power supply configuration. The new transmitter is shown in Figure 21-1.

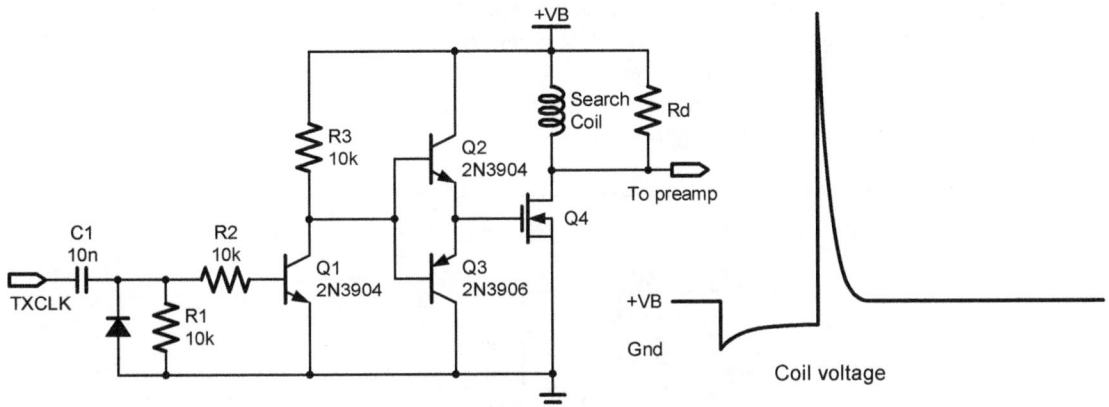

Fig. 21-1: **NMOS Transmitter**

The resulting coil voltage waveform is also shown; compare to Figure 19-9. For Q4 we will start with the popular IRF740 which is rated for 400V. The simple clock drive used in Chapter 20 has been updated with a class-B buffer (Q2/Q3). The buffer provides a lower impedance drive for transient gate currents during both turn-on and (especially) turn-off. A Darlington PNP for Q3 will often improve turn-off speed over a standard PNP.

The TXCLK signal is also AC-coupled to the drive circuit. This is good practice especially when a microcontroller provides timing. Suppose a code bug causes the micro lock to up at the unfortunate instance that Q4 is turned on; this will leave the TX circuit in a state whereby the coil is shorted to the battery and the next likely event is smoke. It is prudent to AC-couple the micro's clock signal to the coil switch so if a clock gets stuck high the switch will turn itself off. The R1-C1 time constant should be selected to work with the desired TX turn-on time.

21.2: Power Supply

The flyback is now positive and the decay settles to +VB instead of ground. This has odd ramifications on the preamp configuration and power supply design. In *Pulse-1* the input and output signals settled to ground (0V) and the preamp required an additional *negative* supply (-VX) to keep the inputs and output in linear operation, as shown in Figure 21-2a. Now the preamp will require an additional *positive* supply (also VX), higher than the battery voltage, in order to do the same. Figure 21-2b illustrates the new power configuration.

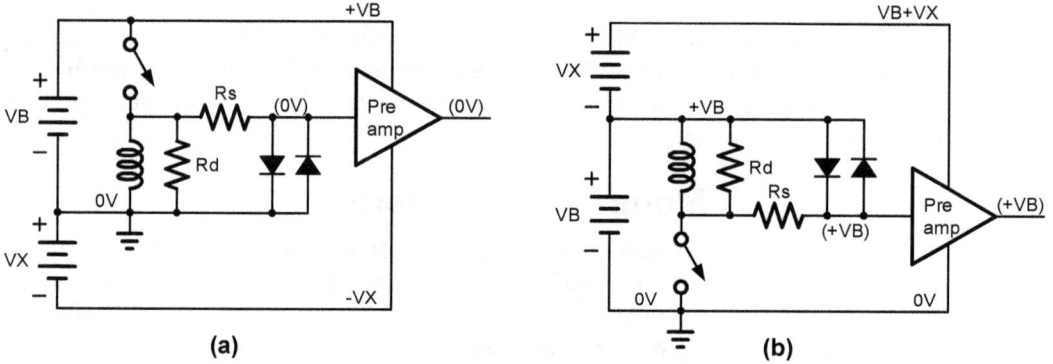

Fig. 21-2: **Power Configuration**

The oddness in this is that the new nominal voltage at both the input and output of the preamp is +VB, and if subsequent stages are DC-coupled this nominal voltage level will persist. It is common practice in PI designs such as this to simply re-label the supply voltages — the high side of the battery becomes 0V, and the low side becomes –VB; see Figure 21-3. This makes things a little easier on the eyes but can create trouble on the test bench. If you are running such a circuit from a bench power supply[1] that shares a ground with an oscilloscope, it is imperative to remember to clip the oscope ground to true ground (-VB) instead of the pretend ground. Furthermore, if a micro needs to be programmed grounding issues can arise from the programming interface such as a USB programmer plugged into a grounded computer. When in doubt, program from a laptop running solely on batteries[2].

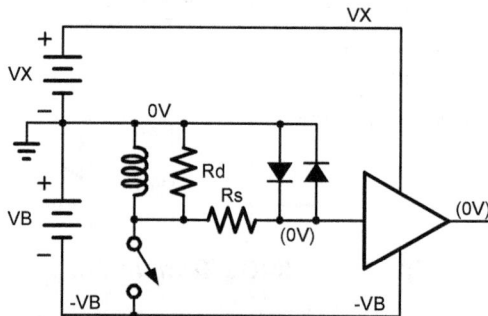

Fig. 21-3: **Convenient Relabeling**

1. Some bench supplies have separate jacks for supply ground and earth ground, and often have a strap connecting them. If so, remove the strap and the supply will be floating.
2. There is also a gadget called a *USB isolator* which plugs inline on a USB cable and effectively AC-couples the signal, preventing damage from mismatched grounds. It costs $20-30 and I found it to be an effective solution after burning up a PI microprocessor.

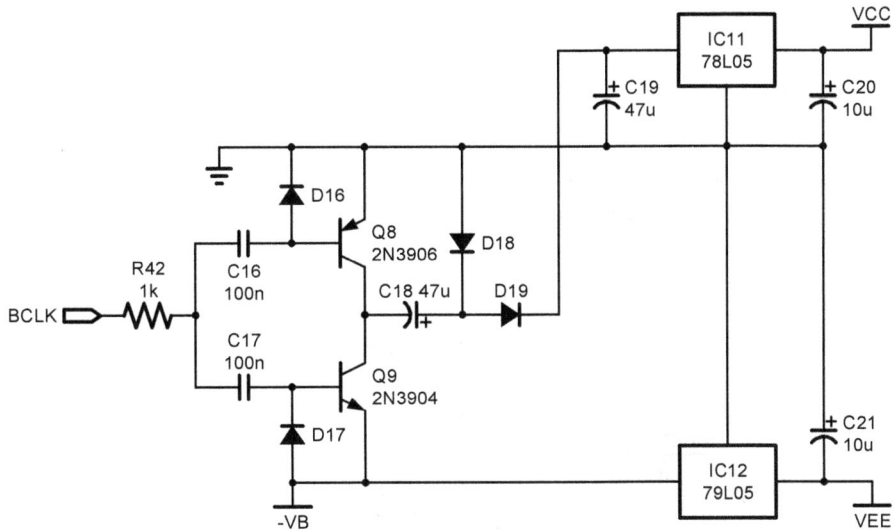

Fig. 21-4: *Pulse-2* Power Supply

Pulse-1 needed an additional negative voltage to power the preamp and this came from an ICL7660 configured as a voltage inverter. Per Figure 21-3 we sort of need the same thing: a voltage that is the negative of −VB. Or, thinking in terms of Figure 21-2b, we need a voltage that is higher than the positive side of the battery. Either way you prefer to think of it, the ICL7660 charge pump could still do the job as it can be configured as a *voltage doubler*. However, we will now switch to a discrete solution to avoid the 10/12V limitation of the 7660. Figure 21-4 shows the new charge pump and, as before, some 5V linear regulators are added to power the analog circuitry. The charge pump requires a clock drive but now a large battery[3] can be used to power the whole thing.

Figure 21-5 shows the newly redesigned front end. Besides the use of a low-side switch, the other major change is that the preamp is now an inverting type instead of non-inverting. This doesn't make a lot of difference in performance but because our new low-side switch produces a positive flyback and therefore a positive-deflecting signal, an inverting preamp restores the output signal polarity to that of *Pulse-1* which allows us to use the remainder of the *Pulse-1* analog circuitry with fewer changes.

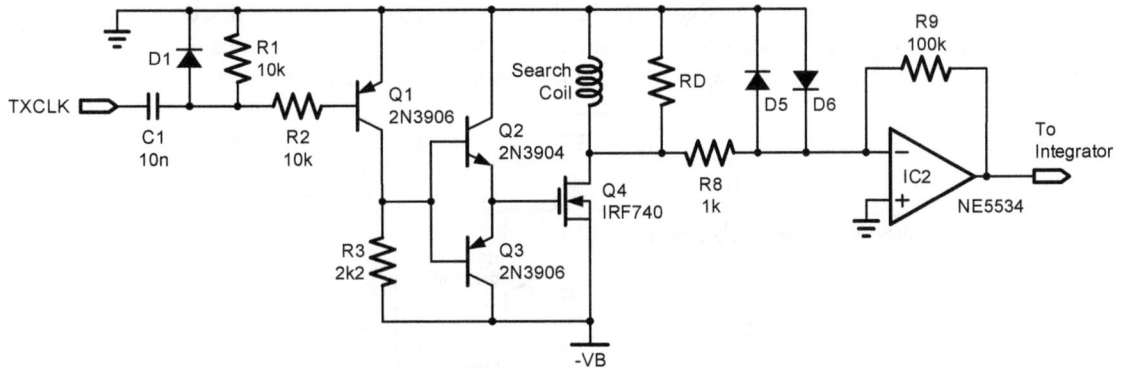

Fig. 21-5: *Pulse-2* Front End

---

3. Generally we're talking about a 12V AA pack or a 14.4V lithium pack. However, you could go higher; the LM78L/79L series have a hard limit of 30V at the input. Don't get carried away, but if you do get carried away mind the voltage ratings on the capacitors. Derating them by 50% is a good rule-of-thumb.

21.3: Receiver Back-End

The next step is to revisit the receiver chain. The sampling integrator used in *Pulse-1* is actually pretty decent, but most newer PI designs have replaced the JFET switch with a MOSFET switch, namely the 4066 "bilateral" switch. These generally have better switch characteristics and are easier to drive. It may seem wasteful to replace a single JFET transistor with a quad switch IC and then use only one switch, but future expansions of this design will need more switches. The SAT/gain stage will remain the same as before; see Figure 21-6.

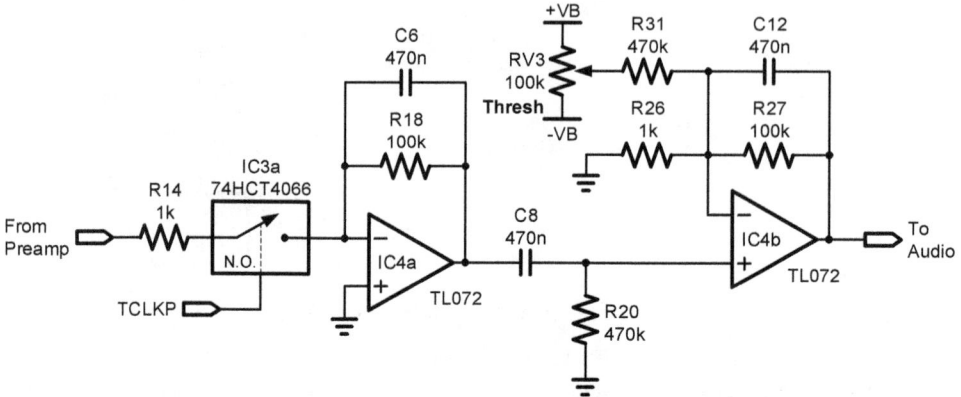

Fig. 21-6: *Pulse-2* **Receiver**

The last part of the receiver chain is the audio circuit. In the *Pulse-1* design the output of IC4b is nominally a little above ground, set by the threshold offset. A target produces a positive voltage followed by a negative "bounce-back" voltage (Figure 20-9d); the audio stage is designed to respond to the positive signal while the negative part of the signal is simply truncated (Figure 20-11c). Because the audio stage can require high transient currents it is desirable to run it directly off the battery, and this was a simple matter in *Pulse-1* where the output signal range of VCC/Ground from the final gain amp (IC4b) needed to drive an audio stage powered from +VB/Ground.

Pulse-2 is basically an upside-down version of *Pulse-1*, except that the output of the final gain amp still produces a positive-going target signal. If we still want to power the audio stage directly from the battery then the signal level needs to be converted from a VCC/Ground range of the gain stage to a Ground/-VB range of the audio amp. Figure 21-7a shows the basic circuit that will accomplish this. Q7 is a common-base amplifier which acts as a level converter. Positive values (above 0.7V) at the input will drive the collector of Q7 positive whereas negative input voltages are truncated.

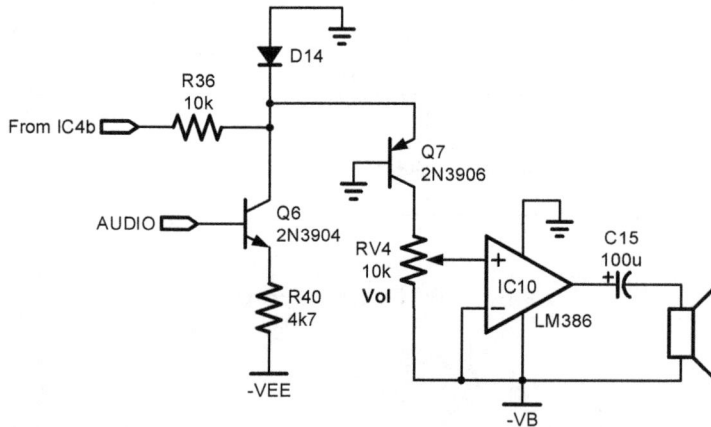

Fig. 21-7: **Audio Solution**

As before, there needs to be a way to chop the input voltage to produce an audible waveform. Q6 does this, with D14 acting as a ground clamp. This is also a good time to move from the simple transistor speaker driver to a better sounding LM386 audio amplifier. This also allows the use of a volume control pot.

21.4: Micro

The clocking circuit in the *Pulse-1* design was a 555 timer and a couple of monostable multivibrators, which is not a bad solution. It is simple and doesn't require programming. Replacing resistors with potentiometers easily gives you control over the pulse rate, TX turn-on time, sample delay, and sample width. Being able to vary those parameters is a useful exercise in learning how a PI detector functions[4].

The *Pulse-2* design will replace all of that with a microcontroller and place all of the timing under the control of firmware. You can still set up any or all of the timing parameters so they can be manually adjusted by reading potentiometers with an on-board ADC and varying the timing as desired. For now, it is limited to just the RX sample delay.

In this design we will choose a simple 8-pin PIC processor, the PIC12F1840; see Figure 21-8. The micro controls not only the transmitter and demodulator clocks but also generates a square wave for both the audio gating and the clock for the charge pump. As with the *Pulse-1* design, everything runs synchronously to minimize clocking noise; Figure 21-9 shows the timing diagram.

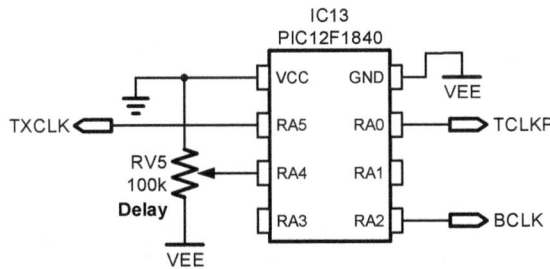

Fig. 21-8: *Pulse-2* **Timing Controller**

Although Figure 21-8 specifies the exact pin connections for the micro, there is some flexibility in this. Three GPIO pins are available as ADC inputs, and all but RA3 can be used as clock outputs. In general, choose the pin connections for the most efficient PCB layout. RA1 is not used in the *Pulse-2* design but it will be used in the *Pulse-3* design.

Finally, the PIC can be powered by a 5V supply connected to the VCC and GND pins. In our case, we will power it from the −5V side so that the VCC pin is grounded and the GND pin is connected to VEE (−5V). This will produce clock pulses with high/low logic levels of 0/−5V which will properly drive the TX circuit and the demod switches.

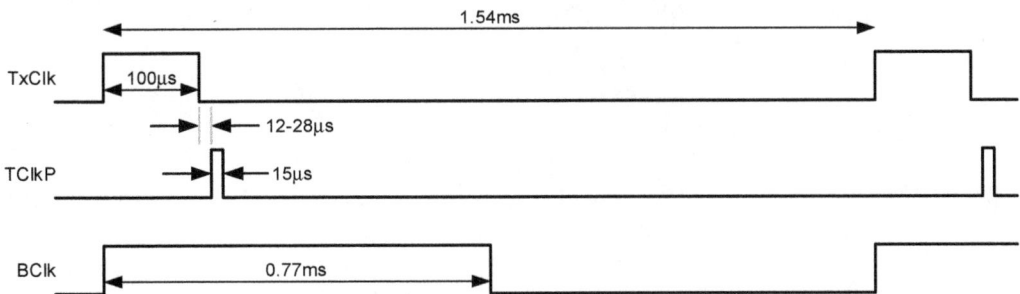

Fig. 21-9: *Pulse-2* **Timing Diagram**

---

4. I designed the *Hammerhead PI* project with lots of variability, specifically intended as a learning platform. See the *Geotech* web site for more info.

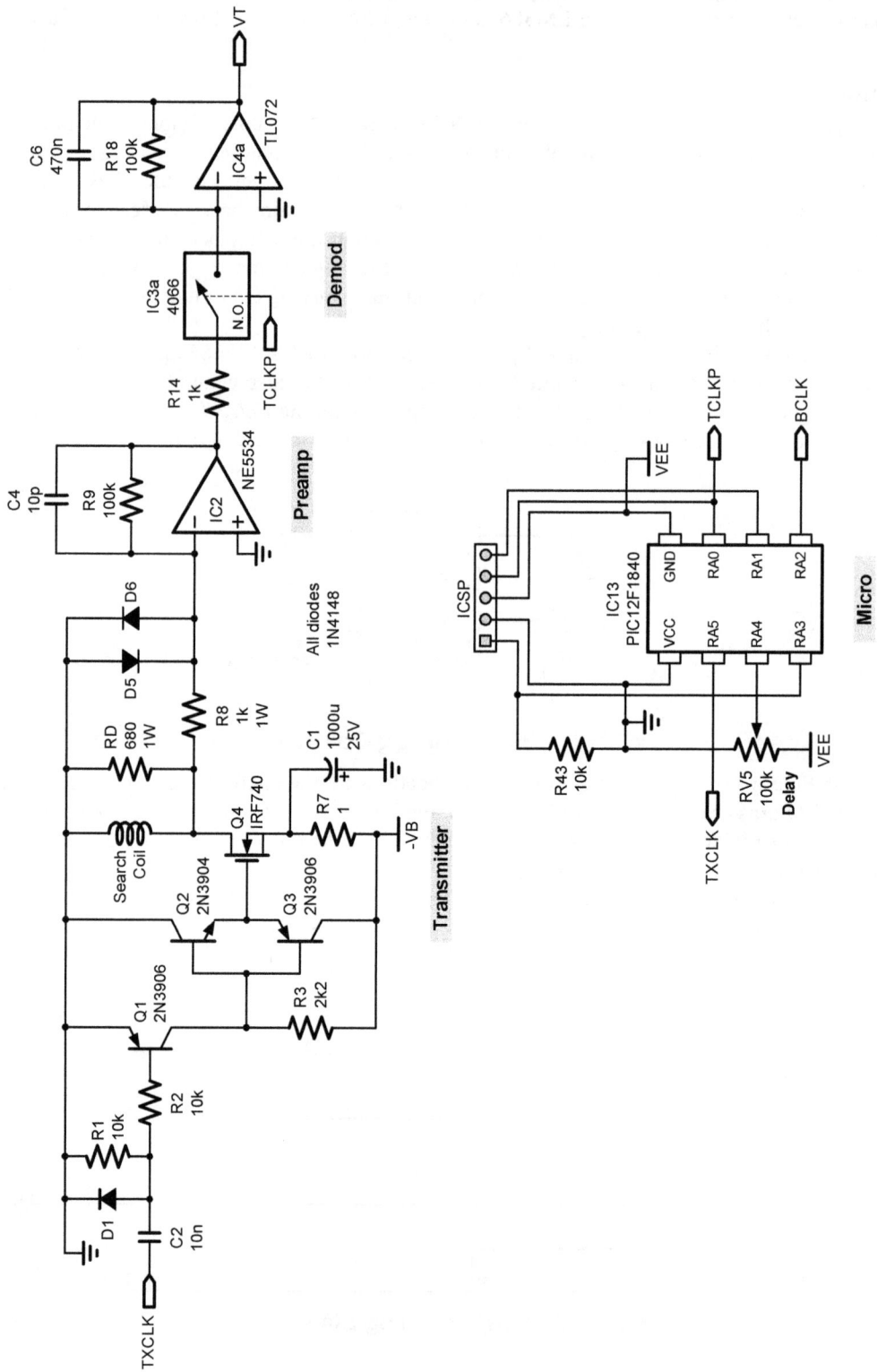

Fig. 21-10: **Pulse-2 — Main Circuit**

Inside the Metal Detector

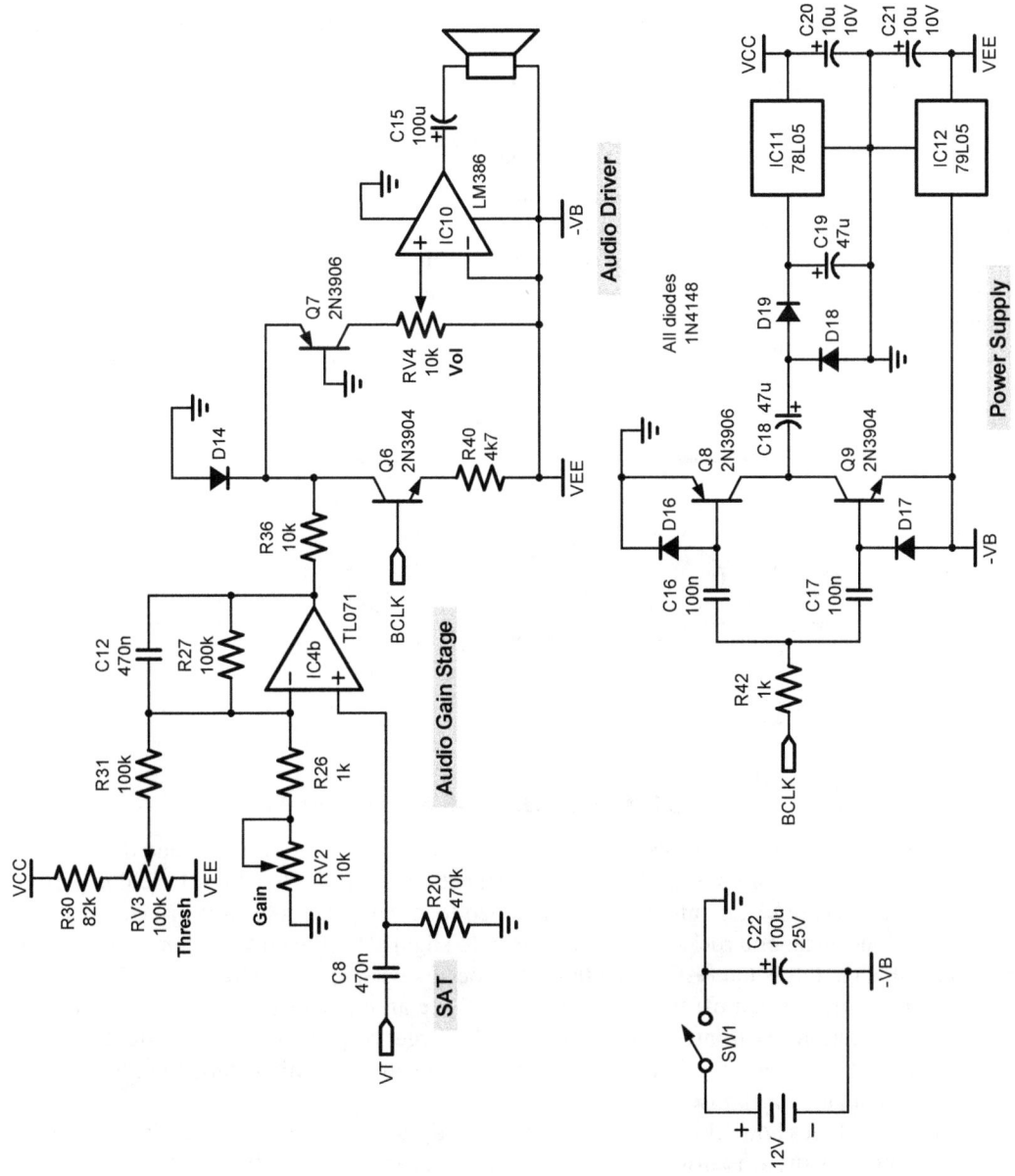

Fig. 21-11: **Pulse-2 — Power & Timing**

Audio Gain Stage

Audio Driver

Power Supply

All diodes
1N4148

Design: Improved PI (Pulse-2)

The final schematic for our improved PI platform design is shown in Figure 21-10 and Figure 21-11. While not a whole lot better than *Pulse-1*, it is a little simpler and may perform slightly better due to the NMOS transmit switch, but that can depend on the quality of the coil. The main advantage of the *Pulse-2* design is it gives us something to build on, and that begins in the next section.

You will notice with this design the component numbering is seemingly haphazard, with many numbers skipped over like R5, R6, and R7. As we further develop our PI knowledge, subsequent example designs will begin to fill in these missing components, making it much easier to compare schematics and apply the prior (and future) subcircuit discussions across all the designs. The same thing was done with the TR and VLF designs.

Earth Field Effect

Basic PI detectors employ a monopolar transmitter; that is, the transmitted signal pulse is of a single polarity. The received signal is likewise monopolar and, being a wideband system, is usually filtered only minimally. Because of these two design characteristics, an odd thing happens with PI detectors: as the coil is swung back and forth, it cuts through the Earth's magnetic field and, from Magnetics 101, a coil in motion through a magnetic field will induce an EMF in the coil. This can give rise to false signals that are generated merely from swinging the coil. In PI detectors this is known as *Earth field effect* (EFE).

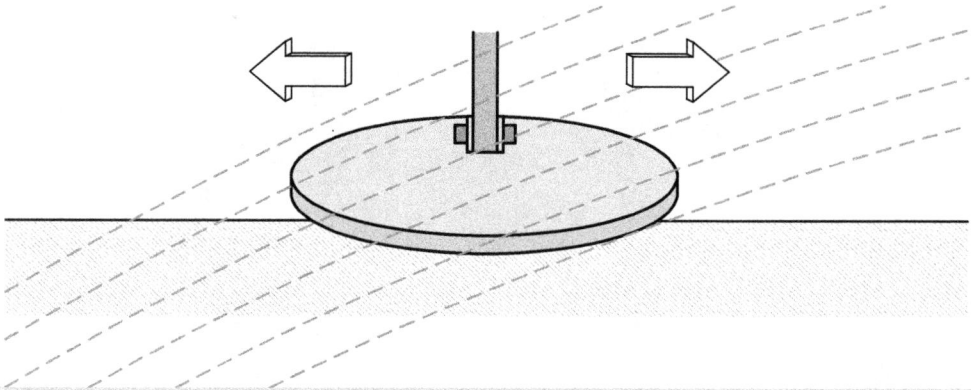

Fig. 21-12: **Coil Moving Through a Magnetic Field**

The severity of the problem depends on the local homogeneity of the Earth field and the exact way the coil is swung, but it can turn a PI detector into a magnetometer. If the local Earth field is perfectly uniform and the coil is carefully swung in a level manner, then the EFE problem is minimal. But local Earth field distortions, magnetic rocks, and the tendency to slightly lift the coil at the ends of the swing will create an EFE signal. It is interesting that this effect occurs whether the PI transmitter is running or not. With the transmitter turned off the receiver will still see an induced signal from the Earth field when the coil is in motion. Also interesting is that, unlike target responses, the Earth field effect can produce both positive and negative signals. A positive signal can sound like a target while a negative signal can mask out a target response.

Figure 21-13 illustrates how EFE affects the flyback decay, as seen at the coil. This is for an NMOS switch and so continues the waveform conventions of this chapter. Normally the flyback decays to some "decay ground" (a) and a target produces a short-lived perturbation of the decay (b). The target response lasts for maybe a few 10s or 100s of microseconds. Earth field effect produces a very long perturbation (c) and the response can last for 10s of milliseconds, appearing to be an overall offset of the entire curve; it can be either positive or negative, depending on the sweep direction of the coil through the Earth field. A target then shows up as a normal perturbation of the offset decay (d).

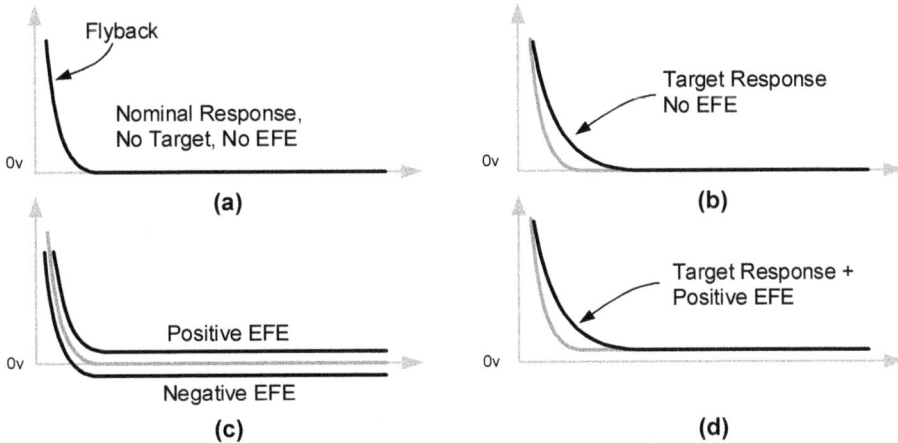

Fig. 21-13: **Target vs Earth Field Effect**

It might be obvious to wonder why EFE does not show up in VLF detectors. It does, but VLF detectors do two things to mitigate the problem. First, most are narrowband systems that high-pass filter the signal before demodulation. Earth field effect is a low frequency phenomenon — the frequency is equal to that of the swing, whereas targets respond in a small portion of the swing — and filtering knocks most of the EFE out. Second, most modern VLFs use full-wave demodulators that subtract equal portions of preamp signal, thereby canceling low frequency signal components. The end result is that Earth field effect is never noticed. While these techniques are used in VLFs for purposes other than Earth field suppression, the fact that they do a good job at EFE suppression suggests two directions we could consider for PI.

21.5: High-Pass Filtering

A first way of dealing with EFE is to high-pass filter the coil response either at the input or the output of the preamp, or between a 2-stage preamp. While it is possible to high-pass filter at the input[5] it is much easier to do so at the output or between preamp stages. A simple RC filter provides modest rejection of EFE (Figure 21-14).

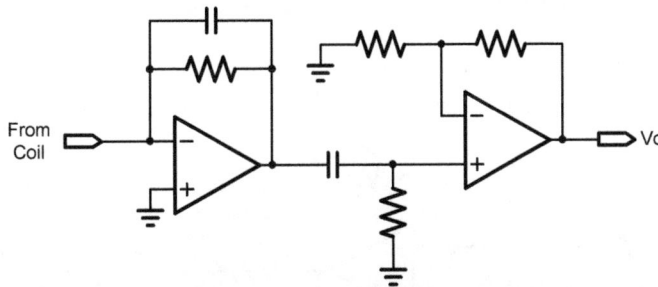

Fig. 21-14: **High-Pass Filtering**

Generally both the target and EFE responses are proportional to the swing speed of the coil so the ratio of their frequency responses is fairly consistent at perhaps 20:1 or so. At this ratio a single-pole high-pass filter would provide, at best, 26dB of attenuation of the EFE signal, which is not all that good. Cascading two stages of high-pass filters would improve the situation.

21.6: Correlated Double Sampling

A more prevalent way of removing Earth field is by subtracting a late sample of the signal decay, a process called *correlated double sampling* because we are sampling the same signal twice. Figure 21-15

---

5. This was done for the PI circuits in *ITMD-1|2*.

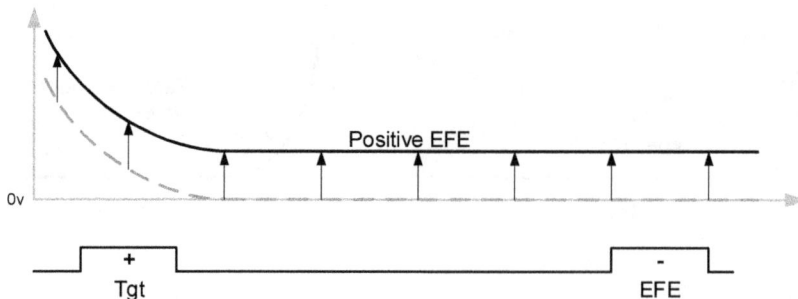

Fig. 21-15: **Subtracting a Late Sample**

shows again the preamp output with no EFE (dashed line) and with a positive Earth field perturbation. The normal target sample is taken soon after the preamp is out of saturation and will see a combined target response and Earth field offset. If a late sample is taken beyond any expected target response, then it will only see the Earth field offset.

The Earth field signal is fairly uniform over a few hundred microseconds, so both the early and late samples will consist of largely equal amounts of Earth field signal. In the case of low conductors the target signal is mostly isolated to the early sample so if the late sample is subtracted from the early sample, then Earth field is removed and the low-conductor target signal is largely unchanged. This is a significant improvement over the high-pass filter solution. For high conductors that persist much longer, there is some modest cancelation of target signal as well, thus some loss of depth. This is a popular technique in commercial PI detectors and was pioneered by Eric Foster.

Figure 21-16 shows a practical implementation of EFE cancelation. The preamp signal is sampled by switch S1 in the usual way. Switch S2 samples an inverted version of the preamp signal — via an inverting amplifier which is configured with a gain of −1 — at the late sample point. The effect is that the S2 sample is subtracted from the S1 sample and the output of the demodulator will be substantially free of EFE.

If correlated double sampling cancels EFE, might it also help with other undesirable low frequency signals? The answer is *yes*; it is a rare win-win situation. Low frequency components like external EMI and internal flicker noise will also be attenuated. The separation of the main sample and the EFE sample determines the amount low-frequency attenuation. Closer spacing will increase the amount of can-

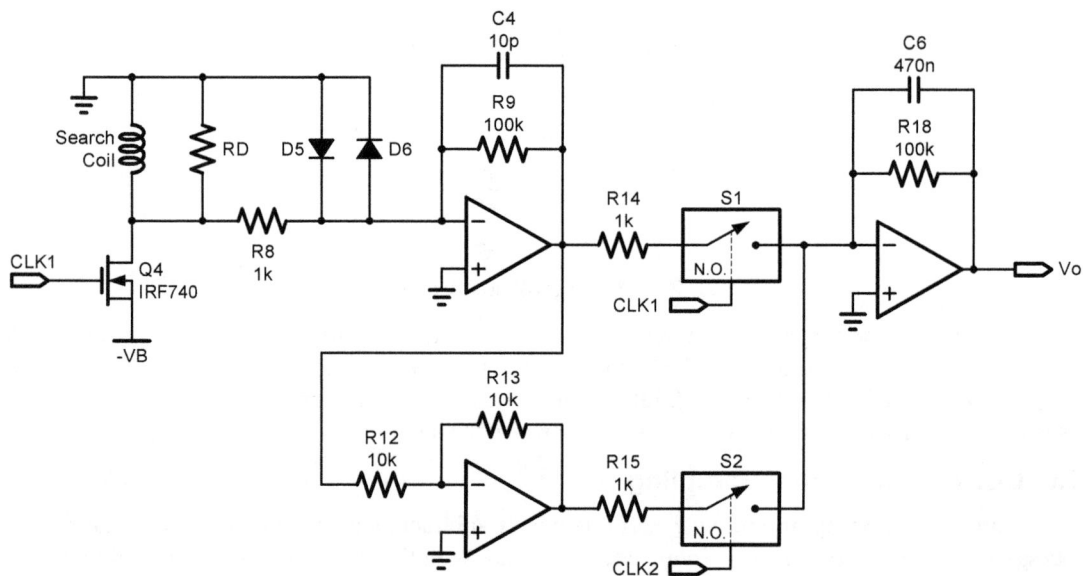

Fig. 21-16: **Practical EFE Subtraction**

celation but also increases the range of conductivities that suffer from loss-of-depth due to their decay falling into the EFE subtraction sample. The details of the math are covered in Appendix C, but for a separation of 200μs the EFE suppression is a decent 58dB and EMI from 50/60Hz mains will be attenuated by over 20dB, a bonus we will gladly accept.

It is useful to pause and consider an alternative method of subtracting the EFE sample, a method which often shows up in real products. In Figure 21-17 the inversion amplifier is removed so that S2 samples the same preamp signal as S1. The subtraction instead takes place at the input of the demod opamp, where the S1 sample is applied to the inverting side and the S2 sample to the non-inverting side. The sampling integrator becomes a *differential sampling integrator*[6]. The overall result is the same as for Figure 21-16 and seems simpler, with the elimination of the inversion amplifier stage and the addition of a resistor and capacitor. A drawback to this approach is that the two caps used in the demod stage must be well-matched[7] in value or there will be an error between the S1 and S2 integration time constants, resulting in poorer cancelation of EFE. Matching the caps to 1% is usually sufficient. The approach in Figure 21-16 uses the same cap for both samples so there is no issue as far as the capacitor is concerned, but it should be noted that the feedforward resistors (R14 & R15) must match to 1% or so. In SMT, 1% resistor tolerance is the norm and in thru-hole 1% is readily available, whereas high-precision capacitors demand a premium or must be hand-matched. Even with matched capacitors, Figure 21-17 will still be more prone to temperature or long-term aging drifts.

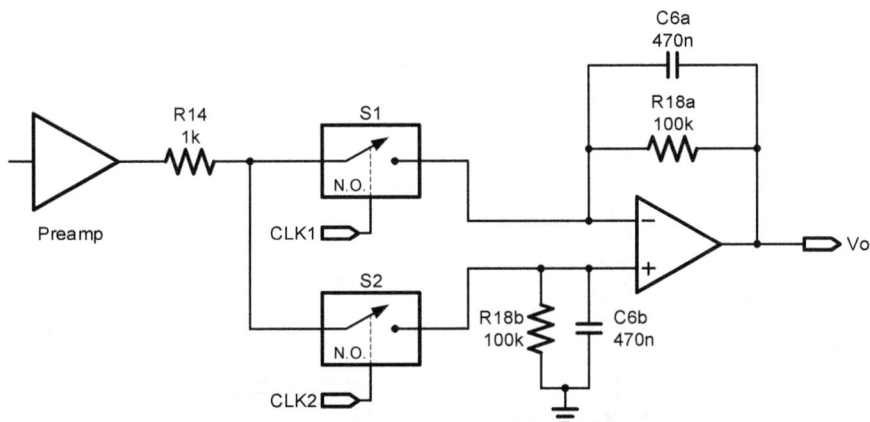

Fig. 21-17: **Alternative EFE Subtraction**

Design: *Improved PI (Pulse-3)*

Pulse-3 is our improved design updated to include EFE cancelation, and the complete schematic is shown in Figure 21-18 and Figure 21-19. Besides adding EFE compensation, there is a minor improvement in the audio stage. A drawback of the circuit in Figure 21-7 is the temperature drift imposed by the V_{BE} of Q7 on the audio level. This will especially cause changes in the threshold as the temperature varies. The solution is to add D15 and R41 to establish a temperature-stable virtual ground on the emitter of Q7 instead of hard-grounding its base.

---

6. We've seen this before; see Figure 16-40.

7. Much the same as with VLF demods.

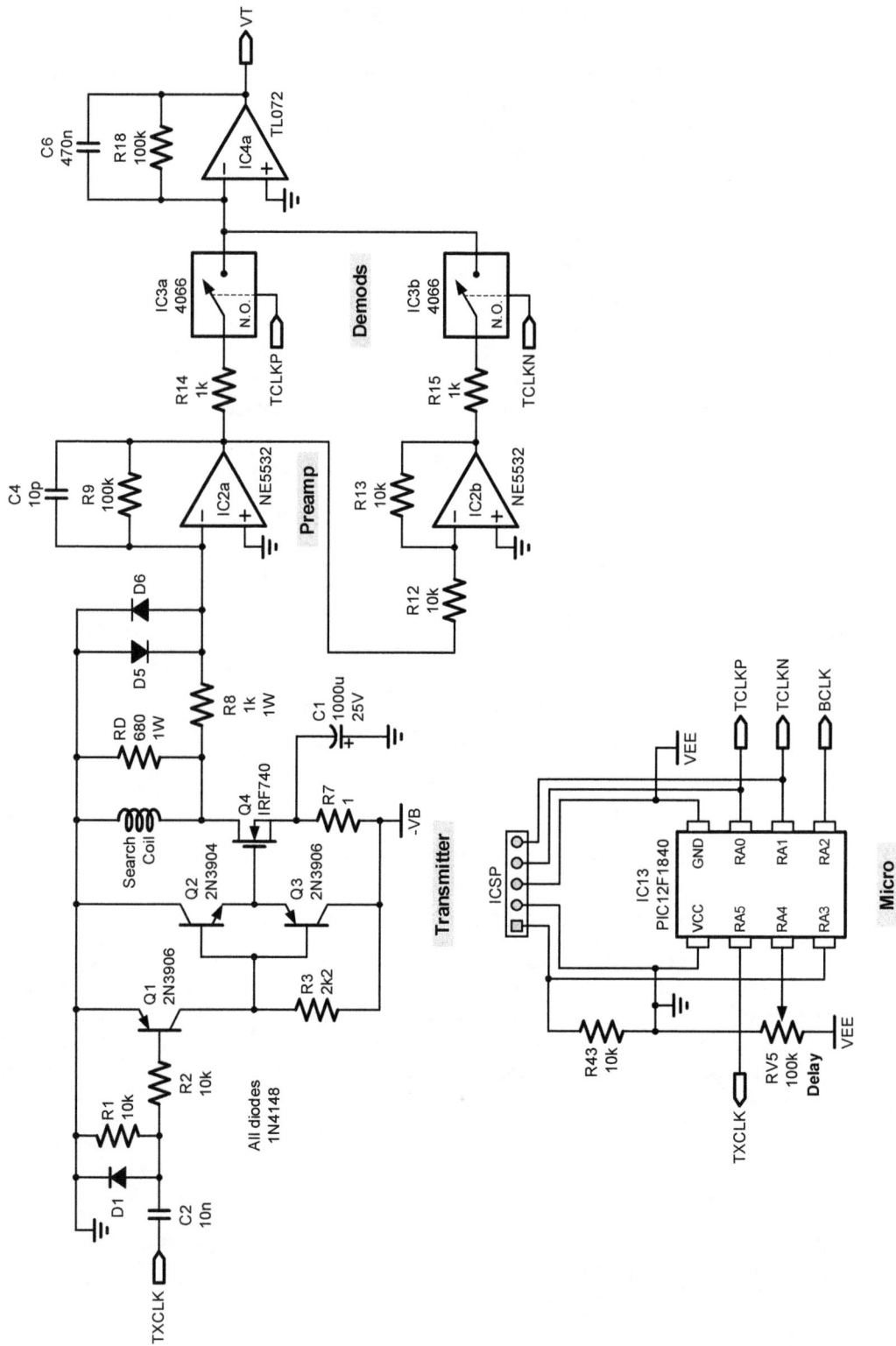

Fig. 21-18: **Pulse-3 — Main Circuit**

Inside the Metal Detector

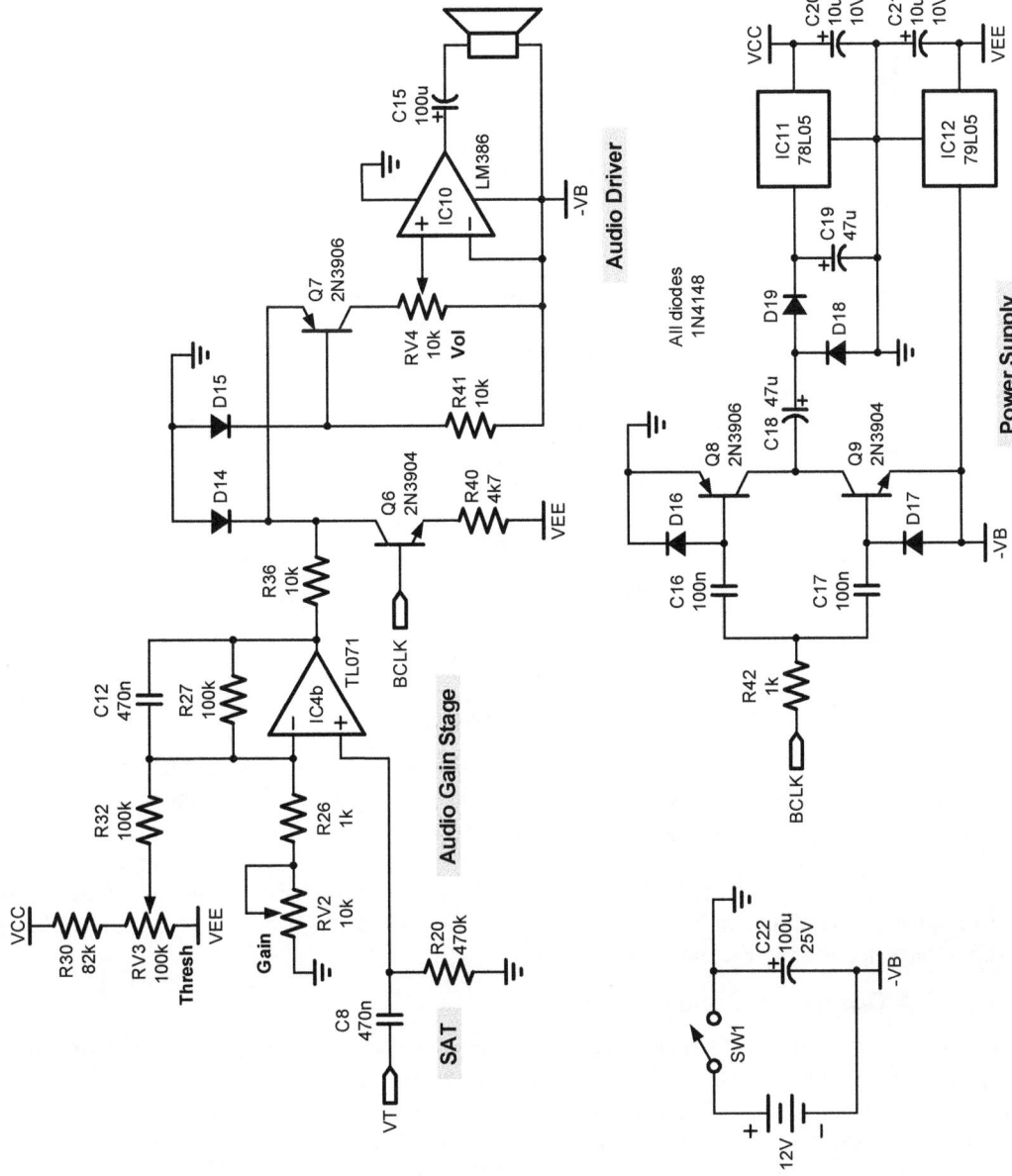

Fig. 21-19: Pulse-3 — Power & Timing

Optimizing

The *Pulse-X* designs have so far evolved without much in-depth consideration for particular component values or pulse timings. Before we continue with more developments to the platform, it is a good time to look at the choices which determine performance.

21.7: Decay Speed

In most cases, the goal of a PI design is to sample as early in the decay as possible and to maximize the target signal to noise signal ratio (SNR). Early sampling is largely controlled by the front end — the coil, the TX circuit, and the preamp. Starting with the coil, it should be designed with low parasitic capacitance. One approach to this is to use a low inductance coil; fewer windings mean less interwinding capacitance. The drawback to this is reduced sensitivity on the RX side of the equation due, again, to the fewer turns. Another technique is to use wire with heavier insulation than normal magnet wire so that the wire-to-wire spacing is increased. In particular, insulation with a lower dielectric constant — such as Teflon — can reduce capacitance. Teflon-insulated wire is, in fact, sometimes used in commercial PI detectors.

As an example, a 10" (25cm) coil with 22 turns of 24AWG Teflon wire comes out to 312µH. With a spiral wrap spacer and shielding added the SRF is 821kHz, resulting in a parasitic C of 120pF. This does not include the coil cable. RG58 coaxial cable has a capacitance specified at 85pF/m. Assuming 58" (1.5m) this adds about 128pF. Finally, for the TX switch we'll assume an IRF740 NMOS transistor, which has an "output capacitance" (C_{oss}) of 330pF[8]. C_{oss} is the combination of C_{DG} and C_{DB} and is a fair measure of the total drain capacitance, which is what we are interested in. The total capacitance is now

	C
Coil	120pF
Cable	128pF
Switch	330pF
Total	578pF

TABLE 1. TX Capacitances

For these choices the switch dominates the overall capacitance and therefore any effort to improve the speed should start with a search for a better NMOS device. It is fortunate that motor control applications are driving better MOSFET designs so improving on the IRF740 is not difficult[9]. Likewise, it is possible to find lower capacitance cable, but keep in mind the current capacity requirement which is likely several amps. For the coil, once Teflon wire is employed one of the last remaining techniques for speed is to reduce the windings, although it is possible to explore some esoteric winding techniques such as spiral, basket, and spider wound coils.

21.8: Other TX Considerations

A fast decay is a primary goal but other considerations must be kept in mind. For the switch, both current and voltage ratings are important. PI designs have a pulsed current and most MOSFETs are rated for continuous current[10] so it is permissible to use the "time-averaged" current which means that even a 1 amp device is usually more than sufficient. A device with a lower current rating generally also has a higher channel resistance which may or may not be important. Likewise, the resistance of the

---

8. Per the Vishay/Siliconix data sheet.

9. Even the IRF740 has improvements. Compare the original IRF740 (C_{OSS} = 330pF) to the IRF740A (170pF) or the IRF740B (59pF).

10. Some data sheets include a "pulsed current" spec in addition to a "continuous current" spec. When available, use the pulsed current value.

cable and the coil windings should also be considered. In detector designs resistance in the TX path always produces an ohmic power loss and we almost always try to minimize it. But in some PI designs, additional resistance is intentionally added so that the TX current pulse flat-tops faster (see the *Explore* section).

For the switch, the voltage rating we are concerned with is V_{DSS} which is the drain-source break-down voltage. Ideally, this should be selected to be higher than the maximum flyback voltage so that the MOSFET never avalanches. If it does avalanche, then this will extend the recovery speed and move the decay curve out in time. It also creates a brief time where there is both a high voltage and a high current in the FET and this will heat up the FET, possibly requiring a heat sink. FETs that do not avalanche tend to run fairly cool[11].

The breakdown voltage of the cable is also important as it sees the full flyback peak. Many coaxial cables that are useful for homebrew projects (RG58, RG6, RG180, etc.) tend to have very high break-down voltage ratings, often above 1kV.

21.9: TX Pulse Rate & Width

Our examples thus far have used a TX pulse width of 100µs and a pulse rate of 1kHz. These are "middle-of-the-road" choices and any optimization will depend heavily on the application.

The TX current is a rising exponential with a fairly slow time constant so the peak transmit current is very dependent on the TX pulse width, up until the point the current reaches its exponential maximum. After that, additional pulse width mostly wastes power except for the rare case of detecting targets with exceptionally long time constants[12]. Most PI detectors terminate the TX pulse long before the exponential maximum as the "return on investment" of depth versus power consumed is not very good. We'll cover this some more in the *Explore* section topic called "Flat-topping."

A faster pulse rate is usually better because it produces more integrator averaging but it also increases power consumption, given the same pulse width. The maximum pulse rate will be limited by the need to keep the EFE sample sufficiently far away from the main (target) sample. "Sufficiently" can depend on whether you are hunting for small gold nuggets or Atocha bars; small nuggets decay quickly so the EFE pulse spacing can be greatly reduced and allow for a higher frequency. At the same time, a wide TX pulse width is not needed for nuggets so power consumption can be kept reasonable.

21.10: Sample Pulse

The sample pulse has two parameters: pulse delay, and pulse width. We normally focus on the delay and emphasize the shortest delay possible. A middle-of-the-road delay might be 15µs which is good for hunting larger jewelry in a salt water environment. 10µs is a bit more aggressive and would be useful for detecting gold nuggets down to a few grains or smaller jewelry like stud earrings, but begins to see salt water as well. 5-7µs is exceptionally difficult to achieve and can pick up gold at 1 grain or less. Extremely low sample delays require a coil and front-end design which are extraordinarily fast and is beyond the scope of this book[13]. However, higher sample delays are quite easy and can be useful in certain circumstances. As the preceding sentences allude to, higher sample delays progressively knock out more targets. If, for example, you are hunting American Civil War relics in an area littered with bird shot from hunters, you will not want to detect the bird shot. A higher delay can be used as a crude form of discrimination.

The sample pulse width is rarely discussed; we usually set it to some nominal value (10µs perhaps) and move on. However, if the sample switch feeds a traditional sampling integrator (as used in many PI

---

11. Remember that traditional PI designs use a "total loss" transmitter so if the coil energy is not dissipated in the MOSFET then it is dissipated elsewhere, namely the flyback resistor and the preamp clamp resistor.

12. Eric Foster designed a PI tow sled for Mel Fisher's search for Atocha silver bars. Those bars had a time constant of several milliseconds, requiring Foster to use a longer than normal TX pulse width coupled with a low pulse rate. The peak current was around 50 amps.

13. Some of it because it gets into proprietary knowledge, but also because such advanced design methods could easily fill another book. However, this book provides all the clues for achieving the next level of performance.

designs) its value determines the effective gain of the sampling integrator (see Appendix C) so it would seem that a wider pulse width would be beneficial. As with the pulse delay, it depends on the desired targets. If the targets have a fast tau they will decay to zero quickly and any additional sample pulse width is only sampling noise, not target signal. Generally we want to sample as much of a changing target response as possible, but no more.

Figure 21-20 shows the step response for two examples: a fast target (perhaps a gold nugget) and a slower target, like a 58-caliber Minie ball. The fast target decays quickly; we want to sample only the region of deviation to get the maximum delta. Anything more and we will sample noise and not signal, resulting in a diminishing SNR. For the slow target, a wider sample width ensures that available target signal is not wasted. As mentioned previously, suppose that while hunting for Minies we also find there is a lot of bird shot in the area (dashed response). We could then increase the sample delay from 10μs to 20μs and set the sample width to 30μs. This will largely eliminate the bird shot while still detecting Minies, although with reduced sensitivity since some of the early (and strongest) signal is being ignored.

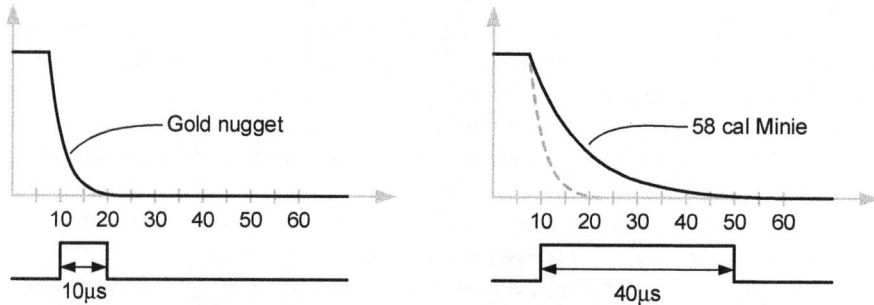

Fig. 21-20: **Optimal Sample Widths**

In practice, the sample delay is usually set to the minimum to maximize sensitivity to all targets. Sample width is often chosen to maximize the most difficult targets — usually small gold — and 10μs is common. If you choose to make the sample delay variable, it would be practical to make the sample width proportional to the delay since an increasing delay progressively ignores fast targets anyway. There is more discussion of proportional timing in Chapter 25.

21.11: Supply Voltage

When we think of optimizing the supply voltage for a PI design we generally think of the power supply that drives the transmitter. And we generally think that "bigger is better." However, it is again a function of what you are searching for. I have designed a PI pinpointer in which the TX supply was 3V and it worked quite well for what it was supposed to do.

A higher TX supply voltage is a good way to compensate for a high coil resistance, and a high coil resistance might be a way to achieve faster flat-topping. If flat-topping is not a goal, then a lower supply voltage coupled with a low resistance coil (using larger gauge wire) can achieve the same peak current as a high supply voltage with a high resistance coil. Don't haul around a huge battery pack if it's not necessary.

Besides the TX supply there is also the RX supply. The designs in this book use ±5V to power the RX analog and it is possible to increase or reduce this. Generally a larger RX voltage increases the dynamic range of the RX circuit; the maximum signal is proportional to the supply voltage and the minimum signal is limited by the noise floor. Some designs may not require a lot of dynamic range; in the PI pinpointer mentioned above the entire circuit — the TX, the RX, and the micro — all ran on 3V.

Explore

21.12: Flat-topping

When the TX switch turns on, the coil current energizes exponentially with a tau of L/R_{eq}, where R_{eq} is the total series resistance. A typical value might be $300\mu H/2\Omega = 150\mu s$. It takes 3τ to reach 95% (5τ for 99%) which is $450\mu s$, but for our project the coil turn-on time is only $100\mu s$. This means the coil current is still ramping up at turn-off. In fact, for the numbers just presented, the coil is only 49% charged and the coil current waveform resembles more of a sawtooth:

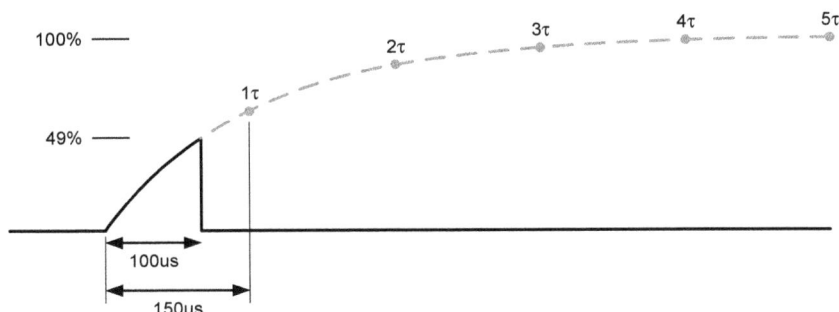

Fig. 21-21: **Coil Turn-On Current**

The exponentially increasing turn-on current generates an increasing magnetic field which induces target eddy currents in a counter-clockwise (CCW) direction (looking down from above). At turn-off, the magnetic field collapses and induces eddy currents in clockwise (CW) direction. If, at turn-off, the turn-on coil current is still increasing then the target will still have CCW eddy currents flowing at the instance the collapsing field starts inducing CW eddy currents. This causes a net reduction in the CW eddies which makes the target signal weaker. Figure 21-22 compares how the eddies behave with a fully-settled TX current (gray) versus a still-rising TX current (black). Assuming both cases have reached the same peak TX current, at turn-off they will both have the same magnitude current step, but in the still-rising case some of that transition is wasted on stopping the residual CCW eddies, resulting in a weaker starting point for the off-time decay.

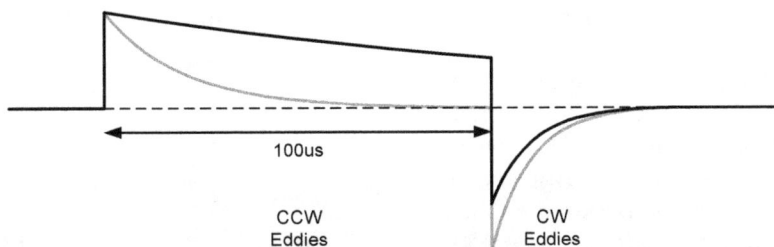

Fig. 21-22: **Effect on Eddy Currents**

Ideally, the turn-on current should be fully settled at turn-off so that the CCW target eddies have fully decayed to zero[14] and don't affect the turn-off eddies. This is called *flat-topping*, and there are two ways to achieve this. The first way is to increase the TX pulse width until the turn-on current has settled. In the previous example, we would increase the pulse width from $100\mu s$ to $450\mu s$ (or more, if 95% isn't sufficient). The average TX current at $100\mu s$ is 135mA (see Appendix C) and will increase to 1.54A at $450\mu s$ which is quite substantial. When the pulse width is increased the pulse rate is often reduced to maintain an overall reasonable average current consumption.

---

14. This assumes the target tau is less than the turn-on time, otherwise we also have to wait for long-tau target eddies to decay beyond the time the coil current has stabilized.

The other way is to decrease the turn-on tau by increasing the coil path resistance. This can be done by simply adding a series resistor to the coil. In our example, we want to be 95% charged[15] in 100μs so the tau needs to be 33μs, therefore the total R_{eq} should be about 9Ω. We assumed 2Ω for the parasitic R_{eq} above, so a 7Ω resistor should be added. Even though the resistor sees a 10% duty cycle, the time-averaged current is still substantial and the resistor will need to have a decent capacity for power dissipation, perhaps 2 watts or more.

Another option to increase path resistance is to simply use a smaller wire gauge for the coil. This moves power dissipation out of the main circuitry and into the coil, so it becomes important to construct the coil in a way that heat can dissipate reasonably well. If flexibility is important the additional discrete series resistor has an advantage in that it can be bypassed if desired, whereas the high resistance of a coil cannot.

The drawback to all of this is that the peak coil current is reduced. Our starting premise — a 100μs turn-on, $\tau = 300ms$, $V_{batt} = 9V$, $R_{eq} = 2\Omega$ — means that the "100% current level" would be 4.5A and the 49% current level is therefore 2.2A. Adding the 7Ω resistor now places the "100% current level" at 1A and so our new peak current is 950mA[16]. This is a far cry from 2.2A and even though it mostly eliminates eddy cancelation it is still a net loss.

So why go through this exercise? First, maybe we don't add enough resistance to achieve 3τ settling; maybe we go for 2τ instead. Perhaps there is a "golden tau" where target sensitivity is maximized. That's why this discussion is in the "*Explore*" section; go explore! Second, it is an introduction for a future discussion where we will use similar methods to greater effect. Be patient, there is much to cover.

21.13: Avalanching

A primary reason for changing to an NMOS switch is to take advantage of better and cheaper devices. PMOS breakdown voltages above 400V are scarce and pricey, but more common for NMOS. The venerable IRF740 — cheap and easy to find — is 400V, but even then with a well-designed coil the flyback will exceed this and avalanche the FET.

Avalanching has two primary drawbacks. First, it causes the FET to dissipate significant power as it experiences both a high voltage (say, 400V) and a high current (upwards of an amp or more). This happens only for a brief instant, perhaps 1μs out of a period of 1ms, so the time-averaged power for this event is maybe less than 1 watt, but it is added to the power dissipated during the exponential turn-on event. The bottom line is, the FET can get quite warm and require heat-sinking. Second, the FET takes a little longer to recover from avalanching, which delays the start of the decay curve and reduces the minimum sample delay.

Figure 21-23 shows[17] the flyback for the IRF740 (400V) versus the IPA95R450P7 (950V). The IRF740 actually breaks down at 450V while the IPA95R450P7 peaks at around 720V but never reaches avalanche. The improvement in the decay curve is not extraordinary — about 250ns — but when trying to achieve an aggressively low sample delay everything matters. A bonus is that the IPA95R450P7 runs cool to the touch, eliminating the need for a heat sink.

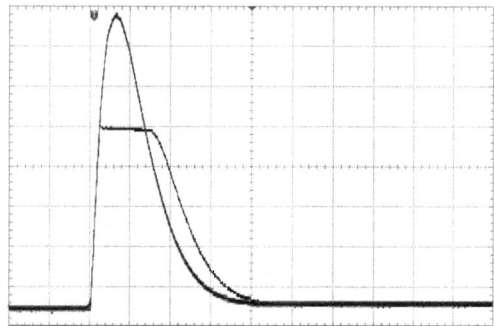

Fig. 21-23: **Avalanching**

---

15. This clearly isn't 100% or even 99% (5τ) but 3τ is often "good enough." Trying to do better will tend to consume gobs of power with little performance benefit.

16. Assuming the current is a perfect 10% pulse, the RMS value is 300mA so the 7Ω resistor will dissipate 2.1W. It's not quite that bad because the current isn't a perfect pulse, so the RMS is a bit lower. But it will get warm.

17. The vertical resolution is 100V/div, horizontal is 500ns/div.

21.14: Capacitance Reduction

There are four primary sources of capacitance in the TX circuitry which limit decay speed:

- Interwinding capacitance in the coil itself
- Coil-to-shield capacitance
- Cable capacitance
- MOSFET capacitance

Interwinding capacitance has been discussed and can be minimized by using thicker insulation, lower permittivity insulation (like Teflon), fewer turns, or special winding techniques like spiral or basket weave. Coil-to-shield capacitance is best minimized with physical spacing between the coil and shield. Cable capacitance is a matter of cable selection — higher impedance cables have lower capacitance per meter so try to avoid the common 50Ω cables[18].

MOSFET capacitance is also normally a matter of good selection. The parameter we're concerned with is C_{OSS}. Newer devices often have superior C_{OSS} and even older devices like the popular IRF740 can be found in newer versions like the IRF740B with reduced C_{OSS}. But there is an additional design trick that can be used to reduce MOSFET C_{OSS}: add a series diode as in Figure 21-24a.

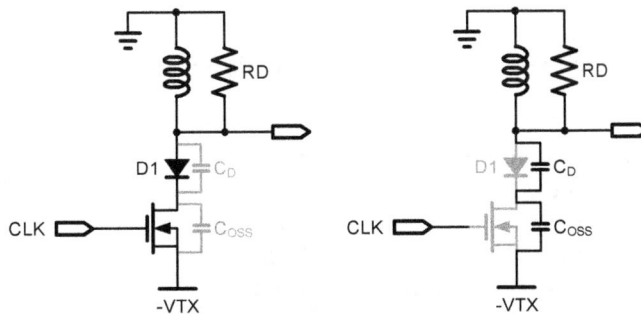

Fig. 21-24: **Series Diode**

Figure 21-24b illustrates the mechanism at play. During flyback the diode conducts and allows C_{OSS} to charge to the high voltage as usual. As soon as the inductor's flyback peaks and begins to decrease, the diode turns off and isolates C_{OSS} from the flyback decay. However, when the diode turns off its depletion capacitance (C_D) prevents total isolation and allows C_{OSS} to discharge somewhat. The equivalent capacitance now seen by the inductor is the series of C_D and C_{OSS} which is:

$$C_{eq} = \frac{C_D \cdot C_{OSS}}{C_D + C_{OSS}}$$

Eq 21-1

If C_D is equal to C_{OSS} this results in half the capacitance, a nice improvement. If C_D is even smaller than C_{OSS} then C_D begins to dominate. So the obvious goal is to choose a diode with a very low capacitance. Because depletion capacitance varies with reverse voltage many diode data sheets show depletion capacitance in a plot versus reverse voltage. Figure 21-25 shows such a plot for the STTH1L06, a 600V diode. At 600V the capacitance drops to as low as 2pF; this means as the flyback decays, the reverse voltage on the diode increases and C_{eq} gets even lower. It is important to choose a diode with a sufficient breakdown voltage to handle the peak flyback.

At the end of the decay the FET's C_{OSS} will still be charged to a high voltage. This is not a problem as the next subsequent TX turn-on event will short the FET and discharge C_{OSS}. However, this can cause a high transient current and while it generally occurs at an unimportant time we prefer to minimize transient noise in PI designs. A resistor placed across the drain/source can dissipate the energy in C_{OSS} over a broader time. Note that this trick only decreases the capacitance on the decay side of the flyback; the rise time of the flyback is unaffected.

---

18. RG58 (50Ω) is typically 85pF/m, whereas RG180 (95Ω) is typically 49pF/m.

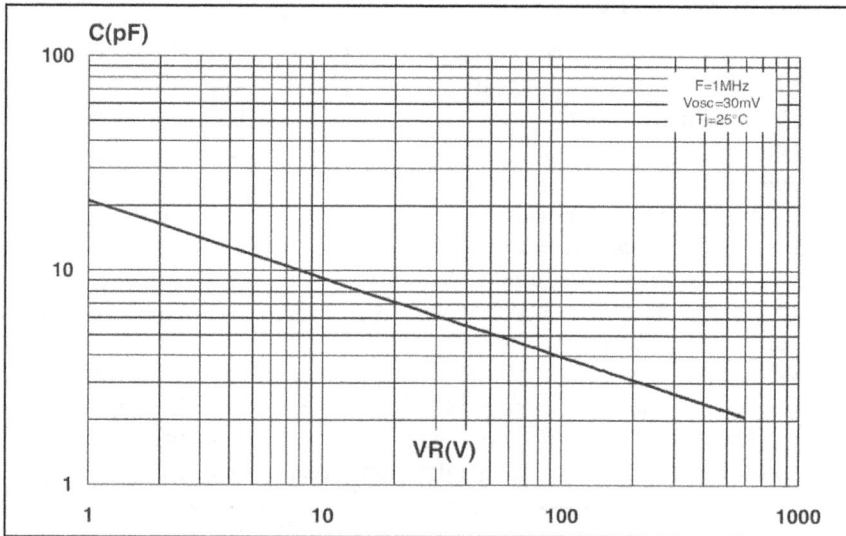

Fig. 21-25: **STTH1L06 Depletion Capacitance**

21.15: Snubbing

In preventing avalanche, an alternative solution is to increase the capacitance of the coil. Besides reducing the peak flyback this will also slow down the flyback, completely counter to our efforts to make the flyback as fast as possible. Instead of just adding a capacitor to the coil, we can instead add what's called a *snubber* circuit. This is a diode-connected capacitor that absorbs the initial flyback energy but is then disconnected for the remainder of the flyback decay. Figure 21-26 illustrates.

When Q1 turns off, the flyback voltage begins to rapidly increase, causing D1 to conduct. This transfers coil energy to the snubbing capacitor CS but also slows down the slew of the increasing flyback voltage. As soon as the flyback voltage peaks and begins to decrease, D1 turns off and CS is removed

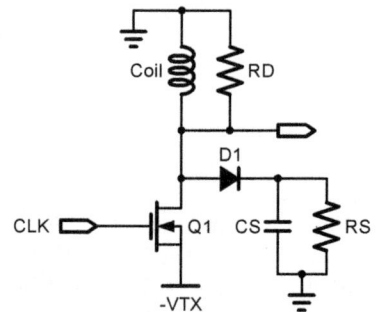

Fig. 21-26: **Snubber Circuit**

from the remain flyback curve, such that CS only slows down the rising flyback, not the decaying flyback. RS is used to drain CS during the TX off time.

The value of CS should be calculated according to what peak flyback is desired. This can be found by equating the energy stored in the coil at turn-off with the energy delivered to the capacitor:

$$\frac{1}{2} \cdot LI^2 = \frac{1}{2} \cdot CV^2 \Rightarrow C = L\left(\frac{I}{V}\right)^2 \qquad \text{Eq 21-2}$$

Suppose we have a 300µH coil with a peak current of 2A and we want to limit the flyback to 400V in order to avoid avalanching an IRF740 NMOS; the total capacitance needed is 7.5nF. This, however, includes the capacitance of the coil, the FET switch, and D1, plus any other parasitics. So CS will need to be somewhat less than 7.5nF.

Figure 21-27 shows the effect of the snubbing circuit. In the lower plot, the flyback is reduced from 800V to 400V. Note that the rising voltage is substantially slower but the speed of the falling slope appears unchanged. The decay settling is delayed by about 1µs. The upper plot is the coil current, which shows that the snubber reduces the slew rate. This can reduce target eddies and lower sensitivity, so be cautious in using a snubber in a PI design and always design for an optimum capacitance. Making it smaller than necessary will cause the flyback to exceed the goal. Making it larger than necessary will

Inside the Metal Detector

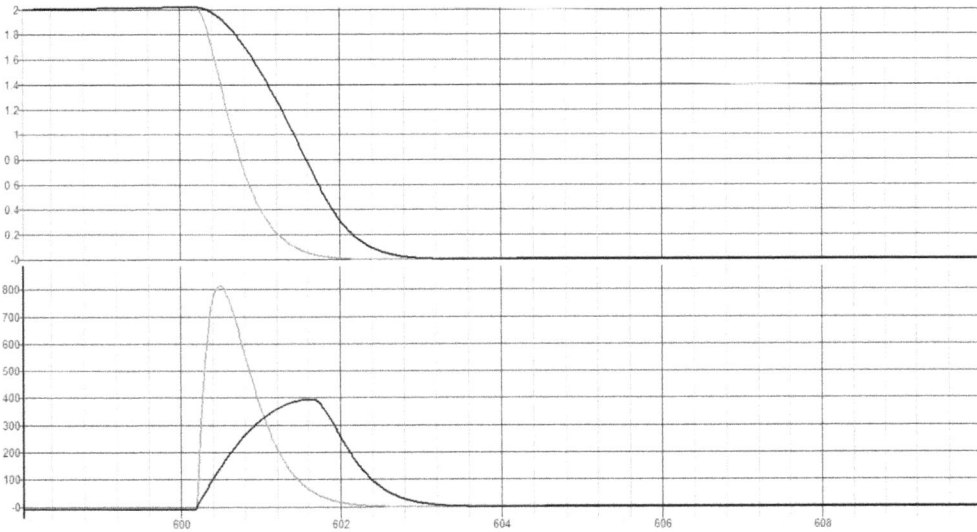

Fig. 21-27: **Snubber Effect**

further slow down the slew rate and increase the decay settling. Even though the snubber increases the minimum attainable sample delay, sometimes it may be a better choice than allowing the MOSFET to avalanche. Compare with Figure 21-23.

It is possible to add the snubbing capacitor (and resistor) to the cathode side of D1 in the circuit of Figure 21-24. In fact, D1 in Figure 21-24 is technically being used as a snubber.

21.16: Improved TX Switch Drive

The NMOS driver used in Pulse2 consists of a PNP inverter followed by a class-B stage. The purpose of the class-B stage is to provide a low-impedance output capable of handling the high transient gate current. This scheme works well but can be improved. The highest transient current occurs at turn-off so replacing the 2N3906 with a Darlington PNP can improve turn-off performance (Figure 21-28). A Darlington PNP, however, begins turning off at a higher output voltage and runs the risk of "running out of gas." That is, it can leave the NMOS gate not fully pulled down. A simple solution is to place a pull-down resistor at the output of the class-b stage to ensure complete turn-off. This is also a good idea even when using a regular PNP transistor.

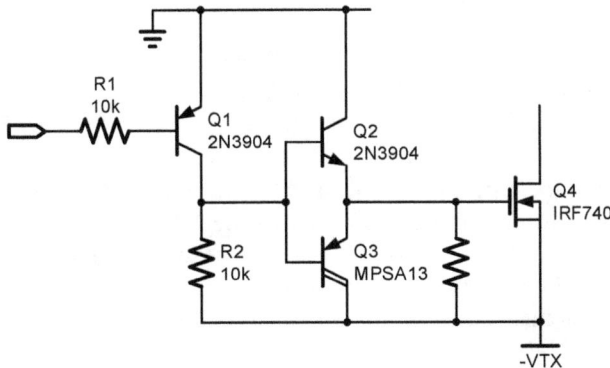

Fig. 21-28: **Improved TX Switch Drive**

21.17: Alternate TX Switch Drive

The *Pulse-X* designs so far have used bipolar transistors and resistors to drive the MOSFET switch. In particular, a class-B stage is needed to quickly remove the gate charge during turn-off. An alternative is to use a SPDT analog switch to drive the MOSFET (Figure 21-29). This approach has several advantages; first, it is simpler. Second, it has logic levels that are compatible with micros even down to 3V. Third, it has a low impedance rail-to-rail drive for the MOSFET gate. Fourth, it draws less current.

A common SPDT switch IC is the 4053 but there are other choices like Vishay's DG419. Important parameters in selecting a switch are its maximum supply voltage, logic drive, and turn-on resistance. In Figure 21-1 the switch is powered directly from the TX supply. However, the 74HCT4053 is limited to a supply of 10V and the DG419 is limited to 12V. It is not uncommon to power a PI transmitter from 12V or more so be mindful of these limits. If you want to use a higher battery voltage than the switch can tolerate then simply add a voltage regulator to supply the switch.

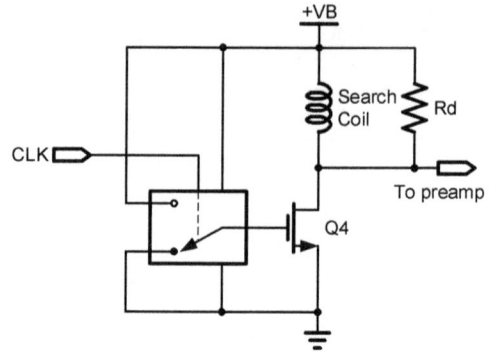

Fig. 21-29: **Alternate Gate Drive**

Some switches like the DG419 and 74HCT4053 will work with 3V logic regardless of the supply voltage. Others, like the 74HC4053, have logic levels that depend on the supply voltage and may not work with a 3V micro, or even a 5V micro. Switch resistance can vary quite a bit. The 4053-type switch may be as high as 75Ω while the DG419 is 14Ω. Analog Devices' ADG419 has a maximum 25V single supply, a 2.1Ω switch resistance, and 3V logic compatibility; however, it is pricier. Multiple switches (such as in the triple 4053 package) can be paralleled to reduce switch resistance.

21.18: MOSFET Drive voltage

A final consideration in the TX drive circuitry is the amount of voltage (V_{GS}) to apply to the MOS switch. For turn-off it is obvious that V_{GS} should be zero but driving a slightly negative V_{GS} could improve the turn-off speed. On the turn-on side we want to apply enough V_{GS} to ensure a low enough R_{DS}(on). However, once that point has been reached additional V_{GS} does not help and can even degrade the turn-off speed.

The reason for this is because, once the MOSFET has reached saturation, additional V_{GS} increases the channel charge and this charge must be removed at turn-off. The more you overdrive the turn-on voltage, the longer it will take to turn off. This effect will vary with MOSFETs with some being worse than others. You should also be aware of the

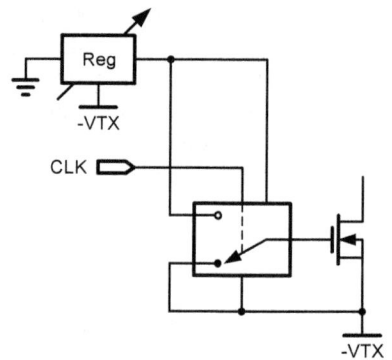

Fig. 21-30: **Variable Gate Drive**

absolute min/max gate voltage that a MOSFET will tolerate, often around ±20V. Also be aware that some of the newer SiC MOSFETs require a higher gate drive to achieve saturation. If you are building a platform for experimentation then a good solution is to make the gate drive voltage variable by using a linear regulator. See Figure 21-30.

Another consideration for driving the FET is a gate driver chip specifically made for this purpose. Due to the explosive growth in MOSFET motor drives there are many to choose from, and inexpensive.

21.19: Clamp Performance

The basic resistive-diode clamp in front of the preamp is simple and robust but not very performance-oriented. In *Pulse-3*, R8 determines the peak current through the diodes and, for R8 = 1kΩ, a

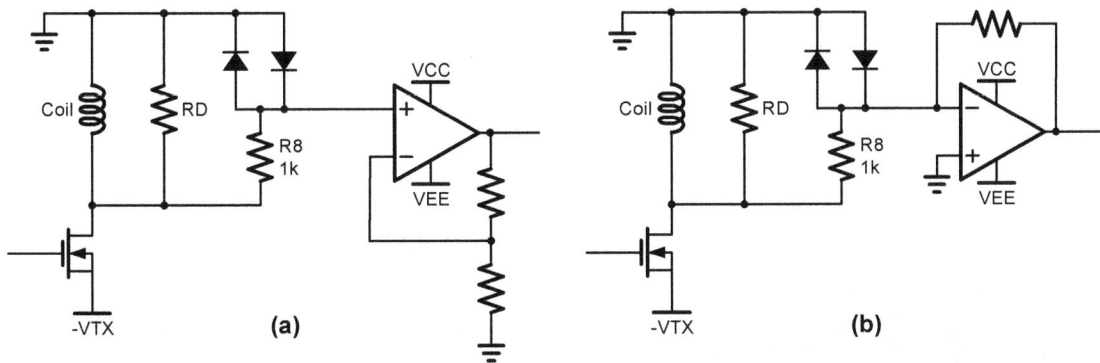

Fig. 21-31: **Clamping Effects**

400V flyback produces about 400mA[19]. The speed at which the diode can turn off depends a lot on the amount of peak current it starts with[20]. We can increase R8 and thereby decrease the peak current and speed up the overall settling, assuming the overall settling isn't limited by something else. This suggests that a higher value for R8 would be preferable, but by now you've come to expect the lunch ain't free.

R8 also directly contributes to (and possibly dominates) the thermal noise of the front-end. Our $1k\Omega$ resistor produces $4nV/\sqrt{Hz}$ which is slightly higher than the NE5534's input noise of $3.5nV/\sqrt{Hz}$. Increasing R8 only makes this worse and even though the sampling integrator averages out random noise its effectiveness is limited. But it may be a worthwhile trade-off; being able to operate at a lower sample delay may be worth a little more noise.

Clamp performance also depends on the preamp configuration. In *Pulse-1* we started with a non-inverting amplifier and in *Pulse-2* changed that to an inverting type. Let's consider the non-inverting preamp, redrawn in Figure 21-31a. When the clamp is on, R8 is effectively in parallel with RD so the overall damping resistance is reduced. This overdamps the response and slows down the decay. Near the end of the decay the diode turns off and R8 is removed from the damping action. Optimal damping is normally achieved through trial-and-error and includes the effect of R8, but preferably we would like R8 to be high to get the fastest possible damping throughout the decay.

In Figure 21-31b R8 is also effectively in parallel with RD during clamping. When the diodes turn off and the opamp is in its linear operation, R8 sees virtual ground at the opamp pin so it is still effectively in parallel with RD. Therefore, unlike the non-inverting preamp, R8 in the inverting topology is in parallel with RD for the entire decay.

21.20: Two-Stage Preamp

The PI designs have so far incorporated a single-stage preamp with a typical gain between 100 and 1000. The speed of the preamp response depends on two mechanisms: overdrive recovery[21] and bandwidth. The overvoltage recovery determines how quickly the opamp comes out of saturation; after that, the bandwidth determines how quickly the opamp can settle to its proper value.

A single preamp with, say, a gain of 900 can be replaced with a two-stage preamp, each running a gain of 30. For the same type of opamp this improves the bandwidth by a factor of 30 and, even though there are two stages that must settle, the bandwidth improvement can often exceed the delay of the additional stage. Splitting the gain evenly between the two stages results in maximum bandwidths for each stage but may not result in an overall minimum settling time. You may find that setting the gain of the first stage a bit lower and the second stage proportionally higher better optimizes the overall settling.

---

19. This suggests a peak power of 160 watts, or a time-averaged power of perhaps 1 watt or so. This is why R8 needs to have a decent power rating.

20. Because diffusion capacitance in a PN junction increases with current.

21. Briefly mentioned in the *Explore* section of Chapter 20.

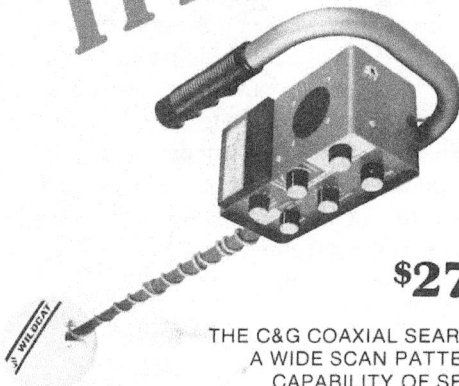

22 Ground Balanced PI

"The key to keeping your balance is knowing when you've lost it."

— *Unknown*

A traditional PI design only looks at target responses during the decay period after the TX signal is turned off. Therefore it sees no reactive signals, only resistive. This provides a mixed advantage over VLF in that ferrite responses usually vanish so quickly that they are not detected[1], hence there is some natural rejection of ground mineralization. But with no reactive response, this comes at the expense of iron discrimination.

If PI detectors do a good job of ground rejection, it might seem odd that there is a need for ground balance. But a lot of ground minerals have a viscous magnetic response: a short-term magnetic memory that results in a non-instantaneous decay of ground. Therefore, some ground can seemingly look like a normal target. Well-behaved ground is somewhat homogeneous so if you can maintain a constant coil height above the ground this ground signal will be constant, whereas a target has a faster transient response. This is the same way BFO detectors had to be used; lower the coil to the ground, retune the detector, and carefully maintain the coil height. The problem is that ground is often not very well behaved, and we certainly don't want to rely on painful practices of the 1960s.

22.1: Viscous Remanence Magnetization

In Chapter 4 we briefly considered the viscous magnetic response. The proper name for this is *viscous remanence magnetization*, or VRM. To recap, VRM is characterized by an ability to be temporarily magnetized by a driving field, but when the field is removed the material's magnetization decays back to zero or some low-level static value. That is, it tends to hold the magnetization for a very brief time before returning to a (largely) demagnetized state.

In a PI detector, the driving field is removed in a (nearly) step response, and the step response of VRM is a decay with a power law response[2], commonly assumed to be close to a 1/t curve. As we've seen, most eddy targets are largely exponential in their step response so there is hope that the ground signal can be distinguished from targets using the temporal nature of the responses.

Figure 22-1 shows a 1/t viscous ground response plotted in a log-linear graph, along with representative low and high conductor exponential responses[3]. Suppose we take our typical early sample for target detection (t_1) and then take a slightly delayed sample (t_2). Assuming the usual integration, there is a precise amount of gain that can be applied to the late sample such that when it is subtracted from the early sample, the net output signal due to the ground response will be zero.

Once the proper gain has been established in the late channel to achieve cancelation of the ground signal what, then, happens to our target signals? When the low conductor waveform is processed identically, the late sample is grabbing a greatly diminished signal so that the amplitude of the early sample

---

1. It is often said that PI does not "see" ground mineralization and that is somewhat true for ground that is similar to pure ferrite. The response of pure ferrite is very fast and is usually diminished by the time the RX sample pulse is fired, although an incredibly fast PI design can begin to encroach upon the ferrite response.

2. A power law response is $f(t) = t^\alpha$, whereas an exponential response is $f(t) = \alpha^t$.

3. Time is linear; the log of the response (i.e., dB) converts the normally exponential decays into straight lines. The log of the 1/t ground response still has a "decayish-looking" curvature.

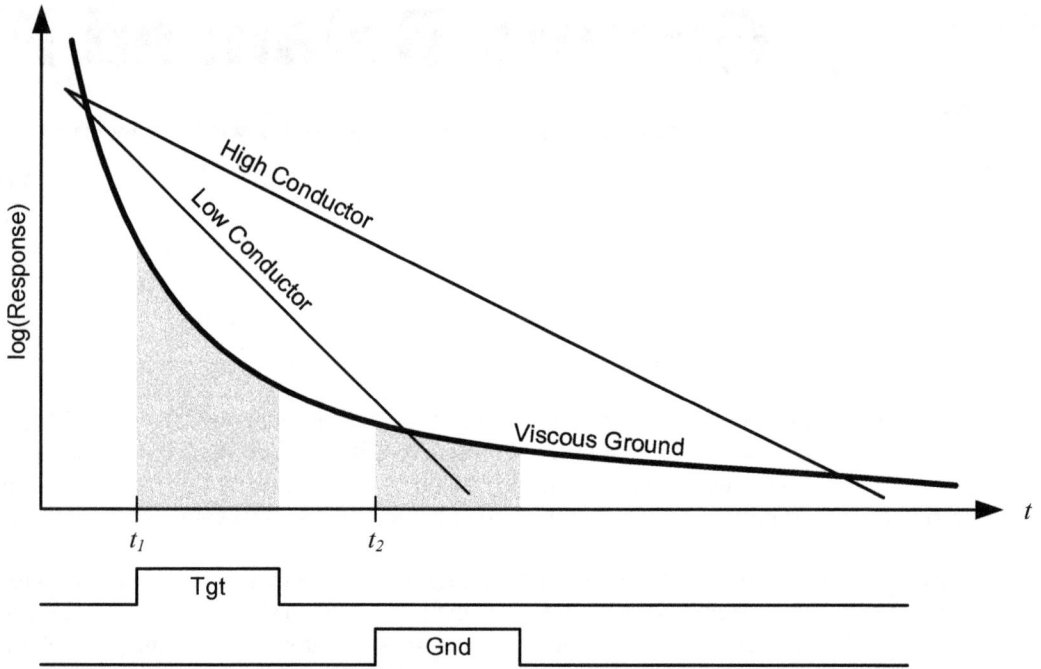

Fig. 22-1: **Ground vs Target Decays**

still exceeds the gained amplitude of the late sample and its response is preserved, though slightly diminished. This is not true for the high conductor signal, however. The amplified late sample exceeds the early sample such that the subtraction results in a *negative* signal.

So far we have three cases: a low conductor produces a positive signal, a high conductor produces a negative signal, and ground produces no signal at all. The positive-versus-negative signal is, in itself, not a big deal and can be applied to a VCO audio stage to produce, say, a higher tone for positive responses and a lower tone for negative responses, with ground producing no change at all from the threshold tone.

From this it is apparent that somewhere in the range of target conductivities there will be a particular conductivity for which the amplified late sample subtracted from early sample exactly cancels and therefore matches the response of ground even though one response is exponential and the other is $1/t$. That particular target conductivity will generate no output signal and will be ignored, just like ground. In ground balanced PI detectors this is called the *target hole*. It goes beyond ignoring a particular conductivity; the result is that conductivities approaching the ground balance point are increasingly diminished. Figure 22-2 shows the relative responses of a White's TDI in all-metal mode (ground balance

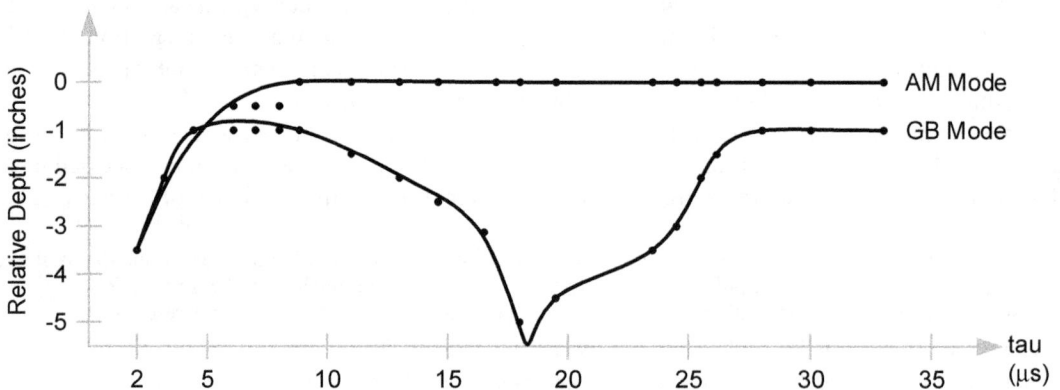

Fig. 22-2: **White's TDI: Target Response vs Tau**

disabled) and in ground balance mode. The result is overall diminished sensitivity of about 1 inch, plus a target hole at $\tau = 18\mu s$. This is the main drawback of the two-point subtractive ground balance scheme that was pioneered by Eric Foster in his *Goldscan* series and used in various other PI models.

22.2: Implementation

Implementation in hardware can be achieved with a similar method as we used for EFE cancelation in the last chapter. The normal early target sample is taken, and a secondary target sample is taken but with a very short delay after the early sample. This is unlike the EFE sample which is usually taken as far away as possible from the early sample to minimize residual target signal. The secondary sample is then amplified to exactly equal the amount of ground response seen in the early sample so that when they are subtracted the ground response vanishes. Figure 22-3 illustrates.

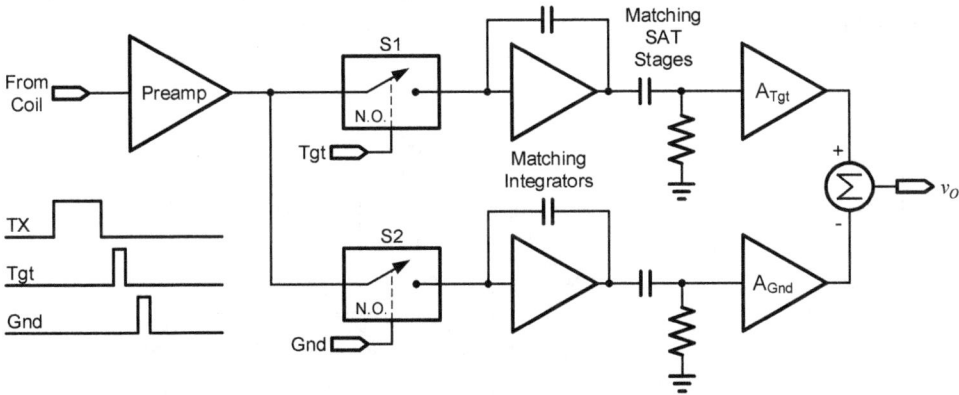

Fig. 22-3: **Ground Balance Scheme**

The primary sample channel is called the *target channel*[4] and the secondary sample channel is called the *ground channel*. In most ground balanced PIs the ground channel can be disabled which results in an "all-metal" mode of operation. This is a bit of a misnomer as traditional PI has no discrimination and is always all-metal. However, subtractive ground balance as described here results in a break point between low and high conductivities and can be considered a crude form of discrimination. The break point, though, is a result of the ground balance and cannot be used to independently control a discrimination level as well. That is to say, you get what you get.

Achieving ground balance is accomplished by adjusting the channel gains until the subtractive output is nulled for ground signals. In some designs A_{Tgt} is fixed while A_{Gnd} is variable and the subtraction is fixed. In other designs the gains are fixed and the subtraction is variable. In either case, the ground channel is sampled later than the target channel and therefore has a weaker signal. The additional gain needed in this channel to achieve balance can be somewhat large and therefore the ground channel is usually noisier than the target channel, resulting in an overall performance that is noisier than with the ground balance disabled. So noisier operation, an overall reduction in depth, and a target hole are all characteristic drawbacks of subtractive ground balancing. But without ground balancing there are some ground conditions that will overwhelm even PI detectors, and certainly VLF detectors. So it's a price worth paying to still be able to hunt.

Subtractive ground balance does nothing for our previous problem, that of Earth field effect. Achieving both ground balance *and* EFE cancelation requires expanding the normal late EFE sample into two late EFE samples, one applied to the target channel and one applied to the ground channel. Figure 22-4 illustrates. It is worth noting the polarities of the ground sample pulses: the Gnd sample is negative while the Gnd-EFE sample is positive. Selecting these polarities at the demod sampling switches inverts the signal polarity for the entire ground channel, which means that instead of subtracting the Ground channel from the Target channel to cancel ground we can instead add them: Tgt + (−Gnd).

---

4. In VLF designs we also call the R channel the "target channel." The PI "target channel" is unrelated.

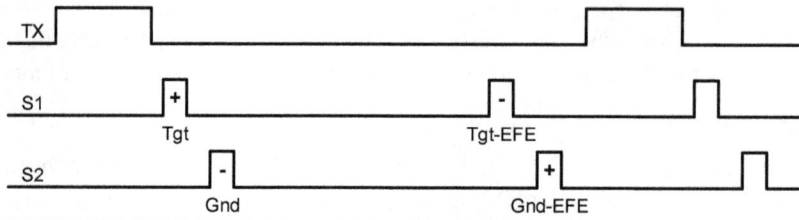

Fig. 22-4: **Ground Balance + EFE**

An implementation of all of this is shown in Figure 22-5. The two channels each have two sampling switches, one for the early sample and one for the EFE sample. The EFE sample is taken from a negated input signal so that it is subtracted from the main sample. Therefore, the outputs of each demodulator should be substantially immune to EFE. The clocks needed for the additional demodulators are provided by the PIC micro, which requires selecting a larger micro than that used in the *Pulse-3* design.

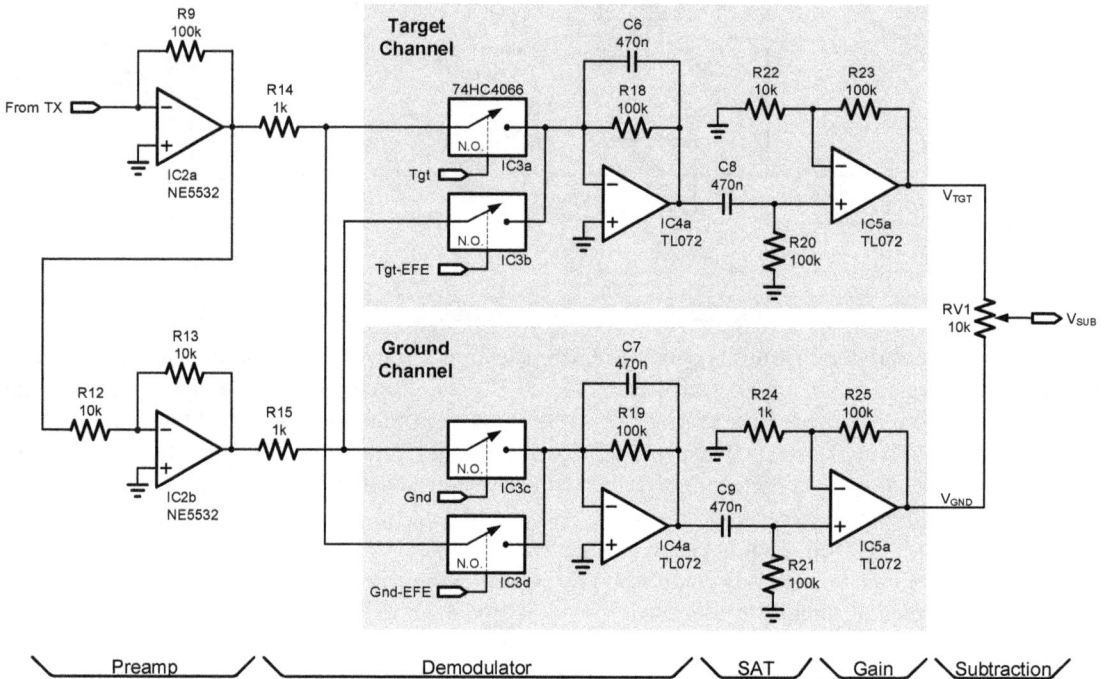

Fig. 22-5: **Practical Ground Balance Circuit**

SAT stages in each channel remove unwanted DC offsets. After the SAT stages, the target signal is applied to a nominal fixed-gain amplifier while the weaker ground signal feeds an amplifier with an overall higher gain. Because the ground channel was inverted via the demod switch polarities we can apply both signals to a simple potentiometer to effect variable subtraction.

22.3: Signal Responses

Each demodulator feeds a SAT stage, which performs the same function as in previous PI circuits. Recall (in Figure 20-9c) that a positive target transient at the input of the SAT stage results in an output with both a positive lobe and a negative "bounce-back" lobe. Figure 22-6 shows the resultant signals for four conditions: no target, ground[5], low conductor, and high conductor. The no-target condition needs

---

5. 'Ground' is shown as a fast transient response (as if it were a target) for comparative purposes. Mineralized soil tends to have a broad continuous response as the coil is swept over it, but viscous hot rocks can easily behave like a target.

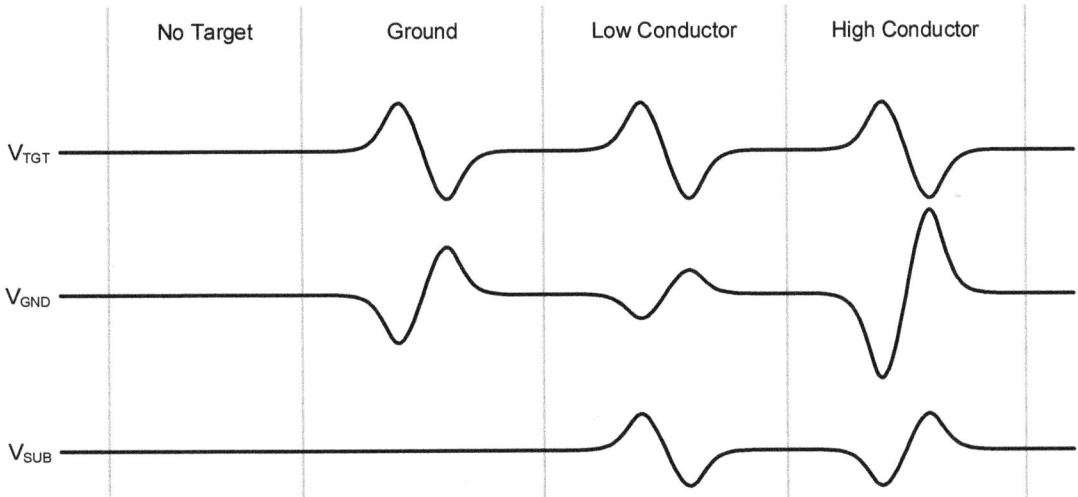

Fig. 22-6: **Signals**

no explanation. With the gains adjusted properly, a ground signal in both the target channel and ground channel have the same amplitude so they exactly cancel. For low conductors the ground signal is weak so the subtraction signal is slightly attenuated but remains in-phase. For high conductors the ground signal is stronger than the target signal so the subtraction signal is phase-inverted so that the leading lobe is now negative.

As with VLF demodulators, matching of the two channels is critical to the performance of this circuit. If there is a difference in the channel delays then the subtraction circuit will not see time-aligned signals; the result is poor ground subtraction and the detector will be noisy in bad ground. Thus R18 should be matched with R19 and C6 with C7, to within 1% of each other[6]. (Briefly returning to Figure 21-17, the use of a differential demodulator requires the simultaneous matching of 4 capacitors, which makes that particular approach even less appealing.) The resistance matching in the switches is also important, which illustrates the preference of using a 4066-type quad analog switch where the switch-to-switch matching within a chip is inherently good. It is also preferable to use dual opamp packages where the two demod opamps share a package. Finally, these opamps should have low IB currents to avoid imposing charge currents on the integration caps during hold times.

The same care extends to the SAT and gain stages. As with the demodulators, it is important that C8 and C9 are closely matched to each other (1% or better), and that R20 and R21 are likewise matched. It is also best to use a dual opamp for the gain stages and that they have ultra-low IB currents to avoid offsetting the SAT level. This is especially critical for a super-slow SAT speed or if SAT is disabled for a non-motion pinpoint mode.

22.4: Back-End & Audio

The final gain stage is imported from the *Pulse-3* project. In the *Pulse-3* design, all targets produce the same response: a positive lobe directly followed by a negative lobe, and the negative lobe gets truncated in the audio stage (see Figure 20-11). The resultant audio is a constant tone that is amplitude-modulated by the strength of the target response.

This design will add a VCO to the audio so that a rising signal not only produces a louder response but also at a higher pitch. Figure 22-7 shows the conceptual back-end design and Figure 22-8 shows the waveforms at points A, B, and C.

As with *Pulse-3*, the negative lobes are truncated and both targets seemingly produce similar audio responses. However, the low-conductor target has an audio report followed by a null, whereas the high-conductor has a null followed by an audio report. We've seen a similar response in 17.10 in the discus-

---

6. The absolute values are not as critical as the pair matching.

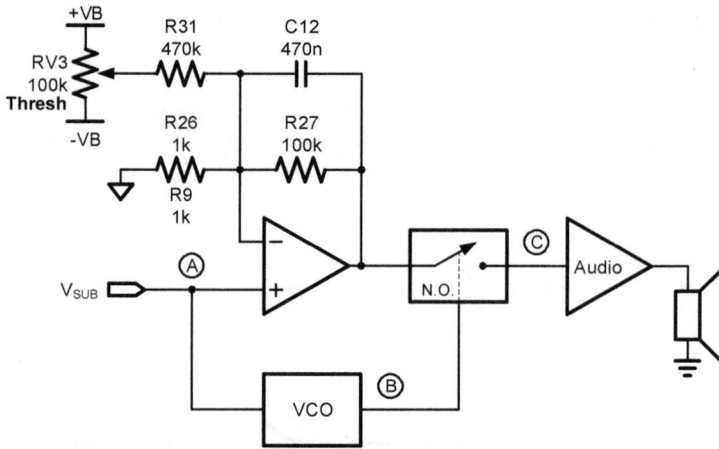

Fig. 22-7: **Back-End Design**

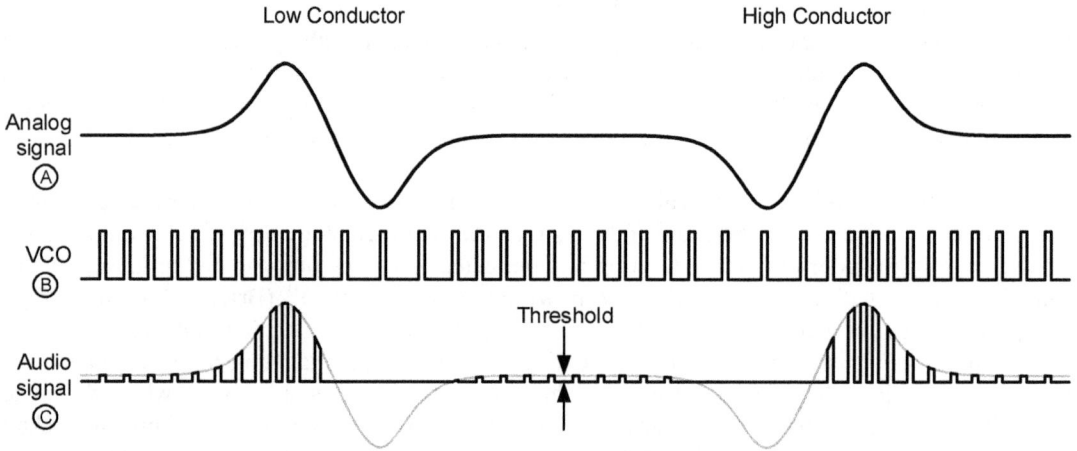

Fig. 22-8: **Audio Signals**

sion of the audio characteristics of a first derivative VLF gold detector design. The low-conductor response is a "zzzip" sound whereas the high-conductor is more of a "puzzz" response.

If the goal is to have all targets sound alike, this is pretty close. If, instead, the goal is to have an easily discernible sound for low and high conductors, this leaves much to be desired. As mentioned early in the chapter, the addition of a ground channel creates the possibility of producing separate audio responses for targets on either side of the ground balance point. However, such an audio design is rather complex and to avoid introducing too many major design changes at once this feature will be held for the next chapter.

Figure 22-7 shows a box for the VCO and, as we know, there are many ways to implement a VCO. This design will use a 555 timer as shown in Figure 22-9. In most 555 applications pin 5 (Control Voltage) is left open or with a cap to ground, but this pin can be used to override the bias voltages applied to the trigger comparators in the 555. Using the output pin to drive the external RC network creates an astable circuit, now with a voltage-controlled frequency. However, the 555 frequency increases as the control voltage decreases so we will need to invert the signal before applying it to the Control Voltage pin. A final audio circuit design is presented in Figure 22-10.

The constant-tone amplitude-modulated audio of *Pulse-2* & *Pulse-3* can be synchronized to the TX pulse rate to minimize incoherent interference. Performance purists should realize that the VCO audio cannot be synchronized and runs the risk of creating more internal noise. Ideally, if you could synchro-

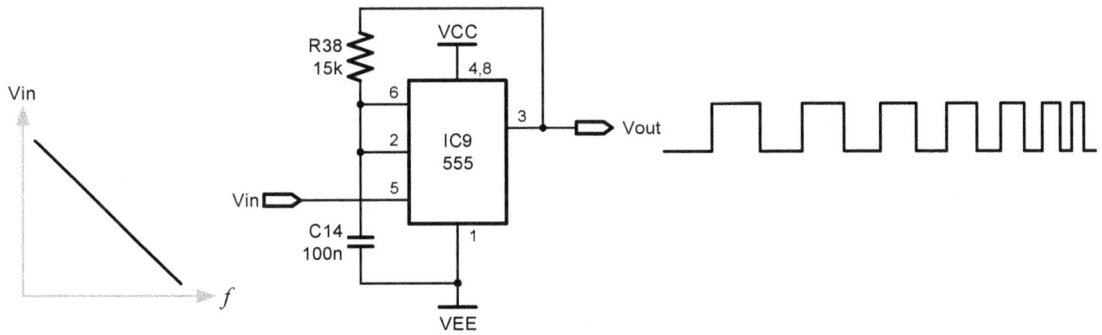

Fig. 22-9: **VCO Details**

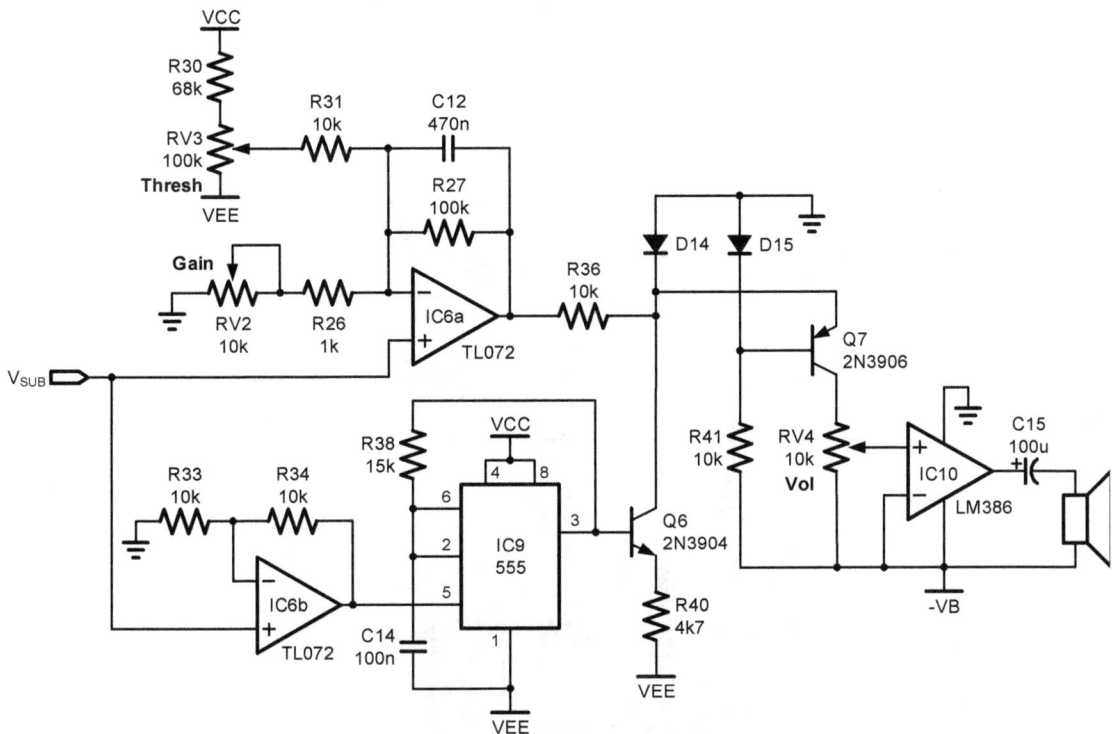

Fig. 22-10: **Final Audio Design**

nize the threshold frequency to the TX pulse rate then audio interference during the most critical time — when there is no target — would be minimized. This would probably require moving the VCO into the microcontroller and performing this task in code and is left to the reader to explore. Even then, actual performance benefits may be very minimal.

Design: *Ground Balanced PI (Pulse-4)*

The *Pulse-4* project adds the ground balance techniques described in this chapter to the *Pulse-3* project from the previous chapter. Schematics are shown in Figures 22-11 through 21-13.

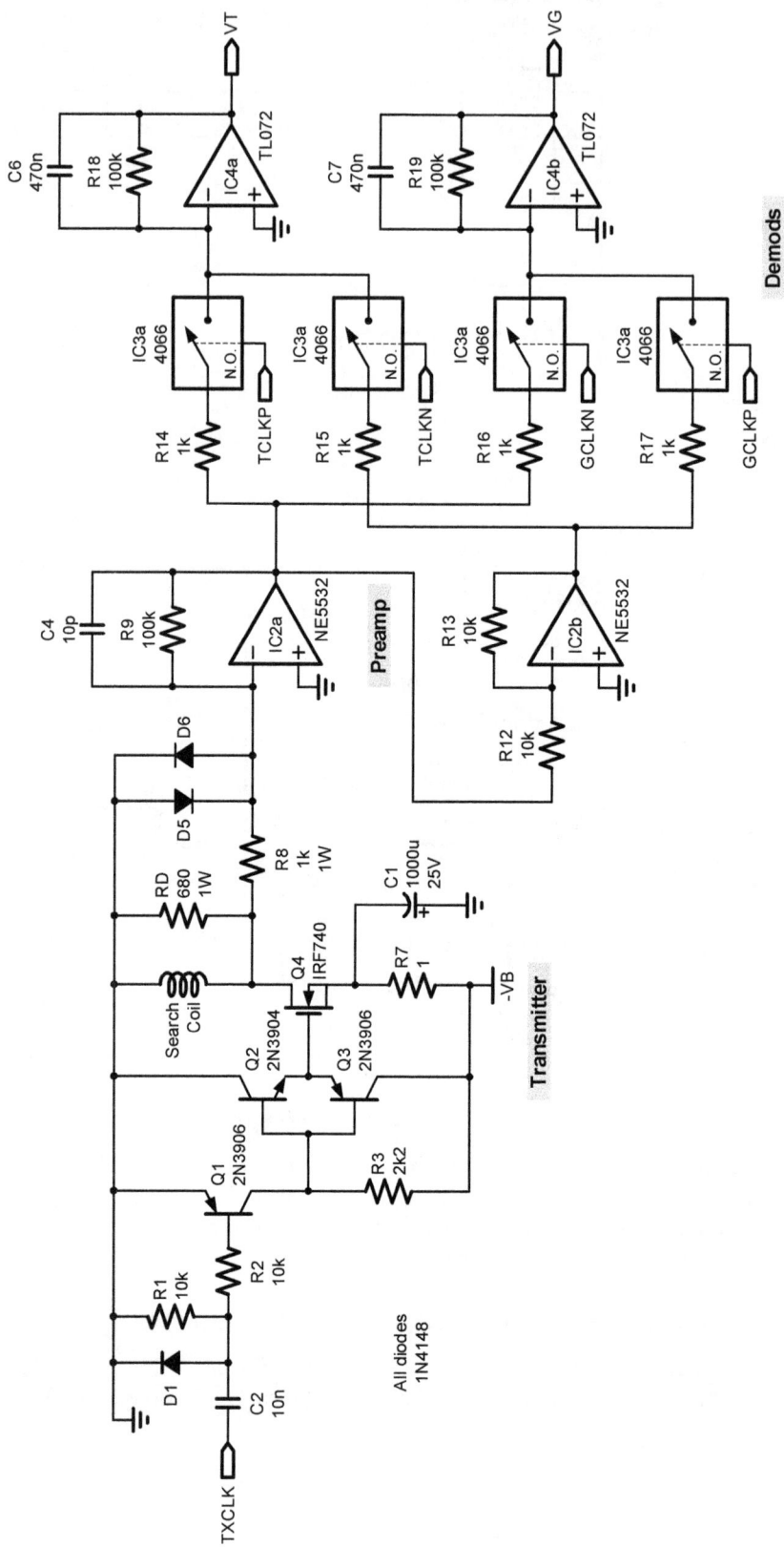

Fig. 22-11: **Pulse-4 — TX & RX Front-End**

Inside the Metal Detector

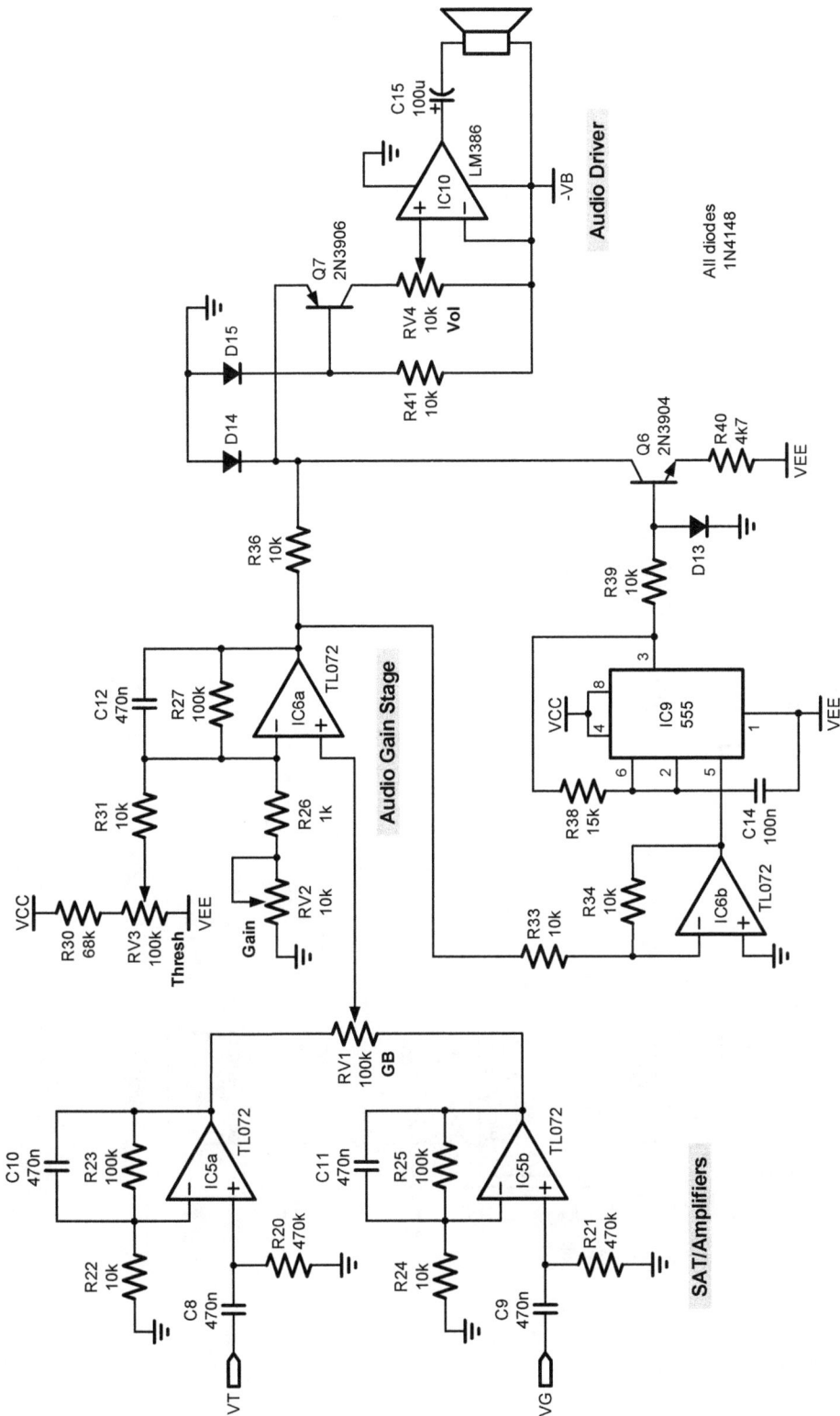

Fig. 22-12: Pulse-4 — RX Back-End & Audio

All diodes
1N4148

Power Supply

Micro

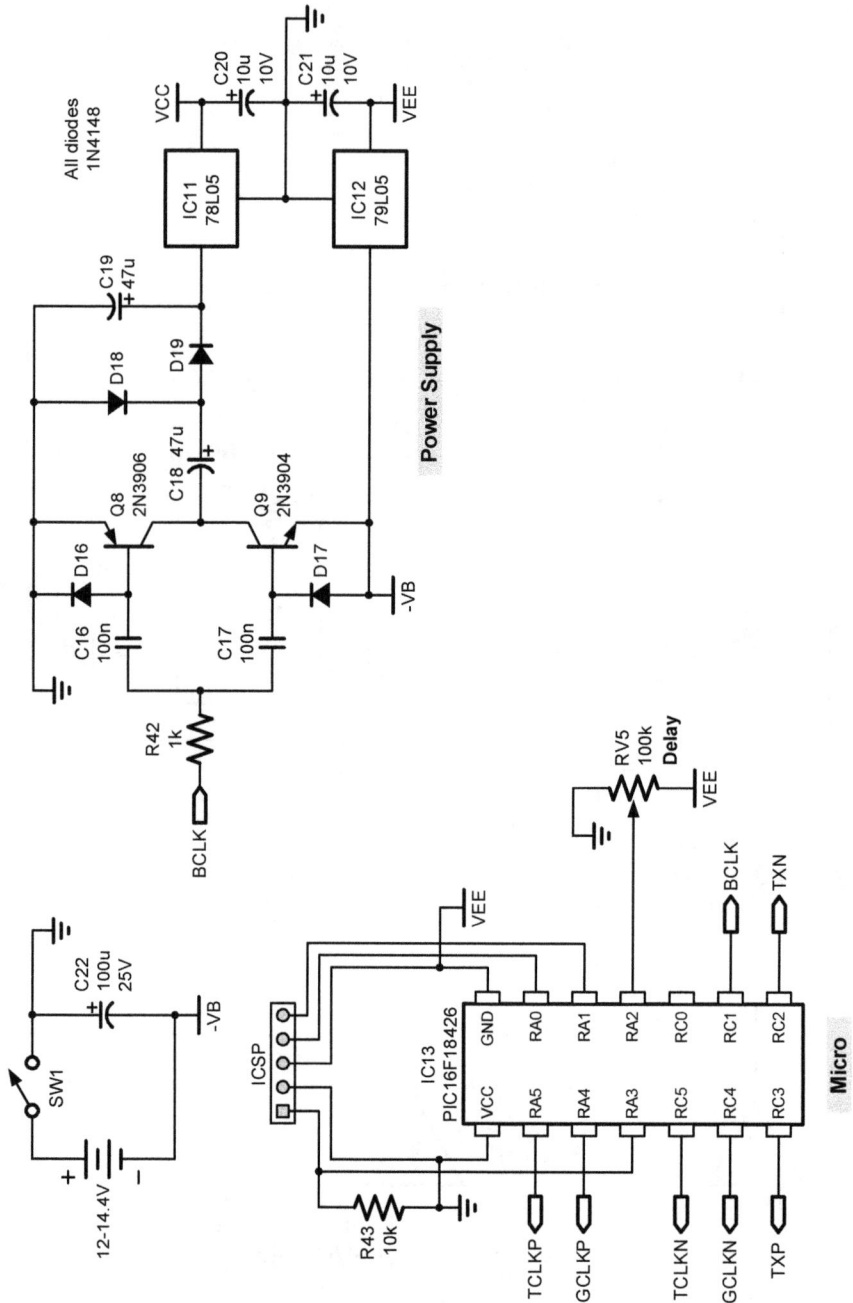

Fig. 22-13: Pulse-4 — Power & Timing

Explore

22.5: PWM SAT Control

The SAT circuit consists of a series capacitor and shunt resistor, the same as used in *Pulse-3*. It was mentioned that a variable speed SAT (VSAT) could be implemented in *Pulse-3* by simply using a potentiometer for the shunt resistor. The same could be done here, except that both SAT circuits need to continuously match each other. A ganged potentiometer could be used, but those will probably match to only 5-10%, which is far from the 1% (or better) we'd like to see.

Another option is to use a fixed shunt resistor and add a series "chopping" switch and put the switch under PWM control. Figure 22-14 illustrates using a NMOS device as a switch. When the switch is open, the SAT has no "return to zero" shunt element so the retune speed is infinite. When closed, the retune speed is set by the usual RC time constant. If the switch alternates between open & closed, then the retune speed will be the average, meaning that it will be inversely proportional to the duty cycle of the switch. If the switches are implemented with e.g. a 4066 chip then they will closely match each other, and therefore the SAT speeds of the two channels will also closely match.

The chopping action of the switch produces switching noise in the channels, so a high switching speed like 100kHz-1MHz will keep this noise well above the audible range.

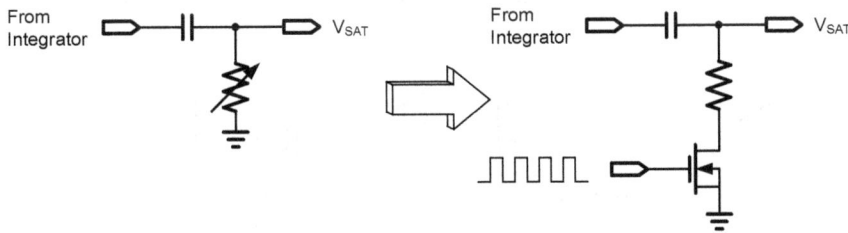

Fig. 22-14: **PWM SAT Control**

22.6: Target Hole Selection

The 2-point subtractive ground balance method results in a target hole at some particular conductivity, but we have not considered what controls that particular conductivity point. It is mostly controlled by the timing scheme used in the demodulators: the pulse widths of the target and ground samples, and the spacing between them. Other influences include the transmit pulse width and the bandwidth of the preamp.

It is possible to change these parameters and move the target hole. The target hole is (currently) unavoidable so we want it in the least obnoxious place possible. If the detector is being used for prospecting and the vast majority of nuggets are expected to be under, say, 10 grams then a target hole above 10 grams makes sense. If the detector is being used to hunt American Civil War relics, then the target hole placement becomes more problematic as cuff buttons, jacket buttons, Minie balls, and other items tend to span quite a range of conductivities. The best target hole position might depend on the hunt site or what relics you prefer to find.

Currently no commercial PI detector allows movement of the target hole; you get what you get. Yet it would not be difficult to add this control as it is easily accomplished with no more than a change in pulse timing, and timing is preferably controlled by a micro. It is, however, far more difficult to explain to users exactly what this control does, and users who don't understand a function control tend to misuse it and then complain that it's a bad feature.

22.7: 4-Channel "no hole" PI

If the target hole location can be controlled, then there is a way to eliminate it completely: run two complete receiver channels in parallel, each with a different target hole, and combine their results. The concept is shown in Figure 22-15. Each receiver is a complete ground-balance circuit, essentially every-

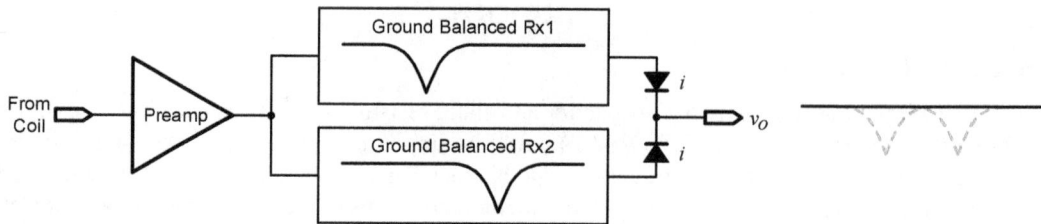

Fig. 22-15: **Dual GB PI Concept**

thing in Figure 22-5 except for the preamp and preamp inverter, and the receiver outputs are combined. In this scheme, the hole in Rx1 is filled in by Rx2, and vice versa. The resulting response versus target tau is flat[7] with no hole.

This significantly increases the design complexity and also demands good delay matching between four channels instead of two. Notice that the two receiver outputs are diode-connected[8] which is a simplified "OR" function. Since this design now has two (albeit invisible) target holes, it can conceptually provide separate audio responses for three target conductivity regions instead of two. Figure 22-16 shows how this scheme might be implemented. The timing pulses shown are for illustrative purposes only, and EFE samples are omitted. Furthermore, it may be advantageous to use a separate preamp for the second receiver.

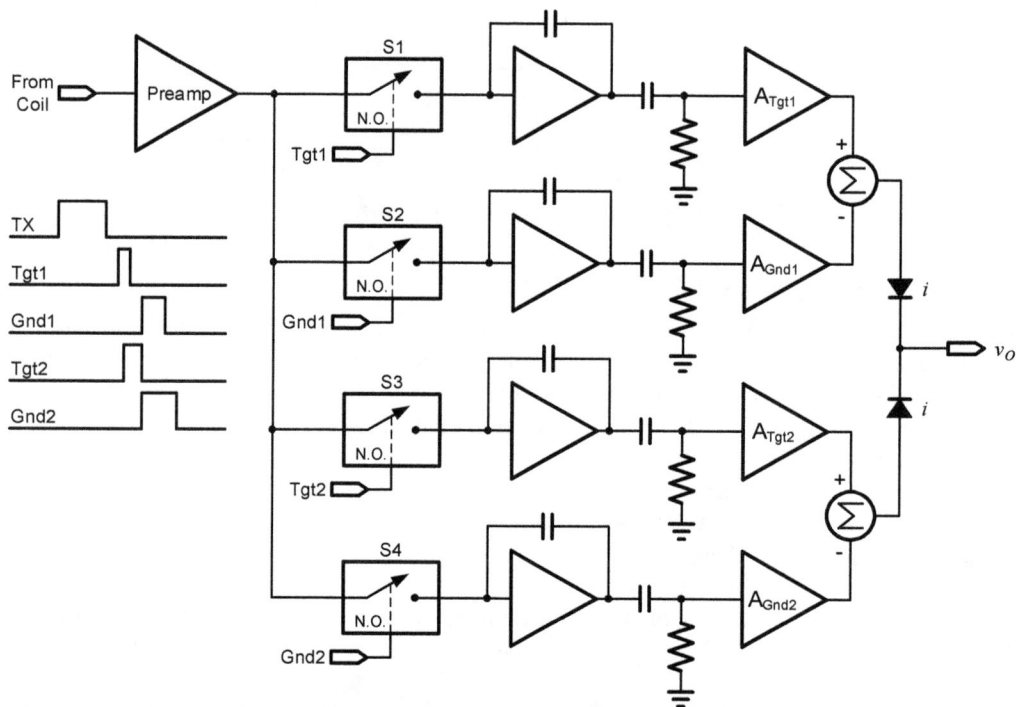

Fig. 22-16: **Dual GB Block Diagram**

A major issue with this approach is that each receiver must be independently ground balanced. Since the outputs are blended together this might require disabling one receiver while ground balancing the other receiver. In short, it could require extreme user skill to operate this kind of design unless there is a way to synchronize the balancing of the two receivers. That is, they are calibrated so that a single control can balance both receivers simultaneously.

---

7. Probably not completely flat, but at least there are no invisible targets.

8. The *i* means ideal diode, which can be implemented with an opamp.

22.8: PWM Ground Balance

In Figure 22-3 the ground channel demodulator output is amplified by a higher (and variable) voltage gain compared to the target channel. Instead of using voltage gain to control the ground balance, the ground channel demodulator can be clocked with a wider and variable sample pulse width, which effectively increases its "gain" by integrating more of the input signal. Variable ground balance control is now implemented by adjusting the width of the ground channel integration[9]. Figure 22-17 shows the timing for this.

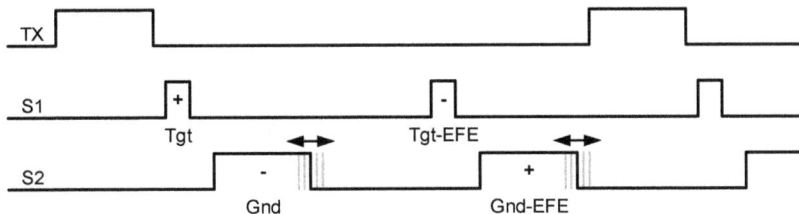

Fig. 22-17: **Alternate Ground Balance Method**

Since this method uses wider integration instead of additional voltage gain it has the potential of reducing the ground noise through more averaging. However, if the clock pulse is provided by a micro then it may be difficult to achieve the needed adjustment resolution for accurate ground balance control. Another drawback is that unless the target channel sample width is fairly narrow (say, 5µs) it can become difficult to get the needed integration width on the ground channel to achieve ground balance. That is to say, the options available for sample timing may be more limited. This can be mitigated by giving the ground channel some additional (but fixed) gain while still using the pulse width as the GB control. Alternatively, a wider (but fixed) sample width could be applied to the ground channel while using a variable (but lower) gain to adjust the balance. In either case, the EFE pulse width on the ground channel still needs to match the ground sample pulse width, which further eats into the available timing budget.

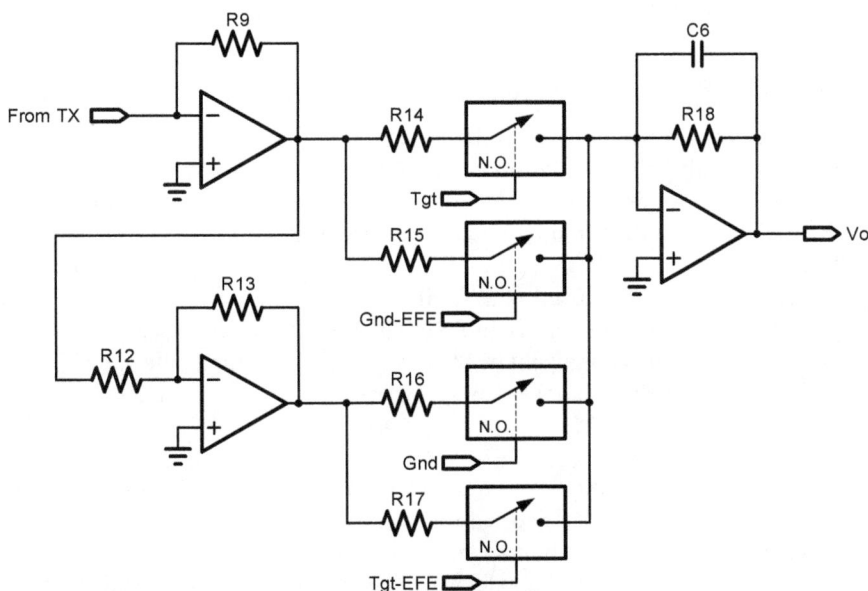

Fig. 22-18: **Integrated PWM Ground Balance**

---

9. It is also possible to vary the GB by using a wider fixed-width ground sample and varying its delay relative to the target pulse.

If ground balance can be achieved in the sampling integrator itself then the need for a separate ground channel integrator is no longer absolute, and everything could be combined into a single integrator stage which eliminates matched capacitors. In Figure 22-18, note that the preamp signal is used for the target and ground-EFE samples, while an inverted signal is used for ground and target-EFE samples. Also, each sample switch has its own feedforward resistor, allowing the gains to be individually adjusted if needed. This again would permit the ground samples to use a higher gain to augment the pulse width control. A final simplification is that a single EFE sample can be used: eliminate the Tgt-EFE sample and reduce the Gnd-EFE sample by the width of the Tgt pulse (assuming all switch channels use the same resistor values).

PWM ground control can also be applied to the aforementioned 4-channel no-hole scheme with the same caveat previously mentioned in that section. Each ground channel will require a different GND pulse and it is likely that they will also need to be uniquely pulse-width varied to achieve ground balance. Therefore, this could mean placing them on separate controls and balancing each receiver separately unless they can be synchronized.

22.9: Digital Design

A rather obvious topic of exploration is to digitize most of the design and implement it in DSP. As with VLF, the most practical approach is to digitize after the demodulators to take advantage of the integration and gain they provide. Figure 22-19 illustrates.

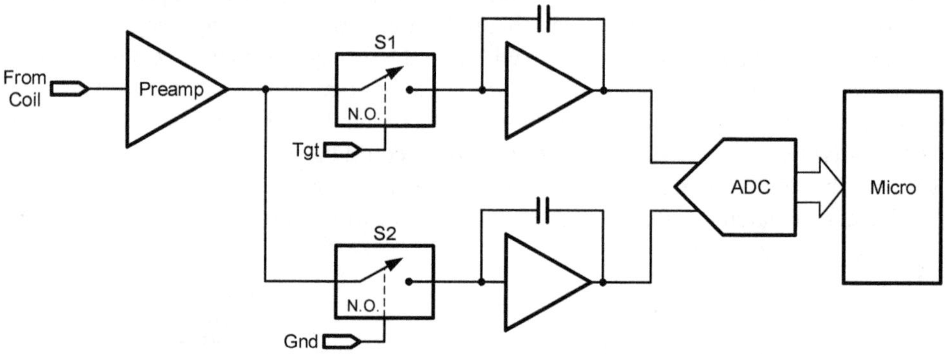

Fig. 22-19: **Practical Digital PI**

One advantage to learning analog metal detector design is that you now know exactly what to implement in DSP. Translating the analog functionality to DSP isn't difficult, but it can be difficult to equal the behavior and performance of an analog system. Most PI detectors produce a smooth audio with tonal subtleties that help in determining good targets from, say, hot rocks. It can be difficult to replicate that in DSP.

Directly replicating an analog design in DSP also imposes the limitations of the analog design. There are things that can be done quite easily in DSP that are difficult or impossible in analog-land, so exactly replicating the analog design may not be the best path. For example, the signal processing in the 4-channel no-hole concept introduced above may be far easier in DSP than analog.

22.10: Direct Sampling

The digital design can be simplified even more by eliminating the analog demodulators and using direct sampling. This was previously discussed for VLF designs in Chapter 18. In Figure 22-20 the ADC samples directly off the preamp output and demodulation is achieved by the sampling itself. That is, if the ADC samples every pulse decay at a delay of 10μs and keeps a windowed average of the samples then that is roughly equivalent to a demodulator sampling at 10μs. What the ADC cannot do is an extended integrating sample window, for example, where a demodulator can integrate a 5μs-wide portion of the decay curve. But the ADC can repeatedly sample multiple different points of the decay curve which could prove more useful in terms of ground balance or target identification.

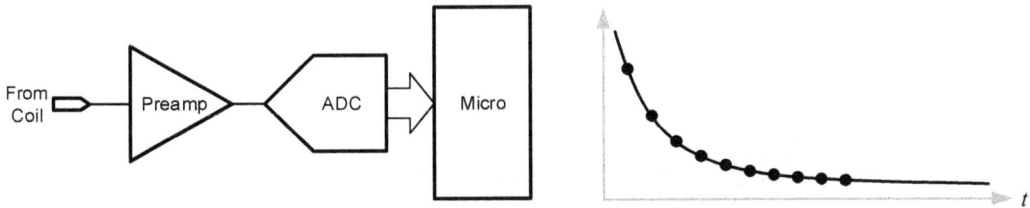

Fig. 22-20: **Direct Sampling PI**

The direct sampling approach requires a higher precision ADC than for sampling the demod outputs. For example, it may be possible to get reasonable performance using a 12-bit ADC when sampling the demod outputs whereas direct sampling may require 16 bits or more. If you want to quickly sample multiple decay points (such as the 10 samples shown) then moderately high speed is also needed, perhaps 100kHz or more. Transferring and processing all that data also becomes a challenge and a high-speed micro (100+ MHz) may be required. Although Figure 22-20 shows 10 sample points, in reality 3 or 4 sample points may be all that are needed.

22.11: Split Coil

All the PI circuits so far have been presented with a mono coil. It is an easy matter to alter the front end to accept a coil with separate TX and RX windings, as shown in Figure 22-21. Such a coil can either be induction-balanced (such as a DD) or not. If not, the coils can be the same diameter and coaxial so that the assembly appears to be a mono.

The advantage of this scheme is that it allows the TX and RX coils to be separately optimized for inductance and resistance. For example, fewer turns on the TX might produce a stronger TX field, and more turns on the RX produces a stronger induced target signal. If the coil is induction-balanced, then the value of R8 can be reduced since the flyback is much lower and this can reduce thermal noise. Plus, if we can avoid a high-current conduction of D6 then this will make flyback recovery faster. An IB coil also allows sampling during the TX turn-on time which can be used to attempt iron ID (see sections 25.14 and 26.12).

It is also an easy matter to still use a mono coil with such a setup. A split coil (IB or not) will need at least 3 pins for the connector. A mono coil assembly can bridge the TX and RX pins inside the coil connector to maintain plug-in compatibility. This is, in fact, what Minelab did with the SD/GP/GPX series of coils.

Finally, note that with a split coil it is no longer necessary to run the TX driver from Ground/–VB. It can instead be run from +VB/Ground which eliminates the weird backwards supply notation. That is, the top of the TX coil can be connected to +VB while the RX coil remains grounded. However, doing so also eliminates the possibility of using the mono coil option.

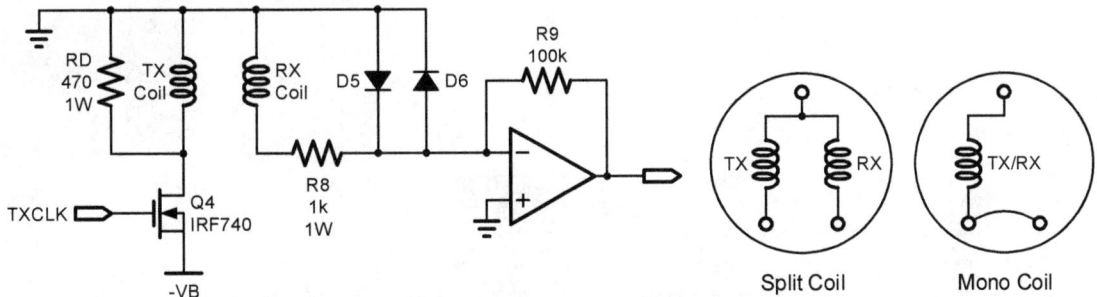

Fig. 22-21: **Split Coil Front-end**

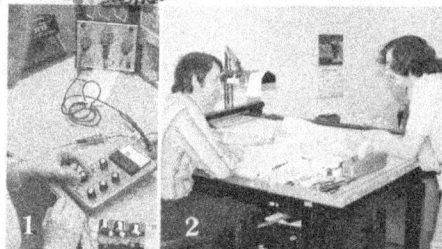

C	
H	
A	**23**
P	
T	
E	
R	

Bipolar PI

"I hate being bipolar. It's awesome."

— *Kanye West*

So far our PI investigations have involved *monopolar* pulsing whereby all TX pulses[1] have the same polarity. Such designs offer simplicity and speed. Simplicity because it uses a mono coil and a single FET switch; speed because it can be designed with low parasitic capacitance to achieve a very low sample delay. The lower the sample delay, the better you can detect low-conductivity targets like small gold.

There are cases where we don't need the raw speed, or we are willing to add complexity to further improve performance in some other way. The use of bipolar pulsing probably began in mine detection where certain types of mines are triggered not by pressure but by a change in the net magnetic field. An example is an anti-tank mine which is triggered by a tank passing over it and distorting the Earth field. Monopolar PI detectors produce a net magnetic field and can potentially trigger such a mine. With bipolar pulsing, the coil current is pulsed in both directions so that there is an alternating positive and negative magnetic field, resulting in a zero net field.

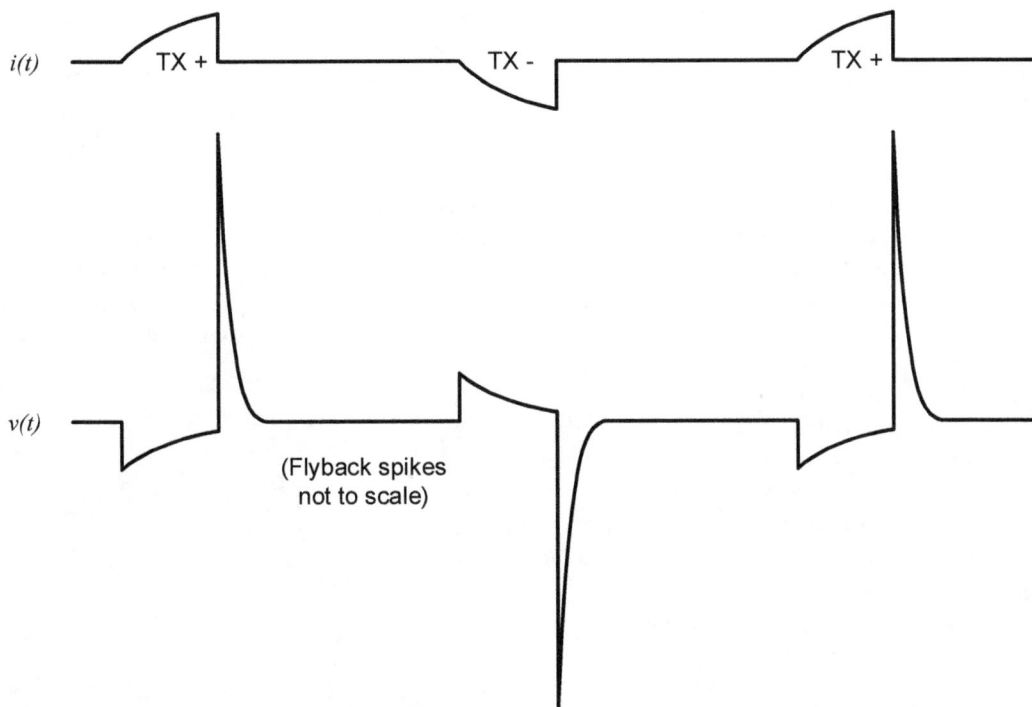

Fig. 23-1: **Bipolar Pulsing — Coil Current & Voltage**

---

1. Whether considering voltage, current, or magnetic field.

Transmitter

There are two predominant transmitter designs for bipolar pulsing. The first is to pulse a mono coil with a positive and negative current. This can be done with either a half-bridge or a full-bridge driver circuit, as shown in Figure 23-2:

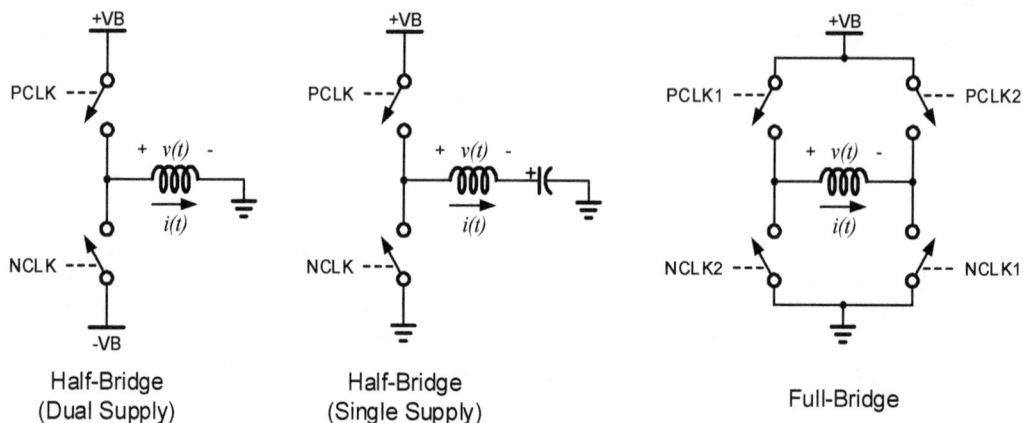

Half-Bridge
(Dual Supply)

Half-Bridge
(Single Supply)

Full-Bridge

Fig. 23-2: **Bridge-Driven Mono Coil**

The half-bridge solution requires a dual power supply[2] if the coil is directly grounded. However, a single supply can be used if a large cap is added in series with the coil. Over time, the capacitor will achieve an average charge of VB/2. The full-bridge option can also be designed with a dual supply but, because the coil is floating during the "off" time, there is no particular need to do so.

The use of a mono coil is an advantage with these circuits. The half-bridge driver is simpler but requires a large coupling capacitor unless a bipolar power supply is available. The full-bridge driver requires more switches with more clock signals but, because both current polarities are driven with the full supply voltage, it has roughly twice the current drive of the single-supply half-bridge circuit. It also has better waveform symmetry than the half-bridge approach when real switches are implemented.

The real switches are assumed to be MOSFETs, although bipolar transistors will also work. Regardless, both options have higher capacitive parasitics than a simple monopolar driver and will therefore have a slower flyback decay. Although we have generally emphasized the need for speed, there are cases where it is not as important. A good example is a walk-through security detector where the primary targets — guns, knives, and other weapons — are somewhat large and have large time constants. In fact, walk-throughs are often set up to ignore smaller items such as coins, watches, and eyeglasses. Both Garrett and Fisher produce PI walk-throughs that use full-bridge coil drivers.

A second way to produce a bipolar current drive is to split the TX coil into two halves and drive each half with opposite currents. Figure 23-3 shows this scenario. The coil is a center-tapped winding resulting in opposing fields (note the polarity dots) and is ideally constructed with bifilar wire to optimize their matching. Assuming the switches are realized in the same way as a monopolar design, this approach maintains the low parasitic capacitance and speed of the monopolar driver at the expense of a more complicated coil design.

Besides the avoidance of magnetic triggers in mine detection, bipolar pulsing offers other benefits. One

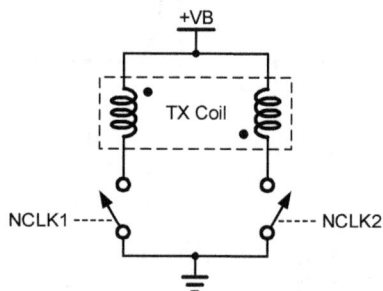

Fig. 23-3: **Differential Coil Driver**

---

2. A dual power supply is often referred to as a "bipolar supply" which is not to be confused with the topic of discussion, bipolar transmitters. A bipolar transmitter can be designed with a positive supply, a negative supply, or a bipolar supply.

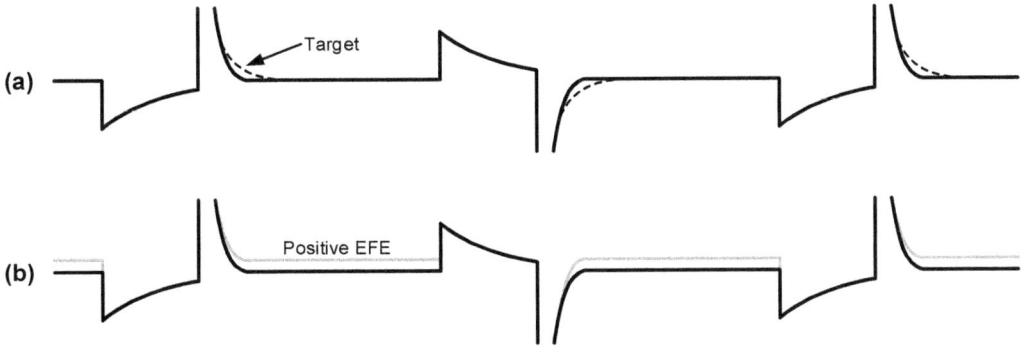

Fig. 23-4: **Inherent EFE Rejection**

related to the discussions in Chapter 23 is the inherent rejection of Earth field effect (EFE). While target responses follow the polarity of the TX signal, the offset produced by EFE has the same short-term polarity[3] regardless of the TX polarity, as shown in Figure 23-4. Waveform (a) shows a target perturbation only; the receiver inverts one of the RX polarities before sampling, effectively subtracting the negative pulse cycle from the positive cycle and resulting in a coherently-summed signal. Waveform (b) shows an Earth field perturbation only; because the EFE offset is always positive (or always negative[4]) the same subtraction process that results in an additive target signal largely cancels the Earth field signal. Thus there is no need for explicit EFE samples which, in turn, enables a faster overall pulse rate since there is no need to budget any time for late samples. And, unlike the EFE sample-subtraction method presented in Chapter 21, bipolar pulsing cancels EFE with no detrimental effect on the target signal[5].

The same rejection mechanism for EFE also applies to any other low frequency EMI. "Low frequency," in this case, means "lower than the pulse rate," generally by an order of magnitude[6]. If the pulse rate is 1kHz, then a bipolar design will do a pretty good job of rejecting EMI below 100Hz. It will likewise cancel flicker noise generated in the preamp. Without the need for EFE samples, bipolar permits faster pulse rates than monopolar, and increasing the pulse rate increases the integrated target signal and improves SNR. It's a rare win-win situation, but the *no-free-lunch* constraint still applies: power consumption increases with pulse rate.

23.1: H-Bridge Design

In this section we'll look more closely at the H-bridge approach. Though we'll focus on the full-bridge driver, most of the discussion also applies to the half-bridge driver. Figure 23-5 shows a more realistic schematic for an H-bridge with MOSFET devices replacing the ideal switches and the requisite damping resistor added. If both opposing switches turn off at the same time then the coil is left floating and since neither side is nailed to any voltage rail (or ground) both sides of the coil will attempt to "fly back" in opposite directions. For example, suppose Q1 and Q4 are turned on and produce a coil current with the direction shown by $i_L(t)$. At turn-off the positive side of the coil ($v1$) will attempt to flyback in the negative direction and the negative side of the coil ($v2$) will attempt to flyback in the positive direction. Because the MOSFETs have parasitic body diodes, the negative flyback on $v1$ will get clamped by Q3 and the positive flyback on $v2$ will get clamped by Q2. This clamping action will dramatically slow down the discharge of the inductor.

---

3. Over a single sweep, which spans perhaps 100s or even 1000s of samples.

4. The "always" qualifier is, at a minimum, for two adjacent cycles. In reality, the EFE offset changes in both strength and polarity as the coil is swept, but changes little across two cycles. This assumes a decent pulse rate and a reasonable coil sweep speed.

5. This "natural" cancellation of EFE due to the bipolar nature of the signal is exactly the same mechanism that suppresses EFE in VLF designs.

6. See Appendix C for the math.

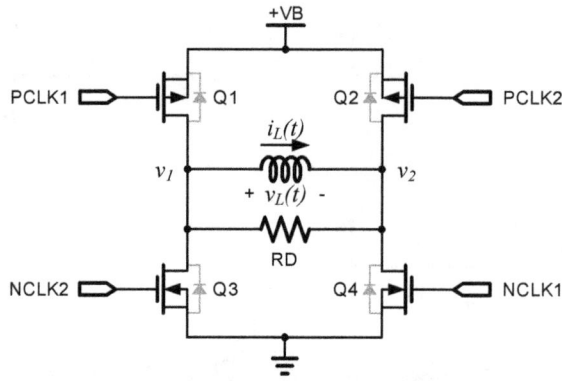

Fig. 23-5: **Full-Bridge Driver (Almost)**

One solution is to place flyback-blocking diodes on all four FETs which will allow unimpeded fly-back action on both sides. That leaves the coil completely floating which is fine if the design uses a separate RX coil. This actually provides the best performance because the four diodes help reduce overall parasitic capacitance by isolating the MOSFET C_{OSS}, as we saw in the *Explore* section of Chapter 21.

If the design instead relies on using the same coil as both TX and RX, then it needs to end up at a predictable and repeatable level. If blocking diodes are applied only to either the high-side switches or the low-side switches, then the unblocked switches will clamp the inductor via their body diodes. For the sake of economics the diodes should be applied to the PMOS devices as shown in Figure 23-6.

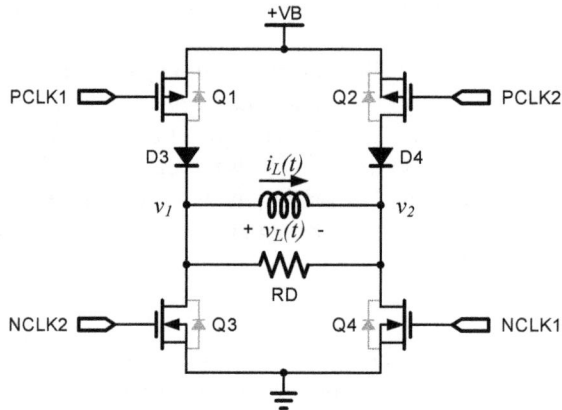

Fig. 23-6: **Full-Bridge Driver**

Returning to the previous supposition, if Q1 and Q4 are conducting current $i_L(t)$ as shown, when the switches turn off $v1$ will attempt to flyback in the negative direction but the body diode of Q3 will clamp it to 1 diode below ground. Meanwhile, the high-voltage blocking diode D4 preserves the positive flyback action on $v2$ by preventing conduction via the body diode on Q2. In this case both $v1$ and $v2$ are properly referenced to true ground and there is no need in this design to go through the power supply re-labeling gymnastics done in the *Pulse-2, 3,* and *4* designs.

The diodes must be rated for the full expected flyback voltage, and so must the NMOS devices unless they are allowed to avalanche. The PMOS devices never experience the flyback voltage and therefore may be low-voltage devices. This keeps the highest demands on the NMOS devices where we are more likely to find a good variety to choose from at lower prices. It is equally possible to place the blocking diodes on the NMOS devices. Then the PMOS body diode will clamp the coil to the positive rail, flyback action will be negative, and the PMOS devices will need to have a high breakdown voltage; but this is the harder (and more expensive) way to do it. And since this approach also clamps the signals to +VB there will be the need to relabel +VB as ground and ground as −VB, as explained in *Pulse-2.*

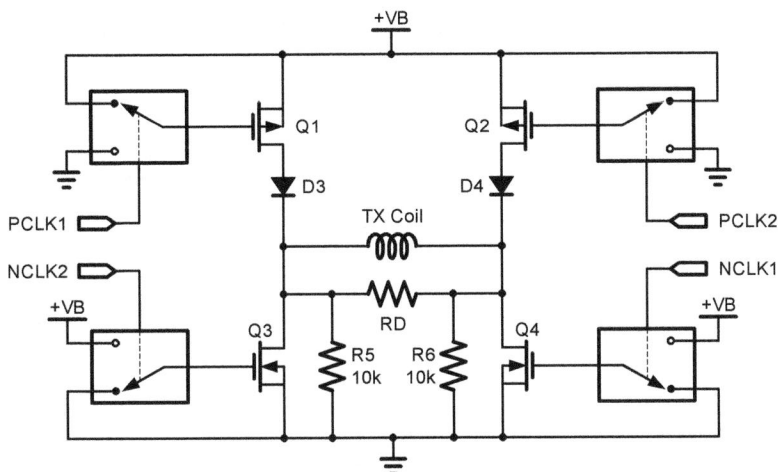

Fig. 23-7: **Complete Full-Bridge Driver**

The clamping of the negative flyback by the NMOS body diodes is a temporary action lasting only as long as there is current flow. Near the end of the flyback decay the NMOS body diode turns off and the coil is left floating. If the coil is to directly drive a receiver preamp then resistors should be added across the NMOS devices to ensure that the coil remains referenced to ground. Finally, all the MOS switches need some kind of clock voltage drive. The clocks are provided by a micro (3V or 5V logic) so we would need to design level shifter and translator circuitry to provide a +VB/ground voltage swing on the gates as was done in previous *Pulse-X* designs. An easier solution is to use analog switches which can often take a reduced clock swing but still switch a high voltage.

The switches in Figure 23-7 can be "4053" type. It is important to choose the right 4053 series; the 74HCT4053 is a "high-speed CMOS" family (the "HC" part) but with a TTL logic interface (the "T" part). The TTL interface ensures that it can be clocked from 3V or 5V logic even if the chip is powered from, say, 12V. Other series chips (H, C, HC, LS, etc.) will not have the logic voltage margins to work properly.

There is a drawback to using analog switches. The switch resistance in a vanilla 4053 can be as high as 150Ω or so and this can slow down the MOSFET drives, especially with turn-off. There are a number of proprietary SPDT analog switches with a much lower switch resistance; for example, the ADG1636 has $R_{ON} = 1.1\Omega$ while still maintaining TTL logic compatibility up to a supply of 16V. Other devices can be found with slightly higher resistances but up to 44V supply capability, giving some flexibility should you want to experiment with a higher TX voltage. Many of these improved switches are only available in SMT and are sometimes aggressively small, making them difficult to work with. Another option is to place a Class-B driver between the switch and the MOSFET gate. Yet another option is to simply stack multiple 4053 chips on top of each other to reduce R_{ON}.

Timing for the H-bridge can be simple, with each diagonal PMOS/NMOS pair clocked simultaneously as shown in Figure 23-8. There is a way to eliminate the NMOS drive switches by tying the NMOS gates to the PMOS drains resulting in the need for only two clock signals and two analog switches. Instead of simultaneous clocking, it is also possible to hold the NMOS devices on for an extended time after the TX pulse (alternate timing), in which case the hold-down resistors R5 & R6 are not needed. However, this approach demands separate switches and separate clock signals for each MOSFET.

The design for this chapter will opt for the simultaneous timing for overall simplicity. Figure 23-9 shows the final TX circuit we will be using. Again, clocks are AC-coupled to the PMOS devices for safety. The NMOS devices are self-driven from the PMOS drains.

Advantages of the H-bridge approach start with the use of a simple mono coil. The choice of inductance value is much the same as with a monopolar design: it depends on the application. But typical values are similar, in the range of 100-500µH.

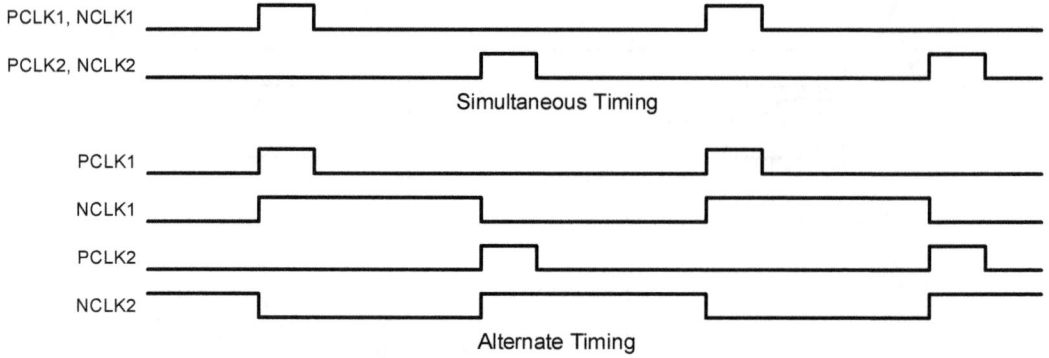

Fig. 23-8: **Full-Bridge Timing**

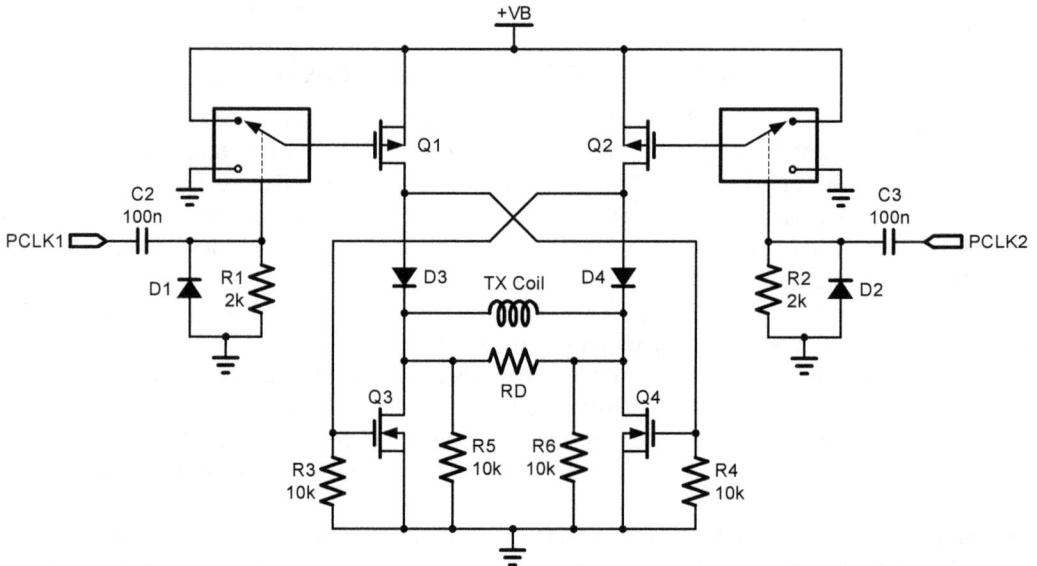

Fig. 23-9: **Final Full-Bridge Driver**

Assuming the PMOS/NMOS pairs (Q1/Q2 and Q3/Q4) are closely matched in their performance, then the forward and reverse turn-on currents through the coil will be closely matched. This makes the overall operation very symmetrical. This is generally not the case with the half-bridge driver because forward and reverse symmetry rely on matching the operation of a PMOS with an NMOS. Better half-bridge symmetry can be achieved if the high-side switch is a matching NMOS device instead of a PMOS device. The high-side NMOS switch will need a more complex bootstrapped driver, and the proliferation of solid-state motor control has provided a wide variety to choose from. This same thing could be done to the full-bridge transmitter: use four matched NMOS devices, with bootstrapped drivers for the high-side switches. High-side NMOS devices still need blocking diodes for the body diode conductance problem so there is little advantage in using NMOS transistors on the high side.

As previously stated, the overall parasitics are increased in an H-bridge driver. The coil current now sees two FET channel resistances, although it is fairly easy to find low-resistance FETs such that the coil is still the dominant resistive element. However, capacitance is more difficult to avoid. Recalling the capacitance table on page 412, the FET capacitance was already the dominant piece. It is now increased by at least the capacitance of the high-side diode on the active flyback node. Even the capacitance of the non-flyback side can slow things down during the body diode clamping. All of this forces a reduction in the damping resistor and slows down the flyback decay. Even though it is possible to find lower capacitance devices such that the TX circuit no longer dominates, the same is true for a simple monopolar design. In short, an H-bridge will always be at least a little slower than a monopolar design.

23.2: Differential Coil Design

The differential coil design trades off coil simplicity for performance. You can think of this as two mono transmitters with tightly coupled coils. Each coil is energized with its own FET[7] to create a monopolar current. The coil that is not being energized is magnetically coupled but, being in a relatively high-impedance state (having only the damping resistor), the induced current is low. In that respect, the coil behaves like an autotransformer. However, when the energized coil is turned off and flyback ensues, the coupled coil is also induced with a reverse-polarity flyback voltage, and both flyback responses get altered by a nearby target.

Because this approach is so close to the NMOS monopolar driver thoroughly covered in the last two chapters, we'll jump right to the full design. Figure 23-10 shows a basic differential TX circuit and the clocking waveforms. With the monopolar design there is a slight advantage in adding a series diode to each NMOS to reduce the effects of drain capacitance (*Explore* in Chapter 21); in this design, we'll see that they are mandatory.

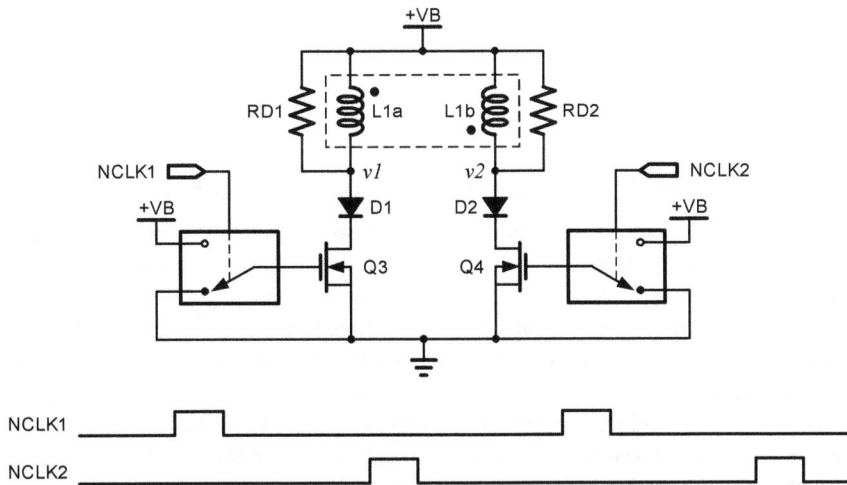

Fig. 23-10: **Differential Driver**

The voltages at *v1* and *v2* are shown in Figure 23-11. Some discussion of what is happening is useful. When Q3 is turned on a step voltage is applied to L1a resulting in an exponentially rising current. The tau of this exponential is L1a/RL, where RL is the total parasitic series resistance, exactly the same as with the monopolar driver. Because RL is small (a few ohms at most) this tau is somewhat slow (see Figure 21-21).

L1a and L1b are tightly coupled so the step voltage on L1a is also induced onto L1b. However, its tau is set by 0.5*L1b/R2 where R2 is hundreds of ohms, so this tau is 100 times faster than that of L1a. Thus whatever current is induced in L1b quickly decays to zero. This explains a minor glitch seen on the non-energizing coil during turn-on.

At turn-off things get interesting (Figure 23-11). The energized coil (we'll stick with L1a as the example) experiences the now-familiar stepped current which produces a positive flyback voltage of several hundred volts and decays at a tau of $0.5 \times L1a/R1$. This voltage is induced into L1b and is seen as a negative flyback at *v2*, and decays at a tau of $0.5 \times L1b/R2$ which should be identical to the L1a tau. That is to say, both coils experience the same flyback but with opposite polarities. A target signal also distorts both flybacks, and also in opposite directions; the positive flyback (*v1*) is distorted in the positive direction, and the negative flyback (*v2*) is distorted in the negative direction. The opposite of course happens when L1b is energized, and the induced negative flybacks are why the previously optional diodes are now mandatory; without the diodes the negative flybacks will be instantly shunted by the NMOS body diodes.

---

7. We will stick with NMOS for best practice, although PMOS would also work. Or NPN, or PNP.

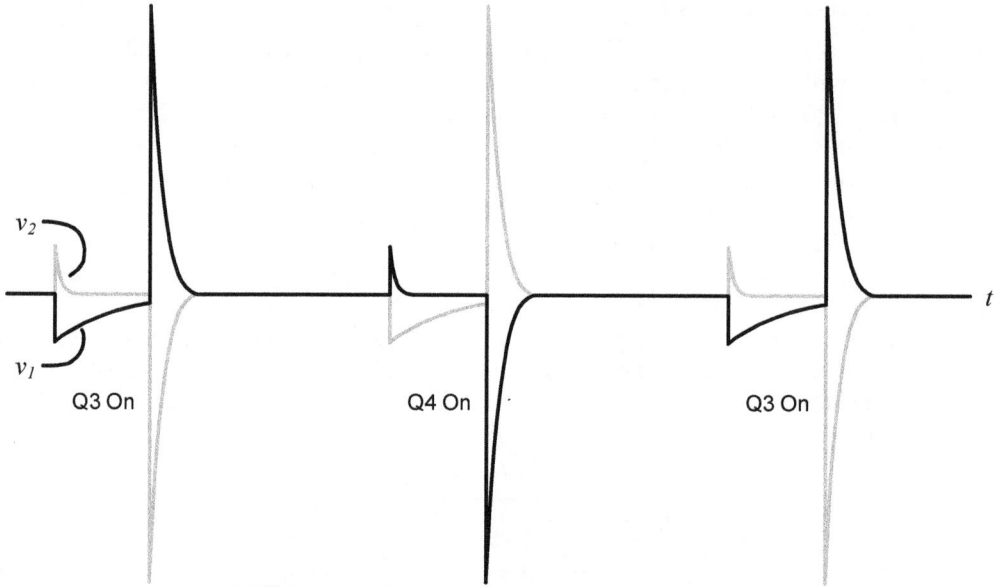

Fig. 23-11: **Differential Driver Voltage Waveforms**

The apparent result of all this is that during transmit only one-half of the coil is active, but during receive both halves are equally active. It is the equivalent of having a TX coil inductance of L1 and an RX coil inductance of 4·L1. It is therefore necessary to reconsider our choice of coil inductance; if we choose the normal L1 = 300µH then the apparent inductance on the RX side is 1200µH and that may prove excessively slow. However, achieving 300µH on the RX side means 75µH on the TX and that may not be optimal, either.

Because each half of the center-tapped coil is pulsed at separate times and the received results are subtracted there is no need to inductively balance the coil halves, although doing so will work just fine. With the differential TX the normal approach to winding the coil is to wind both coil halves together. The use of bifilar magnet wire produces the best coil matching and therefore the best bipolar pulse matching. However, bifilar wire is not the best choice for low capacitance. Carefully co-winding insulated (PVC or Teflon) stranded wire or perhaps using twisted pair may offer better overall results.

Although we won't make use of the differential coil in any projects it has some very interesting possibilities that will be covered in the *Explore* section.

Receiver

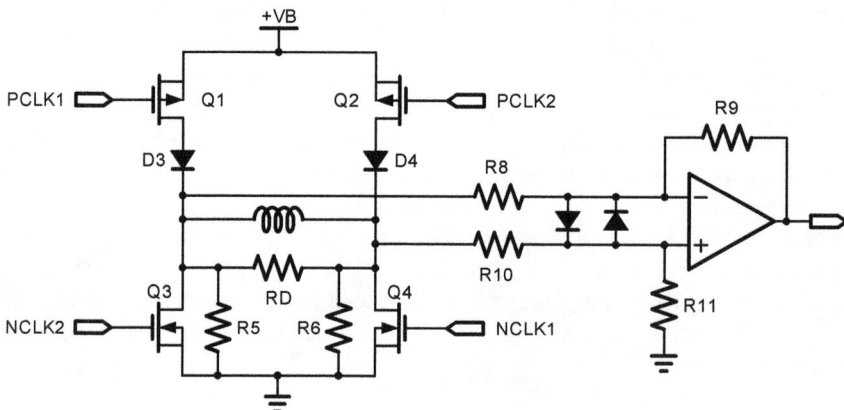

Fig. 23-12: **Typical Bipolar Receiver**

Regardless of the design used, a bipolar transmitter requires some kind of bipolar receiver. Any of these circuits can be used with a separate RX coil, or the TX coil(s) can be direct-coupled to the preamp as we have been doing in the earlier *Pulse-X* circuits. Figure 23-12 shows how a preamp can be differentially coupled to an H-bridge circuit. The same approach will work with a differential TX driver. R11 provides a reference point to ground so R5 and R6 in Figure 23-9 may not be needed. However, the use of R5 and R6 may improve settling.

23.3: Demods & GB

The bulk of the RX circuitry from the preamp to the ground balance subtraction is identical to that in *Pulse-4*. There is no longer a need for EFE samples but there is now a need to have separate samples for the positive and negative pulse polarities. Figure 23-13 shows the same demod scheme as in Figure 22-5 but with the demod clock signals relabeled.

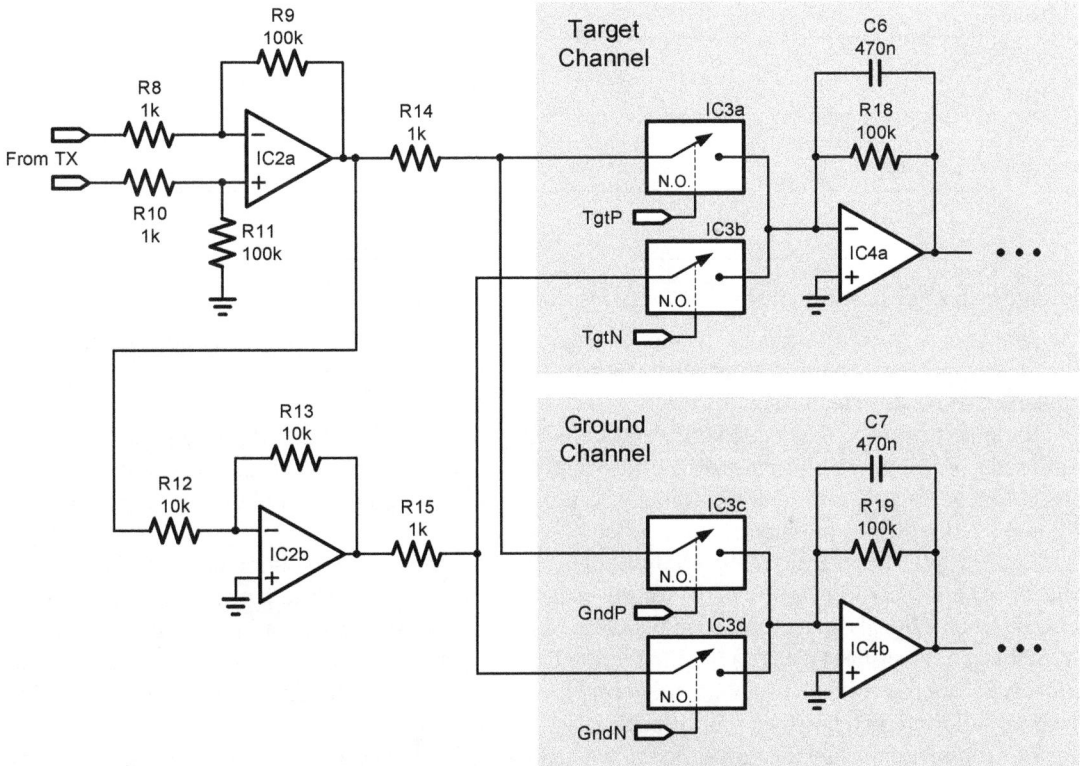

Fig. 23-13: **Bipolar Demodulators**

The demod timing for this scheme is slightly more complicated than for the *Pulse-4* monopolar design. The TX timing adds additional complexity. The complete TX/RX timing is shown in Figure 23-14. The opposing TX polarities combined with the alternating RX polarities produce additive polarities going into the integrator stages.

23.4: Audio Stage

The *Pulse-X* designs have progressed from a simple constant-tone audio with modulated volume to a tone- and volume-modulated audio using a VCO. The *Pulse-4* VCO audio gave a rising pitch for all target responses, though with a slightly nuanced audio null before or after the response, depending on the conductivity. It is possible to differentiate the low conductor side of ground balance from the high conductor side using different tones. For example, the White's *TDI* produces a high tone for low conductors and a low tone for high conductors. This seems logically backwards but gold nuggets are typically low conductors and that is what the TDI was designed for. With the Minelab series of gold PI

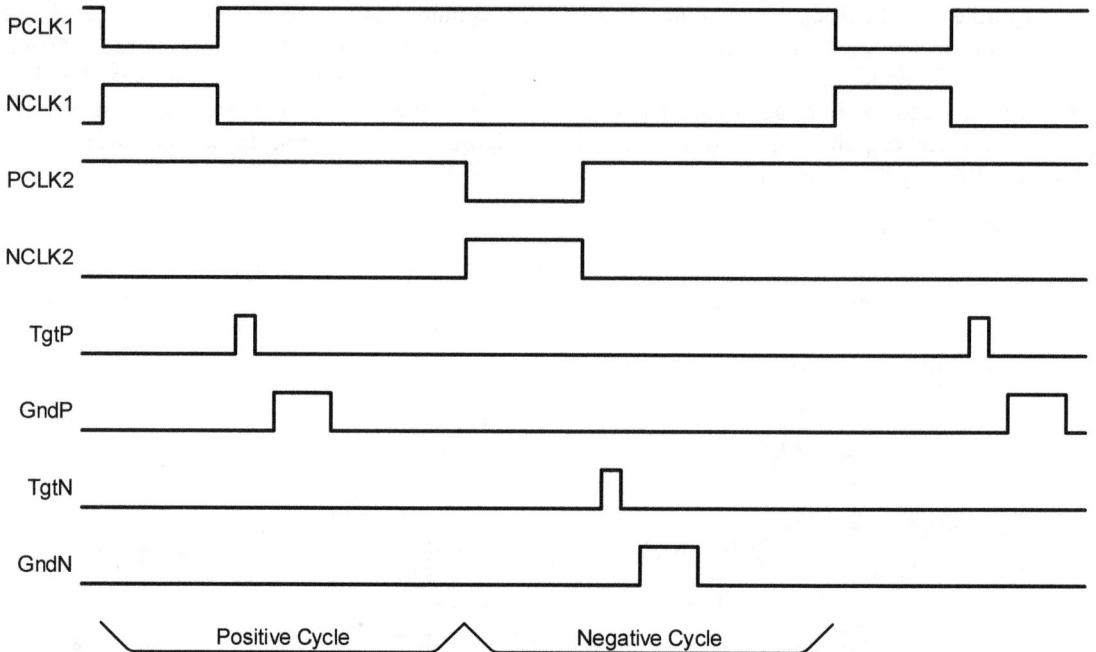

Fig. 23-14: **Bipolar Timing**

detectors a low conductor produces a high tone followed by a low tone, what many describe as a "wee-woo" response. High conductors produce an opposite "woo-wee" response. *Pulse-5* will show how to implement these same responses.

To do this we need to produce audio on both the positive lobe and the negative lobe, but to do so differently. For this we will employ a pitch change that is proportional to the signal level: the idle threshold level is assigned a certain frequency, a rising signal produces a higher pitch than the threshold, and a falling signal produces a lower pitch.

Figure 23-15 shows a block diagram of how a 2-tone system might be implemented. The signals at Points A-D are shown in Figure 23-16 for the cases of a low conductor and a high conductor. The Target and Ground signals are combined at the Ground Balance pot and feed a variable gain stage. The output of the gain amplifier (A) has a nominal voltage in the absence of targets. A low-conductor produces a positive response followed by a negative bounce-back, while a high-conductor produces a negative response followed by a positive bounce-back.

This bidirectional signal A is applied to a VCO stage to produce a frequency[8] (B) which is proportional to voltage. For no targets (or balanced ground) the frequency is chosen to be some nominal idle

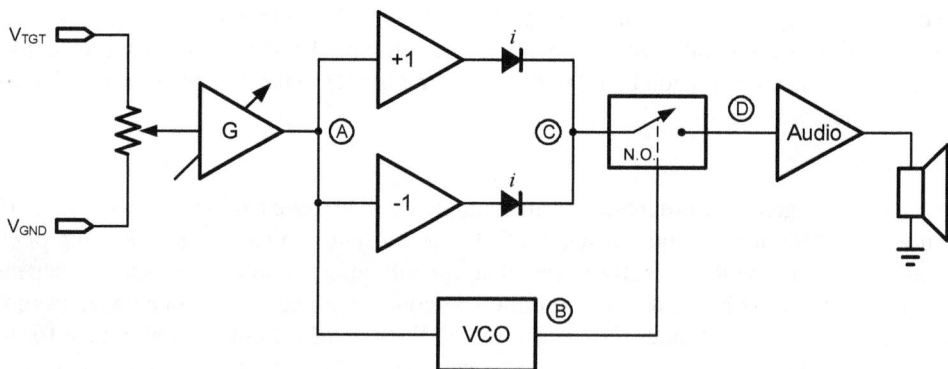

Fig. 23-15: **Audio Block Diagram**

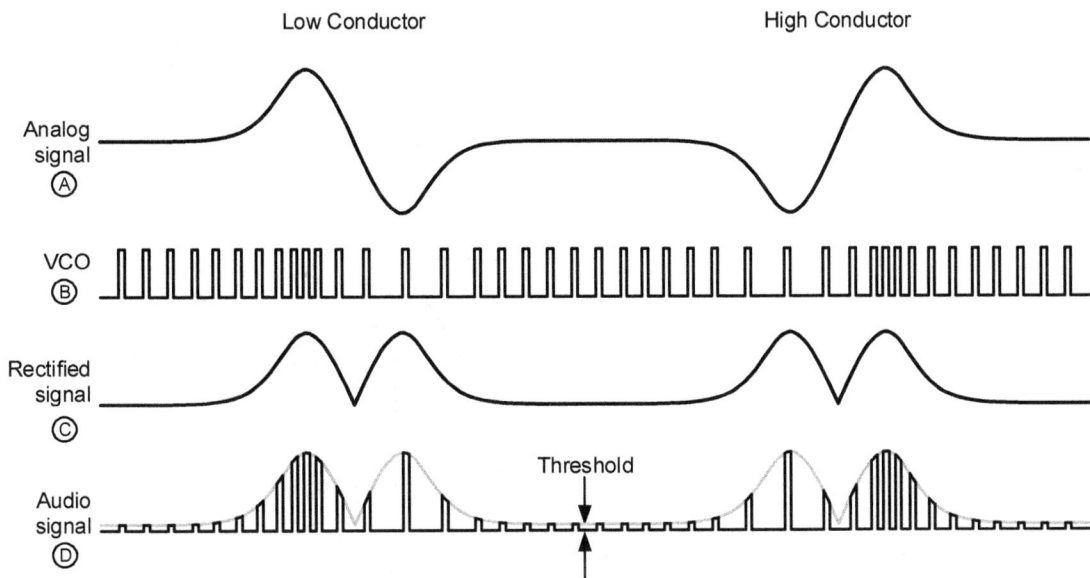

Fig. 23-16: **Audio Signals**

value. An initial higher voltage from low-conductors increases the frequency and its bounce-back signal decreases the frequency. An initial lower voltage from high-conductors decreases the frequency and its bounce-back signal increases the frequency. The bidirectional signal A is also applied to a buffer stage and an inverting stage, each of which is followed by a rectifier. Each positive-negative signal is rectified into a double-humped positive signal and recombined to produce the response at C.

Finally, this signal is chopped by the VCO output (B) to produce the final audio signal (D) which drives the audio amplifier. A low conductor will create a "wee-woo" response and a high conductor will create a "woo-wee" response. The amplitude of the input signal (A) determines the amount of frequency change and also the change in loudness.

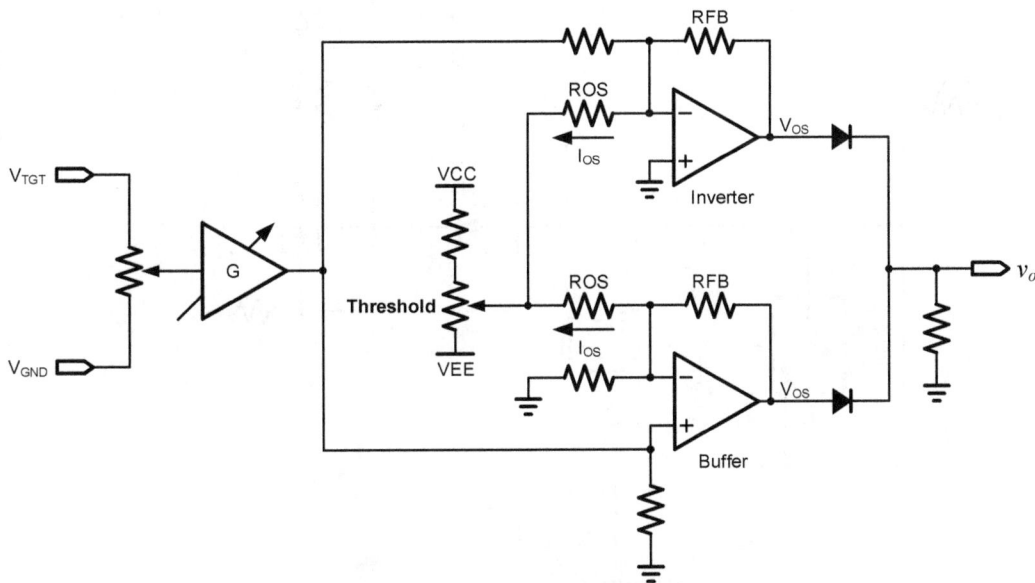

Fig. 23-17: **Threshold Details**

8. The VCO signal is shown as a variable frequency pulse train but could also be a variable frequency square wave.

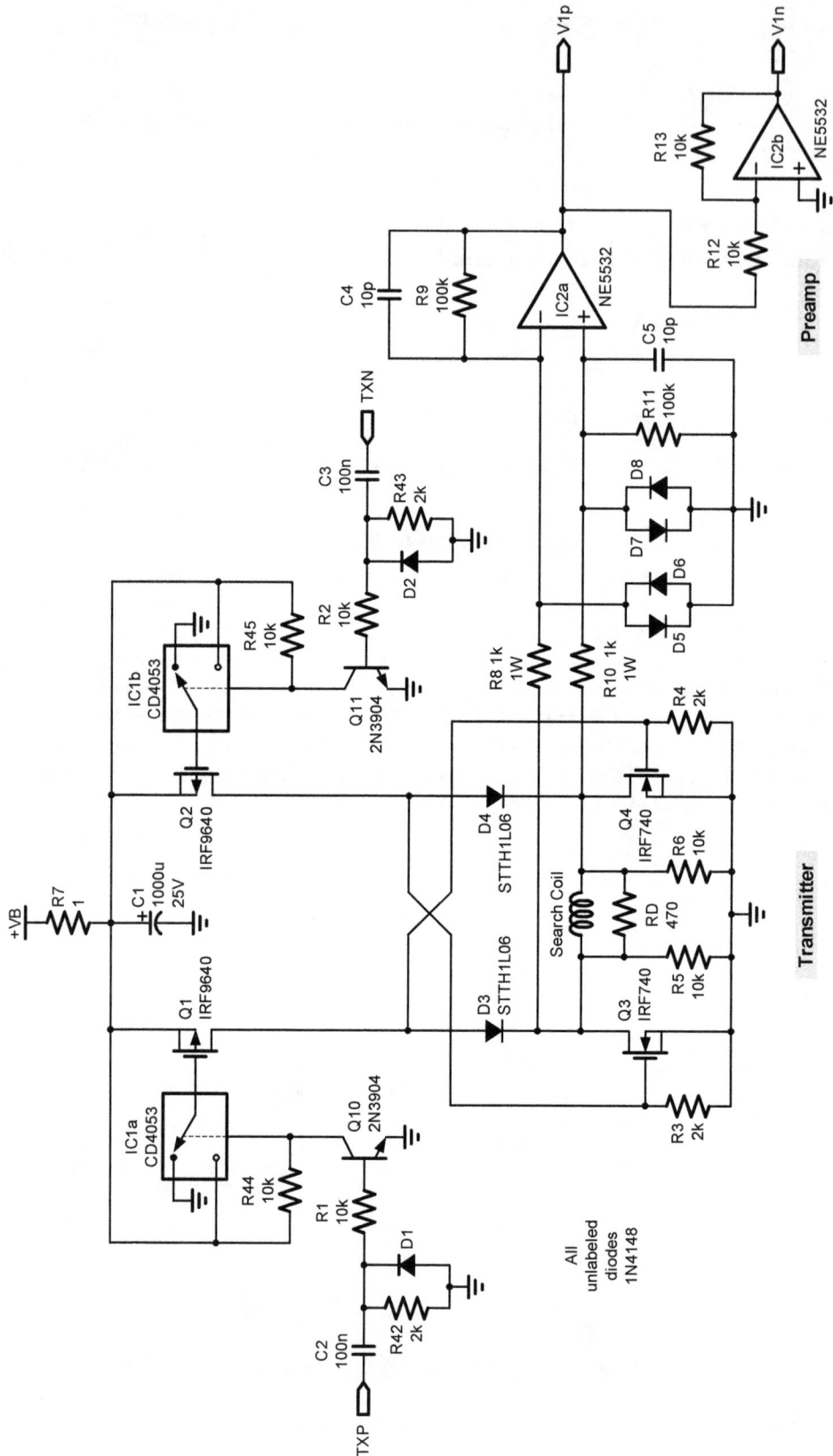

Fig. 23-18: Pulse-5 — TX and Front-End

Inside the Metal Detector

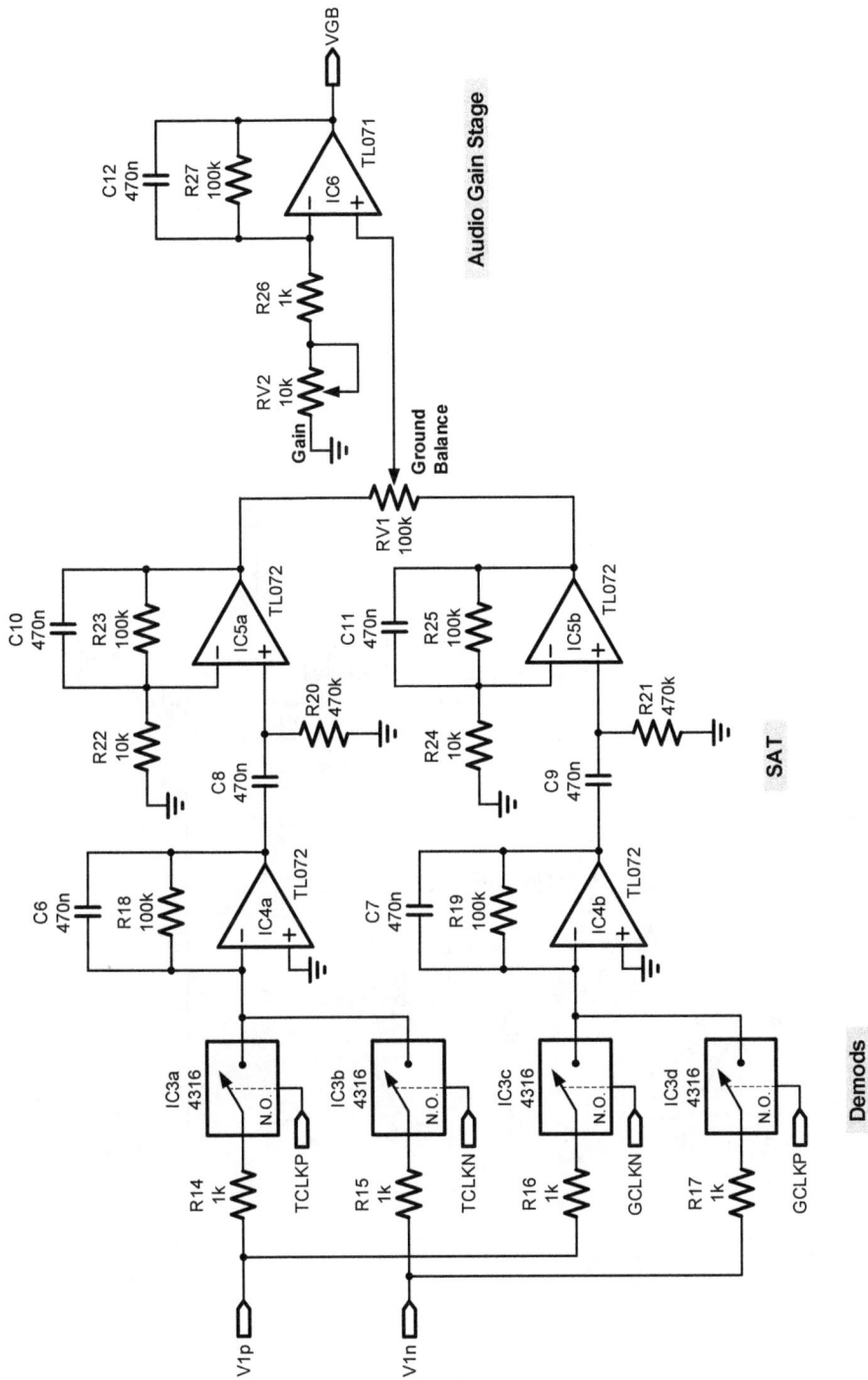

Fig. 23-19: **Pulse-5 — RX**

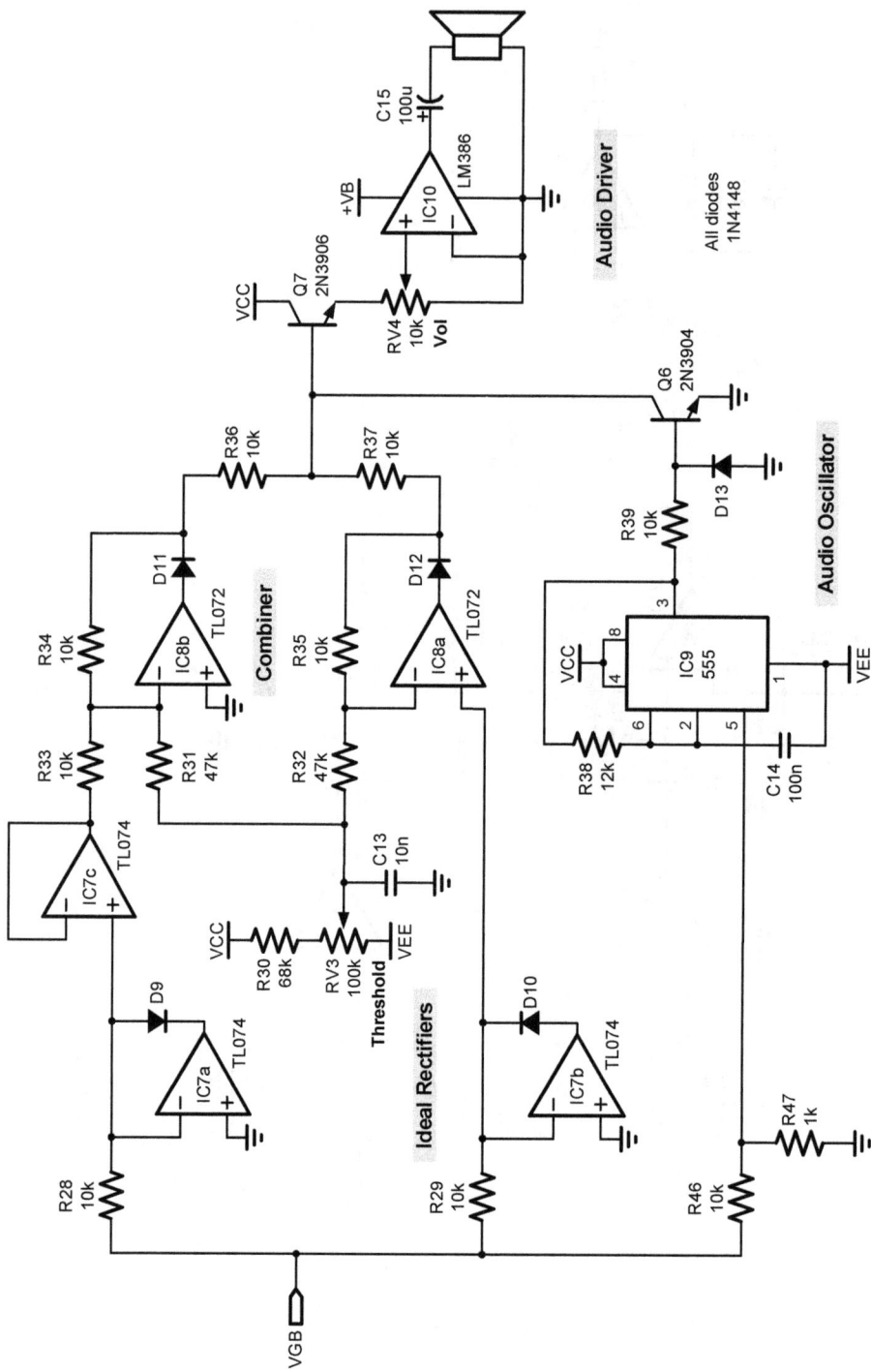

Fig. 23-20: **Pulse-5 — Audio**

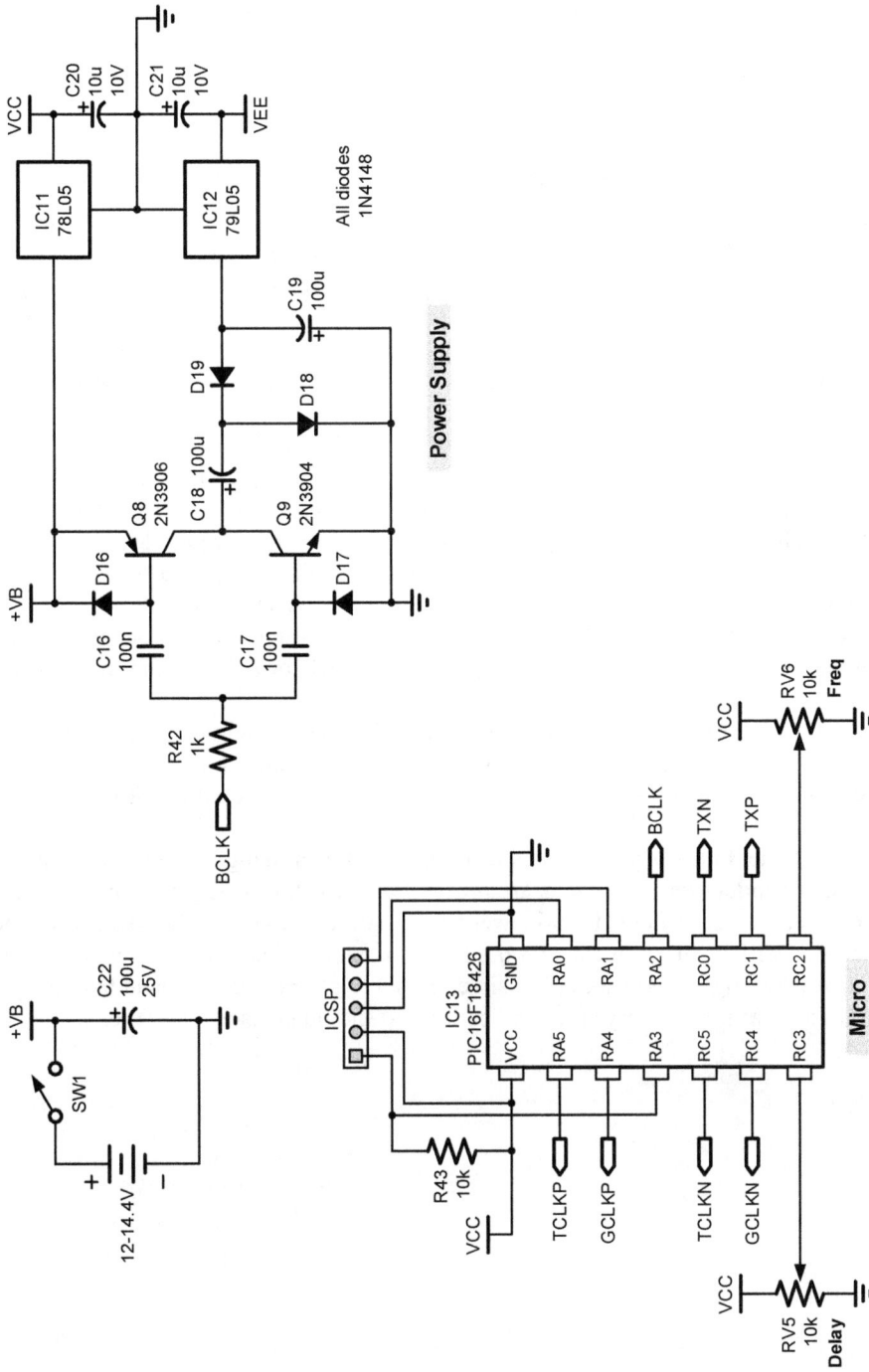

Fig. 23-21: **Pulse-5 — Micro & Power**

Audio signal D shows an idle threshold level but Figure 23-15 does not suggest where the threshold might be applied. In the previous *Pulse-X* designs the threshold is applied at the gain-adjust stage. In this case that will not work as the threshold would be inverted by the inverting rectifier stage. Instead, the threshold is applied equally to both the buffer and inverting rectifier stages.

Figure 23-17 shows how this is done. If the inverter and buffer stages have the same feedback resistor (RFB) and offset resistor (ROS) then the offset currents (I_{OS}) will be identical and so therefore will the output offset voltages (V_{OS}). So no matter what the audio response is doing it will always recover to the same threshold level.

Design: *Bipolar PI (Pulse-5)*

The final schematic for the Bipolar PI design is shown in Figures 23-18 to 22-21 and closely follows everything presented so far. In this design the negative side of the battery is ground so the charge pump is configured as a *voltage inverter* to generate the -5V rail (compare to Figure 21-4). Since the micro clocks the charge pump, it must now be powered from VCC and ground. This creates a problem with 4066 switches (IC3) used in the demods; the clocks are no longer referred to the lowest voltage rail of the 4066 so they will not work. Instead, we use the 4316 (specifically, the 74HCT4316) which includes a logic ground pin.

In the transmitter, the 4053 switches (IC1) that drive the PMOS devices could be 74HCT4053 but they are limited to 10V max which would then limit the overall battery voltage to 10V. To use a higher battery, we need to instead use a CD4053 (20V max) but that requires a higher clock swing. This is accomplished by adding a level translator (Q10,Q11) to boost the clock amplitude.

The micro includes a control for the sample delay as seen in *Pulse-4*, plus a second pot is added for an addition user control. This could be used for either the TX frequency or the TX pulse width with care taken not to exceed power dissipation limits. Or, better yet, adjusting both frequency and pulse width while maintaining a fairly consistent power dissipation. See 25.1 for a discussion of proportional timing.

It is sometimes preferable to run a ground-balance PI detector with the GB turned off. This is sometimes referred to as all-metal mode[9]. By adjusting the GB control all the way to "zero" (completely towards the output of IC5a) no ground channel signal is subtracted and it is effectively in all-metal mode.

As mentioned in the *Audio Stage* section, the usual PI audio has a high tone for low conductors and a low tone for high conductors. PI designs are more often used for prospecting and beach detecting where the more desired targets (nuggets and jewelry) fall on the low conductive side. A high tone indicates "good target" while iron often falls on the high-conductor side where it generates a low tone. Keep in mind that other high-conductors (like silver coins) might be desirable and that many low conductors (like foil) are trash. It's a very crude indication and should be used with care.

---

9. A bit of a misnomer, since most PI designs are always all-metal. In this case, it means that all metals have exactly the same wee-woo response.

Explore

23.5: Differential Dual Coil Variants

The differential coil driver (Figure 23-10) requires a center-tapped coil which effectively means there are now two coils. For a direct comparison with the bipolar H-bridge design the coils are wound together as a single entity. There is no reason why the coils cannot be wound separately and placed non-concentrically, for example, in a DD arrangement, and can even be induction-balanced.

Another possibility is to place the coils concentrically but make them different sizes. As an example, make one coil 12"/300mm and the other coil 6"/150mm (Figure 23-22). Having two different sized coils running simultaneously is similar to the White's "Dual Field" coil[10] that was standard on the *TDI* models except that this approach is perhaps superior. Each coil alone performs as well as it would in a mono coil system but is enhanced by the receive signal of the non-energized coil.

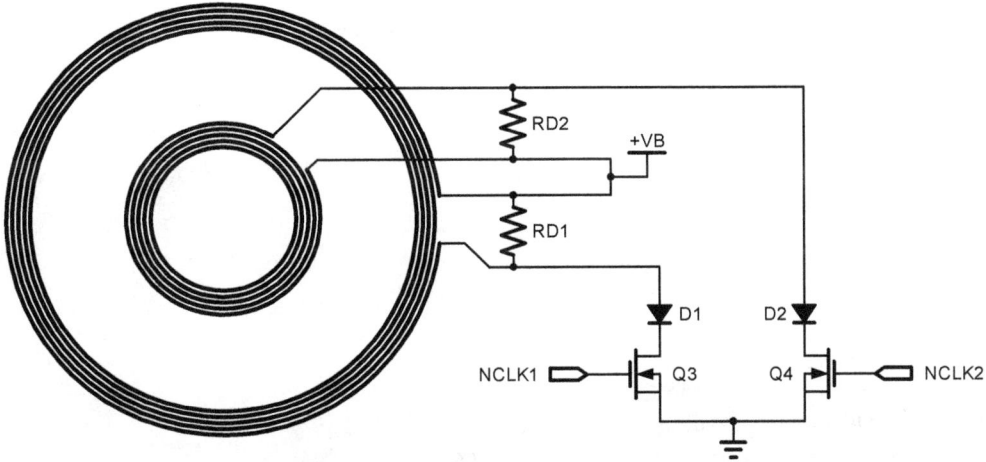

Fig. 23-22: **Differential "Dual-Field" Coil**

23.6: Simultaneous Differential Pulsing

In the *Twin Loop Treasure Seeker*[11] project two non-concentric coils are concurrently pulsed but then subtracted to cancel ground effects and EMI. The differential coil driver in Figure 23-10 could be used to drive the coils simultaneously but the coils will need to be connected in-phase. Interestingly, besides canceling ground effects and EMI this approach also cancels EFE.

To use *Pulse-5* in this way may require using the circuit with ground balance disabled; that is, RV1 adjusted all the way to IC5a. Even then the audio response from targets is likely to be very strange and will sound different when sweeping left-to-right versus right-to-left.

23.7: *TDI* Audio

The audio design in *Pulse-5* mimics the Minelab PI audio response. It was mentioned that the White's *TDI* produces a simpler "wee" response for low conductors and a "woo" response for high conductors without the secondary signal. The secondary signal is caused by the SAT bounceback so to achieve the TDI response it is necessary to eliminate the bounceback part of the signals before the Target and Ground signals are combined at the GB pot. The details will not be covered but the concept is shown in Figure 23-23. The diodes following the SAT stages are elements that eliminate the bounceback signals and must be "ideal rectifiers."

---

10. Patent US7994789

11. By Robert & David Crone, ETI Magazine Sept. 1989

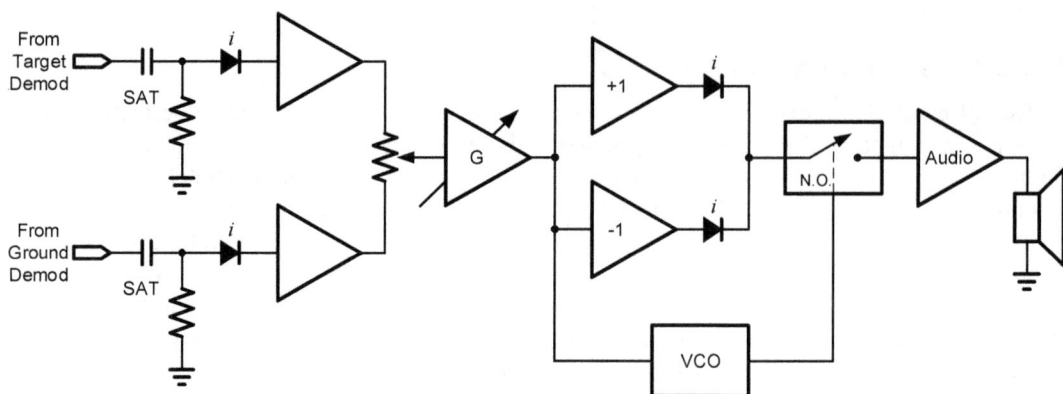

Fig. 23-23: **TDI Audio Concept**

23.8: Security Walk-Through

The bipolar PI is an excellent candidate for a security walk-through detector. A walk-through consists of two vertical panels which contain the TX & RX coils. The circuitry is normally in a *bridge* between the two panels at the top. The area between the panels where the subject walks through is called the *gate*. Figure 23-24 shows a Fisher model I designed[12].

The usual approach is to place a TX coil in one panel and the corresponding RX coils in the opposing panel. The TX coil may be one large coil, as large as 2 feet by 7 feet. The RX side is often broken up into smaller coil areas called *zones*; this allows the walk-through to designate roughly where the target is located. Commercial walk-throughs have anywhere from 1 zone up to 66 zones. A 33-zone walk-through[13] would have 11 vertical zones by 3 horizontal zones. This does not necessarily mean there are 11 RX coils as it is possible to interpolate between a pair of coils; six RX coils would suffice. A horizontal middle zone is always interpolated.

There are three major design challenges in a walk-through:

1. Good sensitivity horizontally across the gate
2. Immunity to mains (50/60Hz) interference
3. Immunity to large metal objects or noisy electronics placed next to the walk-through (like an X-ray machine)

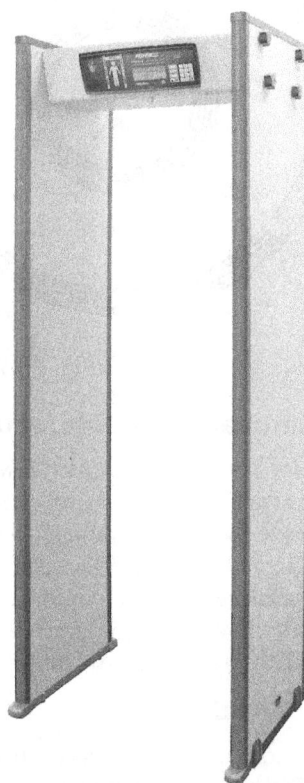

Fig. 23-24: **Fisher Walk-Through**

In a typical ground search metal detector where the TX and RX coils are coplanar sensitivity drops very rapidly with distance from the coils. In a walk-through, the TX coil is in one panel and the corresponding RX coils are in the opposing panels so the sensitivity is better behaved. A target near the TX panel is more strongly energized and easier to see by the RX coil. As the target moves toward the RX side of

---

12. I also designed a walk-through for White's. Even though it performed very well they decided not to enter that market so it was never released.

13. The Garrett *PD6500i* is an example.

Inside the Metal Detector

the gate it is less strongly energized by the TX field but much closer to the RX coil. The result is a modest drop in detectability from the TX side to the RX side.

Normally this would make the walk-through a little weaker on the RX side of the gate. However, most walk-throughs have a TX coil in the left panel with corresponding RX coils in the right panel, plus a TX coil in the right panel with RX coils in the left panel. They are alternately pulsed and their results combined to produce excellent sensitivity near both panels with a reduced sensitivity at the very center of the gate. Therefore the real design challenge is sufficient sensitivity at the center of the gate.

Immunity to mains interference is best dealt with through synchronization. If the TX pulse rate is an exact multiple of the mains frequency then the interference signal is coherently sampled and appears as an offset instead of a signal. Since most walk-throughs are plugged into a wall outlet the mains frequency is conveniently available in the power supply circuitry. It can be extracted and fed to a PLL (hardware or software) to create a higher frequency master clock that is exactly synchronized.

Walk-through detectors typically have to deal with large metal objects placed next to them. It could be a table used for handbag inspection or a large X-ray machine with a motorized conveyer belt. As we've seen, many metal detectors are motion-oriented so that a static object quickly gets tuned out. But that static object is still seen by the preamp and demods and can easily cause a signal overload.

Most walk-through designs use butterfly coils (sideways figure-8) on the RX side. Figure 23-25 shows how the coils might be configured in a cross-fired multi-zone walk-through. When the TXR coil is energized the RXRn coils are used for detection; likewise, the TXL and RXLn coils work together. This might be considered to be a 6-zone walk-through: 3 vertical zones, plus left and right. However, it is possible to interpolate both horizontally and vertically so this coil design would also support a 15-zone walk-through: 5 vertical zones, plus left, middle, and right.

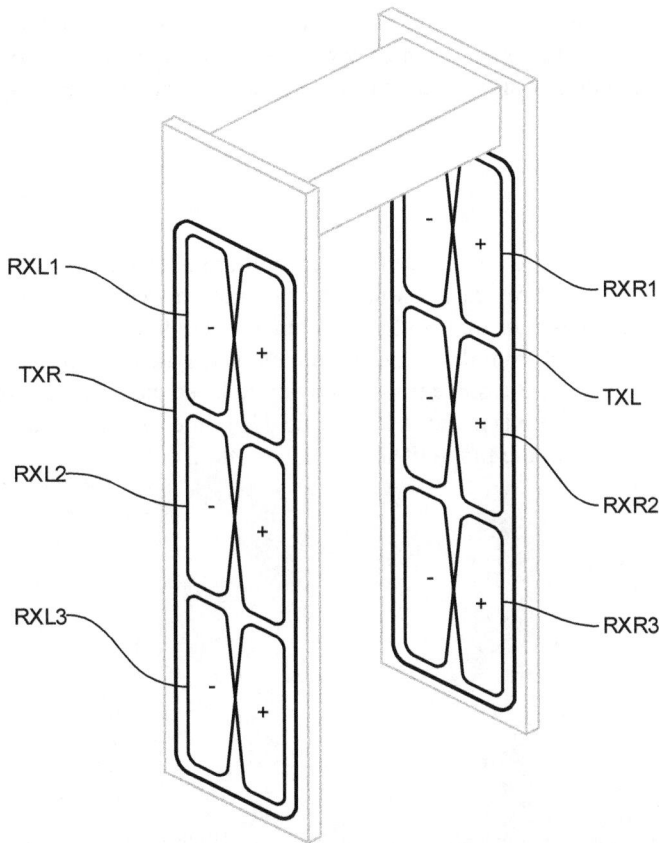

Fig. 23-25: **Walk-Through Coil Configuration**

The sideways figure-8 RX coils have two benefits. First, as we've discussed about figure-8 coils in general, they do a good job at canceling far-field interference. But they also help cancel large static targets, or at least static targets that have a reasonably equal presence on both lobes of the coil. This is because one lobe exhibits a positive detection and the other lobe a negative detection. When they both see an equal target signal at the same time, they roughly cancel[14].

This does not cause a problem for normal target detection. As a person walks through the detection gate the leading lobe of an RX coil will see the target first, and then the trailing lobe. This means that as a target passes through the gate the overall result at the demod output is a positive-negative signal as shown in Figure 23-26. This looks very much like a first derivative signal in a VLF, or the post-SAT signal in a PI. If we take the (negative) derivative of this signal then we end up with a peak response as the target is passing the center of the panel. Everything else is the same signal processing implemented in any other detector.

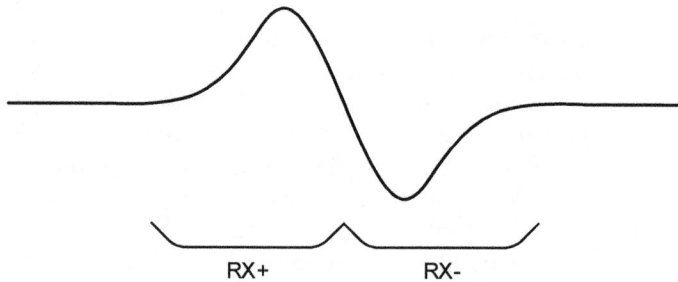

Fig. 23-26: **Raw Target Response**

This all seems simple and it is, conceptually. For best mains and interferer cancelation a bipolar pulse design should be used. This means not only that the panels have to be fired alternately, but that each panel must be fired with a positive and a negative pulse. For example, a firing sequence might be:

- Left-Positive
- Left-Negative
- Right-Positive
- Right-Negative

All while synchronized to the AC mains. Furthermore, there is often a need to run multiple walk-throughs next to each other. To prevent cross-talk interference between walk-throughs it is preferable that only one unit at a time pulses its transmitters; the other units should remain "quiet."

If all the units are powered from the same AC mains then synchronizing multiple units for good behavior is not terribly difficult. But since this effectively reduces the pulse rate of each unit there is a sensitivity penalty. Normally there is no need to provide dead time for more than four units; a fifth unit should be far enough away from the first unit so they don't interfere.

---

14. It is possible to place a large X-ray machine within inches of a walk-through panel without overloading the RX signals. However, an X-ray machine also emits a lot of EMI and that might not be as well-balanced. One of the worst culprits is the motor for the conveyor belt.

PART 6
Advanced Methods

Part 6 presents concepts that are often covered by active patents. US patent law generally permits building a patented circuit for the purpose of evaluating the claims of the patent only. Anything beyond that — even building a circuit for personal use only — is infringement. If you reside outside of the US, a US patent may not apply to you but there may be WIPO patents that do. It is your responsibility to understand and follow all applicable laws regarding patents. Furthermore, any material presented in this section under an active patent is done so merely to explain concepts. It is not intended to be an inducement to infringe. This is why there are no complete buildable circuits included in this section.

Inside the Metal Detector

24

Multifrequency

"Some problems are so complex that you have to be highly
intelligent and well informed just to be undecided about them."

— *Laurence J. Peter*

This section of the book covers "Advanced Methods" which is, of course, somewhat relative. Multifrequency, for example, has been around for a while now but is still considered advanced, at least compared to standard VLF. Although multifrequency didn't arrive on the hobby market until 1991, the technique has been around longer than that. Patent US3012190[1] describes a 2-frequency mine detector. Westinghouse was granted a patent (US3686564) in 1972 for a 2-frequency IB-style metal detector. Bob Gardiner was known to have designed a 2-frequency detector in the 1970s (see US3986104) though it seems he never produced it.

It was Minelab who first brought a multifrequency design to the hobby market masses in 1991 with the *Sovereign*. Fisher also landed their initial multifrequency model, the *CZ6*, later in 1991. In the years since then the multifrequency offerings have slowly expanded, with a more rapid pace of products beginning in 2018 by more companies. In the same 30-year span hundreds of single-frequency VLF models have been released by even more companies so obviously multifrequency is more challenging. A radically new platform like multifrequency takes quite some time to develop and since Minelab and Fisher released their initial products within months of each other, both companies had independently developed their technologies simultaneously. And their approaches could not have been more different.

It is a valuable exercise to investigate these and other designs to see the various ways multifrequency can be achieved, as well as potential trade-offs. We'll start with that, then look at frequency responses of some example targets, and finally cover some of the design methods.

24.1: Overview

A modern single-frequency detector might be described by the block diagram in Figure 24-1. A sinusoidal transmitter (either self-driven or clocked) drives the TX coil, and the RX coil feeds a narrowband preamp followed by quadrature demodulators. The demodulator clocks are typically derived from the TX voltage drive but can also be driven by the micro. The analog signals are digitized and further processing is done in DSP.

Single-frequency designs use a sinusoidal transmitter because it is both simple and power-efficient to resonate a coil with a capacitor, and because the receive signal can easily be demodulated to extract a target phase response. In a multifrequency design, it is possible to time-sequentially resonate a coil with two capacitors[2] but it is exceptionally difficult. Instead, all MF designs to date drive the coil with a digital voltage waveform which results in a coil current (and B-field) consisting of linear ramps. Raw target responses would normally be processed as time domain signals (see the *Ramp Response* section in Chapter 4) but the use of narrowband channel filters and/or narrowband demods converts these signals back to sinusoids, suitable for standard VLF processing.

---

1. This patent was applied for in 1946 but not awarded until 1961; an amazing 15 year-long examination.

2. There is the curious case of patents US5642050 & US5654638 which describe a sequential 4-frequency detector using switched resonant capacitors in the transmitter for maintaining sinusoidal currents. Lore has it that a prototype was built and when it worked it worked amazingly well, but it didn't work very often. The project (called *East Coast Beach*) was canceled because of this. During my time at White's I saw the full *ECB* schematic and most likely it was simply too far ahead of what micros could do at the time.

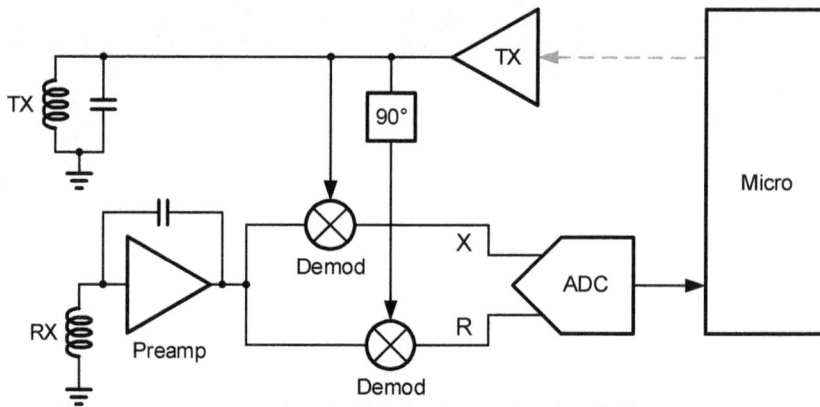

Fig. 24-1: **Single-Frequency Block Diagram**

So a typical two-frequency detector is similar to a single frequency design but with separate analog channels for each frequency. In Figure 24-2 the sinusoidal transmitter is replaced by some kind of 2-frequency transmitter. As we will see later, this is usually achieved with a digital voltage drive clocked directly from the micro. The RX coil again drives a preamp but the preamp must now be wideband. The preamp signal is then split into two channels, one for each frequency. Each channel is first band-pass filtered[3], followed by the normal quadrature demodulators and ADC. The demod clocks are directly driven from the micro instead of being derived from the TX drive.

Because there are now two channels, both channels will need to be calibrated for ground (ferrite). Additionally, specific targets give different phase (and amplitude) responses at different frequencies, so both channels will need to be calibrated so that their responses can be combined (or correlated) in some useful way. Finally, using two frequencies (with enough spread) allows you to use a subtraction method to notch out salt responses. That is, you can simultaneously ground balance to ferrite *and* eliminate salt, which is why multifrequency detectors are preferred when salt water discrimination is needed.

Before diving into any details let's take a look at how some multifrequency designs have been done so far. When looking at a multifrequency design many people make the mistake of probing the TX volt-

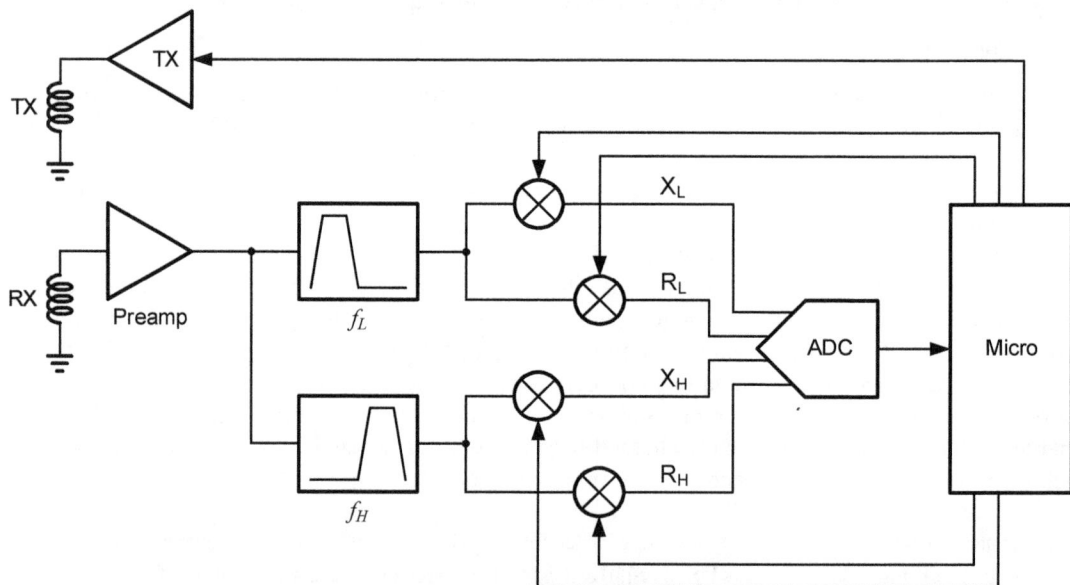

Fig. 24-2: **Two-Frequency Block Diagram**

---

3. We'll see later that some MF designs don't need channel filters.

age waveform and trying to analyze the operation from what they see on the oscilloscope. It is far more useful to look at the TX *current* waveform. Figure 24-3 shows the voltage and current waveforms for the White's *DFX*; the frequencies are clearly easier to visualize when looking at the coil current. There is obviously an overall low frequency, and there are five smaller "ramplet" cycles for each low-frequency cycle, suggesting that the high frequency is five times the low frequency.

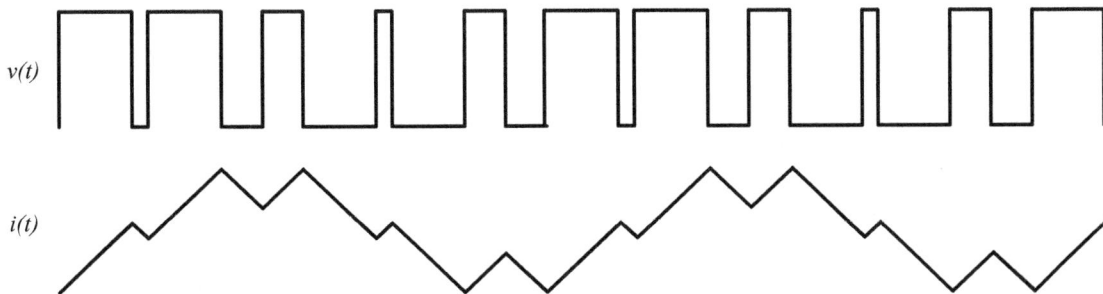

Fig. 24-3: **TX Current vs Voltage (DFX)**

24.2: Case Studies

24.2.1: Fisher *CZ*

The simplest approach to analyze is the Fisher *CZ* platform; see Figure 24-4 for the basic architecture of the *CZ*. The coil is driven with a 5kHz square wave which produces a 5kHz triangle wave current (Figure 24-5). The 5kHz square wave has enough energy at the 3rd harmonic (15kHz) to be a useful second frequency. The spectrum of a square wave is shown in Figure 24-6; the 3rd harmonic is 1/3rd the amplitude of the fundamental.

The received signal is band-pass filtered into the two constituent frequencies (5kHz and 15kHz) which are then demodulated as if they were each single frequency sinusoids (which, at that point, they mostly are). Thus the *CZ* has a continuous-time transmitter, and both frequency channels use continuous-time demodulators to extract amplitude and phase. The whole design is a clear-cut frequency domain approach.

The first *CZ* models (*CZ5, CZ6, CZ20*) were based on a common all-analog design (no microprocessor) and — along with the later *CZ3D* — have been the only all-analog multifrequency detectors ever produced. The *CZ7* added a microprocessor and, in 1997, became the first multifrequency model with a display and multi-notch discrimination.

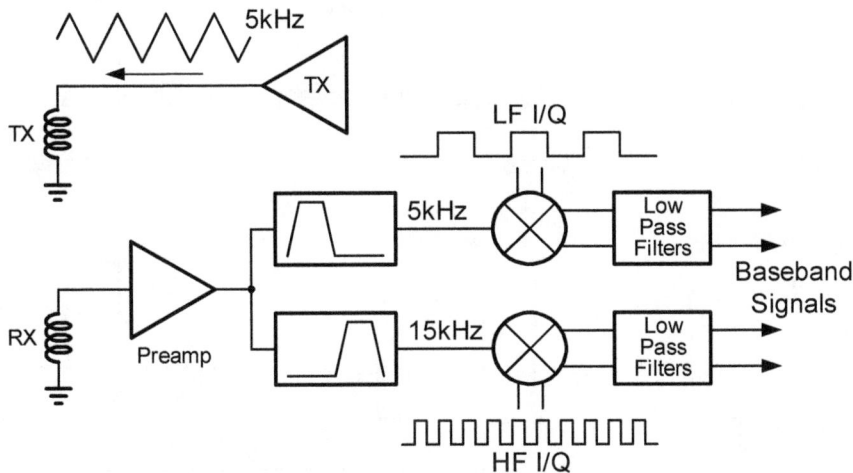

Fig. 24-4: **Fisher CZ**

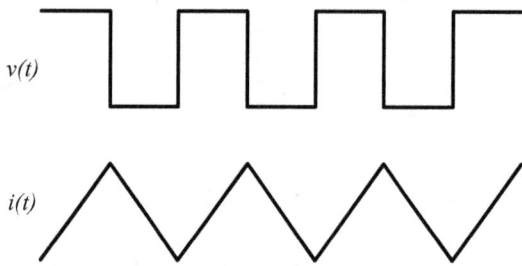

Fig. 24-5: **Fisher *CZ* TX Waveforms**

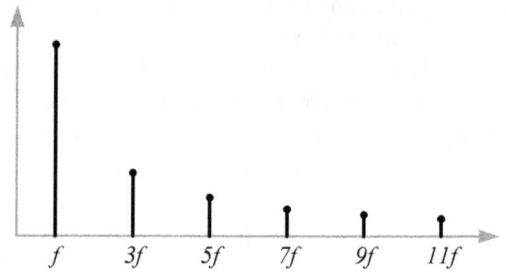

Fig. 24-6: **Square Wave Spectrum**

24.2.2: White's *DFX*

The White's *DFX* approach (Figure 24-7) is similar to the *CZ* except that the transmitted signal (Figure 24-3) is more balanced between the two operating frequencies (3 & 15 kHz); the frequency spectrum is shown in Figure 24-8. The additional high-frequency energy "ramplets" don't make the TX waveform any less continuous-time than with the Fisher and, again, the received signal is frequency-filtered and demodulated to extract amplitude and phase. In all of the examples so far the frequencies are simultaneously transmitted in a continuous manner, so we call this *Simultaneous Multifrequency* (SMF).

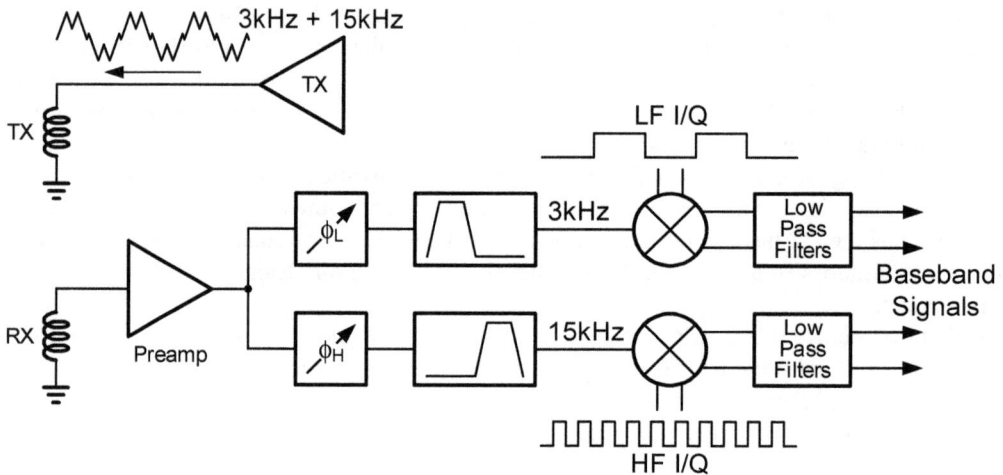

Fig. 24-7: **White's *DFX***

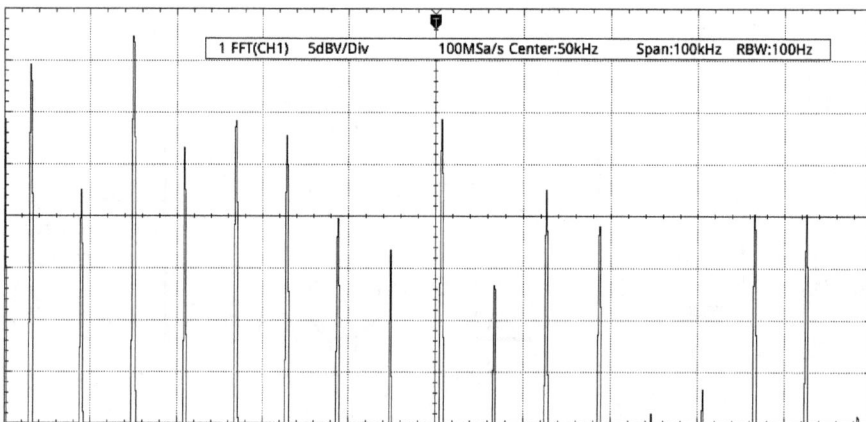

Fig. 24-8: ***DFX* Spectrum**

Single frequency modes are available in the *DFX* (3 or 15 kHz) but the TX continues to use the same multifrequency waveform regardless of mode and the single frequency modes simply ignore the results from the unused frequency. That is, the single frequency modes are not optimized.

A different element of the design shown in Figure 24-7 is the phase rotation technique used to get the demod clock phases correct. In single frequency VLF designs this is accomplished by adjusting the phases of the demod clock drivers (Figure 16-12) or by creating a mathematical G channel. As with any 2-frequency design, the *DFX* has four demod clocks — X and R for the low frequency and X and R for the high frequency. Instead of adjusting all of these, the RX signal in each channel filter is phase adjusted and the demod clock phases remain fixed.

24.2.3: Minelab BBS/FBS

A different MF technique is found in Minelab BBS & FBS[4] detectors whereby two frequencies are transmitted in a time-sequential manner. The TX produces a single cycle of 3.125kHz followed by eight cycles of 25kHz; this pattern is then repeated (Figure 24-9). It is useless to look at the spectrum of this waveform because it is time-sequential; we will instead treat it as two ordinary triangle waves. Because both frequencies have the same slew rate, the amplitude of the 25kHz cycles is 1/8th that of the 3.125kHz cycle. This reduces sensitivity so to help make up for that the 25kHz cycle is repeated eight times. This certainly improves the situation but it is not a complete fix, and while the BBS/FBS models (*Sovereign, Explorer*) were widely known for their excellent sensitivity to deep silver (due to the dominant 3.125kHz cycle) they were equally known for being somewhat lackluster on small jewelry because of the weaker 25kHz cycle. The later *E-Trac* model (called "FBS2") did show some improvement to small gold sensitivity, still using the same TX waveform.

Since the frequencies are time-sequential the demodulators must also be clocked in a time-sequential manner (Figure 24-10). Although the TX waveform is clearly continuous-time, because of its time-sequential nature the demods cannot possibly run in a continuous mode. Instead, the demods are operated in discrete time. Temporal separation of the demodulation eliminates the need for band-pass channel filtering. So BBS/FBS uses a continuous-time transmitter, a discrete-time receiver, and (likely) time-domain signal processing. Because the Minelab approach transmits the frequencies sequentially, we call this *Sequential Multifrequency*[5].

The overall architecture of BBS/FBS is shown in Figure 24-11. The elimination of channel filtering makes the circuitry simpler than the frequency domain methods of Fisher and White's but the timing (for both TX and demods) is more difficult. Since timing is provided by a timer on a micro this is really not much of an issue.

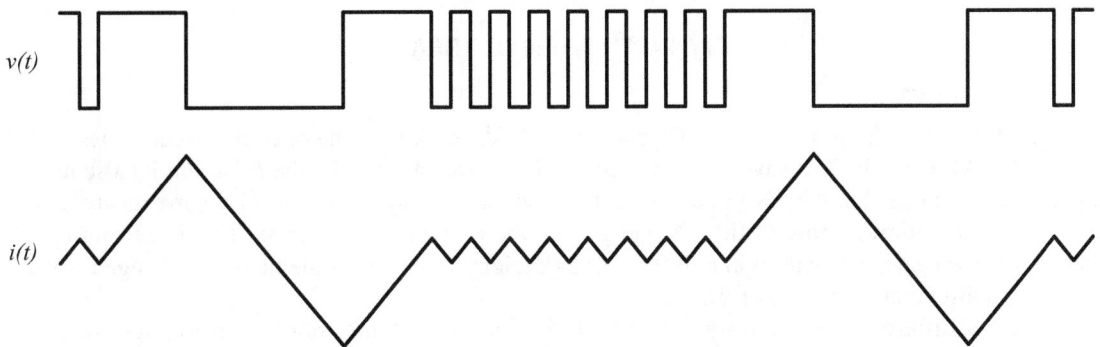

Fig. 24-9: **BBS/FBS TX Voltage & Current Waveforms**

4. BBS & FBS are *broad band spectrum* and *full band spectrum,* Minelab's marketing terms for what they described as 17 frequency and 28 frequency designs. Ironically, both methods use the exact same 2-frequency transmit signal of 3.125kHz and 25kHz.

5. Unfortunately, sequential multifrequency and simultaneous multifrequency have the same acronym, SMF. Most people already use SMF to mean simultaneous multifrequency so for sequential multifrequency we will use the acronym SQMF.

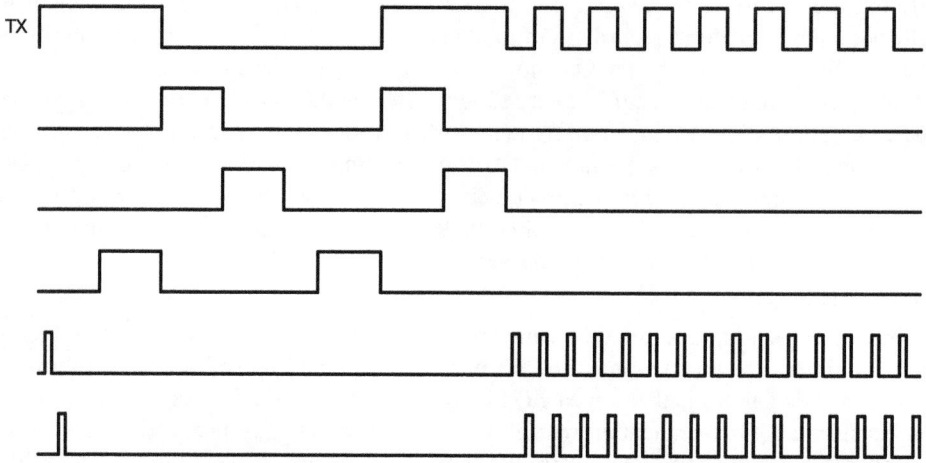

Fig. 24-10: **BBS (*Sovereign*) Demod Timing**

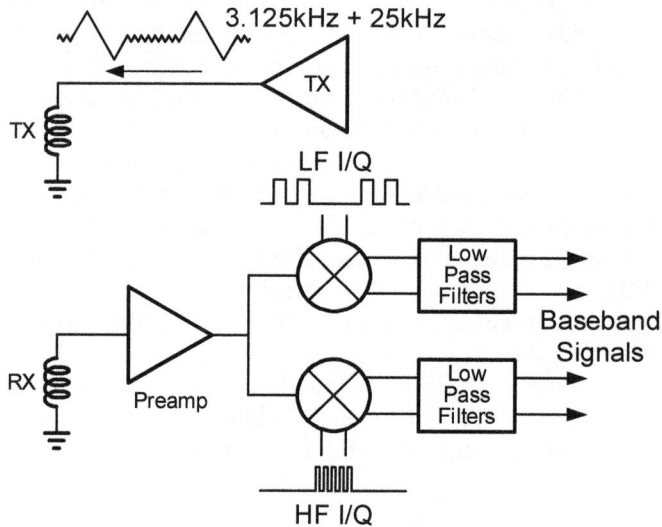

Fig. 24-11: **Minelab BBS/FBS**

24.2.4: White's *V3*

The White's *V3* is similar to the *DFX* but adds a third frequency. The operating frequencies are 2.5, 7.5, & 22.5 kHz and the TX waveforms[6] are shown in Figure 24-12. Like the *DFX*, the *V3* also has single frequency modes but the TX voltage waveforms are simple square waves (TX currents are triangle waves) which optimize more of the TX energy into the frequency of interest. This is a benefit of the digital TX drive commonly used in multifrequency designs — it is a simple matter to change the waveform and substantially alter the operation.

Another difference between the *V3* and *DFX* is that the *V3* does not bother to phase-rotate the demod clocks to align them to the analog signal phases. Rather, phase rotation is done digitally after the X and R signals are digitized and sent to the micro. The digital rotation constants are calibrated in the factory. There is a drawback to this approach — if the R channel demod clocks are very far from the ideal ground null then the R channels can easily overload in hot ground. Therefore it is best to get the demod clock phases at least somewhat close to ground null and then use digital rotation for the remainder.

---

6. Note the pronounced curvature of the TX current waveform. This is due to TX coil resistance.

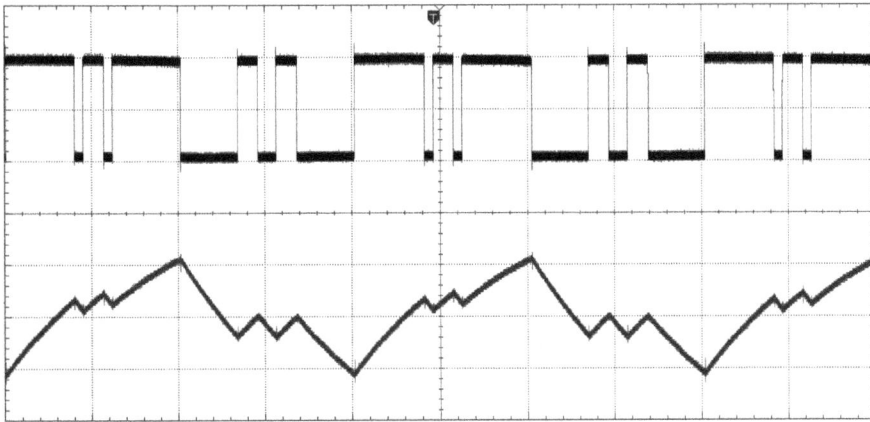

Fig. 24-12: *V3* TX Voltage & Current Waveforms

24.2.5: Minelab *Equinox*

The Minelab *Equinox* transmits a waveform similar to the *V3* but eliminates the analog filtering & demodulation. Instead, it uses a fast precision ADC to directly sample the preamp output, and demodulation and filtering are performed in the microprocessor DSP. This offers more flexibility in the choice of frequencies, filtering, and demodulation schemes.

The Park-mode TX waveforms are shown in Figure 24-14. It consists of 2.6kHz, 7.8kHz, and 39kHz for 3-frequency operation with a 1:3:15 ratio. There are also Beach and Prospecting multifrequency modes with different frequencies and weightings. The *Equinox* also includes single frequency modes (5, 10, 20, & 40 kHz) which, like the White's *V3*, use more optimized square wave voltages. The *Equinox* goes a step further in that the TX voltage is progressively boosted for higher frequencies resulting in a peak coil current that is fairly consistent, at least up to 20kHz. The 40kHz mode is not boosted as much.

On the receive side, the design is hidden in firmware making it difficult to offer any more details, but we can certainly speculate. As with an analog multifrequency design, the Equinox must separate the frequencies and demodulate them. This could be done by using a digital band-pass channel filter followed by a wideband demodulator using +1/−1 multiplication. However, in moving the demods to the digital domain it makes more sense to take advantage of techniques that are better suited for software methods.

One of these is a narrowband demodulator[7] which eliminates the need for channelization filters. Instead of multiplying the input signal by +1/−1, the narrowband demodulator multiplies the input sig-

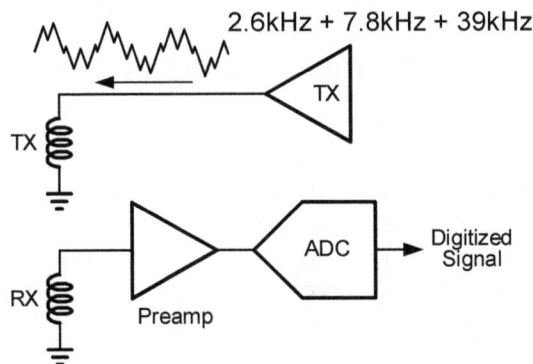

Fig. 24-13: **Minelab Equinox**

7. The assumption that Equinox uses narrowband demods is predicated on patent US7579839. It's also likely that the Minelab X-Terra direct-sampling single frequency designs did the same.

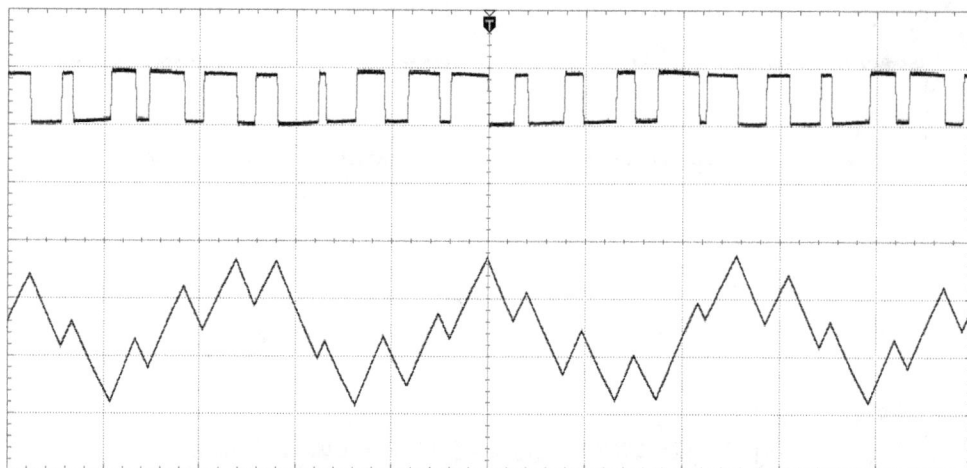

Fig. 24-14: *Equinox* **Park Waveform**

nal by $\sin(2\pi f t)$ where f is the channel frequency of interest. In metal detectors we always have a pair of quadrature demodulators for extracting target phase. In using narrowband demodulators we clock one demodulator with $\sin(2\pi f t)$ and the other with $\cos(2\pi f t)$ to give us the desired quadrature outputs. These outputs are then low-pass filtered (most likely with IIR filters) and sent for further processing. Figure 24-15[8] shows how this might look for the Park program.

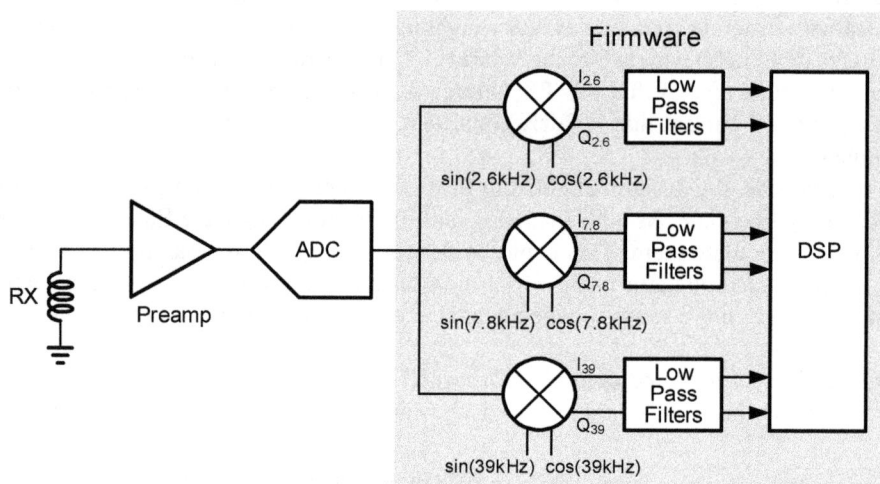

Fig. 24-15: **Equinox RX**

24.2.6: Boosted Sequential Multifrequency

Patent US11022712 offers a glimpse at an improvement to sequential multifrequency. A drawback of all the MF methods mentioned so far is that the transmitter is driven off a single fixed voltage[9]. This limits the triangle/ramp sections to a single slew rate. Target response depends on dB/dt so if the same slew rate always occurs even in a multi-frequency detector then the target response variation is limited. The multifrequency method in US11022712 uses a multi-slew rate transmitter with the different slew rates transmitted in a time-sequential manner. So, like Minelab's BBS/FBS, this method is also sequential multifrequency.

---

8. A little shorthand was used, where $\sin(2.6\text{kHz})$ really means $\sin(2\pi \cdot 2600t)$.

9. The single frequency modes of the Equinox do scale the TX voltage, but not the multifrequency modes.

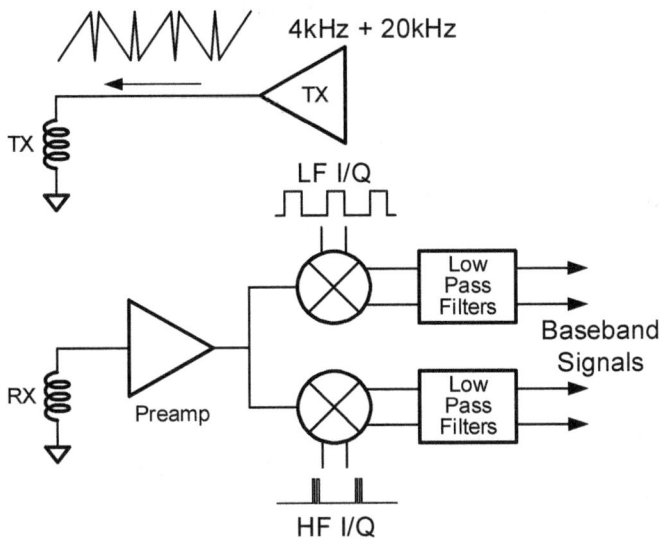

Fig. 24-16: **Boosted SQMF**

The challenge of this approach is in the design of the transmitter. Preferably, all frequencies should have the same peak-to-peak current which means that the TX supply voltage must be proportional to the frequency. Suppose that you want to transmit 4kHz, 16kHz, and 64kHz, or a 1:4:16 frequency ratio. This means that the supply voltage ratio must also be 1:4:16; for example, 4V, 16V, and 64V. Fig. 12 in the patent suggests a way to do this but the devil is in the details, and the patent devilishly does not offer those details.

Besides an improved target response, boosted SQMF also improves depth at higher frequencies because the transmitted peak current remains constant[10]. Another improvement is that it is much easier to alter the weighting of the frequencies for particular types of hunting (Fig. 7 in the patent). For example, for nugget hunting you may want to transmit 64kHz most of the time, with an occasional 4kHz ramp thrown in for ground (or iron) analysis.

24.3: Multifrequency Signals

In order to design a multifrequency detector it is important to first understand how target responses behave for different frequencies. This is especially important for ground signals, including salt. This section will explore those responses.

We will consider only the case of frequency domain multifrequency designs, whereby targets are characterized by amplitude and phase. Chapter 4 briefly covered alternative time-domain responses and an MF design can certainly use that approach, where understanding the design requirements would require characterizing target responses in the time domain. For brevity we will focus on frequency domain and realize that a time-domain design would be of a similar nature.

24.3.1: Measuring MF Target Responses

The frequency response of a target is not a straightforward request. If a fixed-inductance TX coil is driven with a fixed-amplitude sinusoidal voltage then the transmitted field strength will decrease as frequency increases because the reactance of the coil increases. This could affect the target response, especially with matrix targets like ground or salt which can have a strong three-dimensional effect. A better solution is to drive the TX coil with a fixed current amplitude over frequency to ensure the target always sees the same field strength. This can be accomplished either by increasing the TX voltage drive with increasing frequency or by driving the coil with a constant-current drive circuit such as a Howland current pump.

---

10. This puts extreme demands on the high frequency performance of the search coil.

At the same time, a fixed-amplitude target response imposed on a fixed-inductance RX coil results in an induced voltage that increases with frequency due to the derivative function (Equation 7-1). We would prefer an RX signal that is directly proportional to the target signal regardless of frequency. This can be achieved by using a current-mode preamp as is used in the Magnetic Field Probe in Appendix A. This, along with a constant-current drive, produces target RX signals which have no first-order circuit-induced frequency variations.

Once we have a preamp signal, it can be quadrature-demodulated in the usual VLF manner to extract a reactive (X) and resistive (R) signal for calculating amplitude and phase:

$$A = \sqrt{X^2 + R^2}$$

$$\phi = \tan^{-1}\left(\frac{R}{X}\right)$$

In a real metal detector design careful attention is given to the gains and bandwidths of the various stages to ensure they don't overload and to obtain good target response speed. In a test circuit like this, we can make the demod circuitry intentionally slow with far more gain to better see small signal responses with less noise.

24.3.2: Ground Response

The ground response is due to magnetic minerals in the soil which, for most average soils, are non-viscous (basically magnetite). Some extreme soils can have viscous minerals and some soils can have conductive salts (notably beaches[11]). The salt response is considered separately in the next section.

Table 24-1 shows the responses[12] of ferrite (magnetite) and a viscous "hot rock." The ferrite response over frequency is very consistent, both in amplitude and phase, whereas the viscous hot rock shows a slightly increasing phase angle but a slightly decreasing amplitude.

f	Ferrite		Hot Rock	
	Amp	Phase	Amp	Phase
1kHz	1.01	0°	1.20	3.7°
2kHz	1.03	0°	1.13	4.1°
5kHz	1.02	0°	1.03	5.0°
10kHz	1.00	0°	1.00	5.9°
20kHz	1.02	0°	0.93	6.6°
50kHz	1.02	0°	0.91	7.8°
100kHz	1.03	0.2°	0.94	8.9°

Table 24-1: **Ground Responses**

24.3.3: Salt

Like ferrite mineralization, salt water has almost no phase change over frequency, remaining very close to 90° from 1kHz to 100kHz. However, the amplitude response is roughly proportional to frequency. Figure 24-17 shows the amplitude response versus frequency (all phases are approximately 90°). This is why an extremely low single-frequency VLF (like the 2.4kHz Fisher *1280X*) can work reasonably well in salt water; the low frequency is less sensitive to the salt.

24.3.4: Iron Responses

An iron target response can vary quite a bit depending on whether it is a small piece that is predominantly magnetic in nature or whether it is larger and has a significant eddy component. Figure 24-18 shows the frequency vectors[13] for a square nail with vertical and horizontal orientations. A 3/8-inch

---

11. Although heavily fertilized soils can contain conductive ionic compounds, i.e., salts. Many desert soils are also high in concentrated salts due to evaporation.

12. Normalized to the 10kHz response.

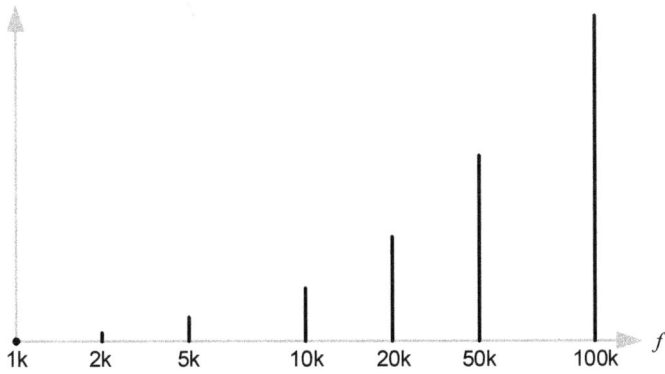

Fig. 24-17: **Salt Response**

steel washer is shown in Figure 24-19. The vertical nail is mostly magnetic with little phase change but a decreasing amplitude as frequency increases, opposite that of pure eddy targets. The horizontal nail also has a decreasing amplitude as frequency increases but with far more phase change due to the larger eddy component. The steel washer has an increasing amplitude with frequency (owing to its large eddy component) and phase increases with increasing frequency much like that of a pure eddy target but enters into the ferrous quadrant at the lowest frequencies.

The responses shown in these plots were all measured with the target statically placed in the center of the coil. In Chapter 5 it was also shown that ferrous target responses can vary positionally under the coil, not only with target strength (as shown, for example, in Figures 5-6 and 5-7) but also with phase. In other words, the amplitude and phase of a ferrous target not only varies with frequency, but even positionally as the coil sweeps over it. All of this shows why multifrequency is so valuable in identifying ferrous targets as their behavior is often completely at odds with well-behaved eddy targets.

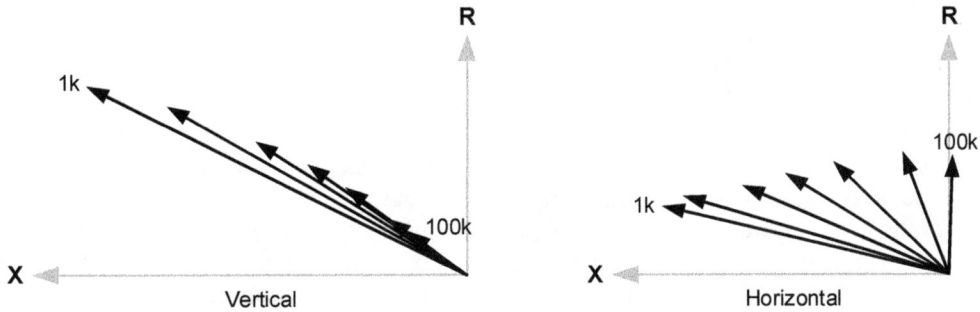

Fig. 24-18: **Iron Response: Square Nail**

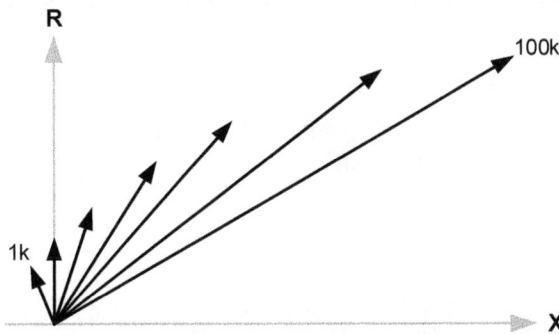

Fig. 24-19: **Iron Response: 3/8-inch Steel Washer**

---

13. The next several vector plots include frequencies of 1, 2, 5, 10, 20, 50, and 100 kHz.

24.3.5: Eddy Responses

Single-domain eddy targets (most coins and rings) follow a fairly consistent frequency response whereby both the amplitude and the phase asymptotically increase toward a limit at 180°. This was discussed and illustrated in Section 16.11. Figure 24-20 shows the measured responses[14] of a US nickel and quarter with the trend lines added. The nickel (cuponickel) is considered a "low conductor" coin while the quarter (silver) is a "high conductor" and this is reflected in the fact that the 1kHz quarter response has a higher phase shift than the 10kHz nickel response[15]. But in both cases, the frequency responses trace out a semicircle from (0,0) for a DC response to some maximum amplitude that approaches 180° at very high frequencies.

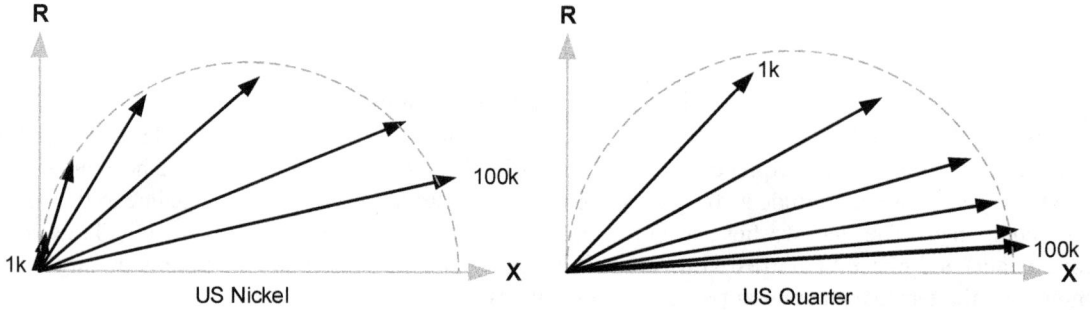

Fig. 24-20: **Measured Eddy Responses**

24.4: Multifrequency Processing

This section will discuss methods and characteristics of multifrequency processing. It will assume the use of two frequencies and in the examples a simultaneous 3kHz and 15kHz will be used. However, the discussions are equally valid for other frequencies, for more than two frequencies, and for sequential frequencies.

24.4.1: Correlated Processing

Suppose we have a two-frequency design as in Figure 24-2. Each frequency is separately demodulated, resulting in its own X and R signal components. Therefore, each frequency channel could be treated as a separate single frequency detector; each can be separately ground-balanced and the X,R results can be processed to produce a target amplitude and phase.

The results from the two frequencies can then be compared to determine if a target produces results that are consistent with some kind of expected behavior. For example, we know that well-behaved eddy targets (such as most coins and rings) have a phase shift that increases with frequency:

$$\phi = \tan(2\pi f \tau) \qquad \text{Eq 24-1}$$

If, for example, we use 3kHz and 15kHz and a target produces the following phase shifts:

$$\phi_{3k} = 160°$$

$$\phi_{15k} = 176°$$

then we can conclude that the responses are likely from an eddy target with a tau of 146µs. In the US this might be a silver or clad quarter. On the other hand, suppose the target phase shifts are:

$$\phi_{3k} = 160°$$

$$\phi_{15k} = 161°$$

---

14. Measuring the frequency responses of targets is a tricky job, and there can be a fair amount of measurement error which is why the responses don't exactly follow the trend lines.

15. It should also be noted that these plots have been scaled to approximately the same size. In reality, if both coins are at the same depth, the quarter responses are somewhat stronger. See Figure 16-30 for a more realistic comparison of signal strengths.

A 3kHz single frequency detector would still assume this is a US quarter and a 15kHz detector might call it a zinc cent but a multifrequency design should recognize it as more likely a ferrous-eddy target since the phases don't follow a typical eddy curve.

This type of processing is called *correlation* because we are correlating the results of two or more frequencies. Other types of responses can also produce correlated results that are distinguishable. Normal ground tends to have a magnetic response with little amplitude or phase change over frequency. Viscous ground and hot rocks have little amplitude variation and phase that is constant or increases slightly with frequency. Salt tends to have almost the same phase response versus frequency but with an amplitude that is close to being proportional to frequency.

Iron responses vary depending on the nature of the iron; that is, whether it is primarily magnetic, primarily eddy, or a combination. Primarily magnetic iron targets tend to have little variation in phase response versus frequency but the response amplitude drops with increasing frequency. Eddy iron targets (especially steel bottle caps) can behave more like coins. Combination magnetic-eddy iron targets can be all over the place but often exhibit a noticeably non-eddy behavior like the washer response in Figure 24-18. And combination responses can often vary with depth as the magnetic and eddy proportions vary, or as the reactive ground signal overwhelms the reactive target signal. See Chapter 5 for further discussion.

Correlated homogeneous processing is, conceptually, the simplest approach in multifrequency design because it directly builds on what we already know about single frequency designs. However, it requires simultaneously ground balancing two channels and in modern designs this includes ground tracking. But an advantage of this is that there are two all-metal G signals to consider, assuming salt notching is not needed. In some cases a particular target may have little or no response at one frequency but a good response at the other frequency. For example, small jewelry might respond well at 15kHz but not at 3kHz. So two frequencies offer a better response to a broader range of conductors. And for use in saltwater environments, a simple weighted subtraction of the G signals nulls the response of salt.

24.4.2: Salt Notching

Salt has almost the same phase response versus frequency (about 90°) but an amplitude that is roughly proportional to frequency. In a system where each frequency channel has a ground-nulled G signal we can notch out salt with some simple mathematics, either in hardware or software. In Figure 24-21, a 15kHz G signal is subtracted from a 3kHz G signal (with the signals weighted 5:1, same as the frequency ratio) to produce a G_{Salt} signal that is substantially free of a salt response. Each channel's G signal is ground balanced prior to subtraction by either adjusting the RCLK phase (Figure 16-25) or doing some mathematical adjusting (Figure 16-26, and implied in Figure 24-21) to null ground. The resulting G_{Salt} signal is simultaneously balanced to ground and to salt and this G_{Salt} signal is now used for targeting.

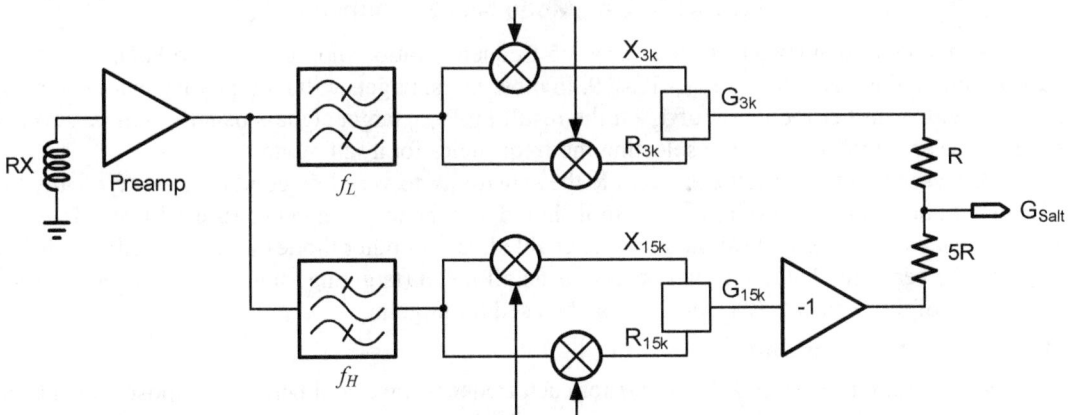

Fig. 24-21: **Salt Subtraction**

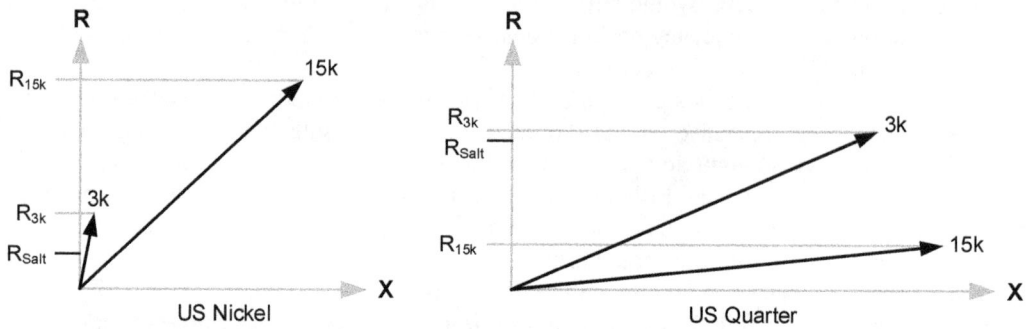

Fig. 24-22: **Salt Mode Eddy Responses**

The processing that creates G_{Salt} also affects how targets are seen. Figure 24-22 shows the individual frequency vectors for a US nickel and quarter with their respective R values. If we pretend for a moment that it is the target's R signals that are subtracted (very close to reality) then the composite R value that is our new targeting signal is $R_{Salt} = R_{3k} - R_{15k}/5$ and is marked on the R-axis. From this we can see that the quarter salt response is very close to the 3kHz response, but the nickel salt response is greatly attenuated. This is true for all low conductors where the f_L phase response approaches 90°; that is, notching salt at 90° also attenuates the composite responses of any target that is close to 90°. This is similar to the effect of how TR discrimination attenuates targets near the discrimination point; see Figure 14-17.

Ferrous target behave similarly; the square nail salt mode responses are given in Figure 24-23. Ferrous targets with a dominant magnetic response (like nails) will have a smaller high frequency magnitude so the composite salt response is usually a little lower than the f_L response. Ferrous targets with a large eddy component can have significant variability in the composite salt response.

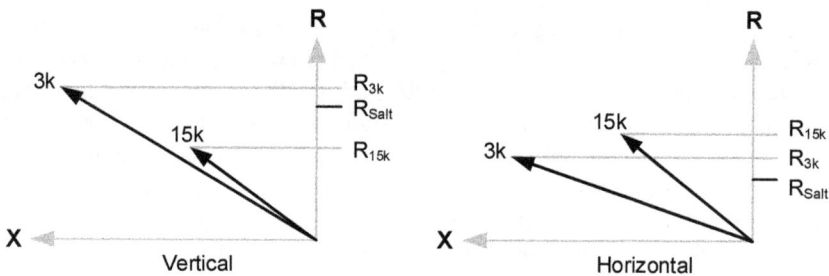

Fig. 24-23: **Salt Mode Nail Responses**

This example used 3kHz and 15kHz for a 1:5 frequency ratio. Suppose we used 3kHz and 27kHz for a 1:9 ratio. In this case, $R_{Salt} = R_{3k} - R_{27k}/9$. In most cases, targets will naturally have a weaker R_{27k} signal and with a further reduction of 1/9th the resulting R_{Salt} target signals usually will be stronger. This might be an important factor in selecting the frequencies for a salt mode.

As we've seen, notching salt can degrade the sensitivity to very low conductors — typically thin foil trash (☺), but also tiny jewelry (☹) — so it should only be used when necessary. Most MF models have a "beach" mode whereby salt notching is enabled, and in other modes it is disabled. If salt is an issue, then the beach mode G_{Salt} signal is used as an all-metal (targeting) signal. If salt is not an issue, either or both of the G signals (G_L and G_H) can be used for targeting.

24.4.3: Composite Processing

Instead of creating separate X,R signals for each frequency, we could create composite signals that combine the demodulator results of each frequency. In the previous section a G_{Salt} signal was created by mathematically combining the G signals from each frequency channel. This is an example of a composite signal. Even the G signal itself (as created in Figure 16-26) is a composite signal.

Suppose we have well-behaved ground that has a consistent amplitude and phase at our two operating frequencies. We simply subtract the demodulator outputs like this:

$$X_C = X_{15k} - X_{3k}$$
$$R_C = R_{15k} - R_{3k}$$

where X_C and R_C are now *composite* signals. Because ground has the same X and R response at both frequencies, both of these composite signals will be zero and the result should be a ground-free signal on both the X and R channel. If the ground responses are not exactly the same then a small multiplier can be computed to produce nulled composite signals.

The subtraction required to null the ground response also affects all targets. Eddy targets have a larger magnitude and greater phase shift at the higher frequency so subtracting the lower frequency from it results in a composite vector that will always have a positive X_C, but R_C can be positive or negative. Figure 24-24 shows the results for a US nickel and quarter. Note that the composite quarter response is in the third quadrant which is normally devoid of target responses. This means that we will need to process responses throughout the second and third quadrants, but since processing is normally done in software this is a simple matter.

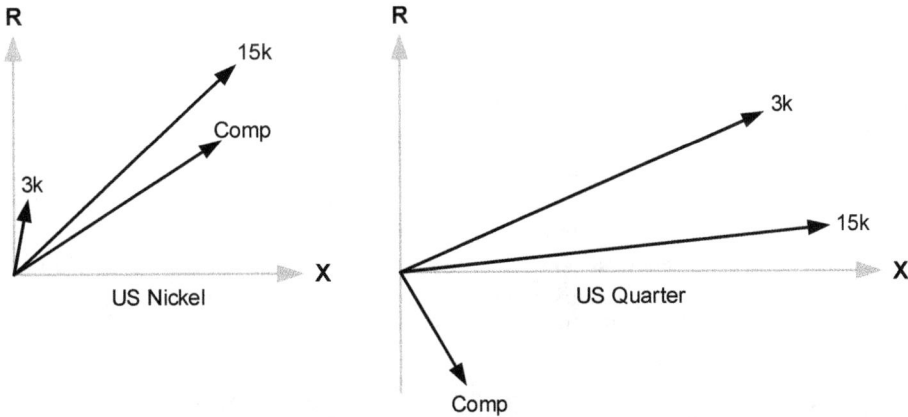

Fig. 24-24: **Typical Composite Eddy Responses**

Composite ferrous target responses are similar in nature but with a twist. Since the ferrous response magnitude typically decreases with increasing frequency, it means that the composite response will usually end up in the second or third quadrants, right along with non-ferrous targets making it impossible to distinguish non-ferrous from ferrous using only the peak composite response. However, we still have the responses of the individual frequencies to work with, although they can be contaminated with ground signal. For example, we could compare the second derivative X responses and negate X_C if $|X_L| > |X_H|$. And we also know that many ferrous targets are ill-behaved as the coil sweeps over them, which allows us to look at composite response consistency on either side of the peak response.

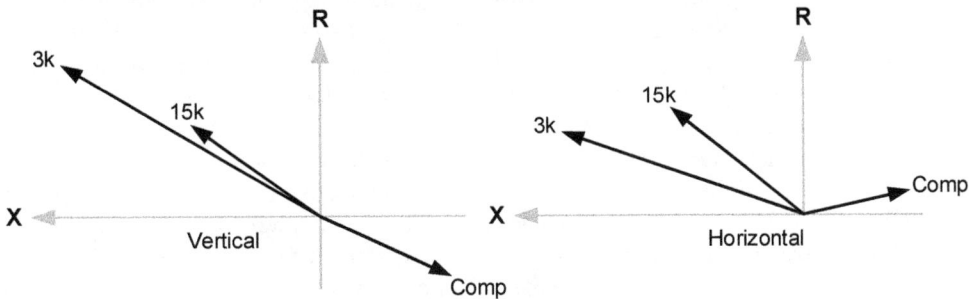

Fig. 24-25: **Typical Composite Nail Responses**

In both the ferrous and non-ferrous cases we see composite signals that can span the second and third quadrants. This suggests that there are targets that can have a purely reactive composite response; that is, exactly on the +X axis. In single-frequency VLF designs, a ground-free R signal (or G signal) is used as the targeting signal. That is, it is the signal we watch to determine when any target — ferrous or non-ferrous — is present. Since it's possible for a composite signal to have a zero or near-zero R_C response this means the R_C signal is no longer a good choice for targeting. The X_C response becomes the better signal to use for targeting because it is ground-free and the only target with a zero X_C response is salt, which we don't want to detect anyway.

There is a major drawback in using X_C for targeting. In more highly mineralized ground the X_L and X_H signals can be large and subject to slight variations in relative amplitudes. Subtracting two large signals makes the result sensitive to those slight variations which can produce a noisy X_C signal. Since it's now the targeting signal, this results in a noisy audio. While this might work in milder soils it is unlikely to be tolerable in more highly mineralized ground.

Another method of composite processing involves only the R signals. From Figure 24-20 we can see that single domain eddy targets (especially coins) have a predictable frequency response that follows a semicircle. We can therefore look only at the ratio of R_{3k} and R_{15k} to determine the tau of the target. The advantage to this is that no ground-contaminated X signals are involved. The drawback is that iron targets may get misidentified as non-ferrous.

24.5: Multifrequency Efficiency

Regardless of which TX method is used, MF has an inherent inefficiency compared to single frequency designs. In MF, the total energy transmitted by the search coil is divided amongst the number of frequencies being transmitted. This is true whether the frequencies are sent sequentially or simultaneously. It is a more obvious case for SQMF because one frequency is completely suppressed while the other frequency is transmitting. If each frequency is transmitted with a 50% duty cycle (as is the case with Minelab's BBS/FBS) then each has one-half the energy that could be transmitted with a single frequency design. Of course, the duty cycles could be skewed so that one frequency is favored over the other, but an individual frequency will still never equal what can be done with a dedicated SF design. In SMF the frequencies are transmitted simultaneously but the total TX energy is still split between the frequencies. Ideally the split is 50-50 but, again, the waveform timing can be skewed to favor one frequency over the other.

Besides the fact that MF divides the total TX energy between the frequencies, there is an additional inefficiency in that the current waveform is a triangle wave (or ramplets) which is rich in harmonics. Some of the TX energy ends up in the harmonics which, in a frequency-domain design, are removed by the channel filters and is therefore wasted. Although a very minor loss, a pure sinusoidal SF design does not share this inefficiency.

It would be easy to say that, apples for apples, a SF design of, say, 15kHz will always beat an MF design of 3kHz+15kHz for all targets that respond best at 15kHz. This seems obvious for a system that uses homogeneous processing where each frequency is processed as if it were a separate metal detector. With composite processing, the act of creating a ground-free composite signal may result in common-mode noise subtraction that produces a higher SNR. It also means that the signal can then be gained up more significantly without the risk of headroom problems. This was also done with the G signal in Chapter 16. The same is true with the creation of the salt-free composite signal; with the (potentially) large salt signal removed, and assuming that in most beach conditions the mineralized ground signal is also not significant, the composite salt signal can run at a higher gain.

If the composite signals can be run at a higher gain without overloading then it's possible that the overall sensitivity will be higher than a detector in which either ground or salt signals limit the maximum gain. Whether this is true depends on the design details. For example, in a direct sampling design the RX signal is digitized before demodulation so that the creation of composite signals does not offer any headroom advantages. That is, once the signals are digitized the SNR is pretty much set, therefore a single frequency mode should beat a multifrequency mode for optimal targets, everything else equal.

24.6: Multifrequency Design

A multifrequency design is an ambitious project and a complete working design beyond the scope of this book. However, we will step through the various analog elements of an MF design and try to explain some of the requirements and potential pitfalls. This exercise will assume a traditional SMF approach using two frequencies and analog demodulators, much like the *DFX*. Deviations for other designs will be noted.

24.6.1: Transmitter

Many MF designs use a full H-bridge drive similar to the right-hand circuit in Figure 23-2. Figure 24-26 jumps straight to a useful design. Compare this circuit to the PI full-bridge driver in Figure 23-7. Flyback diodes D1 & D2 are not needed because there are no flybacks; the coil is always driven and is never switched to an 'off' state. With no flyback there is no need for a damping resistor and the FET switches do not need to be rated for high voltages. With no 'off' state the opposing high-side and low-side devices can be directly switched together.

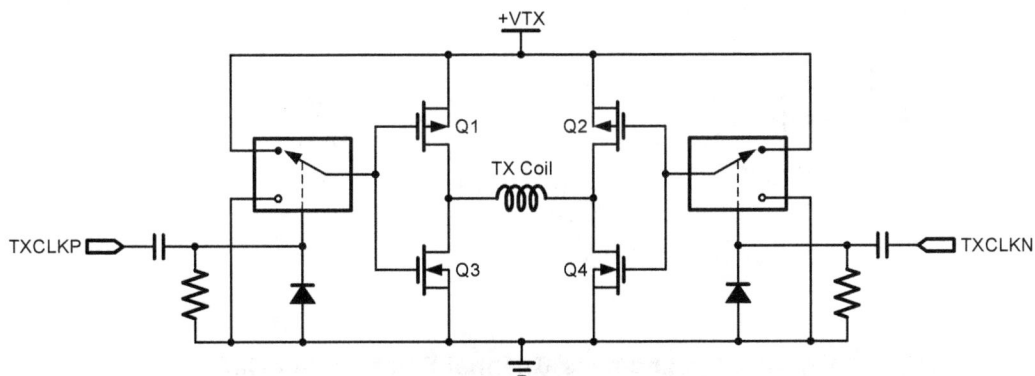

Fig. 24-26: **TX Circuit**

It is assumed that the TX clocks will be provided by a micro and the 4053-type switches provide switching thresholds that will work for 3V logic (with 74HCT technology). In the various PI designs we took care to include a lock-up protection method so if the clocks got stuck the TX circuit would not self-destruct. That is also needed here and component values will depend on the lowest TX frequency used. This same design is also usable for unboosted SQMF (BBS/FBS) but not boosted SQMF.

Choosing the timing for the TX waveform is not so straightforward. It is best done through trial-and-error, either looking at an FFT of the drive waveform (as in Figure 24-8) to determine relative frequency strengths, or by testing the responses of various targets to get a desired balance.

24.6.2: Preamp

The RX coil and preamp design is a simple wideband non-inverting amplifier. The RX coil is shown in Figure 24-27 with load resistor R1. In single-frequency designs this resistor can sometimes be omitted but in multifrequency it acts as a damping resistor to prevent high-frequency ringing from the induced step responses targets can produce (and also an imperfect coil null). Although R1 is assigned a value of 1k it may need to be adjusted depending on the design of the RX coil.

As in a SF design, the gain of the preamp should remain modest to avoid an overload condition from either ground signals or strong surface targets. This will need evaluation at all frequencies. For an SMF design the preamp may either be

Fig. 24-27: **Preamp Circuit**

completely wideband (to the limit of the opamp) or band-pass restricted but with a sufficient bandwidth to pass all desired frequencies. For SQMF — and especially when using time-domain demodulation —

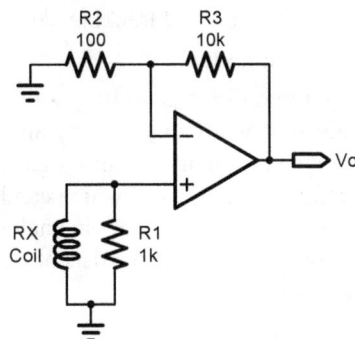

the preamp is normally run wide open to minimize transient artifacts. The opamp will need a sufficient GBW to pass the highest operating frequency with the desired minimum amplitude and phase changes.

24.6.3: Channel Filters

Prior to demodulation the wideband output of the preamp must be filtered into two narrowband channels. This effectively extracts the low and high signal-frequency sinusoidal responses from the single wideband response. The filters obviously need to be band-pass and for best results (with both ground balance and target ID) crosstalk should be minimized. The *DFX* uses a simple single-stage band-pass filter but because the frequencies have a 1:5 ratio this does not provide a lot of isolation. The Fisher *CZ5* uses a more complex multi-stage filter (the 5kHz channel filter[16] is shown in Figure 24-28), including a twin-T notch filter to better suppress the 15kHz component.

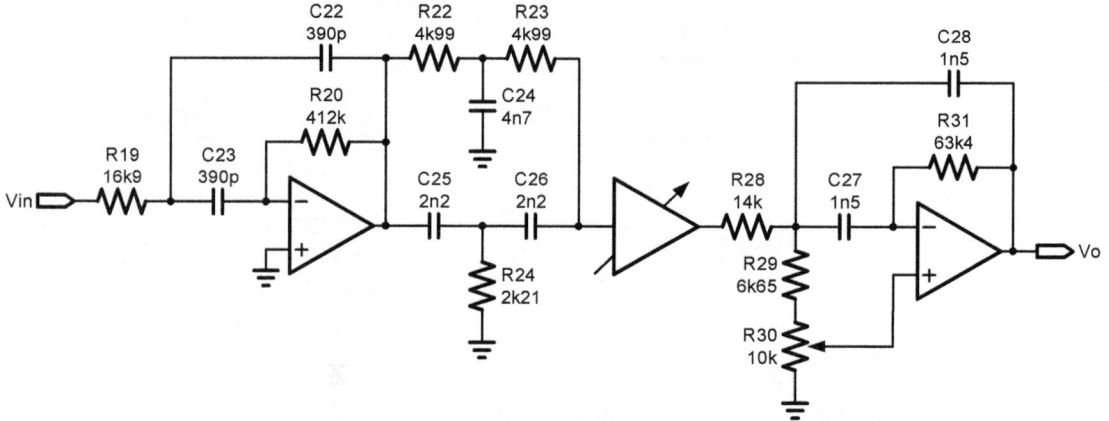

Fig. 24-28: **Band-pass Channel Filter (Fisher CZ5)**

24.6.4: Demods

Typically the demods in a multifrequency design are the same as in a single frequency design. Any of the various demod schemes presented in Part 4 can be used but a normal full-wave SHA-type (Figure 14-40) works as well as anything. A 2F design now requires four demod clocks, an XCLK and RCLK for each frequency channel. Unlike a typical analog SF design, these are not derived from a TX phase reference circuit but instead are driven directly from a micro. For this, a micro that has timer-driven PWM outputs with good phase control makes the job simpler. In some designs (like the White's *DFX* or the Fisher *CZ*) ground phase calibration is done by adjusting the phase of the analog signal rather than the phase of the demod clocks.

Typically the analog circuitry is calibrated for a particular phase references such as ferrite and salt water. Further ground tracking can be done in DSP by software phase rotation.

24.6.5: Remainder

In a non-direct sampling multifrequency design, once we have gotten past the demodulators the signals are usually digitized by an ADC and all further processing is done in software. However, the raw X signals might contain a significant ground component and the R_H signal might have a significant salt component, either of which can limit dynamic range. So it is often advantageous to create whatever composite signals you want in analog circuitry and digitize them to maximize dynamic range.

Beyond that, DSP consists of figuring out the behavior of the signals and determining what the target might be.

---

16. The variable gain stage is used for calibrating (nulling) the X_C signal; R30 is used for phase rotation to null the R_H signal to ferrite.

Explore

24.7: Dual Preamps

Figure 24-2 shows a single wideband preamp driving two frequency channels. If you are going to narrowband-filter the channels for frequency domain processing, then it is possible to use two narrowband preamps instead of a wideband preamp. Besides the opportunity to reduce noise, it also allows for an optimized preamp gain for each frequency.

24.8: Narrowband Demodulators

In the discussion of the Equinox (24.2.5) the concept of narrowband demodulation was introduced. In all of the analog discussions we've only dealt with wideband demodulators whereby the RX signal is multiplied by $+1, -1$ using a clocked analog switch. A narrowband demodulator is one where the RX signal is multiplied by $\cos(\omega t), \sin(\omega t)$. This can be done in analog circuitry but requires a linear demodulator instead of a switch; such a demodulator is found, for example, in the Gilbert cell (Figure 16-38).

Narrowband demodulation is much simpler in a direct-sampling design. The input signal is now a stream of numbers from the ADC and all we have to do is digitally multiply them by numbers from sine/cosine functions, which can be

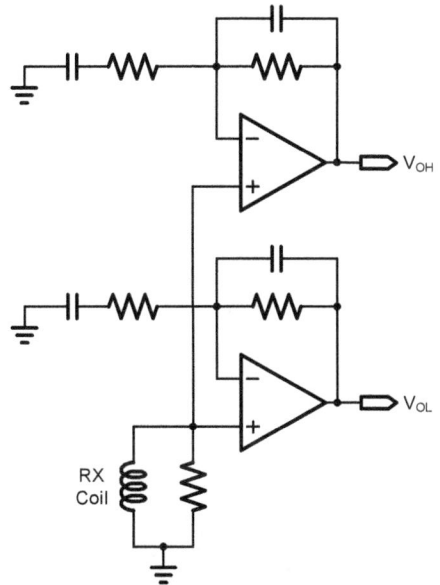

Fig. 24-29: **Dual Preamps**

provided by a lookup table. Figure 24-30 shows an extended frequency simulation response of a simple multiplication; compare this with the wideband analog demod sim in Figure 14-44. Also note the conversion gain is 0.5 instead of 0.636.

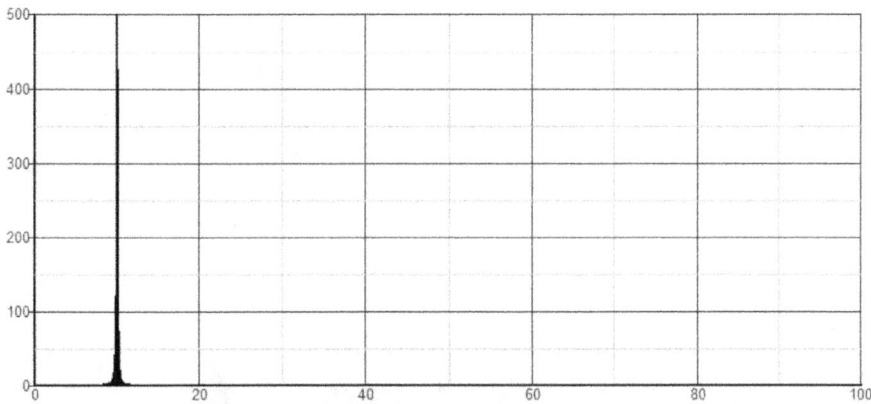

Fig. 24-30: **Narrowband Demod Response**

24.9: Asynchronous Demodulation

Software narrowband demodulation is widely used in radio systems but in metal detectors it is covered by a Minelab patent (US7579839) which is specifically limited to "synchronous" demodulation. To avoid this patent, the Nokta Legend was designed with asynchronous narrowband demodulators and is covered by patent US11914095. Asynchronous demodulation means that the demod outputs are no longer quasi-DC signals, rather they are periodic signals but at a much lower frequency than that of the TX, perhaps as low as several Hz. This results in X and R signals that continuously rotate with time, but this can be further demodulated in software to create the desired quasi-DC signals.

24.10: Moreland-Kelley Filter

Analog simultaneous multifrequency designs typically use channelization filters to split the wideband signal into multiple narrowband signals for demodulation. As an example, Figure 24-31 shows the band-pass filter used in the 7.5kHz channel of the White's *V3*. Because the channel frequencies are as little as three octaves apart it is difficult to do this with just band-pass filters so often you will see a notch filter added in, as with the *CZ* filter in Figure 24-28.

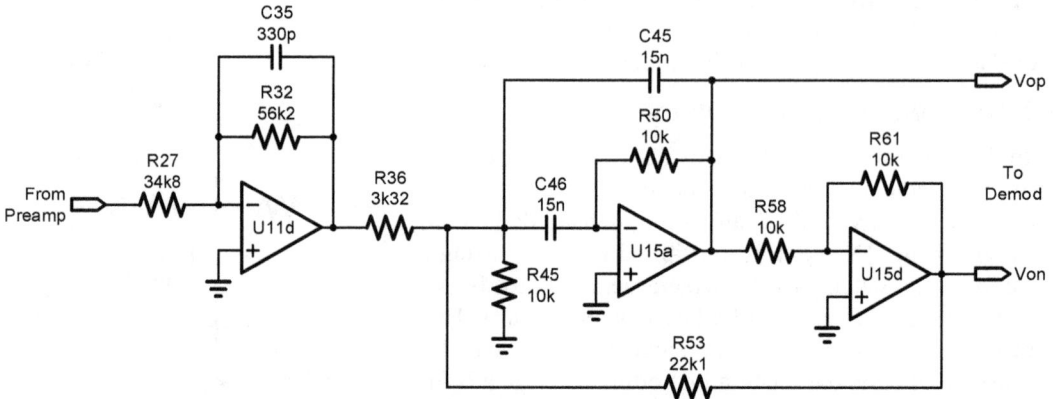

Fig. 24-31: **White's *V3* Channel Filter**

A filter that I designed specifically for this task combines two state-variable filters in such a way that both filters exhibit both the band-pass function for their desired frequency and the notch function for the undesired frequency. My co-worker, Jeff Kelley, recommended making it variable so that it could track, say, frequency offset adjustments. This is done by making the integrator resistors variable using PWM switching.

Figure 24-32 shows a 2-frequency version of the filter (1kHz and 5kHz) . The preamp drives both filters and the outputs of each filter is fed back to the opposite filter, resultingin a subtraction that cre-

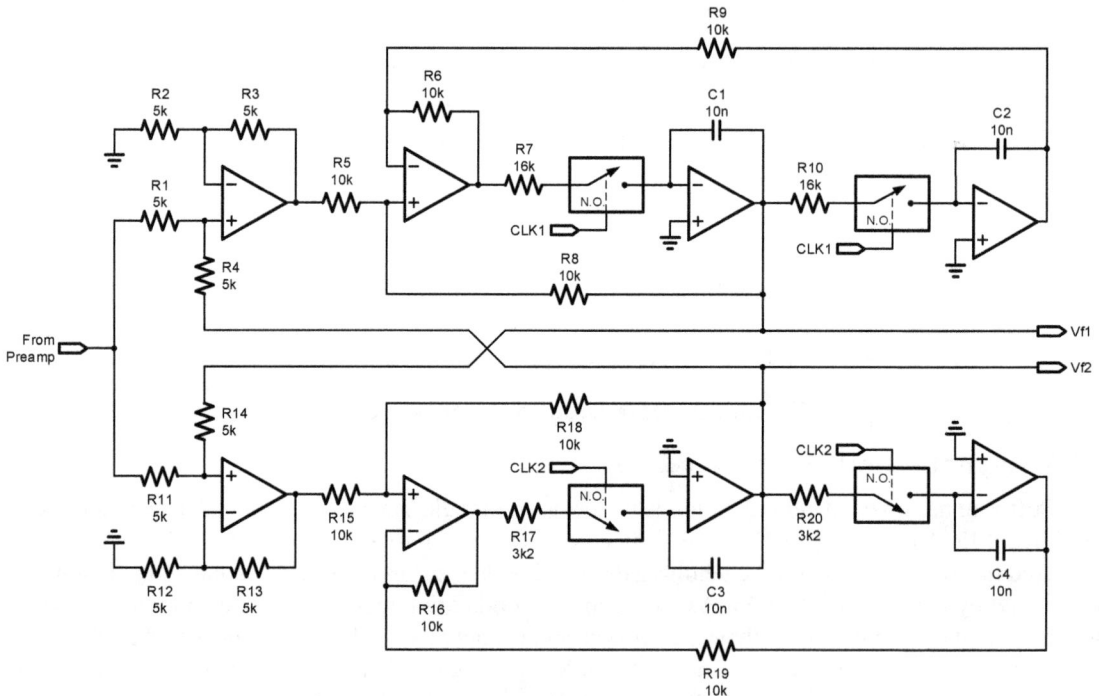

Fig. 24-32: **Moreland-Kelley Filter (2-Frequency)**

ates the notch. The switches in each filter are clocked at a high frequency (sa y, 1MHz) and the duty cycle is varied to create a shift in the center frequency. The filters can be shifted individually using separate clocks, or in unison (CLK1 = CLK2). Figure 24-33 shows the frequency responses of both outputs (CLK1 = CLK2 = 100% duty cycle). It is not difficult to expand this to three or more frequencies.

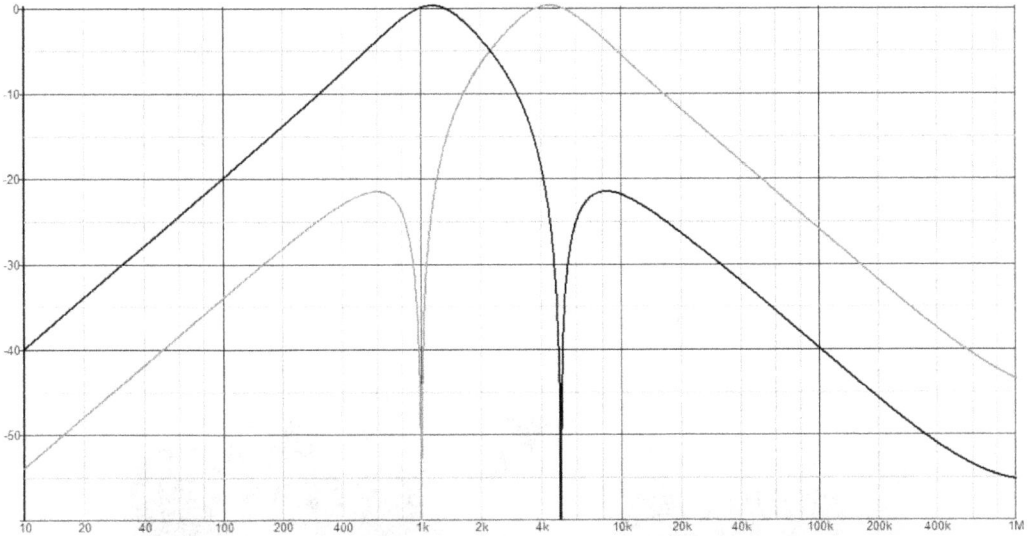

Fig. 24-33: **Moreland-Kelley Filter Response**

25

Advanced PI

"Any sufficiently advanced technology
is indistinguishable from magic."
— *Arthur C. Clarke*

Part 5 of this book on Pulse Induction is limited to a traditional single-pulse design with the coil turn-on being passively energized. That is, a voltage is applied to the coil and its current ramps up in a natural exponential. Those PI discussions also considered the use of single or multiple samples. Adding a sample to cancel Earth Field effect was presented in Chapter 21. Chapter 25 then introduced an additional sample to implement subtractive ground balance thereby giving us a four-sample system, including the two EFE samples. We then noted that the 2-point subtractive ground balance had the unfortunate drawback of a target hole. This chapter will look at more advanced techniques in both the transmitter and receiver designs.

Multi-Period PI

Instead of a single TX pulse width we could transmit two or more different pulse widths, which can be considered analogous to multifrequency in a frequency-domain detector. An early example of a multi-period PI (MP-PI) is described in patent US4717006 (1988) for a vending machine coin detector[1]. Minelab[2] introduced MP-PI to the hobby market in 1995 with the SD2000[3].

The SD2000 provides a good example of multi-period so let's look at the transmit waveforms:

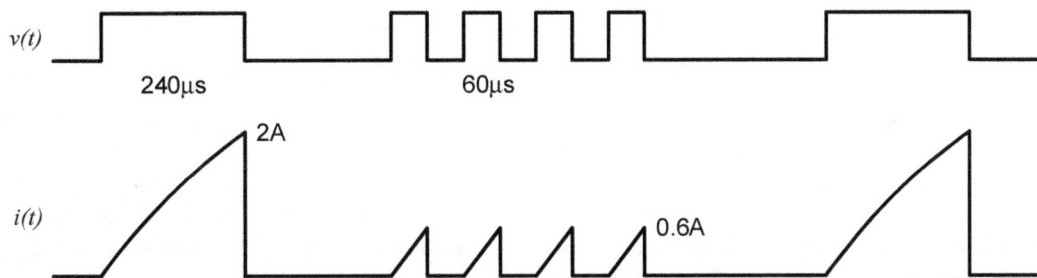

Fig. 25-1: **SD2000 Transmit Waveform**

The pulse train is monopolar and there are four narrow (60µs) pulses for every single wide (240µs) pulse. On the RX side there is a channel that samples the narrow pulses and two channels that sample the wide pulses. It is then possible to combine the results from the three channels to effect ground balance and target detection. To see how this is done let's see how PI timing changes affect the ground response.

---

1. When searching the patent database for metal detector patents, be sure to include similar technologies such as vending machines, coin sorters, and slot machines.

2. Minelab's trademarked name is MPS, or *Multi-Period Sensing*.

3. See US5537041.

25.1: Proportional Timing

Consider a traditional ground-balanced PI design as described in Chapter 22, either monopolar or bipolar. The performance depends greatly on the timing parameters:

- Pulse rate
- Pulse width
- Target delay
- Target width
- Ground delay
- Ground width

If these values are locked in proportion then as they are changed the ground balance point will be roughly maintained. Figure 25-2 shows two sets of timings[4] which produce very different target sensitivities but should have a similar ground balance point. The timings in Figure 25-2a might be selected for small/low conductive targets such as beach jewelry, while the timings in Figure 25-2b (all scaled by 2.5X) might be more suitable for relic hunting. This will allow a detector to have a single control knob for selecting a far greater range of performance than only a sample delay adjustment (common on many PI detectors) can provide[5]. Besides having similar ground balance points, the overall power consumption also remains fairly stable. This technique is called *Proportional Timing*[6].

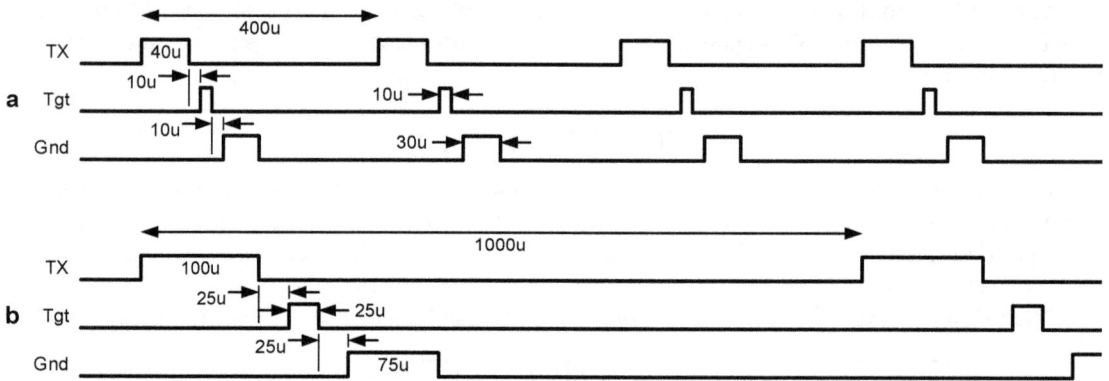

Fig. 25-2: **Proportional Timing**

Although the two timing sets in Figure 25-2 will have similar ground responses, they will have different target responses. Figure 25-2a will be more sensitive to low conductors and less sensitive to high conductors, and Figure 25-2b will be the opposite. Now suppose we combine the two into a multi-period system; we would then simultaneously have a good response to both low and high conductors while being (roughly) ground balanced. However, because both pulse widths would have the same ground balance point, there will still be a target hole.

25.2: Multi-Voltage Drive

In the previous section it was noted more than once that the two timing sets would have *similar* ground balance points. In order to achieve a consistent ground balance setting with proportional timing

---

4. For simplicity, a repeated simple monopolar cycle is shown without the EFE samples; they would need to be added in and must also be proportional. If the design is bipolar then the alternate cycles will be positive/negative TX pulses, also proportional.

5. I have modified a White's TDI to use proportional timing with a main delay of 6μs to 50μs and all the remaining timing following in proportion. It does not, however, have a variable V_{TX} or coil resistance so the ground balance must be adjusted when changing the setting. At 6μs the pulse rate is over 6kHz and easily detects 1-grain nuggets. At 50μs the pulse rate is 750Hz for, perhaps, cache hunting. Power consumption is fairly constant across the adjustment range.

6. See US8749240.

---

there is another critical detail: the peak currents at turn-off should be equal. In the traditional PI transmitter, a single switch and a single power supply result in an exponentially rising current in which a narrow pulse will end up with a lower peak current than a wider pulse, as in Figure 25-1. But a second switch with a second power supply can be used to equalize the peak currents, as in Figure 25-3.

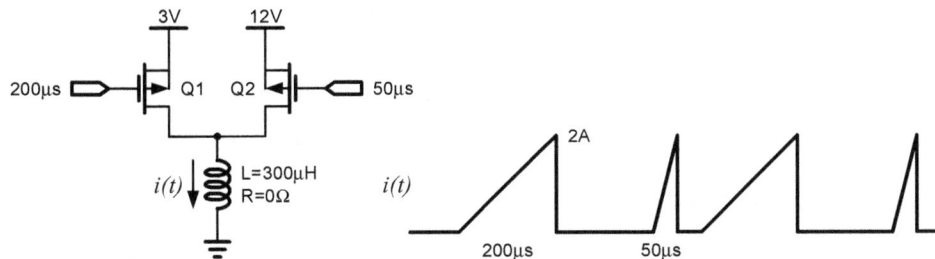

Fig. 25-3: **Dual Voltage TX**

The circuit above uses high-side switches for clarity but this method works with low-side switches as well. It should be noted that, as drawn, the circuit has a serious flaw: when Q2 is energized, the 12V supply will short to the 3V supply via Q1's body diode. An additional blocking diode is needed on the drain of Q1 which will require an upward adjustment of Q1's power supply. The waveform shown has perfectly linear current slopes since the coil is specified with resistance of 0Ω; a real coil with resistance will impose an exponential rise in current which has more effect on the wide pulse width, requiring more adjustment to the power supplies in order to equalize the currents.

This particular technique was introduced in the *GP Extreme* in 2000[7] and Minelab trademarked the name *Dual Voltage Technology* (DVT). It was an improvement to the normal MP-PI method used in the SD series detectors and resulted in more accurate ground balancing. The increased peak current of the short pulse also increases the sensitivity for small gold. It is also possible to balance each channel differently so that they have different target holes, thereby producing no overall target hole.

25.3: Equalized Multi-Voltage Drive

An additional enhancement to the multi-voltage TX circuit is to equalize the *dB/dt* of the two pulse widths just before turn-off[8]. This further improves the matching of the ground responses from each channel. Figure 25-4 shows the TX drive current for two different TX pulse widths; they are aligned to the turn-off edge for easier visual comparison. The wider pulse width gets closer to a flat-topped current and therefore has a lower *dB/dt* at turn-off.

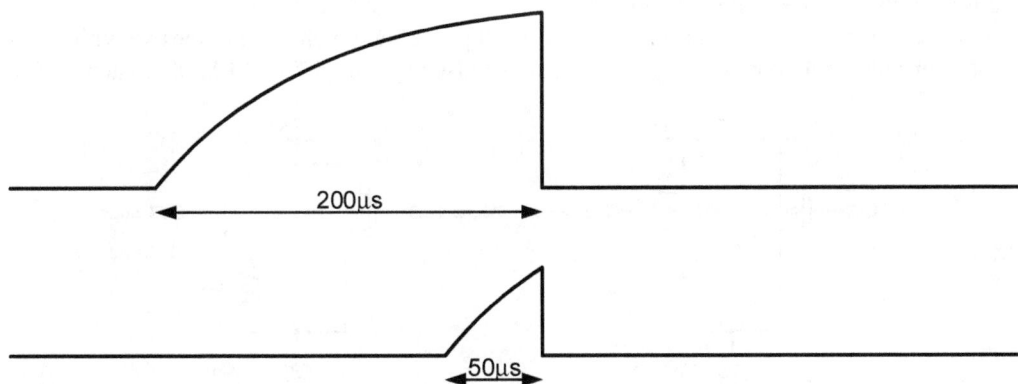

Fig. 25-4: **Current Response for Two Pulse Widths**

---

7. See US6653838.

8. In reality, maintaining the same ending dB/dt means that the curve shapes are overall the same, even if one occurs over a shorter time period and with less peak current.

There are two ways to force the narrow TX pulse to have the same ending dB/dt as the wide pulse. To see what they are, let's take a look at an example circuit with some real numbers. Figure 25-5 shows a basic monopolar TX circuit where the TX coil has $L = 300\mu H$ and $R_L = 3\Omega$. We'll again use a high-side switch for clarity and assume it's ideal. Thus the turn-on tau is $\tau = L/R_L = 100\mu s$.

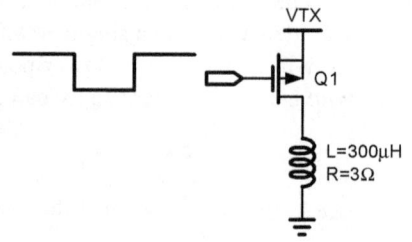

Fig. 25-5: **Example TX Circuit**

Now let's do a little math. The TX current is given by

$$i_L(t) = I_{max}(1 - e^{-t/\tau})$$

where I_{max} is V_{TX}/R_L. The slope of the current is

$$\frac{di_L}{dt} = \frac{V_{TX}}{R_L} \cdot \frac{1}{\tau} \cdot e^{-t/\tau} = \frac{V_{TX}}{L} \cdot e^{-t/\tau} \qquad \text{Eq 25-1}$$

The wide pulse in Figure 25-4 has a turn-on time of 2τ so the ending current slope is

$$\frac{di_L}{dt} = \frac{V_{TX}}{L} \cdot e^{-2}$$

We'll leave it in this form for now.

The ending slope of the narrow pulse can be found the same way. If we assume that the inductance cannot change then the two variables we can modify are the supply voltage V_{TX} and the series resistance of the coil[9] R_L, which alters τ. Let's start with V_{TX}; the narrow pulse has a turn-on time of 0.5τ and the slopes can be equated:

$$\frac{di_L}{dt} = \frac{V_{TXn}}{L} \cdot e^{-0.5} = \frac{V_{TXw}}{L} \cdot e^{-2}$$

where V_{TXn} and V_{TXw} are the supply voltages for the narrow and wide pulses. The solution to the equation above is to set $V_{TXn} = 0.22 \cdot V_{TXw}$.

The other solution is to modify the coil tau by adding a series resistor. The equation then becomes

$$\frac{di_L}{dt} = \frac{V_{TX}}{L} \cdot e^{-0.5\tau w/\tau n} = \frac{V_{TX}}{L} \cdot e^{-2\tau w/\tau w}$$

or $\tau_n = \tau_w/4$. Since $\tau = L/R$, this means the narrow pulse needs 4x the series resistance of the wide pulse, or an additional 9Ω in series with the coil's 3Ω.

How we can achieve these solutions is given in Figure 25-6 which, again, does so with high-side monopolar transmitters. In both cases the same coil is pulsed by two different PMOS switches. The cir-

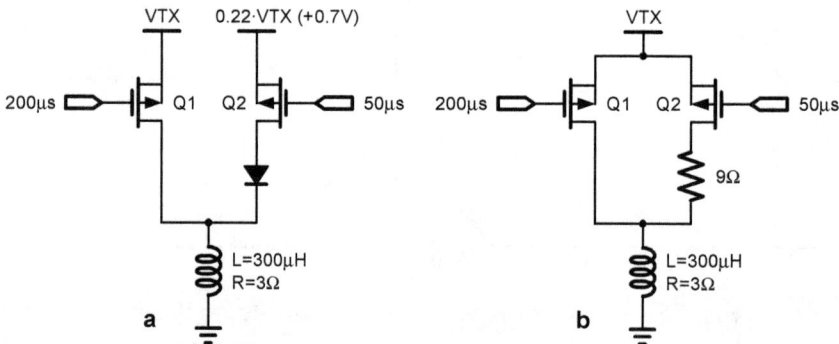

Fig. 25-6: **TX Circuit Solutions**

9. Although we assume we cannot change the coil's inductance, we can always add more series resistance to the coil.

Inside the Metal Detector

cuit on the left uses two different drive voltages while the circuit on the right adds an additional resistor to the narrow pulse switch. The diode in the two-voltage solution is necessary to isolate Q2 during Q1's drive and will require an increase in V_{TXn} by 0.7V or so. These particular solutions are perhaps useful in the situation where you want to alternate between the two pulse widths. In the case where proportional timing is used in a continuously variable manner the V_{TX} solution could be implemented with a variable voltage regulator and using only one PMOS switch.

Either of the two solutions will equalize *dB/dt* but the resulting narrow pulse will have a significantly reduced peak current compared to the wide pulse. We need both pulses to have the same peak current in order to maintain ground balance. It is possible to adjust both V_{TX} and R_L to equalize both *dB/dt* and the peak current. For the example conditions above (plus an arbitrarily chosen wide-pulse V_{TXw} of 10V), a final circuit that works is shown in Figure 25-7. Other solutions can be found for a different wide-pulse V_{TXw} or different search coil parametrics. The optimized values can be calculated but when the diode and damping resistor are added in the math gets rather complicated. Spice simulations are a quicker way to find the optimum values. A plot of the coil current is given in Figure 25-8.

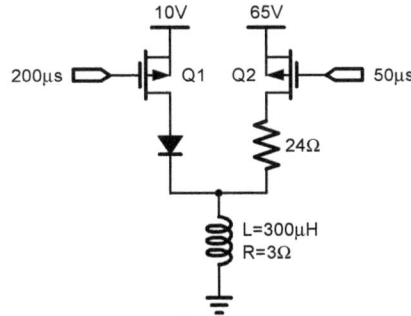

Fig. 25-7: **Optimized TX Driver**

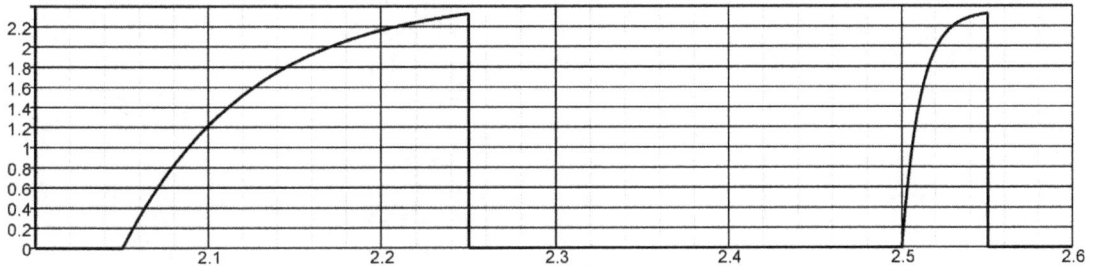

Fig. 25-8: **Optimized TX Current**

25.4: Multi-Step Drive

So far we have only considered cases where each pulse is produced by applying a single voltage for a fixed amount of time and then disconnecting the voltage. It is possible to create an overall pulse whereby one or more drive voltages are applied sequentially with a "dead time" between them.

Consider the waveform shown in Figure 25-9 which shows an overall current pulse being generated in a "stair-step" fashion. This can be accomplished by applying a high drive voltage to the coil during time segments "a" and then applying a low "hold" voltage during time segments "b." The amount of hold voltage required depends on the current level at that point and will vary for each "b" segment. The drive voltage used for each "a" segment can be constant or can vary; in Figure 25-9 the equal "a" slopes imply the drive voltage is increasing for each step. A constant drive voltage will result in an overall rising exponential-looking curve, assuming the drive times and dead times are consistent.

The advantage of this technique is that each of the small "ramplets" in the overall larger pulse will have their own PI response during the dead time. As we will see more clearly in the Constant Current PI section, a PI current pulse does not need to return to zero current in order to see a PI response, it only

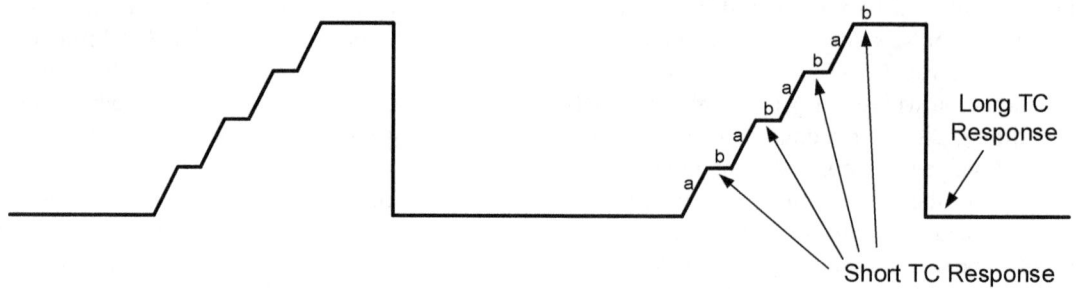

Fig. 25-9: **Multi-Step Current Pulse**

needs to return to a DC (unchanging) current. However, a return-to-zero operation is required for using a mono coil[10] so this technique is only usable with non-mono coils, although they need not be induction-balanced. Therefore, each short ramplet induces an EMF in the target, and during each subsequent dead time the target response can be observed.

The end result is an overall long pulse (useful for energizing high conductors) that includes a number of short pulses (useful for energizing low conductors) at the same time. This is, in fact, the method employed in the Minelab *GPX6000*. The *GPX6000* also includes two medium pulses; the overall current waveform is shown in Figure 25-10 as measured with the Magnetic Field Probe. The "tilt" seen in the zero-current baseline is a result of the bandwidth of the MFP but the tilt seen in the dead time segments ("b") is not. Instead of holding the dead time current at a constant level using a DC voltage, the coil is simply shorted. The low resistances of the coil and shorting switch produces a severely over-damped response that has a lower slope[11]. While not as "clean" as a zero-slope dead time it is still effective in obtaining low conductor responses.

In the case of the *GPX6000* there are effectively five ramplets that are useful for target detection. The last two ramplets at the peak of the overall pulse are likely used to flat-top the pulse so that it resembles something like the flat-top in Figure 25-9. Again, in order to process a response during a non-zero current a true mono coil will not work, so it is likely that the "mono" coils available for the *GPX6000* are not truly mono but actually use separate co-wound TX and RX coils.

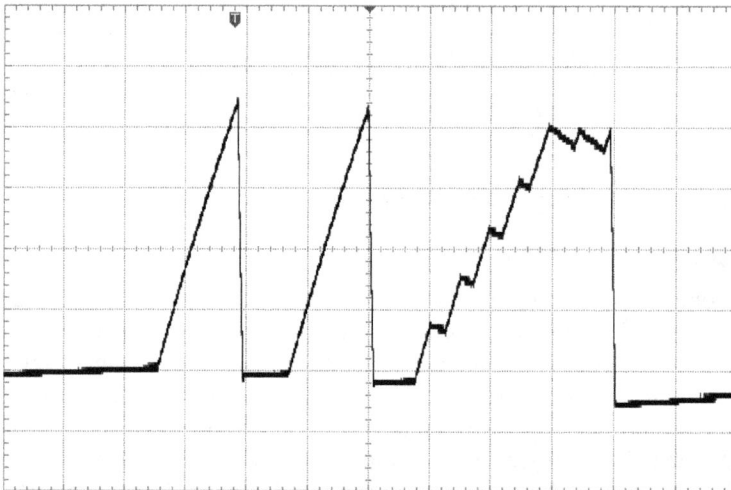

Fig. 25-10: *GPX6000* TX Current

---

10. That is, the coil cannot be driven, it must be effectively open-circuited except for the damping resistor.

11. Note that the dead time slope increases as the current level increases, as you would expect. If you could make the resistances zero, then the coil current would circulate with no loss and the dead time slope would be flat.

25.5: Two-Slope Ground Balance

The PI system considered in Chapter 22 uses two subtracted samples to cancel ground. The fundamental problem with 2-point subtraction is that 2 points create a straight-line approximation of the RX response, and that is not enough to uniquely distinguish an exponential response (eddy target) from a power law response (viscous ground). Figure 22-16 showed a 4-channel design whereby each 2-channel system is calibrated to have a different target hole. Since a target hole corresponds to a particular log-linear decay slope, the 4-channel solution can be considered a two-slope ground balance.

The two-slope ground balance can be simplified. Looking back at the curves of Figure 22-1, consider taking a third sample as shown in Figure 25-11. There are a few ways to process these samples. The first is to use S1-S2 to create a usual ground balanced signal (we'll call this signal G1) and then use S1-S3 to create a second ground balanced signal (G2). G1 and G2 will have different target holes so combining them produces a ground balanced signal with no target holes. A second way is to create the second ground balanced signal (G2) by using S2-S3, with similar effect. Be aware that with this latter method it is possible to select the sample timings (or gains) so that G1 and G2 have identical target holes, but this is fairly easy to avoid. Both of the above methods effectively create two 2-point straight-line approximations, much like the concept in Figure 22-16, but with a little less circuitry.

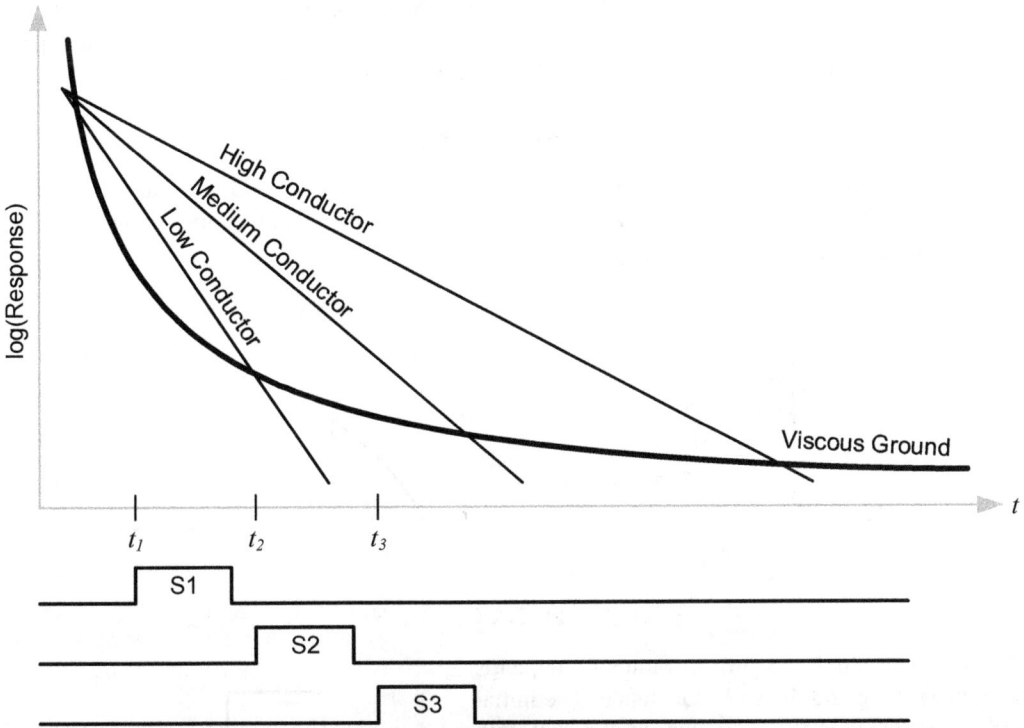

Fig. 25-11: **3-Sample GB**

The two-slope method as shown works with a single pulse width transmission. For multiple pulse widths, each pulse width channel can use two samples to achieve a slope, and the two channels can employ timings that produce different effective slopes. This results in two different target holes and prevents the occurrence of an overall target hole.

In a multi-period pulse design, we've noted that two samples from each pulse width can be combined to produce two separate slopes for ground balance. But samples can also be combined across the different pulse widths to effect ground balance. US6586938 describes several multi-period pulse methods for combining the samples from short and long pulses. In determining exactly how to combine samples, it is useful to measure the responses for a variety of metal targets (both ferrous and non-ferrous) and also a variety of soil or rock samples to see how they behave to varying pulse widths and peak currents.

25.6: Noise

In Chapter 21 an additional sample pulse was added to the basic PI design to cancel Earth Field Effect. EFE is a low-frequency self-induced noise. Additionally, there are other external EMI noise sources from power lines, radio waves, and the like. In the single-sample design of Chapter 21, a single EFE sample was added (Figure 21-15). In the two-sample ground balanced design of Chapter 22, two EFE samples were added, one for each main sample (Figure 25-4).

It is not necessary to have a separate EFE cancelation sample for each of the main samples; EFE cancelation may be done with a single EFE sample. In the case of a two-point subtractive ground balance, we have

$$V_{GB} = S_1 - k \cdot S_2 \qquad\qquad \text{Eq 25-2}$$

where S_1 is the target sample, S_2 is the later ground sample, and k is the multiplier used to null the response to ground[12]. k may be effected by either the sample width of S_2, the amplification of S_2, or a combination. This leaves EFE uncompensated; to cancel EFE, we need only a single additional sample:

$$V_{GB,EFE} = S_1 - k \cdot S_2 + (k-1)S_3 \qquad\qquad \text{Eq 25-3}$$

where S_3 is the EFE sample taken (hopefully) well beyond the normal responses of targets and ground.

Besides canceling EFE, this method also attenuates low frequency EMI. See the math in Appendix C and Table C-4 in particular to see how attenuation is affected by the spacing of S_3.

Constant-Current PI

The traditional PI transmitter is shown in Figure 25-12. It applies a relatively low voltage to the coil and the current exponentially ramps up to some value before the switch is opened and the coil current ceases. This results in the familiar high-voltage flyback.

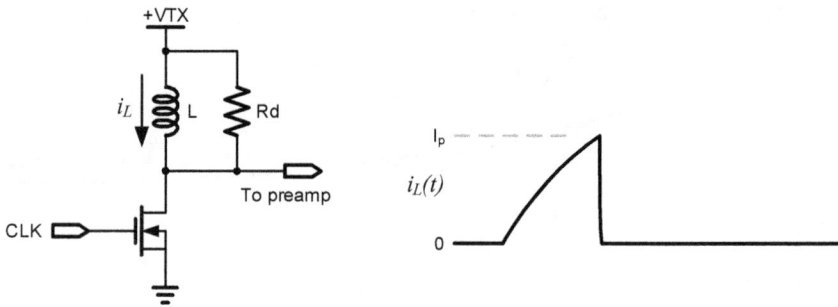

Fig. 25-12: **Traditional PI**

The speed at which the current ramps up depends on the voltage drive and the coil inductance. The initial turn-on current slope is V_{TX}/L which implies that with a very high voltage it is possible to quickly ramp the coil current to a desired value such as 1 amp. Once the coil current reaches that level it can then be held there using a much lower voltage. Figure 25-13 shows how

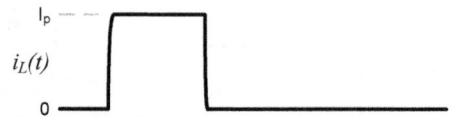

Fig. 25-13: **CCPI Waveform**

this waveform would look. Because the current is quickly flat-topped and held at a constant level through the entire TX pulse it is called *Constant-Current PI* (CCPI).

CCPI can be monopolar or bipolar, and single-period or multi-period. It can have a return-to-zero after each pulse, or it can continuously transition between positive and negative current. Figure 25-14 shows some of the possibilities. One advantage of CCPI is that it minimizes residual reverse eddy cur-

---

12. Appendix C shows the mathematics in calculating k.

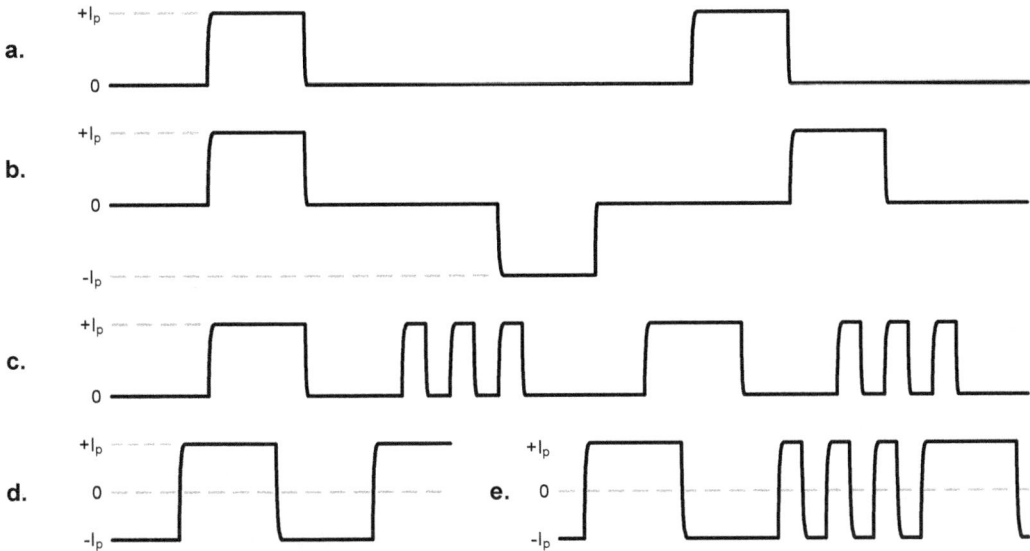

Fig. 25-14: **CCPI Possibilities**

rents at turn-off; see the *Explore* section of Chapter 21. Another advantage is that, once the peak current is achieved, it can be held steady with a low voltage, resulting in a lower power consumption than traditional PI. For example, a coil with a total resistance of 1Ω can be held at 1A with only a 1V power supply.

25.7: Brute-Force Drive

There are several ways to achieve a fast turn-on and we'll start with the brute-force approach shown in Figure 25-15. A high-voltage power supply $(+V_{TXH})$ charges a capacitor during the TX off-time. When the PMOS switches are closed, the high voltage on the capacitor dumps its energy into the coil and the coil current quickly jumps to a high value. This discharges the capacitor and allows the diode to turn on so that the low voltage supply $(+V_{TXL})$ can sustain the coil current at a fixed level. When the PMOS switches turn off, the coil current collapses as usual and the high-voltage capacitor begins charging up again.

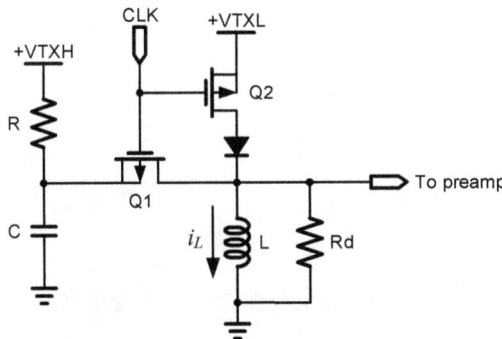

Fig. 25-15: **Brute-Force CCPI TX**

The values of $+V_{TXH}$, $+V_{TXL}$, R, and C all depend on the coil parameters and the desired peak current and turn-on slew rate. For example, suppose the TX coil is $300\mu H$ with a resistance of 1Ω and we want a peak current of 1A with a turn-on slew of under $5\mu s$. If the diode drop is typically 1V then $+V_{TXL}$ will need to be 2V. Because the turn-on action involves transferring energy from a capacitor to an inductor the turn-on curve is actually a quarter-sine wave. A quarter sine wave of $5\mu s$ represents a period of $20\mu s$ or $f = 50kHz$, and from that we can calculate C to be 33.8nF. Let's round that down to 33nF. Finally, the energy in the coil is $\frac{1}{2}\cdot L\cdot I^2 = 150\mu J$ and the energy in the capacitor $(\frac{1}{2}\cdot C\cdot V^2)$ must be the same, so $+V_{TXH}$ must be 95V. R is selected to ensure the capacitor is fully charged during the TX off time. For example, 2kΩ will fully charge the cap in 330μs.

It should be noted that the circuit as shown in Figure 25-15 won't actually work. In order to turn off Q1 and Q2 their gate voltages will need to be roughly equal to $+V_{TXH}$ and $+V_{TXL}$ respectively, and to turn them on they will need to be several volts lower. This can be solved by capacitively coupling the gates separately and using diode clamps to the supply voltages. See previous PI transmitters for examples.

We already know that when the coil is suddenly turned off, a high voltage flyback results. Now we see that in order to suddenly turn the coil on, a high voltage drive[13] is required. The high voltage circuitry that provides the turn-on "kick" is often referred to as a *kickstart* circuit. Like a traditional PI transmitter, the brute-force CCPI is a total-loss system. That is, all the energy put into the coil during turn-on is dissipated as heat at turn-off. Typically, the total power consumption of a brute-force design is not much different than that of a passively driven coil. Next, we'll look at ways to recycle some of the energy and improve efficiency.

25.8: Recycling Kickstart Methods

The brute-force circuit uses a fixed high-voltage power supply to produce the required transient energy for a fast turn-on. Typically this supply will need to be on the order of 100-200 volts, which is not a simple matter. Fortunately, the flyback produced when the coil turns off is a convenient high voltage generator so if we can harness that energy we could eliminate the separate high-voltage power supply and gain some efficiency by recycling energy.

US9348053 and US9366778 both describe PI transmitters that incorporate energy recycling kickstart methods. Another similar design is found in the Garrett *ATX* (and, presumably, in the Garrett *Axiom* as well). The schematic in Figure 25-16 shows the usual TX coil, damping resistor, and main switch Q1. Suppose that Q2 is off and the coil is turned on, with current supplied by the +5V supply via D1. When the coil is turned off, the flyback turns D2 on and most of the flyback energy is dumped onto the 56nF capacitor[14]. An immediate benefit of this is that the energy is not dissipated by the damping resistor, therefore the damping resistor need not be rated for high power dissipation.

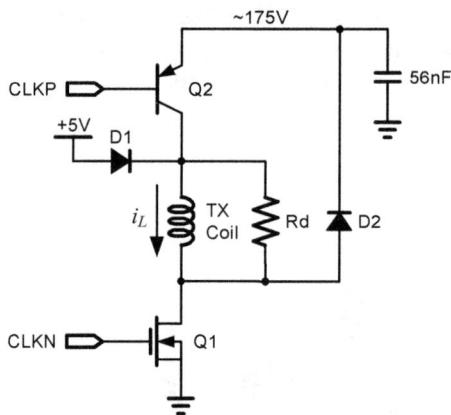

Fig. 25-16: **Garrett *ATX* Transmitter**

In the *ATX*, the flyback energy charges the capacitor to about 175V. At the next turn-on cycle, Q2 is briefly turned on at the same instant as Q1 and the energy in the capacitor is used to quickly charge the coil, resulting in a rapid rise of current to about 5 amps. This completely discharges the capacitor and D1 automatically turns on so that the +5V supply can sustain the coil current.

It is interesting to note that the *ATX* transmits two different pulse widths — 16μs and 250μs — and only the 16μs pulse uses the kickstart circuit. The long pulse is a traditional exponential turn-on that peaks at about 4 amps[15]. See Figure 25-17.

---

13. A "flyforward" voltage?

14. Think you've seen this before? See 21.15.

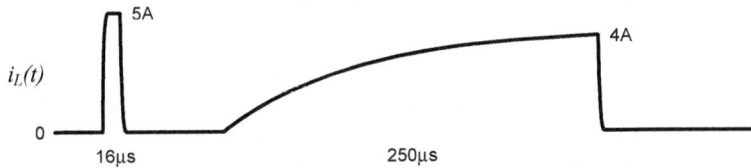

Fig. 25-17: **Garrett ATX Transmit Waveform**

25.9: Square Wave Drive

The previous discussions on CCPI considered a TX pulse that returns to zero between pulses. The return-to-zero is not necessary and we could instead produce a square wave current. Figure 25-14 shows a bipolar pulsed square wave (d) and even a multi-period bipolar waveform (e). Both are considered CCPI because the turn-on current is constant.

The square wave current approach is used in the Minelab *GPZ7000*. Because the current is constant after each transition the resulting *dB/dt* is zero and we can still look at target decay responses even though the TX coil might still be pumping an amp or two of current. An advantage of the square wave method is that, for the same frequency, we now get twice the results (after both transition edges) and the pulse frequency can be much higher, perhaps 20kHz or more[16]. Another advantage is that no damping resistor is needed because there is no "off" time; transition speed is entirely dependent on coil inductance and total parasitic capacitance. A disadvantage is that a separate receive coil is required, although it need not be induction-balanced.

Like the pulsed CCPI method, a square wave drive requires a high voltage kickstart circuit, either brute-force or energy recycling. Figure 25-18 shows a simple transmitter circuit[17] that is both bipolar and uses an energy recycling kickstart. An obvious missing element is the high voltage kickstart capacitor; in this circuit, it is the interwinding capacitance of the coil itself that serves as the kickstart capacitor. That is, the energy from the coil's magnetic field is converted into a high voltage that is temporarily stored on the interwinding capacitance, and when the drive polarity reverses that voltage then kickstarts the coil current in the opposite direction.

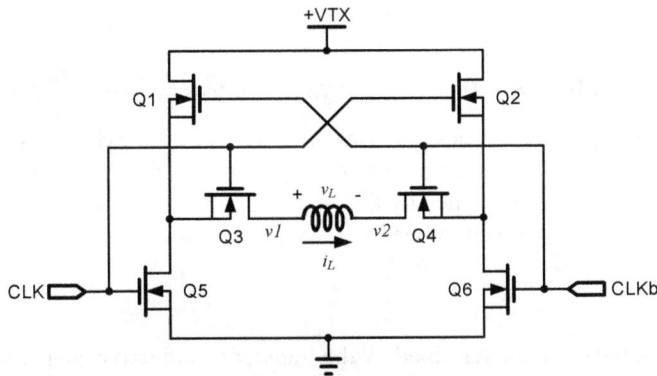

Fig. 25-18: **Square Wave Transmitter**

Q3 and Q4 are the switches which isolate the flyback energy to the coil itself and therefore are the only two MOSFETs that must be rated for the full flyback voltage. The other four MOSFETs can be low voltage, but all six devices should preferably have low ohmic losses. Notice that the high-side devices are NMOS. This works well as long as the V_{OH} of the clock drives is high enough compared to

---

15. The newer Garrett *Axiom* design has identical TX pulses, with a slight change in the dead time spacing.

16. Consider a 25kHz square wave design for prospecting; this gives 50,000 samples/s versus the customary 1000-3000 samples/s.

17. See US9250348.

+V_{TX} so that Q1 and Q2 are solidly turned on. This is not difficult because +V_{TX} will likely be in the range of 2-3 volts.

During one polarity, switches Q2, Q3, and Q5 are turned on by a positive voltage applied to the CLK line. With CLKb low, the remaining switches are off. However, the parasitic body diode of Q4 will conduct current even though Q4 is turned off, and therefore current can flow from +V_{TX} through Q2, Q4, Q3, and Q5. The magnitude of the current depends on +V_{TX} and the combined resistances of the switches and the coil, with the coil resistance typically dominating. As an example, if the coil resistance is 1Ω and the body diode voltage drop of Q4 is 1V, then +V_{TX} = 2V will produce a current of 1A through the coil, assuming the switch resistances are negligible.

When CLK is driven to a low state Q2, Q3, and Q5 are turned off and current through the coil ceases. This produces a high flyback voltage across the coil with *v1* attempting to slew positive and *v2* attempting to slew negative. However, *v2* is clamped to ground via the body diodes of Q4 and Q6, therefore a high positive flyback voltage is seen on *v1*. The energy from this flyback is stored on the parasitic capacitance of the coil. Simultaneously, CLKb transitions to a high state and turns on Q1, Q4, and Q6 and allows the body diode of Q3 to conduct, and the high voltage stored at *v1* kickstarts the current through the coil in the opposite direction. The resulting coil current and flyback voltage waveforms are shown in Figure 25-19.

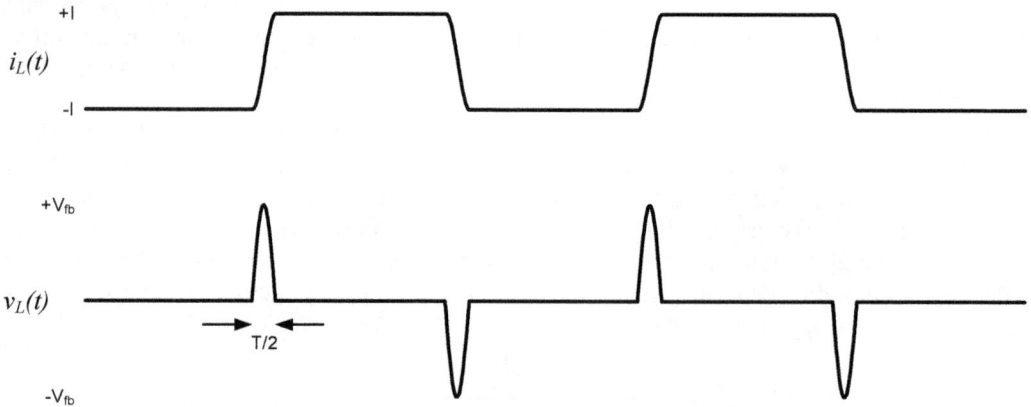

Fig. 25-19: **Square Wave TX Waveforms**

The waveforms show that the flyback voltages have a half-sine waveform and can peak at several hundred volts. The current transitions are a half-cosine waveform followed by a constant DC current. We know that the DC current is determined by the power supply V_{TX} and (mostly) the resistance of the coil. The transition speed is determined by the LC resonance:

$$\frac{T}{2} = \pi\sqrt{LC} \qquad\qquad \text{Eq 25-4}$$

The peak of the flyback pulses are determined by the transfer of inductive energy to capacitive energy:

$$\frac{1}{2}LI_L^2 = \frac{1}{2}CV_{fb}^2 \Rightarrow V_{fb} = I_L\sqrt{\frac{L}{C}} \qquad\qquad \text{Eq 25-5}$$

As an example, suppose we have a TX coil that is 300µH with a parasitic C of 300pF and the peak current is 1 amp. The transition time will be 0.94µs and the peak flyback will be 1000V.

With this circuit it is important that the flyback-blocking devices Q3 and Q4 never avalanche as this will damage other devices. Likewise, the coil wire should be rated for the maximum breakdown, or wound in a way that adjacent windings never see an excessive voltage, such as a spiral winding. If the calculated flyback peaks exceed a breakdown voltage then an additional physical capacitor can be added in parallel with the coil. This will both slow down the transitions and reduce the flyback peaks. In the previous example, adding a 1.5nF capacitor in parallel with the coil will increase the transition to 2.3µs and reduce the flyback to 408V, allowing the use of common IRF740 devices.

This circuit has the ability to produce high pulse currents from a relatively low power supply voltage. A typical design would use a buck regulator to provide V_{TX} to keep overall power consumption as low as possible. A minor drawback of this circuit is the voltage drop incurred in the body diode of Q3 and Q4. It is possible to eliminate this by independently controlling Q3 and Q4 and turning both of them on during the constant current phase of transmission, thereby making the coil current simply V_{TX}/R_L. An aggressively low resistance TX coil could push V_{TX} below 1V. This might also require a coil cable with an aggressively low resistance. Fortunately, an advantage of this circuit is that cable capacitance is less of a concern, making cable selection a bit easier.

25.10: Current Control

In most of the CCPI transmitters the flat-top coil current is heavily dependent on the resistance of the coil. This gives rise to two sources of variation: one due to the coil-to-coil resistance variation, and the other due to the temperature coefficient of the copper wire (+3900ppm/°C). It is therefore useful to have a feedback mechanism to dynamically control the coil current[18].

Figure 25-20 shows how this might be done as applied to the square wave transmitter. A (very) small resistor is used to sense the coil current and feed a control voltage back to the TX voltage regulator. To avoid the transitions a sampling circuit can sample the control voltage only at the flat-top regions.

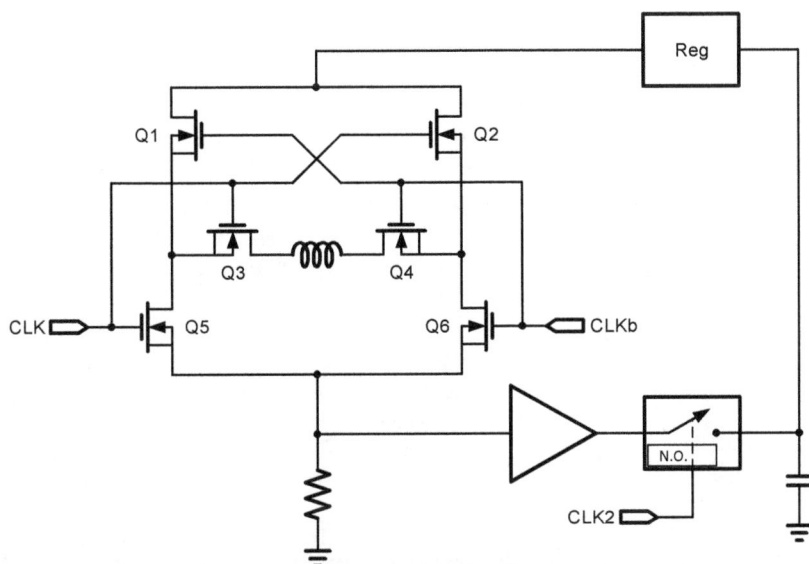

Fig. 25-20: **Current Control**

25.11: Switching Losses

Transmitters which recycle the flyback energy to provide a kickstart voltage are never 100% efficient; there is always some loss in the process. At the very least, the resistance of the coil itself produces an energy loss (in heat) as the coil current transitions to a capacitance voltage, and the voltage back to a current. Other parasitic capacitances (such as with the switching devices) can also cause losses. The result is that the turn-on current is slightly less than the turn-off current of the previous cycle. The resulting current waveform will have a slight "tilt" of the flat-top region as the current attempts to ramp back up to the proper peak value. See Figure 25-21.

The current tilt's *di/dt* can continue to induce signals in both the ground and targets. Whether this causes a problem on the RX side depends on the relative *di/dt* of the tilt compared to the overall peak current. If the effect is small enough it may be "in the mud" compared to the primary driver of target responses, the *di/dt* during the transitions.

---

18. US9250348 covers the use of a feedback mechanism to control the TX current.

$i_L(t)$

Fig. 25-21: **Current Tilt Due to Losses**

Tilt is caused by losses during the current transitions from one polarity to the other. One major cause of this is the resistance of the coil itself. As the current slews from positive to negative there is a small amount of energy loss due to the coil resistance such that the energy transfer according to Equation 25-5 is not ideal. Another cause is the capacitance and transition time of the blocking switches Q3, Q4. Therefore, minimizing coil resistance and using low capacitance switches with fast clocks will help.

It is possible to correct for tilt with an additional slight "kick" from the V_{TX} supply, either just before or just after the transition. This can be placed under feedback control using a process similar to controlling the peak current. In fact, the methods can be combined: instead of taking a single sample of the flat-top region, take two samples: one early and one late. In comparing the two samples, if they are unequal then adjust the pulse width of an extra injected "kick" pulse to eliminate tilt.

Explore

25.12: Flyback Isolation

Some of the more advanced PI designs on the market have replaced the simple resistor-diode clamp at the input of the preamp with an active switch. Figure 25-22 shows how such a switch might be added to the design. Besides eliminating the settling time of the diode clamp it eliminates the noise of the series resistor and also makes the damping response solely dependent on RD.

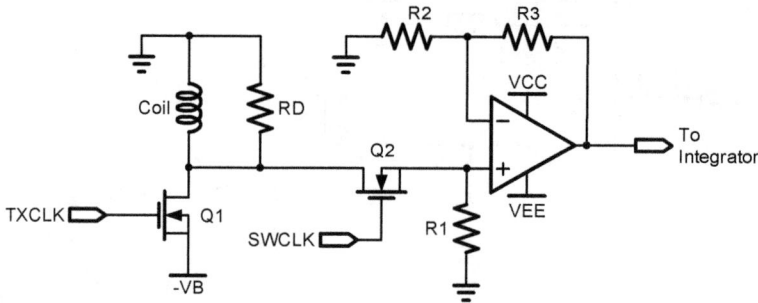

Fig. 25-22: **Active Isolation Switch**

Getting the correct switch timing is difficult. If the switch is turned on too soon, the preamp will still get overloaded, though perhaps not damaged. If turned on too late, fast target signals are lost. Figure 25-23 shows a way to drive the switch using the flyback signal itself. If damping resistor RD is split into a voltage divider then when the following is true

$$v_{flyback} \cdot \frac{RD2}{RD1 + RD2} < V_{TH}$$

the comparator will turn on the switch. RD2 is typically small, say, a few percent of RD1. Note that the comparator will turn the switch on during the TX-on pulse so this will need to be dealt with in some way. In general, this is just a starting concept and is left to the reader to explore.

US10181720 shows an automatic method of isolation using a diode bridge (Figure 25-24). When the flyback is a few 10s of millivolts above ground, D1 turns off and blocks the coil voltage. Likewise, D2 blocks negative coil voltages. When the coil voltage has settled close to 0V, both diodes conduct and allow the coil voltage to reach the preamp. Resistors R1 and R2 set the conduction currents for the diodes and should be chosen such that, when *v1* settles to ground (0V), *v2* is also roughly 0V. Diode D1

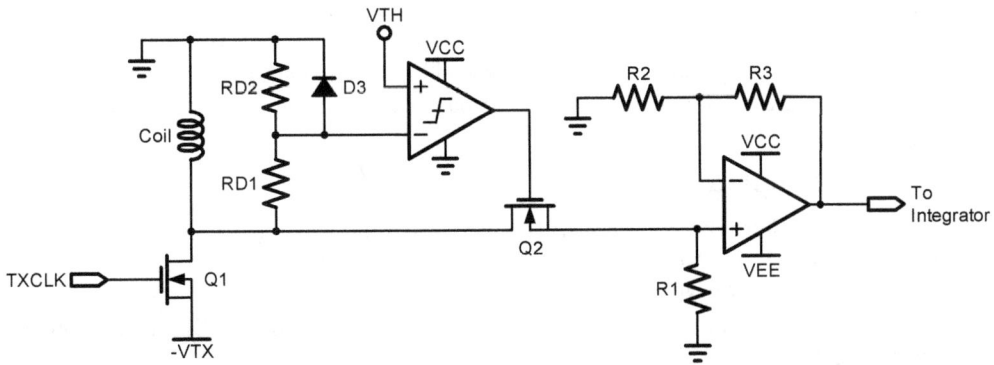

Fig. 25-23: **Flyback-Driven Isolation Switch**

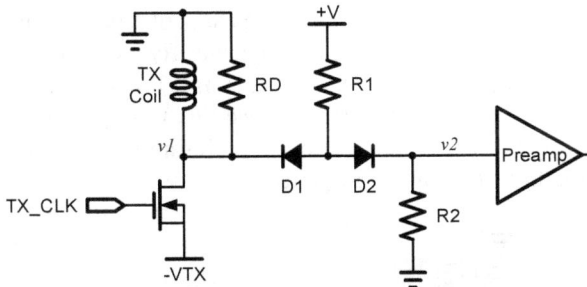

Fig. 25-24: **Bridge Isolation Switch**

must have a breakdown voltage at least as high as the expected flyback; D2 can be a lower voltage diode (enough to block VTX) but for best offset stability should be the same type as D1.

A problem with this circuit is that unequal currents in the diodes give rise to an input offset that, with a high-gain preamp, can create a substantial output offset. As designed with R2 connected to ground, an input offset is unavoidable because D1 is grounded through the resistance of the coil (a few ohms) and D2 is grounded through R2, which is surely to be higher than a few ohms. For example, a 10:1 current imbalance produces a 60mV input offset which, in most PI designs, is enough to rail the preamp. This problem can be partially mitigated by connecting R2 to a negative rail.

Figure 25-25 shows an improved bridge isolation switch where D2 is replaced by transistor Q3 and a diffpair (Q1,Q2) is added to zero the offset[19]. D1 still blocks the flyback, during flyback the diffpair keeps the input of the preamp close to 0V, and during conduction the voltage drops of D1 and Q3 can be

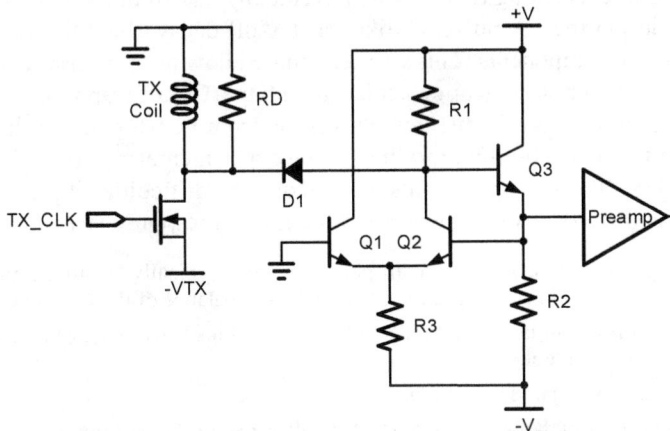

Fig. 25-25: **Improved Bridge Isolation Switch**

almost exactly balanced. A drawback is that there is no negative clamping when the coil switch is energized. Another issue is that the diffpair that maintains a near-zero offset will tend to track out target decays. Both problems are easily solved, the former by adding a base clamp to Q3 and the latter by adding a continuous or sampled low-pass filter to the diffpair. Besides being an improvement, this scheme avoids infringing the previous patent.

25.13: Dynamic Damping

Any PI TX design where the coil current is stopped and the coil left floating needs a method of damping to prevent oscillation. Usually this is done with a physical resistor but it can also be done with a dynamic circuit. US7075304[20] shows a method where the damping resistor is replaced with a MOSFET whose conductance is controlled by a DAC and microcontroller. The MOSFET therefore acts as a voltage-controlled resistor.

Figure 25-26 shows a method that is described by patent AU2013101058. In this scheme the traditional resistor is replaced by a MOSFET which can behave as a variable resistor. Note the similarity with Figure 25-22. In Figure 25-22 Q2 is used strictly as an isolation switch; in Figure 25-26 Q2 performs (along with Q3) both isolation and damping. During flyback, Q3 shorts the preamp input to ground and Q2 is driven by a control voltage to behave as a linear current sink. When the flyback reaches 0V, Q3 turns off and Q2 becomes a low resistance link from the transmitter to the preamp. The control voltage is supplied by the sample/hold circuit which samples the preamp output and automatically adjusts to establish the fastest damping possible.

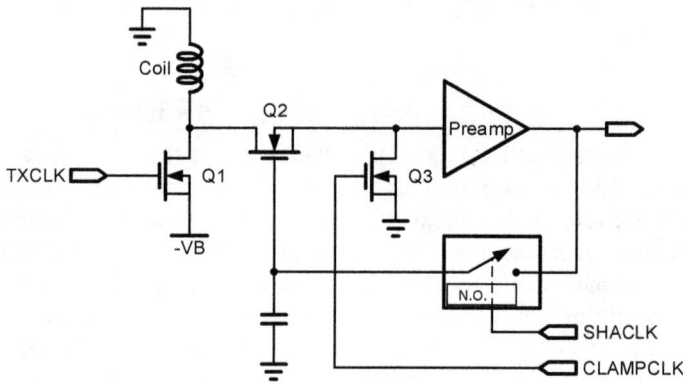

Fig. 25-26: **Dynamic Damping**

25.14: Iron Discrimination

Anyone who experiments enough with PI will eventually use an induction-balanced coil and look at the receive signal during the TX pulse. Unlike the TX-off decay, the TX-on response will include both reactive and resistive components. This will seemingly allow us to discriminate beyond the simple low-conductor/high-conductor classification achieved with the TX-off response.

On the bench, it is quite easy to distinguish ferrous from non-ferrous by looking at the TX-on response[21], and often this is encouraging to the novice experimenter[22]. In the field things are not so easy. Minelab patent US5506506[23] covers this method and the difficulties it presents. Namely, ground mineralization adds its own reactive response, and the ground response strength quickly overwhelms

19. The offset will be improved by the gain of the diffpair, which will typically be around 80 when +V = 5V. It is further improved by selecting resistor values for a good inherent balance of the D1 and Q3 voltage drops.

20. The inventor listed for this patent is Carl V. Nelson of Johns Hopkins University. Mr. Nelson has numerous PI-related patents that are worth a look.

21. See 26.12 for further discussion and signal plots.

22. As stated somewhere earlier in the book, good bench results gets you 10% of the way there.

23. "Metal Detector for Detecting and Discriminating Between Ferrous and Non-Ferrous Targets in Ground"

the iron target's reactive response, making the iron target appear to be mostly (or entirely) resistive, just like a non-ferrous target. That is to say, the method may work for shallow iron targets, or iron targets in mild ground, but not so much for deeper iron in mineralized ground. The same is true for VLF detectors.

Another approach to iron discrimination is to take a more detailed look at the TX-off decay response by taking more samples of the decay. Non-ferrous targets tend to have a fairly straight-line decay in a log-linear plot, except for an initial excitation non-linearity of about one tau (the inertial response). Ferrous targets tend to have a more pronounced non-linear response due to the high permeability of the material. Even then, an exceptionally high conductive target (say, an Atocha bar) might easily be mistaken for an iron target. Like the TX-off analysis, this seems straightforward and will probably work well on the bench. In the field, especially in viscous soils, the composite response may be a significantly muddled mess.

26 Hybrid Methods

> "Any technology distinguishable from
> magic is insufficiently advanced."
>
> — *Barry Gehm*

Modern detectors can be defined almost exclusively as either frequency-domain designs (such as VLF) or time-domain designs (such as PI). Most multifrequency designs fall in the frequency-domain category as they are extracting target phases. What differentiates the two methods is that frequency-domain designs use continuously changing TX waveforms whereas time-domain designs have dead times in the TX waveforms[1]. It is possible to combine the two methods and create a detector that has both a VLF response and a PI response.

Half-Sine

26.1: Transmitter

In VLF detectors the transmitted field is a continuous sinusoid. In PI systems it is a periodic pulse. With half-sine (HS) we combine the two and produce a periodic pulse whereby each pulse is half of a sine wave (Figure 26-1). This method is as old or perhaps older than multifrequency but has been underutilized in metal detector design. It was pioneered by Barringer Research in the 1960s and 70s for aerial detection of ore bodies; see patents US3020471[2] and US4506225.

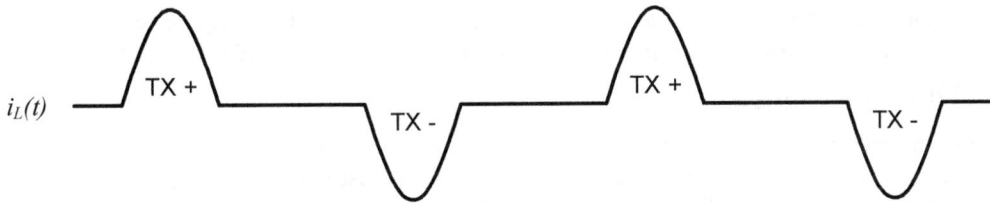

Fig. 26-1: **Half-Sine TX Current**

The half-sine current waveform shown in Figure 26-1 is a bipolar type much like the bipolar PI design in Chapter 23. However, each pulse is instead an alternating half-sine wave. From VLF transmitters (and circuit design in general) a capacitor placed in parallel with the TX coil (parallel resonance) is an effective way to produce a sinusoid. A capacitor can also be connected in series with the coil (series resonance) to do the same, and that is the approach taken here.

Figure 26-2 shows the half-sine transmit circuit. It is an H-bridge design that is a single capacitor more complicated than the bipolar PI H-bridge from Figure 23-6. The only difference is the additional series capacitor. The half-periods of the pulses follow the familiar LC tank frequency equation:

$$f_0 = \frac{1}{2\pi\sqrt{LC}}$$

Eq 26-1

Therefore each half-pulse has a duration of

---

1. We'll ignore for now that there are time-domain detectors that use continuous waveforms, like the Tarsacci.
2. Note Figure 3 in US3020471 which shows a peak half-sine current of 300 amps!

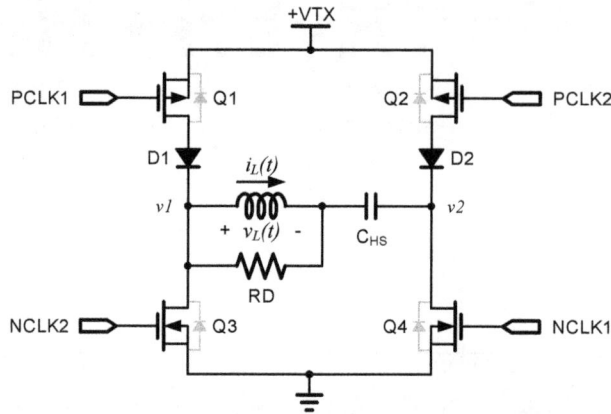

Fig. 26-2: **Half-Sine Transmit Circuit**

$$T_{on} = \pi\sqrt{LC} \qquad\qquad\qquad \text{Eq 26-2}$$

The spacing between the pulses has the same rules as the spacings between PI pulses: basically whatever you want within the constraints of what needs to be done and how big of a battery you can carry.

As an example, suppose we want something similar to the *Pulse-5* design: 100μs pulses at a bipolar pulse rate of 2kHz. Furthermore, we'll specify the same 300μH coil that has been used throughout the *Pulse-X* designs. The required capacitor is therefore

$$C = \left(\frac{T}{\pi}\right)^2 \cdot \frac{1}{L} = \left(\frac{100\mu}{\pi}\right)^2 \cdot \frac{1}{300\mu} = 3.38\mu F \qquad\qquad \text{Eq 26-3}$$

The closest standard value is 3.3μF which results in a pulse width of 98.85μs. The pulse width of the driving clocks needs to be at least this wide but can also be a little bit wider. That is, if PCLK1 and NCLK1 turn on for 100μs the half-sine pulse width will still be 98.85μs because the half-sine current naturally stops when it reaches zero and the diodes prevent reverse current flow. Therefore timing isn't critical except that the clock widths should be at least as wide as the desired half-sine pulse.

26.2: Hybrid Response

This approach produces two distinct regions of operation. There is a sinusoidal region where the TX current behaves exactly like a normal sinusoidal VLF, at least for a half-cycle; we will call that the *VLF region*. There is also an off-region where the TX current has ceased; this mimics the behavior of a PI and we will call that the *PI region*. If we use an IB coil arrangement and sample the RX signal both during the half-sine and also just after cessation (see Figure 26-3) we will get both a VLF-like response and a PI-like response at the same time — a true hybrid detector.

Note that a positive and a negative VLF region effectively equate to a single sinusoidal cycle. We will consider the half-sine's *frequency* to be the frequency of this sinusoidal cycle, regardless of the dead-time spacing. For the previous example where the pulse was 100μs, a positive/negative pulse

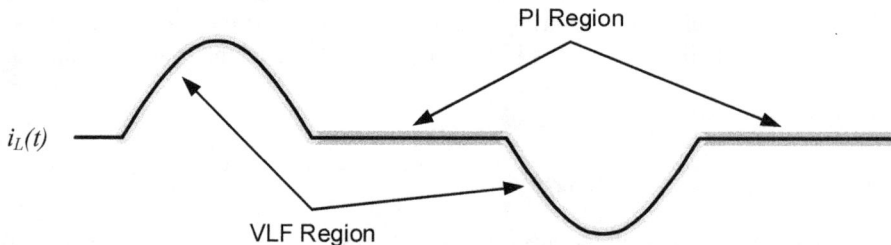

Fig. 26-3: **Hybrid Response**

totals 200μs and so will have a frequency of 5kHz. This frequency is equivalent to a normal VLF running at the same frequency. But the pulses also have dead time between them, so we also define a half-sine *pulse rate* to be the rate of the individual pulses. Again, the prior example stated a pulse rate of 2kHz, which means the 100μs pulses are separated by 400μs of dead time.

26.3: Efficiency

The transmitter for a standard sinusoidal VLF utilizes a parallel capacitor with the coil to create a parallel LC tank. The purpose of this is to recycle the energy used to create the TX signal resulting in low overall power consumption. The half-sine TX circuit also has an LC resonant system but instead of parallel resonance it uses series resonance. It therefore seems that the half-sine transmitter might offer efficiencies similar to a VLF and it can, but we need to understand how this works.

Referring to Figure 26-2, when the coil is initially energized in one direction (via Q1, Q4) current flows from V_{TX} through the coil and capacitor to ground. The current causes the capacitor voltage to charge eventually to V_{TX}; this is the point that current ceases to flow because diode D1 prevents reverse current flow. This is the natural end of the first half-sine cycle.

When the coil is energized in the opposite direction (via Q2, Q3) the same thing happens as well except that the starting voltage of the capacitor now adds to V_{TX}. That is, the energy that passed through the coil in the previous cycle is now stored in the capacitor as a voltage, and that voltage (ideally) doubles the drive voltage ($2V_{TX}$) on the subsequent cycle. The voltage doesn't actually double because there are resistive losses in the TX circuit that consume energy. Through more cycles the peak capacitor voltage will continue to build up and, subsequently, the peak half-sine current will also continue to increase until the resistive energy loss in each cycle equals the potential incremental boost and the whole thing levels out.

What this means is that the effective drive voltage for the coil may be quite a bit higher than V_{TX}, depending on the resistive losses. Depending on the desired peak current, this means that you can possibly run a low-voltage V_{TX}. This is how the half-sine transmitter efficiency is manifested: instead of a higher V_{TX} that needs to provide little current to a parallel resonance system, the series resonance system consumes a higher current but can use a lower V_{TX}. Energy is recycled in a way that is not as intuitive as a parallel resonant system, but knowing how it works is important when it comes time to design a half-sine transmitter.

26.4: Slew Rate

A potential disadvantage of half-sine compared to PI is in the turn-off slew rate. We have generally preached that a faster current slew rate at turn-off evokes a stronger target response, although Figure 5-24 shows the limitations of this gospel. It is obvious that the half-sine slew rate is far less than that of the typical PI. For example, if the previous example 5kHz half-sine has a peak current of 2 amps[3] then the slew rate at turn-off is

$$SR = \omega \cdot I_{peak} = 62.8 mA/\mu s \qquad \text{Eq 26-4}$$

Even a mundane PI detector can achieve a turn-off slew rate of 1A/μs so at 5kHz the half-sine slew rate is around 16 times slower.

But depending on target conductivity and the frequency chosen, the half-sine PI response can perform as well as a traditional PI. For example, Figure 26-4 also shows a 50kHz half-sine which, with the same peak current of 2 amps, has a final turn-off slew rate of 0.63A/μs, much closer to traditional PI. However, the half-sine frequency is usually selected for the desired target conductivities to be detected and we take whatever pulse response we get. For example, 50kHz would most likely end up in a gold nugget design whereas 5kHz would be more suited for deep silver.

---

3. To compare with a traditional PI, which might also have a typical peak current of around 2 amps.

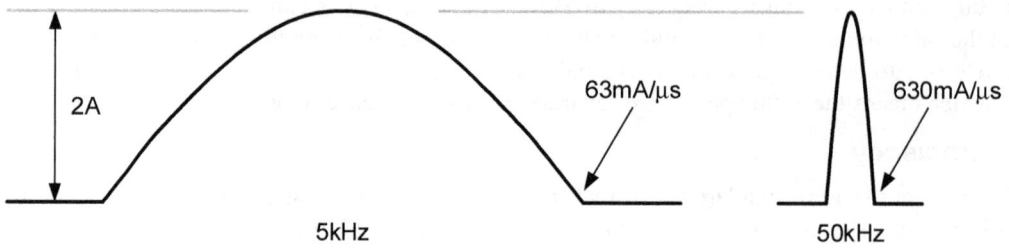

Fig. 26-4: **Low-Slew vs High-Slew**

26.5: Ideal Target Responses

Before a receiver circuit can be designed for a half-sine system we first must determine the behavior of the target responses and how we might sample them. In order to take full advantage of both regions of operation an IB coil is required. If a mono coil is used it will be limited to the off-time PI response which takes all the fun out of it, so we will proceed assuming an IB coil built to the recommended dot-polarity convention in Chapter 8.

As a first-cut attempt at an intuitive analysis consider how targets respond in a continuous sinusoidal system. For this we can simply look back to the waveforms in Chapter 14, keeping in mind that they represent a frequency of 10kHz. Figure 26-5a replicates Figure 14-7b except that the transmit current waveform is added. These are therefore raw voltage waveforms as seen at the RX coil (see Figure 14-6).

Also added are a couple of gray dashed lines which exactly frame a half-sine of the TX current. If we simply crop all the waveforms to these dashed lines then the results are shown in Figure 26-5b. The TX voltage v_L is shown with sudden transitions at the beginning and end of the half-sine current which is very close to reality. In an ideal inductor the voltage is the derivative of the current ($v = L di/dt$) so for a bipolar half-sine current we would expect the inductor voltage to look like that in Figure 26-6. The non-ideal inductor has series resistance and distributed capacitance so the actual voltage waveform

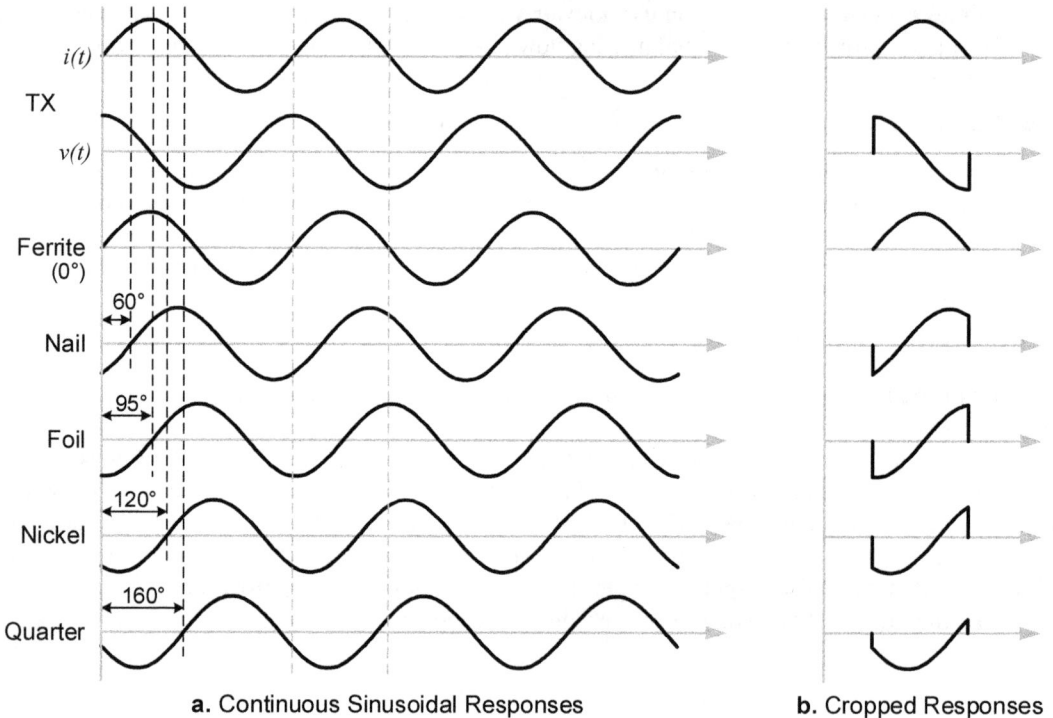

a. Continuous Sinusoidal Responses

b. Cropped Responses

Fig. 26-5: **Continuous vs Cropped Responses**

Inside the Metal Detector

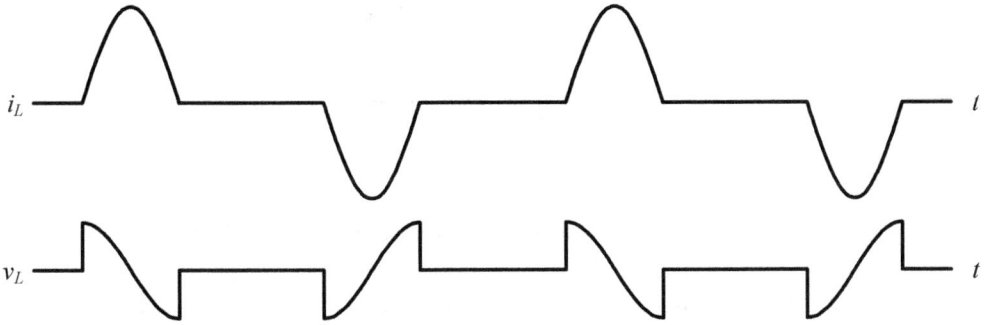

Fig. 26-6: **TX Voltage Waveform**

doesn't have such perfect step transitions, but they are pretty close. Notice that the zero-crossings of v_L occur at the peaks of i_L, a useful aid in determining the locations of the current peaks. In real life the series resistance of the coil will slightly delay the voltage zero crossing but there are ways to manage that.

The remaining responses in Figure 26-5b are targets with varying phase delays from 0° (ferrite) to 160° (silver quarter). The response for ferrite looks identical to the TX voltage waveform and the fast edges remain fairly accurate because the response of ferrite is almost instantaneous. Other targets do not have instantaneous responses so the truncated waveforms are not accurately depicted.

26.6: Actual Target Responses

Like the TX coil voltage, the target responses are shown with sudden transitions at the beginnings and ends of the half-sine current. Magnetic targets like iron exhibit a response delay due to the movement through the B-H curve (see Chapter 4). This delay not only shows up in the phase shift of the sinusoidal response but also as non-ideal rise and fall times of the start and stop transitions.

Eddy targets can be even more pronounced in this respect and are easier to mathematically analyze than the time domain response of a B-H curve. Consider the previous half-sine system with a frequency of 5kHz and a target with a tau of 20µs. Each half-sine pulse is 100µs wide and for this frequency we would expect the target to have a phase shift of

$$\phi = 90° + \text{atan}(\omega\tau) = 90° + \text{atan}(2\pi \cdot 5k \cdot 20\mu s) = 122.1° \qquad \text{Eq 26-5}$$

The ideal truncated response would look like that in Figure 26-7.

As we know from studying PI responses eddy currents do not occur instantaneously; they build up and decay exponentially. This exponential response must be applied to the half-sine case as well. A 20µs target tau takes about 100µs to fully settle and that happens to be the width of the pulse. With the exponential build-up and decay applied the response now appears as in Figure 26-8 along with the ideal truncated response (gray curve).

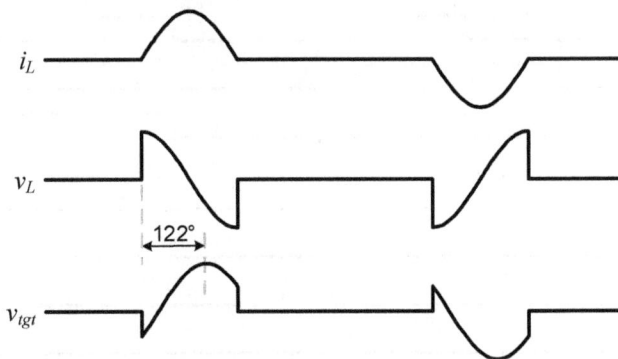

Fig. 26-7: **Ideal Target Response (tau = 20µs @ f = 5kHz)**

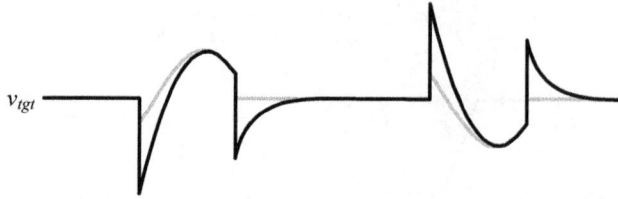

Fig. 26-8: **Target Response With Exponential Build-Up**

The exponential build-up causes both a higher initial transient and a lag in the initial part of the response; as the start-up exponential portion decays away the waveform eventually catches up to the ideal steady-state response. This delays the zero-crossing of the signal and further alters the negative-positive ratio so it will also alter the reactive and resistive demodulator ratio, meaning that calculated phase angle won't be what you'd expect compared to a continuous sinusoidal system. However, it is an easy matter to adjust the math (whether done in analog circuitry or DSP) to align the half-sine responses with the actual targets.

We also see that there is now a response "tail" beyond the half-sine excitation and this becomes our PI response. The PI decays behave exactly the same as with the bipolar pulse design in Chapter 23 except, perhaps, with less kick. But the lack of a several-hundred-volt flyback means we can also sample earlier, even immediately.

26.7: Receiver

The VLF regions can be sampled in a quadrature manner exactly like a normal VLF, the difference being that the VLF positive and negative regions are no longer contiguous. Figure 26-9 shows what the timing will look like. Sampling the PI regions is likewise the same as was done in the bipolar PI design (*Pulse-5*) with an early sample and (potentially) a late sample for subtractive ground balancing.

Both VLF and PI sampling require knowledge of when the half-sine pulse ceases but the end of the pulse is set by the LC resonance, not a timing pulse. Therefore we need some way to detect the end of the pulse. For that we can look at the TX voltage waveform v_L. A simple comparator circuit can be

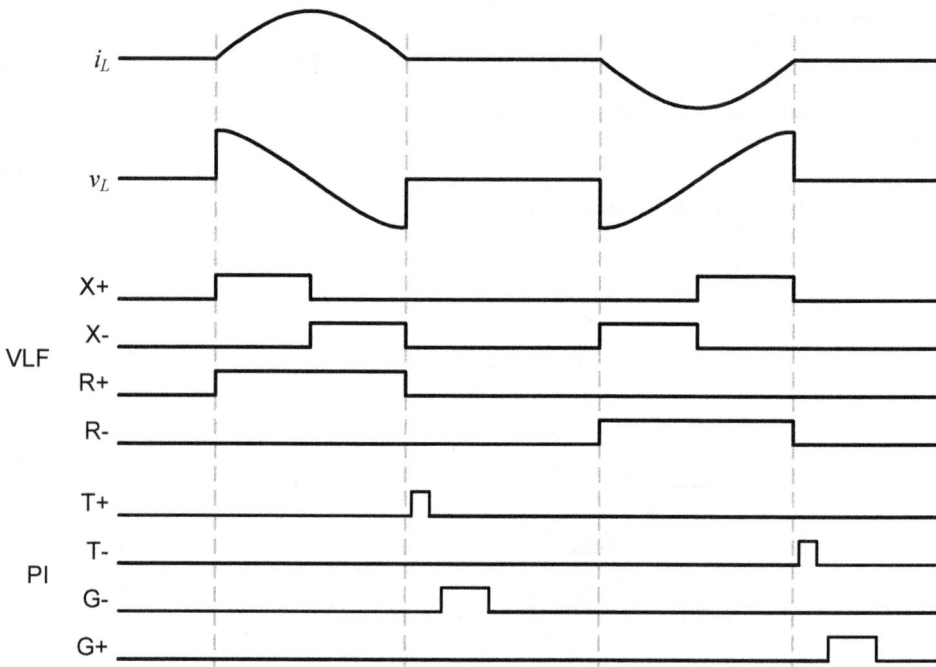

Fig. 26-9: **VLF & PI Timing**

devised to detect the final voltage transition at coil turn-off. It can also be useful to know exactly where the peak of the of the current pulse occurs. Again, a comparator can be used to detect the voltage zero-crossing.

26.8: Conceptual Circuit

Although the *Advanced Methods* section of this book does not include any ready-to-build designs, everything needed to design and build a half-sine project has been presented in earlier chapters. We now present a conceptual circuit which can be filled in with earlier real circuitry.

The transmitter shown in Figure 26-2 is usable as-is but, unlike the bipolar PI transmitter in Figure 23-9, requires four clock signals instead of two. Also, in PI designs it is always a good idea to AC-couple at least some of the TX clock lines so that the transmitter can never be left in a state where DC current can dump through the coil. In the half-sine transmitter the series capacitor naturally blocks DC current so this precaution is not necessary, meaning that you can simply add some 4053-type switches to Figure 26-2 and drive it directly from a micro.

The concept for the RX circuit is shown in Figure 26-10. This only shows the RX coil, preamp, demods, SAT filters, and post-SAT amplifiers. The remainder of the design might consist of analog dis-

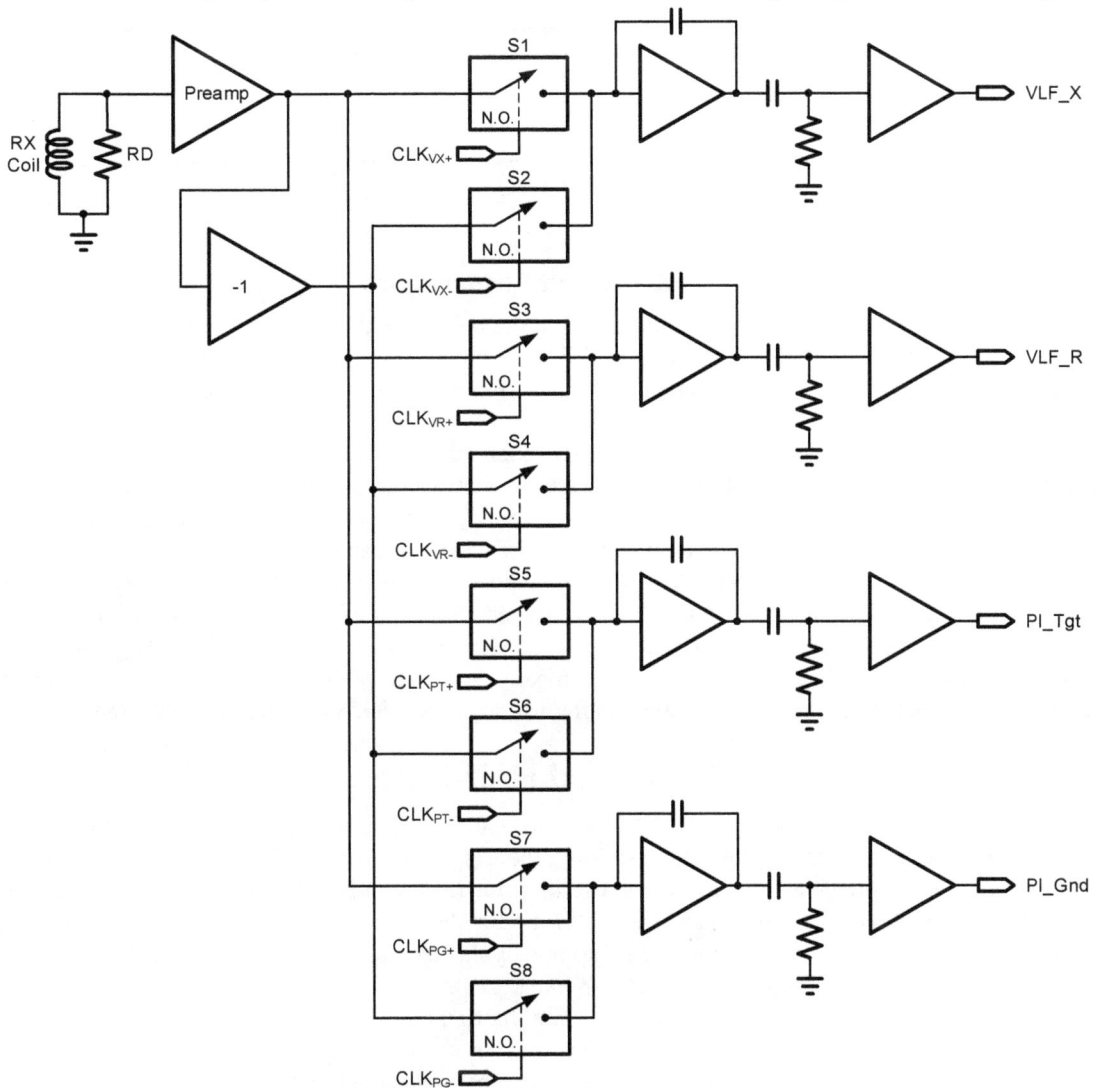

Fig. 26-10: **Half-Sine RX Concept**

crimination and ground balance circuitry and audio, all from previous designs. A modern design would instead use an ADC to digitize the four signals and do everything in DSP. In fact, if the demod offsets are tolerable then even the SAT filters would be moved into DSP.

26.9: Processing

The result of this approach is that there are basically two complete metal detectors running at the same time. An obvious way to deal with this is to treat them separately: ground balance the VLF and PI sides independently, and monitor their responses independently. The PI side would preferably be an all-metal channel that gives the most depth, and the VLF side would be used for target ID within the limitations of its depth.

Another approach is to blend the responses together. Let's look at how this might be done to implement ground balance. In perfect lossless ground the RX response will be the same as $v_{ferrite}$ in Figure 26-5b. If the resistive channel samples (R+ & R-) are taken as in Figure 26-9 then the ferrite response in the resistive channel should be zero. But it's not necessary that the sample widths fill the entire pulse width; we could shorten the samples as shown in Figure 26-11 and still get the same results.

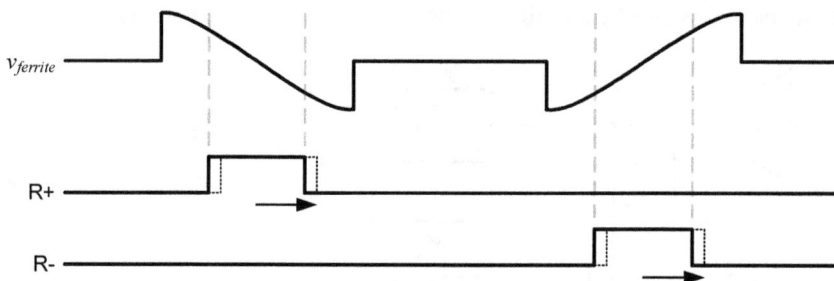

Fig. 26-11: **Shortened Resistive Sampling**

If the R+ and R- pulse widths are maintained but the pulses are then moved a little to the right then the resistive channel will exhibit a slightly negative response to purely reactive ground. Meanwhile, a PI response will exhibit a positive reaction to ground. If the two responses are added in proper proportion then the ground response can be nulled.

Meanwhile, a metal target will cause the resistive channel to react with a positive signal as will the PI channel. Adding the resistive and PI channels therefore cancels the ground response but *increases* the response to targets. And depending on the delay of the PI response the depth increase can be made to favor low conductors or high conductors. It is truly a rare case of a free lunch. It is not so certain that the free lunch will taste all that good as the VLF ground response and the PI ground response may not cancel well in highly variable ground conditions. This was a research project of mine at White's and had reached the very point of implementing this method of ground balance. I had it working and was beginning limited ground testing when I parted ways with the company. Since I developed the technology at White's it was not appropriate to continue investigating it after leaving. See US9285496 for more information.

Truncated Half-Sine

Depending on the frequency of the VLF portion, the Half-Sine approach can leave us with a PI turn-off slew rate that is uninspiring. In describing the transmitter we noted that the TX clock pulses need to be at least as wide as the half-sine pulse in order for the pulse to complete its natural cycle. Extending the clock pulse a little wider does nothing. But shortening the clock pulse will truncate the half-sine before it has a chance to complete its normal cycle. The resulting TX current is shown in Figure 26-12. This approach is called *truncated half-sine* (THS).

The advantage of this approach versus the normal half-sine is that the turn-off slew rate is substantially higher, basically similar to a normal PI. This produces a stronger response for the PI portion of the receive signal and also gives a more consistent PI turn-off regardless of the VLF frequency chosen. This is a system I was developing at White's[4] and had built a dirt-swinging prototype before I left.

Fig. 26-12: **Truncated Half-Sine Current Waveform**

In terms of VLF sampling and signal processing, we don't really need the entire half-sine wave-form to understand what is under the coil. What we need for ground response is the peak of the half-sine where *di/dt* is zero, and a little beyond that to account for ground variations and coil resistance. Practically all other targets, ferrous or non-ferrous, can be determined by processing only the first half of the half-sine pulse. So the cut-off needs to be somewhere after the crest and a good choice is between 2/3rds and 3/4ths of the half-sine pulse. The example waveform in Figure 26-12 is a 5kHz half-sine truncated at the 3/4 point.

26.10: Transmitter

The transmitter for THS is exactly the same as for HS; the only difference is that the timing is changed to turn off the H-bridge before the natural end of the half-sine pulse. A consequence of this is that a large flyback voltage will be developed and the components for the H-bridge must be selected with this in mind. Figure 26-13 somewhat illustrates the coil voltage; it is not to scale as the flyback peaks may be in the hundreds of volts.

One disadvantage of THS versus HS is that the early truncation leaves energy in the coil that must be removed. The remaining energy is dissipated with a damping resistor in the usual way, resulting in a high-voltage flyback. The HS approach allows all the coil energy to discharge back into the series capacitor (minus resistive losses) which recycles the energy. Therefore, THS is less fuel-efficient than HS, and the energy loss increases with earlier truncation.

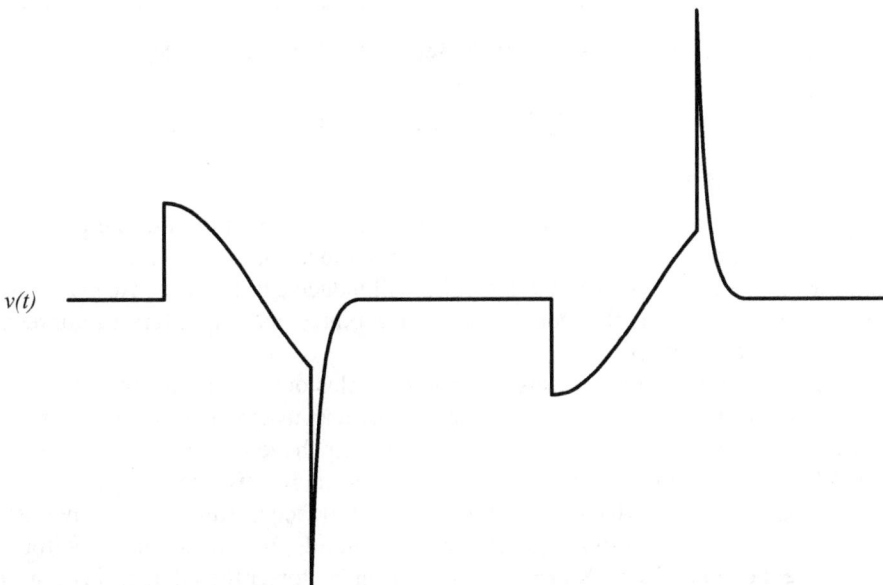

Fig. 26-13: **Truncated Half-Sine Voltage Waveform**

---

4. The same patent (US9285496) mentioned two paragraphs ago. The free-lunch ground balance method men-tioned in the Half-Sine section was actually developed on a Truncated Half-Sine platform; it should work with either one.

26.11: Receiver

The receiver design for THS is exactly the same as for HS. Unlike the normal half-sine system the exact moment of current cessation is controlled by the clock timing so there is no need for a method to detect the end of the current pulse. The timing for THS is more abbreviated than what HS can support as shown in Figure 26-14. Notice that both the reactive and resistive samples are reduced by 50% compared to the half-sine case. Because all the circuitry is basically identical and the main differences are a matter of timing it would not be difficult to design a system that can run as either HS or THS.

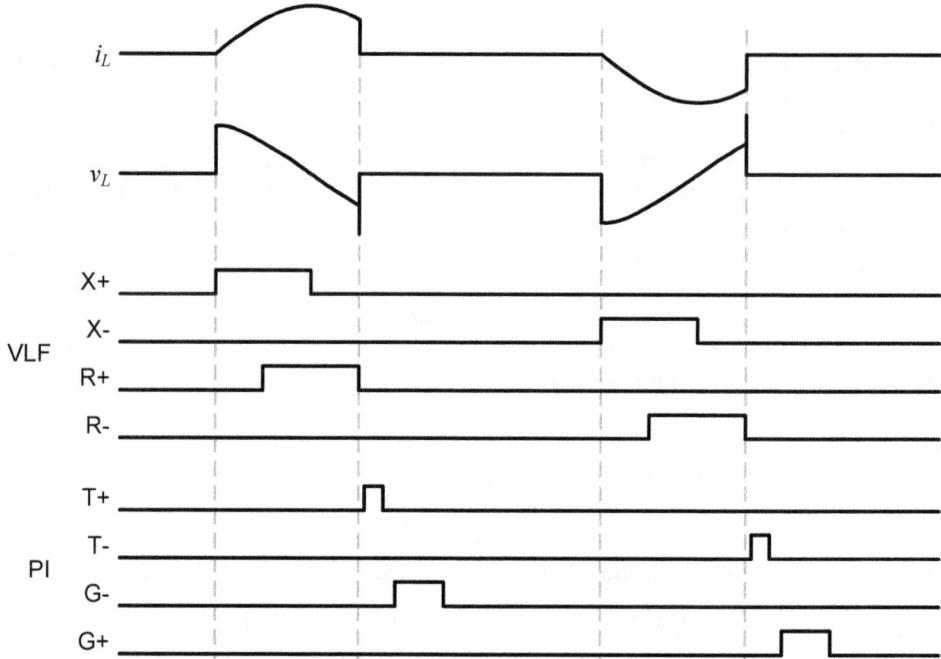

Fig. 26-14: **Truncated Half-Sine Demod Timing**

Other Hybrids

26.12: PI-TX Response

In the *Explore* section of Chapter 25 there was brief mention about monitoring the target response during the TX pulse turn-on for the purpose of PI discrimination. For this, an induction-balanced coil is required. Like any active TX waveform, the TX pulse will induce a target response that consists of both resistive and reactive components that could be used to produce not only a ferrous/nonferrous indication but even a VLF-like target ID.

Figure 26-15 shows a reference (no target) waveform plus our five usual target responses as voltages on an induction-balanced RX coil. It is easy to see that ferrous targets (ferrite, nail) have a positive TX-on response while non-ferrous targets (foil, nickel, quarter) have a negative TX-on response[5]. This is the same as VLF where ferrous and non-ferrous have opposite reactive polarities.

It is not difficult to imagine what demod timing could produce useful results. Obviously, monitoring the response from the TX turn-on requires separate demods from those that monitor the turn-off response. And since the *di/dt* of the TX turn-on is significantly slower than that of the turn-off, it is reasonable to assume that the TX turn-on response will be a weaker signal. That is, the turn-off response offers the most raw depth but with no target ID, and the turn-on response can be used to ID targets but to a lesser depth.

---

5. When you finally get to Appendix C, take a look at Figure C-6 and notice the similarities.

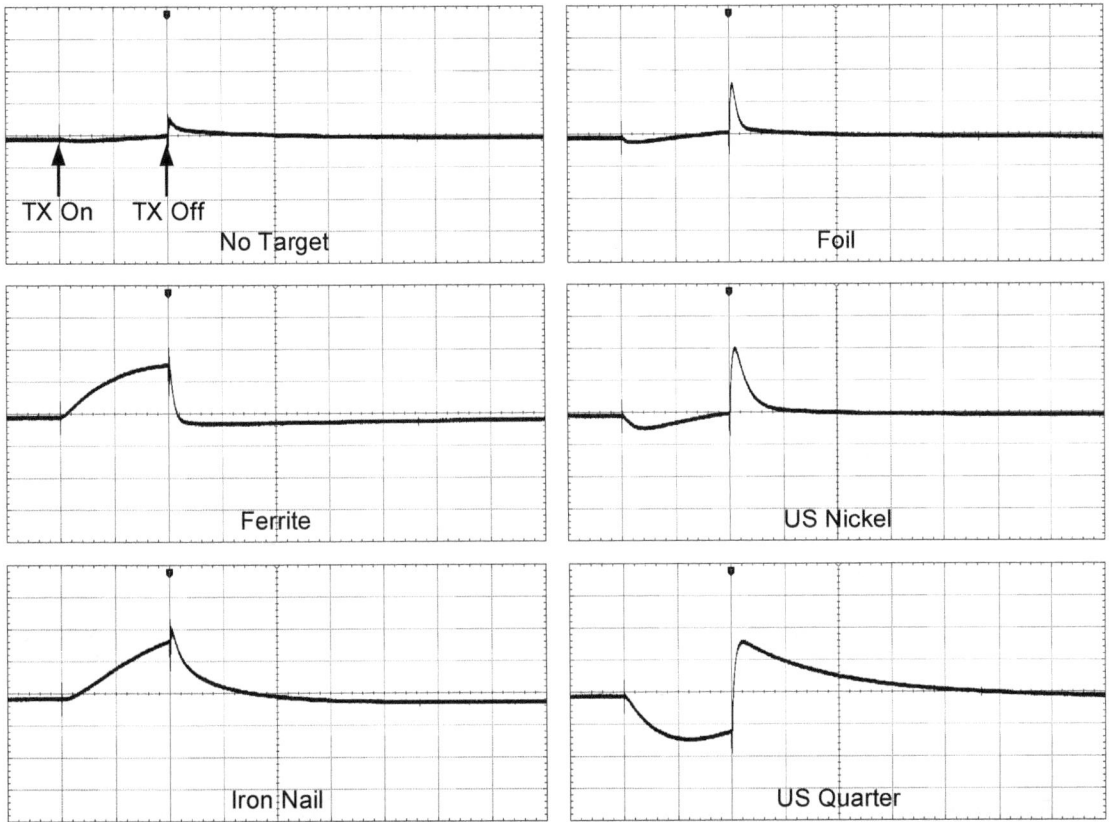

Fig. 26-15: **PI-TX Target Responses**

This is, in fact, exactly the method used in many PI detectors on the market (such as the Minelab *GPX* series) to achieve some level of iron ID. It is why this feature only works when using DD coils and not mono coils. However, as mentioned in the *Explore* portion of Chapter 25, the reactive ground response can quickly overwhelm the reactive target response and make it difficult to get a true reading of the target. Furthermore, small flat pieces of steel (especially from rotting tin cans) have mostly an eddy response and tend to look non-ferrous. These are often prevalent at old mine workings.

26.13: Gold Sweeper

The *Gold Sweeper* is a design developed by Allan Westersten and is described in US7701204. In concept it is similar to the truncated half-sine whereby the TX drive includes a ramping current, followed by a flat-top region, followed by a high-slew turn-off. Westersten suggests sampling during the TX waveform ramp-up (points a & b), during the flat-top (point e), and during the flyback (points c & d). The patent suggests that variable ground is automatically compensated for with no resulting target hole. Westersten claimed to have a working prototype with good performance. He died in 2022 and the patent has lapsed.

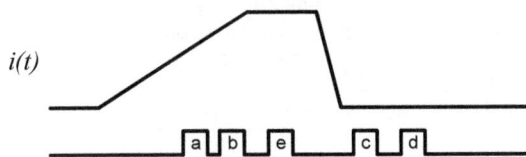

Fig. 26-16: **Gold Sweeper Concept**

26.14: Pulse Devil

The *Pulse Devil* is a design developed by David Emery[6] and is described in patent US7710118[7]. It is a quasi-hybrid in that it uses a PI transmitter to produce a VLF-like response but never really extracts a PI-like response. However, rumors of the design suggest that it had near-PI depth while offering VLF-like discrimination.

Because the RX coil is resonated with a capacitor the pulse response of a target causes a decayed ringing response. Quadrature demodulators are used in the same way as a VLF to extract target amplitude and phase. In the absence of a target, the coil response is a minimal feedthrough spike since it is induction-balanced. Therefore a reference burst (BCLK) is injected on the RX coil in between TX pulses which creates a large ringing response that can be used to established timing for the demods. See Figure 26-17. A metal target creates a larger decaying sinusoid (similar to the reference burst) which can be demodulated to determine the phase response of the target and, at a minimum, determine whether it is ferrous or non-ferrous.

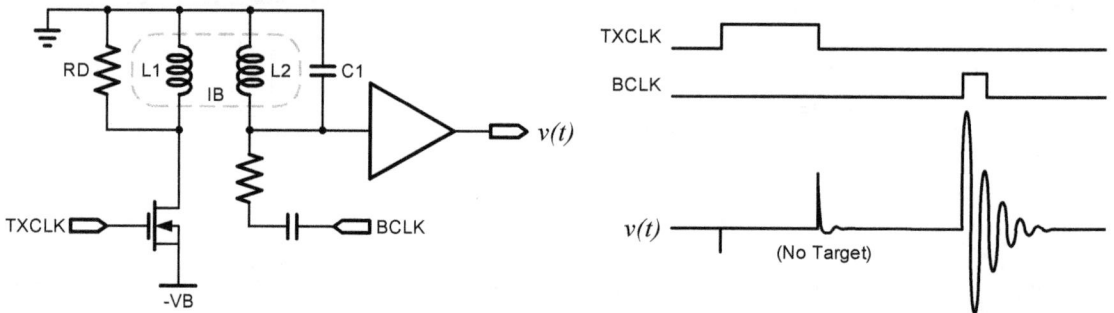

Fig. 26-17: **Pulse Devil Concept**

26.15: Voodoo

Voodoo is a design developed by George Overton[8] and published in full detail in a 2020 book called *The Voodoo Project*[9]. *Voodoo* combines PI with a VLF-ish free-ringing response that has the ability to distinguish ferrous from non-ferrous. Figure 26-18 illustrates the concept.

As with most any hybrid design, *Voodoo* requires an induction-balanced coil. The TX coil is pulsed in the normal way of a PI and is properly damped to produce a classical PI response. The RX coil, however, is not damped and is allowed to ring at a frequency determined by a resonant capacitor (C1). In this respect it operates much the same as the *Pulse Devil* concept. As with the *Pulse Devil*, the VLF response in the absence of a target will be a small feedthrough spike such as that shown in Figure 26-17. Also like the *Pulse Devil*, a target's VLF response will be a large decaying sinusoid that can be demodulated to extract the phase response. Unlike the *Pulse Devil,* there is no reference burst to establish demod timing; the demod timing is simply referenced off the TX pulse by the micro. Overton suggests this works as well and there is no need for a reference burst.

Voodoo therefore has two complete metal detector paths: a PI path with a single demodulator and a VLF path with quadrature demodulators. Since the PI path does not include a ground channel it cannot be properly ground balanced, but the VLF path can be. However, the deepest depth comes from using

---

6. Emery was active on the detector technology forums for several years. At one time he stated he was going to self-produce the *Pulse Devil* and sell it to the public but never did. I met Emery at a GPAA Gold Show in North Carolina and he was to bring the *Pulse Devil* with him to demonstrate it to me. He showed up but without the *Pulse Devil*. A few years later I joined White's Electronics and invited Emery to visit and demonstrate the *Pulse Devil* for a potential licensing deal. He was receptive to the idea but never followed up. See the "Going Solo" section in Chapter 26.

7. Now expired due to non-payment of maintenance fees.

8. George was the co-author of the first two editions of *ITMD*.

9. Found on Amazon, ISBN 979-8690296544. Shameless plug.

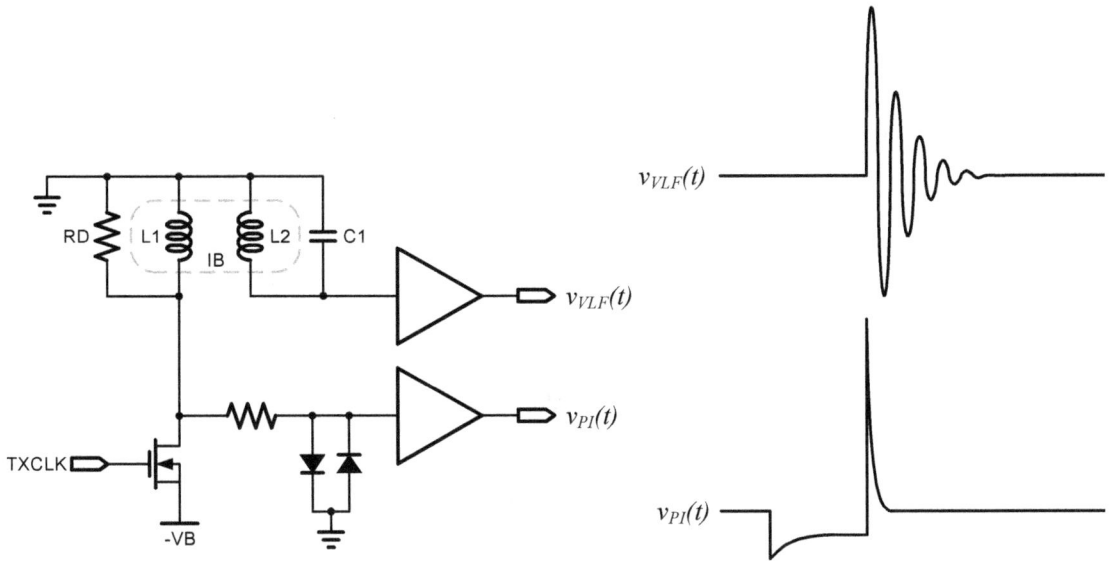

Fig. 26-18: **Voodoo Concept**

the PI side for target indication, with the VLF side used for discrimination up to its useful depth. While the design is more useful for milder grounds, there is no reason why a ground channel cannot be added to the PI side.

Explore

26.16: Multifrequency Half-Sine

In arriving at a usable design for this chapter we simply chose a desired pulse width (100µs) and came up with the required series resonant capacitor (3.3µF). A different pulse width could have been chosen which would have required a different capacitor. This would have produced a different half-sine frequency as far as the VLF side is concerned, and a different turn-off slew rate and likely a different pulse rate as far as the PI side is concerned.

The aforementioned patent US9285496 includes information on how to implement a system that can transmit two (or more) frequencies. Figure 26-19 short-cuts straight to a two-frequency solution — the H-bridge includes a new half-bridge leg for a second capacitor C_H. As an example, if it is 330nF (assuming the same 300µH coil) then it will produce a 10µs pulse width, or an effective 50kHz frequency.

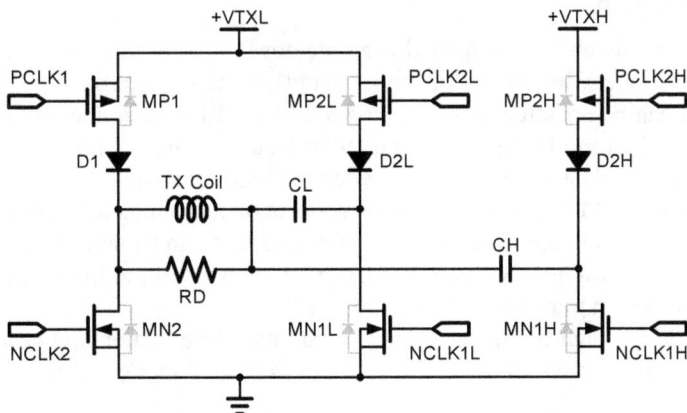

Fig. 26-19: **Two-Frequency Transmitter**

Notice that the high-frequency leg is also powered by a higher voltage V_{TXH}. This is not absolutely necessary but if the same voltage is used to drive two different frequencies then the higher frequency will have a lower current amplitude and an identical turn-off slew rate compared to the lower frequency. Figure 26-20a illustrates using 3:1 frequency ratios. In order to keep the peak half-sine currents equal (Figure 26-20b) the high-frequency half-bridge leg will need a supply voltage higher in the same proportion as the frequencies. That is, if the low frequency is 5kHz powered by 5V, then a high frequency of 50kHz would need 50V.

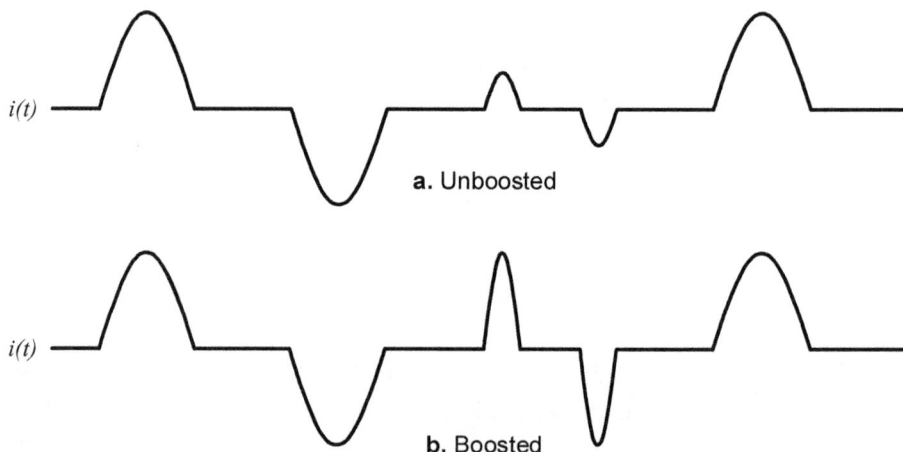

Fig. 26-20: **Boosted vs Non-Boosted**

This sounds like a tall order but it's not so bad. There are numerous switching boost regulators available that will do the job. And while a higher frequency requires a higher voltage supply its current consumption drops compared to the low frequency. This eases the demand on the boost regulator.

The two frequencies can be alternated in sequential time as shown in Figure 26-20b. As we saw in the chapter on multifrequency designs the sequence can be altered to enhance one frequency over the other, for example, to better detect certain desired targets. Or, if the high frequency leg is unboosted then it may be necessary to run several high-frequency cycles for each low-frequency just to better equalize their performances, much like what was done in Minelab's BBS/FBS technique.

The RX design will require a full suite of additional demodulator channels for each additional frequency. A two-frequency design would therefore have eight complete channels with each channel needing two demod clocks. The sheer number of clocks (16) probably limits the design to two frequencies. I built a boosted two-frequency half-sine transmitter at White's as part of my research project and it worked very well, but never married it to a receiver.

26.17: Faking a Hybrid

For the purpose of this chapter, a hybrid is a detector that simultaneously operates in a PI-type mode and a VLF-type mode, whatever those modes entail[10]. Another approach is the dual-mode design whereby the detector can be operated in a PI-type mode or a VLF-type mode but not at the same time. An early example of a dual-mode detector is the 1970s Bounty Hunter *Outlaw* which could be run in either BFO mode or TR mode with the flip of a switch (and maybe some retuning).

Another research project I did at White's was a concept of combining a *Surfmaster PI* with an *MXT* (the *Triton Project*). The challenge in designing a detector that can be switched from a PI mode to a VLF mode using the same coil mostly lies in the design of the transmitter. In Figure 26-21 capacitor C1 can be optionally switched in to resonate the coil for VLF mode.

Unlike a true hybrid, a dual-mode detector can share the demodulators between the two modes. A single pair of demods can be used for the X and R channels in VLF mode and also used for the T and G

10. Or, perhaps, operates simultaneously in a time-domain mode and a frequency-domain mode.

channels in PI mode. All this assumes that the design is heavily based in DSP and not analog circuitry. A flip of a switch changes the CapSEL line, TX and RX timing, the DSP path, and possibly the preamp gain and bandwidth.

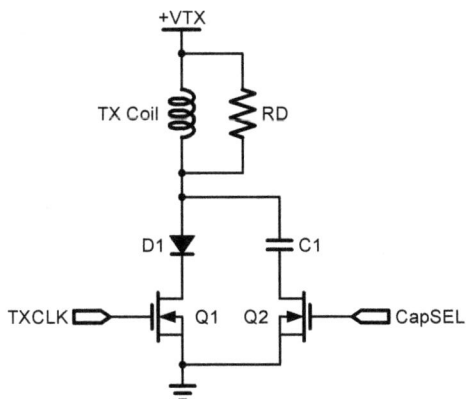

Fig. 26-21: *Triton* **Transmitter**

GARRETT ELECTRONICS

Garrett Electronics Close-Up...

We want you to feel a part of the great Garrett team. We are proud of our company and our leadership role in the industry. Our engineers will continue their exhaustive research and development while maintaining close contact with both the professional and the hobbyist in the field. Our office, production, and customer service personnel are dedicated to providing you with the best possible service and product quality. The Garrett team effort is your assurance of the finest in metal detector instrumentation!

1. Whenever you are in the Garland area (near Dallas), stop by for a visit. You are always welcome!

2. All kinds of "treasure found" are displayed in the museum. In our demonstration area the operating capabilities of all our detection equipment are explained to customers.

3. Bill Bosh is Garrett's Dealer Coordinator. Garrett detectors are marketed in every state in the U.S. and in several foreign countries.

4. On a continual basis, our engineering and design team, led by Bob Podhrasky, improve and update our products, as well as develop new and more efficient detector instrumentation.

5. This is our final assembly area where all Garrett detectors are carefully assembled and tested. Charles Garrett personally oversees all operations.

6. Each Garrett detector receives 300% inspection. Every part and assembly is 100% quality control inspected at least three times. Garrett detectors are built to last!

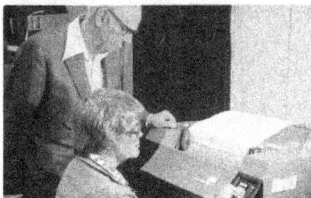

7. Detector orders come in from all over the world. Orders are speeded through our computerized system. Fast, accurate shipping to our customers and dealers, regardless of destination, is our byword.

8. Our shipping personnel make an early start each day to insure that all orders are shipped, usually, by the first day following their receipt.

9. Our customer service and repair departments speedily take care of any problems or service requirements you have.

10. We maintain a continuing "in touch" program of factory seminars, training and informing Garrett dealers of the latest in detectors and products and know-how so that they may serve you better.

11. Garrett detectors have become known world-wide through our international expansion program. Charles Garrett visits Frank Mellish, our United Kingdom representative, at his London treasure hunting supplies shop.

12. Garrett's new and up-dated equipment is demonstrated to factory dealers so that they know how to instruct their customers in the correct use of the newest and best equipment.

PART 7
Leftovers

eti ELECTRONICS TODAY INTERNATIONAL

Dec. 1980 $1.60 NZ$1.75

GOLD DETECTOR
Can discriminate 'trash' from 'treasure'

BUILD A pH METER
Keep your pool or fish tank healthy; lots of other uses etc.

Voltage Regulators –circuits and techniques

Learn BASIC

Pot core coil design

Hi-Fi Reviews:
Marantz ST500 FM/AM tuner
PAS-30 loudspeakers from G.R.D.

Win a ZX80 COMPUTER — WITH SOFTWARE

---

elementary Electronics
SEPTEMBER/OCTOBER 1980

Including Science & Electronics

INFLATION-BUSTING TECHNOLOGY

- Metal Detector makes your fortune
- Energy Sentry checks power savings
- Darkroom Contrast Meter cuts paper losses

...CEPTS IN ...CTRONICS
...omputer Thermostat ...microprocessor
...o Digitizer ...rcuitry aids ...ng

GE's new HELP! —don't leave home without it!

Australia's Top Selling Electronics Magazine

Electronics Australia
DECEMBER 1984
AUST $2.80
NZ $3.30

F/A-18: Australia's new tactical fighter
McDONNELL DOUGLAS

Stand-alone EPROM copier

BEACHCOMBER Metal Detector

Special Subscription Offer: **WIN A PCB ETCHING KIT!**

Realistic CD player review

Printers: new technology, lower prices

EVERYDAY WITH **PRACTICAL** **ELECTRONICS**
JUNE 1994

INCORPORATING ELECTRONICS MONTHLY   FULLY S.O.R. £1.95

SMART SWITCH
REDUCE YOUR ELECTRICITY BILL

ADVANCED TENS UNIT
FOR PAIN RELIEF

DIGITAL WATER METER
KEEP A CHECK ON CONSUMPTION

MICROCONTROLLER
P.I. TREASURE HUNTER

INDEPENDENT MAGAZINE for ELECTRONICS, TECHNOLOGY and COMPUTER PROJECTS

27

Oddball Circuits

"Results! Why, man, I have gotten a lot of results.
I know several thousand things that won't work."

— *Thomas A. Edison*

This book generally classifies detector circuits as proximity, IB, and PI. Sometimes a method (like half-sine) blends technologies but it is still clearly anchored in these three broad methods. Sometimes a circuit simply doesn't seem to fit anywhere at all, or does something in a weird way. This chapter looks at some of the oddballs.

27.1: Thomas Scarborough

Thomas Scarborough is a minister in South Africa who occasionally dabbles in electronic design and has come up with a number of unusual metal detector circuits. Some of his designs attempt to implement a metal detector with the absolute minimum number of components. Over the years his circuits have been published in various electronics magazines.

27.1.1: Beat Balance[1]

This design combines beat frequency with induction balance so Scarborough calls it "Beat Balance." "Induction balance" really refers to the coil design and, as we've seen throughout the book, IB coils can be applied to most any metal detector technology. The Beat Balance approach clearly falls into the proximity genre and operates very much like a BFO.

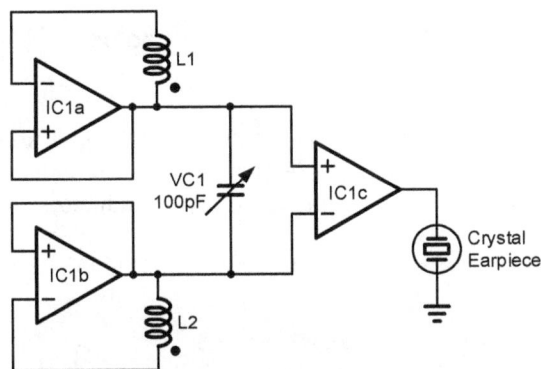

Fig. 27-1: **Beat-Balance**

In the absence of a metal target the matched IB coils run at the same frequency so the output is DC and no tone is heard. When a target enters one of the coil fields its frequency shifts, resulting in a difference tone in the earpiece. A target in the exact center of the coil has equal effect on both oscillators, resulting in an audio null. Therefore, sweeping over any metal target will always produce a double beep.

---

1. *Everyday Practical Electronics*, May 2004.

27.1.2: CCO Metal Detector[2]

The CCO (Coil Coupled Operation) is a variation of the BFO whereby the search oscillator uses a transformer-coupled oscillator (TCO) topology. Figure 27-2 shows the schematic. The search oscillator runs at about 500kHz and its frequency is influenced by metal targets. Potentiometer VR1 provides a means for adjusting the frequency. The reference oscillator (Colpitts type, but using an inverter) is said to run at 2.7MHz which means that the circuit uses harmonic mixing. It is likely that the two oscillators actually run at a nearly exact ratio of 5:1. Opamp IC2 mixes the two frequencies and provides an audio signal to the earpiece.

The search coil is not induction-balanced, if it was the oscillator would not oscillate. But sensitivity to proximate metal likely depends on being fairly close to the IB point. It is also unclear (I have never built this circuit) whether the TCO is sensitive to ground mineralization; I suspect it is.

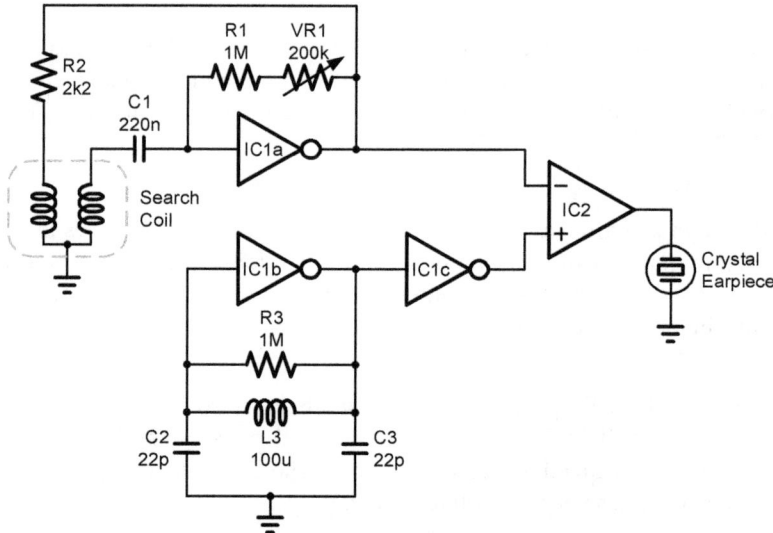

Fig. 27-2: **CCO Metal Detector**

27.2: Balanced Inductor Design

The term "induction balance" traditionally means that the RX coil is arranged so that it is situated in an induction null of the TX field. This design does not do that so, strictly speaking, it is not an induction balance design. However, it does require balanced driven inductors so we'll refer to it as a *balanced inductor* design.

In this design a differential TX circuit drives a pair of coils and the output is taken at the center of the two coils:

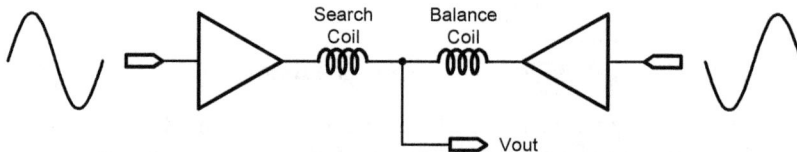

Fig. 27-3: **Balanced-Inductor Transmitter**

If the transmit drive is well-balanced and the coils are matched then the nominal output at the center point is zero. If a target is brought near one of the coils then the balance is upset and a voltage waveform appears at the center point.

---

2. *Everyday Practical Electronics*, Nov 2004.

Assuming the target affects only one of the coils then that coil simultaneously behaves as a transmit coil and a receive coil. As a transmit coil it produces a magnetic field that stimulates the target. As a receive coil, the target produces a field which induces an EMF in the coil and that EMF can be seen at the center point.

A design using this method might place one coil winding in the coil search head and the other coil inside the control box. The sole purpose of coil in the control box is to balance the search coil so that the center-point voltage is well-nulled. This would suggest that the coils should be well-matched but obviously an exact copy of the search coil will be too large for the control box. Also, the control box coil should be designed so that it will not be sensitive to any metal that might come near, like a user's ring or wristwatch. A potential solution is to use a shielded inductor in the control box and to replicate its inductance and resistance when making the search coil winding. Even then, there may be differences that result in a residual signal at the balance point, including imbalances due to coil resistance and the effects of temperature.

The center connection of the inductors can be fed to a preamp and then quadrature demodulated for doing the same kind of processing done in VLF-IB designs. Doing so, in fact, produces the same reactive and resistive baseband signals that can be used to extract target phase. In the end, the balanced-inductor approach gives a full VLF-like response but with what appears to be a mono coil.

27.3: Minelab *Go Find*

In Figure 27-3, if the search coil is resonated with a series capacitor and then driven at its resonant frequency, then its overall impedance will appear to be purely resistive. This means that the balance coil can be replaced with a balance resistor (Figure 27-4a) and the center connection will still be a nulled signal and a target response can be quadrature demodulated.

The Minelab *Go Find* uses this approach. It is critical that the balance resistor matches the resistance of the search coil resonance resistance, which is likely dominated by the wire resistance of the coil. For a copper wire coil, this can be done by using a PTC resistor with a matching tempco. Instead of this, the *Go Find* replaces the balance resistor with two anti-phase coils (Coil1, Coil2 in Figure 27-4b) whose total wire resistance matches that of the search coil. Coil1 and Coil2 are, in fact, wound with bifilar twisted wire to ensure they do not pick up either the transmitted magnetic field or the target field.

This method is explained in patent US9557390. The patent also describes noise reduction methods whereby an additional non-inductive bridge is used to create a noise-only signal that is subtracted from the main signal.

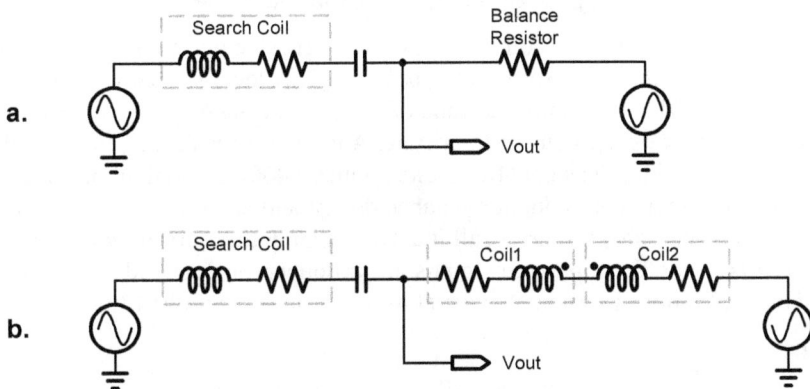

Fig. 27-4: **Minelab Go Find**

27.4: Buckless Concentric Coil

Normally a concentric coil includes a TX bucking coil adjacent to the RX coil for the purpose of nulling the static RX signal. It is possible to create a nulled output without the bucking coil, as shown in Figure 27-5. With the RX coil connected to the driving TX voltage with the right coupling polarity, it is possible for the induced RX voltage to cancel the direct TX voltage. For the diameter of the RX coil (typically half the diameter of the TX coil) this requires getting the number of turns just right. Challenges with this approach include the fact that turns are in integer increments and may not be exact, and parasitic effects (coil resistance and interwinding capacitance).

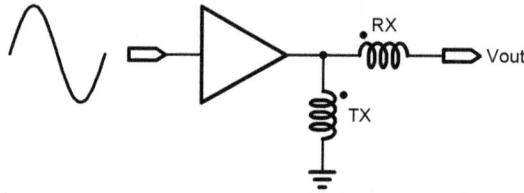

Fig. 27-5: **Buckless Concentric Coil**

27.5: White's Mono-Coil TR

Around 1981 White's produced a basic detector called the *Coinmaster TR* which has a search coil containing a single center-tapped winding. The detection schematic for the *Coinmaster TR* is shown in Figure 27-6.

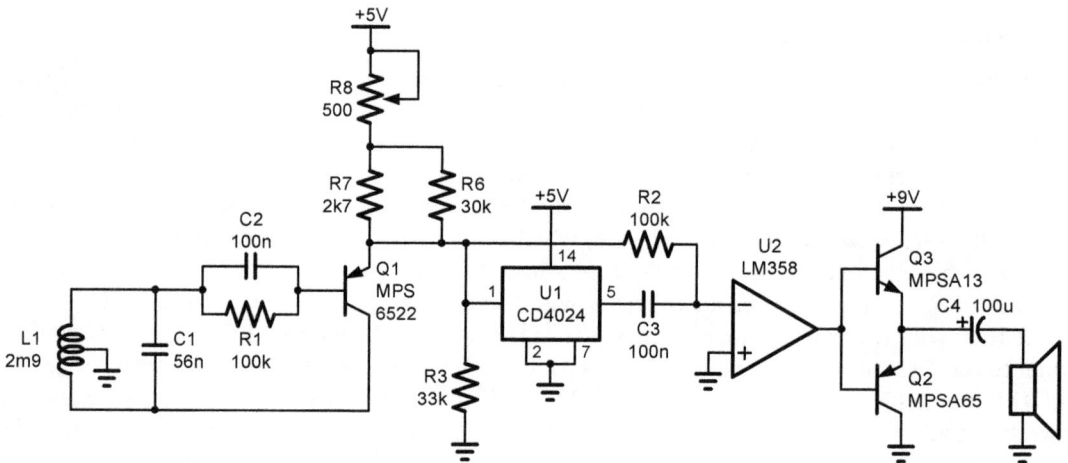

Fig. 27-6: **White's *Coinmaster TR***

In the oscillator the center-tapped coil produces a 12.5kHz sine wave on the collector of Q1 and an identical but 180° phase-shifted sine wave on the base of Q1. A level-shifted version of the base signal appears on the emitter of Q1 and its amplitude and offset are controlled by R8. In the absence of a target the offset is too high for the binary counter to operate. A target causes the amplitude to drop so that the counter operates, which divides the signal by 32 and applies a 400Hz signal to the speaker.

This circuit is certainly not an induction balance design and is more akin to the energy theft technique of Chapter 9. It is probably unfair to call it a TR design in the spirit of what we consider a traditional TR to be. It is ironic that opposite signals are produced on each side of the coil in a similar manner as the balanced-inductor design above, yet the operation is entirely different.

27.6: Tarsacci

In 2018, Dimitar Gargov independently began producing a metal detector called the Tarsacci *MDT 8000*. It is a selectable frequency VLF design yet it does not use the usual frequency domain processing of other single-frequency VLFs. Instead of transmitting a sinusoid, the TX voltage drive is a square wave. The resulting RX signal for eddy targets is an exponential curve, and the Tarsacci analyzes the curvature directly instead of band-pass filtering it into a sinusoid for frequency domain processing. That is to say, the Tarsacci is a time-domain VLF design. US10969512 is a patent that explains the operation in more detail.

27.7: Inductive Sensor Chips

Single-chip metal detector ICs have been produced that allow construction of a basic design with minimal parts. These chips are usually called "inductive proximity sensors" which is a marketing term for "metal detectors." However, because they are intended for close proximity their detection distances are usually quite limited.

27.7.1: CS209A

An early chip was the CS209A from Cherry Semiconductor, a part which is now obsolete. Figure 27-7 shows a typical application circuit which produces an audible beep when metal is detected. Potentiometer RV1 adjusts the sensitivity. Pins 4 and 5 are open-collector outputs, with pin4 normally high and pin 5 normally low. A data sheet can still be obtained through On Semiconductor and lists typical detection distances well under one inch.

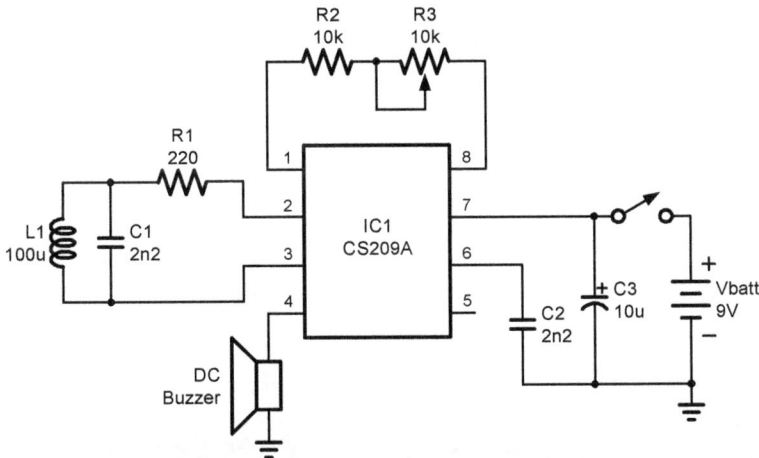

Fig. 27-7: **CS209A Application**

27.7.2: LDC1000

Although the Texas Instruments LDC1000 — which they call an "inductance-to-digital converter" — has been obsoleted, it has spawned a family of similar parts. The single channel LDC1001 has replaced the LDC1000 and other parts have two or four channels. Unlike the CS209A, an LDC chip cannot be used by itself; it requires a microcontroller. The LDC1001 is capable of running from 5kHz to 5MHz. An application circuit is not given here as it mostly consists of the LDC chip and the micro.

THE GROUND EXCLUSION BALANCE

The Ground Exclusion Balance works because each kind of "target" actually has two effects on the field. The field produced by a conducting target weakens the transmitted field, and also delays the detected unbalance by a time that is characteristic of the particular target (whether it is made of nickel, silver, gold, etc., as well as being influenced by its size and shape). The field produced by an artificial magnetic material, such as a bottle cap or nail, strengthens the transmitted field, and also makes the field "seen" by the receive coil earlier than the regular field. It advances the time of the detected field.

The effect produced by magnetite bearing ground differs from either of these. This produces quite a strong unbalanced field, which gets bigger as the loop assembly is lowered closer to the ground. This is what, in earlier types of instruments, completely swamps all but the biggest targets you want to measure, so they pass, undetected.

The Ground Exclusion Balance does its work, because the unbalanced field produced by a mass of ground containing magnetite has unique timing characteristics. It produces an unbalanced pickup very close in time to that of the transmitted field.

Because of this closeness in time, the unique, patented circuitry of the Ground Exclusion Balance instrument can tune it out, so the only field detected is one that **differs** from this precise timing. If the detected field is later in its timing, the object is conducting, if it is earlier in timing, the material is magnetic. An important feature of this method is the search frequency chosen, so as to achieve maximum penetration in the search, along with adequate exclusion capability.

Individual areas where ground contains a form of magnetite have slightly differing characteristics, which the instrument can adjust out so as to completely exclude the local ground. In a locality, concentration of magnetite may also vary, but the instrument excludes such variations in the same way that it ignores variation in height at which you hold the loop—**completely!** Such variations do not call for adjustment.

THE G.E.B. MODELS HAVE A INCREASED DEPTH OF DETECTION IN HIGHLY MINERALIZED SOIL OF APPROXIMATELY 200% OVER CONVENTIONAL BFO AND TR MODELS·.

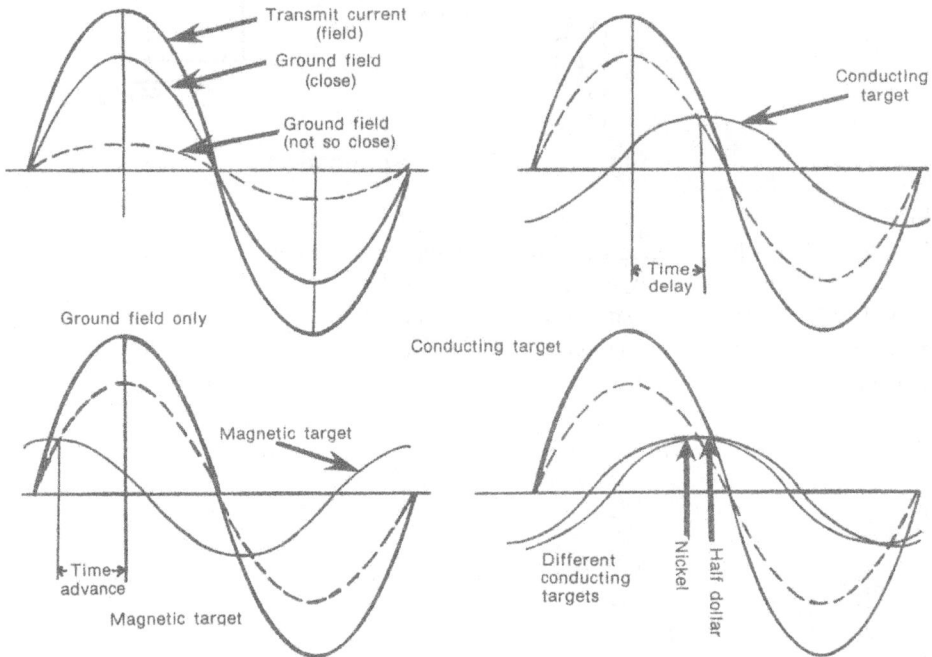

CHAPTER 28 Things & Stuff & Junk

"Engineering is the art of modelling material we do not wholly understand, into shapes we cannot precisely analyse, so as to withstand forces we cannot properly assess, in such a way that the public has no reason to suspect the extent of our ignorance."

— *British Institute Of Structural Engineers, 1976*

The purpose of this chapter is to cover some loose ends, things that did not fit anywhere else in the book. Some of this material covers practical matters, while some is intended to address popular myths and misconceptions.

28.1: Is it an antenna?

Many detectorists refer to the coil as an antenna. It is not, at least in the classical way that an antenna is used. Like an antenna, it converts electrical current into electromagnetic (EM) energy, and vice versa. But the EM energy produced by an antenna behaves differently depending on how far away from the antenna it is. There are two regions: the near-field region and the far-field region. With antennae, effective radiowave propagation occurs in the far-field region, while the near-field region is strongly reactive. The transition between the two regions is gradual, but basically occurs between one and two wavelengths away.

For a typical VLF operating at 10kHz the wavelength is 30,000 meters (roughly 18-1/2 miles); obviously, a metal detector does not operate in the far-field region. In either region the EM energy gets weaker as it travels farther away, because the energy is spreading out into a greater and greater expanse. In the far-field region the EM energy falls off as a function of $1/d^2$ (where d is the distance away), just like you would expect from basic algebra, where the surface area of a sphere is proportional to r^2 (radius). But reactive effects cause the near field energy to fall off as a function of $1/d^3$.

A radio antenna is designed to resonate at a particular frequency such that its capacitive and inductive reactances cancel and it becomes resistive. From classical antenna theory, elements like dipoles and circular loops are most efficient at transferring energy when they have dimensions (length or circumference) equal to a half wavelength. As an example, suppose you want to receive channel 3 (60MHz) on a television; the wavelength $\lambda = c/f$ (where c is the speed of light) so $\lambda = 5m$. A half-wave dipole for channel 3 would therefore be 2.5m long. For our 10kHz metal detector, we would need a loop with a diameter of about 3 miles. Obviously our coils are nowhere near the optimal size for classical radiowave propagation and that is not what we are using them for.

Instead of an antenna, a metal detector coil behaves exactly like a transformer; it transfers energy via induction. You can think of the system as being a double transformer; the TX coil and the target form one transformer, and the target and the RX coil form a second transformer. In fact, for the purposes of analysis and simulation, metal detectors are often modeled exactly this way.

28.2: Transmit Power

This topic often comes up in forum discussions. A common opinion is something like this:

"Metal detector power cannot exceed FCC limitations and is already maxed out."

Often a number is given for that limit, such as 100mW or 1W. Unfortunately, these statements come from people who understand neither metal detector operation nor FCC regulations.

When it comes to radio emissions, most people are familiar with the concept that radios emit power. After all, AM & FM radio stations often tout their power output: "WKLS, comin' at ya with

100,000 watts of pure rock-n-roll!." CB radio users also often understand that their rigs are limited to 4 watts. Metal detectors don't work like radios, and the search coil does not work like an antenna.

Although a circular loop could be used as an effective antenna[1], for the size of the search coil, metal detectors operate at frequencies way too low for this. For example, a 10kHz detector has a wavelength of 30km, yet a typical coil circumference is less than 1m. Instead, the search coil is operated as a purely inductive element that creates a magnetic field. In electronics, an ideal inductor does not dissipate power, so output "power" has no real meaning. Detectors only develop a local magnetic field and, unlike radios, don't propagate electromagnetic energy. Magnetic field strength is determined by the size of the coil and the "ampere-turns" — that is, the number of turns in the coil, and the peak current flowing through those turns.

You might think that a certain amount of "power" from the transmit circuitry is needed to achieve a desired magnetic field strength, but that will depend on the design of the circuitry. In most metal detector designs the search coil is resonated with a capacitor so that the coil current required to create the magnetic field is recycled through the capacitor during successive AC cycles. If the coil and transmitter have no resistance and there is no target in the TX field, then the efficiency is 100% and the coil consumes no power at all, no matter how strong of a magnetic field is produced. However, there are resistive losses in the coil and circuitry resulting in power losses. Furthermore, a metal target in the field "steals" energy to create eddy currents. These power losses are manifested in heat produced, not in radio waves transmitted.

The TX coil does not effectively create electromagnetic[2] radio waves and, in fact, the electric field portion is intentionally suppressed through the use of shielding. So for metal detectors, an AC current is passed through the coil to create an AC magnetic field, and that field does not propagate beyond the near-field region immediately around the coil.

The bottom line is, "output power[3]" has no real meaning when it comes to metal detectors. Instead, we can talk about the field strength of the coil's magnetic field in terms of, say, Teslas. Measuring the "power" of a detector's transmitter (its current consumption at its supplied voltage) tells us nothing about the transmitted field strength. In order to measure this you would need a magnetic field strength meter[4]. See Appendix A for such a design.

28.3: FCC Regulations

Even though a metal detector does not actually transmit electromagnetic power, they still need to meet FCC regulations[5]. The reason is because a poorly designed metal detector (say, with an unshielded coil) could have higher than expected EMI emissions that interfere with other devices. Metal detectors generally fall under the Code of Federal Regulations, Title 47, Part 15, or "47 CFR 15", or just "Part 15." Rule 15.3o specifies that induction devices are treated as intentional radiators:

> (15.3o) Intentional radiator. A device that intentionally generates and emits radio frequency energy by radiation or induction.

Notice that the rule specifically mentions the emission of RF "energy," not "power," and the magnetic field does have energy.

Rule 15.209 gives the radiated emissions limit for intentional radiators over a number of frequency bands. For metal detectors, only the first band is relevant:

1. I once built a Yagi-Uda array of circular loops instead of dipoles.
2. An electromagnetic wave consists of an electric field coupled with a magnetic field, both oscillating in a tightly coupled dance.
3. Anyone who has ever watched Tim "the Toolman" Taylor knows what More Power means. But Tim never discussed what more ampere-turns means, or more B-field strength. This is why even metal detector companies sometimes talk about transmit "power." It's a commonly understood expression, even if it is not technically correct. Transmit "energy" is more accurate.
4. Magnetic field strength meters are often marketed to the pseudoscience fields of ghost hunting and aura health. Be aware that these meters may not have the bandwidth needed to measure metal detector fields.
5. For the USA; or whatever radio emissions rules are in effect in other countries.

Frequency	Field Strength Limit	Distance
9kHz - 490kHz	2400µV/m/kHz	300m

The field strength limit of 2400mV/m/kHz depends on the frequency. For a 10kHz detector, the limit is 240µV/m. For an 80kHz detector, the limit is 30µV/m. These measurements are taken at a distance of 300 meters. Below 9kHz there are no limits, meaning that a metal detector can emit whatever it wants[6].

A field strength measured in µV/m corresponds to an electric field emission, which is normally part of an electromagnetic wave. Recall that metal detector coils are shielded to suppress the electric field, which makes this particular measurement fortuitous for metal detectors as a well-shielded coil with massively strong magnetic field strength could pass.

Just for fun, it is possible to do some calculations assuming radio wave transmission and back into an equivalent isotropic transmit power. Let's assume a detector frequency of 10kHz. The field strength limit is 240µV/m at 300m. The power density at that point is

$$P_D = \frac{E^2}{Z_0} = \frac{(240\mu V/m)^2}{120\pi\Omega} = 152.8 pW/m^2$$

Z_0 is the intrinsic impedance of free space and is roughly 377Ω. This gives us the power density at 300m away. To get the equivalent isotropic transmit power we simply multiply this by the surface area of a sphere with a radius of 300m:

$$P = P_D(4\pi r^2) = 152.8\frac{pW}{m^2}(4\pi \cdot 300^2) = 172.8\mu W$$

Therefore, *if* our 10kHz detector transmitted radio waves, it would be limited to 172.8µW of (isotropic) transmit power[7]. But it doesn't transmit radio waves, so that number is meaningless.

While 15.209 regulates the intentionally transmitted emissions of the detector, there are other regulations that apply to metal detectors. Most detectors now use a microcontroller with digital clocking, so the entire circuitry must be tested as an *unintentional* radiator under 15.109. Whereas 15.209 is concerned with emissions only at the transmit frequency, 15.109 is concerned with emissions at all frequencies beginning at 30MHz:

Frequency Band	Field Strength Limit	Distance
30-88 MHz	100 µV/m	3m
88-216 MHz	150 µV/m	3m
216-960 MHz	200 µV/m	3m
>960 MHz	500 µV/m	3m

Because circuit board EMI can easily contain a strong E-field component, passing 15.109 is usually a bigger challenge than passing 15.209. This is especially true for PI detectors. Sometimes it is necessary to add ferrite beads to cabling, or series resistance to exceptionally fast digital lines, or shield paint to the enclosure.

If the detector is plugged into an AC wall outlet — say, for the purpose of recharging an internal battery — then either 15.107 or 15.207[8] applies. This regulation limits the amount of noise that can be injected into the wall outlet by the detector. If the detector uses a stand-alone battery charger then the detector would not be tested for 15.107 but the charger would be.

---

6. This may not be true elsewhere. For example, the EU regulates emissions all the way down to DC.

7. Power is often specified in dBm. 172.8µW is −7.62dBm.

8. 15.107 is for an unintentional radiator, 15.207 is for an intentional radiator. Usually 15.207 applies, but the use of a battery charger can also be considered "unintentional" radiation. Often the testing lab will use their interpretation of the rules, so different labs may apply different tests.

Finally, some detectors include a wireless transceiver for implementing wireless headphones or, in the case of the XP *Deus*, for basic operation. In that case, the wireless transceiver is treated as a radio and must be tested according to its frequency band. For example, a Bluetooth radio operates in the 2.4GHz ISM band so it would fall under regulation 15.247.

Other regulations that typically apply to metal detectors are:

- 15.203 states that the "antenna" (search coil) must have a unique connector so that only a factory-supplied antenna can be used.
- 15.205 are some keep-out bands, mostly military.
- 15.215 ensures that the detector does not generate interferers in other particular bands.

As footnote 8 mentions, testing labs often determine what tests to apply so it can vary. Testing can be very expensive, often $10,000 - $20,000 just for FCC. EU testing (CE certification) can be even higher. And often, when the rules are changed, you have to resubmit a detector for a complete re-test.

28.4: Power Limits

Metal detector "power" effectively is not limited by 47 CFR 15.209. The practical limit on detector field strength has more to do with signal-to-ground ratio and practical battery life than FCC regulations. Signal-to-ground ratio is the amount of target signal compared to the amount of ground signal. As you crank up the transmitted field strength both go up fairly proportionally. Because deeper targets have more ground volume above them, this is a losing proposition for achieving greater depths. And, eventually, the stronger ground signal will overload the front-end of the detector resulting in zero target detection. The only cases where it can help is in very mild ground mineralization or when ambient EMI forces a reduction of receiver sensitivity. Even then, a stronger TX signal requires a better IB coil null to prevent overload from the residual RX signal.

Even in zero mineralization (air) the depth gained from a stronger TX field is downright depressing. A detector coil operates in the reactive near-field region where signal strength falls off as a function of $1/d^3$. The same is true for the target's returned signal strength, so the round-trip signal falls off as a function of $1/d^6$. What this means is that in order to double our detection depth we would need to increase the transmitted field strength by a factor of 64. Put another way, doubling the field strength will increase detection depth by a mere 12%, so if you could barely detect a target at 8" you would barely detect it at 9" if the coil drive is doubled. While the adage "every little bit helps" applies, at some point your will to carry a massive battery pack will be tested, or you will be swapping batteries often.

28.5: Metal in the Coil

In the 1970s and even into the 1980s it was common to see metal bolts used in the coil pivot mount, and even aluminum lower rods that extended all the way — or almost all the way — to the coil. In most cases static metal in or near the coil may cause a minor error in the IB null but, in motion-mode detectors, won't be detected as a target since it is stationary relative to the coil. And in non-motion-mode detectors the static metal becomes a signal offset that can be tuned out.

However, in the early days of the TR-Discriminator designs the static metal — especially the aluminum lower rod — created an error in the RX phase that affected the discrimination setting. This is when manufacturers began adding a plastic insert to the end of the aluminum lower rod, eventually making the entire lower rod out of plastic or fiberglass. And, for good measure, replaced the steel or brass clevis bolt with a plastic bolt.

But there are still at least two sources of metal in the coil: the coil windings[9], and solder connections. Also, many new models have a small circuit in the coil for a preamp or a microcontroller, so there may be a PCB with components, or perhaps even a battery as with the *Deus* models.

These internal metal components do not affect the coil null because they are present during the factory nulling procedure. However, they can still cause problems by creating ground noise. Ground min-

---

9. The wire used in coil windings can sustain local cross-sectional eddy currents which affect the overall eddy response in the RX coil, or alter the TX current in the TX coil. The problem increases as wire gauge increases.

eralization distorts the TX field so as the search coil is swept over variable ground the TX field distortion will vary. Variations in a magnetic field will induce responses even in an otherwise static piece of metal, and these induced responses will appear to the user as if a like piece of metal was passed over with the coil. In other words, it is possible for ground variations to "light up" a piece of metal in the search coil.

In the cases of the coil windings and the solder joints, these both tend to be very small with low conductivity. Therefore they mostly affect high frequency detectors like VLF prospecting models and high-speed PI models. Ironically, these are also the types of detectors most likely to be used over highly variable mineralization while looking for tiny low conductors. To mitigate the problem, designers often use litz wire when a larger wire gauge is needed, usually in the TX coil. This has a bonus advantage of reducing resistive losses in the coil due to skin effect, which reduces power loss[10]. With the RX coil, resistance usually is not as critical so it can be wound with a small gauge standard magnet wire.

Solder joint problems are mitigated by keeping the joints as small as possible (no large solder blobs) and keeping the joints farther from the ground side of the coil and outside of the RX coil. In WIPO patent 2011/116414 Minelab discloses a method of encasing solder joints inside a ferrite body to prevent the creation of ground-induced eddies.

When a PCB is placed inside a coil care must be taken to orient the PCB in such a way that any eddy current interaction with the RX coil is minimized. That is, place it outside the RX coil, perhaps on-edge, and perhaps in-line with the normal sweep direction. In placing metal (whether a PCB or just solder joints) outside the RX coil any induced eddy signal will present a response that is negative to a normal response; that is, it will end up in the fourth quadrant[11] where it may be easy to ignore.

In recent years many manufacturers have begun making the detector rod system out of carbon fiber for its light weight. However, carbon fiber is conductive so if the lower rod is also carbon fiber this brings us full-circle back to the problems of the aluminum lower rod. Compared to aluminum, carbon fiber is not as conductive and therefore presents a problem mostly to high-frequency VLF and fast PI designs used for prospecting. In fact, carbon fiber has a response almost identical to salt water and, when the lower rod is at a normal angle to the coil, it poses almost no problem for most types of detecting using lower frequencies. But at higher frequencies a carbon fiber lower rod can cause noise issues. When the coil is bumped about in normal use the rod moves slightly relative to the coil and causes falsing, and when the rod is laid against the coil (for example, during target retrieval) it creates a static amplitude and phase offset which makes the R (or G) demod more susceptible to overload or noise.

28.6: Power Line Interference

You might find that, in some instances, it is possible to hunt near power lines without interference and in other instances the same detector near the same power lines is noisy. The demodulators in a metal detector synchronously demodulate the detector's RX signal but also asynchronously demodulate any other incoming signal. Front-end filtering attempts to reduce incoming EMI but invariably some gets in.

Even though a detector might run at, say, 10kHz and the power line frequency is 60Hz (USA), the power line signal that makes it to the demodulator will get aliased into the baseband signal according to the ratio of the demod clock frequency and the power line frequency. If the demods are running at 60Hz, or any multiple of 60Hz, then the ratio is a whole number and a 60Hz input will produce a 0Hz signal at baseband. That is, the 60Hz EMI will look like a DC offset at the demod output. This is an ideal situation because the DC offset is removed by the motion filters. It was mentioned in Chapter 23 that, to minimize power line interference, a walk-through detector should be synchronized to the AC mains. But we can't really do that with a battery-operated detector[12].

---

10. Covered in Chapter 8.

11. See Chapter 18 for discussions on third and fourth quadrant signals.

12. A traditional pulse induction detector has a dead time between pulses. It is possible to use the dead time to monitor for AC mains interference and adjust the pulse rate until the interference is synchronous.

When the demod clock is not an integer multiple of the power line frequency then the residual amount tells us what will be seen at the demod output. For 10kHz and 60Hz, the ratio is 166.67. The fractional part — 0.67 — tells us that we are $0.67 \times 60 = 40$Hz from the next lowest integer multiple but also $0.33 \times 60 = 20$Hz from the next highest integer multiple. The smaller of these two numbers is what matters, and so we will potentially have a 20Hz interferer at the output of the demodulator.

The demod outputs are usually low-pass filtered with just enough bandwidth to pass a target response but, hopefully, low enough to at least somewhat suppress the aliased EMI signal. This is difficult with a 60Hz mains (and even more so with a 50Hz mains) as the highest aliased frequency you can ever hope for is 30Hz. Metal detectors are sometimes designed to use a TX frequency that results in a power line alias of exactly half the power line frequency. So instead of exactly 10kHz, you might choose 9.99kHz or 10.05kHz.

This brings us back to the lead-in sentence where sometimes there is interference and sometimes there is not. As it turns out, the AC mains does not run at an exact frequency; it varies slightly throughout the day, sometimes up to ±0.5%, although it is always adjusted for a long-term average accuracy[13]. For 60Hz, this might be a spread of 59.7Hz to 60.3Hz. If we carefully selected our TX frequency to be 9.99kHz and the local power line has risen to, say, 60.1Hz, then the ratio is now 9.99kHz/60.1 = 166.22 and the aliased frequency is now $0.22 \times 60 = 13.4$Hz. If this is inside the low-pass filter you will hear it.

28.7: Detector Cross-Talk Interference

Besides EMI from power lines and radio transmitters, metal detectors can interfere with each other. This can happen even with detectors that operate at different frequencies. Pulse induction detectors are notorious for interfering with just about anything around and are banned in many competition hunts. Even if you are not hunting near another detectorist, you might find that your metal detector and pinpointer interfere with each other.

It is not uncommon for a given detector model and a given pinpointer model to interfere with each other for one person, but another person using the exact same models does not have interference. The situation is very similar to what was described above for AC mains interference. Instead of a variation in the AC mains frequency, we have variations in the detector frequencies. In the old days, the TX frequency of a metal detector was set by an LC tank circuit which could vary by several percent from one unit to another (otherwise identical) unit. And changing coils could also cause significant shift in frequency. This made the old-style detectors a crap-shoot for, say, working around power lines.

The TX frequency of modern designs is set by a micro which, at best, is crystal controlled to within 25ppm, but often not that tight. 25ppm is 0.0025% so a 10kHz detector would have, at worst, a ±0.25Hz which is quite good. 100ppm is probably more typical for an external crystal which is ±1Hz, still not bad. but when your detector is operating at 50kHz, that becomes a ±5Hz change which could be significant. And many detectors don't use an external crystal, but rather use an oscillator inside the micro which is often ±0.5% for a whopping ±50Hz for a 10kHz TX. Add to that a pinpointer that also has its own tolerance and getting two that don't interfere becomes a statistical gamble.

The point of this section and the last is this: interference can happen with any detector, even one that seemingly has a good reputation for being immune to interference. And interference between a detector and a pinpointer is the same: it can happen for one person, but not another.

28.8: Silent EMI

Even when EMI is not being heard in the audio, it can still cause interference and reduce target depth[14]. The idea is that even though the demodulated EMI itself is out-of-band of the low-pass demod filters and is suppressed for the purpose of audio, it is still present in the demodulators and also in the

---

13. So that AC-motorized clocks will keep time.

14. Silent EMI was first postulated by metal detectorist Tom Dankowski. Tom runs the tech-leaning "Dankowski Detectors" forums and sells advanced detectorist training videos (free plug).

preamp beforehand. EMI in the preamp would only present a problem if it is large enough to overload the preamp, and it would have to be quite large for this to happen.

The demodulators are likely a different issue. Substantial EMI, especially in the presence of a weak target, can alter the amplitude and phase response of the target and mask its true nature. It is possible for EMI to move the phase response of a target into a discriminated area or completely to the ferrous side of the response scale. It is also possible for EMI to suppress the strength of the target. The end result is that the target can be masked and completely disappear.

This can happen sporadically so that, on one pass, you hear the target but on a subsequent pass it vanishes. In short, pay attention to the repeatability of targets, especially weaker ones. If responses are sporadic, you may be experiencing silent EMI.

28.9: The Halo Effect, Part 1

The halo effect is well-known by experienced detectorists and, briefly, is the concept that long-time buried targets can create a stronger response than freshly buried targets. This belief is reinforced by users who seem to detect unusually deep targets that, after recovery, cannot be detected as deep in a simple air test. The predominant theory is that long-time buried targets react with the ground to produce an enlarged detection "halo" area around the target itself. When the target is dug up, the halo area around the target is destroyed and sensitivity to the target returns to normal.

Whether the halo effect is real or not is very much debatable. There is no doubt that iron targets corrode in the ground and often the immediate soil around the target is discolored. Since an iron target can have a magnetic response it is possible that the iron oxide that leaches into the ground creates an enlarged magnetic response that makes the target easier to detect. Therefore, for iron targets the possibility of a halo effect has a plausible explanation.

Non-ferrous targets rely on eddy currents for detection, therefore a halo effect would need to increase the creation of eddies. The only reasonable possibility for this to happen is if target corrosion somehow created conductive ionic salts around the targets which could support eddy currents, much like wet salt sand does at the beach. This is highly unlikely with gold which is a mostly inert metal, and fairly unlikely with silver. Both gold and silver coins that are dug often come out of the ground in a bright clean state. Copper coins are often found in a slightly corroded state (mostly discolored) but, even then, copper is relatively resistant to corrosion.

There is a slim possibility that nitric acid in the soil, perhaps from the long-term use of fertilizer, could attack copper and perhaps even silver coins. With a copper coin this could create copper nitrate, which is an ionic compound and could conduct electricity. Even a small volume of ionic matrix could allow a coin's eddy currents (which are normally pushed to the perimeter of the coin by skin effect) to slightly expand beyond the edge of the coin, making the coin appear larger than it is and therefore detectable at greater depth. It's a theory with deniable plausibility.

If the halo effect is not real then what could account for excessively deep targets? A coin that is tilted or on-edge will pinpoint off to the side of the coil's center. The deeper the coin, the more off-center it will be. During retrieval most people will start with a small hole, especially when detecting on turf. As they dig deeper to find the elusive target, they naturally open the hole wider until, eventually, the off-center coin falls to the bottom of the hole. Thus it appears that the coin was found at the depth of the hole, when in reality it may have been much shallower. This is exacerbated by the fact that a tilted coin will read deeper than it really is on a detector with a depth meter. For example, if a coin that is tilted 45° reads 10 inches on a depth meter, it is really only 7 inches deep.

28.10: The Halo Effect, Part 2

It has been a more common claim with PI users that they are often able to detect items deeper in wet salt sand than in dry sand or air. Eric Foster did some experiments which seemed to confirm this.

As we saw in Chapter 4, skin effect pushes eddy currents to the physical perimeter of a target. In wet salt sand, electrical conductivity no longer stops at the edge of the target. The wet salt matrix has an eddy response of its own, usually at a lower conductivity than whatever target is buried. As the coil passes directly over the target, skin effect pushes the target eddies to the perimeter and they interact with the matrix eddies, which will tend to continue pushing the target eddies beyond the edge of the target. However, these "extended eddies" will likely weaken quickly with distance due to the low conductivity of the matrix. But it may still be enough to make the target look slightly larger than it really is, thereby increasing detection depth.

28.11: Treasure Hunting Methods

People who build their own metal detectors broadly fall into two categories: those who do it for the pure fun of it, and those who do it out of need. The latter tend to be those who either cannot afford a commercially built detector or who live in a country where metal detectors are illegal. Often they have visions of detecting a large treasure and are looking for easy devices to achieve this goal. It is a common occurrence on the *Geotech* forums for a new user to pop up and ask, "Which detector can I build that will detect gold 10 meters deep?" This section is for you.

Building a metal detector is only a single step in a treasure hunt and should probably not be the first step. The first step is the research required to figure out where to hunt, what you are likely looking for, and how deep it might be. These parameters will often dictate what kind of device you will need to succeed, and it may not be a metal detector.

Metal detectors are mostly for near-surface targets. Individual coins are detectable to 25-30cm, caches and hoards as deep as maybe 1 meter if it's big or in a solid metal container. Beyond that there is not much that a metal detector will detect. But there are other tools that might be useful. If iron is prevalent then a magnetometer may be the preferred tool. Some companies advertise magnetometers[15] to treasure hunters with the implicit claim that they will locate buried gold. Mags cannot do that, they can only detect magnetic anomalies like iron.

However, it's possible to carefully make a gridded survey of an area with a magnetometer that can data-log the output and plot the results to see magnetic anomalies in the soil. Such a survey might reveal compacted areas like foundations, underground voids, or where fire pits might have been. Another method that lends itself to gridded surveys is earth resistivity, which can also reveal compacted areas or underground voids.

For those who have never used it, Ground Penetrating Radar (GPR) is often assumed to be a good treasure hunting tool because it literally shows what is underground. However, reading the results takes skill that requires training, and discerning a buried rock from a buried gold bar is not easy. Beyond that, there are some soils in which GPR simply does not work.

It is important to understand the capabilities and limitations of any of these tools. The bottom line is, do your homework; research both the treasure and the instruments you think you might need. And don't pretend that you can build a metal detector for $50 that will locate a gold-filled cavern 10 meters deep. That won't happen.

28.12: Treasure Hunting Mythology

Along with real treasure and real treasure hunting devices are the myths and outright scams. Back in the 1960s and 70s there were a number of "True West" and "True Treasure" type magazines[16] that pitched many lost treasure stories. Problem is, the vast majority of those stories were simply fabricated, made up by imaginative authors, often under fake names, for whatever money the magazines were paying for stories. If you research a treasure story and cannot trace its roots beyond one of these magazine articles, then it's almost certainly a fabrication.

Even then, there are a lot of fake treasure stories that originated outside these magazines. For example, the Oak Island "Money Pit" supposedly originated around 1799 so it certainly has a long lineage. But its longevity doesn't make up for all the gaping holes in the story and the fact that most of the island has been dug up to no effect. A good method is to rate each treasure with a truth probability and then

---

15. Often they don't call them magnetometers, but something like "Ground Penetrating Locator."

16. True West and True Treasure were literally the names of two such magazines. There was also Treasure World, Old West, Golden West, Frontier Times, Argosy, and many others.

ignore anything below your personal threshold for wasting time. Here are some well-known examples with my own truth-o-meter ratings:

- Oak Island   0.001%
- Beale treasure   0.01%
- LUE   0.05%
- Dents Run   0.1%
- KGB treasures   0.2%
- Any Confederate gold treasure   0.5%
- Any iron door treasure   0.5%
- Buried Nazi gold train   1%
- Victorio Peak   2%
- Lost Adams diggings   5%

For me, none of these treasure legends rank high enough to waste any more time on them beyond the effort it took to write the list. They are all almost guaranteed to be fake. I never give a treasure story a 0% rating because it's impossible to prove a negative; however, for my money anything less than, say, 10% is a sure loser. To get above 10% there better be some credible contemporary evidence. And 10% would only be a threshold that got me to start doing some real research. That research would need to get me above 50% to actually bother hunting for a treasure.

If you are going to waste time hunting for a fake treasure then the best way is to use a fake treasure hunting device. There are plenty of people willing to sell such a device and the performance claims can range from the reasonably believable to the outrageously fantastic. Most of these devices are advertised as "long range locators" for gold, silver, and even diamonds[17] and usually include distances of 10s to 100s of meters and depths of meters or 10s of meters. There is no such technology that can actually do what these devices claim to do and they are nothing but money-making scams.

Most devices typically have a handle, a body, and one or more antenna protruding from the front of the box. The handle or the antenna(e) almost always swivel in some fashion and these devices are, in fact, nothing but a dowsing rod significantly dressed up to appear as more than a dowsing rod. The additional window dressing (knobs, switches, lights, meters, etc) do nothing at all. Dowsing itself does not work[18], and making it look high-tech or spending $10,000 on one will not improve the odds.

Dowsing is a mind trick whereby a visual cue triggers an ideomotor action. It can truly feel like the dowsing device is moving on its own, and this mind trick is used to convince a buyer that the long range device can actually locate treasure. And this is why a fake treasure hunting device is the best way to hunt for fake treasures: the device gives you the illusion that you are succeeding in a situation where success is impossible. The treasure you are looking for does not exist, but the long range device just "detected" it and makes you feel like a real treasure hunter. The fact that the treasure is never retrieved can be blamed on any number of circumstances, but you can tell everyone that you found it and you'll probably believe every word.

28.13: Great Ideas

As an engineer who had worked at two metal detector manufacturers, I have often been approached by people who have a great idea for a metal detector, or a detector feature. Quite often they think their idea has value and would like to see that value realized, and at that point they want to sign a contract before they spill the beans. Supposing you have a great idea, here are the gotchas.

First, your idea may not be novel. Metal detector companies have engineers who live and breathe metal detector design and have often come up with way more ideas of their own than the time to develop them. If you disclose your idea, the company may immediately recognize it as prior knowl-

---

17. Any device that claims to locate diamonds is an automatic fake.

18. Probably all dowsers will disagree with this statement. However, when the rigors of scientific test protocols are applied, dowsing very consistently fails to perform any better than guessing. In field use, success relies on intuition (knowing where to look in the first place) and luck. The success rate of actual treasure recoveries using dowsing or long-range locators (LRLs) is pretty dismal.

edge. Then, they will either have to disclose evidence of that prior knowledge, or you will have to accept their word and leave dissatisfied.

Second, an idea is only that: an idea. Developing an idea into a workable solution is 99% of the effort. Companies are reluctant to pay just for an idea because the idea is the easy part. In fact, *ideas are a dime-a-dozen*; bring me a dozen ideas, and I'll pay a dime. If you want to sell a concept, *bring a working prototype*. Even then, that may only be 25% of the effort needed to get it into production, but a hold-in-your-hand working demo is worth far more than a sketch on a piece of paper.

Third, your idea may not be realizable at all. Unless you're proficient at electronics and metal detector design in particular, you may not understand the underlying physics and the brick walls it imposes. This, again, is why a working prototype is far more likely to get you past the front door.

If you are absolutely convinced that you have a huge winner of an idea, you are not proficient enough to build a prototype, and you are convinced that companies are not inclined to deal with you fairly, then write up a provisional patent application. For a micro entity (individual), it costs $60 to file (USPTO) and gives you one year to pitch the idea to companies. At the end of that year, if no one bought the idea then you have a choice: abandon it, or proceed to a full patent application. The latter choice is expensive, but the risk is that a company you pitched the idea to could steal it and take it to market.

All of this suggests it is a no-win situation for pitching an idea, and it is certainly true that you would be at an extreme disadvantage. Short of hiring a lawyer, probably the best you can hope for is to sign a mutual NDA and throw your cards on the table. I believe most detector companies will be honest and fair.

28.14: Going Solo

Over the years of hanging out in detector forums, I've run across several tech-heads who had (seemingly) good ideas but, instead of pitching those ideas to an established company, they decided they wanted to establish their own company. In most cases, those efforts failed. I once wrote an analysis of why that happens, and some rules to follow if you decide to take the dirt road instead of the paved road.

My analysis as to why so many solo efforts fail is:

1. Grossly underestimating the complete picture. An impressive result on the lab bench doesn't equate to impressive results in Australian dirt. Or any dirt. Once you've got good results on the bench, you're 5% of the way there.
2. Ego gets in the way. Wanting to prove yourself before the proof is ready to serve is the part that really damages the effort. Bold performance claims naturally produce skepticism with folks who want to see the evidence. When the evidence doesn't happen, skepticism turns to disdain. And as this process stretches out over years, disdain becomes outright ridicule. The designer then spends an inordinate amount of time responding to the doubts and ridicule. This problem is massively amplified if the hype is coupled with contempt for the competition.
3. Wanting complete control. The development of a new metal detector is not a one-man job, especially if it's your first one. Think you want to also manufacture the detector? This is a purt-near-guaranteed failing proposition. The turf is littered with dead metal detector companies.

Despite the fact that I work for a major manufacturer of beepers, I am personally a proponent for independent efforts. I would love to see new ideas released and be successful. But in most cases, developers take horribly wrong paths. Here are a few suggestions for a better path:

1. Keep your mouth shut. While premature public hyping doesn't necessarily break a project, it does incredible damage to the reputations of the design and the designer when things don't work out, as is the case 95% of the time.
2. Get help. Competent help. Preferably by someone who's been-there-done-that. You can't do it all alone.

3. A day in the field is worth a month in the lab. Get it off the bench ASAP, you will quickly see whether that great idea is worth a crap.

4. Focus on the things that matter the most. Performance is #1. A talking detector is #1000. When you have the level of performance that will definitely sell, **stop the design process** and move on to making it producible. You'll get to implement that Next Great Idea in Version 2.

5. Forget about building detectors. Ain't gonna happen, at least on a scale that will pay the rent. Find someone else to do that. Again, someone who knows how. Also lets you move on to Version 2, which is more fun than potting coils all day.

6. Forget about patents. Unless you get the Partner Production Company (#5) to deal with them. That $5,000 patent ain't worth squat unless you've got the $100,000 to back it up.

7. Respect the competition. Like it or not, they are already successful, have lots of satisfied users, and are making lots of money. Focus on beating them with a better product, not with public displays of contempt. And never ever ever criticize a user of a competitor's product. That's one lost sale, or more.

28.15: Counterfeiting

Over the years a number of popular detector models from different manufacturers have been counterfeited. These counterfeits look almost identical to the real thing and in some cases operate almost identically. During my time at White's, the *GMT* was counterfeited and the only obvious way to tell the copy from the real deal was by the lower fiber rod. The counterfeits used a plastic rod. There were other subtle differences, like the film capacitors used on the PCB and the ejection pin marks on the plastic parts, but you would need to directly compare with a real unit to see this.

In some cases the counterfeits look real but perform very differently. One Minelab PI model was counterfeited but, instead of copying the PI circuitry, they used a poor-performing VLF circuit. But it otherwise looked real. The Fisher *Gold Bug 2* counterfeit has, instead, a digital *Gold Bug*[19] circuit and some of the panel controls are not connected to anything. In Figure 28-1 notice the potentiometer and two toggle switches that have no connections.

Counterfeits are almost always limited to highly successful detector models and mostly those that have strong sales in the African gold market. The counterfeiters tend to be based in China (due to a lack of intellectual property enforcement) and sell through distributors in Turkey and Dubai. I have visited the detector distributors in Dubai and seen genuine and fake product mixed together. But the fakes have begun to commonly show up in the US and Europe as well, often on eBay and even Amazon.

Contrary to popular belief, most counterfeits are not "overruns" or night-shift productions of models already manufactured or even partly manufactured in China, and therefore equivalent in quality to the real deal. For example, the White's

Fig. 28-1: **Fake Gold Bug 2**

GMT was wholly produced in Sweet Home, with PCBs assembled there and plastic that was injection-molded in nearby Albany. All metal parts were fabricated in-house and the lower fiber rods were made in California. By comparison, a counterfeit might have identical-looking plastic parts but you can find subtle differences that show they are made with different molds. In short, counterfeits are made in an independent factory using copycat components and usually to an inferior quality. The GMT coils were especially poorly made.

---

19. An unfortunate naming confusion. The older *Gold Bug 2* is an all-analog 71kHz design, the newer *Gold Bug* is a digital 19kHz design.

Counterfeiting is fairly simple, requiring the purchase of 2 or 3 sample units to strip and duplicate the components. Plastic parts can be 3D-scanned for making new injection molds. A stripped PCB can be delaminated or X-rayed to exactly copy the copper layers. IC components are usually marked with their part number and passives can be measured to determine values. The final requirement is the micro's firmware, and virtually every micro on the market has security holes that allow the firmware to be "ripped," no matter what copy protection is applied. So setting the micro's security bits will stop an amateur but not determined hacker.

Therefore, in producing a metal detector, start with the assumption that the schematic, the PCB layout, and the firmware are all easily extracted by a counterfeiter. If that's so, what can be done to dissuade counterfeiting? In the past, companies[20] potted portions of the circuitry in epoxy. This stops the casual nosy inspection but, with a little effort, epoxy can be removed. Minelab has coated their PI PCBs with a white epoxy paint to pretty much the same effect. Another strategy has been to grind off IC markings to make it more difficult to identify the chips. Again, this is circumvented by decapping the chip[21] and reading the chip ID directly off the die.

These methods certainly add to the work needed to counterfeit the product and may be enough to dissuade a small-scale copycat. It is not likely to stop a more sophisticated operation looking to make a substantial amount of money off a proven successful product. For anyone willing to invest in the tooling required to exactly copy all of the plastic parts[22], stripping epoxy or decapping ICs is a minor additional annoyance, not a show-stopper.

Potentially effective methods involve deception. For example, instead of grinding the markings off of ICs, have them relabeled as something else. Chip companies will do this for a small extra fee. Garrett relabels the microcontrollers in their detectors with the Garrett logo and a Garrett part number. Unfortunately, this is obvious and the counterfeiter only needs to decap the chip to get the real part number. A better method is to relabel the chip with a different but real part number that misleads the counterfeiter. For example, suppose the design requires a high-bandwidth low-noise opamp. Have that opamp relabeled with a part number that represents a low-bandwidth, higher-noise opamp. When the counterfeiter builds the detector they find that it works but not as well, and now they have to figure out why. And if there is one thing a counterfeiter doesn't want to do, it's learn the details of how the product works.

Another possibility is to add firmware traps that are triggered by recognizing that the firmware is a copy. This requires using a micro with a unique embedded serial number and custom-programming each micro to recognize its own serial number. Then, if the firmware is simply cloned to another micro, the serial number won't match. You can then decide whether to crash the program or do some subtly devilish behavior. Again, whatever you do in firmware can be figured out, but it takes a lot of time and this is something counterfeiters don't want to do.

The bottom line is, the harder a counterfeiter has to work, the more likely they will move on to something easier to copy. Subtlety is the key here; obvious barriers are the easiest to circumvent, while random minor problems are tough to track down.

---

20. At least Garrett, Compass, and White's, probably others.

21. Professionally, using fuming nitric acid under a ventilation hood while wearing protective gear. But a homebrew method is to place the IC in boiling pine resin, which will dissolve all the plastic and leave only the die and leadframe. I've used this method to find the real identity of counterfeit chips. *Good ventilation required.*

22. Just the tooling for a coil shell can run $10k-$20k. A complete modern detector will likely run $100k or more.

PART 8
Appendices

The World's Only Modular Metal Detector

Buy a base unit and configure your own custom system. Change your mind, change your module. If we upgrade a module—no problem. It's easy and inexpensive for you to drop in a new chip or board and have the latest technology. TREASURE BARON: The only detector that will never be obsolete.

Discovery Treasure Baron Base Unit-*Released May, 1993*

S-style rod, hip mountable case. 12.5 khz operating frequency, 3.7 lb less batteries, 57" fully extended; 46" collapsed, controls include: power/audio range, mode toggle/all metal, motion discrimination, disc ID/iron/set disc ID tone, push-push iron acceptance or rejection, headphone jack, Thunderhead 8" waterproof loop

Available Modules

Pro Hunter-*Released May, 1993*

Activates GND/SALT toggle on base unit and adds notch discrimination, adjustable manual ground balancing, LED battery power indicator, LED target depth indicator

Deep Hunter-*Released May, 1993*

Adds two power boost levels for better depth perception and definition, allows ni-cad use with monitored onboard recharging (requires AC power adapter)

Ni-Cad-*Released Jan, 1994*

Allows ni-cad use with monitored onboard recharging (requires AC power adapter)

Black Sand-*Released Jan, 1994*

Adds two power levels of penetration for extreme conditions, great for high trash areas, allows ni-cad use with monitored onboard recharging (requires AC power adapter)

Gold Trax (v. 1.0)-*Released May, 1994*

Provides true micro controller ground tracking, fixed GB, iron indicator, slew control to customize the settings, user-friendly touchpad, no need for user to "pump" the unit to set up the ground tracking feature

Gold Trax (v. 2.0)-*Released July, 1994*

User installable upgrade for version 1.0 Gold Trax owners offers greater discrimination, additional specifications, and more! All new Gold Trax modules ship with v. 2.0 circuitry

Coin Trax/ID-*Released Fall 1996*

New! 1996

Adds notch discrimination (9 notches), LED battery power indicator, LED target depth indicator, iron indicator, provides true microprocessor-controlled automatic ground tracking, and our exclusive Turbo Ground Balance and SST SmartScan Technology (LED display scans continuously, locks onto target and identifies target automatically)

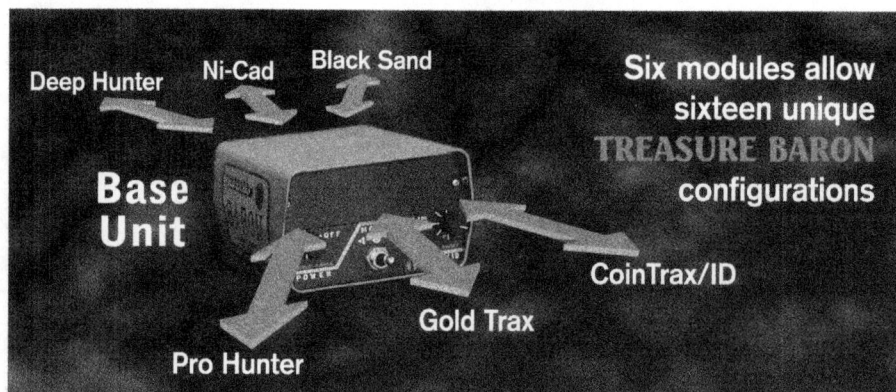

Deep Hunter   Ni-Cad   Black Sand

Base Unit

Pro Hunter

Gold Trax

CoinTrax/ID

Six modules allow sixteen unique TREASURE BARON configurations

A | Magnetic Field Probe

APPENDIX

"Physics is about questioning, studying, probing nature.
You probe, and, if you're lucky, you get strange clues."

— *Lene Hau*

It is a valuable exercise to see how a TX field behaves. Since magnetic fields are invisible we need a way to visualize them. One way is shown in Figure 3-6 whereby a DC current is used to create an alignment pattern in iron filings. Although this offers an excellent visualization it is difficult to set up, has limited distance, and does not give any insight as to the strengths of the field lines. The Magnetic Field Probe (MFP) can measure field strength and also determine flux direction at any point in space.

A.1: Theory

From Equation 4-2 and Equation 4-4 we know that an incident magnetic field induces an EMF in a simple metal target:

$$\varepsilon = -A\cos\theta \cdot \frac{dB}{dt} \qquad\qquad \text{Eq A-1}$$

A simple metal target behaves as a shorted single-turn inductor. A coil placed in an alternating magnetic field also develops an EMF but the coil can have multiple turns of wire. The EMF is now

$$\varepsilon = -N \cdot A\cos\theta \cdot \frac{dB}{dt} \qquad\qquad \text{Eq A-2}$$

where N is the number of turns, A is the area of the coil and θ is the angle of incidence of the flux lines, with 90° resulting in a maximum induced voltage. The EMF developed is a real voltage that, unlike with a target response, can be measured across the terminals of the coil, usually by connecting it to an oscope. If the coil is connected to an amplifier as in Figure A-1 then the probe is usable even for weak signals.

Many people have made such a probe to measure magnetic fields. There are two problems with the voltage-mode approach. First, because the incident field is AC the resulting EMF strength is proportional to frequency due to the derivative term in Equation A-2. Second, if you want to see what the magnetic field waveform actually looks like[1], this method won't do that. Again, you get the derivative of the waveform, but not the waveform itself.

One way to correct these problems is to incorporate an integration function into the amplifier. This might be done by replacing the feedback resistor with a capacitor. An easier way is to connect the probe coil in the current-mode

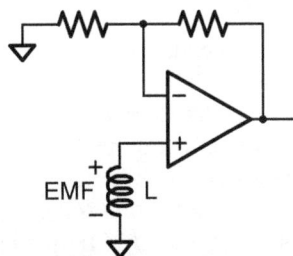

Fig. A-1: **Voltage-Mode Probe**

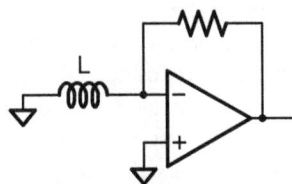

Fig. A-2: **Current-Mode Probe**

1. For example, in a multifrequency detector.

A: Magnetic Field Probe

539

configuration of Figure A-2. This is a curious circuit because, at first glance, it appears the coil is simply shorted with ground on one side and virtual ground on the other side. But the same forces that create eddy currents in a disconnected piece of metal create a coil current here, and the current flows through the feedback resistor to produce an output voltage.

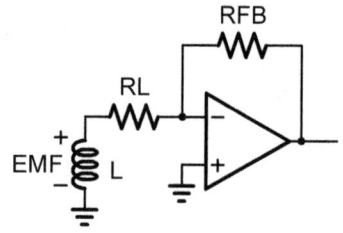

Fig. A-3: **Equivalent Circuit**

To clarify further, consider the coil as a series connection of the inductance (and its EMF) along with its series wire resistance as in Figure A-3. If the coil inductance dominates then it will perform the desired integration function and the coil current will be proportional to the incident field strength but independent of frequency. Temporarily assuming $R_L = 0$, the coil current is given by

$$i(t) = -\frac{1}{L}\int \varepsilon\, dt \qquad\qquad\qquad \text{Eq A-3}$$

Substituting ε from Equation A-2 (with $\theta = 90°$):

$$i(t) = \frac{-NA}{L} B(t) \qquad\qquad\qquad \text{Eq A-4}$$

where A is the cross-sectional area of the probe coil. For a long solenoidal coil, $L = \mu N^2 A / l$ where l is the length of the probe coil[2] so

$$i(t) = -\frac{l}{\mu N} B(t) \qquad\qquad\qquad \text{Eq A-5}$$

An interesting outcome is that the current increases with the length of the coil but decreases with the number of turns. If the windings are a tightly packed single-layer solenoid then length is proportional to N, suggesting that 1000 turns gives the same result as 1 turn. This is not quite true because at 1 turn the solenoidal assumption is not valid and the math changes. But it might be safe to say that 1000 turns is no better than 100 turns. Also note that the cross-sectional area of the coil, A, falls out of the math.

A.2: Sensor

Now that the idea has been laid out, let's convert it into a practical circuit. When measuring a magnetic field it is important that the sensor alters the field as little as possible. Equation A-5 suggests that a high-permeability core will reduce $i(t)$ making an air-core coil a better choice, and an air-core coil will also have little effect on the magnetic field being measured. However, we will see shortly that an air-core probe is not very practical.

Another consideration is the frequency response; the sensor should have enough bandwidth to measure the highest desirable frequency with minimal error. Throughout the book we have consistently considered 1kHz - 100kHz as our potential range of detector frequencies so we want to design for at least that. However, using 1kHz for the low −3dB frequency means we will end up with a gain of 0.707 and we'd like to do better than that. Let's set a target minimum gain of 0.98 (−0.17dB) which will require moving the f_L and f_H roll-off frequencies out by a factor of five times the usual −3dB points. This means we need that $f_L = 200$Hz and $f_H = 500$kHz. See Figure A-4.

For the purpose of deriving the induced current the theoretical math above assumed the coil's resistance R_L is 0Ω. However, the coil resistance is part of the DC gain of the opamp which is R_{FB}/R_L (referring to Figure A-3) so a zero-ohm coil is neither possible nor desirable. The coil resistance also sets the low end of the bandwidth at $f_L = 1/(2\pi L/R_L)$. We want this to be 200Hz which means the required coil tau (L/R_L) is 796μs.

---

2. This assumes the probe will be a long solenoidal coil.

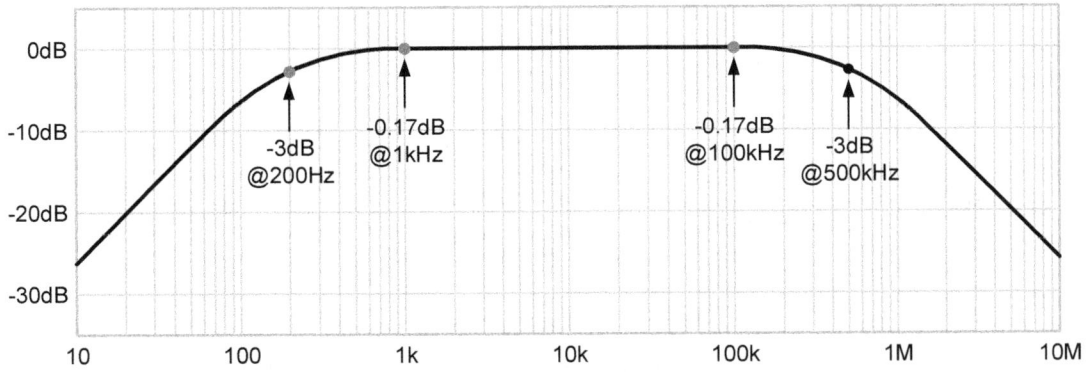

Fig. A-4: **Desired Bandwidth**

Let's do a quick reality-check using an example coil. An air-core solenoidal coil with 50 turns is found to have L = 12.1µH and R_L = 0.315Ω. The low-frequency limit is therefore

$$f_L = \frac{1}{2\pi(12.1\mu H)/0.315\Omega} = 4.14\text{kHz} \qquad\qquad \text{Eq A-6}$$

which is not even close to good enough. Extending the low end argues for more turns as L is proportional to the square of the turns but R_L is only proportional to the turns. This also requires that the overall length of the inductor remains the same which means the additional turns must be layered.

The above inductor has a tau of 38.4µs and for 200Hz we need a tau of 796µs. This argues for 21 times more turns, or a total of 1050 turns. This would be quite a challenge to wind and we would eventually find out that even more turns are needed to compensate for dimensional inefficiencies of the layered windings.

An easier way to improve low frequency performance is to increase L by using a high-permeability core which has no effect on R_L. Let's try another example: a Wurth 7447480331 inductor. At 330µH and 0.3Ω it has a tau of 1100µs which is quite a bit better than we need. The overall diameter is 10mm and the length is 14mm so it is also fairly compact for getting good measurement resolution.

There are other off-the-shelf inductors that will work and you can also wind your own. When considering a purchased inductor make sure it is an unshielded type. Most will list both L and R_L so that you can determine whether the tau meets the design goal. Keep in mind that the ferrite core will affect the magnetic field being measured so keep it as small as reasonably possible. A smaller sensor also improves measurement resolution.

A.3: Remaining Design

Now that we have a probe inductor that meets the low frequency requirement, let's look at the high frequency limitations. f_H is limited by the opamp's gain-bandwidth product or, perhaps, a bandwidth-limiting capacitor added to R_{FB}. The gain is set by the feedback resistor R_{FB} divided by the DC resistance of the inductor so, for example, if the coil resistance is 0.3Ω then R_{FB} = 100Ω results in a gain of 333. Since we want a minimum f_H of 500kHz this suggests we need an opamp with a gain-bandwidth product of 166MHz. The LT6231 (215MHz) will work for this and results in a slightly higher f_H of 645kHz.

The field strengths we want to measure can vary considerably so we would like a probe with considerable dynamic range. For this we'll follow the fixed gain preamp with a second amplifier stage having a selectable gain. The not-so-final circuit design in Figure A-5 shows the second stage with (negative) gains of 1, 10, and 100. Since each stage is inverting the overall sensor-to-output relationship is non-inverting, assuming the coil is connected properly.

The schematic is missing an R2 and R3 which is a clue that there is more to come in the final design. Furthermore, we need a power supply circuit to make it work. The power supply is dependent on the battery(s) to be used and the desired VCC. The LT6231 can run up to 12V so we will choose 10V for our analog voltage. The battery depends on the enclosure so let's take a look at that.

A: Magnetic Field Probe **541**

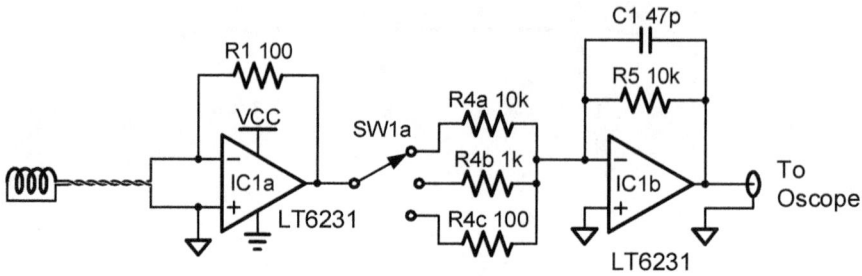

Fig. A-5: **Probe Circuit**

We'll finish off the MFP in a nice enclosure and for this I've chosen Bud Industries' "Grabber Style C" plastic enclosure (HH-3431) which includes a battery compartment. It can be used with either a 9V battery or two AA batteries; we will use the AA batteries which requires the purchase of the battery terminal kit (HH-3450-C). See Figure A-6.

Fig. A-6: **Enclosure**

Creating a 10V supply for the opamps from a 3V battery requires a boost regulator. Again, we will generate a boosted voltage of 12V and then use a linear regulator to drop back down to 10V. A rail splitter is used to create the "analog ground" of 5V. Figure A-7 shows a block diagram of the power supply.

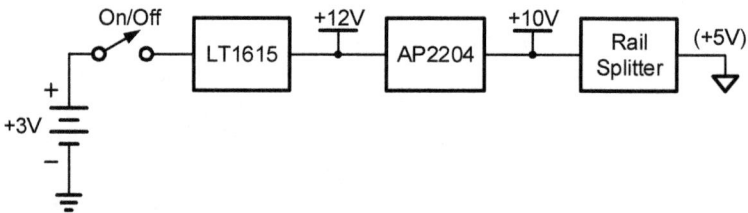

Fig. A-7: **Power Block Diagram**

One problem with the Magnetic Field Probe design is that the gain can be exceptionally high, as much as 10,000. The LT6231 can have an input offset up to 450µV which means the output of the second stage could have an offset as high as 4.5V. The rail splitter uses half of a TL072 so we will use the other half to implement a very slow retune circuit similar to what was used in the TR-Disc design. The difference is that this one does not have a retune switch and runs live all the time. If the integrator speed is well below our 200Hz minimum then the probe bandwidth will not be affected. In this design the integrator tau is 1s which is 0.16Hz.

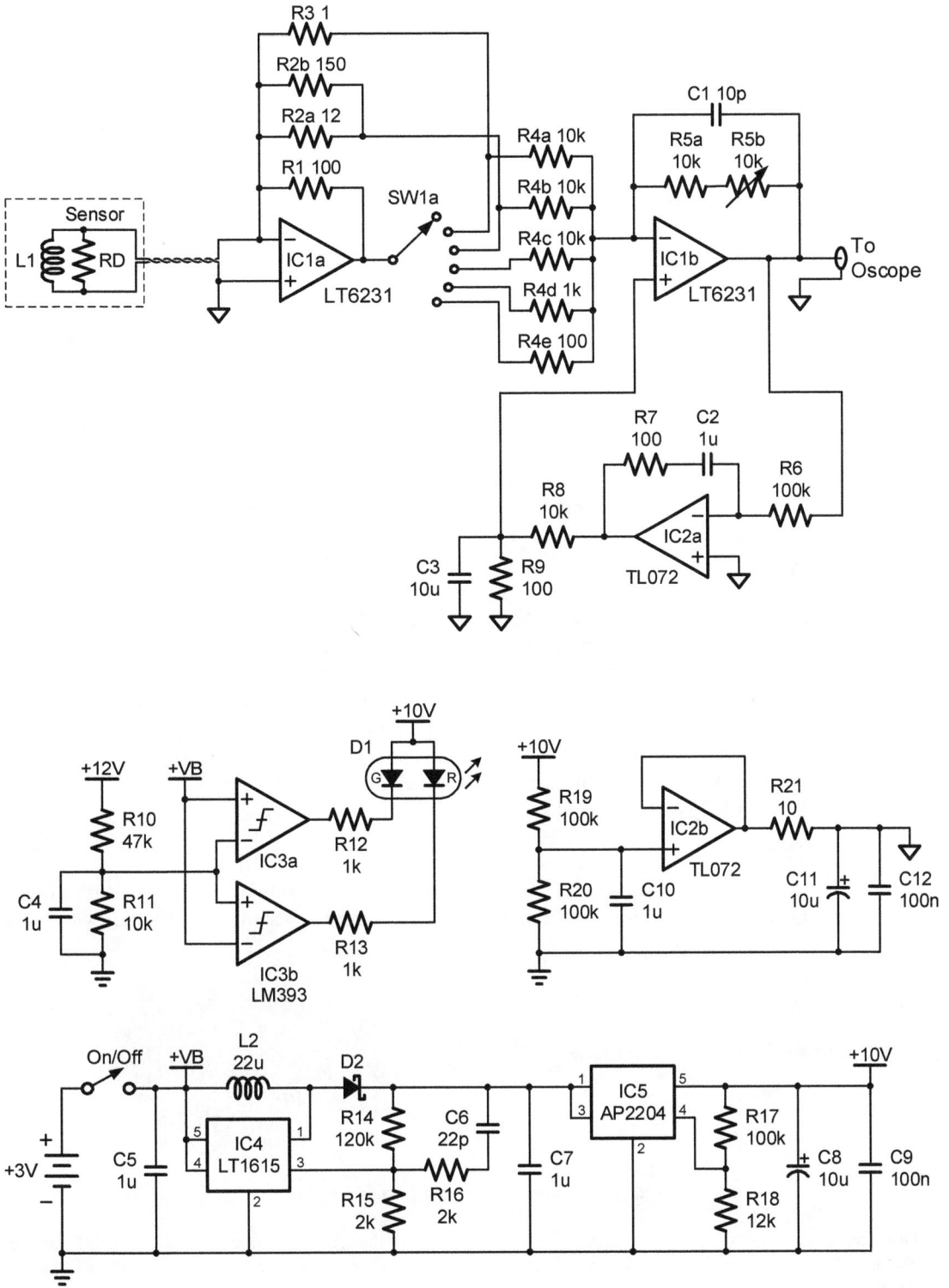

Fig. A-8: **Final MFP Circuit**

Another potential problem is that the first-stage gain is set by the ratio of the feedback resistor to the resistance of the sensor coil. The sensor coil resistance is due to the copper wire which has a temperature coefficient (TC) of approximately +3900ppm/°C, which means it will increase 0.39% for every 1° rise in temperature. Ideally we should use feedback resistors which also have a TC of +3900ppm/°C so that the overall gain is temperature independent. This is especially important if the MFP is used to measure actual field strengths but not so important if it is used to visualize waveforms and measure relative strengths. In any case, in a laboratory environment the ambient temperature will likely vary no more than ±5° so this represents a small ±2% gain variation.

Finally, Figure A-5 shows three gain selections for an overall gain of 100, 1,000, and 10,000. As it turns out the lowest gain is still too high for placing the probe in the middle of a modestly powerful transmit coil. Therefore, we will add two more lower gain settings: 1 and 10. However, these need to reduce the gain of the first stage, not the second stage, so that the first opamp isn't overloaded with a too-strong signal.

Figure A-8 shows the final circuit with the additional gain settings and the retune circuitry. The two new low gain settings (G=1 & G=10) are implemented by reducing the gain of the first stage. For G=10, R2a and R2b are in parallel with R1 resulting in an effective feedback value of 10Ω. For G=1, R3 in parallel with R1 is 0.99Ω, or close enough to 1Ω.

The second stage feedback resistor includes a trim pot for trimming the overall gain so that a particular peak field strength gives a particular peak voltage on the oscope. For this design, at a gain setting of 1 a peak-to-peak field strength of 1mT produces 1V-pp. Since the second stage has a lower gain than the first stage a feedback capacitor is added to moderately reduce the bandwidth and reduce noise. Although we want to probe fields up to 100kHz, those fields might have significant harmonic content and we don't want to restrict the bandwidth more than necessary. A 10pF cap results in a second stage bandwidth of 1.6MHz. Also, at the two lower gain settings the first stage bandwidth is higher than the previously cited 645kHz and at these gain settings non-sinusoidal signal fidelity is slightly improved.

The power switch is combined with the gain selection switch by using a DP6T rotary switch. The LT1615 can boost from a minimum input voltage of 1.2V to a maximum output of 34V[3]. We will set the output to 12V and then use a linear regulator (AP2204[4]) to drop back down to 10V. A rail splitter is devised using one half of a TL072 JFET opamp. As a final touch the power supply circuitry should also have a "power on" indicator and a low battery indicator. These are provided by a red/green LED which is driven by a dual comparator. When power is on, the LED is green; if the battery drops below a set threshold (~2.1V) the LED switches to red.

A.4: Construction

The probe inductor has parasitic capacitance and is susceptible to ringing. To prevent this a damping resistor should be added, preferably right at the inductor. Figure A-9 shows the inductor with a 1kΩ

Fig. A-9: **Sensor Damping Resistor**

3. There are several pin-compatible parts that can be substituted for the LT1615. Some are in a 6-pin SOT package with an internal synchronous isolation switch, and the 6th pin is an output voltage. The PCB for this design will accept these versions as well.

4. As of publication, the AP2204 has been replaced with the AP2205. And, again, there are some other linear regulators that are pin-compatible with the AP2204.

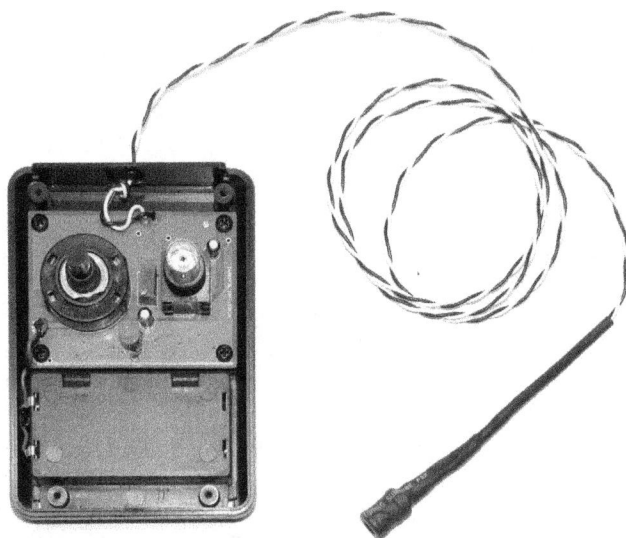

Fig. A-10: **Inside the MFP**

1206 SMT resistor soldered to the pins. Because the resistor is connected across ground and virtual ground it has no effect at all on signal fidelity. For protection the sensor assembly can be enclosed in one or two layers of heat shrink or plastic tubing. The probe leads should be twisted pair or coax; the length can be anything you want but the resistance of the leads adds to that of the inductor to reduce the sensor tau, so keep it reasonable.

An open but assembled MFP is shown in Figure A-10. The PCB is mounted to the back half of the clamshell with wiring to the battery terminals and the sensor wire placed through the front insert piece. With the exact parts specified it is a tight fit in the enclosure; you may find it necessary to slightly reduce the height of the mounting bosses.

Figure A-11 shows the completed MFP. A dimensioned drawing of the label is provided at the end of this addendum along with a detailed parts list. Gerber files for the PCB can be found on the *Geotech* web site.

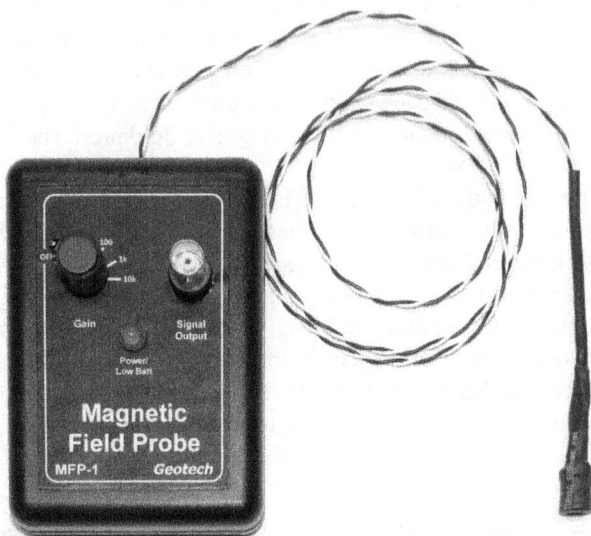

Fig. A-11: **Finished MFP**

A.5: Calibration

If all you want to do is get relative measurements or visually see the TX field waveform then calibration is optional. Wind a moderately large coil (say, 10"/25cm) with a moderate number of turns (say, 50) using average wire (say, 24AWG). Measure the coil's inductance and resistance. Then drive the coil with a 10kHz sinusoid at some nominal peak voltage (say, 5vp). This will establish a current in the coil with the peak value

$$i_p = \frac{v_p}{|Z_L|} = \frac{v_p}{|R_L + jX_L|} = \frac{v_p}{\sqrt{(R_L)^2 + (X_L)^2}}$$

where $X_L = \omega L$.

As an example, a 52-turn 10" coil has L = 1887uH and R = 5.82Ω. X_L is therefore 118.56Ω and $|Z_L|$ is 118.71Ω. The magnetic field density at the center of the coil is given by Equation 3-3:

$$B = \frac{\mu NI}{2r} = \frac{\mu N}{2r} \cdot \frac{v_p}{Z_L} = \frac{(4\pi \cdot 10^{-7}) \cdot 52 \cdot v_p}{25cm \cdot 118.71\Omega} = 2.202 \cdot v_p \ (\mu T)$$

That is, the B field is 2.2μT (peak) per volt (peak) of the applied voltage drive. We can then adjust the peak voltage drive to, say, 4.54v to generate a field of 10μT. Knowing the field at the center of the coil we can place the sensor at that spot and adjust R5b to get a particular reading on the oscope. For the example above and with the gain switch at a setting of 1000, R5b was adjusted so that the 10μT peak waveform produced a 10V peak reading on the oscope. This means that the 1000x switch setting reads out 1μT/V on the oscope. The remaining switch settings can be directly inferred from this:

Gain	Scale
1	1mT/V
10	100μT/V
100	10μT/V
1k	1μT/V
10k	0.1μT/V

A.6: Usage

The MFP is capable of measuring B-fields from about 100nT up to 10s of mT with a flat frequency response from 1kHz to 100kHz. The amplifier can directly drive an oscope or an AC voltmeter, but be aware that the AC-volts setting on many multimeters has a very low bandwidth. Also keep in mind that the MFP requires a changing magnetic field. Although Equation A-5 implies otherwise, the derivative & integral are still embedded in the equation so a DC magnetic field produces no EMF, and therefore no probe response[5].

An added benefit to the current-mode approach is that the waveform presented at the output of the opamp is identical to the magnetic field waveform being measured. This makes it useful for analyzing the transmit waveform itself, for example, a multifrequency waveform. Figure A-12 shows the TX voltage drive waveform for a Minelab *Vanquish* along with the TX field waveform as produced by the MFP. While the voltage waveform is difficult and confusing, the TX field is much easier to analyze.

Notice that in the *Vanquish* waveform there is some slight exponential bowing of the ramp segments. Measuring the TX coil yields L = 151μH and $R_L = 0.58\Omega$, resulting in $\tau = 260\mu s$. This is well below the tau of the sensor coil (1100μs) so it is safe to say that the bowing is a real result of the *Vanquish* search coil and not due to the MFP's sensing coil.

---

5. You may notice, especially at higher gain settings, that waving the sensor around in the air will cause the output level to bounce around. This is normal because moving the sensor through Earth's (relatively static) magnetic field produces the same effect as a changing magnetic field.

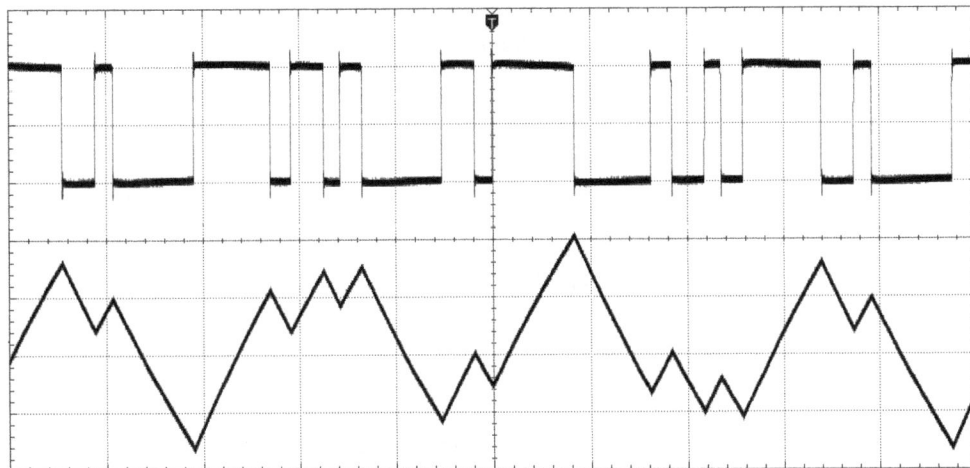

Fig. A-12: **Minelab *Vanquish* B-Field Waveform**

Parts List

Resistors: (5% 0805 unless otherwise noted)

R1, R4e	100 (1%)
R2a	12 (1%)
R2b	150
R3	1 (1%)
R4a, R4b, R4c	10k (1%)
R4d	1k (1%)
R5a	10k
R6, R17, R19, R20	100k
R7	100
R8, R13	1k
R9, R21	10
R10	15k
R11	4k3
R12	470
R14	120k
R15, R16	2k
R18	12k
RD	1k (1206)

Potentiometers:

R5b	10k trimpot

Capacitors: (10% 0805 unless otherwise noted)

C1	10p
C2, C8	10u
C3, C4, C5, C7, C10	1u
C6	22p
C9, C11	10u elect., 16v thru-hole
C12, C13, C14, C15	100n

Inductors:

L1	330uH; Wurth 7447480331

Diodes:

D1	Red-Green LED common anode
D2	Schottky

ICs:

IC1	LT6231 (SOIC)
IC2	TL072 (SOIC)
IC3	LM393 (SOIC)
IC4	LT1615 (SOT23-5)
IC5	AP2204-adj (SOT23-5)

Switches:

S1	Alpha 105-SR2612F-26-21RN (6PDT Rotary)

Connectors:

CONN1	TE Connectivity 1-1634505-0 (BNC)

Misc:

Enclosure	Bud Industries HH-3431 (Grabber Style C)
Battery terminals	Bud Industries HH-3450-C
Control knob	As preferred
LED bezel	Optional
Grommet	Optional; for probe wire through front insert
Coil cable	Coax or twisted pair
Batteries	(2) AA alkaline

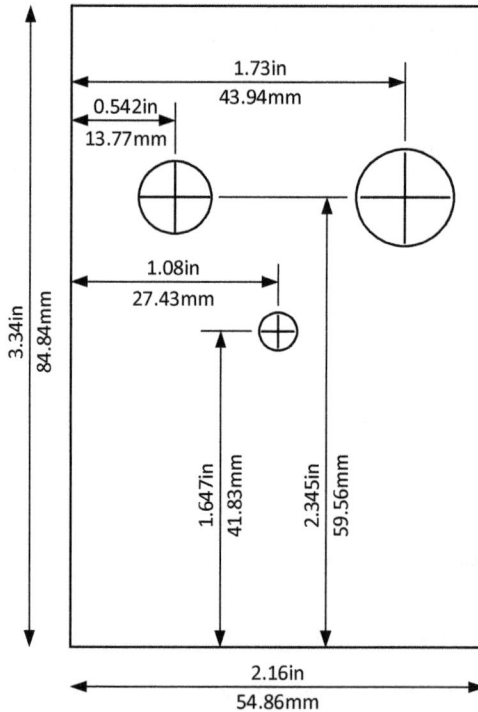

Fig. A-13: **Label Dimensions**

Coil Calculations

"On two occasions I have been asked [by members of Parliament], 'Pray, Mr. Babbage, if you put into the machine wrong figures, will the right answers come out?' I am not able rightly to apprehend the kind of confusion of ideas that could provoke such a question."

— *Charles Babbage*

In designing and building coils it is often useful to predetermine how many turns are needed to achieve a particular inductance. For this a coil calculator is handy. This appendix shows a number of formulae which can be used in an Excel spreadsheet or even developed into a stand-alone app. In cases where both imperial and metric versions of a formula are commonly found on the Internet they are both presented here.

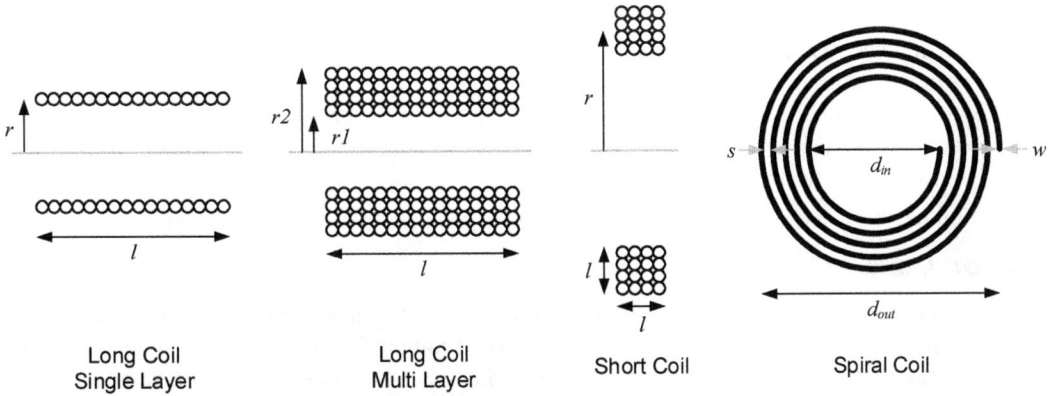

Fig. B-1: **Coil Diagrams**

B.1: Solenoidal Coils

In metal detectors solenoidal coils are mostly found in pinpointer designs. The standard equation for a solenoidal inductor is

$$L = \frac{\mu N^2 A}{l}$$

Eq B-1

where L = inductance in Henries
μ = core permeability
N = number of turns
A = cross-sectional area (meters)
l = length of solenoid (meters)

For an air-core inductor $\mu = \mu_0 = 4\pi \times 10^{-7}$. Pinpointers typically use a ferrite core where $\mu = \mu_0\mu_r$ and μ_r is a "relative permeability" multiplier. μ_r depends on the exact ferrite used and is typically in the hundreds.

Another commonly found equation used for solenoidal coils is an approximation by Harold Wheeler[1]:

---

1. H.A. Wheeler, "Simple Inductance Formulas for Radio Coils," Proc. I.R.E., vol 16, Oct 1928, pp 1398-1400.

549

	Imperial		Metric	

$$\text{Imperial} \qquad\qquad\qquad \text{Metric}$$

$$L = \frac{N^2 r^2}{10l + 9r} \qquad\qquad L = \frac{39.5 \cdot N^2 r^2}{10l + 9r} \qquad\qquad\qquad \text{Eq B-2}$$

where L = inductance in μH
  r = coil radius (inches/meters)
  l = coil length (inches/meters)
  N = number of turns

The Wheeler formula is valid only for an air core winding and is typically accurate to within 1% for $l >$ $0.8r$. This dimensional ratio is what we would call a "long" coil. Note that μ_0 is included in the multiplier constants and that the result is in μH, not H.

Equation B-2 is for a single-layer solenoidal coil but Wheeler[2] also developed a formula for a multi-layer solenoidal coil:

$$\text{Imperial} \qquad\qquad\qquad\qquad \text{Metric}$$

$$L = \frac{0.8 N^2 (r_1)^2}{6r_1 + 9l + 10(r_2 - r_1)} \qquad\qquad L = \frac{31.6 N^2 (r_1)^2}{6r_1 + 9l + 10(r_2 - r_1)} \qquad\qquad \text{Eq B-3}$$

where L = inductance in μH
  r_1 = inner coil radius (inches/meters)
  r_2 = outer coil radius (inches/meters)
  l = coil length (inches/meters)
  N = number of turns

This equation has high accuracy when the three terms in the denominator are about equal.

B.2: Short Coils

Most common metal detector coils are scramble-wound "skinny donuts," what would be called "short" coils by the folks who write coil formulae. Even thought they are scramble-wound the bobbin usually has a rectangular cross section so we will assume the coil cross section is rectangular. For short and medium coils with a square cross section we find an equation derived by Herbert Brooks[3]:

$$L = \mu_0 \cdot r \cdot N^2 \left(\left(0.5 + \frac{l^2}{48r^2}\right) \ln\left(\frac{32r^2}{l^2}\right) - 0.84834 + 0.051\frac{l^2}{r^2} \right) \qquad \text{Eq B-4}$$

where L = inductance in Henries
  μ_0 = permeability of air ($4\pi \times 10^{-7}$ H/m)
  r = mean radius of the winding in meters
  l = coil length (which equals thickness c) in meters
  N = number of turns

This can be rewritten as

$$L = 1.26 \cdot N^2 r \cdot \left(\left(0.5 + \frac{x}{12}\right) \ln\left(\frac{8}{x}\right) - 0.84834 + 0.2041x \right) \qquad \text{Eq B-5}$$

where $x = (l/(2r))^2$ and L is now in μH.

---

2. We also met Mr. Wheeler back in Chapter 6 when we mentioned that he is possibly the inventor of the concentric coil per patent US2451596. He also worked at Hazeltine and was involved in the design of early IB mine detectors during WWII.

3. H.B. Brooks, "Design of Standards of Inductance, and the Proposed Use of Model Reactors in the Design of Air-Core and Iron-Core Reactors," Bureau of Standards Journal of Research, Vol 7, May 12, 1931, pp 289-328.

B.3: General Short/Long Coils

In 1982 Wheeler published his continuous inductance formula[4]:

$$L = 0.002\pi d N^2 \left(\ln\left(1 + \frac{\pi d}{2l}\right) + \frac{1}{2.3004 + 3.2 l/d + 1.7636(l/d)^2} \right)$$

Eq B-6

where   L = inductance in μH
  d = mean diameter of the winding in cm
  l = coil length (which equals thickness c) in cm
  N = number of turns

It is called the continuous inductance formula because it is accurate for both long and short coils.

B.4: Spiral Inductors

A modified version[5] of Wheeler's equation for circular spiral inductors[6] is

$$L = \frac{2.8r \cdot N^2 d_{ave}}{1 + 3.45\rho}$$

Eq B-7

where   L = inductance in μH
  $d_{ave} = 0.5(d_{out} - d_{in})$
  d_{out} = outside diameter (meters)
  d_{in} = inside diameter (meters)
  $\rho = (d_{out} - d_{in})/(d_{out} + d_{in})$
  N = number of turns

d_{ave} is the average diameter of the spiral and ρ is what is called the "fill factor"; that is, how far the spiral extends to the center. In metal detector coils the center is usually left substantially open (a ρ of perhaps 0.1). Figure B-1 shows the spiral diagram with both d_{in} and d_{out} labeled but in practice you might design a PCB coil around, say, d_{out} with a particular trace width (w) and spacing (s) from which d_{in} can be calculated. The reference paper also offers alternative multiplier values for the cases of square, hexagonal, and octagonal spiral coils, which are more prevalent in IC design.

B.5: Accuracy

Any coil formula will only provide an estimate for the inductance. Many of the formulas assume a perfectly uniform current distribution; that is, a *current sheet* result which means the use of square wire[7] with practically zero spacing between the windings. In practice round wire is used with, at a minimum, an enamel coating and perhaps even a thicker insulation.

There is a correction method introduced by E. B. Rosa[8] which can be applied to a current sheet formula to account for round wire and exaggerated wire pitch which can correct the results to high levels of accuracy. For the purpose of metal detectors that level of accuracy is not necessary and the uncorrected formulas usually will be close enough. However, keep in mind that you will experience variation depending on the thickness of the insulation and even how tightly the coil is wound and bundled. Best results are obtained with tightly wound, carefully layered magnet wire that is bonded.

In most cases a formula will overestimate the inductance so that when you build the coil to the physical specs the inductance comes out a little low. It is best to always add, say, 5% additional turns to the coil until you measure it, then back off the turns as needed. Also, all the formulas assume circular

4. Harold A. Wheeler; Inductance Formulas for Circular and Square Coils, Proceedings of the IEEE, Vol. 70, No. 12, December 1982, pp. 1449-1450.

5. S. Mohan, M. Hershenson, S. Boyd, & T. Lee, Simple Accurate Expressions for Planar Spiral Inductance, IEEE J. of Solid State Circuits, Vol 34 No 10, Oct 1999, pp 1419-1424.

6. H.A. Wheeler, "Simple Inductance Formulas for Radio Coils," Proc. I.R.E., vol 16, Oct 1928, pp 1398-1400.

7. Square magnet wire is available from several companies.

8. E.B. Rosa and F.W. Grover, Bulletin of the (American) Bureau of Standards, Vol 8 No 1, 1931, p 122.

coils. Any time you take a circular coil and re-shape it in to any thing else (D, square, elliptical, etc.) the inductance will always drop somewhat so be prepared to add even more spare turns. See B.7 below for information on how to estimate non-circular turns.

B.6: Coil Calculator

The first and second editions of *ITMD* included the source code for a JavaScript-based coil calculator for estimating the inductance of skinny-donut coils. Equation B-5 is what co-author George Overton used as the basis for this calculator, and the source code is available on the *Geotech* web site.

B.7: Non-Circular Coils

All the coil equations presented assume circular windings. Suppose you want D-shaped windings for a DD coil, or skinny rectangular windings for a *Cleansweep*-type coil; how would you calculate the required number of turns? In that case, do the following:

1. Find the perimeter[9] of the winding you want to make.
2. Use that perimeter and calculate the radius of an equivalent circular coil.
3. Calculate the area of the coil you want to wind and the area of its circular equivalent.
4. Take the ratio of those areas: $A_{ratio} = A/A_{eq}$. This will always be < 1.
5. Divide the target inductance by $\sqrt{\sqrt{A_{ratio}}}$. Call this the equivalent circular inductance.
6. Use the equivalent circular coil dimensions and the equivalent circular inductance in a coil calculator to find the number of turns.

Example: Make a 27AWG 2mH RX winding for a *Cleansweep* coil.

A *Cleansweep* winding is a rectangle with the dimensions of 50mm x 440mm. The perimeter is therefore 980mm. A round coil with the same perimeter has a radius of 156mm.

The area of the *Cleansweep* TX is 22000mm^2. The area of the circular equivalent is 76453.8mm^2. The area ratio is 22000/76453.8 = 0.288.

The target equivalent circular inductance should be $2\,\mathrm{mH}/\sqrt{\sqrt{0.288}} = 2.73\mathrm{mH}$. The coil calculator computes 53 turns. It is good practice to add 5% which comes to 56 turns. A 56-turn *Cleansweep* coil was made and the resulting inductance was 2.098mH. One less turn (55) will be as close as you can get to 2mH.

---

9. Most of the line art in this book is drawn in Microsoft Visio, which can report the circumference and area of arbitrary closed shapes. This makes it a good tool for designing coil windings.

PI Math

"I have had my results for a long time,
but I do not yet know how I am to arrive at them"

— *Carl Friedrich Gauss*

Single-frequency sinusoidal systems are fairly intuitive and the underlying math straightforward. Not so with time-domain systems, and often people develop misconceptions about what is happening. The concepts presented in the PI section of this book were done so with light coverage of the math, partly because it was not really necessary to get the main points across, and partly because many readers may not have a background in calculus and it would become a huge distraction. This appendix dives into the gory math and looks at ways to better optimize designs.

The Transmit Side

C.1: TX Turn-On Response

We'll start with the transmit behavior and explore how to optimize the coil. Equation 19-3 shows that the coil current rises exponentially during the turn-on period. The turn-on time is finite and usually shorter than the time needed to reach the maximum current (5τ) so the actual peak current at turn-off depends on the transmit pulse width. The equation for this is

$$I_{peak} = I_{max}(1 - e^{-T_{on}/\tau}) = \frac{V_{TX}}{R_{eq}} \cdot (1 - e^{-T_{on}R_{eq}/L}) \qquad \text{Eq C-1}$$

where T_{on} is the transmit pulse width and R_{eq} is the total series resistance during turn-on. Usually the series resistance is dominated by the resistance of the coil itself, although some designs intentionally add a series resistor to effect faster flat-topping at the expense of peak current. As an example consider the following:

- $L = 300\mu H$
- $R_L = 2\Omega$
- $V_{TX} = 10V$
- $T_{on} = 100\mu s$

The maximum current is $10V/2\Omega = 5A$. The turn-on tau is $300\mu H/2\Omega = 150\mu s$. The peak current is

$$I_{peak} = 5A \cdot (1 - e^{-100\mu s/150\mu s}) = 2.43A$$

Figure C-1 illustrates.

The peak current isn't the only important issue. The transmit magnetic field is proportional to the peak current times the number of turns of wire, the so-called ampere-turns (NI) parameter. In a typical scramble-wound coil the inductance is roughly proportional to N^2 and resistance is proportional to N, therefore the time constant (L/R_L) is roughly proportional to N. This tells us that as we add more turns to the coil two things happen: the time constant increases which slows down the rate of charging, and the increased resistance reduces the maximum current. Either of these alone would reduce the peak current (assuming a consistent TX pulse width) so taken together things are even worse. However, adding more turns does help the other half of the ampere-turns parameter that we really care about, but not enough to overcome the harsh reduction in peak current.

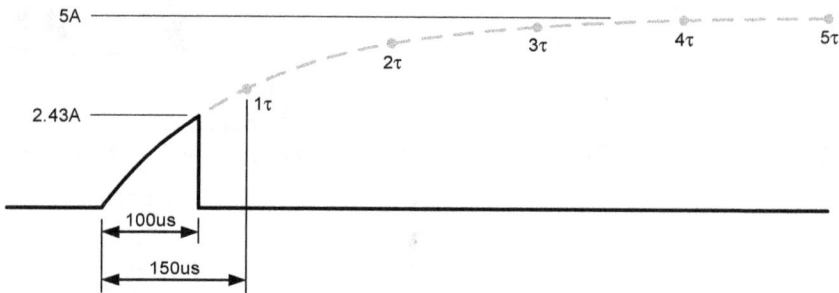

Fig. C-1: **TX Charging**

At this point it is useful to look at a range of coils with actual numbers. For this we'll use a coil diameter of 10" (25cm) and 24AWG PVC-insulated wire. The number of turns range from 15 to 30; the TX supply is assumed to be 10V and the TX turn-on time is 100μs. Switch and battery resistance are ignored. (L and R_L are measured from real coils[1].)

N	L(μH)	$R_L(\Omega)$	tau (μs)	Imax (A)	Ipeak (A)	NI	N^2I	Iave (mA)
15	143.1	1.00	143.1	10	5.03	75.4	1131.4	280.5
20	242.6	1.29	188.1	7.75	3.20	63.9	1278.8	174.0
25	369.0	1.62	227.8	6.17	2.19	54.8	1370.9	117.7
30	519.4	1.94	267.7	5.15	1.61	48.2	1446.0	85.3

Table C-1: **TX Efficiencies vs. Number of Turns**

The table shows how coil selection (in terms of N turns of wire) translates into inductance, series resistance, tau, and the peak current at turn-off. From this we see that, as expected, the peak TX current increases as the number of turns decreases. A single-turn coil would maximize the peak current. However, TX field strength is also dependent on the number of turns and the TX efficiency (NI) still increases with lower N as the increase in current outruns the reduction in turns[2].

This might be the end of the story and the lesson learned is to use fewer turns. But the traditional PI has a mono coil that is used for *both* TX and RX, so now we need to consider the RX side. The magnetic field from a target induces a voltage in the coil that is proportional to the turns as well; recall Equation 7-1. Since it's the same coil the turns are the same and the overall *round-trip efficiency* can be considered to be the ampere-turns-squared (N^2I). The table above includes a column for this conceptual number and now we see that higher N wins. In this case, the increase in current doesn't outrun the *squared* reduction in turns for lower N coils. Had the coils been ideal ($R_L = 0$) then the current would always be a perfect sawtooth and always be exactly inversely proportional to N^2, ampere-turns would be exactly inversely proportional to N, and total round-trip efficiency would be constant regardless of turns.

A final column in Table C-1 shows the average transmit current for each N, calculated by integrating the total TX current during the transmit turn-on time T_{on} and averaging across the overall pulse period T:

$$I_{ave} = \frac{1}{T}\int_0^{T_{on}} \frac{V_{TX}}{R_L}\left(1 - e^{-\frac{t}{\tau_{TX}}}\right) = \frac{T_{on}}{T} \times \frac{V_{TX}}{R_L}\left[1 - \frac{\tau}{T_{on}}(1 - e^{-T_{on}/\tau_{TX}})\right] \qquad \text{Eq C-2}$$

Note that our tau (L/R_L) has been formally designated τ_{TX} because we are looking at the TX side of coil activity. T = 1ms (f = 1kHz) was used to calculate the values in the table. The first term T_{on}/T is an over-

1. Curve-fitting L vs N results in $(0.993*N1/N2)^{1.88}$ instead of the ideal $(N1/N2)^2$. This will vary depending on coil dimensions and choice of wire.

2. This was also seen in Chapter 7, specifically Figure 7-5.

all duty cycle multiplier; the remaining equation gives the total integrated current during the transmit pulse. This assumes a total-loss system (which PI generally is) so that even if $R_L = 0$ the energy stored in the coil at turn-on is completely lost during the turn-off phase. In fact, if $R_L = 0$ then

$$I_{ave} = \frac{T_{on}}{T} \times \frac{1}{2} \cdot \frac{V_{TX}}{L} \cdot T_{on}$$

and current consumption is inversely proportional to N^2. But the coil's resistance also dissipates power during the turn-on phase resulting in even more loss, and the loss is progressively worse at lower N where the turn-on current is also higher. So the round-trip efficiency and power efficiency both favor higher N.

C.2: TX Turn-Off Response

This might be the end of the story and the lesson learned is to use more turns, but by now you have probably guessed this isn't the end of the story. This analysis so far shows how number of turns affects the transmit field and round-trip efficiency. When the transmitter turns off the coil suddenly operates in RX mode. On the RX side the coil's inductance and parasitic capacitance also play a strong role in the minimum achievable sample time which directly affects target sensitivity, especially for small low-conductors like jewelry and nuggets. Let's explore how.

Once a coil is wound its inductance, series resistance, and SRF[3] can all be measured. This tells us everything we need to know about the coil. As we saw, inductance and series resistance control the turn-on behavior which plays into the TX/RX efficiencies. Inductance and SRF control the turn-off behavior which impacts the decay speed and minimum sample time. As with turn-on behavior, it is possible to derive a mathematical equation that describes the turn-off behavior, including flyback and decay.

Coil capacitance is calculated by a slight rearrangement of Equation 19-6:

$$C_L = \frac{1}{(2\pi f_0)^2 \cdot L} \qquad \text{Eq C-3}$$

where f_0 is the measured SRF. R_D is then calculated using Equation 19-8. This gives us values for the three elements of a parallel RLC circuit. The inductor also has series resistance which normally would complicate the solution, but in our case the series resistance is very small compared to the damping resistance and can be ignored on the turn-off side of operation[4].

The currents through the parallel elements of an RLC circuit are described by the differential equation

$$C_L \cdot \frac{d^2v}{dt^2} + \frac{1}{R_D} \cdot \frac{dv}{dt} + \frac{1}{L} \cdot v = 0 \qquad \text{Eq C-4}$$

To completely solve this, we need two initial conditions: the current through the coil and voltage across the capacitor. The coil current is simply the peak current at turn-off, and Table C-1 has those calculated values listed. Since the capacitor is considered to be a lumped element across the coil, its voltage is the same as the coil voltage at the instance just before turn-off, which should be the supply voltage[5], or 10V.

The Laplace transform of Equation C-4 is

$$s^2 + \frac{1}{R_D C_L} \cdot s + \frac{1}{L C_L} = 0 \qquad \text{Eq C-5}$$

and the solution via the quadratic formula is

---

3. Self-resonant frequency; see Chapter 8.

4. Similarly, the damping resistor was ignored on the turn-on side of operation, as it has little effect.

5. This, again, assumes the switch and battery have no resistance. In reality, whether the capacitor's voltage at $t = 0$ is 10V or 6V or 2V has almost no impact on the final response solution. Coil current at $t = 0$ does.

$$s_{1,2} = \frac{1}{2R_DC_L} \pm \sqrt{\left(\frac{1}{2R_DC_L}\right)^2 - \frac{1}{LC_L}} \qquad \text{Eq C-6}$$

We want critical damping which is achieved when the complex conjugate roots are exactly equal. This means the term under the square-root must be zero:

$$\left(\frac{1}{2R_DC_L}\right)^2 = \frac{1}{LC_L} \qquad \text{thus} \qquad L = 4R_D^2C_L \qquad \text{Eq C-7}$$

This can be rearranged to give the required value for the damping resistor that was casually presented in Equation 19-8:

$$R_D = \frac{1}{2}\sqrt{\frac{L}{C_L}} \qquad \text{Eq C-8}$$

We can complete the solution to get a real equation that describes the flyback curve. The response curve at turn-off is not a simple exponential but has the form

$$v(t) = K_1 \cdot e^{-t/\tau_{RX}} + K_2 \cdot t \cdot e^{-t/\tau_{RX}} \qquad \text{Eq C-9}$$

where $\tau_{RX} = 1/s = 2R_DC_L$ and the RX subscript shows that we are looking on the RX side of the coil activity. This is the general solution for a critically-damped parallel RLC circuit found in textbooks and we can use this to calculate settling time.

K_1 and K_2 can be found by applying the initial conditions. At $t = 0$ we have $v(0) = K_1$ which will equal the supply voltage minus the resistive loss of the coil, or $V_{TX} - i(0) \cdot R_L$. The other initial condition is the coil current, which must equal the sum of the currents through the resistor and capacitor:

$$i(t) = C_L \cdot \frac{dv(t)}{dt} + \frac{v(t)}{R_D}\bigg|_{t=0} \qquad \text{thus} \qquad \frac{dv(t)}{dt} = \frac{i(0)}{C_L} - \frac{v(0)}{R_DC_L} \qquad \text{Eq C-10}$$

We know everything except dv/dt, which can be found by taking the derivative of Equation C-9:

$$\frac{dv(t)}{dt} = -\frac{K_1}{\tau_{RX}} \cdot e^{-t/\tau_{RX}} + K_2 \cdot e^{-t/\tau_{RX}} - \frac{K_2}{\tau_{RX}} \cdot t \cdot e^{-t/\tau_{RX}}\bigg|_{t=0} = \frac{-K_1}{\tau_{RX}} + K_2 = \frac{-K_1}{2R_DC_L} + K_2 \quad \text{Eq C-11}$$

Combining Equation C-10 and Equation C-11:

$$\frac{i(0)}{C_L} - \frac{v(0)}{R_DC_L} = \frac{-K_1}{2R_DC_L} + K_2 \qquad \text{Eq C-12}$$

thus

$$K_2 = \frac{i(0)}{C_L} - \frac{v(0)}{R_DC_L} + \frac{K_1}{2R_DC_L} = \frac{1}{C_L}\left(i(0) - \frac{v(0)}{2R_D}\right) \qquad \text{Eq C-13}$$

since $K_1 = v(0)$.

In most cases $i(0)$ is on the order of amps and $v(0)/2R_D$ is perhaps a few milliamps, so we can approximate $K_2 = i(0)/C_L$. The very final solution to the flyback waveform is

$$v(t) = v(0) \cdot e^{-t/\tau_{RX}} + \frac{i(0)}{C_L} \cdot t \cdot e^{-t/\tau_{RX}} \qquad \text{Eq C-14}$$

where, again, $\tau_{RX} = 2R_DC_L$.

Before we try to make good use of all this effort we should check our work through a simple example. For this we'll use real values from an actual built coil:

- $L = 344.3\mu H$
- $C_L = 204.4pF$ (from a measured SRF of 600kHz)
- $R_L = 2.8\Omega$
- $R_D = 648.9\Omega$ (calculated from Equation C-8)

- $V_{TX} = 10V$
- $T_{on} = 100\mu s$
- $i(0) = 1.99A$ (calculated from Equation C-1)

From this $\tau_{RX} = 265.3ns$; the solution is

$$v(t) = 10 \cdot e^{-t/265.3ns} + 9.74 \times 10^9 \cdot t \cdot e^{-t/265.3ns}$$

and when plotted gives the familiar flyback/decay curve in Figure C-2.

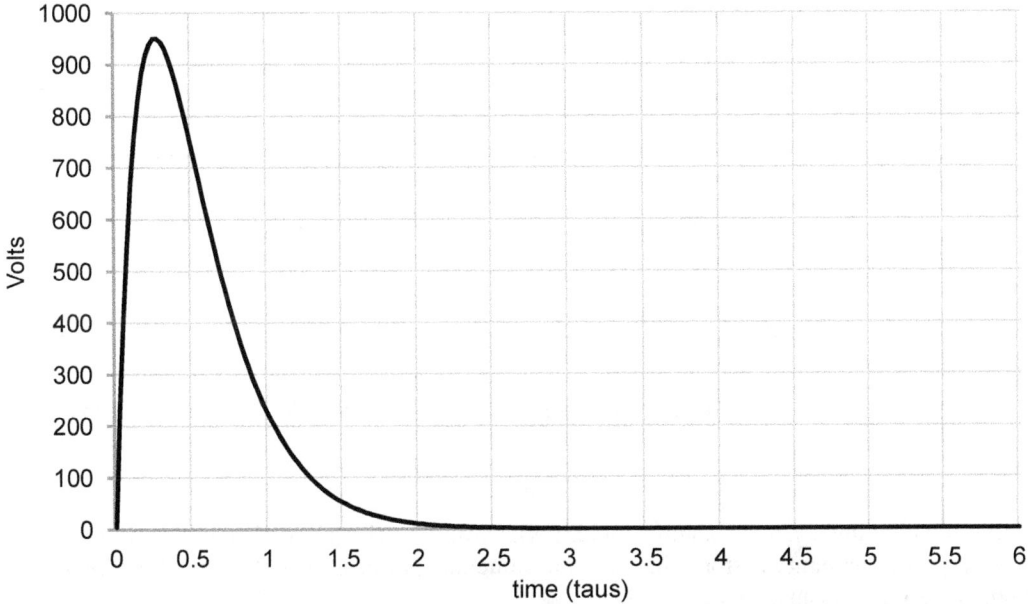

Fig. C-2: **Flyback Curve**

There are two things in the flyback curve that interest us. The first is the peak voltage of the flyback which can be calculated by setting the derivative of Equation C-14 to zero since, at the peak, the slope is zero. The result is

$$\frac{dv(t)}{dt} = \frac{-v(0)}{\tau_{RX}} \cdot e^{-t/\tau_{RX}} + \frac{i(0)}{C_L} \cdot e^{-t/\tau_{RX}} - \frac{i(0)}{C_L} \cdot \frac{t}{\tau_{RX}} \cdot e^{-t/\tau_{RX}} = 0$$

Simplifying gives us

$$t = \tau_{RX} - \frac{C_L}{i(0)} v(0) = 265.3ns - 455ps = 264.8ns$$

From this we can see that the second term is negligible, therefore in a PI system the flyback peak occurs at almost exactly τ_{RX}. Plugging $t = \tau_{RX}$ back into Equation C-14 gives a peak voltage of 951.8V.

The second item of interest is how long it takes to "settle." In an exponential decay we often consider 5τ to be sufficiently settled but that is not good enough here. Let us suppose the coil voltage is connected to a preamp with a gain of 500, and that the preamp output must settle to at least 0.5V or less for the demodulators to function properly. The input to the preamp must therefore settle to 500mV/500 = 1mV for that to happen. To solve, we can set $v(t) = 1mV$ and iteratively solve for t and reach the answer of $4.67\mu s$. In this case, $4.67\mu s$ is slightly more than 17.6τ. So instead of 5τ we need almost 18τ.

Before moving on let's reflect on some practical observations. First, the flyback has a peak of 952 volts, which is substantially more than the 100-400 volts casually mentioned in the PI chapters. The reason for this discrepancy is that our example coil has only 204.4pF of self-capacitance and only includes the raw coil; adding a shield, a cable, and some circuit capacitance will surely increase the total capacitance and produce a somewhat lower flyback peak. The circuits in the PI chapters also use the IRF740 MOSFET which is rated for 400V and typically avalanches at 450V. This limits the flyback, and a higher flyback can likely be achieved with a better MOSFET.

Second, it seems that if the voltage is sufficiently settled by 4.67μs then we have an exceptional design. But, again, all that extra capacitance that is yet to be factored in will quickly dash our hopes. Plus, there is a diode clamp and a preamp stage (or two) to be considered which add even more delay. In general, a sample delay of 15μs is pretty easy; 10μs is slightly difficult; 6μs is a challenge; and less than that requires religious dedication.

Finally, is there any simplification to all the mess we went through? There is. It turns out that the second term of Equation C-14 dominates and we can ignore the first term. This simplifies things to

$$v(t) = \frac{i(0)}{C_L} \cdot t \cdot e^{-t/\tau_{RX}} \qquad\qquad \text{Eq C-15}$$

where $\tau_{RX} = 2R_DC_L$. This certainly looks easier, and so does the derivative at the peak:

$$\frac{dv(t)}{dt} = \frac{i(0)}{C_L} \cdot e^{-t/\tau_{RX}} - \frac{i(0)}{C_L} \cdot \frac{t}{\tau_{RX}} \cdot e^{-t/\tau_{RX}} = 0 \qquad\qquad \text{Eq C-16}$$

thus

$$\frac{i(0)}{C_L} \cdot e^{-t/\tau_{RX}} = \frac{i(0)}{C_L} \cdot \frac{t}{\tau_{RX}} \cdot e^{-t/\tau_{RX}} \qquad\qquad \text{Eq C-17}$$

which occurs when $t = \tau_{RX}$, something we already observed. Using simple substitution we get

$$V_{peak} = \frac{i(0)}{C_L} \cdot \tau_{RX} \cdot e^{-1} = 2 \cdot i(0) \cdot R_D \cdot e^{-1} \qquad\qquad \text{Eq C-18}$$

Using this equation in the above example yields 950.2 volts, which we'll call "close enough." Note that R_D is not just any damping resistor value, it is the value that produces exactly critical damping thereby including the effects of capacitance.

In computing the settling time, there is a shortcut for that, too. Instead of trying to solve for a specific time t at which the decay reaches, say, 1mV, suppose we pose the question as, "How many time constants does it take to reach 1mV?" We can now take the ratio of Equation C-15 and Equation C-18:

$$\frac{1\,mV}{V_{peak}} = \frac{\dfrac{i(0)}{C_L} \cdot t \cdot e^{-t/\tau_{RX}}}{\dfrac{i(0)}{C_L} \cdot \tau_{RX} \cdot e^{-1}} = \frac{t}{\tau_{RX}} \cdot \frac{e^{-t/\tau_{RX}}}{e^{-1}} \qquad\qquad \text{Eq C-19}$$

Taking the natural log of both sides and rearranging, we get

$$t = \tau_{RX}\left[\ln\left(\frac{V_{peak}}{1\,mV}\right) + \ln\left(\frac{t}{\tau_{RX}}\right) + 1 \right] \qquad\qquad \text{Eq C-20}$$

This still has no closed solution, but let's plug in V_{peak} = 950.2V and watch what happens. The first term is 13.76, and we add 1 to get 14.76. If we ignored the middle term then the answer would be t = 14.76τ_{RX}. The iterative solution we got was 17.6τ_{RX}, so we call this "not close enough." But if we plug t/τ_{RX} = 14.76 into the middle term, that term becomes 2.69. When added to 14.76, the result is 17.45τ_{RX} which is within 1% of the iterative answer, and our engineering soul is soaring. So the general (simplified) answer becomes

$$t = \tau_{RX} \cdot (x + \ln(x)) \quad \text{where} \quad x = \ln\left(\frac{V_{peak}}{V_{min}}\right) + 1 \qquad\qquad \text{Eq C-21}$$

where V_{min} is the desired decay voltage value.

This gives us all the information needed for the flyback voltage and its decay (with no target) but, meanwhile, what is happening to the coil current? Using the simplified voltage in Equation C-15 the coil current is now

$$i_L(t) = \frac{1}{L} \cdot \int v(t) \, dt = \frac{i(0)}{LC_L} \cdot \int t \cdot e^{-t/\tau_{RX}}$$

Eq C-22

The solution to this is (using the CRC Math handbook)

$$i_L(t) = -\frac{i(0)}{LC_L} \cdot (\tau_{RX})^2 \cdot e^{-t/\tau_{RX}} \left(\frac{t}{\tau_{RX}} + 1 \right)$$

Eq C-23

Realizing that $LC_L = \tau_{RX}^2$ we get

$$i_L(t) = -i(0) \cdot e^{-t/\tau_{RX}} \left(\frac{t}{\tau_{RX}} + 1 \right)$$

Eq C-24

This describes a modified exponential decay. Figure C-3 shows this (negated) curve plotted versus tau along with a normal exponential curve $(e^{-t/\tau_{RX}})$ for comparison (both normalized, that is, $i(0) = 1$):

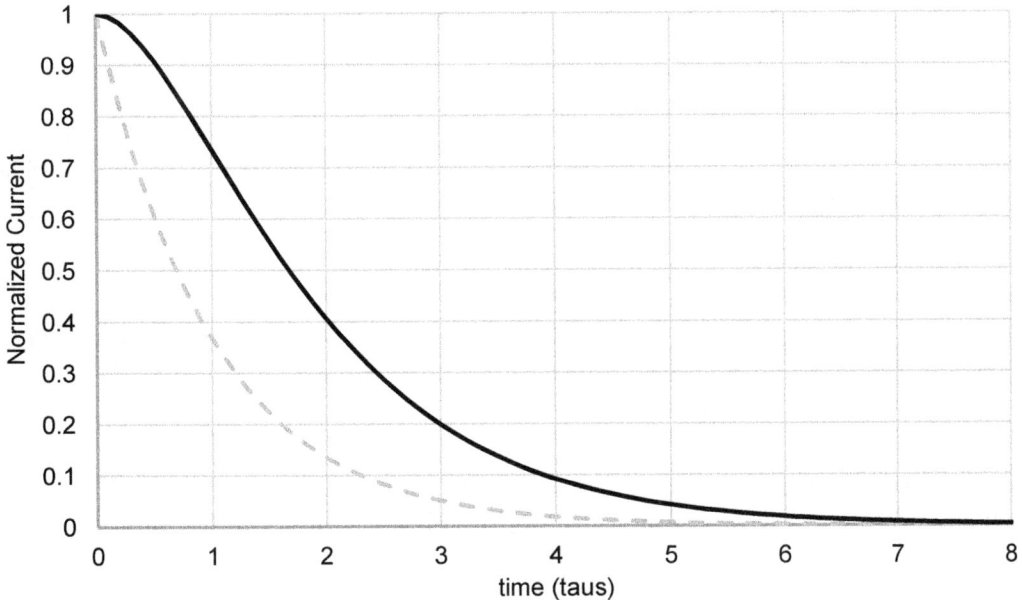

Fig. C-3: **Turn-Off Current vs Std Exponential**

The slope at $t = 0+$ is zero as would be expected for an RLC circuit, and the settling is somewhat extended compared to the normal exponential. It is worth noting because, unlike a simple RL decay, "substantial settling" occurs beyond the normal 5τ.

As a final exercise we'll add a few more columns to our comparison table and see how things stand with the flyback and decay side. Table C-2 shows inductance versus N and adds the measured self-resonant frequency (SRF), plus all the calculated parameters of the response.

N	L (μH)	SRF (kHz)	C_L (pF)	R_D (Ω)	tau (ns)	V_{peak} (V)	Settling (τ)	Settling (μs)
15	143.1	980	184.3	440.6	162.4	1629.9	18.03	2.93
20	242.6	730	195.9	556.4	218.0	1308.7	17.80	3.88
25	369.0	570	211.3	660.8	279.2	1066.4	17.58	4.91
30	519.4	470	220.8	766.9	338.6	906.6	17.41	5.89

Table C-2: **Turn-off Characteristics vs. Number of Turns**

The Target Side

C.3: TX Turn-On Eddies

In the flat-topping discussion it was shown that residual turn-on eddies can reduce the turn-off eddy response (Figure 21-22). We'll now dive into that math and see what really happens. Because we will be working on both sides of the TX turn-off event along with both coil and target response, we'll need to expand our subscripts a little more. 'TX' continues to denote the turn-on (transmit) side, 'RX' denotes the turn-off (receive) side, and 'tgt' will be used for target parameters and responses. Furthermore, there is a reverse target response (tgtr) and a forward target response (tgtf).

The turn-on current from Equation C-1 is restated as

$$i_{TX}(t) = \frac{V_B}{R_L} \cdot (1 - e^{-t/\tau_{TX}}) \qquad \text{Eq C-25}$$

This current creates a magnetic field and along the center axis of the coil as described by Equation 3-4:

$$B_{on}(t) = \frac{1}{2} \cdot \frac{\mu N r^2}{\sqrt{r^2 + d^2}^3} \cdot i_{TX}(t) = k_{TX} \cdot i_{TX}(t) = k_{TX} \cdot \frac{V_B}{R_L} \cdot (1 - e^{-t/\tau_{TX}}) \qquad \text{Eq C-26}$$

k_{TX} replaces a bunch of detail that is inconsequential to our purpose[6]. The magnetic field interacts with a target and induces an electromotive force $\varepsilon = -d\Phi/dt$ as we saw in Equation 4-2. For non-ferrous targets $\Phi = B \cdot A \cos\theta$ where A is the target surface area and θ is the tilt angle, which we'll assume to be $0°$ (i.e., the coin is laying flat):

$$\varepsilon(t) = -A \cdot \frac{dB}{dt} = -k_{TX} \cdot \frac{V_B}{R_L} \cdot \frac{A}{\tau_{TX}} \cdot e^{-t/\tau_{TX}} = -k_{TX} \cdot A \cdot \frac{V_B}{L} \cdot e^{-t/\tau_{TX}} \qquad \text{Eq C-27}$$

We can model a typical eddy target as a lumped RL circuit (see Figure 4-22) where the inductor provides the EMF, and solve for the eddy current:

$$\varepsilon(t) = R_{tgt} \cdot i_{tgtr}(t) + L_{tgt} \cdot \frac{d}{dt} i_{tgtr}(t) \qquad \text{Eq C-28}$$

Notice that the target eddy current has been labeled $i_{tgtr}(t)$. The 'r' denotes the fact that this is a *reverse* eddy current. This will be explained shortly. At this point it is easier to use Laplace transforms to solve:

$$-k'_{TX} \cdot \frac{V_B}{L} \cdot \frac{1}{s + \frac{1}{\tau_{TX}}} = R_{tgt} \cdot I_{tgtr} + s L_{tgt} \cdot I_{tgtr} \qquad \text{Eq C-29}$$

or

$$I_{tgtr}(s) = -k'_{TX} \cdot \frac{V_B}{L \cdot L_{tgt}} \cdot \frac{1}{s + \frac{1}{\tau_{TX}}} \cdot \frac{1}{s + \frac{1}{\tau_{tgt}}} \qquad \text{Eq C-30}$$

where $k'_{TX} = A \cdot k_{TX}$[7]. Using partial fraction expansion this becomes

$$I_{tgtr}(s) = -k'_{TX} \cdot \frac{V_B}{L \cdot L_{tgt}} \cdot \frac{\tau_{TX}\tau_{tgt}}{\tau_{TX} - \tau_{tgt}} \left(\frac{1}{s + \frac{1}{\tau_{TX}}} - \frac{1}{s + \frac{1}{\tau_{tgt}}} \right) \qquad \text{Eq C-31}$$

Converting back to the time domain we get

---

6. However, it is worth noting that k_{TX} has the unit of inductance/area.

7. Which now has the unit of inductance.

$$i_{tgtr}(t) = -\frac{k'_{TX}}{L_{tgt}} \cdot \frac{V_B}{L} \cdot \frac{\tau_{TX}\tau_{tgt}}{\tau_{TX} - \tau_{tgt}} \cdot (e^{-t/\tau_{TX}} - e^{-t/\tau_{tgt}}) \qquad \text{Eq C-32}$$

When $\tau_{tgt} = \tau_{TX}$ there appears to be an anomaly: the tau ratio goes to infinity, but at the same time the exponential terms subtract towards zero[8]. For this case we can return to the s-domain (Equation C-30) and re-evaluate with $\tau_{tgt} = \tau_{TX} = \tau$:

$$I_{tgtr}(s) = -\frac{k'_{TX}}{L_{tgt}} \cdot \frac{V_B}{L} \cdot \frac{1}{(s + \frac{1}{\tau})^2} \qquad \text{Eq C-33}$$

for which the inverse Laplace transform is

$$i_{tgtr}(t) = -\frac{k'_{TX}}{L_{tgt}} \cdot \frac{V_B}{L}(t \cdot e^{-t/\tau}) \qquad \text{Eq C-34}$$

In short, there is no problem and everything is continuous. The result, therefore, is an eddy current with subtractive exponentials of opposite signs. The first exponential is due to the exponential rise of the transmitted magnetic field and the second exponential is due to the natural eddy current decay in the target. When the target decay is much faster than the field time constant ($\tau_{tgt} \ll \tau_{TX}$) then first exponential dominates and the eddy current decay closely follows the exponential of the incident field:

$$i_{tgtr}(t) = -\frac{k'_{TX}}{L_{tgt}} \cdot \frac{V_B}{L} \cdot \tau_{tgt} \cdot e^{-t/\tau_{TX}} \qquad \text{Eq C-35}$$

When the target has a much slower decay than the applied field ($\tau_{tgt} \gg \tau_{TX}$) then second exponential dominates and the eddy current decay rate is dominated by the target itself:

$$i_{tgtr}(t) = \frac{k'_{TX}}{L_{tgt}} \cdot \frac{V_B}{L} \cdot \tau_{TX} \cdot e^{-t/\tau_{tgt}} \qquad \text{Eq C-36}$$

Figure C-4 shows three responses for $\tau_{tgt} = 5\mu s$, $50\mu s$, and $500\mu s$ with $\tau_{TX} = 150\mu s$. The target taus

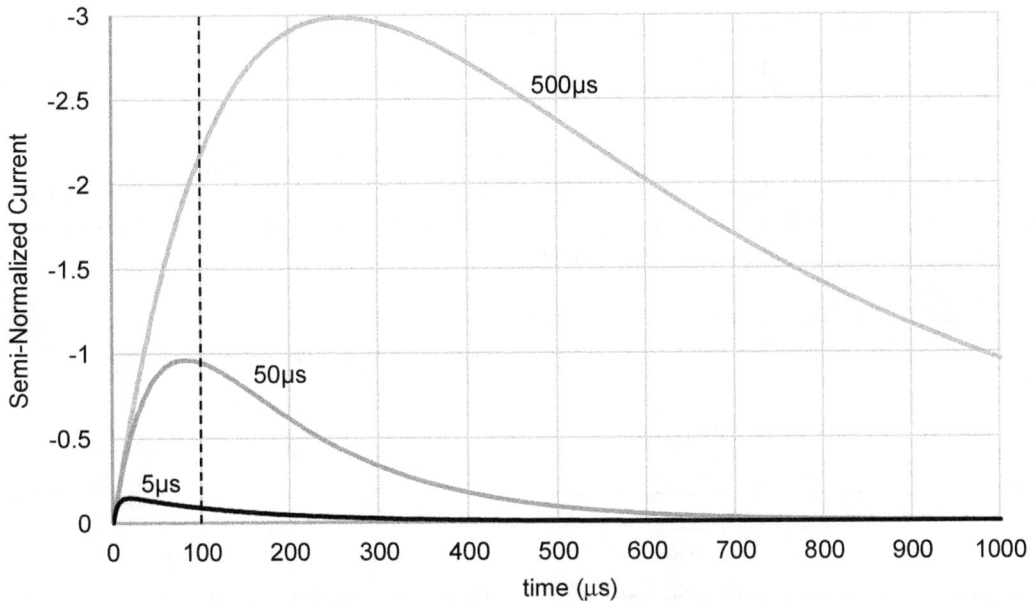

Fig. C-4: **Target TX Eddy Response**

---

8. To see this, plug in $\tau_{tgt} = 0.99\tau_{TX}$.

---

represent fast, medium, and slow targets and the TX tau represents the typical result of a 300µH coil with 2Ω of resistance. The curves are normalized with $k'_{TX}/L_{tgt} = 1$ [9] and use the same example values as before:

- L = 300µH
- R_L = 2Ω
- V_{TX} = 10V

Notice that the response (y) axis is negative; that's because the reverse eddies flow in a (relatively) negative direction but it's useful to draw them with the same orientation as positive eddies for easy visual comparison.

The time axis represents the TX pulse width and a pulse width of 1000µs is unreasonable for most PI designs. If the TX pulse width is, say, 100µs (dashed line) then the low conductor target has peaked and is closely following the exponential response of the TX; the medium conductor target has only just reached its peak current; and the sluggish high conductor target still has an increasing eddy current. In the end what we are interested in is the residual current level at the turn-off event. It is not possible to derive an absolute numerical value for this eddy current because the elements of the k'_{TX} variable are unknown, as are the tau components (R & L) of the target. For now we will leave this solution in a vague state and head over to the RX side of the target response.

C.4: TX Turn-Off Eddies

Recall that the turn-off coil current from Equation C-24 is

$$i_L(t) = -i(0) \cdot e^{-t/\tau_{RX}}\left(\frac{t}{\tau_{RX}} + 1\right)$$

Eq C-37

where $i(0)$ is the initial current (which is the same as the peak turn-on current) and τ_{RX} is ½·L/R_D. As with the TX turn-on current, this produces a magnetic field

$$B_{off}(t) = \frac{1}{2} \cdot \frac{\mu N r^2}{\sqrt{r^2 + d^2}^3} \cdot i_{RX}(t) = k_{TX} \cdot i_{TX}(t) = -k_{TX} \cdot i(0) \cdot e^{-t/\tau_{RX}}\left(\frac{t}{\tau_{RX}} + 1\right)$$

Eq C-38

k_{TX} is exactly the same as before since none of the relevant parameters have changed. The target EMF is

$$\varepsilon(t) = -A \cdot \frac{dB}{dt} = \frac{A \cdot k_{TX} \cdot i(0)}{(\tau_{RX})^2} \cdot t \cdot e^{-t/\tau_{RX}}$$

Eq C-39

Notice that the EMF has a form of $t \cdot e^{-t/\tau_{RX}}$ exactly like the flyback voltage that was derived in Equation C-15. This is fortuitous because — as we've seen before — the target EMF should always have the same waveform as the coil drive voltage waveform, but negated. It is a clue that we have not made a math blunder. The Laplace equation for this situation is:

$$\frac{k'_{TX} \cdot i(0)}{(\tau_{RX})^2} \cdot \frac{1}{\left(s + \frac{1}{\tau_{RX}}\right)^2} = I_{tgtf}(sL_{tgt} + R_{tgt})$$

Eq C-40

where, again, $k'_{TX} = A \cdot k_{TX}$. Thus

$$I_{tgtf}(s) = \frac{k'_{TX} \cdot i(0)}{L_{tgt} \cdot (\tau_{RX})^2} \cdot \frac{1}{\left(s + \frac{1}{\tau_{RX}}\right)^2} \cdot \frac{1}{s + \frac{1}{\tau_{tgt}}}$$

Eq C-41

I_{tgtf} denotes a *forward* eddy current as opposed (literally) to the previous reverse eddy current. Partial

---

9. The reason k'_{TX}/L_{tgt} is normalized will become apparent later. Remember, though, that k'_{TX} has dimensions of inductance which means that k'_{TX}/L_{tgt} is dimensionless.

fraction expansion yields

$$I_{tgtf}(s) = \frac{k'_{TX} \cdot i(0)}{L_{tgt} \cdot (\tau_{RX})^2} \cdot \left[\frac{\tau_{RX}\tau_{tgt}}{\tau_{RX} - \tau_{tgt}} \cdot \frac{1}{\left(s + \frac{1}{\tau_{RX}}\right)^2} + \left(\frac{\tau_{RX}\tau_{tgt}}{\tau_{RX} - \tau_{tgt}}\right)^2 \cdot \left(\frac{1}{s + \frac{1}{\tau_{tgt}}} - \frac{1}{s + \frac{1}{\tau_{RX}}}\right) \right]$$

And the time domain solution is

$$i_{tgtf}(t) = \frac{k'_{TX} \cdot i(0)}{L_{tgt}} \cdot \left(\frac{\tau_{tgt}}{\tau_{RX} - \tau_{tgt}} \cdot \frac{t}{\tau_{RX}} \cdot e^{-\frac{t}{\tau_{RX}}} + \left(\frac{\tau_{tgt}}{\tau_{RX} - \tau_{tgt}}\right)^2 \cdot \left(e^{-\frac{t}{\tau_{tgt}}} - e^{-\frac{t}{\tau_{RX}}}\right) \right)$$

It is useful again to pause and see what the equation is really saying. First, unlike the case where τ_{TX} and τ_{tgt} are very close (or even equal) in value, τ_{RX} is invariably smaller than τ_{tgt}. A moderately mundane τ_{RX} might be $0.5 \cdot 300\mu H/680\Omega = 220ns$ with more aggressive designs even smaller, while τ_{tgt} is generally no less than 1μs and often larger. The term $\tau_{tgt}/(\tau_{RX} - \tau_{tgt})$ will therefore almost always approach the value of −1. Assuming that's the general case the simplified eddy current solution is:

$$i_{tgtf}(t) = \frac{k'_{TX} \cdot i(0)}{L_{tgt}} \cdot \left(e^{-\frac{t}{\tau_{tgt}}} - \left(1 + \frac{t}{\tau_{RX}}\right) \cdot e^{-\frac{t}{\tau_{RX}}} \right)$$

Eq C-42

Since we previously recognized that k'_{TX} has the unit of inductance, Equation C-42 says that the target eddy current response is the peak transmit current $i(0)$ (at TX turn-off), scaled by k'_{TX}/L_{tgt}, and modified by two exponential responses. Figure C-5 shows the responses again for τ_{tgt} = 5μs, 50μs, and 500μs using the same TX values as the previous example, and with $R_D = 680\Omega$ so that $\tau_{RX} = 220ns$. The curves are again normalized with $k'_{TX}/L_{tgt} = 1$ so actual amplitudes will vary. Note that the same variables (k'_{TX}/L_{tgt}) were normalized in Equation C-32 to produce the plot in Figure C-4, therefore this plot represents the same relative forward eddy response as the reverse eddies in Figure C-4.

Contrasting Figure C-5 with Figure C-4 we see that the near-step response of the TX turn-off event results in both high and low target conductances having equally fast initial eddy build-up, compared to

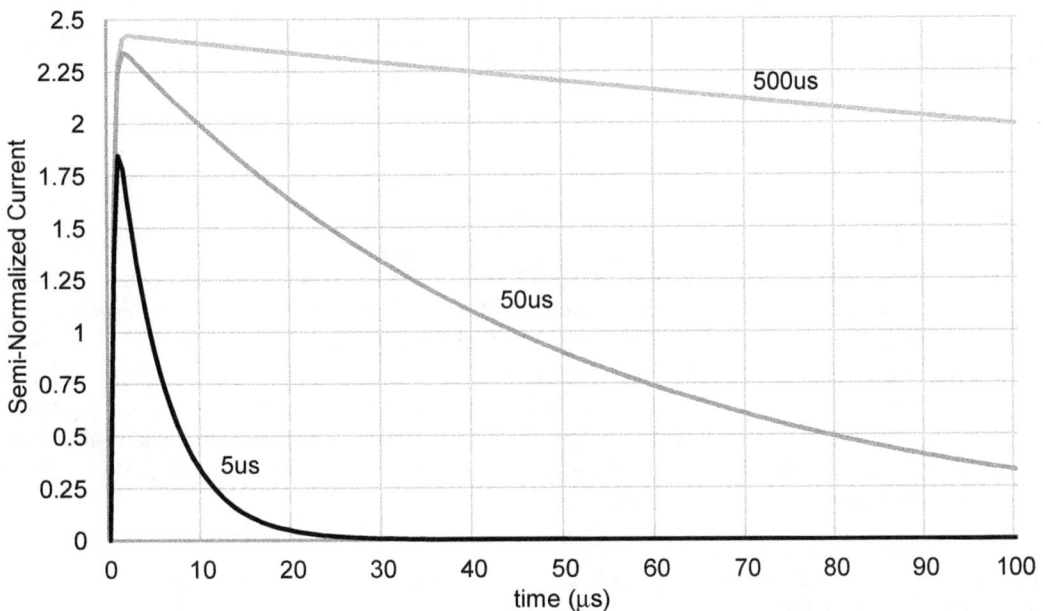

Fig. C-5: **Target Eddy Response**

the highly variable responses when the TX current is exponentially ramping up. This suggests that sampling after the turn-off event should always be done as early as possible to maximize sensitivity to any target conductivity. Figure C-5 also suggests that it is possible to sample too soon, before the eddy currents have reached their peak. In the plot above this is roughly 1-2μs, which corresponds to roughly $5\tau_{RX}$. This seemingly places another boundary on what we can ever hope to achieve, but we've already seen that the coil response alone exceeds this by a wide margin[10]. See Equation C-21 for a refresher.

C.5: Complete Eddy Response

We have derived the equation for a reverse eddy current and another equation for the forward eddy current. Reverse eddies occur during TX turn-on and flow in one direction; forward eddies occur during the turn-off event and flow in the opposite direction. It is the forward eddy response we are interested in but residual reverse eddies left over from the TX turn-on event end up in the forward response and not in a good way.

Because the two eddy equations represent activity on either side of a step event they cannot be directly combined. Rather, Equation C-32 is used to calculate the magnitude of the reverse eddy current at turn-off, and that eddy current then becomes an initial condition for the turn-off side of the action. Per superposition, this initial condition current has the target tau applied to it and it decays to zero. The *total* target eddy current (i_{tgt}) is now:

$$i_{tgt}(t) = i_{tgtf}(t) + i_{tgtr}(t_{TXW}) \cdot e^{-\frac{t}{\tau_{tgt}}}$$

Eq C-43

where

$$i_{tgtr}(t_{TXW}) = \frac{-k'_{TX}}{L} \cdot \frac{V_B}{R_{tgt}} \cdot \frac{\tau_{TX}}{\tau_{TX} - \tau_{tgt}} \cdot (e^{-t/\tau_{TX}} - e^{-t/\tau_{tgt}})\Big|_{t = t_{TXW}}$$

Eq C-44

and t_{TXW} is the transmit pulse width. This shows that the desired target eddy response is diminished by the residual reverse eddies left over from the TX turn-on. This is why it is generally desirable to allow the TX current to flat-top before the turn-off event; a flat-topped current results in a dB/dt = 0 which induces no EMF in the target. However, even allowing enough time to flat-top may not be enough. You also want whatever reverse eddies that were induced before flat-topping to have time to decay away. The curves in Figure C-4 suggest that this could take a while: five times the worst-case target tau.

As a final example to wrap things up, the two sides can be plotted together to give a clarifying picture of what is happening. As with the previous examples, we will again use the following:

- L = 300μH
- R_L = 2Ω
- R_D = 680Ω
- t_{on} = 100μs
- τ_{TGT} = 5μs, 50μs, 500μs

This gives us τ_{TX} = 150μs, τ_{RX} = 0.22μs, and $i_{tgtr}(100\mu s)$ = 2.34A, and results in the curves shown in Figure C-6. The reverse eddies occur during the turn-on time from 0 to 100μs and are shown as negative currents. Note that these portions of the curves are identical to the first 100μs of the curves in Figure C-4. The TX coil turns off at 100μs and the forward eddy responses begin there. These portions of the curves are the same as those in Figure C-5 except that they do not begin at i = 0, they begin where the reverse eddies stopped[11]. In the case of the 5μs target there is little degradation because the residual reverse eddy current is relatively small. But for the 500μs target the reverse eddy current significantly degrades (by almost 90%) the forward eddy response.

---

10. This assumes a mono coil. In a system with separate TX & RX coils that are induction-balanced, it is possible to sample almost down to zero.

11. Compare these curves to the Foil, Nickel, and Quarter responses shown in Figure 26-15.

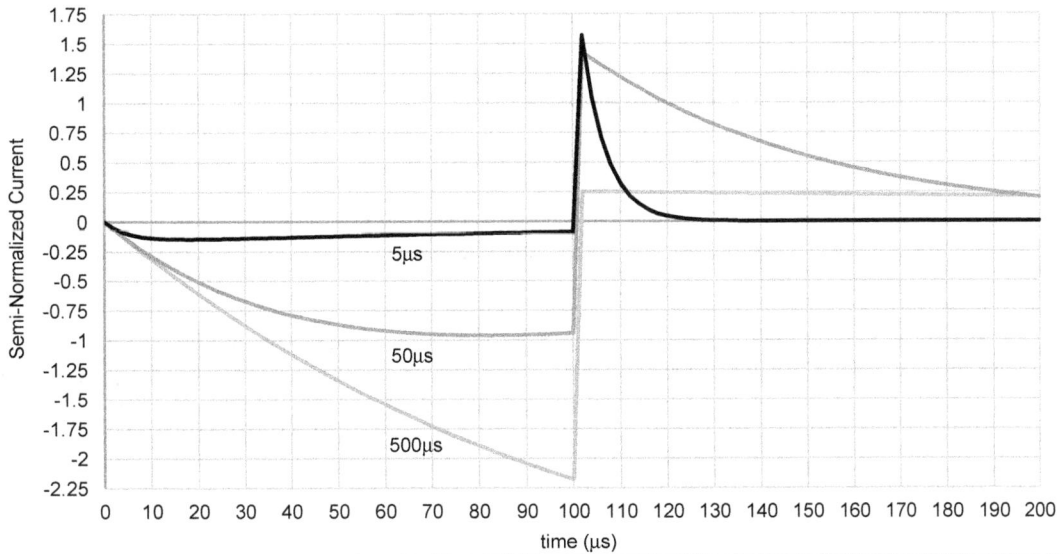

Fig. C-6: **Complete Target Eddy Response**

This explains why many PI detectors can detect a US nickel deeper than a silver US quarter. It is also why it is important to design the PI for the intended targets. If you want to detect Atocha bars, then the TX turn-on pulse width should be at least five times the tau of a silver bar which can be many milliseconds. If you are nugget hunting then the turn-on time makes very little difference; from Figure C-6 you can see that a 5μs target will show little additional degradation when the pulse width is 10μs versus 100μs.

Another item worth noting is that for high-tau targets the reverse eddy currents are often a greater magnitude than the forward eddies, even when the reverse eddies are allowed to settle out. For example, the 500μs target has a peak reverse eddy (Figure C-4) of "3" (keeping in mind these are relative numbers, not actual currents) but has a peak forward eddy of about "2.4" (Figure C-5). When the TX pulse width is shortened, the difference becomes even greater. This suggests that detecting high conductors will benefit from using an IB coil and sampling the turn-on response. The same is not true for low conductors as their turn-off response is always greater than their turn-on response.

C.6: Target Response vs Turn-Off Slew Rate

Before leaving the target side of the math let's look at one more element of the design. We've generally preached that a faster turn-off slew rate will produce a stronger target "kick." This is true but it does so with diminishing returns. We know that the di/dt[12] of the pulse turn-off determines the amount of induced EMF in a target, which in turn determines the strength of the eddy currents. If we want a stronger target response then we can either increase the coil's ampere-turns at turn-off or decrease the turn-off time, or both. Increasing the ampere-turns is a no-brainer; you can increase the turns, reduce the amount of parasitic series resistance, increase the drive voltage, or some combination; these have their own trade-offs. Reducing the turn-off time is a bit more difficult; maybe a faster MOSFET, faster MOSFET driver, or lower coil parasitic C.

It is tempting to reduce the turn-off time to as close to zero as possible but the additional effort quickly outruns the additional benefit. Let's suppose we have a given coil with a peak current I_p at turn-off and that the turn-off is a linear ramp that occurs in time Δt. The induced target EMF is therefore a step voltage with an amplitude proportional to the turn-off slew rate. If we decrease the turn-off time by half ($\Delta t/2$) this will double the slew rate and double the EMF step voltage, but the step will last half as long. Figure C-7 illustrates.

---

12. Assuming a constant ampere-turns.

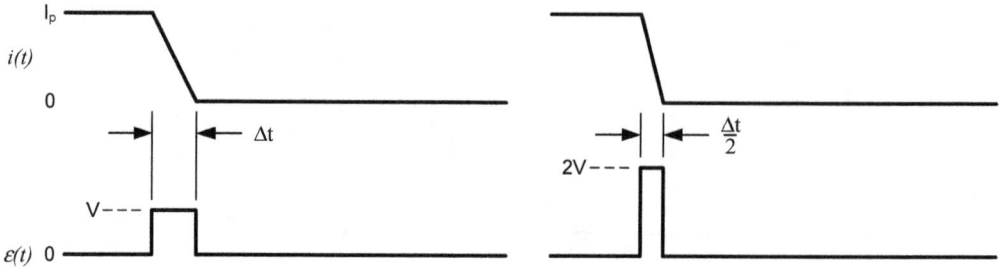

Fig. C-7: **EMF vs Slew rate**

The EMF step drives the creation of eddy currents which increase exponentially per the tau of the target with an initial rate proportional to the EMF. Therefore, the total peak amplitude of the eddy currents depends on both the amplitude and duration of the EMF. A shorter, higher EMF produced by a faster turn-off slew rate produces a higher peak eddy current, but not by a lot. Figure C-8 shows the eddy current waveforms for the same Δt and $\Delta t/2$ transitions above. I_{max} and $2I_{max}$ are the maximum eddy drives that would eventually occur if the EMFs were sustained.

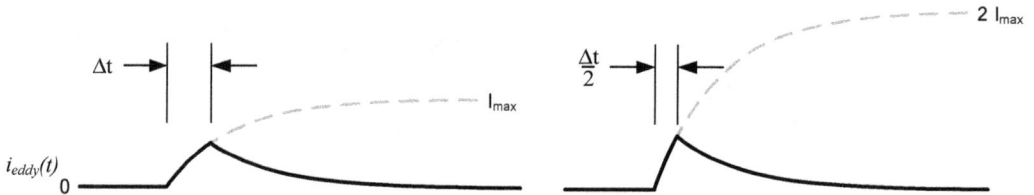

Fig. C-8: **Eddy Current vs Slew rate**

It is apparent that doubling the turn-off slew rate has a non-linear (and rather marginal) effect on the peak eddy current. The target tau plays a strong role in the relative effect and it is useful to calculate the improvement as a function of the target tau:

$$I_{eddy} = I_{max} \cdot (1 - e^{-\Delta t/\tau})$$
Eq C-45

By plugging in Δt as a function of τ and scaling I_{max} accordingly we can get a clearer picture of the relationship:

$$I_{eddy}(\Delta t = \tau) = I_{max} \cdot (1 - e^{-\tau/\tau}) = 0.632 \cdot I_{max}$$

$$I_{eddy}\left(\Delta t = \frac{\tau}{2}\right) = 2I_{max} \cdot (1 - e^{-\tau/2\tau}) = 0.787 \cdot I_{max}$$

etc...

SlewRate	Efficiency
τ	63.2%
$\tau/2$	78.7%
$\tau/3$	85.0%
$\tau/4$	88.5%
$\tau/5$	90.6%
$\tau/10$	95.2%
$\tau/20$	97.5%
$\tau/50$	99.0%
τ/∞	100%

Table C-3: **Eddy Efficiency vs Turn-Off Slew**

Given a fixed peak coil current, the peak eddy efficiency versus slew rate is shown in Table C-3 where the slew rate is stated in terms of the target tau. What this means is if a desired target tau is 10μs then a perfect turn-off step of 0μs will produce the maximum eddy response, but a turn-off slew of 2μs (τ/5) will produce 90.6% of the maximum. From this we can see that the effort in improving turn-off slew depends on what we are wanting to detect. If our goal is a silver cache detector or a meteorite detector where taus are typically a few 100μs then turn-off slew is not so important. For a sub-gram nugget hunter, more effort may be well worth it.

In reality the TX turn-off is not linear, it is a modified exponential as given by Equation C-24. Therefore the induced EMF is not a step response but rather a transient exponential as given by Equation C-39. However, those equations always assumed an instantaneous step and things get quite a bit more complicated with a slower step and also considering that the TX switch (usually an NMOS device) is also non-linear. Actual results from simulations suggest that switching speed is even less critical than Table C-3 suggests.

The Receive Side

C.7: RX Coil Response

A target in the presence of a metal detector forms a double transformer. The previous section dealt with the first half of that system: the transmitter-target transformer. Now we look at the second half: the target-receiver transformer. The eddy currents generated in the target form the primary-side signal for the receiver. We now know the mathematical behavior of the eddy currents and can continue the analysis back to the metal detector.

The B-field emanated by the target is

$$B_{tgt}(t) = \frac{1}{2} \cdot \frac{\mu r^2}{\sqrt{r^2 + d^2}^3} \cdot i_{tgt}(t) = k_{tgt} \cdot i_{tgt}(t) \qquad\qquad \text{Eq C-46}$$

where r in this case is the radius of the target[13] and noting that the target is a single-turn coil (N=1). Otherwise, the components of k_{tgt} are the same as for k_{TX}. $i_{tgt}(t)$ is generally given by Equation C-43 and will certainly include the forward eddy component that could be modified by a reverse eddy component. For slight simplicity we'll only consider low-tau targets where the reverse eddy effect can be ignored (Equation C-42). It is not a terribly difficult exercise to expand these calculations to include the reverse eddy effect. Also, if an IB coil is used then the entire reverse and forward target responses can be considered in the RX EMF; again, this is not difficult to calculate but we will not do so here.

B_{tgt} reaches the coil which, by now, has transformed into a receiver coil. Obviously it has the same diameter and number of turns as the transmitter coil we started with[14]. The induced EMF in the coil is

$$\varepsilon(t) = -A_{RX} \cdot N \cdot \frac{dB_{tgt}}{dt} = -A_{RX} \cdot N \cdot k_{tgt} \cdot \frac{d}{dt} i_{tgt}(t) \qquad\qquad \text{Eq C-47}$$

where A_{RX} is the area of the coil and N is the number of turns. The derivative of the forward current given in Equation C-42 is

$$\frac{di_{tgtf}(t)}{dt} = -\frac{k'_{TX}}{L_{tgt}} \cdot i(0) \cdot \left(\frac{1}{\tau_{tgt}} \cdot e^{-\frac{t}{\tau_{tgt}}} - \frac{t}{(\tau_{RX})^2} \cdot e^{-\frac{t}{\tau_{RX}}} \right) \qquad\qquad \text{Eq C-48}$$

This results in an induced EMF of

---

13. We'll assume a well-behaved round disc like a coin.

14. This, of course, assumes a mono coil. The RX coil could be a separate coil with any diameter and number of turns with only a slight complication in the math.

$$\varepsilon(t) = A_{RX} \cdot N \cdot k_{tgt} \cdot \frac{k'_{TX}}{L_{tgt}} \cdot i(0) \cdot \left(\frac{1}{\tau_{tgt}} \cdot e^{-\frac{t}{\tau_{tgt}}} - \frac{t}{(\tau_{RX})^2} \cdot e^{-\frac{t}{\tau_{RX}}} \right) \qquad \text{Eq C-49}$$

In the equations for the TX side of the target response we normalized the k'_{TX}/L_{tgt} term to 1 to obtain relative responses that could be compared. Figure C-9 has three more unknown terms — A_{RX}, N, and k_{tgt} — so it will be impossible to produce a numerically comparable result. The best we can do is see what the waveform shape looks like. Figure C-9 shows each of the exponential terms individually, plus the composite waveform, for a 1μs target.

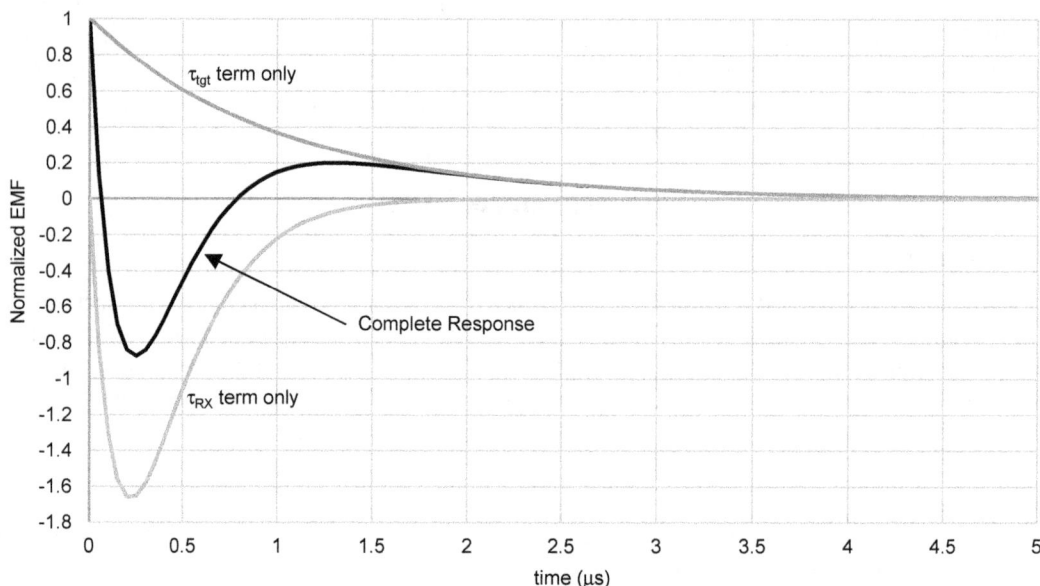

Fig. C-9: **RX Response Waveform**

The τ_{RX} term has the form of a negative $t \cdot e^{-\alpha t}$ and dies out at about 2μs. The tau of this term is due to the RX coil which is $\tau = 0.5 \sqrt{L/R_D}$ and is independent of the target being detected. That is, this part of the response will not change for different targets. The τ_{tgt} term is a simple exponential with a tau due only to the target. After a time of about $10 \cdot \tau_{RX}$ the composite response very closely tracks the τ_{tgt} term and we can assume it is purely target. A 1μs target was chosen for Figure C-9 to better illustrate the effect of the τ_{RX} term. With higher target taus the τ_{RX} term vanishes relatively quickly compared to the target response.

It is important to remember that Equation C-49 represents the EMF that is induced in the coil by the target response only. It does not include the coil's flyback response as shown in Figure C-2. The flyback can peak at 100s of volts while the target EMF may be in the millivolts. We therefore have to wait for the flyback to decay to a low enough level before we can try to discern the target response.

C.8: EFE Sample

Most PI designs utilize a late sample for Earth field cancelation. The late sample is subtracted from the early sample which not only cancels the slow Earth field signal but cancels any sufficiently low frequency signal. In digital signal processing, the act of subtracting (or adding) a later sample to an earlier sample forms what is called a *comb filter*.

A subtraction comb filter has a frequency response described by the transfer function

$$H(f) = 2|\sin(\pi f t_d)| \qquad \text{Eq C-50}$$

where t_d is the delay between the first sample and the subtractive sample[15]. An example value for t_d in a

---

15. An "addition" comb filter has a transfer function $2|\cos(\pi f t_d)|$ which begins with a peak at DC instead of a null.

PI detector might be 100µs. This would make the first peak occur at 5kHz and effect a high-pass filter between DC and 5kHz (Figure C-10).

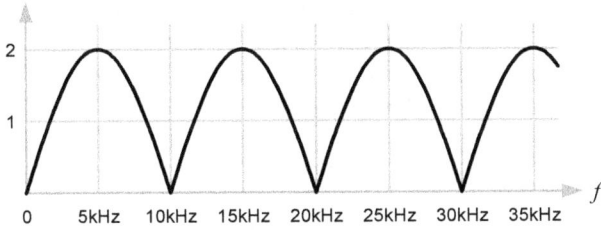

Fig. C-10: **Comb Filter Response**

The frequency of Earth field depends on how fast the coil is being swung, with 1Hz being a practical nominal rate. At $t_d = 100$µs and f = 1Hz the gain is

$$H(1\text{Hz}) = 2|\sin(\pi \cdot 100\mu)| = 0.000628 = -64\text{dB}$$

While not perfect cancelation, it is a reasonably good. The same can be done for power line interference:

$$H(50\text{Hz}) = 2|\sin(\pi \cdot 50 \cdot 100\mu)| = 0.0314 = -30\text{dB}$$

or

$$H(60\text{Hz}) = 2|\sin(\pi \cdot 60 \cdot 100\mu)| = 0.0377 = -28.5\text{dB}$$

This is a bonus and we'll take whatever we can get.

The question then arises, what is the "frequency" of the target response? Both Earth field and power line EMI continuously span the sampling operation of the demod, but each target response is a exponential decay that only spans a single sampling pair. If the early sample is V_1 and the late sample is V_2 then the "gain" is

$$h(t) = \frac{V_1 - V_2}{V_1} = \frac{e^{\frac{-t_s}{\tau}} - e^{\frac{-(t_s + t_d)}{\tau}}}{e^{\frac{-t_s}{\tau}}} = 1 - e^{\frac{-t_d}{\tau}}$$

where t_s is the sample delay of the first sample, t_d is the amount of delay from the first sample to the EFE sample, and τ is the time constant of the target. The amount of attenuation depends on sample spacing relative to the target tau. A (relatively) fast target has little or no attenuation, while a slow target does.

It's useful to look at the attenuations of EFE, EMI, and targets for a few different sample spacings:

Spacing →	50µs	100µs	500µs
$\tau = 5$µs	0.9999	1	1
10µs	0.9933	0.9999	1
20µs	0.9179	0.9933	1
50µs	0.6321	0.8647	0.9999
100µs	0.3935	0.6321	0.9933
200µs	0.2212	0.3935	0.9179
500µs	0.0952	0.1813	0.6321
60Hz	0.0188	0.0377	0.1882
50Hz	0.0157	0.0314	0.1569
EFE (1Hz)	0.0003	0.0006	0.0031

Table C-4: **Comb Filter Signal Attenuation**

Clearly a shorter spacing better suppresses EFE and EMI but also attenuates high conductors. In general, a sample spacing 3-5 times longer than highest desired target tau should be the goal. A spacing of 50µs might be useful for a high-frequency nugget detector where the only nuggets are expected to be on the order of a few grams. But a cache detector would prefer a much wider spacing, 500µs or more.

Bipolar TX designs do not employ a second sample for EFE cancelation but rely on the bipolar action to do the same job. The subtraction of the positive and negative cycles occurs at the pulse frequency and also results in the exact same comb filter response as described previously, where t_d is replaced with the transmit period, or rather the reciprocal of the TX frequency:

$$H(f) = 2\left|\sin\left(\pi\frac{f}{f_{TX}}\right)\right|$$

However, there is no target attenuation with this method. That is, you get attenuation of EFE and EMI with no loss in target signal[16]. The drawback is you have less control over the effective spacing. To achieve the same EFE/EMI attenuation of 100µs (as above) you will need a TX frequency of 10kHz which is rather high but possible in a nugget hunter. In applications where high conductor targets are desired a wider TX pulse is needed and, to keep power consumption in check, is coupled with a lower pulse rate frequency which degrades EFE/EMI suppression.

C.9: Ground Balance

Subtractive ground balance creates a "hole" in normal target responses. This section will look at the math behind the problem. Consider that the PI ground balance mostly deals with a viscous ground response and that response (as a voltage at the RX coil) is ideally a function of 1/t:

$$v_{gnd} = k_{gnd} \cdot t^{-\alpha} \qquad\qquad \text{Eq C-51}$$

where k_{gnd} is some multiplier and α is typically close to 1. Also consider that a single domain target has a response that is dominated by a single exponential:

$$v_{tgt} = k_{tgt} \cdot e^{-t/\tau} \qquad\qquad \text{Eq C-52}$$

where k_{tgt} is, again, some multiplier and τ is the tau of the target. In both cases the k multipliers have units of volts and we will consider them constant over the course of a single pulse.

In a subtractive ground balance, a late amplified[17] sample (usually in a ground channel) is subtracted from an early sample (usually in a target channel) and the channel gains adjusted so that they exactly cancel for a ground signal. For this exercise we will consider the samples to be point samples, therefore for a ground signal the two samples result in

$$v_{GND} = (k_{gnd} \cdot t_1^{-\alpha}) - k_{amp} \cdot (k_{gnd} \cdot t_2^{-\alpha}) = 0 \qquad\qquad \text{Eq C-53}$$

or

$$k_{amp} \cdot k_{gnd} \cdot t_2^{-\alpha} = k_{gnd} \cdot t_1^{-\alpha}$$

where k_{amp} is the excess ground channel gain. This is satisfied when $k_{amp} = (t_2/t_1)^\alpha$. Suppose the early sample is $t_1 = 10$µs and the late sample is at $t_2 = 25$µs, and $\alpha = 1.1$; k_{amp} would need to be 2.74. We can see from this that k_{amp} depends on the relative spacing of samples and also on the exact response of the ground.

The same subtraction is also applied to the target eddy response in Equation C-52. This results in

$$v_{TGT} = (k_{tgt} \cdot e^{-t_1/\tau}) - k_{amp} \cdot (k_{tgt} \cdot e^{-t_2/\tau})$$

We cannot solve this for an absolute response because the k_{tgt} and τ values are arbitrary. But we can get a relative response like this:

---

16. Finally, the free lunch we've been yearning for?

17. This can be done by literally amplifying the sample, or by integrating a wider sample, or a combination.

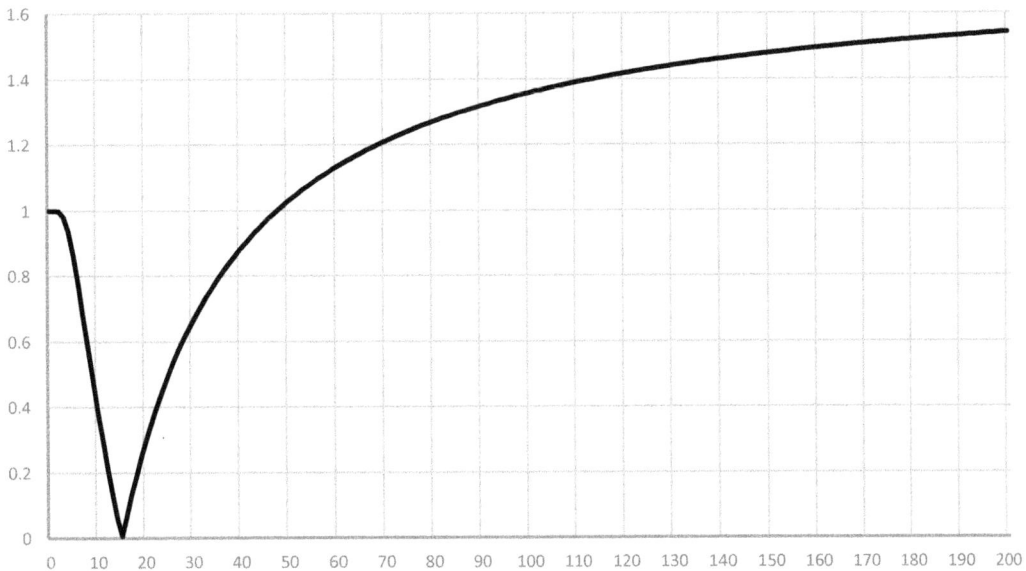

Fig. C-11: **Subtractive GB Relative Response vs Tau**

$$Tgt_{rel} = \frac{v_{TGT}}{v_{tgt}} = 1 - k_{amp} \cdot e^{-(t_2 - t_1)/\tau} \qquad\qquad \text{Eq C-54}$$

Using the sample times and k_{amp} from the ground balance example above we now have

$$Tgt_{rel} = 1 - 2.74 \cdot e^{-(25 - 10)\mu s/\tau}$$

Suppose we have a target whose tau is 5μs. This tells us that relative target response for the subtractive ground balance is 86% of the case where ground balance is disabled. That is, the ground balance subtraction causes a 14% reduction in target response for the particular numbers in this example.

From Equation C-54 it is also apparent that some combination of ground balance and target tau will result in a complete subtraction of the target signal. The target tau for which that occurs can be found by setting Tgt_{rel} to zero:

$$1 - k_{amp} \cdot e^{-(t_2 - t_1)/\tau} = 0$$

or

$$\tau = \frac{t_2 - t_1}{\ln(k_{amp})}$$

Again, using the example numbers above, this is true when $\tau = 14.88\mu s$. Because this particular target tau produces no response at all, we call this the *target hole*. A target with a lower tau would produce a relatively positive response and a target with a higher tau would produce a relatively negative response. In gold PI detectors these are often coupled to a VCO audio so that targets with a tau lower than the target hole produce a high tone and those with a higher tau produce a low tone.

The use of subtractive ground balance will always cause a reduction in detection depth for targets below the ground balance point, and the reduction gets progressively worse for taus closer to the ground balance point. Above the ground balance point, the subtractive target strength can eventually exceed that of a non-ground balanced response, depending on the parameters. A response plot for the example numbers used above is shown in Figure C-11. The target hole is at 14.88μs and targets above 47μs benefit from the higher ground channel gain. Compare this plot to Figure 22-2.

C.10: Power Consumption

For a traditional PI design with a monopolar TX and a total energy loss every cycle, the power consumption can be estimated as follows. First, assume that the TX current approximates a linear ramp which is often the case in beach and gold models. The peak current at the end of the TX pulse is

$$I_{peak} = \frac{V_{TX}}{L} \cdot t_{on} \quad \text{(amps)} \qquad \text{Eq C-55}$$

where V_{TX} is the TX drive voltage, L is the search coil inductance, and t_{on} is the TX turn-on time. The energy stored in the coil at the end of the TX pulse is

$$E = \frac{1}{2} \cdot L I^2 = \frac{1}{2} \cdot \frac{V_{TX}^2}{L} \cdot t_{on}^2 \quad \text{(joules)} \qquad \text{Eq C-56}$$

Power consumption is just energy per unit of time; in this case, Equation C-56 is the energy lost in every transmit period T so that power consumption is

$$P = \frac{1}{T} \cdot \int_0^t E(t)\, dt$$

$$= \frac{1}{2} \cdot \frac{V_{TX}^2}{L} \cdot \frac{t_{on}^2}{T} \quad \text{(watts)} \qquad \text{Eq C-57}$$

As an example, suppose we have the following:

- $L = 300\mu H$
- $V_{TX} = 12V$
- $t_{on} = 100\mu s$
- $T = 1ms$ (1kHz)

The estimated power consumption will be

$$P = \frac{1}{2} \cdot \frac{12^2}{300\mu} \cdot \frac{100\mu^2}{1m} = 2.4 \text{ watts}$$

Keep in mind this estimate is only for the TX circuit and will be on the low side. Series resistance (not included above) will increase the power somewhat. It also does not include all the remaining circuitry, but it is usually the TX circuit that dominates power consumption. It is useful to calculate power consumption in order to select the proper battery capacity and to also to plan for thermal management.

<table>
<tr><td>A
P
P
E
N
D
I
X</td><td># D</td><td># **Resources**</td></tr>
</table>

A
P
P
E
N
D
I
X

D

Resources

"Don't start vast projects with half-vast resources."

— Bruce Pittman

In the Introduction we mentioned that information on metal detector technology is remarkably difficult to find, especially compared to other electronic devices. Metal detectors represent a very small market with only a handful of companies who actively design them, so there is a natural tendency to closely guard intellectual property.

However, over the years there have been a few books that cover detector technology, plus a number of detector designs presented in electronic hobby magazines. For the most part these have dealt with only simplistic designs, often basic BFO and TR circuits. A far better resource for advanced developments is the patent office (See Appendix E).

This chapter presents a number of resources and references for the curious who want to dig deeper into detector technology.

Web Sites

Geotech — https://geotech1.com — This web site is run by the author of this book and contains the single largest collection of metal detector information anywhere. Support files (Gerbers, source code, and the like) can be found at https://geotech1.com/itmd, for this edition and previous editions, and errata and updates to the book are also posted. *Geotech* also hosts a large discussion forum on a variety of technical topics and projects at https://geotech1.com/forums.

Findmall Tech forum — http://www.findmall.com/list.php?34 — This is a tech forum for discussion of PI detectors, established by PI guru Eric Foster.

THunting.com Tech forums — http://www.thunting.com/smf/index.php?c=15 — These are the tech-related forums of the larger THunting.com web site.

MD4U — http://www.md4u.ru — These forums are in Russian.

Coil64 — https://coil32.net/online-calculators.html — The web site has multiple coil calculators that cover different winding topologies.

Accel Instruments — https://www.accelinstruments.com/Magnetic/Coil-Calculator.html — This is a magnetic field calculator for circular coils and can calculate the field strength anywhere in the 3D space around the coil. The calculated field strength appears to be z-component (aligned with the coil's axis). The coil is described in terms of physical parameters (dimensions, wire size, etc.) and it also gives the resulting electrical parameters (inductance, DC & AC resistance, capacitance). Litz wire is supported.

Books

Title	Author(s)	Date
Modern Divining Rods	R.J. Santschi	1927
How to Build Proximity Detectors and Metal Locators	John Potter Shields	1965
The Electronic Metal Detector Handbook	E.S. "Rocky" LeGaye	1969
Official Handbook of Metal Detectors	Dr. Arnold Kortejarvi	1969
How to Build Your Own Metal & Treasure Locators	F.G. Rayer	1976
How to Build Metal/Treasure Locators	Robert Traister John Traister	1977
The Complete VLF-TR Metal Detector Handbook	Roy Lagal Charles Garrett	1979
How to Build Gold & Treasure Detectors (Collection of former magazine projects.)	ETI Magazine	1981
The Advanced Handbook of Metal Detectors	Charles Garrett	1985
Building Metal Locators: A Treasure Hunter's Project Book	Charles D. Rakes	1986
Schatzsuche mit Metalldetektoren (German)	A. Gatowski	1990
Treasure: The Business and Technology (Chapter 3: Tools for Archaeology covers some advanced methods)	Phillip S. Olin	1991
Inside Treasure Hunting (Chapters 1-5 cover history & technology)	Ty Brook	1999
Ortungstechnik für Profis (German)	Wolfgang Schüler	1999
Das Kompendium der Metalldetektoren (German)	Rolf Wilhelm	2001
Inside the Metal Detector	George Overton Carl Moreland	2012
Los Detectores de Metales en Áreas Históricas: Metal Detectors in Historical Areas (both Spanish & English)	Jose Antonio Agraz Sandoval	2012
Inside the Metal Detector 2nd Ed	George Overton Carl Moreland	2015
The Voodoo Project	George Overton	2020
Arduino Nano Pulse Induction Metal Detector Project	George Overton	2021
Arduino Nano VLF Metal Detector Project	George Overton	2024

Research Papers

Title & Author(s)	Publication	Date
An Experiment with the Induction Balance	Scientific American	Jul 16 1881
New Use for the Induction Balance	Scientific American	Jul 30 1881
On the Theory of Electro-Magnetic Induction — *W. Thomson*	British Association for the Advancement of Science	1848
Induction-Balance and Experimental Researches Therewith — *D. E. Hughes*	Procedings of the Physical Society of London	1879

Upon New Methods of Exploring the Field of Induction of Flat Spirals — *Alexander Graham Bell*	Publication unknown (possibly Telegraphic Journal and Electrical Review)	1879 (presented)
Note on the Theory of the Induction Balance — *Lord Rayleigh*	British Association for the Advancement of Science	1880
On Intermittent Currents and the Theory of the Induction-Balance — *Oliver Joseph Lodge*	Procedings of the Physical Society of London	1880
Upon the Electrical Experiments - Part 1 — *Alexander Graham Bell*	Telegraphic Journal and Electrical Review	Dec 23, 1882
Upon the Electrical Experiments - Part 2 — *Alexander Graham Bell*	Telegraphic Journal and Electrical Review	Dec 30, 1882
Note on the Theory of the Magnetic Balance of Hughes — *Silvanus P. Thompson*	Proceedings of the Royal Society of London	1883
The Induction Balance and Telephonic Probe	Journal of the American Medical Association	1887
Note on Induction Coils or "Transformers" — *John Hopkinson*	Proceedings of the Royal Society of London	1887
Simple Inductance Formulas for Radio Coils — H.A. Wheeler	Proceedings of the I.R.E.	Oct 1928
A Sensitive Induction Balance for the Purpose of Detecting Unexploded Bombs — *Theodore Theodorsen*	Proceedings of the National Academy of Sciences	Nov 1930
Design of Standards of Inductance, and the Proposed Use of Model Reactors in the Design of Air-Core and Iron-Core Reactors — H.B. Brooks	Bureau of Standards Journal of Research	May 12, 1931
Measurement of magnetic susceptibility with the Hughes induction balance — *RJ Duffin*	Journal of Geophysical Research	Sept 1946
Instrument for Detecting Metallic Bodies Buried in the Earth — *Theodore Theodorsen*	Journal of Applied Physics	Nov 1950
The Magnetic Field of a Plane Circular Loop — *C.L. Bartburger*	Journal of Applied Physics	Nov 1950
A Conducting Sphere in a Time-Varying Magnetic Field — *J.R. Wait*	Geophysics	Oct 1951
A Pulsed Bomb Locator — *F.B. Johnson*	Ministry of Supply (Internal memorandum)	1956
Eddy Currents — *Beam, DeBlois, & Nesbitt*	Journal of Applied Physics	Dec 1959
Further Developments of the Pulse Induction Metal Detector — *E.J. Foster*	Prospezioni Archaeologiche	1969
A Diver-Operated Underwater Metal Detector — *E.J. Foster*	Archaeometry	1970
Magnetic Viscosity, Quadrature Susceptability, and Frequency Dependence of Susceptibility in Single-Domain Assemblies of Magnetite and Maghemite — *Mullins & Tite*	Journal of Geophysical Research	Feb 1973
Inductance Formulas for Circular and Square Coils — H.A. Wheeler	Proceedings of the IEEE	Dec 1982
Magnetic Susceptibility and Viscosity of Soils in a Weak Time Varying Field — *Dabas, Jolivet and Tabbagh*	Geophysics Journal International	1992
Time Domain Magnetization of Soils — *Dabas & Skinner*	Geophysics	Mar 1993

Simple Accurate Expressions for Planar Spiral Inductance — S. Mohan, M. Hershenson, S. Boyd, & T. Lee	IEEE Journal of Solid State Circuits	Oct 1999
Induction Balance Study for Metal Detection — *Wirjawan, Zollman, and Mau*	American Institute of Physics Conference Proceedings	Oct 2005
Induction Coil Sensors - A Review — Slawomir Tumanski	Journal of Measurement and Science Technology	Jan 2007
Multi-Transmitter Multi-Receiver Null Coupled Systems for Inductive Detection and Characterization of Metallic Objects — Smith, Morrison, Doolittle, Tseng	Journal of Applied Geophysics	Mar 19, 2007
A Hybrid Method for UXO vs. Non-UXO Discrimination — *Kappler & Gasperikova*	Journal of Environmental & Engineering Geophysics	Dec 2011
Design and Verification of Search Coil Inductance for Pulse Induction Metal Detection — *David Desrochers*	Undergraduate Thesis for the University of Arkansas	May 2020

Periodical Articles

Title & Author(s)	Publication	Date
An Experiment with the Induction Balance	Scientific American	Jul 16 1881
New Use for the Induction Balance — *George M Hopkin*	Scientific American	Jul 30 1881
Searching for the Bullet — *Bell & Garfield*	Harper's Weekly	Aug 13 1881
Bell Induction Balance	Lancet	Jan 20 1883
The Induction Balance — *Alexander Graham Bell*	The Engineer	Feb 2 1883
A Successful Form of Induction Balance for the Painless Detection of Metallic Masses in the Human Body — *Alexander Graham Bell*	American Journal of Science	1883
Electrical Ore Finder	Scientific American	Aug 1892
Unscientific & Scientific Divining Rods — *George M Hopkins*	Scientific American (Supplement)	Aug 20 1892
Unscientific & Scientific Divining Rods	Scientific American	Jan 20 1906
Using an Induction Balance	Scientific American	Nov 13 1915
Buried Shells Found By Induction Balance	Popular Mechanics	Feb 1916
How to Build a Buried Treasure Finder	Science & Invention	Aug 1921
Detection of Minerals by Electric Methods	Science (Supplement)	Jun 6 1924
Buried Treasure - Apparatus for Locating Minerals — *Albert G. Ingalls*	Scientific American	Dec 1925
Radio Gold Explorer — *Joseph H. Krause*	Science & Invention	Mar 1926
The Radio Watchman at the Gate — *K. Schett*	Radio News	Apr 1926
Strange Radio Devices Locate Buried Treasure — *Albert Henry Kingerly*	Popular Science Monthly	Apr 1928
How Radio Prospecting Takes the Gamble Out of Mining — *C.S. Gleason*	Radio News	Feb 1929
Certain Aspects of Henry's Experiments on Electromagnetic Induction — *Joseph S. Ames*	Science	Jan 22 1932

How to Build the Radio "Treasure" Finder — *Clyde J Fitch*	Radio Craft	June 1932
Science Aids Quest for Gold	Scientific American	Oct 1932
How to Build the New Treasure Finder — *E.F. Sarver*	Radio Craft	July 1933 Nov 1933 Apr 1934
Another "Treasure Finder"	Scientific American	Sept 1933
Latest in Radio Treasure Finders	Modern Mechanix	Nov 1933
An Improved Treasure Locator — *C.W. Palmer*	Radio Craft	Aug 1934
New Device Locates Buried Metal	Modern Mechanix	Oct 1934
How to Build an Inexpensive Metal Locator — *Ben A. Elliott*	Popular Mechanics	Sept 1935
Metallascope (update)	Scientific American	Apr 1 1936
Dowsing for Cable — *R.I. Crisfield*	Radio World	Sept 1936
Treasure-Seeking Circuits — *J.E. Anderson*	Radio World	Sept 1936
Hunting Riches by Radio — *J.E. Anderson*	Radio World	Oct 1936
Radio Circuits for Treasure Seeking — *J.E. Anderson*	Radio World	Nov 1936
Power Crux of Treasure Quest — *J.E. Anderson*	Radio World	Dec 1936
Newest in Treasure Locators — *Gerhard Fisher*	Radio Craft	Dec 1936
Prison Gun Detector — *David Luck & Charles Young*	Radio World	Dec 1936
The Earth as a Treasure Chest Explored with Radio Devices by Those Seeking Riches — *J.E. Anderson*	Radio World	Jan 1938
Science Seek the Treasure Trove — *Jerry Brown*	Science & Mechanics	Feb 1938
A Practical Metal Detector — *WC Broekhuysen*	Electronics	Apr 1938
Finding Hidden Treasure — *Maxwell Reid Grant*	Radio News	May 1938
Build Your Own Treasure Hunter — *Charles E. Chapel*	Radio News	Sept 1938
Plans for a Radio Treasure Finder — *Gerhard Fisher*	Science & Mechanics	Feb 1939
How to Make a Modern Radio Treasure Locator — *Allan Stuart*	Radio Craft	Sept 1939
How to Build an Inexpensive Metal Finder — *Harry A. Fore*	Popular Mechanics	Jan 1940
Building a Modern Miniature-Tube Metal-Treasure Locator — *G.M. Bettis*	Radio Craft	Dec 1940
FM Metal Treasure Locator — *G.M. Bettis*	Radio News	Oct 1942
Metal Locator — *G.M. Bettis*	Radio Craft	Apr 1943
Metal Detectors — *W.H. Blankmeyer*	Electronics	Dec 1943
Hughes Balance Metal Detectors — *Eric Leslie*	Radio Craft	Jan 1944
A Precise "Treasure" Locator — *MC Greenridge*	Radio Craft	Mar 1944
Mine Locators — *Connery Chappell*	Radio News	Mar 1944
Treasure Locators — *John Haynes*	Radio Craft	July 1944
Locating Land Mines — *Paul Horni*	Electronic Industries	Jan 1945
Non-metallic Mine Locator — *T.E. Stewart*	Electronics	Nov 1945
Vehicular Mounted Mine Detectors — *H.G. Doll et al*	Electronics	Jan 1946

How Mine Detectors Work — *Eric Leslie*	Radio Craft	July 1946
Electron-coupled Zero-beat Metal Detector — *Harry A. Fore*	Popular Mechanics	Aug 1946
Treasure Finding Modernized —*W.E. Osborne*	Radio News	Sept 1946
Try This 1946 Treasure Finder — *W.E. Osborne*	Radio News	Nov 1946
The Modern Divining Rod — *Alvin B Kaufman*	Radio News	Apr 1947
Small-space Metal Locator — *John Haynes*	Radio Craft	Sept 1948
Industrial Metal Detector Design	Electronics	Nov 1951
A Modern Metal Locator — *Rufus P Turner*	Radio & Television News	Sept 1954
Hidden Metal Detector — *Thomas A. Blanchard*	Radio-TV Experimenter	Vol 3 1955
Two Transistorized Metal Locators — *Edwin Bohr*	Radio Electronics	Mar 1955
Portable Metal Locator — *Harvey Pollack*	Popular Electronics	June 1955
How a Metal Detector Works — *Unk*	Popular Science	Aug 1955
Metal Detector Finds Ducts & Pipes — *Carl David Todd*	Electronics	Jan 1957
Transistorized Metal Detector — *Rudolf Graf*	Popular Electronics	June 1959
Underwater Metal Detector — *Elgin Ciampi*	Electronics Illustrated	July 1959
Modern Methods of Metal Hunting	Science Digest	Aug 1959
Underwater Metal Detector — *C.L. Henry*	Science & Mechanics	Aug 1960
Transitone Locates Hidden Wiring — *Harry Parker*	Radio Electronics	Dec 1960
Simple Metal Locator — *Frederick H. Calvert*	Electronics World	July 1961
Underwater Metal Detector	Skin Diver	Aug 1961
The Radio Gizmo — *Fred Maynard*	Electronics Illustrated	Apr 1961
Transistorized Metal Detector — *W.E. Osborne*	Electronics World	Mar 1962
Lodestar Metal Detector — *Charles Caringella*	Popular Electronics	Sept 1962
Simplified Metal Locator — *B.F. Miessner*	Radio Electronics	Sept 1962
Underwater Metal Detector — *James E. Pugh*	Radio-TV Experimenter	Spring 1963
Finding Buried Stuff — *C. Beeler*	Radio Electronics	Apr 1966
Underwater Metal Hunting for Fun or Profit — *Olle Klippberg*	Radio Electronics	June 1966
A Treasure Finder You can Build — *Ronald Benry*	Popular Science	July 1966
Summer Fun with a Sensitive Metal Finder — *GH Gill*	Radio Electronics	July 1966
Electronic Metal Locators: Basic Types & Design Factors — *Don Lancaster*	Electronics World	Dec 1966
Build the IC-67 Metal Locator — *Don Lancaster*	Popular Electronics	Jan 1967
Underwater Metal Detector	Skin Diver	May 1967
Build the Beachcomber — *D Meyer*	Popular Electronics	July 1967
Build a Treasure Finder — *Charles Rakes*	Radio Electronics	Nov 1967
Gold Grabber — *Charles Green*	Radio-TV Experimenter	Aug/Sept 1968
Metal Detector — *D. Bollen*	Practical Electronics	Jan 1969

Build a "Different" Metal Locator — *Leslie Huggard*	Popular Electronics	Feb 1969
Metal Detector — *F.G. Rayer*	Practical Electronics	Jan 1970
Metal Locator — *Halvor Moorshead*	Practical Wireless	Mar 1970
Treasure Witcher — *Charles Rakes*	Science & Electronics	Apr/May 1970
New Approach for the Metal Locator — *Irving Gottlieb*	73 Magazine	Feb 1971
The Induction Coil — *George Shiers*	Scientific American	May 1971
Treasure Tracer — *Halvor Moorshead*	Practical Wireless	Aug 1971
Metal Detector — *D. Bollen*	Everyday Electronics	May 1972
Treasure Tracer Mk2 — *Halvor Moorshead*	Practical Wireless	May 1972
Treasure Detectors for Land Use — *L George Lawrence*	Popular Electronics	Sept 1972
Coin Collector (Project 531) — *Unk*	ETI	July 1973
Ferret Metal Detector — *R.P. Perry*	Practical Wireless	Nov 1973
PLL Metal Detector — *Unk*	Practical Electronics	Dec 1973
Coin Collector (Project 531) — *Unk*	ETI (Aus)	Dec 1974
Sensitive Metal Detector — *Unk*	Elektor	Nov 1976
Metal Pipe or Wiring Locator — *C. Whitehead*	Practical Electronics	Dec 1976
Induction Balance Metal Locator — *Unk*	ETI	Feb 1977
BFO Metal Detector — *D Waddington*	Wireless World	Apr 1977
Induction Balance Metal Detector (Project 549) — *Unk*	ETI (Aus)	May 1977
Seekit — *W. Opel*	Practical Wireless	May 1977 June 1977
Treasure Locator — *Unk*	Everyday Electronics	Oct 1977
Metal Detector — *J.P. McCaulay*	ETI	Nov 1977
CMOS Twin Oscillator Detector — *Mark Anglin*	Electronics	Dec 22, 1977
Metal Locators Are Fun — *W. Neville*	Electronics Australia	Dec 1977
Induction Balance Metal Locator — *Unk*	ETI	Feb 1978
Balance Circuit for ETI Metal Locator — *C. Bray*	ETI	July 1978
Discriminitive Metal Detector — *RCV Macario*	Wireless World	July 1978
Treasure Hunter — *N. Hunter*	Everyday Electronics	Oct 1978
Metal Locators — *Unk*	Hobby Electronics	Nov 1978
Detecteur de Metaux Sensible — *Unk*	Elektor (France	Jan 1979
Metal Detection — *Unk*	Practical Wireless	Jan 1979
Sandbanks Metal Detector — *PJ Wales*	Practical Wireless	Jan 1979
Treasure Hunter Sound Adapter— *N. Hunter*	Everyday Electronics	Feb 1979
Low Cost Metal Locator — *Robert Penfold*	Everyday Electronics	June 1979
Prospector Metal Locator — *Ron De Jong*	Electronics Australia	Nov 1979
Build a Metal Locator — *Robert Krieger*	Popular Electronics	Jan 1980

Pulse Induction Metal Detector — *JA Corbyn*	Wireless World	Mar 1980 Apr 1980 May 1980
ETI Metal Locator — *Unk*	ETI	Mar 1980
Simple Sensitive Metal Detector — *Phil Wait*	ETI (Aus)	Mar 1980
Pipe & Metal Locator (Project 566) — *Phil Wait*	ETI (Aus)	Apr 1980 Oct 1980
Off-Resonance Metal Detector — *G Wareham*	Wireless World	June 1980
Eddy Currents — *Unk*	ETI	Aug 1980
Magnum Metal Locator — *Andy Flind*	Practical Electronics	Aug 1980 Sept 1980
Houndog IB Metal Locator — *Leslie Huggard*	Elementary Electronics	Sept/Oct 1980
All About Metal Detectors — *R Gallagher*	Radio Electronics	Nov 1980
A Discriminating Metal Detector (Project 1500) — *Phil Wait*	ETI (Aus)	Dec 1980
Metal Detector — *Jennings Kimberley*	Elektor	July 1981 Aug 1981
Coinshooter — *William Lahr*	Popular Electronics	Aug 1981
HE "Diana" Metal Detector — *Unk*	Hobby Electronics	Sept 1981
Metal Detector — *Unk*	Elektor	Nov 1981
VCO Add-On for Diana — *Unk*	Hobby Electronics	Nov 1982
The Art and Science of Metal Detection — *Dick Turner*	Hobby Electronics	May 1984
Principles of Metal Detection — *Dick Turner*	ETI (Can)	July 1984
Beachcomber Metal Detector — *Colin Dawson*	Electronics Australia	Dec 1984
Absorption Metal Detector — *Unk*	Elektor	July/Aug 1985
Metal Detector — *Unk*	Elektor	July/Aug 1985
Metal Pipe Detector — *Unk*	Elektor	July/Aug 1985
Metal Detecting for Fun & Profit — *Gerald Pattee*	Modern Electronics	July 1985 Aug 1985
Metal Detector — *Unk*	Elektor	May 1987
EE Buccaneer IB Metal Detector — *Andy Flind*	Everyday Electronics	July 1987
Mini Metal Detector	Electronics & Beyond	Dec 1987
Pipe & Cable Locator — *Robert Penfold*	Everyday Electronics	Apr 1988
Phaser TR Metal Detector — *Unk*	Practical Electronics	Oct 1988
Metal Detector (Project 1539) — *Keith Bindley*	ETI	Jan 1989
Metal Detector — *Robert Penfold*	Everyday Electronics	May 1989
EE Treasure Hunter — *Mark Stuart*	Everyday Electronics	Aug 1989
Twin Loop Treasure Seeker — *Robert and David Crone*	ETI	Sept 1989

Probe Pocket Treasure Finder — *Andy Flind*	Everyday Electronics	Sept 1989
Simple Metal Detector (Circuit Notebook) — *Darren Yates*	Silicon Chip	Mar 1991
Circuit Circus — *Charles Rakes*	Popular Electronics	Mar 1991 June 1991 July 1991 Aug 1991
Metal Detector *Kamil Kraus*	Elektor	July 1992
Metal Detector — *Kamil Kraus*	Electronics World	Sept 1992
Metal Detector — *Kamil Kraus*	Wireless World	Sept 1992
David Hughes: Electromagnetic Pioneer — *James P Rybak*	Popular Electronics	Nov 1992
Induction Balance Metal Locator	Silicon Chip	May 1994 June1994
Microcontroller PI Treasure Hunter — *Mark Stuart*	Everyday with Practical Electronics	June 1994
Mini Metal Detector — *Z. Kaszta*	Elektor	Oct 1996
Simple Metal Detector — *Robert Penfold*	Everyday with Practical Electronics	Apr 1998
A Detector for Metal Objects — *John Clarke*	Silicon Chip	May 1998
A Century of Detector Designs — *Dick Turner*	Treasure Hunting	July 1998
BFO Metal Detector — *Rachel & Steve Hageman*	EDN	Sept 1998
Metal Detectors — *Gavin Cheeseman*	Electronics & Beyond	May 1999 June 1999
Fortune Finder — *J. Clarke*	Silicon Chip	Dec 1999
Frequency Meter Metal Detector — *Andrei Chtchedrine & Yuri Kolokolov*	Circuit Cellar	May 2001
Metal Detection — *Charles Rakes*	Poptronics	Aug 2001
Metal Detecting Circuits II — *Charles Rakes*	Poptronics	Sept 2001
Matchless Metal Locator — *Thomas Scarborough*	Silicon Chip	June 2002
Bounty Treasure Hunter — *Thomas Scarborough*	Everyday with Practical Electronics	Oct 2002
Build the Frisker — *William Sheets & Rudolf Graf*	Poptronics	Nov 2002
Simple BFO Metal Locator — *Thomas Scarborough*	Silicon Chip	Dec 2002
Back-to-Basics Metal Detector — *Bart Trepak*	Everyday with Practical Electronics	Mar 2003
Minimalist Induction-Balance Metal Detector — *Thomas Scarborough*	Elektor	Oct 2003
Beat Balance Metal Detector — *Thomas Scarborough*	Everyday with Practical Electronics	May 2004
Poor Man's Metal Locator — *Thomas Scarborough*	Silicon Chip	May 2004
The Metal Detector and Faraday's Law — *J.A. McNeil*	The Physics Teacher	Sept 2004
One-Component Metal Detector — *Thomas Scarborough*	Elektor	Oct 2004

CCO Metal Detector — *Thomas Scarborough*	Everyday with Practical Electronics	Nov 2004
CCO Metal Detector — *Thomas Scarborough*	Elektor	July 2005
Poor Man's Metal Locator — *Thomas Scarborough*	Everyday with Practical Electronics	June 2006
Frequency Meter Metal Detector — *Andrei Chtchedrine & Yuri Kolokolov*	Circuit Cellar	May 2007
Back-to-Basics Metal Detector — *Bart Trepak*	Everyday with Practical Electronics	Mar 2007
EPE Mini Metal Detector — *Thomas Scarborough*	Everyday with Practical Electronics	July 2007
Build a Four Transistor Metal Detector — *Paul Florian*	Nuts & Volts	Feb 2008
Poor Man's Metal Detector — *Thomas Scarborough*	Elektor	Dec 2009
Hand-Held Metal Locator — *John Clarke*	Silicon Chip	July 2009
Metal Locator — *John Clarke*	Everyday with Practical Electronics	July 2011
LRC Beat Balance Metal Detector — *Thomas Scarborough*	Everyday with Practical Electronics	Feb 2012
A Really Simple Metal Detector — *Mahmood Alimohammadi*	Silicon Chip	July 2012
Simple Metal Locator uses Overlapping Coils — *Mahmood Alimohammadi*	Silicon Chip	Sept 2012
Metal Detector uses a TL074 Quad Op Amp — *Mahmood Alimohammadi*	Silicon Chip	June 2013
Simple 2-coil VLF metal detector — *Mahmood Alimohammadi*	Silicon Chip	Dec 2013
ATmega-Based Metal Detector with Stepped Frequency Indication — *Mahmood Alimohammadi*	Silicon Chip	Mar 2017
Drift-free Induction Balance Metal Detector — *Mahmood Alimohammadi*	Silicon Chip	Mar 2018
Regenerative BFO Metal Detector — *Thomas Scarborough*	Silicon Chip	Mar 2023
'Huygens Beam' BFO Metal Detector — *Thomas Scarborough*	Silicon Chip	Sept 2023

Other

Title & Author(s)	Source
How Metal Detectors Work — *Mark Rowan & Bill Lahr*	White's Electronics booklet
Metal Detector Basics and Theory — *Bruce Candy*	Minelab white paper
Metal Detecting Terminology	Minelab white paper
Ground Mineralisation	Minelab white paper
Basic Metal Detector Operation	Minelab white paper
MPF Technology	Minelab white paper
ZVT Technology — *Bruce Candy*	Minelab white paper
DIF Technology	Minelab white paper

Patents

E

> "People equate patents with secrecy; that secrecy is what patents were designed
> to overcome. That's why the formula for Coca-Cola was never patented. They
> kept it as a trade secret, and they've outlasted patent laws by 80 years* or more."
>
> — *Craig Venter*
>
> *Now 138 years.

For advanced technical information an excellent resource is the patent office. While patents can offer insight to the latest developments they are sometimes written in an (often intentionally) obscure manner, on top of the legalese. This appendix presents a number of resources and references for the curious who want to dig deeper into detector technology.

E.1: The Patent Minefield

Patents are often poorly understood by lay people. They are at best difficult to read, even if you are well-versed in the topic of the patent. Besides the content of the patent, there is also the question of exactly what a patent is good for. We'll take a look at both aspects.

A patent is a way to protect an invention from copycats. Most of the world patent laws are very similar but for this discussion we'll assume U.S. law. We'll also only consider utility patents[1]. A patent filed in one country only provides protection for that country. However, it provides two channels of protection. It prevents anyone from selling an infringing device in that country no matter where the device came from. It also prevents anyone from making an infringing device in that country no matter where the device is sold. What it does not prevent is a foreign company making an infringing device and selling it everywhere else. To obtain patent protection in other countries, you need to file in those countries.

Patents are granted on a "first-to-file" basis[2]. Suppose you invent the World's Best Ground Balance technique and you wait around and file for a patent two years later; you find that someone else filed a patent on the same technique 6 months prior. Even though you might have documented evidence that you were the first to invent the idea, it doesn't matter; the first person to the patent office wins.

Patents protect a device for 20 years from the filing date. During that time, no one in the country of the patent is allowed to make, buy, sell, or use an infringing device. However, an exception is generally granted for building the patented device for the purpose of evaluating the claims of the patent. Beyond that, permission or licensing is needed. Because the patent only covers the country of the patent, it's important to consider where your device might be used. For example, if your World's Best Ground Balance gets used in an American-made detector that is overwhelmingly popular in England, then a US patent does not prevent a foreign company from copying your WBGB and selling it into England. They just cannot make or sell it in the US. Patents can be filed in multiple countries. The World Intellectual Property Organization (WIPO) allows a single patent filing to include coverage in up to 157 countries.

Getting a patent will typically run from $5,000 to $20,000 depending on the patent and how much legal help you need. Having a patent does not automatically stop other people from infringing, it only gives you the right to sue them if they do. Enforcing a patent in the face of infringement can easily cost $100,000, far more than that if there is a trial, and even way more if it involves foreign countries. Sure, if you win your legal costs will be covered, assuming the infringer can afford to pay the damages. If

---

1. There is also a "design" patent, which covers the look-and-feel of a device. These usually involve just drawings.

2. The US switched from "first-to-invent" to "first-to-file" on 16 March 2013.

you don't defend a patent against infringement then you can lose your patent rights. So the device better be worth both the cost of getting the patent and the cost of defending it.

For an individual, an undefended patent is only good for PR and a plaque to hang on the wall. For a corporation, patents can be a valuable asset; they can be important if the company is sold to or merged with another company, and they can also be used in brokering "patent swaps" where each company licenses each other's patents. This is commonly done in the electronics industry, especially in IC design.

Finally, getting a patent is no guarantee that you will have anything worthwhile for long. It is often quite easy to work around a patent and devise a scheme that is just as good (or maybe better) but does not infringe. A good example is US10181720 which is shown in Figure 25-24. It is a passive diode switch to prevent the flyback from damaging the preamp. In about 30 minutes, I devised an alternate switch (Figure 25-25) which is an improvement and does not infringe. This is a case where the very narrow wording of the patent claims makes it easy to circumvent, and renders the patent worthless.

E.2: Patent Requirements

For an invention to be patentable it must meet three requirements:

1. It must be "useful." That is, it must be realizable and usable; you cannot patent an idea or concept. If you can't build it and make it work, then it is not useful and not patentable.
2. It must be "novel." That is, it must be new, something no one else has thought of, at least to the point where it has been made public. Many patents fail this requirement by failing to find relevant prior art. Often the patent examiner[3] will also miss the prior art and grant the patent. This is usually the easiest way to invalidate a patent.
3. It must be "non-obvious." That is, it cannot be a simple application of common knowledge to a device where that common knowledge had not yet been applied[4]. This is another case where patents will often be granted that fail this requirement but then get invalidated in an infringement case.

Patents generally consist of two parts: the body, and the claims. The body is where the claimant is supposed to teach what the patent is covering. It usually includes prior art plus a detailed description of what is being patented including how and why it is an improvement over the prior art.

The claims are where the specifics of the patent are listed. There are independent claims — those which stand alone — and there are dependent claims — those which are extensions of independent claims. Often a patent will have only one or two independent claims and many dependent claims. Wording of the claims is of paramount importance and poorly worded claims may be completely ineffective at protecting the device. Wording is often a trade-off between being too narrow — and therefore being easy to circumvent — and being too broad — and therefore being easy to challenge. Claims are often written to protect both the 'device' itself and also 'methods' used in the device, and often these claims appear very closely worded.

A good patent is not easy to write and a bad patent is easy to circumvent. Given the cost of a patent (as of this writing, about \$5,000 - \$20,000 USD), the do-it-yourself route is not recommended; get a patent attorney. Challenging a patent you think is bogus is equally expensive and also requires an expert level of patent knowledge.

Circumventing a patent requires an equal level of patent knowledge as writing or challenging one. Often people believe that if they make an improvement to a device, they can then make that device (and even patent it) without infringing a prior patent. As an example, suppose there is a patent on a PI detec-

---

3. Having traversed the US patent system many times, I am of the opinion that patent examiners are being asked to judge technology that they simply don't understand. The current method of having young, inexperienced examiners review what should be cutting-edge technology is guaranteed to produce sub-mediocre patents.

4. A favorite example of mine is US4890064, Minelab's very first patent. Minelab managed to get a patent on the use of litz wire in metal detector coils to reduce self-induced eddy currents in the windings, which is the very reason that litz wire was created in the first place, and is widely used elsewhere exactly for that purpose.

tor which uses 2 or more simultaneously transmitted pulse widths to improve ground balance and the patent includes a claim which protects *"the use of 2 or more simultaneously transmitted pulse widths."* You discover a different way to use 2 pulse widths to implement an even better ground balance. Even though you might be doing something different with the 2 pulse widths, the mere use of 2 pulse widths for any purpose may infringe the original patent. Furthermore, the patent office may grant you a patent on your improvement but you still can't legally use it without permission from the original patentee.

All said, patents are like a minefield; difficult to maneuver through, often with unpleasant surprises lurking beneath the surface. For the individual, it is usually best to just avoid the minefield. If you decide you want to step in, a patent attorney is overwhelmingly recommended.

Patents

E.3: Proximity Methods

Patent No.	Title	Comments
US3355658	Differentiating Metal Detector for Detecting Metal Objects and Distinguishing Between Detected Diamagnetic and Non-Dia-magnetic Objects	Gardiner, VLF BFO
US3467855	Object Detector and Method for Distinguishing Between Objects Detected Including a Pair of Radio Oscillators	Basic BFO
US3519919	Frequency Stabilizing Element for Metal Detectors	
US3546628	Oscillating Metal Object Detector	Simplest detector on earth
US3601691	Metal Detector Responsive to Small Metallic Objects for Dif-ferentiating Between Ferrous and Non-Ferrous Objects	Gardiner, Z-response
US3626279	Metal Detector Utilyzing Radio Receiver and Harmonic Signal Generator	Harmonic technique increases sensitivity; also printed spiral coil
US3662255	Apparatus for Locating Concealed or Buried Metal Bodies and a Stable Inductor Usable in Such Detectors	Garrett Zero-Drift BFO
US3742341	Inductively Coupled Metal Detector Arrangement	Appears to be an off-resonance technique
US3823365	Metal Detecting Apparatus Having Improved Ground-Effect Immunity	Uses induction balance
US3875498	Metal Detector for Distinquishing Between Precious Metal Objects and Other Metals	D-Tex, your real basic BFO
US3896371	Metal Detector With a Resonating Circuit Being Driven by a Frequency Higher Than Its Natural Resonance Frequency	A.H. Electronics, off-resonance type
US3961238	Selective Metal Detector Circuit Having Dual Tuned Resonant Circuits	Gardiner, very basic Z-response
US3986104	Dual Frequency Metal Detector System	Gardiner
US4130792	Metal Detector With Feedback Tuning	Z-response with feedback
US4196391	Metal Locator with Stereotonic Indication of Translateral Posi-tion Stereo	
US4204160	Metal Detector With Automatic Optimum Sensitivity Adjust-ment	
US4255710	Plural Search Frequency Directional Metal Detector Apparatus Having Enhanced Sensitivity	Uses two gated oscillator fre-quencies

US4263553	Discriminating Metal Detector With Compensation for Ground Minerals	A.H. Electronics, off-resonance detector
US4321539	Digital BFO Metal Detecting Device with Improved Sensitivities at Near-Zero Beat Frequencies	
US4439734	Metal Object Locator Including Frequency Shift Detector	Induction balance *and* ground balance
US4678992	Electronic Metal Detector	A.H. Electronics, off-resonance type
US5025227	Metal Detection Circuit	Appears to be off-resonance
US7068028	Method and Apparatus for Metal Target Proximity Detection at Long Distances	Appears to be a variation of energy theft

E.4: Induction Balance (TR/VLF) Methods

Patent No.	Title	Comments
US3405354	Apparatus for Limiting Phase-Angle Response Range, Particularly in Eddy Current Testing Apparatus	Uses synchronous demodulation to determine target phase, see 3848182
US3471772	Instrument for Measuring the Range and Approximate Size of Buried or Hidden Metal Objects	Synchronous detectors, lots of fundamental theory & equations
US3471773	Metal Detecting Device with Inductively Coupled Coaxial Transmitter and Receiver Coils	Basic TR with 4B-style coil (NOT coaxial)
US3826973	Electronic Gradiometer	Technos PRG (Phase Readout Gradiometer)
US3835371	Apparatus for Detecting the Presence of Electrically Conductive Material Within a Given Sensing Area	Focuses on submersible probe & cabling method
US3848182	Apparatus for Limiting Phase-Angle Response Range, Particularly in Eddy Current Testing Apparatus	See 3405354
US3872380	Metal Detector Distinguishing Between Different Metals by Using a Bias Circuit Actuated by the Phase Shifts Caused by the Metals	Gardiner, basic phase response IB
US4016486	Land Mine Detector with Pulse Slope, Width and Amplitude Determination Channels	
US4024468	Induction Balance Metal Detector with Inverse Discrimination	White's TR Discriminator (1975), includes good description with phase diagrams and circuitry with component values
US4030026	Sampling Metal Detector	White's, the basis of most of their early-80s analog detectors
US4053828	Metal Detector with First and Second Nested Rectangular Coils	Not too practical for a handheld
US4096432	Metal Detector for Discriminatory Detection of Buried Metal Objects	Basic phase response IB
US4099116	Metal Detector With Phase Related Selective Discrimination Circuit	Nautilus, feedback method for ground balance and discrimination
US4110679	Ferrous/Non-ferrous Metal Detector Using Sampling	White's

US4128803	Metal Detector System With Ground Effect Rejection	PNI (the old Bounty Hunter), possibly the Red Baron series (RB3/5/7)
US4249128	Wide Pulse Gated Metal Detector With Improved Noise Rejection	White's
US4263551	Method and Apparatus for Identifying Conductive Objects by Monitoring the True Resistive Component of Impedance Change in a coil System Caused by the Object	Title says it all, looks at target response vs. frequency
US4300097	Induction Balance Metal Detector with Ferrous and Non-ferrous Metal Identification	Techna, now First Texas Mfg. (Bounty Hunter)
US4303879	Metal Detector Circuit with Mode Selection and Automatic Tuning	Garrett (ADS?)
US4325027	Metal Detector for Locating Objects with Full Sensitivity in the Presence of Distributed Mineral Material	Compass
US4334191	Metal Detector Circuit Having Momentary Disabled Output	Garrett
US4334192	Metal Detector Circuit Having Automatic Tuning With Multiple Rates	Garrett, probably their Master Hunter VLF
US4344034	Selective Ground Neutralizing Metal Detector	Gardiner patent for phase discriminator
US4348639	Transmitter-Receiver Loop Buried Metal Object Locator with Switch Controlled Reference Voltage	Discovery Electronics' two-box detector
US4423377	Compact Metal Detector of the Balanced Induction Type	Garrett, handheld with integrated double-D coil, see also 4488115
US4470015	Metal Detector System With Undesirable Target and Mineralized Ground Discrimination	Teknetics, lots of diagrams and waveforms
US4486712	Frequency Dependent Pulsed Gain Modulated Metallic Object Detector	
US4486713	Metal Detector Apparatus Utilizing Controlled Phase Response to Reject Ground Effects and to Discriminate Between Different Types of Metals	Tesoro, lots of signal & phase diagrams
US4488115	Low Battery Voltage Indicator Circuit for a Metal Detector	Garrett, see also 4423377
US4507612	Metal Detector Systems for Identifying Targets in Mineralized Ground	Teknetics, tons of info
US4514692	Metal Detector and Discriminator Using Differentiation for Background Signal Suppression	Fisher, basis for early X models
US4563645	Inductively Balanced Oscillatory Coil Current for Metal Detection	
US4628265	Metal Detector and Classifier with Automatic Compensation for Soil Magnetic Minerals and Sensor Misalignment	Fisher
US4659989	Inductively Balanced Metal Detector Circuit with Orthogonal Balancing Signals and Including Phase and Polarity Detection	
US4677384	Target-Identifying Metal Detector	Teknetics
US4700139	Metal Detector Circuit Having Selectable Exclusion Range For Unwanted Objects	Garrett patent with a pretty good explanation of I&Q signal processing

US4709213	Metal Detector Having Digital Signal Processing	Garrett
US4783630	Metal Detector With Circuits for Automatically Screening Out the Effects of Offset and Mineralized Ground	White's
US4868910	Metal Detector with Microprocessor Control and Analysis	White's, ton's of info on their target ID
US4881036	Phase Shift Compensation for Metal Detection Apparatus	
US4894618	Metal Detector Using Cross-Correlation Between Components of Received Signals	Minelab, good explanatory text
US4912414	Induction-Type Metal Detector with Increased Scanning Area Capability	Describes a large array for use with a submersible, includes IB and PI methods
US5148151	Metal Detector Having Target Characterization and Search Classification	Garrett patent on VDI, explanation on phase response and VDI flowchart
US5523690	Metal Detector with Bivariate Display	White's patent detailing VDI, includes 6805 assembly code (see also 5596277)
US5596277	Method and Apparatus for Displaying Signal Information from a Detector	White's patent detailing VDI, includes 6805 assembly code (see also 5523690)
US5691640	Forced Balance Metal Detector	
US5721489	Metal Detector Method for Identifying Target Size	Garrett's target ID patent, lots of signal analysis and flow charts, see also 5786696
US5729143	Metal Detector With Nulling of Imbalance	Bucking signal calibrated for amplitude & phase
US5786696	Metal Detector for Identifying Target Electrical Characteristics, Depth and Size	Garrett's target ID patent, lots of signal analysis and flow charts, see also 5721489
US5969528	Dual Field Metal Detector	Garrett's two-box add-on
US6172504	Metal Detector Target Identification Using Flash Phase Analysis	White's, applies phase info to what's essentially a flash ADC; see 6421621
US6421621	Metal Detector Target Identification Using Flash Phase Analysis	White's, applies phase info to what's essentially a flash ADC; see 6172504
US6583625	Metal Detector and Method in Which Mineralization Effects Are Eliminated	
US6911823	Metal Detector Employing Static Discrimination	White's
US7078906	Simultaneous Time-Domain and Frequency-Domain Metal Detector	Combo PI & phase sampling
US7088103	Metal Detector Having a Plurality of Phase Discrimination Regions with Corresponding Selectable Exception Spaces Therein	White's
US7126323	Systems and Methods for Synchronous Detection of Signals	
US7432715	Method and Apparatus for Metal Detection Employing Digital Signal Processing	Minelab

Patent No.	Title	Comments
US11598897	Method for Operating a Metal Detector and Metal Detector	Series resonant TX driver
US11619758	Method for Operating a Metal Detector and Metal Detector	See also US11598897
EP0580396	Metal Detector with Display	White's VDI

E.5: Multifrequency

Patent No.	Title	Comments
US3012190	Multiple Frequency Alternating Current Network	1961 (filed in 1946!), an early multi-frequency detector
US3686564	Multiple Frequency Magnetic Field Technique for Differentiating Between Classes of Metal Objects	Westinghouse MF detector from 1972
US4942360	A Method and Apparatus of Discrimination Detection Using Multiple Frequencies to Determine a Recognizable Profile of an Undesirable Substance	Minelab, multiple frequency
US4975646	Detector System for Recognizing a Magnetic Material	Multi-frequency
US5537041	Discriminating Time Domain Conducting Metal Detector Utilizing Multi-Period Rectangular Transmitted Pulses	Minelab BBS
US5642050	Plural Frequency Method and System for Identifying Metal Objects in a Background Environment Using a Target Model	White's multi-frequency detector (see 5654638)
US5654638	Plural Frequency Method and System for Identifying Metal Objects in a Background Environment	White's multi-frequency detector (see 5642050)
US6879161	Method and Apparatus for Distinguishing Metal Objects and Employing Multiple Frequency Interrogation	White's
US8063777	Real-Time Rectangular-Wave Transmitting Metal Detector Platform with User Selectable Transmission and Reception Properties	Minelab PI, see US8237560
US8159225	Multi-Frequency Transmitter for a Metal Detector	Minelab, method of creating MF TX signals
US8237560	Real-Time Rectangular-Wave Transmitting Metal Detector Platform with User Selectable Transmission and Reception Properties	Minelab PI see US8063777
US10989829	Method for Operating a Multi-frequency Metal Detector and Multi-frequency Metal Detector	See also US11598897

E.6: Pulse Induction/Time Domain

Patent No.	Title	Comments
US3707672	Weapon Detector Utilyzing the Pulsed Field Technique to Detect Weapons on the Basis of Weapons Thickness	
US4157579	Pulse Generation Employing Parallel Resonant LC Circuit for Energizing a Coil with a Square Wave	Barringer, CCPI
US4506225	Method for Remote Measurement of Anomalous Complex Variations of a Predetermined Electrical Parameter in a Target Zone	Barringer, half-sine
US4868504	Apparatus and Method for Locating Metal Objects and Minerals in the Ground with Return of Energy from Transmitter Coil to Power Supply	Fisher Impulse PI, recycles power from the coil
US4894619	Impulse Induced Eddy Current Type Detector Using Plural Measuring Sequences in Detecting Metal Objects	

US5047718	Improving the Discrimination of an Impulse Technique Metal Detector by Correlating Responses Inside and Outside of a Cut-Off Peak Area	
US5414411	Pulse Induction Metal Detector	White's, block level
US5506506	Metal Detector for Detecting and Discriminating Between Ferrous and Non-ferrous Targets in Ground	Minelab, , compare with US5576624
US5576624	Pulse Induction Time Domain Metal Detector	Minelab, compare with US5506506
US6326790	Ground Piercing Metal Detector Having Range, Bearing, and Metal-Type Discrimination	A PI probe
US6326791	Discrimination of Metallic Targets in Magnetically Susceptible Soil	
US6452396	Method for Detecting the Metal Type of a Buried Metal Target	"Periscope" probe
US6452397	Ground Piercing Metal Detector Method for Detecting the Location of a Buried Metal Object	"Periscope" probe
US6456079	Circuit for Detecting the Metal Type of a Metal Target Object	"Periscope" probe
US6529007	Temperature Compensation for Ground Piercing Metal Detector	"Periscope" probe
US6586938	Metal Detector Method and Apparatus	
US6636044	Ground Mineralization Rejecting Metal Detector (Receive Signal Weighting)	Minelab, multi-width pulse
US6653838	Ground Mineralization Rejecting Metal Detector (Transmit Signal)	Minelab, multi-width + multi-voltage drive (DVT)
US6686742	Ground Mineralization Rejecting Metal Detector (Power Saving)	Minelab, energy recycling
US6690169	Interference Cancelling Metal Detector Including Electronic Selection of Effective Sensing Coil Arrangement	Minelab
US6724305	Pulse Induction Silverware Detector	
US6853194	Electromagnetic Target Discriminator Sensor System and Method for Detecting and Identifying Metal Targets	
US6927577	Digital Nulling Pulse Inductive Metal Detector	
US6967574	Multi-Mode Electromagnetic Target Discriminator Sensor System and Method of Operation Thereof	
US7078906	Simultaneous Time-Domain and Frequency-Domain Metal Detector	Combo PI & phase sampling
US7148691	Step Current Inductive Antenna for Pulse Inductive Metal Detector	
US7474102	Rectangular-Wave Transmitting Metal Detector	Minelab PI, see 7791345
US7652477	Multi-Frequency Metal Detector Having Constant Reactive Transmit Voltage Applied to a Transmit Coil	Minelab
US7791345	Rectangular-Wave Transmitting Metal Detector	Minelab PI, see 7474102
US7924012	Metal Detector Having Constant Reactive Transmit Voltage Applied to a Transmit Coil	Minelab, see US8614576
US8106770	Metal Detector with Improved Magnetic Soil Response Cancellation	Minelab PI
US8614576	Metal Detector Having Constant Reactive Transmit Voltage Applied to a Transmit Coil	Minelab, see US7924012

US8749240	Time Domain Method and Apparatus for Metal Detectors	White's
US8878515	Constant Current Metal Detector	White's
US8988070	Metal Detector for Use with Conductive Media	
US9250348	Transmit Signal of a Metal Detector Controlled by Feedback Loops	Minelab CCPI TX
US9348053	Metal Detector with at Least One Transmit/Receive Switch	Minelab, PI TX methods
US9366779	Signal Processing Technique for a Metal Detector	Minelab, PI multisampling
US9429674	Discrimination Method of a Metal Detector Based on a Time Constant Spectrum	Minelab
US9547065	Method for Detecting Fast Time Constant Targets using a Metal Detector	Minelab, negative capacitance
US9829598	Metal Detector	Minelab
US10181720	Dual Polarity High Voltage Blocking Circuit for a Pulse Induction Metal Detector	Preamp blocking circuit using a diode bridge
US10228481	Ground Eliminating Metal Detector	White's square wave PI
US10969511	Signal Processing Technique for a Metal Detector	Minelab, PI damping method
US11899156	Metal Detector	Minelab, PI with dual RX
EP0654685	Arrangement and Method for Detecting Metal Objects	PI with DD coil
EP0732600	Active Impulse Magnetometer	Looks more like a PI method

E.7: Hybrid/Other Methods

Patent No.	Title	Comments
US7649356	Pulse Induction Metal Detector Having High Energy Efficiency and Sensitivity	White's, half-sine
US7701204	Metal Detector with Reliable Identification of Ferrous and Non-Ferrous Metals in Soils with Varying Mineral Content	Discriminating PI
US7701337	Hybrid Tecnology Metal Detector	
US7710118	Resonant Pulse Induction Metal Detector that Transmits Energy from High Voltage Flyback Pulses	Discriminating PI
US8629677	Hybrid Induction Balance/Pulse Induction Metal Detector	White's, truncated half-sine
US9285496	Truncated Half-Sine Methods for Metal Detectors	White's

E.8: Coils

Patent No.	Title	Comments
US2451596	Unitary Balanced Inductor System	1948, Wheeler concentric coil
US3549985	Metal Detector Device Having a Disk-Shaped Head for Housing a Coil System	Seems to mostly cover plastics & fillers
US3753185	Metal Detector Search Coil	Bill Mahan, D-Tex
US3882374	Transmitting-Receiving Coil Configuration	IB
US4255711	Coil Arrangement for Search Head of a Metal Detector	Compass, concentric IB
US4276484	Method and Apparatus for Controlling Current in Inductive Loads Such as Large Diameter Coils	Pulse method
US4293816	Balanced Search Loop for Metal Detector	White's concentric loop

US4345208	Anti-falsing and Zero Nulling Search Head for a Metal Detector	Daytona search coil
US4552134	Equipment for Determining the Position of a Metal Body in a Medium With Low Electric Conductivity	Particular IB coil arrangement
US4862316	Static Charge Dissipating Housing for Metal Detector Search Loop Assembly	White's concentric loop
US4890064	Metal Detector Sensing Head with Reduced Eddy Current Coils	Minelab
US5038106	Detector of Metalliferous Objects Having Two Pairs of Receiving Loops Symmetrical and Orthogonal to a Driving Loop	see US5039946
US5039946	Metalliferous Objects Detector Having a Pair of Angularly Positioned Driving Loops and a Pair of Parallel, Coaxial Receiving Loops	see US5038106
US5245307	Search Coil Assembly for Electrically Conductive Object Detection	
US5498959	Metal Detector With Multipolar Windings Shaped So As To Eliminate the Neutralizing Effects When Several Metal Masses Are Passing Through Simultaneously	Walk-through type coil arrangement
US5859532	Method of and Measuring Arrangement for Metal Detection With a Coil Device Having Several Separately Controllable Regions	Coil arrangements for walk-through type
US5863445	Etched Coil Unibody Digital Detector	Handheld wand, EP0249110
US6791329	Portable Metal Detection and Classification System	Coil methods for a powered rover
US6822429	Inductive Sensor Arrangement Comprising Three Sense Coil Cooperating with Said Three Field Coils to Perform Three Field/Sense Coil Pairs and Method for Detecting of Ferrous Metal Objects	Say that 3 times fast
US7075304	Variable Damping Induction Coil Metal Detection	Uses a MOSFET for the damping resistor
US7157913	Reconfigurable Induction Coil for Metal Detection	
US7176691	Switched Coil Receiver Antenna for Metal Detector	
US7994789	Dual Field Search Coil for Pulse Induction Metal Detectors	for White's Surfmaster-DF and TDI
US9557390	Noise Reduction Circuitry for a Metal Detector	Minelab Go Find coil
US9989663	Auto Nulling of Induction Balance Metal Detector Coils	White's
US11454736	Metal Detector	Minelab
US11474274	Metal Detector	Minelab
US11658416	Antenna of a Metal Detector	Minelab
EP0249110	Sensors for Metal Detectors	Some various coil arrangements
EP0764856	Sensor for a Metal Detector	Several IB coil arrangements
WIPO 2011/116414	Improvements In Metal Detector Sensor Head	Minelab; addition of ferrite beads to shield solder joints

E.9: Eddy techniques

Patent No.	Title	Comments
US3337796	Eddy Current Testing Device with Means for Sampling the Output Signal to Provide a Signal Proportional to the Instantaneous Value of Said Output Signal at a Particular Phase	Helluva title, the basis for most modern sampled discriminators
US3478263	Wide Frequency Range Eddy Current Testing Instrument	
US4006407	Non-destructive Testing Systems Having Automatic Balance and Sample and Hold Operational Modes	
US4095180	Method and Apparatus for Testing Conductivity using Eddy Currents	
US4188577	Pulse Eddy Current Testing Apparatus for Magnetic Materials, Particularly Tubes	
US4191922	Electromagnetic Flaw Detection System and Method Incorporating Improved Automatic Coil Error Signal Compensation	
US4230987	Digital Eddy Current Apparatus for Generating Metallurgical Signatures and Monitoring Metallurgical Contents of an Electrically Conductive Material	
US4303885	Digitally Controlled Multifrequency Eddy Current Test Apparatus and Method	
US4424486	Phase Rotation Circuit for an Eddy Current Tester	
US5508610	Electrical Conductivity Tester and Methods Thereof for Accurately Measuring Time-Varying and Steady State Conductivity Using Phase Shift Detection	
US5952879	Device for the Simultaneous Demodulation of a Multifrequency Signal, Particularly for an Eddy Current Measurement	

E.10: DSP Methods

Patent No.	Title	Comments
US7579839	Metal Detector	Minelab, narrowband demod
US9239400	Method for Separating Target Signals from Unwanted Signals in a Metal Detector	Minelab
US10078148	Metal Detector	Minelab
US10838103	Effective Target Detection Depth Information for Metal Detectors	Minelab
US11067715	Signal Processing Technique for a Metal Detector	Minelab
US11333785	Metal Detector	Minelab
US11914095	Asynchronous Method for Sampling Signals in Metal Detectors	Nokta, narrowband demod

E.11: Security

Patent No.	Title	Comments
US3676772	Metallic Intrusion Detector System	Pass-through type
US3758849	Metal Detector System Having Identical Balanced Field Coil System on Opposite Sides of a Detection Zone	Walk-through type
US3950696	Trapezoidal Coil Configuration for Metal Detector in the Shape of an Inverted U	Walk-through type

US4012690	Device for Selectively Detecting Different Kinds and Sizes of Metals	Walk-through type
US4605898	Pulse Field Metal Detector with Spaced, Dual Coil Transmitter and Receiver Systems	Walkthrough PI
US4779048	Metal Detector for Detecting Metal Objects	Pass-through type
US4821023	Walk-Through Metal Detector	
US4866424	Metal Detector Coil	Walk-through type
US4906973	Walk-Through Metal Detector	White's
US5121105	Metal Detector	Walk-through type
US5521583	Metal Detection System	Walk-through type, see US5680103
US5680103	Metal Detection System	Walk-through type, see US5521583
US5726628	Metal Detector System	Walk-through type
US5790685	Apparatus and Method for Detecting and Imaging Metal	Pass-through type
US6133829	Walk-through Metal Detector System and Method	Fisher
US6696947	Metal Detector	Foldable security walk-through
US6970086	Wide Area Metal Detection (WAMD) System and Method for Security Screening Crowds	
US7592907	Metal Detector Presenting High Performance	Ceia walk-through with elliptical coils
EP0611970	Multiple Aerial for Metal Detector	Cylindrical coil for walk-through

E.12: Mine Detection

Patent No.	Title	Comments
US5307272	Minefield Reconnaissance and Detector System	Pulsed radar
US5680048	Mine Detecting Device Having a Housing Containing Metal Detector Coils and an Antenna	Appears to combine a metal detector and GPR in one search head
US7310060	Multi-Mode Landmine Detector	Metal detector and GPR in one search head
US7532127	Motion and Position Measuring for Buried Object Detection	Mine detector
US8174429	Mine Detection	Metal detector and GPR in one search head
US8854247	Metal Detector and Ground-Penetrating Radar Hybrid Head and Manufacturing Method Thereof	

E.13: Misc.

Patent No.	Title	Comments
US3836960	Sensor System	VHF/UHF
US3976564	Combination Digger and Sifter for Use With Metal Detector	

US4006481	Underground, Time Domain, Electromagnetic Reflectometry for Digging Apparatus	High frequency wide spectrum detector mounted to a digging tool
US4529937	Metal Detector With Spring Loaded Hinged Support	
US4540943	Belt-Supported Swingable Metal Detector	
US4560935	Remote Actuator for Metal Detector Discriminating Adjust Switch	
US4594559	Metal Detector Audio Amplifier	
US4641091	Device for Testing and Calibrating Treasure Hunting Metal Detectors	
US4644290	Metal Detector Audio Amplifier	
US4719421	Metal Detector for Detecting Product Impurities	Auto phase adjustment
US4779777	Support Bracket for Metal Detector	
US4797618	Caddy for Metal Detector	
US4983281	Metal Detector Scoop Sifter	
US5045789	Detector for Detecting Foreign Matter in Object by Using Discriminant Electromagnetic Parameters	
US5138262	Metal Detector Having Detachable Battery and Speaker Housing	Garrett
US5247257	Electronic Metal Detector Return Signal Phase Changer	A nail and two coils, weird
US5501283	Hole Cutting Device for Recovering Targets Located with a Metal Detector Audio Amplifier	
US5696490	FM (VHF) Infrared Wireless Digital Metal Detector	
US5896031	Quad Coil Vibration Cancelling Metal Detector	
US5963035	Electromagnetic Induction Spectroscopy for Identifying Hidden Objects	Multi-frequency method, interesting data
US5994897	Frequency Optimizing Metal Detector	
US6791329	Portable Metal Detection and Classification System	Detector on wheels
US6838886	Method and Apparatus for Measuring Inductance	Vehicle method
US6870370	Electromagnetic Induction Detection System	Airborne method
US7081754	Metal Detection System With a Magnetometer Head Coupleable to Conventional Footwear and Method of Use	Metal detector on your shoe
US7123016	Systems and Methods Useful for Detecting Presence and/or Location of Various Materials	A mish-mash of techniques, see 6724191
US7132943	Moving Belt Sensor	
US7288927	Remote Substance Identification and Location Method and System	Supposedly an infrared molecular locator (LRL?)
US7310586	Metal Detector with Data Transfer	Minelab, transfer of operating parametrics
US7575065	Metal Detector with Excavation Tool	Garrett pinpointer
US7940049	Portable Wireless Metal Detector	XP Deus
US8854043	Method for Displaying Metal Detection Information	Minelab CTX
US9151863	Method of and Apparatus for a Metal Detection System	Synchronizing multiple detectors

| US11513252 | Metal Detector | Minelab, accelerometer |
| EP0790507 | Metal Detector with Pivoting Detector Coil | Lockable coil pivot |

E.14: Pre-1970

Patent No.	Title	Comments
US269439	Apparatus for Finding Torpedoes	1882, C.A. McEvoy (First US detector patent?)
US1126027	Apparatus for Detecting Pipe Leads or Other Metallic Masses Embedded in Masonry	1915, Dr. Max Jüllig (Figure 8 coil)
US1812392	Method of and Apparatus for Locating Terrestrial Conducting Bodies	1931
US1890786	Radio Distance or Location Finder	1932
US2066135	Apparatus for Locating Bodies Having Anomalous Electrical Admittances	1936, 2-box locator, much like Fisher's
US2066561	Metalloscope	1937, Fisher 2-box
US2129058	Transformer for a Metal Locator	1936, describes an adjustable DD coil
US2139460	Means and Methods for Geophysical Prospecting	1938
US2160356	Geophysical Instrument	1939, another 2-box
US2167630	Electrical Prospecting Method and Apparatus	1938, bore hole unit
US2179240	Metal Detection Device	1939, looks like an early walk-through
US2201256	Electrical Apparatus and Method for Locating Minerals	1940, 2-box variation
US2220070	Method and Apparatus for Magnetically Exploring Earth Strata	1940, bore hole instrument
US2268106	Radiowave Prospecting	1941
US2278506	Apparatus for Geophysical Prospecting	1942, first PI patent?
US2408029	Electrical Prospecting Apparatus	1946
US2447316	Variable Frequency Oscillatory System	1948
US2608602	Detecting Device	1952, bore hole unit, describes phase discrimination
US3015060	Method and Means of Prospecting for Electrically Conducting Bodies	1961
US3020470	Submerged Body Detection System	1962
US3105934	Method and Apparatus for the Remote Detection of Ore Bodies Utilizing Pulses of Short Duration to Induce Transient Polarization in the Ore Bodies	1963
US3471773	Metal Detecting Device with Inductively Coupled Coaxial Transmitter and Receiver Coils	1969

Glossary

"It's like deja-vu, all over again."

— *Yogi Berra*

Throughout the book we've covered a lot of terms and thrown around acronyms like a NASA manual[1], so it's probably a good idea to wrap up with a short review. Most of the terms listed below have been generically applied across brands and models. Some terms are specific to brands, but have occasionally achieved near-generic status, much like "I need a Kleenex" doesn't necessarily mean you want to wait for the actual brand-name tissue.

- ADC — analog-to-digital converter — A device that digitizes an analog signal.
- AFE — analog front end — The analog input circuitry for the receiver, consisting at least of the preamp stage(s) and possibly the demodulator(s).
- AGT — automatic ground tracking
- AWG — American Wire Gauge — The American standard for wire diameter.
- BBS — Broad Band Spectrum — Minelab's term for their multifrequency technology.
- BFO — beat frequency oscillator — A proximity type of detector that compares two oscillators to determine frequency shift effect.
- CCPI — constant current PI — A type of PI whereby the transmit current is a pulse that approximates a rectangular wave (can be CT or DT).
- CDS — correlated double sampling — A method of sampling a signal twice, commonly used in PI designs.
- CE — Conformitè Europëenne (European Conformity) — Regulatory board for the EU.
- CT — continuous time — In a transmitter, a signal that does not linger in an off-state (zero current) but is continuously varying. In demodulators, the act of continuously sampling the received waveform with no "dead time."
- DD — A type of coil using overlapped "D"-shaped TX & RX windings.
- DOD — A type of coil using a central "O"-shaped TX coil with two overlapping "D"-shaped RX coils.
- DSP — digital signal processing — Any signal processing applied to digital signals.
- DSR — direct sampling receiver — Receiver whereby the ADC directly digitizes the RF signal.
- DT — discrete time — In a transmitter, a signal that consists of a repeating waveform that is punctuated by regions of zero current. In demodulators, the act of sampling the received waveform in discrete time windows, with "off" time in between.
- DVT — Dual Voltage Technology — Minelab's term for their PI transmitter technology where different pulse widths have different voltage drives.

---

1. When I worked for NASA, I got a 300-page dictionary of nothing but NASA acronyms, mostly TLAs.

- EFE — Earth field effect — The signal induced when a coil is moved through the Earth's magnetic field.

- EMF — electromotive force — The signal induced in a coil from a changing magnetic field.

- EMI — electromagnetic interference — External RF signals that couple into a metal detector and create havoc.

- FBS — Full Band Spectrum — Another Minelab term for their multifrequency technology.

- FCC — Federal Communications Commission — Regulatory board for the US.

- FD — frequency domain — Signal processing that determines target amplitude and phase.

- FEM — finite element modeling — A software method for modeling magnetic fields.

- GB — ground balance — Generic term for ground balance.

- GEB — Ground Exclusion Balance — White's term for their VLF ground balance.

- GNC — Ground Neutralizing Circuit — Fisher's original term for their VLF ground balance.

- HS — half-sine — A transmitter which produces disjointed (DT) current pulses that approximate half-sinusoids.

- IB — induction balance — A type of coil arrangement, and also a term that encompasses detectors that use such a coil, primarily TR and VLF.

- MF — multifrequency — A detector which transmits and processes two or more frequencies (or equivalents), usually in the frequency domain.

- Multi-IQ — Minelab's term for simultaneous multifrequency with direct sampling.

- NCO — numerically-controlled oscillator — A digital method for generating an AC signal, generally with fine frequency resolution.

- OO, 00 — A type of coil using overlapped "O"-shaped TX & RX windings.

- PI — pulse induction — A type of detector which uses a time-domain step response.

- PLL — phase-locked loop — A circuit that is used for frequency manipulation, and also a proximity type of detector which uses that circuit to determine frequency shift effect.

- Q — quality factor — A measure of the efficiency of a narrowband circuit.

- RF — radio frequency — In general, any radiowave signal; specific to metal detectors, a term commonly used for two-box locators.

- RX — receive or receiver.

- SAT — self-adjusting threshold — A method of AC-coupling the baseband signal so it maintains a reference level. Coined by White's but now universal.

- SF — single frequency — A detector which transmits and processes one frequency.

- SHA — sample-and-hold amplifier — A circuit that samples and holds an applied voltage, commonly used in demodulators.

- SMF — Simultaneous multifrequency — Simultaneous continuous transmission of multiple frequencies.

- SNR — signal-to-noise ratio — The ratio of signal energy to noise energy, used as a figure of merit for circuit design.

- SPD — Synchronous Phase Demodulation — Bounty Hunter's original term for their VLF motion discrimination.

- SQMF — sequential multifrequency — Time-sequential transmission of multiple frequencies.

- SRF — self-resonant frequency — The frequency at which a coil winding has a peak impedance.

- TC — temperature coefficient — The amount by which a circuit or an element varies over temperature, often specified in PPM (parts per million).

- TID — target identification — A method of identifying a probable target. On a display, it might be an icon, a number, or a spectral graph. With audio, it is usually a tone.
- TD — time domain — Signal processing that determines target amplitude and time constant.
- THS — truncated half-sine — A variation of half-sine in which a trailing portion of the half-sine pulse is suddenly truncated.
- TLA — Three Letter Acronym
- TR — transmit-receive — A type of detector which uses an IB coil to transmit a signal while simultaneously receiving a target signal.
- TX — transmit or transmitter.
- VCO — voltage-controlled oscillator — A circuit that produces an AC signal whose frequency depends on an applied voltage.
- VDI — visual display indicator — Generally a 2-digit display number that identifies the probable target (see TID). Coined by White's but now universal.
- VLF — very low frequency — Technically, the 3-30 kHz frequency band, but also a TR type detector which has ground cancel ability.
- VSAT — variable self-adjusting threshold — An SAT circuit which has a variable retune speed. Coined by White's but now universal.
- WT — walk-through — A type of security metal detector.

Index

"The covers of this book are too far apart."

— Ambrose Bierce

Qualcomm 350

R

R channel 278, 297–298, 301, 313, 317–320, 329, 333, 339, 348, 359, 372, 466, 508
R signal 291, 299–300, 314, 317, 332, 357, 377, 472–473
Radio Shack 137, 230
rail splitter 236, 385, 542
Rakes, Charles 2, 219
ramp response 88–92
Raptor 235, 317
ratcheting 344
Rayer, F. G. 2
reactive channel - see X channel
reactive signal - see X signal
recovery speed 83–84, 94, 96
recovery speed test jig 95
Relco 16
resistive channel - see R channel
resistive signal - see R signal
response
    composite 79, 81, 83–84, 86, 91, 94, 96, 294, 499, 568
    eddy 56–58, 73, 79, 81–83, 89–91, 96, 128, 173, 191, 355, 472, 511, 531, 560–561, 563–564, 567, 570
    ground 251, 272, 293–294, 296–297, 423, 425, 470, 483–485, 498, 508–509, 511, 570
    iron 470
    magnetic 51, 53–54, 56, 79, 81–83, 91, 178, 272, 355, 531
    multiple domain 85–86, 96
    single domain 85, 96, 570
    viscous 56, 423, 489, 570
response speed 55, 84, 264, 270
retune circuit 214, 254–255, 257, 262–263, 273, 300, 323, 542, 544
RF detector 30, 107, 238

S

saline soil 193
salt cancel 32, 170, 300–301, 358, 462, 473–474, 476
salt water 19, 70, 168, 249, 470
sample delay 93, 374, 376, 395, 403, 413–414, 416, 419, 421, 439, 454, 484, 558, 569
sample-and-hold 246, 265, 374–375
sampling 246
Santschi, R. J. 2
SAT 372, 376, 386, 427, 433, 507
Scarborough, Thomas 519
SCR-625 14

secondary coil 46–47, 56
security detector 15–16, 20–21, 28, 85, 105, 170, 345, 351, 361, 375, 440, 456–457
self-adjusting threshold - see SAT
self-capacitance 116, 133–135, 137, 139, 149, 152, 196–197, 257, 369, 382, 394, 557
self-resonant frequency (SRF) 116, 135, 138, 149, 367
series resonance 501, 503
SHA 248, 257, 264, 284, 304, 311
shielding
    connection 146
    drain wire 156, 158, 164–165
    paint 145, 156, 164–165
    tape 145, 167
Signetics 219
silent EMI 530
skin depth 62–66, 70, 86
skin effect 62–63, 172, 274, 301, 529, 531
SNR 103, 149, 186, 313, 333–334, 359, 412, 414, 441, 476
snubber 418–419
software-defined metal detector 361
SpectreRF 264
Spice 67, 182, 185, 264–265, 305, 487
step response 75, 91–93, 269, 414, 423, 563, 567
Sunray 167
superconductor 57, 65, 291

T

target channel (PI) 285, 425, 427, 435, 570
target channel (VLF) 119, 298–299, 329, 343
target hole 424–425, 433–434, 483–485, 489, 511, 571
target ID 80–81, 86, 90, 96, 104, 245, 348, 352–353, 356–357, 478, 508, 510
target magnitude 245, 291, 340, 355, 357
target model 60, 67–68, 90
target phase 58–59, 105, 245, 248, 250, 264, 273, 283, 291, 299, 301, 317–318, 340, 343, 348, 352–353, 355–357, 461, 468, 521
target separation 84, 96, 123, 127, 270
target tau 68, 75, 90, 92–93, 415, 434, 505, 561, 564, 566–569, 571
target vector plot 252, 291, 341–342

targeting signal 299, 301
Tarsacci
    MDT 8000 36, 91, 501, 522
Tayloe, Dan 35
Technos 16
    Phase Readout Gradiometer 14, 16, 106, 241
Teknetics 14, 21, 312, 352
    8000 348
    8500 245, 352
    9000 345
    Omega 355, 358
    T2 182
temperature coefficient 149, 178, 183, 185, 495, 544
temperature compensation 181, 185, 193, 204, 409
temperature drift 178, 181, 183, 185, 193, 213, 255, 409
Tesoro 17, 22, 103, 263, 330, 347
    Cortés 352
    Sandshark 137, 168, 377
thermometer code 329
threshold 107, 181, 183, 189, 213–214, 216, 232–233, 236, 238–239, 254, 256–257, 263, 285, 361, 376–377, 387–388, 409, 449, 454
TID - see target ID
Tim "the Toolman" Taylor 526
time domain 27, 33, 36, 51, 88–93, 265, 461, 469, 505, 560, 563
Tinker-Raser Model 505 239
TR 16–17, 27, 29–31, 33, 84, 229–239
Traister and Traister 2
transformer 46–47, 57, 117, 239, 525, 567
transformer gain 117, 149, 233
transmit power 129, 525–527
Transmit-Receive - see TR
TR-Disc 241–260
truncated half-sine 34, 508–510
two-box detector 2, 12–13, 18, 30, 107–109, 232, 238
two-slope ground balance 489

U

US coinage 58

V

varactor 194–195, 197–198, 200, 225
VCO 29, 182–183, 220–222, 224–225, 428–429
VCO audio 447
VCO pinpoint 347–348
VDI 66, 314, 340, 348, 352
vector response 118, 293
Veroboard 3

Bandido II µMax

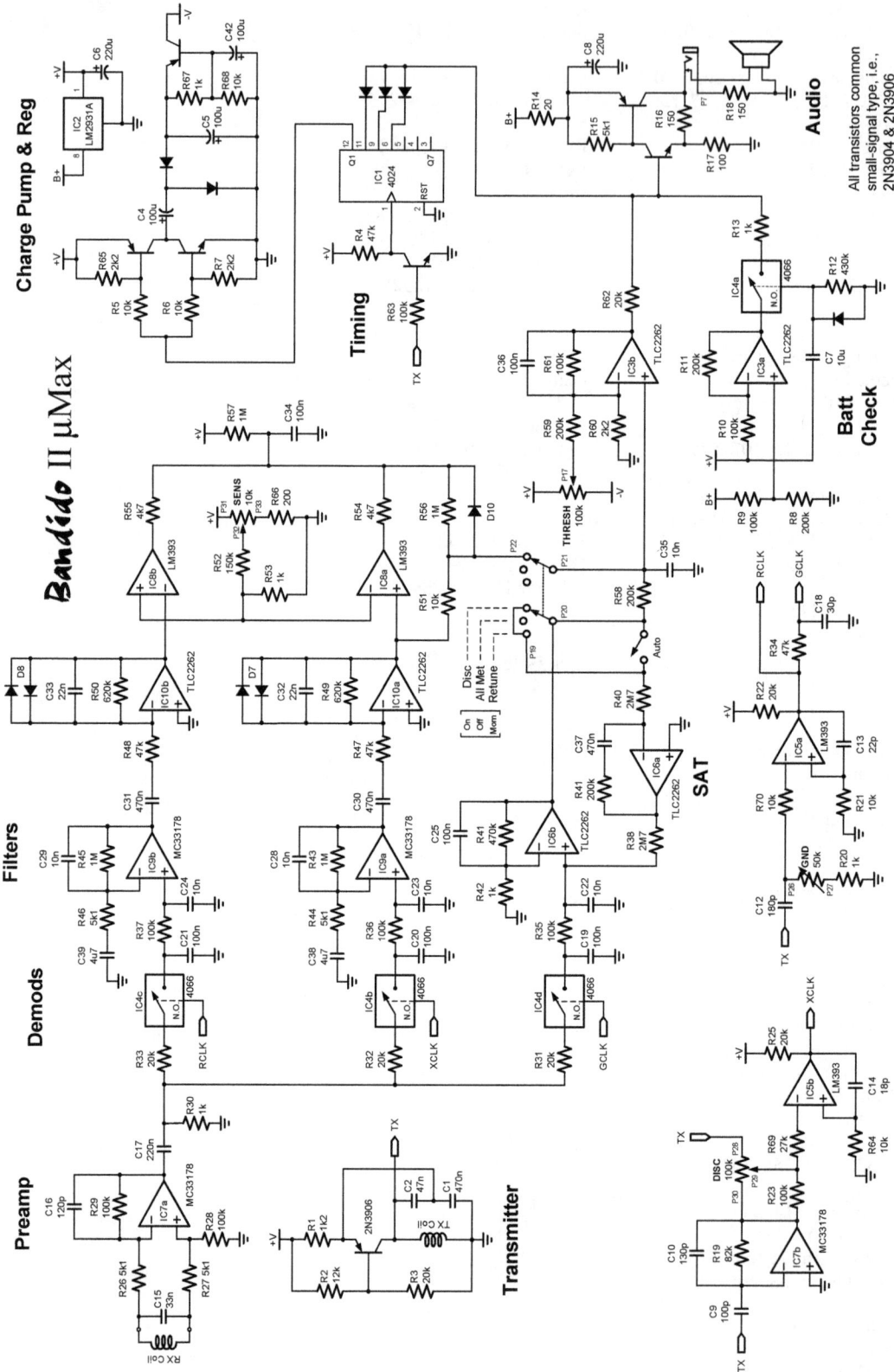

All transistors common small-signal type, i.e., 2N3904 & 2N3906

All diodes 1N4148

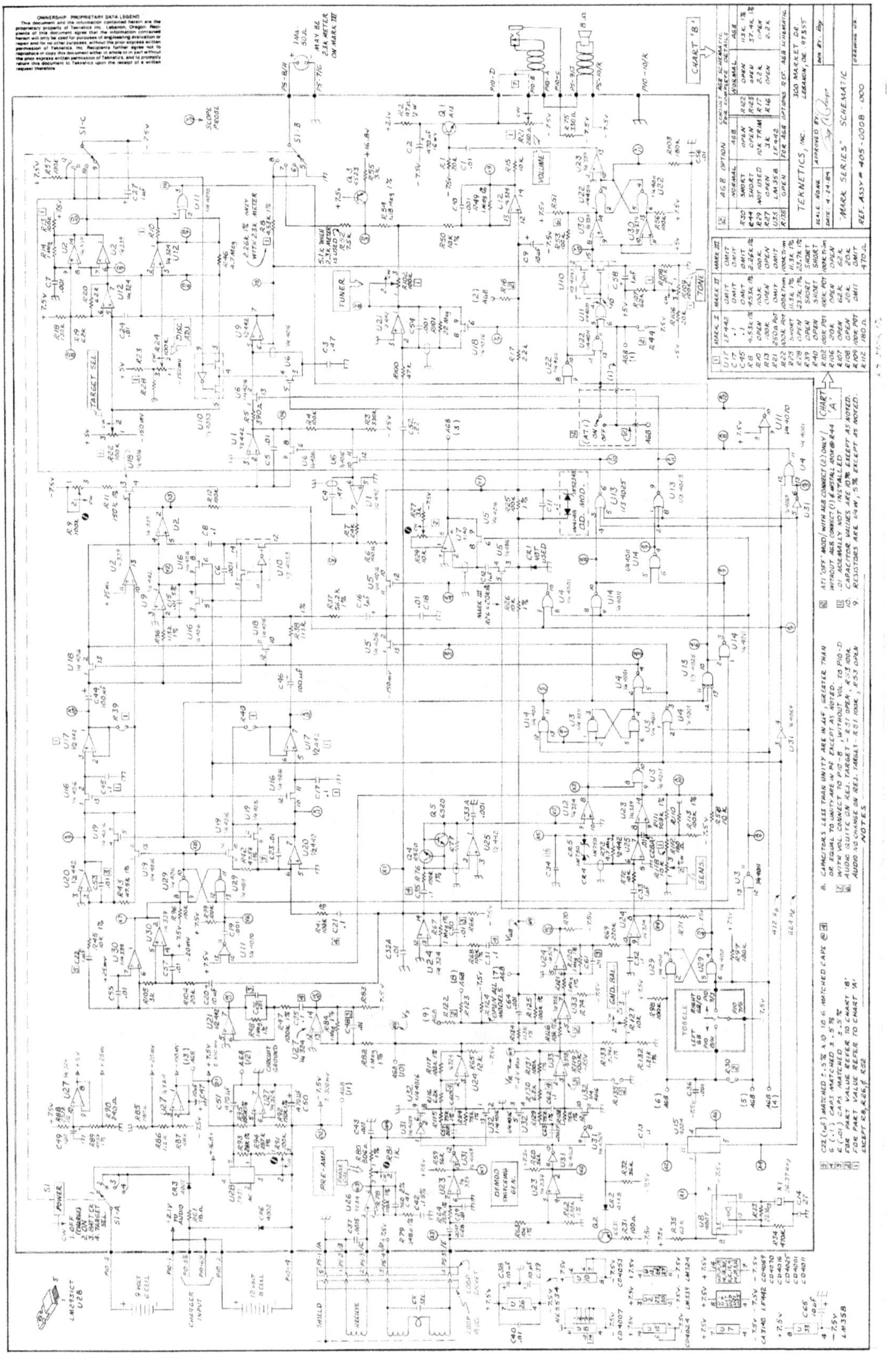
TEKNETICS, INC.
100 MARKET DR.
LEBANON, OR. 97355
"MARK SERIES" SCHEMATIC
REF. ASSY # 405-0008-000

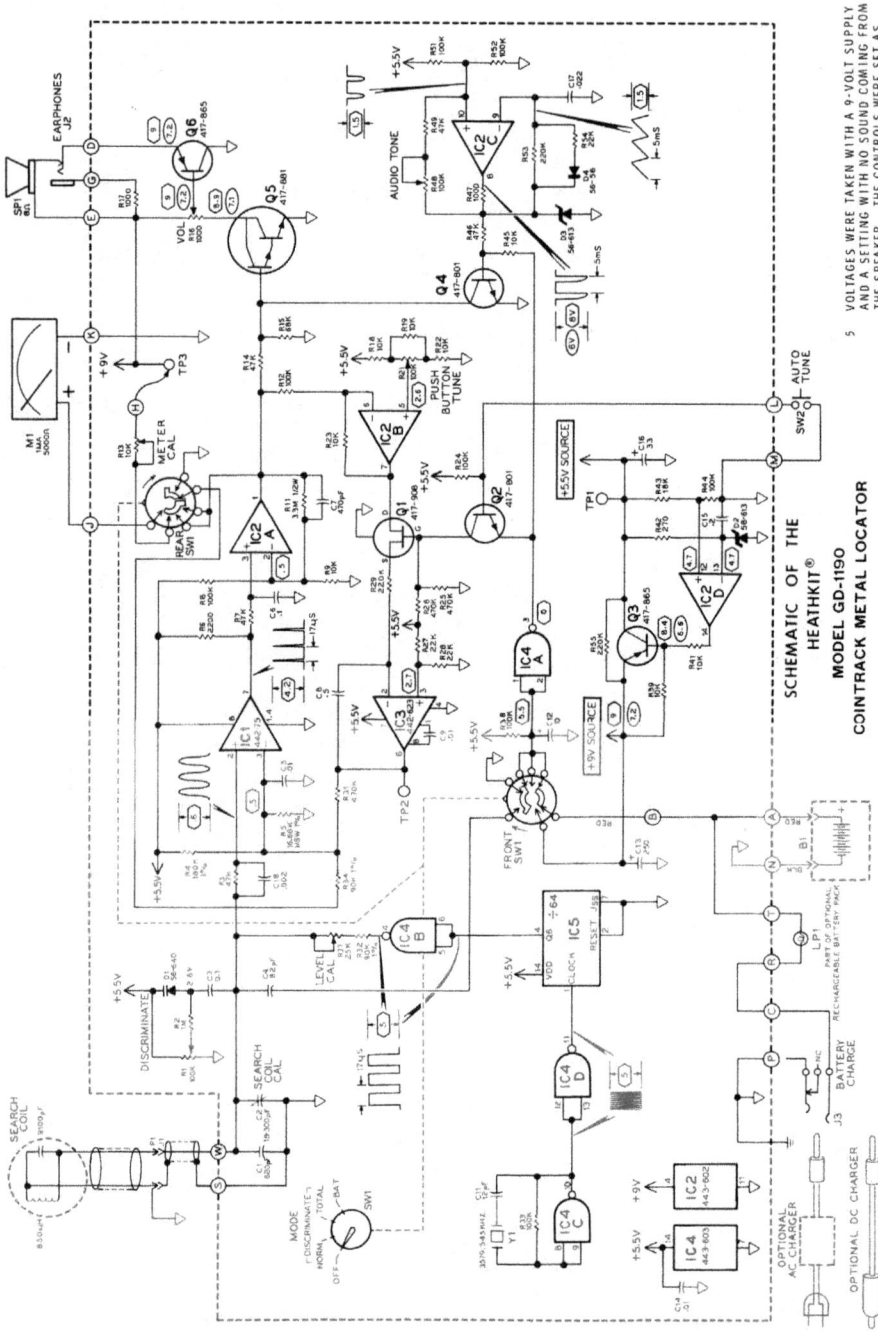

Schematic of the Heathkit® Model GD-1190 Cointrack Metal Locator

NOTES

1. ALL RESISTORS ARE 1/4-WATT, 5% UNLESS MARKED OTHERWISE. RESISTOR VALUES ARE IN OHMS (K=1000; M=1,000,000).

2. ALL CAPACITORS ARE IN µF(MICROFARADS) UNLESS MARKED OTHERWISE.

3. ▽ THIS SYMBOL INDICATES CIRCUIT BOARD GROUND.

4. ⏚ THIS SYMBOL INDICATES CHASSIS GROUND.

5. VOLTAGES WERE TAKEN WITH A 9-VOLT SUPPLY AND A SETTING WITH NO SOUND COMING FROM THE SPEAKER. THE CONTROLS WERE SET AS FOLLOWS:

   VOL...............FULLY COUNTERCLOCKWISE
   PUSHBUTTON TUNE.......CENTER OF ROTATION
   DISCRIMINATE..........CENTER OF ROTATION
   AUDIO TUNE(ON PCB)...CENTER OF ROTATION

6. ⬡ THIS SYMBOL INDICATES VOLTAGES PRODUCED BY REGULAR BATTERIES.

7. ⬯ THIS SYMBOL INDICATES VOLTAGES PRODUCED BY NICKEL-CADMIUM BATTERIES.

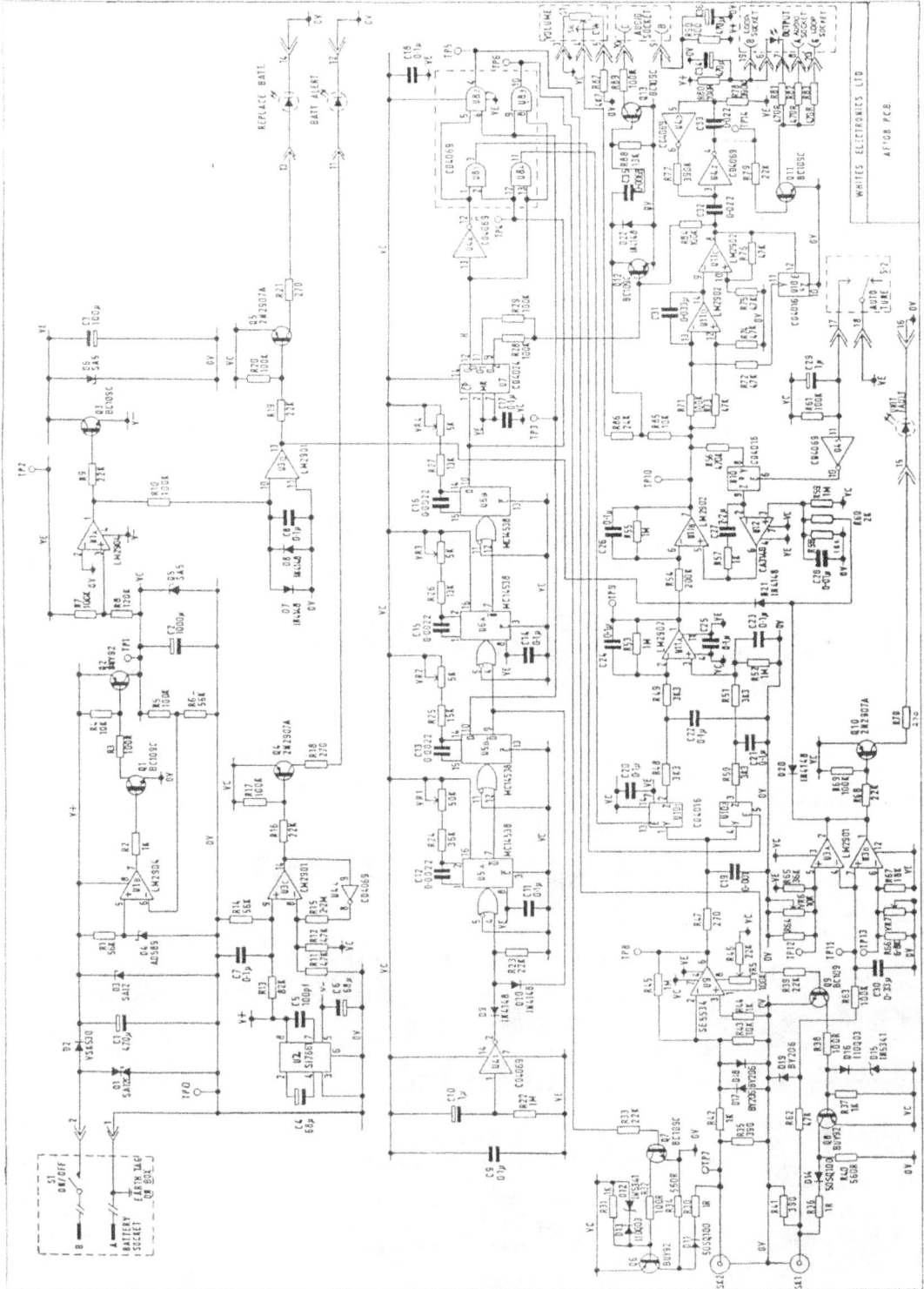

Nokta DETECTION TECHNOLOGIES

GARRETT ELECTRONICS Trademark of Quality

WE WHITE'S ELECTRONICS

Treasure Ray

C&G TECHNOLOGY INC.

COMPASS ELECTRONICS CORP.

RUTUS

DetectorPro

QUEST METAL DETECTORS

a.h. pro line professional metal detectors

Jetco

Troy CUSTOM DETECTORS INC.

WILSON NEUMAN

TeKnetics

nexus

Discovery Since 1981

NAUTILUS METAL DETECTORS

LORENZ DEEPMAX.com

Fisher Labs

white's

ReLCO

XP

CSCOPE

Pillar ELECTRONICS, INC.

The GOLDAK Co. Leading Electronic Manufacturers Since 1933

BOUNTY HUNTER METAL DETECTORS

TREASURE INDUSTRIES

GARRETT METAL DETECTORS

TEKNETICS

GOLD MOUNTAIN

MINELAB

RC RAYSCOPE COMPANY THE DETECTOR CENTER

JW FISHERS

the Detectron corporation

PULSE TECHNOLOGY

VIKING METAL DETECTORS

FISHER m-scope

HEATHKIT

Tesoro

www.ingramcontent.com/pod-product-compliance
Lightning Source LLC
Chambersburg PA
CBHW062010190326
41458CB00009B/3026